Methods in Cell Biology

The Zebrafish: Genetics, Genomics, and Transcriptomics

Volume 135

Series Editors

Leslie Wilson
Department of Molecular, Cellular and Developmental Biology
University of California
Santa Barbara, California

Phong Tran
University of Pennsylvania
Philadelphia, USA &
Institut Curie, Paris, France

Methods in Cell Biology

The Zebrafish: Genetics, Genomics, and Transcriptomics

Volume 135

Edited by

H. William Detrich, III
Northeastern University Marine Science Center, Nahant, MA, United States

Monte Westerfield
University of Oregon, Eugene, OR, United States

Leonard I. Zon
Harvard University, Boston, MA, United States

ELSEVIER

AMSTERDAM • BOSTON • HEIDELBERG • LONDON
NEW YORK • OXFORD • PARIS • SAN DIEGO
SAN FRANCISCO • SINGAPORE • SYDNEY • TOKYO

Academic Press is an imprint of Elsevier

AP

Academic Press is an imprint of Elsevier
50 Hampshire Street, 5th Floor, Cambridge, MA 02139, USA
525 B Street, Suite 1800, San Diego, CA 92101-4495, USA
125 London Wall, London EC2Y 5AS, UK
The Boulevard, Langford Lane, Kidlington, Oxford OX5 1GB, UK

Fourth edition 2016

Notices

Knowledge and best practice in this field are constantly changing. As new research and experience broaden our understanding, changes in research methods, professional practices, or medical treatment may become necessary.

Practitioners and researchers must always rely on their own experience and knowledge in evaluating and using any information, methods, compounds, or experiments described herein. In using such information or methods they should be mindful of their own safety and the safety of others, including parties for whom they have a professional responsibility.

To the fullest extent of the law, neither the Publisher nor the authors, contributors, or editors, assume any liability for any injury and/or damage to persons or property as a matter of products liability, negligence or otherwise, or from any use or operation of any methods, products, instructions, or ideas contained in the material herein.

ISBN: 978-0-12-803474-3
ISSN: 0091-679X

Publisher: Zoe Kruze
Acquisition Editor: Zoe Kruze
Editorial Project Manager: Sarah Lay
Production Project Manager: Malathi Samayan
Designer: Victoria Pearson

Typeset by TNQ Books and Journals
Transferred to Digital Printing in 2017

Len, Monte, and I dedicate the 4th *Edition of* Methods in Cell Biology: The Zebrafish *to the postdoctoral fellows and graduate students who conducted the genetic screens that established the zebrafish as a preeminent vertebrate model system for analysis of development.*

Contents

PART II TRANSGENESIS AND FUNCTIONAL GENOMICS

PART V ZEBRAFISH INFRASTRUCTURE

Contributors

J. Ablain
Howard Hughes Medical Institute and Harvard Medical School, Boston, MA, United States

K. Asakawa
National Institute of Genetics, Mishima, Shizuoka, Japan

H. Ata
Mayo Clinic, Rochester, MN, United States

J. Bakkers
Hubrecht Institute and University Medical Centre Utrecht, Utrecht, The Netherlands

J. Bessa
IBMC-Instituto de Biologia Molecular e Celular, Porto, Portugal; Universidade do Porto, Porto, Portugal

Y.M. Bradford
University of Oregon, Eugene, OR, United States

B.R. Cairns
University of Utah School of Medicine, Salt Lake City, UT, United States

W. Chen
Vanderbilt University School of Medicine, Nashville, TN, United States

J. Cibelli
Michigan State University, East Lansing, MI, United States; BIONAND, Andalucía, Spain

K.J. Clark
Mayo Clinic, Rochester, MN, United States

P. Coucke
Ghent University, Ghent, Belgium

V.T. Cunliffe
University of Sheffield, Sheffield, United Kingdom

F. Del Bene
PSL Research University, Paris, France

A. De Paepe
Ghent University, Ghent, Belgium

F. De Santis
PSL Research University, Paris, France

V. Di Donato
PSL Research University, Paris, France

A. Eagle
University of Oregon, Eugene, OR, United States

S.C. Ekker
Mayo Clinic, Rochester, MN, United States

T. Erickson
Oregon Health & Science University, Portland, OR, United States

T. Evans
Weill Cornell Medical College, New York, NY, United States

D. Fashena
University of Oregon, Eugene, OR, United States

A. Felker
University of Zürich, Zürich, Switzerland

A. Fernández-Miñán
Centro Andaluz de Biología del Desarrollo (CABD), Consejo Superior de Investigaciones Científicas/Universidad Pablo de Olavide/Junta de Andalucía, Sevilla, Spain

K. Frazer
University of Oregon, Eugene, OR, United States

A. Ghosh[a]
Carnegie Institution for Science, Baltimore, MD, United States

M.G. Goll
Memorial Sloan Kettering Cancer Center, New York, NY, United States

J.L. Gómez-Skarmeta
Centro Andaluz de Biología del Desarrollo (CABD), Consejo Superior de Investigaciones Científicas/Universidad Pablo de Olavide/Junta de Andalucía, Sevilla, Spain

[a]Current address: Tata Institute of Fundamental Research, Mumbai, India

D.J. Grunwald
University of Utah, Salt Lake City, UT, United States

M.E. Halpern
Carnegie Institution for Science, Baltimore, MD, United States

J.K. Heath
Walter and Eliza Hall Institute of Medical Research, Parkville, VIC, Australia

K. Hoshijima
University of Utah, Salt Lake City, UT, United States

D.G. Howe
University of Oregon, Eugene, OR, United States

H. Huang
University of California Los Angeles, Los Angeles, CA, United States

J.P. Junker
Hubrecht Institute and University Medical Centre Utrecht, Utrecht, The Netherlands

M.J. Jurynec
University of Utah, Salt Lake City, UT, United States

P. Kalita
University of Oregon, Eugene, OR, United States

K. Kawakami
National Institute of Genetics, Mishima, Shizuoka, Japan

M.C. Keightley
Australian Regenerative Medicine Institute, Clayton, VIC, Australia; Monash University, Clayton, VIC, Australia

F. Kruse
Hubrecht Institute and University Medical Centre Utrecht, Utrecht, The Netherlands

C. Lawrence
Boston Children's Hospital, Boston, MA, United States

S. Lefever
Ghent University, Ghent, Belgium

C. Li
Memorial Sloan Kettering Cancer Center, New York, NY, United States; Weill Cornell Graduate School of Medical Sciences, New York, NY, United States

G.J. Lieschke
Australian Regenerative Medicine Institute, Clayton, VIC, Australia; Monash University, Clayton, VIC, Australia

S. Lin
University of California Los Angeles, Los Angeles, CA, United States

C.G. Love
Walter and Eliza Hall Institute of Medical Research, Parkville, VIC, Australia; University of Melbourne, Parkville, VIC, Australia

L.A. Maddison
Vanderbilt University School of Medicine, Nashville, TN, United States

P. Mani
University of Oregon, Eugene, OR, United States

S. Markmiller
University of California San Diego, La Jolla, CA, United States

R. Martin
University of Oregon, Eugene, OR, United States

S. Masuda
Tokyo Institute of Technology, Yokohama, Japan

A.C. Miller
University of Oregon, Eugene, OR, United States

C.B. Moens
Fred Hutchinson Cancer Research Center, Seattle, WA, United States

C. Mosimann
University of Zürich, Zürich, Switzerland

S.T. Moxon
University of Oregon, Eugene, OR, United States

M.C. Mullins
University of Pennsylvania Perelman School of Medicine, Philadelphia, PA, United States

P.J. Murphy
University of Utah School of Medicine, Salt Lake City, UT, United States

K.N. Murray
University of Oregon, Eugene, OR, United States

A. Muto
National Institute of Genetics, Mishima, Shizuoka, Japan

T. Nicolson
Oregon Health & Science University, Portland, OR, United States

C.J. Ott
Dana Farber Cancer Institute, Harvard Stem Cell Institute, Boston, MA, United States

H. Paddock
University of Oregon, Eugene, OR, United States

F. Pelegri
University of Wisconsin—Madison, Madison, WI, United States

C. Pich
University of Oregon, Eugene, OR, United States

S. Prukudom
Kasetsart University, Bangkok, Thailand; Center of Excellence on Agricultural Biotechnology: (AG-BIO/PERDO-CHE), Bangkok, Thailand

S. Ramachandran
University of Oregon, Eugene, OR, United States

J.E.J. Rasko
Centenary Institute, Camperdown, NSW, Australia; University of Sydney, Sydney, NSW, Australia; Royal Prince Alfred Hospital, Newtown, NSW, Australia

M.P. Rossmann
Harvard University, Harvard, Cambridge, MA, United States

L. Ruzicka
University of Oregon, Eugene, OR, United States

K. Schaper
University of Oregon, Eugene, OR, United States

A.N. Shah
Fred Hutchinson Cancer Research Center, Seattle, WA, United States

X. Shao
University of Oregon, Eugene, OR, United States

A. Singer
University of Oregon, Eugene, OR, United States

K. Siripattarapravat
Kasetsart University, Bangkok, Thailand; Center of Excellence on Agricultural Biotechnology: (AG-BIO/PERDO-CHE), Bangkok, Thailand

F. Speleman
Ghent University, Ghent, Belgium

M. Superdock
Boston Children's Hospital, Boston, MA, United States; Dana Farber Cancer Institute, Harvard Stem Cell Institute, Boston, MA, United States; Harvard Medical School and Howard Hughes Medical Institute, Boston, MA, United States

M. Tanaka
Tokyo Institute of Technology, Yokohama, Japan

J.J. Tena
Centro Andaluz de Biología del Desarrollo (CABD), Consejo Superior de Investigaciones Científicas/Universidad Pablo de Olavide/Junta de Andalucía, Sevilla, Spain

S. Toro
University of Oregon, Eugene, OR, United States

A. van Oudenaarden
Hubrecht Institute and University Medical Centre Utrecht, Utrecht, The Netherlands

J. Vandesompele
Ghent University, Ghent, Belgium

S. Vanhauwaert
Ghent University, Ghent, Belgium

C. Van Slyke
University of Oregon, Eugene, OR, United States

Z.M. Varga
University of Oregon, Eugene, OR, United States

H. Wada
National Institute of Genetics, Mishima, Shizuoka, Japan

M. Westerfield
University of Oregon, Eugene, OR, United States

A. Willaert
Ghent University, Ghent, Belgium

S. Yang
Boston Children's Hospital, Boston, MA, United States; Dana Farber Cancer Institute, Harvard Stem Cell Institute, Boston, MA, United States; Harvard Medical School and Howard Hughes Medical Institute, Boston, MA, United States

L. Yin
Vanderbilt University School of Medicine, Nashville, TN, United States

B. Zhang
Peking University, Beijing, People's Republic of China

Y. Zhang
Peking University Shenzhen Graduate School, Shenzhen, China; University of California Los Angeles, Los Angeles, CA, United States

Y. Zhou
Boston Children's Hospital, Boston, MA, United States; Dana Farber Cancer Institute, Harvard Stem Cell Institute, Boston, MA, United States; Harvard Medical School and Howard Hughes Medical Institute, Boston, MA, United States; Harvard University, Harvard, Cambridge, MA, United States

L.I. Zon
Boston Children's Hospital, Boston, MA, United States; Dana Farber Cancer Institute, Harvard Stem Cell Institute, Boston, MA, United States; Harvard Medical School and Howard Hughes Medical Institute, Boston, MA, United States; Harvard University, Harvard, Cambridge, MA, United States

Preface

Len, Monte, and I are pleased to introduce the fourth edition of *Methods in Cell Biology: The Zebrafish*. The advantages of the zebrafish, *Danio rerio*, are numerous, including its short generation time and high fecundity, external fertilization, and the optical transparency of the embryo. The ease of conducting forward genetic screens in the zebrafish, based on the pioneering work of George Streisinger, culminated in screens from the laboratories of Wolfgang Driever, Mark C. Fishman, and Christiane Nüsslein-Volhard, published in a seminal volume of *Development* (volume 123, December 1, 1996) that described a "candy store" of mutants whose phenotypes spanned the gamut of developmental processes and mechanisms. Life for geneticists who study vertebrate development became *really* fine.

Statistics derived from ZFIN (the Zebrafish Model Organism Database; http://zfin.org) illustrate the dramatic growth of research involving zebrafish. The zebrafish genome has been sequenced, and as of 2014, more than 25,000 genes have been placed on the assembly. Greater than 15,500 of these genes have been established as orthologs of human genes. The zebrafish community has grown from ~1,400 researchers in 190 laboratories as of 1998 to ~7,000 in 930 laboratories in 2014. The annual number of publications based on the zebrafish has risen from 1,913 to 21,995 in the same timeframe. Clearly, the zebrafish has arrived as a vertebrate biomedical model system *par excellence*.

When we published the first edition (volumes 59 and 60) in 1998, our goal was to encourage biologists to adopt the zebrafish as a genetically tractable model organism for studying biological phenomena from the cellular through the organismal. Our goal today remains unchanged, but the range of subjects and the suite of methods have expanded rapidly and significantly in sophistication over the years. With the second and third editions of *MCB: The Zebrafish* (volumes 76 and 77 in 2004; volumes 100, 101, 104, and 105 in 2010–11), we documented this extraordinary growth, again relying on the excellent chapters contributed by our generous colleagues in the zebrafish research community.

When Len, Monte, and I began planning the fourth edition, we found that the zebrafish community had once more developed and refined novel experimental systems and technologies to tackle challenging biological problems across the spectrum of the biosciences. We present these methods following the organizational structure of the third edition, with volumes devoted to *Cellular and Developmental Biology*, to *Genetics, Genomics, and Transcriptomics*, and to *Disease Models and Chemical Screens*. Here we introduce the third volume, *Genetics, Genomics, and Transcriptomics.*

Genetics, Genomics, and *Transcriptomics* is divided into five sections that cover genetic and genomics techniques. Part 1 covers forward and reverse genetics in nine chapters, many of which employ the revolutionary CRISPR/Cas9 technology in novel ways. Precision editing of the zebrafish genome through homologous recombination has now become a reality. Part 2 contains five chapters that describe advances in transgenesis and functional genomics approaches. Spatially resolved transcriptomics at the organismal level, cell type-specific transcriptomics, and the important companion technology, RT-qPCR, are presented in Part 3. We devote Part 4 to five chapters on the emerging analysis of epigenetic regulation of gene expression in the zebrafish. Part 5 concludes the volume with four important chapters on zebrafish husbandry, health monitoring, disease prevention, and information technology.

We anticipate that you, our readership, will apply these methods successfully in your own zebrafish research programs and will develop your own technical advances that may be considered for a future edition of *Methods in Cell Biology: The Zebrafish*. The zebrafish is a remarkable experimental system—the preeminent vertebrate model for mechanistic studies of cellular and developmental processes in vivo.

We thank the series editors, Leslie Wilson and Phong Tran, and the staff of Elsevier/Academic Press, especially Zoe Kruze and Sarah Lay, for their enthusiastic support of our fourth edition. Their help, patience, and encouragement are profoundly appreciated.

H. William Detrich, III
Monte Westerfield
Leonard I. Zon

PART 1

Forward and Reverse Genetics

CHAPTER

Multiplex conditional mutagenesis in zebrafish using the CRISPR/Cas system

1

L. Yin[a], L.A. Maddison[a], W. Chen[1]

Vanderbilt University School of Medicine, Nashville, TN, United States

[1]Corresponding author: E-mail: wenbiao.chen@vanderbilt.edu

CHAPTER OUTLINE

Abstract

The clustered regularly interspaced short palindromic repeat (CRISPR)/CRISPR-associated protein (Cas) system is a powerful tool for genome editing in numerous organisms. However, the system is typically used for gene editing throughout the entire organism. Tissue and temporal specific mutagenesis is often desirable to determine gene function in a specific stage or tissue and to bypass undesired consequences of global mutations. We have developed the CRISPR/Cas system for conditional mutagenesis in transgenic zebrafish using tissue-specific and/or inducible expression of Cas9 and U6-driven expression of sgRNA. To allow mutagenesis of multiple targets, we have isolated four distinct U6 promoters and designed Golden Gate vectors to easily assemble transgenes with multiple sgRNAs. We provide experimental details on the reagents and applications for multiplex conditional mutagenesis in zebrafish.

[a]These authors contributed equally.

Methods in Cell Biology, Volume 135, ISSN 0091-679X, http://dx.doi.org/10.1016/bs.mcb.2016.04.018

INTRODUCTION

The Cas9-based CRISPR system has been widely used to generate mutations in many organisms including zebrafish (Doudna & Charpentier, 2014; Hsu, Lander, & Zhang, 2014). The fusion of the native crRNA and tracrRNA as a single guide RNA (sgRNA) has simplified this three-component system to a two-component system (Hwang et al., 2013; Jinek et al., 2012). The two components can be delivered as synthetic RNAs, expression plasmids, or sgRNA-Cas9 protein complexes. Both the sgRNA and Cas9 RNA can be easily synthesized using in vitro transcription. By injecting the two components into zebrafish embryos, a target gene is recognized by sgRNA and a double-strand break (DSB) is then created by Cas9 endonuclease (Chang et al., 2013; Hwang et al., 2013; Jao, Wente, & Chen, 2013). Repair of the DSB by error-prone nonhomologous end joining or microhomology-mediated end joining results in small indels (Doudna & Charpentier, 2014; Hsu et al., 2014). Exogenous DNA can also be integrated at the DSB through homology-dependent and homology-independent repairs (Doudna & Charpentier, 2014; Hsu et al., 2014).

Conditional gene inactivation is critical to study gene function in particular stages or tissues. It is especially necessary when conventional mutations are embryonic lethal or have defects in multiple organ systems. Conditional inactivation can elucidate the function of genes more precisely. In zebrafish, Cre and Flp approaches have been used to facilitate conditional manipulation of gene expression through integration of gene-trapping cassettes (Floss & Schnutgen, 2008; Ni et al., 2012; Schnutgen et al., 2003).

Considering the high mutagenesis efficiency by the CRISPR/Cas9 system in somatic cells (Jao et al., 2013), we developed a transgenic CRISPR/Cas9 system in zebrafish to allow for target gene mutagenesis in a conditional manner. Transgenic expression of sgRNA allows for longer expression and at later stages than injection of in vitro synthesized RNA can achieve. The wide range of tissue-specific promoters, the temporal control of heat shock induction or tetracycline/ecdysone-based methods, or a combination of different systems such as the HOTCre system can provide a broad potential for tissue and temporal restricted Cas9 expression (Halloran et al., 2000; Hesselson, Anderson, Beinat, & Stainier, 2009; Huang et al., 2005; Knopf et al., 2010; Li, Maddison, Page-McCaw, & Chen, 2014). Because the CRISPR/Cas9 system has only two components and biallelic mutation is achievable, mutant phenotypes can be determined in a single generation. Simultaneous expression of multiple sgRNAs targeting the same gene should increase the likelihood of achieving high degrees of biallelic inactivation. By crossing Cas9 and sgRNA transgenic fish, double positive transgenic fish are putative mutants and can be used for functional studies. Because a substantial number of genes in zebrafish are duplicated (Howe et al., 2013), targeting both genes at the same time can bypass functional redundancy.

1. METHODS

1.1 ASSEMBLY OF U6-BASED sgRNA TRANSGENIC CONSTRUCTS

Expression of multiple sgRNAs provides advantages for studying gene function. First, the same gene can be targeted at multiple sites to increase mutagenesis. Second, gene interaction can be studied by using sgRNAs against candidate genes in a pathway of interest. Third, targeting duplicated genes can overcome redundancy and compensation. To facilitate these outcomes we isolated U6 promoters to drive expression of individual sgRNAs in a transgenic construct. Individual U6 promoters are used for each sgRNA to minimize potential instability of expressing multiple sgRNAs in tandem. Four high efficiency U6 promoters were isolated: U6a (chromosome21), U6b (chromosome9), U6c (chromosome11), and U6d (chromosome6) (Clarke, Cummins, McColl, Ward, & Doran, 2013; Yin, Maddison, et al., 2015). They have equivalent promoter activity in transgenic fish (Yin, Maddison, et al., 2015).

1. Generation of U6-based expression vectors.
   A series of U6 promoter–based expression cassettes were developed, which contain the different U6 promoters and the sgRNA$^{(F + E)}$ scaffold (Chen et al., 2013) (Fig. 1). They are available through Addgene (Addgene plasmid # 64245, 64246, 64247, 64248, 64249). To design sgRNA targeting oligos, we recommend the CRISPRscan tool http://www.crisprscan.org/ that predicts efficient sgRNA with offtarget information (Moreno-Mateos et al., 2015). Addition of the linker sequences outlined in Fig. 2 facilitates cloning using BsmBI into either the U6-based expression vector or into the pT7-sgRNA vector to allow for in vitro transcribed sgRNA (Addgene plasmid #46759) (Jao et al., 2013).
   a. Annealing of targeting oligonucleotides.
      i. Add 2 μL of 100 μM for each oligo, 2 μL of 10× NEB Buffer 2.1, and 14 μL of distilled H_2O for a total reaction volume of 20 μL.
      ii. Incubate the mixture at 95°C for 5 min, decrease to 50°C at 0.1°C/s, incubate at 50°C for 10 min, and chill to 4°C at normal ramp speed.
   b. Ligation to the U6 vector.
      i. The choice of vector depends on the end goal of the transgenic construct. For example, for a single sgRNA, use the U6a vector. If the goal is to express four different sgRNAs, each annealed pair of oligos should be placed in a different vector: pair 1 in U6a, pair 2 in U6b, pair 3 in U6c, and pair 4 in U6d. Fig. 3 outlines the assembly of the transgenic constructs and can be used as a guide for vector choice.
      ii. Mix together 1 μL 10× NEB CutSmart buffer, 1 μL T4 DNA ligase buffer, 0.25 μL U6 plasmid (about 100 ng), 1 μL annealed oligos, 0.3 μL T4 DNA ligase, 0.3 μL BsmBI, 0.2 μL PstI, and 0.2 μL SalI, and adjust with distilled H_2O for a total of 10 μL.

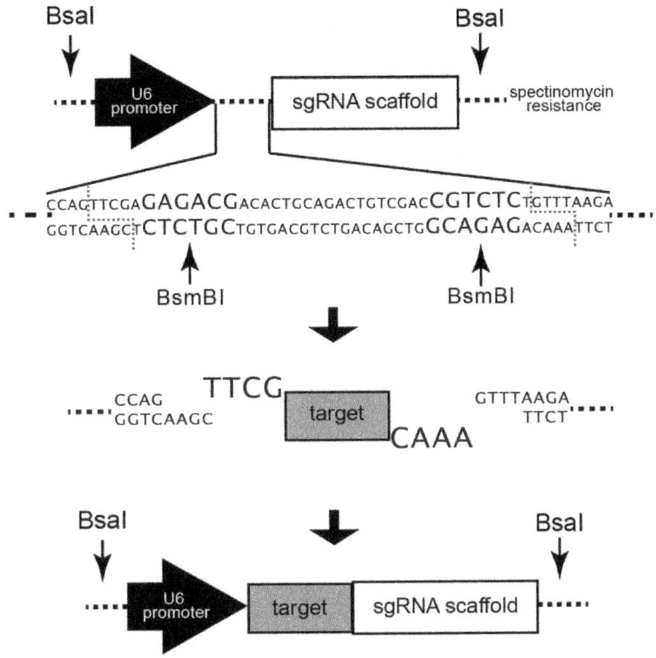

FIGURE 1 Schematics and cloning of the sgRNA expression vector.

Each vector contains a U6 promoter and an sgRNA scaffold. The inclusion of BsmBI and BsaI sites simplifies the insertion of the sgRNA target oligo and subsequent Golden Gate cloning. Digestion with BsmBI (*gray dashes*) leaves specific overhangs that will recognize the linkers on the annealed primer pair for the target.

iii. Incubate for three cycles of 37°C for 20 min and 16°C for 15 min. Follow this with 37°C for 10 min, 55°C for 15 min, and 80°C for 15 min.

iv. Transform chemically competent *Escherichia coli* such as Top10 (Thermo Fisher) with 2 μL of the ligation. Plate 10% transformants onto spectinomycin plates (50 μg/mL). Incubate the plates overnight at 37°C.
   - To increase the number of transformants, either increase the number of cycles during the digestion/ligation step or plate a larger volume of the transformation.

v. Pick single colonies and grow in Luria Broth (LB) medium with 50 μg/mL spectinomycin overnight at 37°C. Prepare plasmid DNA using standard protocols and confirm sgRNA insertion by sequencing with primer pCR8 R1.

2. Construction of the sgRNA expressing transgenes via Golden Gate cloning. To orderly assemble sgRNA cassettes with one, two, or more U6-driven sgRNAs, we developed a Golden Gate strategy (Yin, Maddison, et al., 2015). We generated a series of Tol2-based destination vectors, pGGDestTol2LC, all containing *cryaa:cerulean* (LC) (Hesselson et al., 2009) for lens-specific

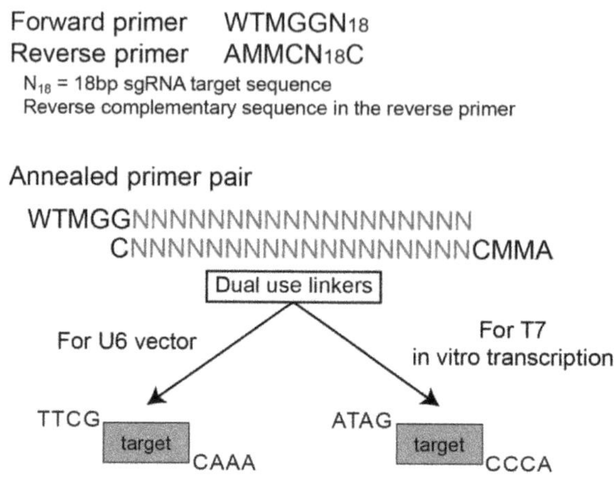

FIGURE 2 Dual use linker sequences for sgRNA targets.

The relevant linker sequences for the forward and reverse primers are indicated. After annealing of the oligos, the linkers allow cloning into either the U6 expression vectors or into the vector allowing T7 RNA polymerase based in vitro transcription.

cerulean expression as a selection marker for positive transgenesis. Each destination vector is designed to receive 1, 2, 3, 4, or 5 *U6:sgRNA* cassettes (Addgene plasmid # 64239, 64240, 64241, 64242, 64243). Golden Gate cloning is facilitated by BsaI sites in the *U6:sgRNA* vectors and for proper assembly the correct combination of plasmid vectors must be used (Fig. 3). A *U6a:sgRNA* for *tyrosinase* is available to use as the first cassette (Addgene plasmid # 64250). This allows an easily identifiable pigmentation phenotype as an indication of mutagenesis (Jao et al., 2013; Yin, Maddison, et al., 2015).

a. Golden gate assembly.
   i. Choose the appropriate destination vector for the number of U6:sgRNA cassettes to be assembled.
   ii. Mix 50 ng of the pGGDestTol2LC vector and 100 ng of each pU6$_x$-sgRNA vector with 2 μL 10× NEB CutSmart buffer, 2 μL T4 DNA ligase buffer, 1 μL BsaI, and 1 μL T4 DNA ligase, and adjust with distilled H_2O for a total volume of 20 μL.
   iii. Incubate for three cycles of 37°C for 20 min and 16°C for 15 min. Follow this with 80°C for 15 min, and cool to room temperature.
   iv. Use 10 μL of the ligation for the transformation and plate 50% of the transformants on ampicillin (100 μg/mL) plates. Pick single clones and grow in LB medium with 100 μg/mL ampicillin. Prepare plasmid DNA using standard protocols.
   v. Verify the multiplexed sgRNA vectors by PCR or sequencing. Each sgRNA element can be verified using a U6 forward primer (Table 1) and the corresponding sgRNA reverse primer (AMMCN_{18}C).

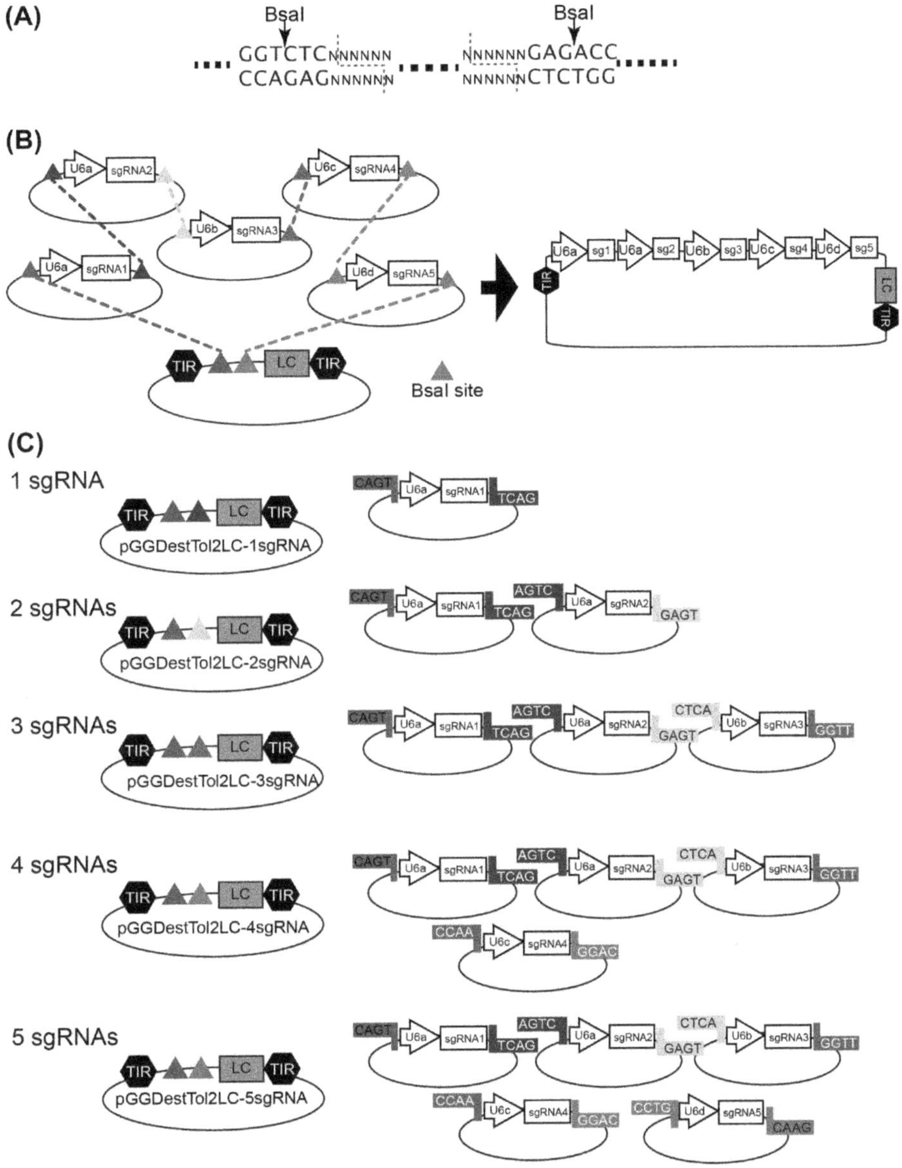

FIGURE 3 Golden Gate cloning of U6-based expression transgene.

(A) Golden Gate cloning is facilitated by the use of BsaI where the overhang following digestion is specifically designed for each component. (B) Example showing the progressive pairing of components in a five sgRNA vectors. Each BsaI site (*triangle*) is designed so that ligation occurs in a specific order (*dashed line*) with the destination vector containing the Tol2 repeats (TIR) and the cryaa:cerulean (LC) marker. (C) The choice of vectors is dependent on the number of sgRNAs to be expressed. The overhang sequence produced by BsaI digestion is specific for each vector and is indicated and color coded for visual simplicity. For the successful production of the sgRNA expression transgenic construct, the correct combination of destination vector and U6-based vector(s) needs to be used. (See color plate)

Table 1 Primers for Verifying Multiplex sgRNA Vectors

Primer Name	Primer Sequence 5′–3′
U6aF	TTTCTCCAGCCTCGGTCATT
U6bF	CTCATTACCCTCCACGTGTCTGTC
U6cF	CCAATCCGAGAGTCTGTGAATGTT
U6dF	CCTGTGATTTGGTGGTTGTGAAAG

b. The confirmed plasmids can then be injected into one-cell stage embryos with Tol2 transposase RNA using standard methods. Embryos that exhibit lens cerulean expression can be selected and raised to maturity.

1.2 CONSTRUCTION OF Cas9 EXPRESSION VECTORS

To achieve conditional control of CRISPR mutagenesis, the expression of at least one component of the CRISPR system needs to be spatially and/or temporally regulated. Although Pol III promoter-driven sgRNA expression can be made to be dependent of Cre activity or tetracycline as has been done for shRNA expression (Tiscornia, Tergaonkar, Galimi, & Verma, 2004; van de Wetering et al., 2003), this will increase the complexity of its implementation since additional components are necessary to achieve the regulation. In contrast, Pol II promoter-driven Cas9 expression can be easily regulated using various tissue-specific promoters and inducible promoters.

1. The Tol2-based multisite Gateway system is used to prepare the conditional Cas9-expression vector (Kwan et al., 2007). A codon-optimized version of Cas9 (Jao et al., 2013) was cloned into a middle-entry vector to generate a universal pME-Cas9 (Addgene #64237). A destination vector containing a fluorescent marker for simplified identification of transgenic carriers is recommended for simple identification of transgenesis, as long as it does not obscure the lens-cerulean expression used for the U6-sgRNA constructs. In combination with a 5′ entry vector containing the promoter of interest, the Cas9 middle-entry vector, and the 3′ entry vector containing a poly A using standard multisite Gateway reactions, the transgenic construct can be easily generated.
2. Once constructed, the Cas9 expression vector can then be injected into one-cell stage embryos with Tol2 transposase RNA using standard methods. Embryos that exhibit expression of the marker, if used, can be selected and raised to maturity.

1.3 SCREENING AND EVALUATION OF STABLE sgRNA or Cas9 TRANSGENIC FISH

1. Evaluation of sgRNA transgenic lines.
   We have found that multiple transgenic lines need to be evaluated to produce those that have the most robust expression of the sgRNA(s). Two rounds of

embryo production are needed to determine the optimal transgenic line (Fig. 4). In the first round, founders are crossed to wild-type fish and germline integration of the transgene evaluated by marker gene expression in F1 embryos. In the second round, the founders with germline integration are crossed to a stable transgenic line with ubiquitous expression of Cas9, and the degree of mutagenesis of the target gene is evaluated in the F1 embryos. This saves both time and resources in that only the optimal sgRNA transgenic lines are raised. However, the second phase of evaluation can be done in subsequent generations where the positive F1 carriers are raised to maturity and then crossed to the ubiquitously expressed Cas9 transgenic line. This increases the number of embryos that carry both transgenes but will take additional time and resources to raise the F1 fish.

a. Cross individual injected founders with wild-type fish. Collect the embryos from successful matings and hold the founder fish in a separate small tank.
   i. At least 30 embryos are needed for screening. If fewer embryos have been produced, return the injected fish to the unscreened tank for additional mating.
   ii. At 3–5 dpf, evaluate lens cerulean expression using a fluorescence microscope with a CFP filter.
      - Positive F1 embryos can be raised to maturity for additional evaluation if desired.

b. Cross the founder fish that exhibit germline integration to a stable transgenic line with ubiquitous Cas9 expression.
   i. We have developed two transgenic lines *Tg(ubi:cas9;CG)* and *Tg(actb2:cas9:LR)* that have been fully characterized and are efficient in producing mutagenesis in conjunction with transgenic sgRNA expression (Yin, Maddison, et al., 2015) (Fig. 5).
   ii. At 3–5 dpf, select embryos that have both the lens cerulean expression and the marker gene for the Cas9 transgenic line.
   iii. If the sgRNA for *tyrosinase* was included in the transgenic construct, the degree of pigmentation in the double transgenic embryos can be easily evaluated (Fig. 5). We have found that this is a good predictor of the efficiency of the other sgRNAs within the construct (Yin, Maddison, et al., 2015).
   iv. If the sgRNA for *tyrosinase* was not included in the construct, the efficiency for one or multiple sgRNAs can be evaluated by other methods. We routinely use a heteroduplex mobility shift assay (HMA) although other methods are available including sequencing and PCR-based approaches (Yin, Jao, & Chen, 2015; Yu, Zhang, Yao, & Wei, 2014). The procedure for the HMA evaluation will be detailed here.
      - Isolate genomic DNA by placing double transgenic embryos with one embryo per well in PCR tubes. Include at least one embryo that is either nontransgenic or is single transgenic as a negative control. Incubate on ice to euthanize the embryos, then remove all water, and

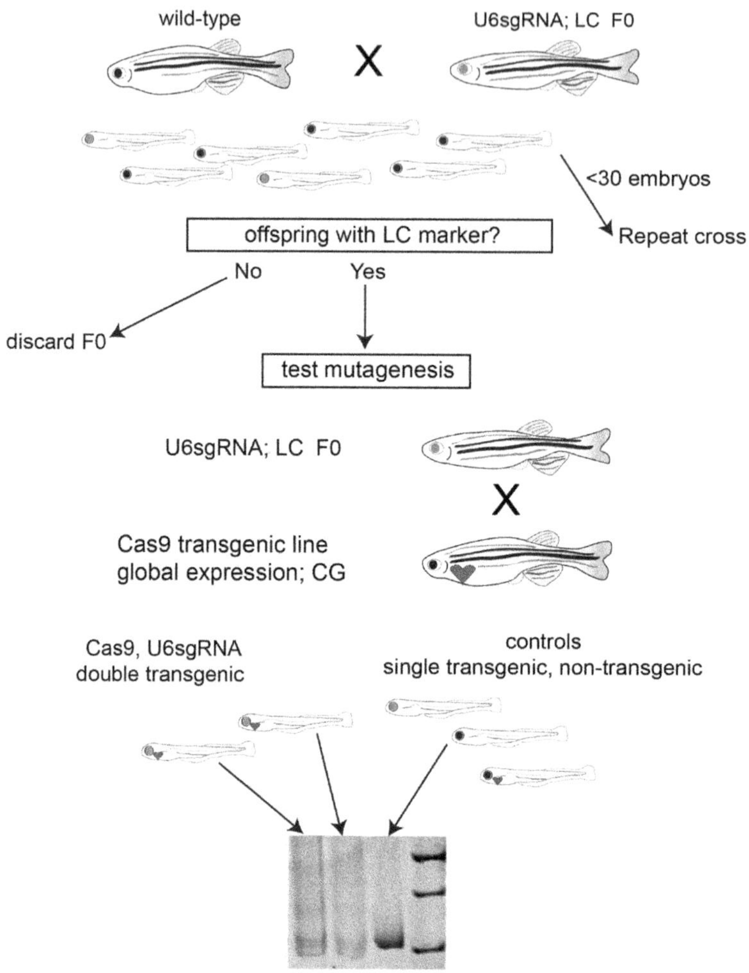

FIGURE 4 Evaluation of efficiency of U6-based sgRNA transgenic lines.

Two rounds of embryo production are needed to evaluate each transgenic line. In the first round, germline transmission is evaluated by crossing the F0 fish to a wild-type fish. In case there is a low germline transmission rate, more than 30 embryos should be collected from each mating. If there are embryos that have cerulean expression in the lens, the founder can be evaluated for efficiency of mutagenesis. In this second round of embryo production, the U6sgRNA F0 fish is crossed to a transgenic line with a high level of Cas9 expression throughout the fish. These Cas9 fish also contain a fluorescent marker such as heart GFP expression. Embryos that have both the lens cerulean expression and the heart GFP expression are used to determine mutagenesis of the gene target(s) using assays such as the heteroduplex mobility shift assay. Mutagenesis is indicated by a reduction or shifting of the PCR product, compared to single or nontransgenic siblings.

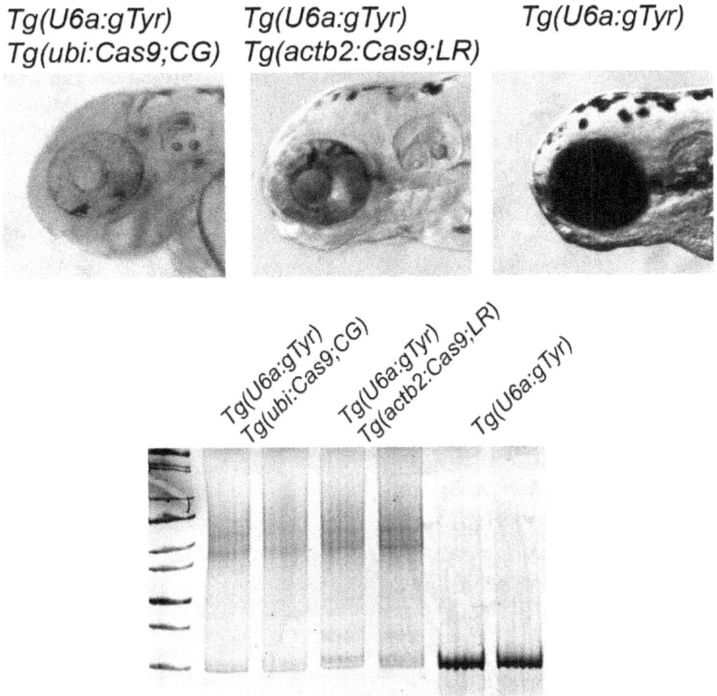

FIGURE 5 Global expression of Cas9 and *tyrosinase* mutagenesis.

Tg(ubi:cas9;CG) and *Tg(actb2:cas9;LR)* are two fully characterized transgenic lines with global expression of Cas9. The *Tg(U6a:gTyr)* has expression of an sgRNA against *tyrosinase*. In combination with either Cas9 transgenic line, a defect in pigmentation can be easily observed in double transgenic embryos but not in single transgenic embryos. Analysis of the *tyrosinase* gene also indicates efficient mutagenesis.

add 30 μL of 20 mM NaOH. Incubate samples at 95°C for 20 min and then cool to 4°C. Add 6 μL of 1 M Tris-HCl (not pH adjusted) to neutralize the samples. Add 164 μL of 10 mM Tris-Cl, pH 8.5 to make the final volume to 200 μL.

- Amplify the target region by PCR from 1 μL of the genomic DNA solution using the primers that flank the predicted cleavage site in the genome. Amplicons that are between 150 and 400 bp are desired for this approach. Run 2 μL of the PCR product on a 0.8% agarose gel to be sure that a single species of expected size is amplified.
- Add 10× stop solution to the PCR reaction for final concentration of 10 mM EDTA and 0.1% SDS.

  Optional: The PCR product can be column purified or ethanol precipitated. If the product is column purified, elute in 50 μL of

10 mM Tris-Cl, pH 8.5. If ethanol precipitated, resuspend the pellet in 50 μL of 10 mM Tris-Cl, pH 8.5. Then mix 200 ng of the purified product, 2 μL of 10× NEB buffer 2.1, and nuclease-free water to a total volume of 20 μL.

- Melt and reanneal the product by incubation at 95°C for 5 min, decrease to 85°C at −2.0°C/s, decrease to 25°C at −0.1°C/s, and hold at 16°C until use.
- Run the products on a 10% polyacrylamide (29:1) TBE gel. If using purified products, load 10 μL of the reaction. If using the PCR reaction directly, load 5–10 μL of the reaction. Run at 120 V for 2–3 h depending on the amplicon size.
- Stain the gel in 1× TBE buffer containing 0.5 μg/mL of ethidium bromide 5 min before imaging.
- The presence of slow-migrating bands is indicative of DNA heteroduplexes (Fig. 5).

  Caution: Presence of multiple bands in a known wild-type sample can indicate polymorphisms present in the amplicon. If present, primers may need to be redesigned to limit the inclusion of these regions.

2. Evaluation of Cas9 transgenic lines.

As with the sgRNA transgenic lines, multiple lines need to be evaluated before choosing the one that drives the most robust mutagenesis. Again, two rounds of embryo production are needed to determine the most useful transgenic line. In the first round, founders are crossed to wild-type fish and germline integration of the transgene evaluated by marker gene expression in F1 embryos. In the second round of evaluation, the founders with germline integration are crossed to a characterized, stable transgenic line with efficient sgRNA expression, such as *tyrosinase*, and the degree of mutagenesis of the target gene evaluated in the F1 embryos.

a. Cross individual injected founders with wild-type fish. Collect the embryos from successful matings and hold the founder fish in a separate small tank.

   i. At least 30 embryos are needed for screening. If fewer embryos have been produced, return the injected fish to the unscreened tank for additional mating.

   ii. At 3–5 dpf, marker expression can be evaluated using a fluorescence microscope with the appropriate filter. Positive F1 embryos can be raised to maturity for additional evaluation if desired.

b. Cross the founder fish with germline integration to a stable transgenic line with efficient expression of sgRNA such as *tyrosinase*.

   i. Degree of mutagenesis can be evaluated using the HMA method presented earlier as long as the population of mutated DNA is sufficiently large.

   ii. If the expression pattern of the Cas9 is limited to a specific tissue, it may be more useful to use a double transgenic line with GFP expression in the tissue of interest and an sgRNA against GFP. Efficiency of the Cas9 transgenic line being tested can be evaluated by examining the degree of EGFP fluorescence. Reduced or absent EGFP would be an indication of functional Cas9 expression. Alternatively, a reporter line similar to the traffic light reporter (Chu et al., 2015; Kuhar et al., 2014), in which expression of a fluorescent protein is activated by Cas9 activity, may be generated and used for evaluating the activity and tissue specificity of Cas9.

2. DISCUSSION

Conditional alleles have been instrumental for functional analysis in mice and will likely be so in zebrafish. Previously we have generated conditional alleles using gene-trap mutagenesis (Maddison, Li, & Chen, 2014; Maddison, Lu, & Chen, 2011; Ni et al., 2012). However, this approach relies on random integration of a conditional gene-trap cassette, and its broad application requires the generation of a large collection of such alleles. In contrast, the transgenic CRISPR approach described in this chapter allows targeted conditional inactivation. Further, it allows simultaneous inactivation of multiple genes, overcoming functional redundancy or compensation of duplicated genes, and facilitating gene–gene interaction studies.

Successful implementation of this approach of conditional mutagenesis requires efficient sgRNAs and robust tissue-specific expression of Cas9. Although a number of studies have identified features of active sgRNA and have incorporated these features into algorithms for identification of active and specific sgRNAs (Chari, Mali, Moosburner, & Church, 2015; Doench et al., 2014; Gagnon et al., 2014; Moreno-Mateos et al., 2015; Varshney et al., 2015; Wang, Wei, Sabatini, & Lander, 2014; Wong, Liu, & Wang, 2015), these designing tools cannot substitute for empirical testing. We recommend testing selected sgRNA using RNA injection into zygotes and using multiple validated sgRNAs for each target gene. Identification of robust and tissue-specific Cas9 drivers is also critical for the success of this approach. In this regard, a reporter line for evaluating Cas9 function in a tissue restricted manner is lacking. However, Cas9 expression may be determined by in situ hybridization and/or immunofluorescence.

A potential concern of transgenic CRISPR mutagenesis is off-target effects. Careful selection of specific sgRNA should largely mitigate this concern. However, long-term coexpression of Cas9 and sgRNA could exacerbate the off-target effect. In this regard, control of Cas9 expression using the HOTCre approach is advantageous (Hesselson et al., 2009; Yin, Maddisson, et al., 2015), although its implementation requires one additional transgene that confers tissue-specific Cre expression. Alternatively, replacing Cas9 with one of the developed split-Cas9 systems may also achieve temporal control of Cas9 activity (Davis, Pattanayak, Thompson, Zuris, & Liu, 2015; Nihongaki, Kawano, Nakajima, & Sato, 2015; Zetsche, Volz, & Zhang, 2015).

SUMMARY

We have presented here an approach to generate conditional mutations in zebrafish. This CRISPR-based approach requires a transgenic line expressing sgRNA targeting the gene of interest and a transgenic line expressing Cas9 in the desired spatial/temporal pattern. Crossing the two transgenic lines allows CRISPR mutagenesis of the target gene in the desired cell type at the desired time. In additional to zebrafish, this approach should also be applicable to other genetically amenable organisms.

ACKNOWLEDGMENTS

We thank members in the Chen laboratory for discussions. The work is supported a grant from National Institute Diabetes and Digestive and Kidney Diseases at NIH (DK088686) and American Diabetes Association (1-13-BS-027).

REFERENCES

Chang, N., Sun, C., Gao, L., Zhu, D., Xu, X., Zhu, X., … Xi, J. J. (2013). Genome editing with RNA-guided Cas9 nuclease in zebrafish embryos. *Cell Research, 23*, 465–472.

Chari, R., Mali, P., Moosburner, M., & Church, G. M. (2015). Unraveling CRISPR-Cas9 genome engineering parameters via a library-on-library approach. *Nature Methods, 12*, 823–826.

Chen, B., Gilbert, L. A., Cimini, B. A., Schnitzbauer, J., Zhang, W., Li, G. W., … Huang, B. (2013). Dynamic imaging of genomic loci in living human cells by an optimized CRISPR/Cas system. *Cell, 155*, 1479–1491.

Chu, V. T., Weber, T., Wefers, B., Wurst, W., Sander, S., Rajewsky, K., & Kuhn, R. (2015). Increasing the efficiency of homology-directed repair for CRISPR-Cas9-induced precise gene editing in mammalian cells. *Nature Biotechnology, 33*, 543–548.

Clarke, B. D., Cummins, D. M., McColl, K. A., Ward, A. C., & Doran, T. J. (2013). Characterization of zebrafish polymerase III promoters for the expression of short-hairpin RNA interference molecules. *Zebrafish, 10*, 472–479.

Davis, K. M., Pattanayak, V., Thompson, D. B., Zuris, J. A., & Liu, D. R. (2015). Small molecule-triggered Cas9 protein with improved genome-editing specificity. *Nature Chemical Biology, 11*, 316–318.

Doench, J. G., Hartenian, E., Graham, D. B., Tothova, Z., Hegde, M., Smith, I., … Root, D. E. (2014). Rational design of highly active sgRNAs for CRISPR-Cas9-mediated gene inactivation. *Nature Biotechnology, 32*, 1262–1267.

Doudna, J. A., & Charpentier, E. (2014). Genome editing. The new frontier of genome engineering with CRISPR-Cas9. *Science, 346*, 1258096.

Floss, T., & Schnutgen, F. (2008). Conditional gene trapping using the FLEx system. *Methods in Molecular Biology, 435*, 127–138.

Gagnon, J. A., Valen, E., Thyme, S. B., Huang, P., Akhmetova, L., Pauli, A., … Schier, A. F. (2014). Efficient mutagenesis by Cas9 protein-mediated oligonucleotide insertion and large-scale assessment of single-guide RNAs. *PLoS One, 9*, e98186.

Halloran, M. C., Sato-Maeda, M., Warren, J. T., Su, F., Lele, Z., Krone, P. H., … Shoji, W. (2000). Laser-induced gene expression in specific cells of transgenic zebrafish. *Development, 127*, 1953–1960.

Hesselson, D., Anderson, R. M., Beinat, M., & Stainier, D. Y. (2009). Distinct populations of quiescent and proliferative pancreatic beta-cells identified by HOTcre mediated labeling. *Proceedings of the National Academy of Sciences of the United States of America, 106*, 14896–14901.

Howe, K., Clark, M. D., Torroja, C. F., Torrance, J., Berthelot, C., Muffato, M., … Stemple, D. L. (2013). The zebrafish reference genome sequence and its relationship to the human genome. *Nature, 496*, 498–503.

Hsu, P. D., Lander, E. S., & Zhang, F. (2014). Development and applications of CRISPR-Cas9 for genome engineering. *Cell, 157*, 1262–1278.

Huang, C. J., Jou, T. S., Ho, Y. L., Lee, W. H., Jeng, Y. T., Hsieh, F. J., & Tsai, H. J. (2005). Conditional expression of a myocardium-specific transgene in zebrafish transgenic lines. *Developmental Dynamics: An Official Publication of the American Association of Anatomists, 233*, 1294–1303.

Hwang, W. Y., Fu, Y., Reyon, D., Maeder, M. L., Tsai, S. Q., Sander, J. D., … Joung, J. K. (2013). Efficient genome editing in zebrafish using a CRISPR-Cas system. *Nature Biotechnology, 31*, 227–229.

Jao, L. E., Wente, S. R., & Chen, W. (2013). Efficient multiplex biallelic zebrafish genome editing using a CRISPR nuclease system. *Proceedings of the National Academy of Sciences of the United States of America, 110*, 13904–13909.

Jinek, M., Chylinski, K., Fonfara, I., Hauer, M., Doudna, J. A., & Charpentier, E. (2012). A programmable dual-RNA-guided DNA endonuclease in adaptive bacterial immunity. *Science, 337*, 816–821.

Knopf, F., Schnabel, K., Haase, C., Pfeifer, K., Anastassiadis, K., & Weidinger, G. (2010). Dually inducible TetON systems for tissue-specific conditional gene expression in zebrafish. *Proceedings of the National Academy of Sciences of the United States of America, 107*, 19933–19938.

Kuhar, R., Gwiazda, K. S., Humbert, O., Mandt, T., Pangallo, J., Brault, M., … Certo, M. T. (2014). Novel fluorescent genome editing reporters for monitoring DNA repair pathway utilization at endonuclease-induced breaks. *Nucleic Acids Research, 42*, e4.

Kwan, K. M., Fujimoto, E., Grabher, C., Mangum, B. D., Hardy, M. E., Campbell, D. S., … Chien, C. B. (2007). The Tol2kit: a multisite gateway-based construction kit for Tol2 transposon transgenesis constructs. *Developmental Dynamics: An Official Publication of the American Association of Anatomists, 236*, 3088–3099.

Li, M., Maddison, L. A., Page-McCaw, P., & Chen, W. (2014). Overnutrition induces beta-cell differentiation through prolonged activation of beta-cells in zebrafish larvae. *American Journal of Physiology Endocrinology and Metabolism, 306*, E799–E807.

Maddison, L. A., Li, M., & Chen, W. (2014). Conditional gene-trap mutagenesis in zebrafish. *Methods in Molecular Biology, 1101*, 393–411.

Maddison, L. A., Lu, J., & Chen, W. (2011). Generating conditional mutations in zebrafish using gene-trap mutagenesis. *Methods in Cell Biology, 104*, 1–22.

Moreno-Mateos, M. A., Vejnar, C. E., Beaudoin, J. D., Fernandez, J. P., Mis, E. K., Khokha, M. K., & Giraldez, A. J. (2015). CRISPRscan: designing highly efficient sgRNAs for CRISPR-Cas9 targeting in vivo. *Nature Methods, 12*, 982–988.

Ni, T. T., Lu, J., Zhu, M., Maddison, L. A., Boyd, K. L., Huskey, L., … Chen, W. (2012). Conditional control of gene function by an invertible gene trap in zebrafish. *Proceedings of the National Academy of Sciences of the United States of America, 109*, 15389–15394.

Nihongaki, Y., Kawano, F., Nakajima, T., & Sato, M. (2015). Photoactivatable CRISPR-Cas9 for optogenetic genome editing. *Nature Biotechnology, 33*, 755–760.

Schnutgen, F., Doerflinger, N., Calleja, C., Wendling, O., Chambon, P., & Ghyselinck, N. B. (2003). A directional strategy for monitoring Cre-mediated recombination at the cellular level in the mouse. *Nature Biotechnology, 21*, 562–565.

Tiscornia, G., Tergaonkar, V., Galimi, F., & Verma, I. M. (2004). CRE recombinase-inducible RNA interference mediated by lentiviral vectors. *Proceedings of the National Academy of Sciences of the United States of America, 101*, 7347–7351.

Varshney, G. K., Pei, W., LaFave, M. C., Idol, J., Xu, L., Gallardo, V., ... Burgess, S. M. (2015). High-throughput gene targeting and phenotyping in zebrafish using CRISPR/Cas9. *Genome Research, 25*, 1030–1042.

van de Wetering, M., Oving, I., Muncan, V., Pon Fong, M. T., Brantjes, H., van Leenen, D., ... Clevers, H. (2003). Specific inhibition of gene expression using a stably integrated, inducible small-interfering-RNA vector. *EMBO Reports, 4*, 609–615.

Wang, T., Wei, J. J., Sabatini, D. M., & Lander, E. S. (2014). Genetic screens in human cells using the CRISPR-Cas9 system. *Science, 343*, 80–84.

Wong, N., Liu, W., & Wang, X. (2015). WU-CRISPR: characteristics of functional guide RNAs for the CRISPR/Cas9 system. *Genome Biology, 16*, 218.

Yin, L., Jao, L. E., & Chen, W. (2015). Generation of targeted mutations in zebrafish using the CRISPR/Cas system. *Methods in Molecular Biology, 1332*, 205–217.

Yin, L., Maddison, L. A., Li, M., Kara, N., LaFave, M. C., Varshney, G. K., ... Chen, W. (2015). Multiplex conditional mutagenesis using transgenic expression of Cas9 and sgRNAs. *Genetics, 200*, 431–441.

Yu, C., Zhang, Y., Yao, S., & Wei, Y. (2014). A PCR based protocol for detecting indel mutations induced by TALENs and CRISPR/Cas9 in zebrafish. *PLoS One, 9*, e98282.

Zetsche, B., Volz, S. E., & Zhang, F. (2015). A split-Cas9 architecture for inducible genome editing and transcription modulation. *Nature Biotechnology, 33*, 139–142.

CHAPTER

Tol2-mediated transgenesis, gene trapping, enhancer trapping, and Gal4-UAS system

2

K. Kawakami[1], K. Asakawa, A. Muto, H. Wada

National Institute of Genetics, Mishima, Shizuoka, Japan

[1]*Corresponding author: E-mail: kokawaka@nig.ac.jp*

CHAPTER OUTLINE

Methods in Cell Biology, Volume 135, ISSN 0091-679X, http://dx.doi.org/10.1016/bs.mcb.2016.01.011

Abstract

The *Tol2* element is an active transposon that was found from the genome of the Japanese medaka fish. Since the *Tol2* transposition system is active in all vertebrate cells tested so far, it has been applied to germ line transgenesis in various model animals including fish, frog, chicken, and mouse, and to gene transfer in culture cells. In zebrafish, the *Tol2* system consists of the transposase mRNA and a *Tol2* transposon-donor plasmid, and is introduced into fertilized eggs by microinjection. Thus genomic integrations of the *Tol2* construct are generated in the germ lineage and transmitted to the offspring very efficiently. By using the *Tol2* transposition system, we have developed important genetic methods, such as transgenesis, gene trapping, enhancer trapping, and the Gal4-UAS system in zebrafish and applied to many aspects of biological studies. In this chapter, we describe how these methods are performed.

INTRODUCTION

The *Tol2* transposable element was identified from the genome of the Japanese medaka fish, and its DNA sequence is similar to those of transposons of the hAT family (Koga, Suzuki, Inagaki, Bessho, & Hori, 1996). It was shown that the *Tol2* element carries a gene encoding a fully functional transposase (Kawakami, Koga, Hori, & Shima, 1998; Kawakami & Shima, 1999) (Fig. 1A). Thus the *Tol2* element is the first active transposon identified from a vertebrate genome. The *Tol2* element also contains DNA sequences that are recognized by the transposase. The minimal *cis*-sequences essential for transposition were analyzed, and it was shown that 200-bp from the left end and 150-bp DNA from the right end of the *Tol2* element are necessary and sufficient (Fig. 1A) (Urasaki, Morvan, & Kawakami, 2006). Any DNA fragment can be cloned between these *cis*-sequences.

The *Tol2* transposition system consists of two components, a transposon-donor plasmid carrying a *Tol2* construct and the transposase activity supplied as a form of mRNA or an expression plasmid. It has been shown that the *Tol2* system is active in all vertebrate cells tested so far (Kawakami, 2007). In zebrafish, a transposon-donor plasmid and mRNA synthesized in vitro by using the transposase cDNA as a template are injected into fertilized eggs. The *Tol2* construct is excised from the donor plasmid and integrated into the genome of the germ lineage during embryonic development, and the transposon insertions are transmitted to the next generation very efficiently (Kawakami, Shima, & Kawakami, 2000; Kawakami et al., 2004) (Fig. 2B). Thus the *Tol2* system has served as an essential tool for transgenesis in zebrafish. Furthermore, important genetic methods, such as gene trapping, enhancer trapping, and the Gal4-UAS system (Asakawa et al., 2008; Davison et al., 2007; Kawakami et al., 2004; Nagayoshi et al., 2008; Parinov, Kondrichin, Korzh, & Emelyanov, 2004; Scott et al., 2007) were developed by using the *Tol2* system. Transgenic zebrafish generated by these methods that express GFP or Gal4 in spatially and temporally restricted fashions have been powerful tools for the study of developmental biology, organogenesis, and neuroscience.

1. TRANSGENESIS BY USING THE *Tol2* TRANSPOSABLE ELEMENT

1.1 RATIONALE

A *Tol2*-donor plasmid DNA and the transposase mRNA (Fig. 1A) are introduced into zebrafish fertilized eggs by microinjection. In the injected embryos, the transposase protein is synthesized and catalyzes excision of the *Tol2* construct from the donor plasmid. The excised *Tol2* construct is integrated into the genome during embryonic development, in the future germ cells. Thus transgenic fish will be obtained in the progeny from the injected fish (Fig. 2B).

1.2 METHODS

1.2.1 Synthesis of transposase mRNA in vitro

1. Linearize pCS-zT2TP (Fig. 1A) by digestion with *Not*I and synthesize mRNA using mMESSAGE mMACHINE SP6 Kit (Ambion Inc.).
2. Purify the transposase mRNA by using "Quick Spin Columns for radiolabeled RNA purification" (Roche), then precipitate the mRNA and resuspend it in nuclease-free water at 250 ng/μL.
3. Analyze the product by gel electrophoresis. For electrophoresis of RNA, a denaturing gel is preferable, but, alternatively a standard agarose/TAE gel can be used (Fig. 2A).

1.2.2 Preparation of a Tol2*-donor plasmid*

1. Clone the desired DNA fragment into an appropriate *Tol2* vector, for instance, either by using the *Xho*I and *Bgl*II sites on T2AL200R150G (Urasaki et al., 2006) (Fig. 1A) or by using the *Tol2* vectors with the Gateway system (Kwan et al., 2007; Villefranc, Amigo, & Lawson, 2007).

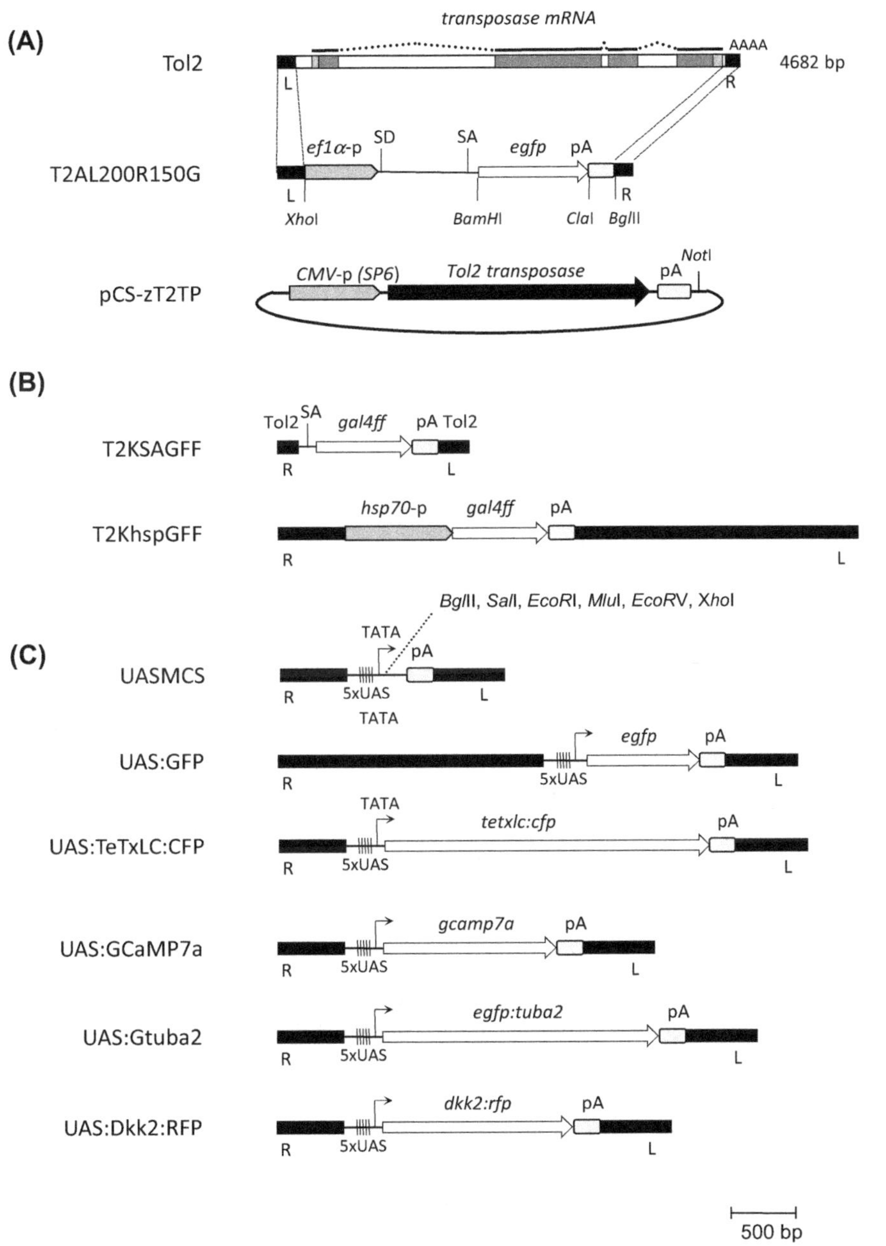

FIGURE 1

The structures of *Tol2* vectors used in zebrafish. (A) The full-length *Tol2* element, the minimal *Tol2* vector T2AL200R150G, and pCS-zT2TP. *Tol2* is 4682 bp in length and encodes mRNA for the transposase (*dotted lines* indicate introns). T2AL200R150G contains 200-bp and 150-bp DNA from the left (L) and right (R) terminals of *Tol2*, the *Xenopus* EF1α promoter (ef1α-p), the rabbit-β-globin intron (from SD to SA), the egfp gene, and the SV40

2. Prepare the transposon-donor plasmid DNA using QIAfilter Plasmid Maxi Kit (QIAGEN), purify the recombinant plasmid once by phenol/chloroform extraction, precipitate it with ethanol, and suspend the plasmid in nuclease-free water at 250 ng/μL.

1.2.3 Microinjection

1. Set up male and female adult zebrafish in a mating box in the evening and collect fertilized eggs in the next morning (Fig. 3A). Microinjection should be carried out at the one-cell stage within 30 min postfertilization.
2. Make an injection ramp by using 1% agarose, a glass plate, and a 6-cm plastic dish (Fig. 3B). Create fine needles for microinjection by using a glass capillary (GC-1, Narishige, Japan) and a puller (PC-10, Narishige, Japan). Cut the tip with a surgical blade (No. 11, Akiyama MEDICAL MFG. CO., Japan).
3. Prepare DNA/RNA solution by mixing the following components; 10 μL of 0.4 M KCl, 2 μL of phenol red solution (Sigma−Aldrich), 2 μL of 250 ng/μL transposase mRNA, 2 μL of 250 ng/μL *Tol2*-donor plasmid DNA, and 4 μL of nuclease-free water (final volume 20 μL). Before injection, centrifuge the mixture at the maximum speed for 1 min to precipitate and remove debris that may clog the injection needle. Transfer the upper 18 μL to a new tube.
4. Fill the DNA/RNA solution into the glass capillary from the backside by using a Microloader tip (Eppendorf, Germany). Attach the filled capillary to a holder (No. 11520145, Leica, Germany) and connect the holder to a 10-mL syringe via a Teflon tube (inner diameter: 0.56 mm, Chukoh Chemical Industries, Japan) (Fig. 3C).
5. Inject ~1 nL of the DNA/RNA solution (the approximate volume can be measured by observing the diameter of the injected bolus by eye) into the cytoplasm of fertilized eggs (Fig. 3D). Incubate the injected embryos in a plastic dish at 28°C.

1.2.4 Excision assay

1. To confirm the transposition reaction, the excision assay should be performed in a subsample of embryos (Kawakami & Shima, 1999) (Fig. 2A and B).

polyA signal (pA). pCS-zT2TP contains the codon-optimized transposase coding sequence downstream of the CMV and SP6 promoters. Unique restriction enzyme sites are indicated. (B) The gene and enhancer trap constructs. T2KSAGFF contains the splice acceptor (SA) and the *gal4ff* gene. T2KhspGFF contains the *hsp70* promoter and the *gal4ff* gene. (C) The UAS-reporter and effector constructs. UASMCS is a cloning vector that contains five tandem repeats of the Gal4-target sequence (5xUAS), followed by a minimal TATA sequence, the multicloning site, and polyA. UAS:GFP, UAS:TeTxLC:CFP, UAS:GCaMP7a, UAS:Gtuba2, and UAS:Dkk2:RFP contain *egfp*, the tetanus toxin light chain gene fused to CFP, the calcium indicator *gcamp7a* gene, the α-tubulin gene fused to *egfp*, and the *dickkopf2* gene fused to *rfp* downstream of UAS, respectively. Note that UAS:GFP was created by using an old version of *Tol2* vectors with longer arms.

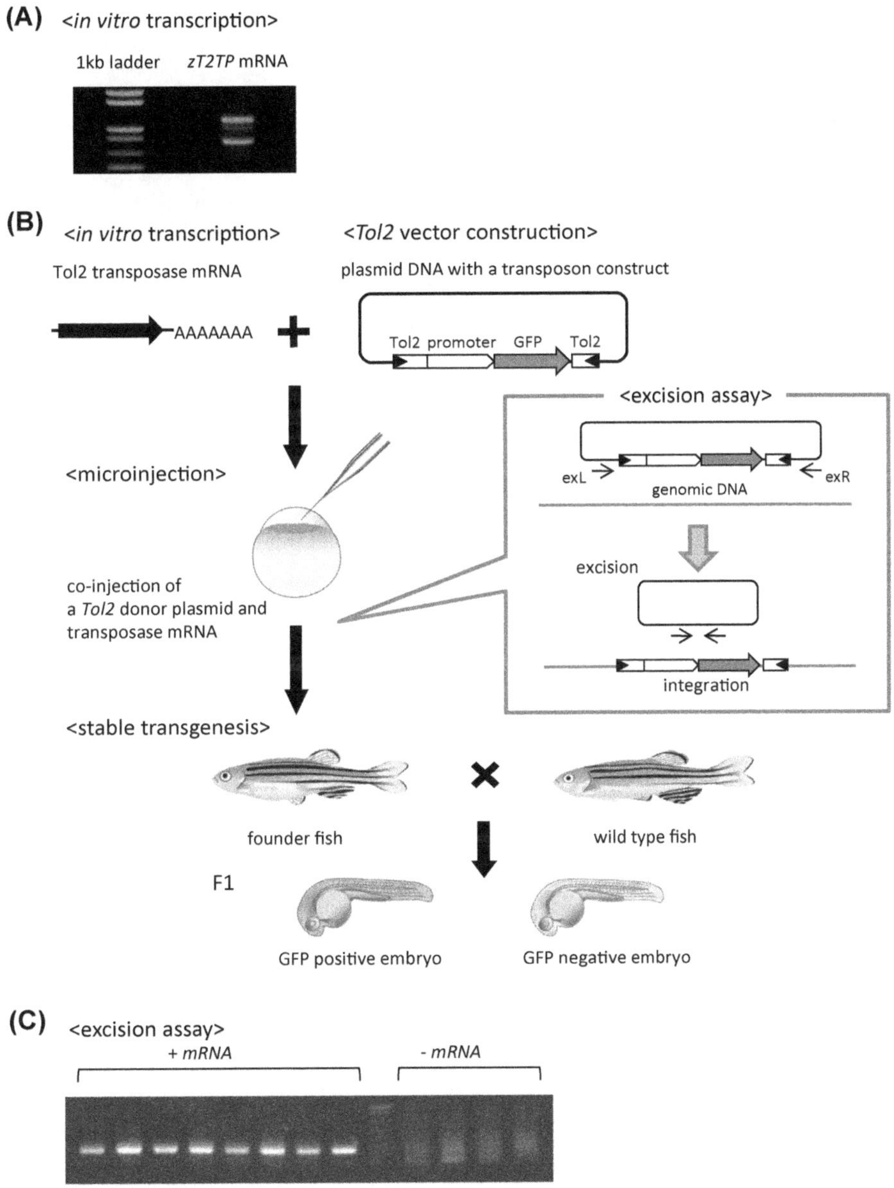

FIGURE 2

Transgenesis by using the *Tol2* transposable element. (A) Electrophoresis of the transposase mRNA synthesized in vitro on a standard agarose/TAE gel. Two bands are detected presumably due to its higher-order structure. (B) A scheme for transgenesis in zebrafish. The transposase mRNA synthesized in vitro and a *Tol2*-donor plasmid DNA are coinjected into fertilized eggs. The transposase protein produced in the injected embryos catalyzes excision

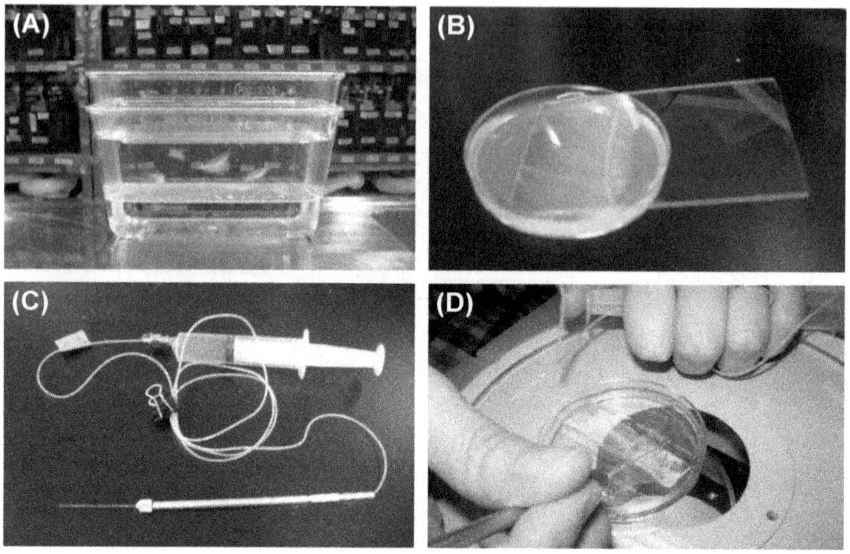

FIGURE 3

Apparatus and tools for microinjection. (A) A zebrafish mating box (Aquaschwarz, Germany). Male/female pairs of adult zebrafish are placed in a plastic tank with a sieve insert. (B) An agarose ramp for microinjection. Melted 1% agarose is poured in a 60-mm petri dish, and then a glass plate is placed to create a slant. (C) Microinjection apparatus. A glass capillary is attached to a holder and connected to a syringe through a Teflon tube. The DNA/RNA mixture is backloaded into the capillary prior to attachment to the holder. (D) Microinjection is performed under a stereoscope. The left hand provides air pressure to the capillary.

2. About 10 h after microinjection, transfer several embryos one by one into 0.2-mL strip tubes (eight tubes per strip). Remove water and add 50 μL of lysis buffer (10-mM Tris-HCl pH 8.0, 10-mM EDTA, 200-μg/mL proteinase K). Incubate the sample at 50°C for 2 h to overnight.
3. Inactivate the proteinase K by heating at 95°C for 5 min. Prepare PCR reaction mixture containing 1-μM primers (exL and exR), buffer, Hi-Fi taq (Roche), and 1 μL of the sample. Perform PCR; 35 cycles of 94°C for 30 s; 55°C for 30 s;

of the *Tol2* construct from the plasmid and integration into the genome. The injected embryos are raised and crossed with wild-type fish. The integrated *Tol2* construct is transmitted to the F1 generation. (C) Electrophoresis of PCR products generated by excision assay. After excision of the *Tol2* construct, the double strand break on the donor plasmid is repaired and re-ligated. Therefore PCR using the exL and exR primers located at both sides of the *Tol2* construct generates short PCR products from the backbone plasmid in embryos injected with both the transposase mRNA and the donor plasmid but not in embryos injected only with the donor plasmid.

72°C for 30 s, and analyze the PCR product on 1.5% gel electrophoresis. When the *Tol2* portion is excised from the donor plasmid, the backbone plasmid is religated and DNA without *Tol2* will be amplified (Fig. 2C).

exL: 5′-ACCCTCACTAAAGGGAACAAAAG-3′
exR: 5′-CAAGGCGATTAAGTTGGGTAAC-3′

1.2.5 Identification of transgenic fish

1. Raise the remaining injected embryos to the sexual maturity (~3 months).
2. Cross the injected fish with wild-type fish and analyze the offspring. When the construct contains a fluorescent marker such as GFP, observe embryos by a fluorescent stereoscope and select GFP-positive embryos.
3. Alternatively, collect a subsample (~50) of the day-1 embryos for PCR analysis. Place embryos in a microtube, add 250-μL DNA extraction buffer (10-mM Tris-HCl pH 8.2, 10-mM EDTA, 200-mM NaCl, 0.5% SDS, 200-μg/mL proteinase K) and incubate them at 50°C overnight. Purify embryonic DNA by phenol/chloroform extraction, precipitate with ethanol, and resuspend in 50-μL TE. Use 1 μL of the DNA sample for PCR (35 cycles of 94°C for 30 s; 55°C for 30 s; 72°C for 30 s) using transgene specific primers. When a PCR-positive F1 pool is found, raise their siblings and analyze them individually at the adult stage for the presence of the transgene by PCR of caudal fin clips.
4. We highly recommend analyzing the F1 fish by Southern blot hybridization to identify fish with single *Tol2* insertions. F1 fish often carry multiple insertions, and this may complicate further analyses. When the F1 fish of interest carry multiple insertions, cross the F1 fish with the smallest number of insertions to wild-type fish, raise F2 offspring, and analyze F2 fish again by Southern blot hybridization.

1.3 MATERIALS

pCS-zT2TP: contains a codon-optimized version of the transposase cDNA downstream of the CMV and SP6 promoters (Fig. 1A).
pT2AL200R150G: contains 200-bp and 150-bp DNA from the left and right ends of *Tol2*, respectively (Fig. 1A) (Urasaki et al., 2006). A DNA fragment can be cloned between unique *BglII* and *XhoI* sites.

1.4 DISCUSSION

Transgenesis using the *Tol2* transposon system is highly efficient. 50–70% of fish injected with the *Tol2* system at the one-cell stage and grown up to the adulthood will become germ line–transmitting founder fish that transmit transgenes to the offspring. From such germ line–transmitting founder fish, six to seven insertions are transmitted on average (Kawakami et al., 2004; Urasaki et al., 2006). This feature, together with high germ line transmission rates, enables generation of thousands of transposon insertions in a mid-scale laboratory. *Tol2* transposon–mediated transgenesis has the following merits. First, since a transposon construct is integrated as a single copy, the expression of the transgene on the construct is less sensitive to silencing in comparison to multimeric or concatemeric transgene integrations. Second, since a transposon vector functions as a cassette, end-to-end integration of a transgene is guaranteed. Third, the transposon insertion does not cause unwanted rearrangements at the integration locus. Forth, the *Tol2* vector has fairly large cargo capacity. 10-kb DNA can be cloned without reducing the transpositional activity (Urasaki et al., 2006), and, furthermore, a BAC-size DNA, namely 100–200 kb DNA, can be cloned into the *Tol2* vector (Suster, Sumiyama, & Kawakami, 2009).

2. GENE TRAP, ENHANCER TRAP METHODS FOR THE Gal4FF-UAS METHODS

2.1 RATIONALE

The Gal4-UAS system allows targeted expression of any desired gene in the Gal4-expressing cells. We employed Gal4FF, a modified version of the Gal4 yeast transcription activator, that has the Gal4 DNA-binding domain and two short transcription activator segments from the herpes simplex viral protein VP16 (Asakawa et al., 2008). To generate transgenic fish with various different patterns of Gal4FF expression, we constructed a gene trap construct T2KSAGFF that contains a splice acceptor and the *gal4ff* gene, and an enhancer trap construct T2KhspGFF that contains the zebrafish *hsp70* promoter and the *gal4ff* gene (Fig. 1B). When the gene trap construct was integrated within a gene and the splice acceptor "trapped" its endogenous transcript, the *gal4ff* gene is expressed under the control of the promoter activity, and, when the enhancer trap construct was integrated in the genome and the *hsp70* promoter was influenced by a nearby enhancer, the *gal4ff* gene is expressed in a pattern dictated by the trapped enhancer.

2.2 METHODS

2.2.1 Gene trap and enhancer trap screens

1. Coinject a plasmid containing harboring T2KSAGFF or T2KhspGFF (Fig. 1B) and the transposase mRNA to fertilized eggs. Raise the injected fish.
2. Cross the injected fish (founder fish) with UAS:GFP reporter fish (Fig. 2B). GFP is expressed where Gal4FF is expressed. Collect GFP-positive F1 embryos by observing them under a fluorescent microscope at different developmental stages (eg, 24 hpf, 48 hpf, 72 hpf, and 5 dpf). Raise the F1 embryos to adulthood (Fig. 4A).
3. To identify a gene or an enhancer trapped by the insertion, analyze genomic DNA surrounding the insertion by Southern blot hybridization and inverse PCR.

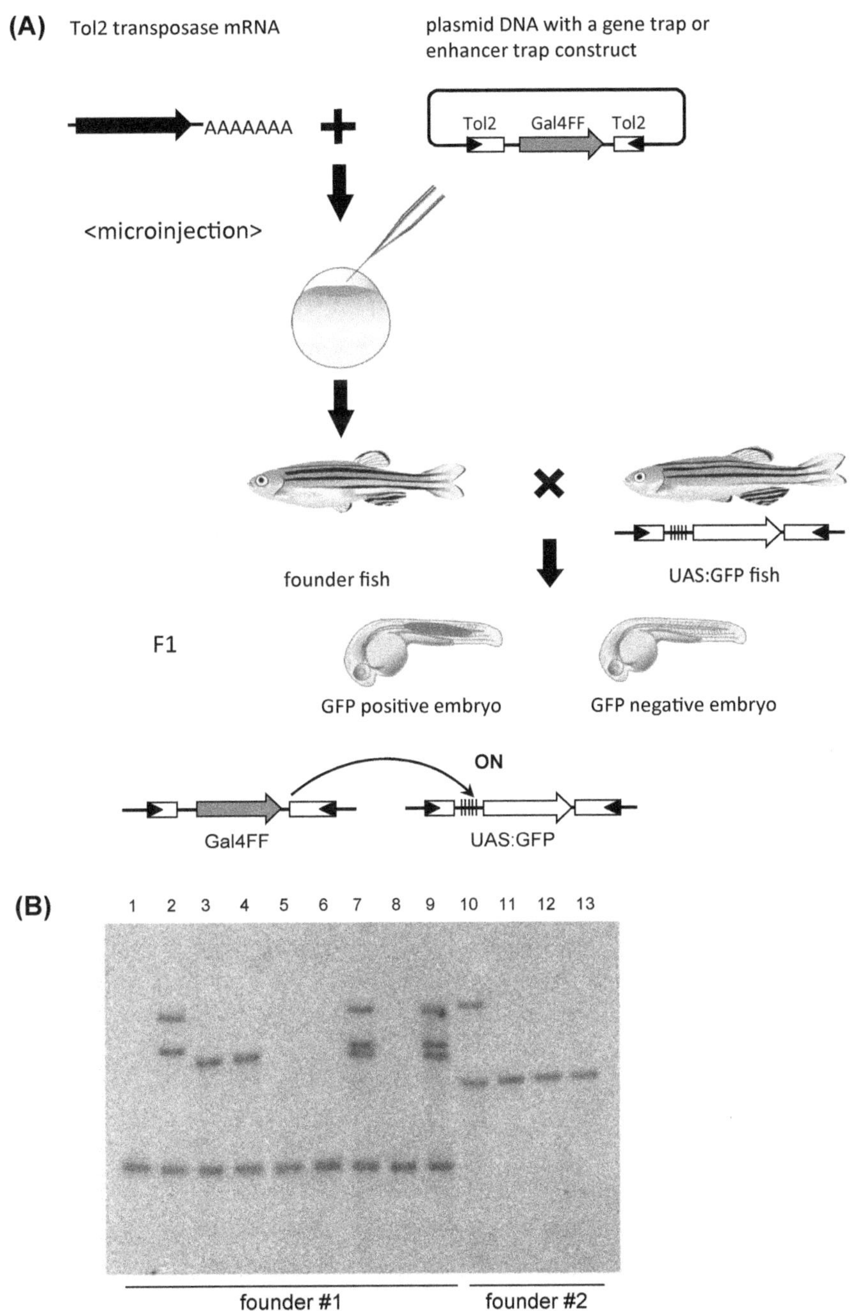
(A)
Tol2 transposase mRNA
plasmid DNA with a gene trap or enhancer trap construct
AAAAAAA
Tol2
Gal4FF
Tol2
<microinjection>
founder fish
UAS:GFP fish
F1
GFP positive embryo
GFP negative embryo
ON
Gal4FF
UAS:GFP
(B)
1 2 3 4 5 6 7 8 9 10 11 12 13
founder #1
founder #2

FIGURE 4

2.2.2 *Analysis of* Tol2 *insertions by Southern blot hybridization*

1. Because transgenic fish often carry multiple insertions, we recommend Southern blot analyses to identify an insertion responsible for the observed expression pattern.
2. Clip caudal fins of the F1 fish and lyse the tissue in 200 μL of DNA extraction buffer (10-mM Tris-HCl pH 8.2, 10-mM EDTA, 200-mM NaCl, 0.5% SDS, 200-μg/mL proteinase K) at 50°C for 3 h to overnight. Purify DNA by phenol/chloroform extraction, precipitate with ethanol, and suspend in 50-μL TE. Approximately 20–30 μg DNA will be obtained.
3. Digest 5 μg of the genomic DNA with *Bgl*II, which cuts most of our transposon constructs once. Perform electrophoresis by using 1% TAE-agarose gel.
4. Soak the gel in 0.25-N HCl for 15 min. Rinse with deionized water, soak in 0.25-N NaOH for 30 min, rinse with water, and transfer in 10X SSC (1X SSC: 0.15-M NaCl, 0.015-M sodium citrate).
5. Place the gel in a vacuum transfer apparatus (BS-31, BIO CRAFT, Japan) with Hybond-XL (15 × 15 cm; GE Healthcare, England) presoaked in 10X SSC. Perform transfer according to the manufacturer's instructions. After transfer, rinse the membrane in 1X SSC and dry completely at 50°C for 2 h to overnight.
6. Make a DIG (digoxigenin)-labeled probe using DIG probe synthesis KIT (Roche). Perform PCR by using the Gal4FF-f and Gal4FF-r primers and the *gal4ff* gene.
   Gal4FF-f: 5′-ATGAAGCTACTGTCTTCT-3′
   Gal4FF-r: 5′-TCTAGATTAGTTACCCGG-3′
7. Place the membrane into a hybridization bag, and add prewarmed 22.5-mL DIG Easy Hyb (Roche) to the bag. Incubate at 42°C for more than 30 min.
8. Replace the prehybridization solution with the hybridization buffer (8 mL of DIG Easy Hyb with 7 μL of a denatured DIG-labeled probe). Incubate at 42°C overnight.
9. Rinse the membrane with 2X washing buffer (2X SSC, 0.1% SDS) twice and then wash with 0.5X washing buffer (0.5X SSC, 0.1% SDS) at 65°C for 15 min twice.
10. Transfer the membrane to a new plastic container containing 100-mL MABT (0.1-M maleic acid, 0.15-M NaCl, 0.3% Tween). Incubate at room temperature for 2 min with shaking.

Gene trapping and enhancer trapping for the Gal4-UAS system. (A) A scheme for gene trapping and enhancer trapping. A trap construct containing gal4ff is injected into fertilized eggs with the transposase mRNA. Injected fish are raised and mated with homozygous UAS:GFP reporter fish. Doubly transgenic F1 embryos express GFP in regions where Gal4FF is expressed. (B) Southern blot hybridization analysis. Lanes 1–9 and 10–13 represent F1 fish from two different founder fish. Fish carrying a single *Tol2* insertion are identified.

11. Discard MABT. Add 150-mL blocking solution, 1% blocking reagent (Roche) in MAB (0.1-M maleic acid, 0.15-M NaCl). Incubate at room temperature for 2 min with shaking.
12. Discard blocking solution. Add 50-mL antibody solution (1/10,000 of anti-digoxigenin-AP in blocking solution). Incubate at room temperature for 30 min with shaking.
13. Discard the antibody solution. Wash the membrane twice with 100-mL MABT.
14. Soak the membrane in 20-mL detection buffer (0.1-M Tris-HCl pH 9.5, 0.1-M NaCl) and transfer into a hybridization bag.
15. Add 2–3 mL CDP-Star (Roche). Soak the membrane evenly in the solution. Incubate at room temperature for 5 min. Analyze the membrane using LAS-1000 (Fuji Film). Exposure time longer than 10 min is recommended (Fig. 4B).

2.2.3 Identification of Tol2 integration sites by inverse PCR

1. Digest 1 μg of genomic DNA with *Mbo*I in 10 μL of reaction buffer at 37°C for 1 h. Incubate the sample at 70°C for 15 min. Add 430-μL H_2O to the sample, incubate at 70°C for 10 min, and cool to 16°C.
2. Add 50 μL 10X T4 DNA ligation buffer (TAKARA, Japan) and 2-μL T4 DNA ligase, then incubate the sample at 16°C for 3 h to overnight.
3. Add 50 μL of 3 M sodium acetate and 1 mL ethanol to the sample. Chill the sample at −20°C for 30 min. Centrifuge the sample at 15,000 rpm at 4°C for 20 min. Rinse once with 70% ethanol and suspend in 20-μL H_2O.
4. First PCR: using 10 μL of the ligation sample, perform the first PCR (30 cycles of 94°C for 30 s; 57°C for 30 s; 72°C for 1 min) using *Tol2*-5′inv-f1 and *Tol2*-5′inv-r1 primers for the 5′ junction or *Tol2*-3′inv-f1 and *Tol2*-3′inv-r1 primers for the 3′ junction.
   Tol2-5′inv-f1: 5′-GTCATGTCACATCTATTACCAC-3′
   Tol2-5′inv-r1: 5′-CTCAAGTAAAGTAAAAATCC-3′
   Tol2-3′inv-f1: 5′-AGTACAATTTTAATGGAGTACT-3′
   Tol2-3′inv-r1: 5′-TGAGTATTAAGGAAGTAAAAGT-3′
5. Second PCR: using 2 μL of the first PCR product, perform the second PCR (30 cycles of 94°C for 30 s; 57°C for 30 s; 72°C for 1 min) using *Tol2*-5′inv-f2 and *Tol2*-5′inv-r2 primers for the 5′ junction or *Tol2*-3′inv-f2 and *Tol2*-3′inv-r2 primers for the 3′ junction.
   Tol2-5′inv-f2: 5′-AATGCACAGCACCTTGACCTGG-3′
   Tol2-5′inv-r2: 5′-CAGTAATCAAGTAAAATTACTC-3′
   Tol2-3′inv-f2: 5′-TTTACTCAAGTAAGATTCTAG-3′
   Tol2-3′inv-r2: 5′-AAAGCAAGAAAGAAAACTAGAG-3′
4. Analyze the PCR product on a 1.5% TAE-agarose gel, then purify and sequence using primers L100-out for the 5′ junction, and R100-out for the 3′ junction.
   L100-out: 5′-AGTATTGATTTTTAATTGTA-3′
   R100-out: 5′-AGATTCTAGCCAGATACT-3′

2.2.4 Search for useful gene trap and enhancer trap fish using the zTrap database

We developed a database *z*Trap (zebrafish gene trap and enhancer trap database; http://kawakami.lab.nig.ac.jp/ztrap/) that contains the data for expression patterns and transposon integration sites (Kawakami et al., 2010) (Fig. 5). Transgenic fish that express Gal4FF (visualized by UAS:GFP expression) in specific cells, tissues, and organs can be searched using the *z*Trap database. For instance, when you click "heart" on the "by region" column, transgenic fish lines that show Gal4FF (GFP) expression in the heart appear on the space in the right. Then by clicking the

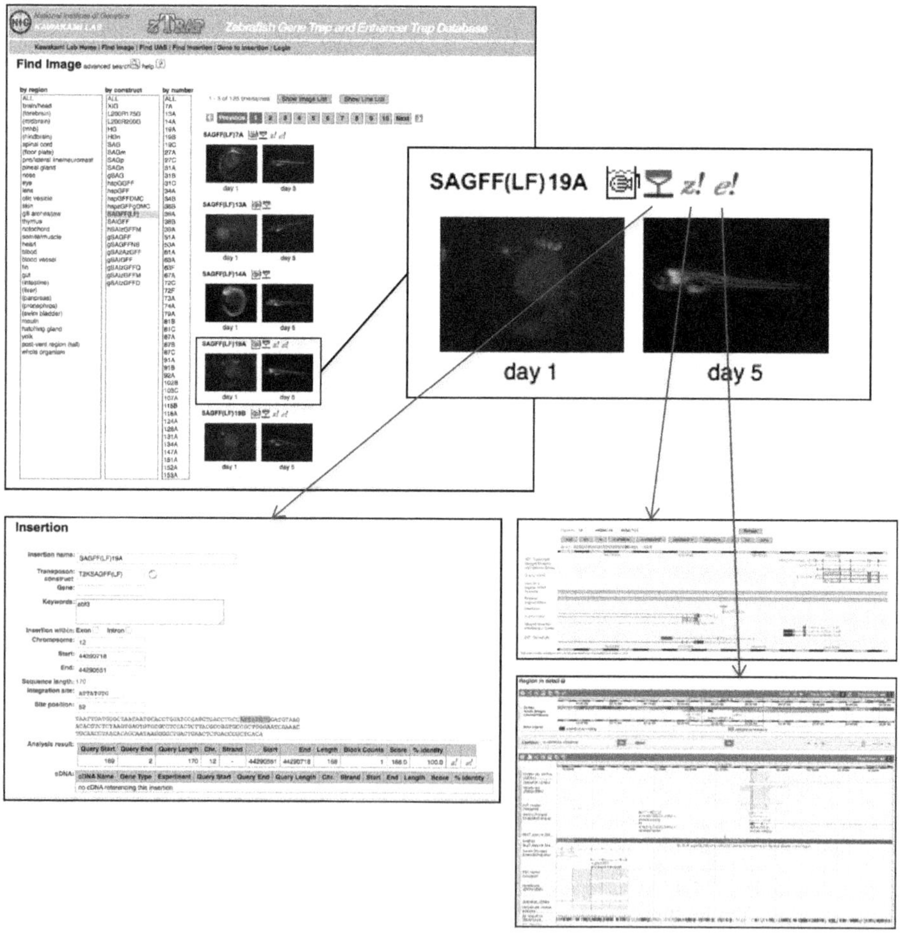

FIGURE 5

The gene and enhancer trap database zTrap (http://kawakami.lab.nig.ac.jp/ztrap/). Transgenic lines that express GFP and Gal4FF in regions of interest are shown by clicking a region name in the left column. By clicking icons, information of the transposon insertion site and links to the internal (*z!*) and ensemble (*e!*) browsers are seen. (See color plate) "transposon icon," the information about the transposon integration site can be seen as a new tub. "*z!*" and "*e!*" icons are link to internal genome browser and ensemble genome browser, respectively.

2.3 MATERIALS

T2KSAGFF: a gene trap construct containing the *gal4ff* gene downstream of a rabbit β-globin splice acceptor (Fig. 1B).
T2KhspGFF: an enhancer trap construct containing the *gal4ff* gene downstream of the zebrafish *hsp70* promoter (Fig. 1B).
UAS:GFP reporter fish: transgenic fish line carrying the *egfp* gene downstream of UAS (Fig. 1C).

2.4 DISCUSSION

We employed Gal4FF to develop the Gal4-UAS system in zebrafish. In previous studies, full-length Gal4 and Gal4-VP16 were used in zebrafish. These showed some disadvantages; namely, the transcriptional activity of the full-length Gal4 was not strong, and Gal4-VP16, which contained a strong transcription activator domain, showed some developmental toxicity (Koster & Fraser, 2001; Scheer & Campos-Ortega, 1999). We found Gal4FF could reliably induce transcription from UAS and showed no obvious toxicity (Asakawa et al., 2008). We demonstrate that the gene trapping and enhancer trapping are powerful to generate transgenic fish expressing Gal4FF in various specific cells, tissues, and organs. Further, analysis of the genomic DNA surrounding the transposon insertions identifies genes expressed in such specific patterns.

3. TARGETED GENE EXPRESSION WITH THE Gal4-UAS SYSTEM

3.1 RATIONALE

The Gal4-expressing transgenic fish are powerful tools to visualize and manipulate specific cell types. For this purpose, desired reporter or effector genes should be cloned into the multicloning site (MCS) of T2MUASMCS (Fig. 1C), and transgenic fish carrying these genes downstream of the Gal4-binding sequence (5xUAS) and the E1b TATA sequence should be generated. The Gal4-expressing fish and UAS-reporter and UAS-effector fish are kept independently, and phenotypes can be analyzed in double transgenic offspring obtained by the cross of these fish.

3.2 METHODS

3.2.1 Inhibition of neuronal activities via the Gal4-UAS system

1. The UAS:TeTxLC:CFP transgenic fish carries a gene encoding the tetanus toxin light chain fused to the CFP gene downstream of UAS (Fig. 1C) and is used to

inhibit neuronal functions in the Gal4-expressing neurons (Asakawa et al., 2008). The UAS:TeTxLC:CFP fish is crossed with transgenic lines that express Gal4FF in specific subpopulations of neurons. Selection of CFP-positive offspring will be easier if the Gal4FF-expressing fish are maintained as double transgenic with UAS:RFP.

2. The UAS:TeTxLC:CFP fish is crossed with SAGFF36B or SAGFF31B fish that expresses Gal4FF in the sensory neurons or interneurons in the spinal cord, respectively. At 2 dpf, touch the tails of the embryos gently with a needle, and take images by using a high-speed digital video camera (FASTCAM-512PC1, Photoron, Japan). Wild-type embryos respond to the gentle touch to the tail and swim rapidly away from the stimulus. However, the SAGFF36B;UAS:TeTxLC:CFP embryos do not respond to the touch, and the SAGFF31B;UAS:TeTxLC:CFP embryos show abnormal escape behaviors (Fig. 6A) (Asakawa et al., 2008).

3.2.2 Visualization of neuronal activities via the Gal4-UAS system

1. The UAS:GCaMP7a transgenic fish carries the codon-optimized GCaMP7a gene, encoding an improved version of the calcium indicator GCaMP, downstream of UAS (Fig. 1C), and is used to monitor cellular Ca^{2+} concentration. When the UAS:GCaMP7a fish is crossed with appropriate Gal4FF lines, neuronal activities are visualized (Muto, Ohkura, Abe, Nakai, & Kawakami, 2013).
2. To analyze neuronal activities in the optic tectum, the UAS:GCaMP7a fish is crossed with the gSA2AzGFF49A fish (tectum-gal4 driver). A gSA2AzGFF49A;UAS:GCaMP7a larva at 6 dpf and a paramecium are put in a recording chamber under an epifluorescent microscope Imager.Z1 (Zeiss). Images are taken by using a scientific CMOS camera (ORCA-Flash 4.0 V2, Hamamatsu Photonics) or a cooled CCD camera (ORCA-R2, Model, Hamamatsu Photonics) at 10 fps (Fig. 6B, left). Alternatively, the larva is immobilized in agarose, and a paramecium is put in a space in front of the larva (Fig. 6B, right). The data are analyzed with ImageJ. To quantify changes in fluorescence intensity, divide the individual frames by a reference image (an averaged image over all frames or an averaged image over a period with no calcium signals).

3.2.3 Visualization of in vivo microtubule structures via the Gal4-UAS system

1. The UAS:Gtuba2 transgenic fish carries the *tuba2* gene encoding an α-tubulin fused to the *egfp* gene downstream of UAS (Fig. 1C). The GFP-Tuba2 fusion protein (Gtuba2) forms heterodimer with β-tubulin and is incorporated into microtubule (MT) filament. Thus, the UAS:Gtuba2 fish is used to visualize MT-based cytoskeletal structures in vivo (Asakawa & Kawakami, 2010).

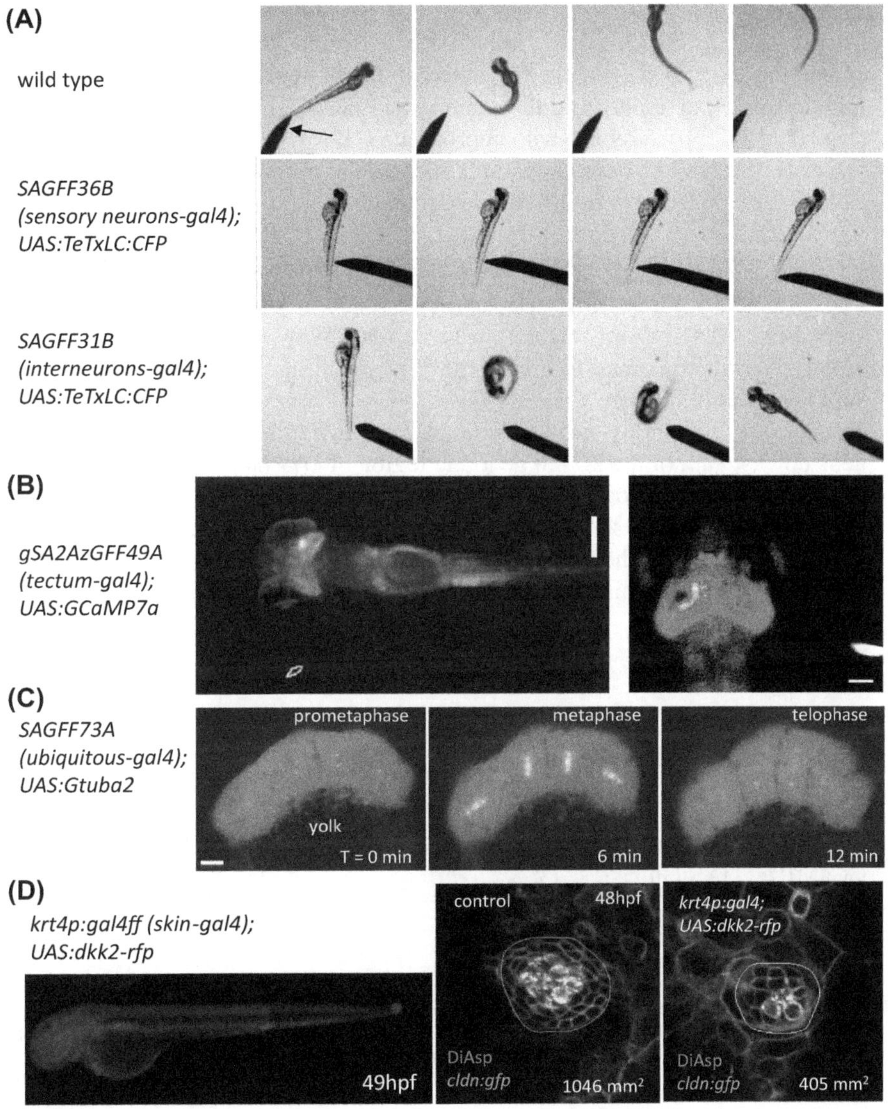

FIGURE 6

Targeted expression via the Gal4FF-UAS system. (A) Inhibition of neuronal activities via the Gal4-UAS system. A wild-type embryo at 2 dpf rapidly escapes from a gentle touch to the tail. The SAGFF36B;UAS:TeTxLC:CFP fish, which expresses TeTxLC:CFP in the sensory neurons, does not respond to the touch. The SAGFF31B;UAS:TeTxLC:CFP fish, which expresses TeTxLC:CFP in subsets of interneurons and motor neurons, responds to the touch but shows abnormal escape swimming. (B) Visualization of neuronal activities via the Gal4-UAS system. The gSA2AzGFF49A;UAS:GCaMP7a larva at 6 dpf expresses the calcium indicator GCaMP7a in the optic tectum. (Left) Calcium signal in the right tectum of a freely behaving

2. The UAS:Gtuba2 fish is crossed with the SAGFF73A fish (ubiquitous-gal4 driver). The SAGFF73A;UAS:Gtuba2 female is used for mating to obtain eggs containing the Gtuba2 protein via maternal expression. In such embryos, the mitotic spindle assembly and disassembly are observable during the cleavage stage (Fig. 6C). By crossing with appropriate Gal4FF drivers, UAS:Gtuba2 can visualize various MT-based apparatus such as the mitotic spindle in dividing cells and the MT arrays in differentiated cells.

3.2.4 Inhibition of the Wnt-signaling pathway via the Gal4-UAS system

1. The *UAS:dkk2-rfp* transgenic fish carries the *dkk2-rfp* fusion gene downstream of UAS (Fig. 1C). Dickkopf2 (Dkk2) is the secreted Wnt inhibitory factor, and the *UAS:dkk2-rfp* fish is used to inhibit the Wnt-signaling pathway in desired tissues (Wada et al., 2013).
2. The *UAS:dkk2-rfp* fish is crossed with the *krt4p:gal4ff* fish that carries the *gal4ff* gene downstream of the 2.2-kb genomic region of *krt4* (*keratin4*). In the *krt4p:gal4ff;UAS:dkk2-rfp* double transgenic embryos, Dkk2-RFP is expressed in the epidermal cells. In the zebrafish lateral line sensory organ, the Wnt-signaling activity is detected in the proliferating neuromasts. The *krt4p:gal4ff;UAS:dkk2-rfp* embryos show significant reduction in the neuromast size (Fig. 6D).

3.3 MATERIALS

T2MUASMCS: contains an MCS downstream of 5xUAS and the E1b promoter (Fig. 1C).
Gal4FF drivers: SAGFF36B, SAGFF31B, gSA2AzGFF49A, SAGFF73A, the *krt4p:gal4ff* fish.
UAS-reporter and effector fish: UAS:GFP, UAS:RFP, UAS:TeTxLC:CFP, UAS:GCaMP7a, UAS:Gtuba2, *UAS:dkk2-rfp*

3.4 DISCUSSION

We employed 5xUAS as a Gal4-binding sequence, which is less sensitive to silencing effects than 14xUAS (Goll, Anderson, Stainier, Spradling, & Halpern, 2009).

larva when the paramecia locate on the left side. Scale bar: 250 μm. (Right) Calcium signal in the left tectum of an immobilized larva when the paramecia locate on the right side (pseudocolored). Scale bar: 100 μm. (C) Visualization of in vivo microtubule structures via the Gal4-UAS system. Mitotic spindles are observed in an 8-cell stage embryo dividing into 16 cells. Scale bar: 50 μm. (D) Inhibition of the Wnt-signaling pathway via the Gal4-UAS system. (Left) A fluorescence image of krt4p:gal4ff;UAS:dkk2-rfp. (Right) Images of a neuromast in wild type and the krt4p:gal4ff;UAS:dkk2-rfp fish. Green: cldn:gfp, epidermal cells and neuromasts. Red: DiAsp, hair cells. Overexpression of dkk2-rfp in the skin reduces the neuromast size. (See color plate)

We have observed different levels of reporter and effector gene expression when fish injected with the same UAS construct were crossed with the same Gal4FF fish. In these transgenic fish, the UAS construct is integrated at different chromosomal loci, and its expression is influenced by positional effects. Therefore it is important to obtain many UAS transgenic F1 fish and select one that works best with a Gal4FF line of interest. Indeed, when we established the UAS:GFP reporter line, we selected the one that showed strongest expression when crossed with a Gal4FF tester fish out of ~30 different UAS:GFP transgenic F1 fish.

ACKNOWLEDGMENTS

This work was partly supported by National BioResource Project (KK), PRESTO program of the Japan Science and Technology Agency (JST) (HW), and Grant-in-Aids (15H02370 to KK; 25830020 to KA; 22500299 and 24120521 to AM; 19770204 and 25440118 to HW) from the Ministry of Education, Culture, Sports, Science, and Technology of Japan.

REFERENCES

Asakawa, K., & Kawakami, K. (2010). A transgenic zebrafish for monitoring in vivo microtubule structures. *Developmental Dynamics, 239*, 2695–2699.

Asakawa, K., Suster, M. L., Mizusawa, K., Nagayoshi, S., Kotani, T., Urasaki, A. … Kawakami, K. (2008). Genetic dissection of neural circuits by *Tol2* transposon-mediated Gal4 gene and enhancer trapping in zebrafish. *Proceedings of the National Academy of Sciences of the United States of America, 105*, 1255–1260.

Davison, J. M., Akitake, C. M., Goll, M. G., Rhee, J. M., Gosse, N., Baier, H. … Parsons, M. J. (2007). Transactivation from Gal4-VP16 transgenic insertions for tissue-specific cell labeling and ablation in zebrafish. *Developmental Biology, 304*, 811–824.

Goll, M. G., Anderson, R., Stainier, D. Y., Spradling, A. C., & Halpern, M. E. (2009). Transcriptional silencing and reactivation in transgenic zebrafish. *Genetics, 182*, 747–755.

Kawakami, K. (2007). *Tol2*: a versatile gene transfer vector in vertebrates. *Genome Biology, 8*(Suppl. 1), S7.

Kawakami, K., Abe, G., Asada, T., Asakawa, K., Fukuda, R., Ito, A. … Yoshida, M. (2010). zTrap: zebrafish gene trap and enhancer trap database. *BMC Developmental Biology, 10*(1), 105.

Kawakami, K., Koga, A., Hori, H., & Shima, A. (1998). Excision of the *Tol2* transposable element of the medaka fish, *Oryzias latipes*, in zebrafish, *Danio rerio*. *Gene, 225*, 17–22.

Kawakami, K., & Shima, A. (1999). Identification of the *Tol2* transposase of the medaka fish *Oryzias latipes* that catalyzes excision of a nonautonomous *Tol2* element in zebrafish *Danio rerio*. *Gene, 240*, 239–244.

Kawakami, K., Shima, A., & Kawakami, N. (2000). Identification of a functional transposase of the *Tol2* element, an Ac-like element from the Japanese medaka fish, and its transposition in the zebrafish germ lineage. *Proceedings of the National Academy of Sciences of the United States of America, 97*, 11403–11408.

Kawakami, K., Takeda, H., Kawakami, N., Kobayashi, M., Matsuda, N., & Mishina, M. (2004). A transposon-mediated gene trap approach identifies developmentally regulated genes in zebrafish. *Developmental Cell, 7*, 133–144.

Koga, A., Suzuki, M., Inagaki, H., Bessho, Y., & Hori, H. (1996). Transposable element in fish. *Nature, 383*, 30.

Koster, R. W., & Fraser, S. E. (2001). Tracing transgene expression in living zebrafish embryos. *Developmental Biology, 233*, 329–346.

Kwan, K. M., Fujimoto, E., Grabher, C., Mangum, B. D., Hardy, M. E., Campbell, D. S. … Chien, C. B. (2007). The Tol2kit: a multisite gateway-based construction kit for *Tol2* transposon transgenesis constructs. *Developmental Dynamics, 236*, 3088–3099.

Muto, A., Ohkura, M., Abe, G., Nakai, J., & Kawakami, K. (2013). Real-time visualization of neuronal activity during perception. *Current Biology, 23*, 307–311.

Nagayoshi, S., Hayashi, E., Abe, G., Osato, N., Asakawa, K., Urasaki, A. … Kawakami, K. (2008). Insertional mutagenesis by the *Tol2* transposon-mediated enhancer trap approach generated mutations in two developmental genes: *tcf7* and *synembryn-like*. *Development, 135*, 159–169.

Parinov, S., Kondrichin, I., Korzh, V., & Emelyanov, A. (2004). *Tol2* transposon-mediated enhancer trap to identify developmentally regulated zebrafish genes in vivo. *Developmental Dynamics, 231*, 449–459.

Scheer, N., & Campos-Ortega, J. A. (1999). Use of the Gal4-UAS technique for targeted gene expression in the zebrafish. *Mechanisms of Development, 80*, 153–158.

Scott, E. K., Mason, L., Arrenberg, A. B., Ziv, L., Gosse, N. J., Xiao, T. … Baier, H. (2007). Targeting neural circuitry in zebrafish using GAL4 enhancer trapping. *Nature Methods, 4*, 323–326.

Suster, M. L., Sumiyama, K., & Kawakami, K. (2009). Transposon-mediated BAC transgenesis in zebrafish and mice. *BMC Genomics, 10*, 477.

Urasaki, A., Morvan, G., & Kawakami, K. (2006). Functional dissection of the *Tol2* transposable element identified the minimal *cis*-sequence and a highly repetitive sequence in the subterminal region essential for transposition. *Genetics, 174*, 639–649.

Villefranc, J. A., Amigo, J., & Lawson, N. D. (2007). Gateway compatible vectors for analysis of gene function in the zebrafish. *Developmental Dynamics, 236*, 3077–3087.

Wada, H., Ghysen, A., Asakawa, K., Abe, G., Ishitani, T., & Kawakami, K. (2013). Wnt/Dkk negative feedback regulates sensory organ size in zebrafish. *Current Biology, 23*, 1559–1565.

CHAPTER

Genetic screens for mutations affecting adult traits and parental-effect genes

3

F. Pelegri*, M.C. Mullins§,[1]

*University of Wisconsin–Madison, Madison, WI, United States

§University of Pennsylvania Perelman School of Medicine, Philadelphia, PA, United States

[1]Corresponding author: E-mail: mullins@mail.med.upenn.edu

CHAPTER OUTLINE

Methods in Cell Biology, Volume 135, ISSN 0091-679X, http://dx.doi.org/10.1016/bs.mcb.2016.05.006

Abstract

Forward genetics remains an important approach for the unbiased identification of factors involved in biological pathways. Forward genetic analysis in the zebrafish has until now largely been restricted to the developmental period from zygotic genome activation through the end of embryogenesis. However, the use of the zebrafish as a model system for the analysis of late larval, juvenile and adult traits, including fertility and maternal and paternal effects, continues to gain momentum. Here, we describe two approaches, based on an F_3-extended family and gynogenetic methods, that allow genetic screening for, and recovery of mutations affecting post-embryonic stages, including adult traits, fertility, and parental effects. For each approach, we also describe strategies to maintain, map, and molecularly clone the identified mutations.

INTRODUCTION

The zebrafish *Danio rerio* continues to gain momentum as a model system to understand a wide range of biological processes, spanning the developmental cycle from oocyte formation and the egg-to-embryo transition (Elkouby, Jamieson-Lucy, & Mullins, 2016; Lindeman & Pelegri, 2012) to aging (Carneiro et al., 2016). Recently, advances in transcriptomic analysis have allowed to better define the dynamic representation of maternal RNAs and the transition to zygotic control in the early embryo (Harvey et al., 2013; Lee et al., 2013; Mathavan et al., 2005), crucial aspects of this developmental cycle. These analyses show that, surprisingly, 95% of genes expressed during the first day of embryonic development are maternally contributed, with about 1/3 of all genes only expressed maternally and 2/3 expressed both maternally and zygotically. Other studies increasingly show the importance of paternally inherited chromatin marks in embryonic development (Jiang et al., 2013; Potok, Nix, Parnell, & Cairns, 2013). These studies highlight the importance of maternally and paternally inherited information in early embryogenesis and exemplify the need for methods to identify and analyze the function of parentally derived products in processes involving intergenerational transmission. The increased technical difficulties of studying later traits in addition to fertility, such as those involved in juvenile development and adult homeostasis and function, are of great medical relevance yet remain as a whole poorly understood.

A major recent technical advance has been the development of increasingly effective reverse-genetic approaches, including the generation of specific gene knock outs with zinc finger nucleases (ZFNs) (Doyon et al., 2008; Foley et al., 2009; Meng, Noyes, Zhu, Lawson, & Wolfe, 2008), transcription activator-like effector nucleases (Cade et al., 2012; Dahlem et al., 2012; Huang et al., 2011) and, more recently, clustered, regularly interspaced short palindromic repeats (CRISPR)/Cas9 (Hruscha et al., 2013; Hwang et al., 2013; Irion, Krauss, & Nüsslein-Volhard, 2014; Jao, Wente, & Chen, 2013; Moreno-Mateos et al., 2015; Ota, Hisano, Ikawa, & Kawahara, 2014). The latter method is able to effectively target specific sequences using a simple, nucleic-acid–based method, providing a "magic bullet" to generate targeted gene mutations. These reverse genetic methods are beginning to be used to complement ongoing genetic analysis such as with the generation of new alleles of existing mutations (Eno, Solanki, & Pelegri, 2016) and the generation of mutations in genes known to interact with already known factors (Campbell, Heim, Smith, & Marlow, 2015), and provide the means for the systematic testing of gene function at both the organismal and tissue levels (Ablain, Durand, Yang, Zhou, & Zon, 2015; Varshney et al., 2015).

Despite the simplicity and efficiency of current reverse-genetic technologies, forward genetic approaches will remain an important avenue to identify factors and pathways involved in specific biological processes. This is because they provide the ability to pinpoint genes acting in such processes without any prior knowledge or assumptions regarding the molecular components of the system. In addition, as our knowledge of biological systems has increased, two themes, pathway reutilization and robustness, have become increasingly apparent. Reutilization of pathways during development leads to pleiotropy in phenotypes associated with mutation of pathway components, and such pleiotropy may interfere with the study of gene function at later stages of development. Robustness in systems is in part underlain by redundancy in gene function, and redundancy can obscure gene function. Both of these themes add complexity to genetic analysis, and forward genetic approaches provide unique advantages to address these challenges. This is because forward genetics selects for genes that exhibit a readily observable effect when mutant, revealing exceptions to pleiotropy and redundancy within gene networks. Thus, forward genetics can help reveal unique aspects of late gene functions that may escape an earlier developmental role or functional duplication. Moreover, forward genetics is unbiased and will identify any type of mutation, including unusual alleles that may bypass both pleiotropy and redundancy and result in an observable phenotype.

Thus, forward genetics remains an effective means to identify new biological pathway components and is an important complement to biological analysis. This approach is of particular relevance to the analysis of adult, parental, and other intergenerational traits for a number of reasons, such as the paucity in knowledge in these processes, obstacles presented by pleiotropy to assess function at late stages of development, and technical difficulties implementing genetic screens designed to identify relevant genes. Fortunately, gene identification after mutagenesis, a traditional bottleneck for forward genetics approaches, is enormously facilitated by current advances in high-throughput sequencing and bioinformatics (Henke, Bowen, & Harris, 2013a, 2013b; Langdon et al., 2016). Altogether, the stage is set for the continued importance of forward genetic approaches as an unbiased means to identify genes involved in adult traits with relevance to biomedicine and reproductive biology.

The genetic attributes of the zebrafish allow the powerful method of forward genetics, so effective in other model systems, to be applied in a vertebrate to identify and study the functions of genes through loss-of-function analysis. Such an approach can identify in a systematic and unbiased manner a myriad of genes that are essential for a given process, establishing pathways of genes, a molecular framework, or filling gaps in our knowledge in particular processes, often providing insights in unpredicted ways. The methods presented in this chapter were originally developed and carried out primarily to identify maternal-effect genes, and subsequently shown also to allow the identification of paternal-effect genes and adult traits. However, these methods can be readily extended to include the genetic analysis of juvenile and adult traits, because the basic approaches are similar. Indeed, maternal and paternal effects, as well as defects during gametogenesis also leading to sterility, can be regarded as a specific type of adult trait (involved in the production of viable progeny). In this chapter, we present a unified set of methods for the forward genetic analysis of juvenile, adult, and maternal- and paternal-effect traits.

EXTENDING THE RANGE OF FORWARD GENETICS ANALYSIS IN ZEBRAFISH SCREENS

In spite of this potential range, zebrafish genetic research has largely focused on the identification and analysis of zygotically driven genetic networks involved in embryonic development. The temporal range of this more thoroughly studied time period has been determined by both biological and practical reasons: the earlier time point in this range corresponds to about 3 hours post fertilization (hpf), the time at which embryos begin to rely on zygotic gene expression. The latter limit of this range is at ~5 days of development, the latest point that the resulting larva is nutritionally self-sufficient and does not require an externally supplied food source, which facilitates analysis by the investigator. Intense studies of developmental processes in this limited developmental range has generated an enormous amount of crucial information to understand key biological questions, including aspects of cell fate diversification, the behavior and movement of cells and tissues during normal development, and the intricate interactions between these two types of processes during the development of the body plan and organ systems. However, development encompasses a much broader period that extends after the embryo exhausts its yolk supply at 5 days post fertilization (dpf) and enters larval, juvenile, and adult stages (Fig. 1) (Gupta et al., 2010). Moreover, development of gonads and gametogenesis in adults is essential for the generation of functional gametes and to provide products essential in the next generation for the cleavage stage embryo prior to the activation of the zygotic genome, closing full circle the generational life cycle.

Because of the above-mentioned repeated deployment of developmental programs throughout the lifespan of an organism, much of the knowledge gained from the genetic analysis of zygotic genes in embryonic development has shed light on understanding related processes at other developmental stages. Thus, zygotically driven signaling pathways involved in early development have been shown in various model organisms to be reused during subsequent growth, in adulthood and aging, and the same pathways are affected in disease conditions such as cancer. However, a large gap remains in understanding the genetic circuitry acting at these other stages of development. The direct genetic dissection of juvenile, adult, and parental-effect traits through forward genetics promises to provide important new clues to these processes.

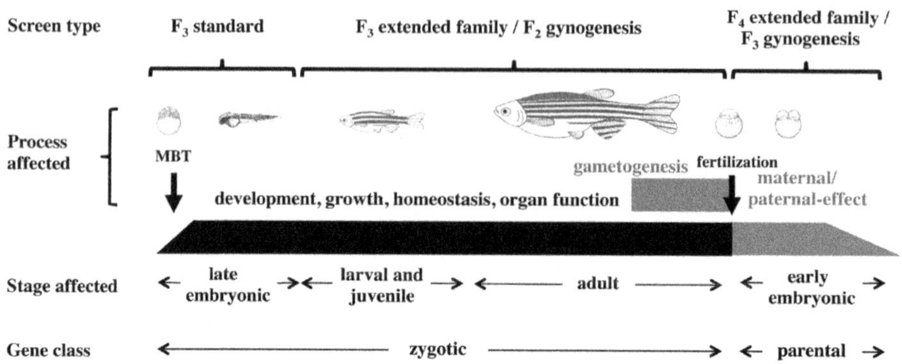

FIGURE 1

Use of various genetic schemes to target different developmental stages in the zebrafish. Standard inbreeding approaches have been used to identify recessive zygotic mutations affecting the period between the activation of the zygotic genome at MBT (ca. 3 hours post fertilization) until day 5 of development. The identification of recessive mutations in stages past day 5 of development and into adulthood has been facilitated by the use of strategies described in this chapter, such as extended family- and gynogenesis-based approaches. The identification of parental-effect genes, which act prior to the MBT, requires the generation of an additional generation. Mutations affecting gametogenesis can be identified using strategies targeting adult or parental-effect genes. See text for details.

APPLICATION OF GENETIC SCREENS BEYOND THE EMBRYONIC LETHAL RANGE

Screening for juvenile and adult phenotypic traits requires screening during the growth and/or aging of the potentially affected clutches (Fig. 1). In addition, and somewhat counterintuitively, screens for the earliest stages of embryonic development must be carried out through genetic schemes that utilize an additional generation of crosses (Fig. 1). This is because the earliest stages of embryonic development rely on maternally and paternallyderived products, and homozygous mutant adults must be generated to observe the effect of mutations in these inherited products. We note that mutations causing defects in gametogenesis that lead to sterility can be screened for either directly in the homozygous mutant individuals (as a bona fide adult trait) or after breeding, preselecting mutant individuals that fail to produce viable offspring (in a maternal- or paternal-effect screen) followed by the analysis of the gonads of the affected parents.

POST-EMBRYONIC DEVELOPMENT: JUVENILE AND ADULTS STAGES

Depletion of nutritional sources in the yolk cell and the development of the digestive system, swim bladder inflation, and the appearance of foraging behavior are synchronized in the zebrafish embryo, such that by day 5 of development larvae no longer rely on maternal food resources and must find and process extraneous nutrients (Kimmel, Ballard, Kimmel, Ullmann, & Schilling, 1995; Parichy, Elizondo, Mills, Gordon, & Engeszer, 2009). Genetic screens probing stages beyond 5 dpf necessitate the raising of the clutch through a full schedule of feeding in an aquarium system. Since genome-wide screening in the zebrafish is carried out by random inbreeding of a sufficiently large number of crosses to maximally homozygose the mutagenized genome, this extra generation and rearing requirements generate a challenge, as now a large number of clutches for each mutagenized genome needs to be maintained and grown until testing.

Nevertheless many of the phenotypes expected to be of high biomedical relevance (eg, propensity to disease, organ function, and physiology) will be manifested in juveniles and adults (for simplicity, hereafter referred to as adults) (Gupta et al., 2010). Thus it would be advantageous to dissect such traits using forward genetics. A handful of genetic screens have identified recessive mutations affecting adult traits, including body shape, pigmentation, and eye and fin morphology (Haffter et al., 1996; Johnson, Africa, Walker, & Weston, 1995; Lang, Patterson, Gordon, Johnson, & Parichy, 2009; Parichy & Turner, 2003b; Wagner, Dosch, Mintzer, Wiemelt, & Mullins, 2004), skeletal structures (Andreeva et al., 2011; Harris et al., 2008), ocular structures (Tschopp et al., 2010), fin regeneration (Johnson & Weston, 1995), scale formation (Rohner et al., 2009), gonadal development (Bauer & Goetz, 2001), and other adult morphological defects (Wagner et al., 2004).

PRIOR TO THE MIDBLASTULA TRANSITION: MATERNAL- AND PATERNAL-EFFECT GENES

In all animals development from fertilization to the activation of the zygotic genome at the midblastula transition (MBT, Signoret & Lefresne, 1971; Newport & Kirschner, 1982a, 1982b) depends on maternal factors made during oogenesis and activated upon fertilization. By necessity, all cellular and developmental processes acting during this time window are carried out solely by such maternal factors. Although the activation of widespread zygotic gene expression at the MBT marks the beginning of zygotic gene control during development, it does not imply an absolute shift between the use of maternal and zygotic products. Rather, in many instances perduring maternal products interact with newly expressed zygotic products to control developmental processes even after the activation of the zygotic genome. Maternal products are instrumental for the generation of the embryonic body plan, for example by establishing both the animal-vegetal and dorsal-ventral axes (Fig. 2). The animal-vegetal axis is established during oogenesis and marks the prospective anterior—posterior axis of the embryo, whereas the dorsal-ventral axis forms during the early cleavage stages prior to the MBT.

Genetic analysis in invertebrate model organisms, such as *Drosophila* and *Caenorhabditis elegans*, has revealed networks of maternal factors involved in basic cellular functions, establishment of egg polarity, and the regulation of cell fates (Kemphues & Strome, 1997; Schnabel & Priess, 1997; St. Johnston &

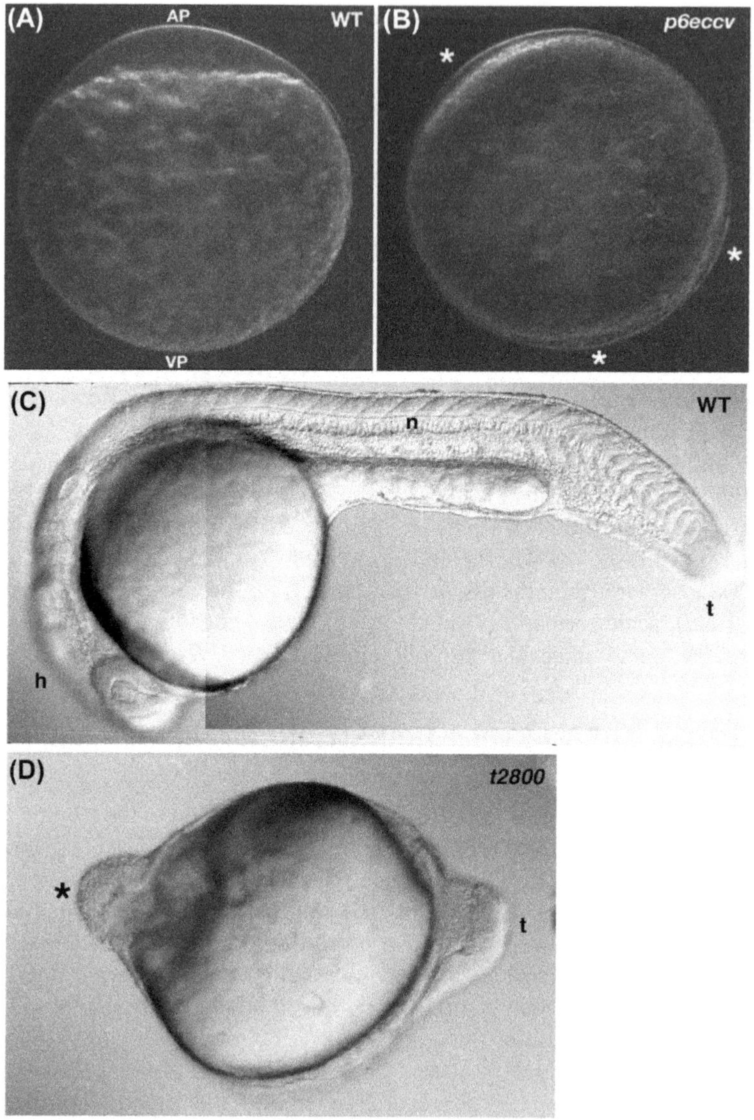

FIGURE 2

Two mutants identified and recovered in systematic screens for maternal-effect mutants. (A and B) The animal-vegetal polarity mutant *magellan*^{p6cv} *(p6eccv)* was identified in an F_4 natural crosses screen (Dosch et al., 2004; Gupta et al., 2010). (A) A wild-type egg shortly after fertilization displaying the blastodisc prominently at the animal pole. (B) In contrast, in the p6eccv mutant the cytoplasm segregates to multiple locations around the circumference of the egg (*asterisks*). AP and VP are the animal and vegetal poles, respectively *(A, B photos courtesy of Florence Marlow)*. (C,D) The mutation *hecate*^{t2800}

Nüsslein-Volhard, 1992). In vertebrates, a much more limited knowledge of maternal gene function has been acquired in model organisms such as *Xenopus* through embryological and molecular approaches, and in mouse through targeted gene mutation approaches. Studies in teleost fish including the zebrafish have also begun to address the requirement of maternally driven genes in early development. Such maternal processes span basic cellular functions such as fertilization, egg activation, and the early cellular and nuclear divisions, as well as the induction of embryonic cell fates and the execution of morphogenetic movements during gastrulation (reviewed in Abrams & Mullins, 2009; Lindeman & Pelegri, 2010). Genetically, mutations in such maternal-effect genes manifest themselves in the embryos from females homozygous mutant for these genes (for recessive alleles), regardless of the genetic composition of the sperm. The genotype of the sperm is not important because the function of the gene in the embryo depends on its expression in the mother during oogenesis and in this way the gene function is supplied to the egg. For this same reason, the mutant phenotype is expressed in 100% of the progeny of mutant females (in cases of fully penetrant mutations), because all eggs are made from the same mutant ovarian germ line. The first forward genetic screens to isolate recessive maternal-effect mutations in vertebrate species were performed in the zebrafish (Dosch et al., 2004; Kishimoto, Koshida, Furutani-Seiki, & Kondoh, 2004; Pelegri et al., 2004; Pelegri & Schulte-Merker, 1999; Wagner et al., 2004) and led to a number of important new findings (reviewed in Abrams & Mullins, 2009; Lindeman & Pelegri, 2010).

Testing of putative homozygous males in the Mullins screen also led to the identification of a surprisingly large number of paternal-effect mutations (Dosch et al., 2004; Wagner et al., 2004; Yabe, Ge, & Pelegri, 2007). Paternal-effect genes were manifested in the progeny of homozygous mutant males, irrespective of the genotype of the mother. The identification of paternal-effect genes was somewhat unexpected considering the well-established view that most of the material inherited by the embryo from its parents is stored in the large, sessile oocyte, as opposed to the much smaller, motile sperm cell. However, the identification of these mutations is consistent with the expected requirement for male germ line genes in chromosomal segregation during meiosis in spermatogenesis (Poss, Nechiporuk, Stringer, Lee, & Keating, 2004), as well as the inheritance of a pair of centrioles in the sperm cells for the centriole-less oocyte (Delattre & Gönczy,

(*t2800*), recovered in an EP-based screen (Pelegri et al., 2004), results in defects in the induction of dorsoanterior cell fates. (C) A wild-type embryo 24 hours after fertilization shows the normal body plan, including the head (h) and the notochord (n), a dorsal mesoderm derivative. (D) An embryo from *hecate* mutant mothers lacks anterodorsal structures and is radially symmetric (Lyman-Gingerich, Lindeman, Putiri, Stolzmann, & Pelegri, 2006; Lyman-Gingerich, Westfall, Slusarski, & Pelegri, 2005). (t) indicates the tail region in both wild-type and mutant embryos, which is less extended and contains multiple folds in the mutant. The *asterisk* indicates a group of cells that accumulates at the anterior of the embryo.

2004; Yabe et al., 2007), and the isolation of paternal-effect mutants in *Drosophila* (Wakimoto, Lindsley, & Herrera, 2004).

Despite this progress, our knowledge of the functions of maternal- and paternal-effect genes (which for simplicity we will refer to as parental-effect) in early development in zebrafish and vertebrates in general remains superficial and disconnected. Ongoing and future screens in zebrafish, together with studies of the current mutants in zebrafish and mouse, and further molecular investigations in *Xenopus* should help fill these gaps of knowledge.

1. STRATEGIES FOR ADULT TRAIT AND PARENTAL-EFFECT SCREENS

There are two approaches that have been used to identify de novo recessive mutations affecting adult and parental-effect genes in the zebrafish: (1) F_3 screen for adult traits and F_4 for parental-effect traits based solely on natural crosses; and (2) an F_2 screen for adult and F_3 for parental-effect based on gynogenesis, specifically the technique of early-pressure (EP). These alternatives differ in various important ways, which are summarized in Table 1 and described throughout this chapter. Here, we discuss these alternatives and provide detailed protocols to implement them.

1.1 F_3 SCREEN FOR ADULT TRAITS AND F_4 FOR PARENTAL-EFFECT TRAITS BASED ON NATURAL CROSSES THAT INTEGRATES A MAPPING STRATEGY

One method to produce recessive, homozygous maternal-effect mutants is through a three-generation inbreeding strategy. Although at first glance this method is expected to occupy an enormous amount of tank space, consolidating the F_3 generation into a single tank comprised of an "F_3-extended family" makes this approach considerably more practical. This scheme begins similarly to F_3 zygotic screens (Driever et al., 1996; Haffter et al., 1996; Mullins, Hammerschmidt, Haffter, & Nüsslein-Volhard, 1994). As shown in Fig. 3A, G_0 male fish are mutagenized with 3 or 3.3 mM ENU as described (Mullins et al., 1994), crossed to wild-type females, and their F_1 progeny raised. Each F_1 fish carries a different set of mutagenized genes derived from the independently mutagenized spermatogonial cells of their fathers. F_1 fish are interbred and the F_2 progeny, referred to as an F_2 family, are raised to adulthood. Each F_2 family contains two mutagenized genomes, one from its mother and one from its father. In a zygotic screen, the fish of a given F_2 family are intercrossed, and the F_3-embryos are screened for recessive mutant defects. To identify adult and parental-effect mutations, the F_3-progeny are instead raised to adulthood. As with a zygotic mutation, for an adult or parental-effect mutation that is present in an F_2 family, one quarter of the F_2 intercrosses will yield F_3 families containing 25% mutant adults or parental-effect mutants.

Table 1 Summary Between an F_4 Screen Based Solely on Natural Crosses and an F_3 Screen Based on Gynogenesis (EP)

	Genetic Approach	
	F_4 Based on Natural Crosses, (F_3-Extended Family)	**F_3 Based on Gynogenesis**
Background strain used in screen	Lethal- and sterile free	Lethal- and sterile-free. Amenable to IVF and gynogenesis
Basic methodology	Sibling pair matings	Induction of gynogenetic clutches using EP or HS2
Number of generations needed	Three, plus maternal-effect test	Two, plus maternal-effect test
Amount of space needed	Large to moderate	Moderate to small
Fraction of females within a family expected to exhibit maternal-effect phenotype	1/16	EP: Variable, between 50% towards 0%, depending on distance between the locus to the centromere HS2: fixed at 50%
Incorporation of mapping scheme	Feasible	More difficult, due to low number of surviving gynogenotes
Identification of maternal-zygotic and paternal-effect mutations	Feasible	More difficult due to lower number of siblings and reduced fertility of gynogenotes

Similar to a zygotic mutation, if a recessive, adult or parental-effect mutation exists in an F_2 family, the probability of identifying it directly depends on the number of F_3 families generated from F_2 intercrosses and the number of F_3 females screened from each F_3 family (see Section 1.1.3 for formulas). To obtain a 90% probability of making a mutation homozygous, eight F_2-intercrosses are required, corresponding to eight F_3 families raised from each F_2 family. If these eight F_3 families each occupy a separate tank, then an eight-fold increase in tank space is required beyond that needed for an F_3 zygotic screen. To make such a maternal-effect screen practical for the moderately sized fish facility, multiple F_3 crosses from a single F_2 family are pooled into an "F_3-extended family," comprised of siblings and cousins, and raised in a single tank. Specifically, if eight crosses from each F_2 family are desired, then equal numbers of each cross are pooled and raised together. If it is necessary to set up the F_2 family more than once to obtain the desired eight crosses, then separate pools are generated on different days, which are then pooled at a later point.

In an F_3-extended family, recessive adult or parental-effect mutants represent 1/16 of the total adults or males or females, rather than one-quarter because F_2

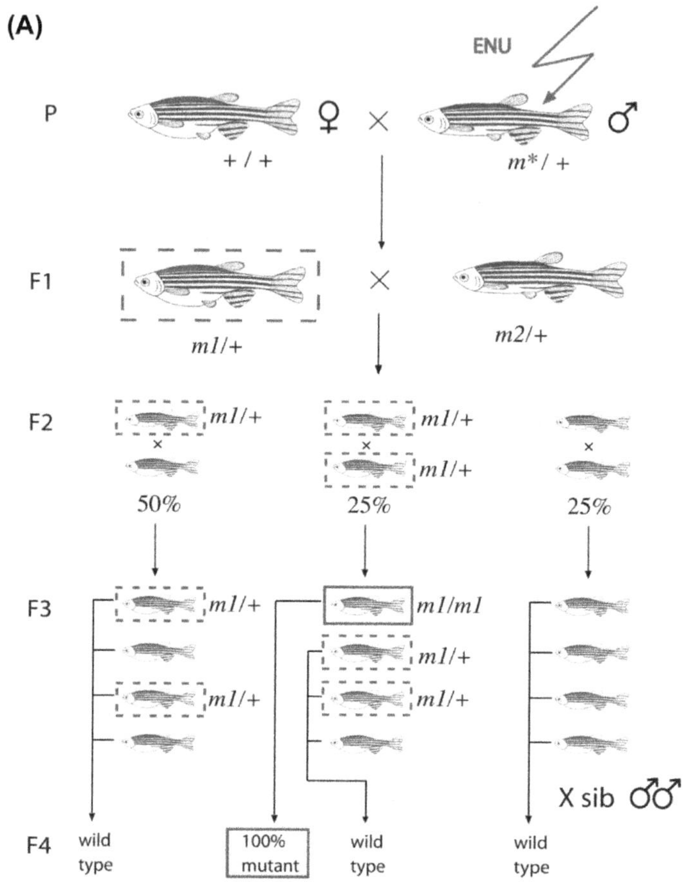

FIGURE 3

(A) F_4 natural crosses screen strategy. Males of the parental generation (P) are mutagenized with ethylnitrosourea (ENU), to induce new mutations *(m*)* and crossed to wild-type females. F_1 fish are raised, each of which carries a different set of mutagenized genes. Two mutations are shown, *m1* and *m2*, each carried by one of the two F_1 fish. Only mutation *m1* is followed in subsequent generations for simplicity. Two F_1 fish are intercrossed and an F_2 family raised. Half of the individuals of the F_2 family are heterozygous for *m1*. The F_2 fish are intercrossed to make an F_3-extended family (see text), composed of equal numbers of F_3 fish from each of the F_2 intercrosses of one family. One quarter of the F_2 intercrosses are between *m1* heterozygotes, producing *m1* homozygotes in 25% of their F_3 progeny. F_3 individuals can be tested for larval, juvenile, and adult phenotypes, as well as for parental effects in the F_4 generation. (B) EP-based screen strategy. F_1 heterozygous females carrying newly induced mutations (*boxed with hatched lines*; *m** and *m1* as in (A)) are treated to induce gynogenetic F_2 clutches, which can contain homozygotes for maternal-effect mutations (*boxed with solid lines*). F_2 individuals can be tested for larval, juvenile, and adult phenotypes, as well as for parental effects in the F_3 generation. A fraction of EP-derived progeny will be heterozygous for the mutation. Hypothetical results are shown using a gene with an average centromere-locus distance (see text).

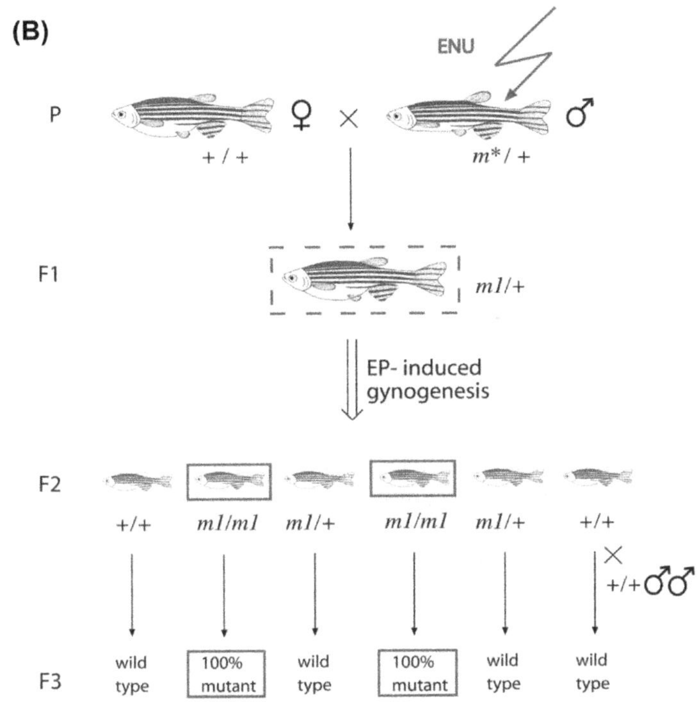

FIGURE 3

Continued

intercrosses generating mutant and nonmutant progeny are pooled. The advantage is that only one tank is occupied and screened, rather than eight, which is a significant saving in space. However, the same number of adults is screened whether they are in eight separate or one consolidated tank. To obtain an 80% probability of identifying a mutant if it exists in an F_3-extended family (see also Section 1.1.3), 25 F_3 individuals must be screened. F_3 individuals are screened for phenotypes using the desired assays (adult phenotypes) or by crossing them to sibling or wild-type individuals and examining their F_4 progeny for defects (parental-effect phenotypes). Far fewer F_4 embryos are examined for defects in a parental-effect screen than a zygotic screen, because all or nearly all embryos are affected, in contrast to 25% for a zygotic mutant phenotype.

1.1.1 An F_3-extended family approach with integrated mapping

A chromosomal mapping cross can be integrated into an F_3 adult or F_4 parental-effect screen using a natural crosses approach, which allows one to map the mutation to a chromosomal position in the F_3 generation. This integrated mapping strategy is particularly beneficial in the case of parental-effect mutations, which are difficult to propagate because they typically produce all nonviable progeny. However, a similar difficulty would arise if the mutations are detrimental during the juvenile or adult stages to the extent that mutant carriers cannot reproduce. In these cases it is necessary to identify heterozygous females and males. This can be greatly facilitated by mapping the mutation to a chromosomal position, which has the additional value of initiating the molecular isolation of the mutated gene. The mapping strategy is discussed further in Section 4.2.

1.1.2 Identification of maternal-zygotic, male sterile, and paternal-effect mutations

Crossing F_3 females to F_3 sibling/cousin males permits the simultaneous isolation of maternal-zygotic, zygotic, as well as male sterile and paternal-effect mutations. F_3 intercrosses provide a ~50% probability of detecting a mutant that requires loss of both maternal and zygotic gene activity, which is not possible if the F_3 female is crossed to a wild-type male. All zygotic mutations in the F_2 family are still present in the F_3 family, so the F_4 embryos from F_3 intercrosses can be screened not only for maternal-zygotic, but also for zygotic mutants, if desired. In F_3 sibling/cousin crosses, paternal-effect and male sterile mutations are also revealed.

If a mutant is identified in the F_4 embryos, the type of mutation induced can be distinguished by crossing the F_3 female and male parental fish separately to wild-type fish and examining the progeny for defects. If the mutation is a maternal- or paternal-effect or a female or male sterile mutation, then the defect is evident in the F_4 embryos, even when the F_3 fish is crossed to wild type. For a maternal-effect or female sterile mutation, the F_3 female will be the cause of the defective embryos, whereas for a paternal-effect or male-sterile mutation, the F_3 male parent will be the sole cause of the defect. For recessive, zygotic and maternal-zygotic mutations, the F_4 embryonic defect will depend on both F_3 parents. A different fraction of affected F_4 embryos is expected for fully penetrant maternal-zygotic versus zygotic mutations. For a maternal-zygotic mutant, the F_3 female is homozygous and the F_3 male heterozygous for the mutation resulting in 50% mutant progeny, whereas a zygotic mutation yields 25% F_4 mutants.

1.1.3 Assessment of the F_4 natural crosses screen for maternal-effect mutations

Typical values for several parameters of an F_4 natural crosses screen are shown in Table 2. The results of an F_4 natural crosses strategy, using an F_3-extended family and incorporating a mapping cross, have been published (Dosch et al., 2004; Wagner et al., 2004). In such an F_4 screen the number of mutagenized genomes (G) screened contributed by a given F_3-extended family is determined with the following formula: $G = (1 - 0.9375^n) \times 2 \times (1 - 0.75^m)$. The term $(1 - 0.9375^n)$ is the probability of identifying an F_3, recessive maternal-effect mutant female present in an F_3-extended family if "n" females are screened within that family. The factor "2" represents the two mutagenized genomes derived from the two F_1 fish. The term $(1 - 0.75^m)$ is the fraction of the two mutagenized genomes expected to be homozygous in the F_3 generation, where "m" is the number of F_2-crosses that comprise an F_3-extended family.

Table 2 Statistics in an F_4 Natural Crosses Screen

% F_2 families used to generate F_3 families[a]	85%
#F_2 intercrosses generating an F_3-extended family	>8
Fraction of F_1 mutagenized genomes homozygous in F_3-extended family	>90%
#F_3 females screened/F_3-extended family	24
Fraction of genomes screened in F_3 family	79%
# Haploid genomes screened/#F_3-extended families[b]	1.4
Maternal-effect mutants identified/genome screened	0.11 genomes
Fraction of candidate mutations that are recovered	95%

[a] *Due to the female bias in hybrid strains that are needed to make the mapping cross, a fraction of the F_2 families have three or less males. These families are difficult to work with and so are discarded.*
[b] *Since F_1 fish are interbred to make the F_2 genome, there are two mutagenized genomes present in the F_2 family and therefore greater than one mutagenized genome is ultimately screened in each F_3-extended family.*

1.2 F_2 ADULT OR F_3 PARENTAL-EFFECT TRAIT SCREENS BASED ON EP-INDUCED GYNOGENESIS

Artificially induced gynogenesis in zebrafish involves diploidization of the maternal haploid genome, producing viable offspring with solely a maternal genetic contribution (Streisinger, Walker, Dower, Knauber, & Singer, 1981). The incorporation of gynogenesis into a genetic scheme for adult or parental-effect mutations allows the direct production of homozygotes for induced mutations from a single heterozygous F_1 carrier, bypassing one generation in comparison to a scheme based solely on natural crosses (compare Fig. 3A and B). Due to the large number of chromosomes present in the zebrafish, genetic screens in this organism involve the whole genome and are essentially blind, screening all mutagenized chromosomes simultaneously, rather than individually as done in the fly and worm. Therefore each generation in a screen generates an exponentially increasing number of crosses. Thus, bypassing one generation through the use of EP allows a significant reduction of the time and space required to carry out an adult or parental-effect screen.

In a basic gynogenesis-based scheme (Fig. 3B), mutations are induced in the germ line of parental (P) males by exposing them to the point-mutagen N-ethyl-N-nitroso-urea (ENU) (van Eeden, Granato, Odenthal, & Haffter, 1999; Mullins et al., 1994; Solnica-Krezel, Schier, & Driever, 1994). P males are then crossed to produce F_1 progeny heterozygous for induced mutations. Eggs are extruded from the F_1 females and gynogenesis is induced. This allows newly induced mutations to become homozygous in up to 50% of the gynogenetic F_2 generation (see Section 1.2.1). Adult F_2 individuals can be screened directly for mutations affecting adult traits as well as parental-effects by testing their F_3 progeny for embryonic phenotypes.

In the originally reported EP-based screen (Pelegri et al., 2004; Pelegri & Schulte-Merker, 1999), the production of F_3 clutches to detect parental-effects was carried out by in vitro fertilization (IVF) using wild-type sperm. Using IVF and a reliable source of sperm allowed the production of F_3 clutches that can be immediately observed and followed synchronously, thus facilitating the identification of early phenotypes (see Section 1.3.2). In addition, because all clutches are fertilized by the same batch of sperm solution, IVF allows the rapid identification of maternal mutations that affect the ability of the egg to become fertilized (which requires retesting in a natural crosses screen approach, because unfertilized clutches are sporadically observed in natural matings from wild-type parents).

Originally gynogenetic lines were shown to exhibit strong sex biases, with a large majority of them consisting almost entirely of males (Pelegri & Schulte-Merker, 1999; F.P. unpublished). In the rare female-rich lines amenable to the identification of maternal-effect genes and gynogenesis itself, males were either absent or unable to produce fertilized clutches. Such background semisterility interferes with the isolation of male sterile mutations (eg, resulting from defects in spermatogenesis), as well as with the use of sibling crosses that may reveal maternal-zygotic effects. Current improvements in gynogenetic techniques have allowed the production of robust lines exhibiting even sex ratios with largely fertile males (Heier, Takle, Hasley, & Pelegri, 2015; F.P. unpublished results), and these improvements should allow the incorporation of schemes to identify paternal sterile and maternal-zygotic mutations in gynogenetic screens. The difference between the sex ratio of the early reported and more recently developed gynogenetic lines may relate to differences in the genetic background of the stock used to generate the lines, as well as a greater ability to select for lines with normal sex ratios using improved gynogenetic conditions. Regardless of the degree of background semisterility, a gynogenetic approach is in principle amenable to identifying paternal-effect phenotypes that manifest in fertilized embryos (eg, defects in chromosome segregation during meiosis (Poss et al., 2004), centriole complement defects (Yabe et al., 2007)), since these phenotypes would be readily observable in most if not all fertilized embryos.

The originally reported gynogenesis-based screens obtained a relatively small number of surviving and fertile individuals within F_2 clutches, which precluded the incorporation of a mapping strategy within the genetic scheme, as described above for screens based solely on natural crosses (see Sections 1.1.1 and 4.2). Improvements in gynogenetic methods allow one to readily generate lethal-free lines from multiple polymorphic genetic backgrounds (see also Section 2.2), allowing in principle the incorporation of mapping strategies into the screening strategy. However, due to a large fraction of carrier siblings within gynogenetic families, an effective strategy is to initiate mapping crosses after identification of the mutation (see Section 4.3).

Although the incorporation of gynogenesis can simplify an adult and parental-effect genetic screen, gynogenesis itself is only efficiently induced under specific conditions. In a scheme for such a gynogenesis-based screen, the main goal is the efficient production of viable and fertile gynogenetic F_2 adults that are homozygous for newly induced mutations. A number of variables need to be optimized to carry out this procedure, which we discuss below. First, a suitable method of gynogenesis needs to be selected. Secondly, an appropriate mutagenesis dosage needs to be chosen to induce a reasonably high rate of mutations, while allowing the production of viable homozygous adult mutants. In addition, lines amenable to gynogenetic procedures need to be selected (see Section 2.2).

1.2.1 Choice of gynogenetic method

There are two main techniques for the artificial induction of gynogenesis in the zebrafish: EP and heat shock (HS) (Streisinger et al., 1981), and recently an alternative to the standard HS method, HS2, has allowed increased efficiency (Heier et al., 2015). In both types of methods, eggs are first artificially fertilized with sperm whose genetic material has been inactivated by UV irradiation. In the absence of further treatment, these eggs would develop into haploid embryos that are inviable. Both EP and HS/HS2 lead to the diploidization of the genetic content of the egg, thus producing viable, diploid embryos (Fig. 4).

In EP, diploidization is induced by the application of hydrostatic pressure between minutes 1.33 and 6 after egg activation (see Section 5.3.6). This treatment inhibits completion of the second meiotic division (specifically the expulsion of the 2° polar body), resulting in a diploid egg. HS and HS2, on the other hand, inhibit cytokinesis during one of the embryonic mitotic divisions (see Section 5.3.5). HS is carried out by applying a heat pulse during minutes 13 to 15 after egg activation to inhibit cytokinesis of the first embryonic cell cycle, whereas the heat pulse in HS2 occurs at minutes 22 to 24 after egg activation to inhibit cytokinesis of the second cell cycle. In both cases, the inhibition of cytokinesis during precisely one embryonic cell cycle, coupled to unaffected DNA synthesis during that cycle, results in the transformation of haploid embryos into diploid ones. Both methods act through heat shock inhibition of centriole biogenesis, with HS2 inhibiting centriole duplication during the first cell cycle and standard HS likely interfering with centrosome reconstitution after fertilization (Heier et al., 2015; Vertii, Zimmerman, Ivshina, & Doxsey, 2015). Hydrostatic pressure through the method of late pressure (LP) has also been used as an alternate method to inhibit the first mitosis (Streisinger et al., 1981). The pressure treatment in LP occurs at a developmental time point similar to that for the heat pulse in HS2 and presumably acts via a similar mechanism. However, LP involves a more cumbersome pressurizing step and has not been widely used.

Methods involving inhibition of embryonic mitosis, such as HS or HS2, are more effective than methods that inhibit meiotic division, such as EP, in the direct induction of complete homozygosity, and therefore might be in principle the technique of choice in an adult or parental-effect screen. This is because HS-derived progeny are homozygous at every single locus, and therefore 50% of HS-derived F_2 progeny are homozygous for a mutation present in heterozygous form in the F_1 mother. EP, on the other hand, due to recombination during meiosis, leads to a variable degree of homozygosity ranging from 50% towards 0% depending, respectively, on whether loci are linked to the centromere or are distally located. Thus, HS or HS2 would in principle provide the highest possible yield of homozygous mutant adults for

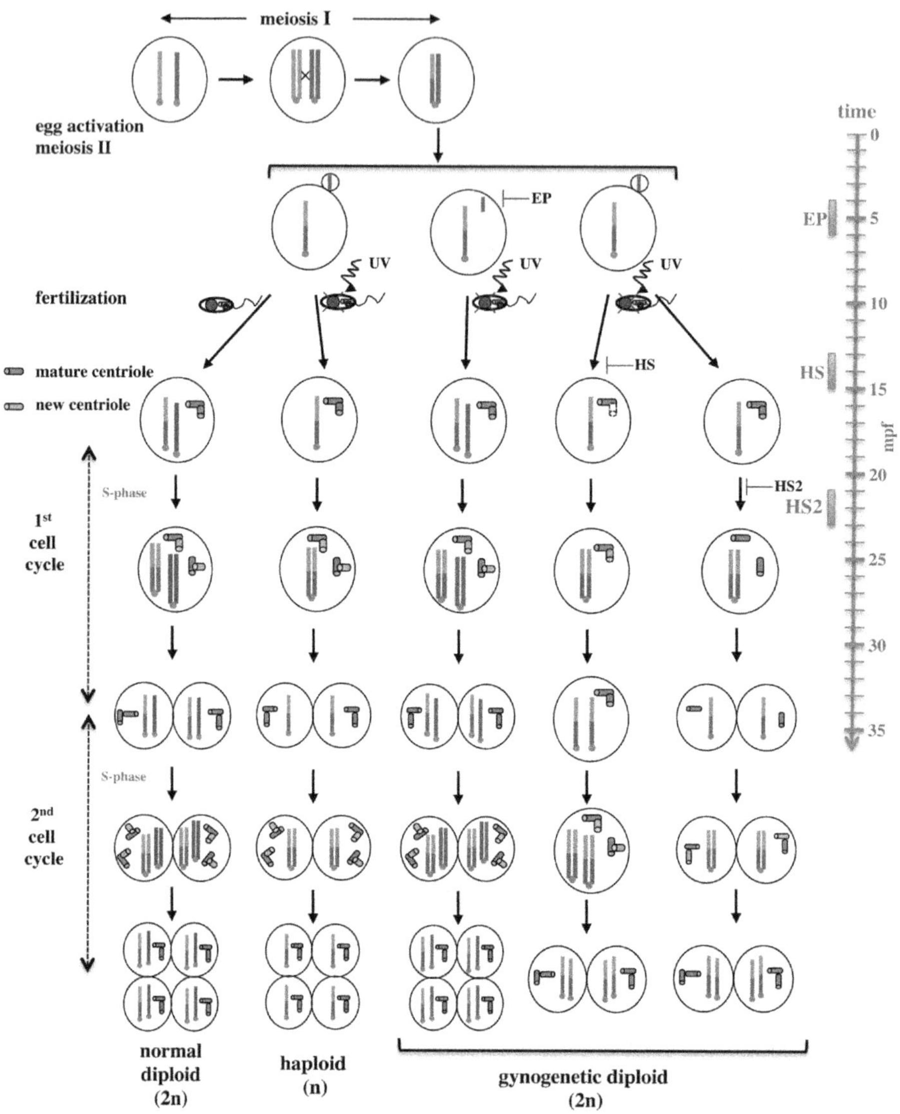

FIGURE 4

Generation of haploid and gynogenetic diploid embryos. Mature eggs are arrested at metaphase of meiosis II. Contact with water results in egg activation and the resumption of meiosis. IVF with UV-treated sperm results in haploid embryos. In EP, hydrostatic pressure at 1.33–6 min after activation (mpa) inhibits polar body extrusion resulting in a diploid oocyte. In the standard HS method, a heat pulse at 13–15 mpa results in a one-cycle stall in cytokinesis for the first cell cycle and the unaffected DNA replication transforms the zygote from haploid to diploid. In HS2, a heat pulse at 22–24 mpa causes a one-cycle stall in cytokinesis for the second cell cycle resulting in the transformation of embryonic blastomeres

Table 3 Comparison of Standard Heat Shock- and Early Pressure-Induced Gynogenesis (Gol-Mix Line)

	Heat Shock	Early Pressure
Viability at d5 (Viable/fertilized eggs)[a]	0.09 n = 3590	0.21 n = 4368
Fraction clutches with >6 viable d5 fish	0.41 n = 29	0.93 n = 29
Clutch size (Viable d5 fish/clutch)	10 n = 29	37 n = 29
Adult viability (Viable at 3 mo./d5 viable)	0.53 n = 324	0.66 n = 218
Fertility (Fertile adults/total adults)	0.23 n = 13	0.65[b] n = 226

[a] *Viable at day 5 are defined as fish that can inflate their swim bladders.*
[b] *Value from F_2 descendants of P males mutagenized with 2 mM.*

all loci regardless of their chromosomal location. Moreover, the expectation of a fixed percentage of homozygous mutant females would aid in the assessment of newly identified phenotypes.

Previous gynogenesis-based screens used EP rather than the standard HS method. This is because, compared to standard HS, EP yields an increased number of gynogenotes at 24 h (Table 3; see also Streisinger et al., 1981), presumably due to a greater ease of inhibiting polar body extrusion during meiosis, as well as higher adult viability and fertility, probably due to higher levels of heterozygosity. The main drawback of the heterozygosity associated with EP gynogenotes is that it leads to an intrinsic bias against the identification of distally located mutations. Measurements of the frequency of homozygosity (F_m) of random zygotic mutations after EP-induced diploidization range from 0.50 to 0.04, with an average value of 0.23 (16 loci; Streisinger, Singer, Walker, Knauber, & Dower, 1986; Neuhauss, 1996). With the assumption that adult and parental genes are similarly distributed

from haploid to diploid. The heat pulse at these stages has been proposed to inhibit cytokinesis through effects on centriole biogenesis. Mature and new centrioles are represented in dark and light green, respectively. Full maturation of the sperm-derived daughter centriole is hypothesized to be affected by standard HS (represented in white). EP results in partial heterozygosity in chromosomal regions distal to recombination events. HS and HS2 produce whole genome duplication and complete homozygosity, regardless of chromosomal location. (See color plate)

Modified, with permission, from Heier, J., Takle, K., Hasley, A., & Pelegri, F. (2015). Ploidy manipulation and induction of alternate cleavage patterns through inhibition of centrosome duplication in the early zebrafish embryo. Developmental Dynamics, 244, *1300–1312.*

throughout the chromosomes, these data suggest that the majority of these genes are sufficiently close to a centromere to be identified through an EP-based screen.

However, the recent finding that HS2 can produce up to four-fold higher yield of gynogenotes compared to standard HS at 24 h (Heier et al., 2015) brings the yield of HS2 to levels comparable to those for EP. This improvement suggests that HS2 can be effectively used for gynogenesis screens, combining acceptable yields with the advantages of full genetic homozygosity and fixed and high rates (50%) of homozygotes for newly induced mutations regardless of chromosomal position.

1.2.2 Mutagenesis dose

In the F_2 gynogenetic generation, homozygosity for mutations in essential zygotic genes will lead to a decreased survival of gynogenotes. For example, the mutagenic dosage used in large-scale zygotic screens (3×1 h treatments with 3 mM ENU) is expected to induce about one embryonic lethal and one larval lethal per haploid genome (Mullins et al., 1994; Solnica-Krezel et al., 1994; Haffter et al., 1996). This implies that under this mutagenic condition, only 59% of what would be otherwise viable EP-derived gynogenotes (25% using HS) would survive to adulthood. Thus, we reduced the ENU dosage in our maternal-effect screen experiments. Similar reductions in the strength of mutagenic treatments were adopted for maternal-effect screens in *Drosophila* and *Caenorhabditis elegans* (see, for example, Kemphues, Kusch, & Wolf, 1988; Lehmann & Nüsslein-Volhard, 1986). We have observed that a mutagenic dosage of 3×1 h 2 mM ENU treatments begins to have a mild effect on the viability of F_2 gynogenetic clutches (not shown). These conditions lead to a mutagenic rate, as assayed by the frequency of newly induced *albino* alleles, estimated to be about one-third of the rate induced by the standard (3 mM ENU) treatment (Mullins et al., 1994; Pelegri & Schulte-Merker, 1999; Solnica-Krezel et al., 1994), or about 0.3–0.4 embryonic lethal mutations per haploid genome. In a previous EP-based screen, an ENU concentration of 2 mM worked well as a compromise between a moderate mutagenic rate and a practical level of viability (Pelegri & Schulte-Merker, 1999).

1.2.3 Assessment of EP-based screens for maternal-effect mutations

The results from genetic screens using an EP-based method have been described elsewhere (Pelegri et al., 2004; Pelegri & Schulte-Merker, 1999). Typical survival and yield values are presented in Table 4.

To estimate the number of genomes screened using an EP-based method, one needs to keep in mind that the number of genomes screened depends on the level of EP-induced homozygosity (F_m), which in turn is inversely related to the centromere to locus distance. Thus, the number of genomes screened will differ according to the position of genes with respect to the centromere. To estimate such values, one can first estimate the critical number of F_2 females that need to be screened in a given family to result in a 90% probability of detecting a newly induced mutation present in that family. For example, with an average F_m value of 0.23

Table 4 Statistics in an F_3 Gynogenesis-Based Screen

% F_2 clutches grown to adulthood[a]	45%
% F_2 clutches with fertile adult females[b]	20%
# Screened haploid genomes/#F_2 clutches with fertile females[c]	0.45 (proximal)–0.27 (average)
Maternal-effect mutants identified/# haploid genomes screened[d]	0.11–0.19
Percentage of candidate mutations that are recovered[e]	44%

[a] *EP-derived F_2 clutches with at least six viable fish on day 6 of development.*
[b] *Fertile females are defined as those that produce normal eggs, which upon activation exhibit the wild-type translucent appearance and can be fertilized to exhibit either a normal or a characteristic abnormality in the early cleavage pattern.*
[c] *The number of genomes screened depends not only on the number of females tested but also on the average distance of the loci to the centromere (Section 1.2.1). The values presented are derived from the number of females tested for each family, so that mutations present in the family have a 0.97 chance of being identified, assuming F_m values of 0.50 and 0.23, respectively, for centromere-linked loci and loci at an average distance to the centromere. F_2 families are considered to carry a candidate mutation when they contain females that produce a phenotype in 100% of the F_3 offspring, and the phenotype appears in more than one independent F_3 clutch.*
[d] *The range given is estimated by assuming that the isolated mutations are all either proximal (0.11) or at an average distance to the centromere (0.19).*
[e] *Mutations in some lines are not recovered due to a variety of reasons: false positives in the original tests, inability to recover the line due to insufficient fish to perform recovery crosses, and variability in the penetrance of the mutation.*

(see Section 1.2.1), nine F_2 tested females per clutch would be needed to detect a newly induced mutation with a 90% probability. This estimate corresponds to an average locus and varies greatly depending on the centromere-locus frequency: to reach a similar frequency of detection for centromere-linked (F_m: 0.5) and distal (F_m: 0.05) loci, the critical number of F_2 females tested per clutch is 4 and 44, respectively. For specific F_m values, each family where the number of tested F_2 females is equal to or larger than the critical number of F_2 females contributes one screened haploid genome. In cases where the number of F_2 females tested per family is less than the critical value for a given F_m, the number of tested F_2 females can be pooled to find a combined number of screened genomes contributed by that pool. The latter value is calculated by dividing the pooled number of tested F_2 females by the critical value needed to screen one haploid genome at a 90% certainty. The total number of haploid genomes screened is the sum of all families with more than the critical number of tested F_2 females and the combined number of genomes calculated from the pooled number of tested F_2 females.

1.3 SCREENING

Adults homozygous for newly induced mutations, either in the F_3 generation in an extended-family screen or the F_2 generation in an EP-based screen, can be screened for adult or parental-effect phenotypes. In practice, both of these types of traits may be screened in tandem within the same screen, first for any number of adult phenotypes and subsequently, upon crossing and creating a subsequent generation, for parental-effect genes in the progeny. This is only applicable, however, when the adult phenotype assays are noninvasive and are not expected to interfere with each other, gametogenesis or mating behavior.

When adult screen assays are not compatible with each other (eg, when adult assays are invasive or involve treatments that may affect another assay), multiple assays can still be combined albeit utilizing different sets of screened fish. However, the screens as normally carried out generate a relatively low number of homozygous mutants: for an F_3 extended family approach 1/16 of a typical size family of 60, or about four (or two if the phenotype is specific to one of the sexes); for an F_2 EP-based gynogenesis approach 1–12 (0–6 if sex-specific) of a clutch of 24 reaching adulthood. Considering that the identification of multiple individuals within the same family can be crucial to distinguish genetically transmitted traits from possible syndromes with a nongenetic basis, screens as typically carried out are therefore reaching the limits of detection. Therefore, in cases of noncompatible assays it is advisable to generate a larger number of adults (ie, an increase in number proportional to the number of noncompatible assays being used). This allows for sufficient screenable adults for reliable detection of new phenotypes while taking advantage of the already generated mutagenized P0 males and heterozygous carrier F_1 individuals.

1.3.1 Screening for adult traits

Adults potentially homozygous for newly induced mutations can be screened for any desired adult phenotype (see, for example, Andreeva et al., 2011; Bauer & Goetz, 2001; Haffter et al., 1996; Johnson & Weston, 1995; Lang et al., 2009; Tschopp et al., 2010). Given the limited number of expected adult homozygous mutants in a given family or clutch (see above), adult trait screens create a particular concern in that the screenable assays need to be robust. One aspect to assay robustness is whether the mutation exhibits the phenotype with a high penetrance when present in homozygous mutant individuals. Ideally, this can be confirmed by the existence of previously isolated, highly penetrant mutations with a similar phenotype. In the absence of such preexisting mutations, a pilot screen may determine whether highly penetrant mutations affecting the trait can be isolated.

Another important variable to assay robustness is the incidence of false positives caused by nongenetic deficiencies and syndromes. This can be determined in practice by determining, prior to initiating the screen, the fraction of individuals that appear to test positive (defective) in a large number of unmutagenized, wild-type adult fish of the same genetic background. Because of sporadic false positive effects, it is highly advisable that, when possible, any putative homozygous mutant individual be tested at least a second time with the primary assay, or tested with a different yet related assay. Such retests or secondary screening tests can eliminate most sporadic effects. Acceptable rates of false positives are therefore determined by the fraction of screened individuals that can be feasibly retested/secondarily tested during the screening process. For example, false-positive incidence rates that result in having to retest up to 10% of all screen families may be acceptable, provided that these subsequent tests can be readily incorporated into the ongoing screen.

Even when a putative mutant individual retests with the same phenotype as in the primary screen, it remains a possibility that the cause is syndromic (nongenetic). In this case, secondary, more precise screens, which may be more involved but have a lower rate of false positive incidence, may allow one to distinguish between nongenetic syndromes and newly identified mutations. Ultimately, however, propagation of the mutation to a subsequent generation in a Mendelian manner is the only fool-proof method to verify the genetic basis for an observed phenotype. Because such a propagation step is labor intensive, the assay(s) used in the screen should attempt to eliminate false positives as much as possible, thus minimizing the fraction of lines that need to be tested for mendelian segregation through additional generations.

1.3.2 Screening embryos for parental-effects

Once adults that may be homozygous for a parental-effect mutation are produced, embryos from those adults are screened for potential defects. For parental-effects, the uncertainty caused by sporadic nongenetic developmental defects is lessened by the fact that parental-effect phenotypes caused by homozygosity in the parent are observed in multiple progeny individuals (ideally most if not all of potentially hundreds of embryos), each of them providing support for the existence of a true genetic mutation. This essentially amplifies the signal from the screen and frequently allows one to infer the presence of a new mutation from a single identified mutant individual (a notable exception to this is the case of axis-deficient or "ventralized" phenotypes, which occur sporadically as a maternal effect at a relatively high frequency in wild-type females; in this case retesting of the female or identification of multiple mutant-affected sibling females with the same phenotype is essential).

To screen for parental-effect phenotypes, embryos are collected as early as possible or within 2 h after fertilization. As mentioned above (see Section 1.2), in EP-based strategies for maternal-effect genes, clutches can be derived by IVF. This allows one both to observe and synchronize their development immediately after fertilization. On the other hand, if the screened embryos are produced through natural crosses, early observation and synchronization of the clutches can be approached by taking advantage of the propensity of zebrafish to lay eggs during the early hours of their daylight cycle. In practice, this is done by setting up crosses towards the end of the light cycle and collecting embryos during the early hours of the following light cycle. Alternatively, pair matings can be set up during the first 2 h of the light cycle, and embryos can be collected in the same day (this prevents the occurrence of eggs being laid during the previous night cycle but has the disadvantage that egg laying is less synchronized, and continuous monitoring for egg laying is needed throughout the first half of the light cycle).

The early collection of eggs allows one to discard clutches that have undergone aberrant ovulation, which can occur in a small fraction of clutches from wild-type females and which if undetected would provide false positives in the screen. This early observation also allows one to determine whether the cleavage and cellularization pattern characteristic of wild-type embryos is observed. Because unfertilized embryos also exhibit a pattern of irregular cleavages (pseudocleavages (Kane & Kimmel, 1993; Kimmel et al., 1995)), it is important that such early embryos are carefully observed to reveal potential differences between pseudocleavage formation and an abnormal early cellular pattern. Once the regular pattern of cellular cleavage characteristic of normal fertilized embryos is detected, fertilized embryos are sorted and transferred to a clean plate at low densities (40 embryos per 10.5 cm dia. plate). Embryos are subsequently screened for deviations from the wild-type developmental pattern (Kimmel et al., 1995; see van Eeden et al., 1999 for a sample scoring chart).

This screening strategy relies on the incorporation of the sperm into the egg, which is necessary for patterns of cleavage distinct from those in unfertilized eggs. Parental-effect mutations acting at earlier steps in oogenesis and egg maturation may also be identified by screening clutches shortly after fertilization; for example, maternal-effect mutations affecting the animal-vegetal axis of the egg manifest their defects in the directional movement of ooplasm during egg activation (Fig. 3B, compare to A; Dosch et al., 2004), and paternal-effect mutations affecting centriole duplication during spermatogenesis exhibit a one-cycle delay in cell division (Yabe et al., 2007). In addition, it is possible to identify mutations affecting gametogenesis by selecting for nonegg laying females or in the case of males, selecting for a failure to fertilize eggs and then screening these individuals for gonadal defects by dissection and/or sectioning (Bauer & Goetz, 2001; M.C.M. unpublished).

2. SELECTION OF LINES FOR GENETIC SCREENS

2.1 SELECTION FOR LETHAL/STERILE-FREE BACKGROUND LINES

An important characteristic desired in a genetic background is the absence of preexisting mutations, whether zygotic or parental. In any kind of screen, whether based on gynogenetic techniques or natural crosses, the use of lines free of preexisting mutations is important for two reasons. First, lines free of lethal mutations diminish unwanted background lethality, which reduces brood sizes and can preclude the isolation of new mutations linked to the background mutation. In addition, the use of lines free of preexisiting mutations eliminates the possibility of isolating multiple copies of a mutant allele already present in the genetic background. Selecting for the absence of lethal or sterile mutations can be carried out in two ways:

2.1.1 Continuous inbreeding

Wild-type stocks free of zygotic lethal and sterile mutations can be obtained by inbreeding individuals for two generations and essentially screening the F_2

generation for lethality and sterility phenotypes. In a stock maintained by mixed breeding of many individuals (to maintain genetic diversity and prevent unhealthy, highly inbred stocks) two generations of inbreeding of several pairs of fish can reduce the likelihood of background mutations being present in the parental generation. Individual pairs of wild-type fish are intercrossed and their respective F_1 progeny raised in separate tanks. The F_1 progeny are then intercrossed and screened for zygotic lethal mutations in the F_2 generation. A reliable indicator of zygotic lethal mutations is the lack of swim bladder inflation at 5 dpf, in addition to obvious defects at earlier stages in 25% of the brood. By examining F_2 embryos from at least 12 intercrosses from one F_1 family, a >95% probability exists that a mutation will be detected, if it exists in that particular family.

If lethality is not observed in any of the 12 crosses, then the F_2 fish from the 12 individual F_1 intercrosses are raised in a separate tank and screened for late lethal mutations, as well as maternal- and paternal-effect, and female and male sterile mutations. The total number of F_2 adults is counted and compared to the number of larvae initially raised. If a late lethal mutation exists, then 25% of the larvae will not survive to adulthood. Several crosses can be raised between unrelated individuals to control for nongenetic lethality associated with normal fish raising. Only lethality significantly beyond that of normal fish husbandry is then considered as a potential late lethal mutation. From each F_2 family 12 males and females are intercrossed and their embryos examined. If the F_3 embryos are normal, then the probability is >95% that maternal- and paternal-effect and sterile mutations do not exist in those F_2 fish. If two such F_2 lines are established from different F_1 fish, then the males can be mutagenized and then interbred to females from another F_2 line in the parental generation of the screen to prevent further inbreeding.

2.1.2 Whole genome homozygosity through gynogenesis

The gynogenetic methods of HS and the higher-efficiency version HS2 induce homozygosity at every locus (see Section 1.2.1), thus allowing one to select in a single generation for fish that lack all background lethal or sterile mutations. HS2-derived gynogenetic clutches can be derived from any desired background line and the population expanded via natural crosses. Previous studies have shown that selection of lines via gynogenetic methods results in improved gynogenetic yields (Streisinger et al., 1981), presumably by the reduction of background detrimental alleles concommitant with the selection of genetic backgrounds that respond favorably to manipulation by the heat pulse treatment. Thus, inclusion of rounds of HS2-derived gynogenesis within the propagation scheme of lethal-free lines allows continued selection for high yields through this method. Lethal-free lines can be derived from as few as a single pair of HS2-derived gynogenotes from a single female, representing less than two haploid genomes, or as a mix of a larger number of HS2-derived gynogenotes to maintain a higher level of genetic diversity.

2.2 SPECIFIC REQUIREMENTS FOR LINES IN GYNOGENESIS-BASED SCREENS

The majority of lines we have examined, including lines recently derived from the wild, tend to produce low yields of fertile gynogenotes (Pelegri & Schulte-Merker, 1999). Selection of appropriate lines is therefore important for an efficient gynogenetic-based maternal screen.

2.2.1 Selection of lines that produce a high yield of gynogenotes

The experimental induction of gynogenesis relies on the manipulation of in vitro fertilized eggs at very early stages. Therefore, it is necessary that females should, as a first requirement, readily yield eggs when manually stripped. In the absence of stimulation by the male during natural mating, fish strains differ greatly in their ability to be manually stripped of eggs (Eaton & Farley, 1974; Pelegri & Schulte-Merker, 1999). This may be related to the fact that, under natural conditions, release of mature oocytes from their follicles into the ovarian lumen requires hormonal stimulation (Selman, Petrino, & Wallace, 1994), which may normally be triggered by vigorous chasing by the males (Eaton & Farley, 1974). Lines most easily stripped of eggs appear to be those that have been propagated by artificial fertilization methods, which also involve stripping of eggs, such as those derived from the AB Oregon line (Streisinger et al., 1981). In contrast, lines from the wild or laboratory lines that have been propagated mostly by natural crosses tend not to be readily stripped of eggs. However, we find that even in unselected lines females can be readily stripped of eggs if they have initiated pair mating via natural crosses. Thus, in preparation for gynogenesis induction we routinely set up females in natural pair matings, and interrupt the mating as soon as the first few eggs are laid. Females from such interrupted natural matings readily yield eggs that are competent for fertilization through the standard manually stripping procedure.

2.2.2 Selection for favorable sex ratios under gynogenetic conditions

Sex determination in fishes varies from species with sex-determining chromosomes to multifactor autosomal ones, and in some cases sex has been shown to be influenced by external factors (reviewed in Chan & Yeung, 1983). The mechanism of sex determination in domesticated zebrafish, although poorly understood, appears to fall into the latter category (Wilson et al., 2014). Domesticated zebrafish lack a single sex chromosome and sex determination is sensitive to growing conditions. Many gynogenetic clutches, after grown to adulthood, exhibit sex ratios that are strongly biased toward males (86%–88% males; see Fig. 4 in Pelegri & Schulte-Merker, 1999). The phenomenon of sex bias in gynogenetic clutches is likely related to the tendency of zebrafish and other teleosts to develop into males under suboptimal conditions, for example, in overcrowded conditions or in subviable genetic backgrounds (see Chan & Yeung, 1983). Presumably, gynogenetic clutches, because of their high degree of inbreeding, also have a suboptimal genetic background that under normal circumstances produces males. Nevertheless, a small fraction of the gynogenetic clutches (5–10%) are composed of a high fraction of females, ranging from 50% to 100% (Pelegri & Schulte-Merker, 1999). Such clutches can be used to generate a line with high rates of gynogenetic females. Treatment of this line with testosterone for the first 14 days of development leads to the production of fertile males for stock propagation and/or mutagenesis. More recent studies involving a larger number of HS2-derived gynogenotes show that gynogenetically derived lines can be robust and of even sex ratio (Heier et al., 2015; F.P. unpublished data). It is likely that underlying genetic variability in the stock used to derive such gynogenetic lines provides genetic constellations that allow normal sex ratios even under gynogenetic conditions.

2.3 A HYBRID/INBRED APPROACH

Selection of lines can increase the frequency of certain desired traits but also leads to inbreeding, the latter of which can result in a reduction in overall robustness and fertility (Thorgard, 1983). Thus, the best lines for gynogenesis-based maternal screens might be hybrids between gynogenetically selected lethal/sterile-free lines. This approach is essential for genetic screens that incorporate a simultaneous mapping strategy. In this case, two separate polymorphic lines can be selected independently via HS2-based gynogenesis. If a single genetic line selected for high yields under EP or HS2 does not exhibit inbreeding depression, it should be also possible to use it as the sole background within a genetic screen scheme, which also allows mapping of the mutation through natural crosses after mutant identification.

3. RECOVERY AND MAINTENANCE OF ADULT AND PARENTAL-EFFECT MUTATIONS

Once individuals are identified as exhibiting an adult and/or parental-effect phenotype, the mutation needs to be recovered. In the case of viable mutations (eg, affecting pigmentation or adult morphology but not viability or fertility), recovery and maintenance of the mutation is straightforward, because new stocks can be initiated from the affected individuals and the mutation can be maintained as a visible marker or even a homozygous stock. Here we present strategies to recover and maintain mutations that are either not viable or not fertile when homozygous. In this case, the observed adult, infertile or parental-effect phenotypes are expected to be caused by homozygosity for recessive mutations, because dominant mutations are unlikely to be propagated through generations that occur prior to screening. Because homozygosity for recessive lethal, infertile, or parental-effect mutations leads to the inability of an affected individual to reproduce, a genetic scheme has to allow the recovery of the mutations through genetically related individuals.

3.1 GENERAL METHODS FOR RECOVERING MUTATIONS

A mutation can be recovered by three general means: through known heterozygous carriers, through siblings of homozygous mutant individuals, and through rare survivor progeny derived from homozygous mutant individuals.

3.1.1 Recovery through known heterozygous carriers

Individuals that produce homozygous affected adults (F_2 parents in an F_3 adult/F_4 parental-effect natural crosses screen, and F_1 individuals in a EP F_2 adult/F_3 parental effects screen) are heterozygous carriers of the mutations. Thus, such fish are stored separately until their progeny reach adulthood and are tested for potential phenotypes. After the 3 to 5 months that are required to grow up and test their progeny, we find that the majority of the separated individual fish are still alive and fertile, and thus can be used to recover the mutation.

3.1.2 Recovery through siblings of homozygous mutant individuals

In some instances a known carrier for a mutation is not available or is not fertile. In these cases, maternal-effect mutations can be recovered by performing crosses between siblings of homozygous mutant individuals (F_3 siblings in an F_3 adult/F_4 parental-effect natural crosses screen and F_2 siblings in an F_2 adult/F_3 parental-effect EP-based screen), a fraction of which are carriers of the mutation. If the mutation is mapped (see Section 4.2), siblings can be selected that are either homozygous or heterozygous for the mutation by genotyping.

If the mutation has not yet been mapped, propagation is ensured by raising progeny from multiple sibling intercrosses. In an F_3-extended family strategy (where families are composed of siblings and cousins), ~50% of the sibling males and females are heterozygous carriers and 6.25% of the males and females are homozygous carriers. Thus, 25% of F_3 sibling/cousin intercrosses are between heterozygotes of the mutation and will yield F_4 homozygous mutant individuals. If the F_3 (or subsequent) generation is made between two heterozygous carriers, then 75% of all F_3 males and females are carriers (50% heterozygous and 25% homozygous for the mutation). In this case, 50% of the intercrosses should yield F_4 homozygous mutants.

In EP-based screens, it is preferable to generate outcrosses, rather than incrosses, for the recovery of mutations. This is because EP-derived fish do not mate as efficiently as wild-type fish. In addition, the background used in an EP-screen results in female-rich tanks, and such abnormal ratios interfere with subsequent propagation of the mutation. We have found that outcrossing to a line such as leopard long fin (also known as TL), which tends to have a slight bias towards maleness, results in hybrid stocks that have normal sex ratios in subsequent generations. Outcrossing also improves the general robustness and fertility of the line.

In the case of maternal-effect mutations isolated in EP-based screens, outcrossing is preferably carried out through sibling F_2 males, rather than sibling females, because females homozygous for a maternal-effect mutation are sterile, whereas homozygous males should be fertile unless the mutated gene also affects male fertility.

This is particularly true in the case of centromere-linked loci (F_m towards 0.5), when most *fertile* females are expected to be homozygous for the wild-type allele (see Section 1.2.1 and below) and therefore cannot transmit the mutation. The reverse logic applies to paternal-effect mutations.

In EP-derived clutches, the frequency of heterozygotes and homozygotes for a given mutation varies depending on the centromere-locus distance. For centromere-linked loci (F_m close to 0.5), 50% of the siblings are homozygous for the mutation. As the centromere-locus distance increases, the fraction of homozygous siblings (F_m) decreases but the fraction of heterozygous siblings increases two times as rapidly. For a distal mutation of $F_m = 0.05$, for example, 5% of the F_2 siblings are homozygous mutant, and 90% are heterozygous carriers. Thus, the overall frequency of F_2 carrier siblings (homozygous or heterozygous), varies from 50% for centromere-linked loci to percentages approaching 100% for distal loci. Therefore, the recovery of mutations through F_2 siblings can also be an efficient strategy in EP-screens. In large F_2 EP-derived clutches, F_m, and therefore the fraction of siblings that are heterozygous or homozygous carriers for the mutation, can be estimated by the proportion of F_2 individuals that exhibit the adult or parental-effect mutant phenotype.

In HS2-based screens, clutches consist solely of fish homozygous for either the mutant or wild-type allele, each at a 50% frequency. Thus, maternal- or paternal-effect mutations need to be recovered by outcrossing F_2 siblings of the opposite sex: ie, males in the case of maternal-effect mutations and females in the case of paternal-effect mutations. When a mutation affects fertility in both sexes, recovery through F_2 siblings is not possible and such mutations need to be recovered through F_1 female carriers or rare F_3 survivors.

3.1.3 Recovery through rare survivors

In cases of mutations that are not 100% lethal or sterile, the mutation may also be recovered by crossing rare homozygous mutant survivors of a generally lethal larval or juvenile defect (F_3 individuals in a natural crosses screen and F_2 individuals in an EP screen), or by raising and crossing rare surviving progeny of a parental-effect or sterile mutant adult (F_4 clutches in a natural crosses screen and F_3 clutches in an EP screen). The presence of such "escapers" may be due to variability in the phenotype caused by residual function of a hypomorphic allele or some degree of redundancy in the affected pathways. Progeny from such subviable/subfertile individuals are expected to be heterozygous carriers for the mutation, and the mutation can be propagated by incrossing fish derived from them.

3.2 MAINTENANCE OF ADULT AND PARENTAL-EFFECT MUTATIONS

Whether a mutation has been mapped or not, it is tempting to maintain stocks carrying it either by repeated inbreeding or through "escaper" embryos (see Section 3.1.3). However, repeated inbreeding eventually generates stocks that are weak and have aberrant sex ratios, typically with an excess of males (see Section 2.2.2), thus interfering with future stock propagation and, in the case of maternal-effect mutations, the identification of homozygous mutant females. Moreover, maintenance of the mutation by repeated propagation through "escaper" embryos may select for genetic modifiers that gradually weaken the mutant phenotype. Lastly, particularly in the case of a higher ENU dose as used in a natural crosses extended family approach, mutagenesis leads to the induction of multiple lethal mutations. Such mutations are typically unlinked to the adult/parental-effect mutation but can nevertheless reduce the size of intercross families and are best crossed out of the mutant background. To address these issues, adult and parental-effect mutations can be routinely propagated through cycles of crosses to a wild-type stock (outcrosses), followed by crosses between siblings (incrosses). Typically, an outcross that is known to carry a mutation can be kept for a period of time and additional incrosses performed from the outcross fish to produce new families containing homozygous individuals. It works well to perform a cycle of one outcross, which can generate several incrosses over a period of a year or more, and then initiate a new cycle by carrying out an outcross from one of the more recent incrosses.

3.2.1 Maintenance of recessive lethal adult or parental-effect mapped mutations

To propagate a mutation that has been mapped to a chromosomal position and/or for which the molecular nature of the mutant gene is known, crosses between known carrier individuals are generated. When only the chromosomal position is available, identification of heterozygous or homozygous carriers is inferred by linkage of markers (eg, Single Sequence Length Polymorphisms), flanking the mutant allele that are polymorphic to those linked to the wild-type allele. For mutations where the molecular lesion is known, methods for single nucleotide polymorphism (SNP) genotyping, for example using PCR-based end-point genotyping (Livak, Flood, Marmaro, Giusti, & Deetz, 1995), can unambiguously identify homozygous or heterozygous carriers, which can be used in propagation crosses. A highly reliable, fast SNP genotyping method recently developed called KASPar (Kettleborough et al., 2013; Nijman, Kuipers, Verheul, Guryev, & Cuppen, 2008; Smith & Maughan, 2015) is also particularly valuable for genotyping point mutations or deletion alleles in zebrafish (M.C.M. unpublished). Outcrossing of genotyped homozygous individuals to wild type produces an F_2 generation consisting solely of heterozygous carriers, which can then be incrossed to produce an F_3 generation containing homozygous mutant individuals, and this breeding cycle can be repeated. To increase efficiency, multiple rounds of F_3 incrosses can be performed from the same F_2 carrier generation before performing another breeding cycle involving a new set of outcrosses.

As an example, in the case of female-sterile or maternal-effect genes, males homozygous for the mutation are selected from siblings of homozygous mutant females (25% of the male siblings are expected to be homozygotes). Outcrosses are then initiated from such homozygous males to wild-type females. The progeny of this cross are all heterozygous carriers and can be interbred to produce a family that contains 25% homozygous mutant females. The reverse strategy can be followed (propagation through homozygous sibling females) to identify sibling carriers and propagate paternal-effect mutations.

3.2.2 Maintenance of recessive lethal adult or parental-effect unmapped mutations

If the mutations are not mapped to a chromosomal location, a similar strategy is followed except through multiple, random crosses. In the case of adult phenotypes, multiple crosses can be initiated through viable, fertile siblings, 66% of which are expected to be heterozygous carriers of the mutation (the expected Mendelian frequency of 75% carriers is modified in this case because homozygous mutant individuals are inviable or not fertile). If the mutations are female-sterile or have a maternal-effect, and similar to when recovering these mutations (see Section 3.1.2), it is more efficient to initiate the crosses through sibling males, 75% of which are expected to be either homozygous or heterozygous carriers for the mutation. The reverse logic would apply to male-sterile and paternal-effect mutations.

Multiple crosses ensure the propagation of the mutation. For example, in the case of a maternal-effect mutation, 75% of outcrosses derived from males that are siblings of homozygous mutant females consist of families of carrier individuals. Thus, five random outcrosses from sibling males of homozygous mutant females ensure a 99.9% probability of transmission of the allele to at least one of the outcrosses. Within such outcrossed-derived families, the percentage of heterozygotes is expected to be 100% or 50%, depending, respectively, on whether the original outcrossed male is homozygous or heterozygous for the mutation. Multiple incrosses from such families allow the recovery of homozygous mutant females in the next generation. For example, eight incrosses from a tank that consists of 50% heterozygous carriers lead to a 90% probability of finding homozygous females (at a 25% frequency) in at least one of the incrossed families.

A variation of these approaches is to outcross identified heterozygous or homozygous carrier individuals. These individuals are identified as carriers by interbreeding them with siblings, raising the progeny and determining if their offspring yields mutant individuals. Parental individuals yielding progeny with mutant adult/parental-effect phenotypes are then outcrossed. These outcross progeny are then inbred to produce a new generation of homozygous mutant individuals, as discussed above. In this modified approach, two blind generations of intercrosses are avoided by first identifying the individual carriers prior to outcrossing them.

During the maintenance of mutations, individual carriers should be outcrossed to fish of the same strain. This avoids increasing the degree of polymorphism in the carrier line, which in turn facilitates the subsequent process of mapping (see Section 4).

4. MAPPING ADULT AND PARENTAL-EFFECT MUTATIONS, AND CLONING THE MUTANT GENE

Mapping a mutation to a chromosomal position can be carried out either simultaneously with an F_3 adult/F_4 parental-effect screen using natural crosses (see Section 1.1), or at any time after the identification and recovery of the mutation. Specific details on mapping protocols have been previously described (Geisler, 2002; Talbot & Schier, 1999). Here we describe the modification of this approach for mapping of adult or parental-effect mutations. Briefly, the approach consists of outcrossing a carrier of the mutation to a polymorphic wild-type stock to yield F_1 hybrid families. Incrosses from the F_1 family in turn allow the production of F_2 adults, which can be tested for homozygosity of the mutation and analyzed for linkage to DNA markers throughout the genome. Prior to reviewing these classic strategies, we will discuss the newest methods to map parental-effect mutations to chromosomal regions and to clone the mutant gene both using Next Generation Sequencing (NGS) technology.

4.1 MAPPING AND MUTANT GENE IDENTIFICATION USING NEXT GENERATION SEQUENCING TECHNOLOGY

With the advent of NGS methods and the availability of the complete sequence of the zebrafish genome (Howe et al., 2013; Kettleborough et al., 2013), it is now possible to resequence the genome of mutant lines at relatively low cost or perform high depth transcript sequencing via RNA-seq at even lower cost (due to its reduced complexity). These methods do not require generation of a map cross to a polymorphic strain, because the inherent sequence variation within the major noninbred zebrafish strains (eg, TU, AB, WIK, TLF) has been found to be sufficient for mapping with both of these methods (Bowen, Henke, Siegfried, Warman, & Harris, 2012; Henke et al., 2013a, 2013b; Hill et al., 2013; Miller, Obholzer, Shah, Megason, & Moens, 2013). Using bioinformatic analysis, linkage of an SNP specifically with high frequency in pooled mutant individuals (discussed below in bulk segregant analysis) and with significantly lower frequency in siblings (heterozygotes and wild-type) is used to determine the chromosomal position of the mutation. Sufficient sequence depth is required to obtain high confidence of linkage of an SNP to the mutation. Confirmation of linkage is then performed in additional mutant and wild-type fish. Specific SNPs with 100% linkage to the mutation can correspond to the mutant lesion. Linkage is insufficient, however, as proof that the SNP is the causative mutation. Traditional methods are still required to prove the final identity of the mutant gene, which can include induction via CRISPR/Cas9 of a second allele or rescue of the mutant phenotype via expression of the wild-type gene. Because of the considerable reduced complexity of the transcriptome versus the genome, RNA-seq is more cost-effective than whole genome sequencing, although it is not certain to identify the causative mutation if the mutation does not lie within the transcript sequence.

These and other NGS methods can also be used to identify the mutant gene when the mutation is already mapped to a chromosomal location. In this case, the analysis is less complex and limited to the defined interval. We recently used RNAseq to identify the mutant gene of a maternal-effect mutant gene split top, which we had previously mapped to a 7.9 Mb region using bulk segregant analysis (see below) (Langdon et al., 2016). Sequencing mRNA from 128- to 256-cell stage mutant embryos, a point prior to the onset of wholesale zygotic transcription, lead to the identification of the mutant gene as encoding the lysosomal endopeptidase, Cathepsin Ba (Langdon et al., 2016). The mutant gene was identified based on the near absence of the transcript in the RNA-seq analysis. The correct identity of the gene was verified by mRNA injection of the wild-type cathepsin Ba gene into embryos from mutant females, which fully rescued them (Langdon et al., 2016).

Sequencing of the defined interval of a mapped parental-effect mutation can also be achieved using targeted pull-down methods (Dapprich, Ferriola, Magira, K, & Monos, 2008; D'Ascenzo et al., 2009; Gupta et al., 2010; Rehman et al., 2010). We cloned two mutant genes via a sequence capture method (Gupta et al., 2010; M.C.M. unpublished). In this method short oligos are made complementary to genomic sequence at 8 kb intervals spanning the region. The oligos are hybridized to the genomic DNA of a wild-type (+/+) and a homozygous mutant fish, extended using biotinylated nucleotides, and then the DNA captured via streptavidin attached to magnetic beads. The DNA is isolated from the beads and prepared for sequencing. Sequence complexity is low, so can be aligned and the candidate causative mutation identified more easily. We performed a 616 kb genomic DNA sequence capture using 77 oligos spanning the region and identified three candidate mutations, one of which was confirmed as the causative one in the *macf1* gene (Gupta et al., 2010). Although it is a relatively quick method, the costs of whole genome or transcriptome analysis are now sufficiently inexpensive and rapid that the additional oligo design for the capture method makes it currently less attractive.

4.2 MAPPING CONCOMITANT WITH F_3 ADULT/F_4 PARENTAL-EFFECT GENETIC SCREENS

A mapping cross can be integrated into an extended-family natural crosses screen strategy (see also Dosch et al., 2004). Two strains that are polymorphic to each other, eg, TU and AB, are mutagenized. The mutagenized males are crossed to females of their respective strain to produce an F_1 generation. F_1 fish are then interbred between the two strains to make a hybrid F_2 generation. F_2 fish are intercrossed to make the mutations homozygous in the F_3 generation and F_3 mutant adults can be used to map the mutation. The F_1 grandparent DNA is crucial in examining linkage using bulk segregant analysis. Thus, the F_1 fish are frozen and kept for mapping purposes, should a mutation be found that one wants to map.

Intercrossing strains to make a map cross gives rise to very robust stocks, so-called hybrid vigor. As a consequence of interbreeding F_1 fish of different strains, we find the F_2 hybrid generation to be particularly healthy and prolific, with increased reproductive longevity, compared to either independent strain. This is advantageous in regenerating the identified mutations to produce additional individuals for mapping (see below). However, a drawback is that hybrid vigor leads to an increased propensity to produce females in the F_2 hybrid generation. Thus, typically 10–20% of F_2 families yield three or fewer males. We assess the sex ratio at about 2 months of age, and discard those with fewer than four males. In the F_3 generation, the sex ratio is not distorted and we rarely find such sex biased families. It is possible that future lines could be developed that do not exhibit the sex bias in the F_2 hybrid generation.

As with zygotic mutations (Geisler, 2002; Talbot & Schier, 1999), bulk segregant analysis can be used to map adult and parental-effect mutations. However, there is considerably more effort to generate adult and parental-effect mutant individuals than zygotic embryonic mutants, which are used in the bulk-segregation protocol. We have found that for maternal-effect mutations, 14 mutant females are sufficient to map a mutation efficiently. In using an extended family screen (see Section 1.1), mutant females represent 1/16 of the total, producing one to five mutant females in a tank of 60 fish. Even five females is insufficient to map a mutation efficiently. Thus in performing such a screen, we keep up to 10 pairs of fish of the F_2 generation in a small 2-L tank. If we are interested in a mutation, we return to the F_2 fish and regenerate the mutant through F_2 pair-wise crosses. Each F_3 family is raised separately and the F_2 fish stored individually. Regenerated F_3 families are screened for mutant females (as described in Section 3.2.2); the F_2 parents of those that yield mutants can then be used to produce more mutant females. The F_3 mutant and nonmutant sibling females are then used for mapping the mutation. We routinely use as few as 14 mutant females to map a mutation, and have mapped mutations with just nine mutant individuals. In the latter cases, considerably more false-positive linkages are detected in bulk segregant analysis. We have reliably regenerated >50 maternal-effect mutants and efficiently mapped most of the mutations using this strategy (Abrams et al., 2012; Bontems et al., 2009; Dosch et al., 2004; Gupta et al., 2010; Holloway et al., 2009; Kapp, Abrams, Marlow, & Mullins, 2013; Langdon et al., 2016; Marlow & Mullins, 2008; Mei, Lee, Marlow, Miller, & Mullins, 2009; Wagner et al., 2004; Yabe et al., 2009; M.C.M., unpublished).

4.3 MAPPING AFTER IDENTIFICATION AND RECOVERY OF MUTATIONS

In the case of viable and fertile adult mutations, homozygous mutant individuals can be directly identified by their phenotype, and can be used to generate mapping crosses by crossing to a polymorphic strain such as WIK. Incrossing of the resulting hybrids leads to a generation with homozygous mutants, which can be used for bulk segregant analysis to find linkage as described above.

In the case of lethal/sterile adult mutations or parental-effect mutations, genetic mapping can be initiated by a process similar to that carried out for the maintenance of unmapped mutations (see Section 3.2.2). Multiple (P) siblings of affected (presumed homozygous) individuals are outcrossed to a polymorphic strain. F_1 hybrid individuals from such outcrosses are incrossed at random to generate F_2 crosses, a fraction of which will yield homozygous mutants.

For adult lethal mutations, 66% of (P) siblings are expected to carry the mutation, and these can be used to initiate the mapping crosses. In the case of female-sterile and maternal-effect mutations, it is most efficient to initiate mapping crosses from homozygous mutant (P) sibling males, because 75% of males are expected to be carriers compared to 66% of sibling females. Among these males, 25% are expected to be homozygous for the mutation. Outcrosses of homozygotes yield F_1 families where all individuals are heterozygous carriers for the mutation, so that all random F_2 incrosses contain homozygous mutant females. Outcrosses of heterozygous carrier males, expected at a 50% frequency amongst the siblings of homozygous mutant females, produce F_1 families where 50% of individuals are heterozygous carriers, so in this case only 25% of random F_2 incrosses yield homozygous mutant F_2 females. Thus, it is more efficient to initiate the mapping strategy using homozygous males, because a much larger fraction of incrosses from the F_1 hybrids (100% compared to 25% if using heterozygous males) yield homozygous F_2 females. Parental males can be identified in advance as homozygotes through genetic crosses, and then the mapping cross initiated with such a male. Alternatively, the mapping strategy can be initiated with eight sibling males of undetermined genotype, which are outcrossed to polymorphic females. This relatively large number of outcrosses increases the probability that at least one of the outcrosses originates from a homozygous male. After the F_1 hybrid fish are incrossed to make an F_2 generation, they are kept separately. Multiple incrosses from F_1 hybrids and testing of F_2 females for maternal-effects allows inferring whether the original P male was homozygous for the mutation, if it was not identified in advance, and which pairs of F_1 hybrids are heterozygous carriers. Such pairs, now identified, can be crossed repeatedly to produce more mutant F_2 females for mapping. This general strategy is reversed in the case of male-sterile or paternal-effect mutations.

F_2 individuals from mapping crosses are separated into two phenotypic classes according to the specific adult/parental-effect assay: phenotypically mutant individuals (ie, individuals homozygous for the mutation) and phenotypically wild-type (ie, siblings that are either heterozygous or homozygous for the wild-type allele). Identified mutant individuals may be retested with the same assay or a different, secondary assay to ensure that they produce the expected phenotype and to check for potential errors in handling during phenotypic identification.

After their classification into phenotypic classes, F_2 individuals are anesthetized, their tail fin clipped and the tail fin DNA is isolated (the remaining part of the body is also frozen and serves as a backup in case additional DNA needs to be isolated). Individual tail fin DNA from 20 individuals of each phenotypic class is used to make two DNA pools. These pools are used to carry out a first-pass mapping (see Section 1.1; Geisler, 2002; Talbot & Schier, 1999) to identify linkage. Once linkage has been found, DNA from single fish is analyzed separately with respect to polymorphisms to markers within the linked region (Geisler, 2002; Talbot & Schier, 1999).

4.4 EFFICIENT FINE MAPPING OF MATERNAL-EFFECT MUTATIONS

Narrowing down the location of a mutation through fine mapping is necessary to identify the molecular nature of the mutated gene through either candidate gene or positional cloning approaches. For viable adult and maternal-effect mutations, a different strategy is used for fine mapping, which can be performed much more efficiently than the initial mapping. Once a viable adult or maternal-effect mutation is mapped, then homozygous male and heterozygous female carriers can be identified by polymorphic markers flanking the mutation. Such F_0 fish are intercrossed to map finely the position of the mutation in the F_1 progeny. In a cross of a heterozygous female to a homozygous male, mapping will be performed with recombinants generated only through meiosis in the F_0 female, because the male is homozygous for the mutation. Thus, each F_1 fish represents a single meiosis, rather than two meioses for progeny from heterozygote intercrosses. However, the loss of recombination events from the homozygous male is offset by the fact that meiotic recombination is suppressed in males compared to females, so that the vast majority of all recombinants generated in intercrosses of heterozygotes are from female, not male meioses (Singer et al., 2002; S.L. Johnson, personal communication; M.C.M. unpublished). Thus, little is lost in crossing heterozygous females to homozygous males, and considerable gain is achieved using this strategy, as discussed below. Because this strategy depends on crosses from one homozygous mutant adult, this approach is not feasible for recessive mutations that result in adult lethality, or sterility in both sexes. Paternal-effect and male sterile mutations are also not effectively mapped by this strategy, given the reduced rate of recombination in male meioses.

Fine mapping of an adult viable or maternal-effect mutation through crosses between heterozygous females and homozygous males is similar to haploid mapping of zygotic mutations (Postlethwait & Talbot, 1997). In both cases, all fish can be examined for recombinants. For such mutations, both the phenotypically mutant and wild-type F_1 adults are examined for recombination between, respectively, the mutation and a wild-type linked flanking marker or the wild-type allele of the gene and a mutant-linked flanking marker. Thus, all the progeny (female progeny in the case of maternal-effect mutations) from a cross between a heterozygous female and a homozygous male are informative, in contrast to only 1/4 of the progeny in intercrosses of heterozygotes. This strategy saves considerable effort and tank space, which is important because it takes significantly more effort to generate adult individuals to map adult or maternal-effect mutations, compared to generating embryos or young larvae to map finely zygotic mutations.

Depending on the complexity of the assay, fine mapping can be made even more efficient in a map cross between a heterozygous female and homozygous male by genotyping all F_1 progeny, rather than phenotyping them. This is the case for maternal-effect mutations, because this strategy avoids having to generate crosses to examine their progeny except for those rare cases where an informational recombination event has occurred. In this case, all F_1 females (and males, see below) are genotyped with the closest polymorphic markers flanking the mutation to determine whether they are nonrecombinant mutants or heterozygotes, or recombinants within the interval of the flanking markers. Only the small subset of recombinants is phenotyped to determine whether they are mutant or wild-type females and thereby establish where the recombination occurred relative to the mutation. As the critical interval is narrowed and closer polymorphic markers are defined, fewer recombinants are identified and consequently fewer females phenotyped in crosses. We typically genotype individual fish at 2 months of age and maintain only the small fraction of recombinants until breeding age to determine their phenotype and the position of the recombination (the recombination "break point") relative to the mutation. The total number of F_1 females genotyped is compared to the number of recombinants to determine the genetic distance between the mutation and the flanking markers.

Once the interval of the maternal-effect mutation is narrowed to a ~0.5 centiMorgan region, we also find it worthwhile genotyping F_1 males from the intercross of the heterozygous female and homozygous male for recombination within the critical interval. The disadvantage of males is that a test cross must be performed between the recombinant male and a heterozygous female to determine whether the male is homozygous or heterozygous for the mutation. Female progeny carrying the recombinant male chromosome of this cross over a mutant chromosome must be tested to determine whether they exhibit a mutant or wild-type phenotype to assess the genotype of the recombinant male chromosome. Although this is a considerable effort, identifying a recombinant that narrows down the interval can be so valuable that we have found it worth the effort.

5. SOLUTIONS, MATERIALS, AND PROTOCOLS

5.1 SOLUTIONS

- MESAB stock solution: 0.2% ethyl-m-aminobenzoate methanesulfonate. Adjust to pH 7.0 with 1 M Tris pH 9.0. Keep at 4°C.
- MESAB working solution: 7 mL stock solution per 100 mL fish water.
- Hank's Solutions: Stock solutions 1, 2, 4, and 5 and Premix can be stored at 4°C. Stock Solution #6 is prepared fresh and added to the premix to form the final Hank's solution.

  Solution #1: 8.0 g NaCl, 0.4 g KCl in 100 mL double distilled (dd) H_2O.
  Solution #2: 0.358 g Na_2HPO_4 Anhydrous, 0.60 g KH_2PO_4 in 100 mL ddH_2O.
  Solution #4: 0.72 g $CaCl_2$ in 50 mL ddH_2O.
  Solution #5: 1.23 g $MgSO_4 \cdot 7H_2O$ in 50 mL ddH_2O.

Hank's Premix:	Combine the following in order: 10.0 mL Solution#1 1.0 mL Solution#2 1.0 mL Solution#4 86.0 mL ddH_2O 1.0 mL Solution#5

Solution #6 (prepare fresh): 0.35 g $NaHCO_3$ in 10 mL ddH_2O

Hank's (Final):	990 μL Hank's Premix 10 μL Solution#6

- E2 saline (used specially during the testosterone treatment because of its higher buffering properties): 15 mM NaCl, 0.5 mM KCl, 1 mM $CaCl_2$, 1 mM $MgSO_4$, 0.15 mM KH_2PO_4, 0.05 mM Na_2HPO_4, 0.7 mM $NaHCO_3$.
- E3 saline (a simpler version of E2 used for routine embryo raising): 5 mM NaCl, 0.17 mM KCl, 0.33 mM $CaCl_2$, 0.33 mM $MgSO_4$, 10^{-5}% Methylene Blue.
- Testosterone stock: 150 mg testosterone in 50 mL absolute ethanol. Store in aliquots at −20°C.
- Testosterone working solution: While stirring, add 10 μL of stock solution per 600 mL of (1) E2 saline, for babies before d6, or (2) fish water supplemented with 3 g/L Red Sea Salt (Read Sea Fish pHarm, Israel), for larvae between d6 and d15. In our hands E2 (instead of E3) and Red Sea Salt in the fish water improve the survival of testosterone-treated larvae. Stir 10 min.

5.2 OTHER MATERIALS

- UV lamp: Sylvania 18 in. 15 W germicidal lamp.
- French Press Cell, 40 mL (SLM-Aminco).
- French Pressure Cell Press (SLM-Aminco) or Hydraulic Laboratory Press (Fisher).
- Heat shock can be applied while embryos are held by a tea strainer similar to that used in embryo collection. Heat shock baskets can also be made by cutting off the bottom of Beckman Ultraclear centrifuge tubes, and heat-sealing a fine wire mesh to the bottom edge of the tube.
- EP vials: Disposable glass scintillation vials, with plastic caps (3.2 cm height and 2.2 cm diameter, Wheaton) or similar vials. The plastic caps are perforated several times with a needle to better allow exposure to the hydrostatic pressure. Only two vials of these type can fit at once in a pressure cell. To fit four vials in one cell, we have custom-built shorter plastic vials (1.8 cm height, including cap, 2.5 cm diameter, 0.3 mm wall thickness), which fit the plastic caps from the scintillation vials.

5.3 PROTOCOLS

5.3.1 Sperm collection (adapted from D. Ransom)

- A sperm solution can be made with testes dissected from 10 males for each 1 mL of Hank's solution. Keep the isolated testes and Hank's solution on ice. Shear the testes with a small spatula and by pipetting up and down with a 1000 μL Pipetteman. Allow debris to settle and transfer supernatant to a new tube. Sperm solution on ice is effective for about 2 h. Sperm still remaining inside the sheared, settled testes can be further collected by adding 300 μL of fresh Hank's and letting the mixture rest for 1/2 h or longer. For more details see (Ransom & Zon, 1999).

5.3.2 UV-inactivation of sperm

- Transfer the sperm solution to a watch glass. Avoid pieces of debris, because they may shield sperm from the UV light. Place the watch glass on ice at a distance of 38 cm (15 in.) directly under the UV lamp. Irradiate for 2 min with gentle stirring every 30 s. Transfer to a new Eppendorf tube with a clean pipette tip. UV-treated sperm solution on ice is effective for about 2 h.

5.3.3 Stripping of eggs

- Females can yield eggs to manual stripping if removed directly from their tank and placed in a clean tank (1–10 females per 2-L tank) the evening before stripping, but the success of this procedure is variable. Significantly more consistent results are obtained through the pre-selection of females, which have initiated natural mating (see also Eaton & Farley, 1974): the day before the IVF or gynogenetic experiment, set up males and females in pair matings in the same tank but with a physical separation (eg, using a transparent plastic divider or, if using mating inserts, with the males underneath the insert and females inside the insert). The morning of the experiment, allow males and females in the same space (lift the divider or place the male together with the female inside the insert). Visually inspect natural pair matings to detect the first signs of egg release. As soon as the first released eggs are observed, interrupt mating by separating females away from males (females pre-selected in this manner can be pooled). Interrupting mating as early as possible during the process allows for the largest number of eggs remaining in the female, resulting in higher IVF/gynogenetic yields. Best stripping and egg clutch quality are obtained during the first 4 h after the start of the light cycle. If kept overnight as pools of fish (as opposed to pair matings), the presence of males together with the separated females does not significantly affect the ability of females to be stripped of eggs in a gol-mix line (a strain of mixed AB/Tübingen background additionally marked by the golden mutation, our observations), although it may have an effect when working with other fish lines (Eaton & Farley, 1974).
- Anesthetize females in MESAB working solution until they reduce their gill movements (2–4 min, MESAB solution may need to be boosted through time

with more stock solution in 0.5–1 mL increments). Overexposure to MESAB will impede the recovery of the female, and fish should be placed in fresh water if they are not going to be used within 1 or 2 min after they stop their movements.

- With the aid of a spoon, rinse a female in fish water and place her on several paper towels to remove excess moisture.
- Place the female on the bottom half of a Petri plate. With a soft tissue, dry further the anal fin area. Excess water may prematurely activate the eggs.
- Slightly moisten the index fingers of both hands (dry hands will stick to the skin of the fish). With one finger support the back of the female, and with the other gently press her belly. Females that can be stripped will release their eggs upon gentle pressure. Healthy eggs have a translucent, yellowish appearance. Separate the eggs from the female with a small, dry spatula. Females can be placed separately in boxes and identifying tags can be attached to the box with the female and the corresponding egg clutch. If necessary, clutches can wait for several minutes before being activated. In this case, we cover the clutches with the Petri plate lid to reduce drying of the clutch. Fertilization can occur after even longer delays (in our hands, up to 6 min), although not in a consistent manner. Egg activation can be delayed for periods of 1.5 h or more with ovarian fluid from the rainbow trout or coho salmon (Corley-Smith, Brandhorst, Walker, & Postlethwait, 1999), or with Hank's saline buffer supplemented with 0.5% BSA (Sakai, Burgess, & Hopkins, 1997).

5.3.4 *In vitro* fertilization

- Add 25 μL of untreated or UV-irradiated sperm to the egg clutch. Mix the sperm and eggs by moving the pipette tip without lifting it from the Petri plate (to minimize damage to the eggs). If desired, proceed at this point to Heat Shock or Early Pressure Protocols. If not, add 1 mL of E3 saline to activate the eggs, and, after 1 min, fill the Petri plate with E3. Incubate at 27–29°C.

5.3.5 Heat shock

We describe below the protocol for HS2, with heat pulse at 22.0–24.0 min post-activation, which results in higher yields than the standard HS (Heier et al., 2015). To implement the standard HS protocol, embryos should be transferred to a fine mesh basket as early as possible for better synchronization and the heat shock pulse should be administered at 13.0–15.0 min post-activation.

- After IVF with UV-treated sperm, add 1 mL of E3 saline to activate the eggs and start the timer.
- After 1 min, add E3 to flood the petri plate.
- At 10 min, transfer the eggs to a tea strainer or heat shock basket. The waiting period allows the egg chorions to expand sufficiently so that they do not sift through the tea strainer mesh. Using a finer mesh size (as with a custom-made heat shock basket) may allow transfer at earlier time point for better

developmental synchronization. Immerse the strainer/basket in a water bath with E3 saline at 28.5°C.

- Heat pulse: at 22 min, blot briefly the bottom of the tea strainer/basket onto a stack of paper towels and transfer to a water bath with E3 saline at 41.4°C.
- At 24 min, blot briefly the bottom of the tea strainer/basket and transfer back to the 28.5°C E3 bath.
- Allow the embryos to rest for about 45 min and transfer to a Petri plate using E3 and a wash bottle. Allow embryos to develop in a 27–29°C incubator (see note under Section 5.3.6).

5.3.6 Early pressure

To maximize the number of clutches produced, one can work on cycles to include up to four clutches in separate vials within the Pressure Cell. For this, 6–12 females are typically anesthetized. Once four healthy-looking clutches are obtained, the females that have not yet been stripped of eggs are transferred to fresh fish water until they completely recuperate. It works well to begin to anesthetize females for the next EP cycle at around 4 min within a current cycle.

- After mixing eggs with UV-treated sperm (see Section 5.3.4), activate up to four clutches simultaneously by adding 1 mL of E3 saline to each clutch and start the timer (at least two people are required to timely manipulate four clutches).
- After 12 s, add more E3. A squirt to the side of the Petri plate will make the fertilized eggs collect in the middle of the plate.
- With a plastic pipette, transfer the fertilized eggs to an EP vial. Fill the vial with E3 and cap it with the perforated plastic lid. Avoid large air bubbles. Place the vials inside the pressure cell, ensuring that no air remains trapped inside it. Record the relative position of the clutch within the pressure cell by placing the tags in the corresponding order on a dry surface. Fill the pressure cell with E3 and close it allowing excess E3 to be released from the side valve. Close the side valve without overtightening. Insert entire assembly on the French Press apparatus and apply pressure to 8000 lbs/sq. in. by time 1 min 20 s after activation. For different strains and/or presses, different pressure values may be optimal (see Gestl, Kauffman, Moore, & Cheng, 1997).
- At 6 min, release the pressure and remove the pressure cell from the French Press apparatus. Maintaining the relative order of the vials, remove the vials from the pressure cell, dry them with a towel and label them with their corresponding number tags. Place the vial in a 27–29°C incubator.
- After all EP cycles have been completed, allow the embryos to rest in the vial for at least 45 min but no more than 4 h. Transfer embryos with their corresponding tags to Petri plates. Let embryos develop in a 27–29°C incubator.

Note: Due to the large amount of embryonic lethality induced by the HS/HS2 and EP procedures, we incubate the embryos at a low density of 80 embryos maximum per 94 mm Petri plate (this is particularly important for the first 24 h of development).

5.3.7 Testosterone treatment

- Before embryos reach 24 hpf, remove the chorions from the embryos, remove as much E3 as possible and replace with testosterone/E2 working solution.
- Each consecutive day, replace 1/2 of the testosterone/E2 with fresh testosterone/E2.
- On day 6, transfer the embryos to mouse cages with 1 L of testosterone solution in fish water supplemented with 3 g/L Coral Reef Salt. Start feeding as normally. Continue replacing 1/2 of the solution every day by carefully aspirating the solution and refilling with fresh testosterone solution.
- On day 15, remove the testosterone by aspirating most of the solution and refilling with fresh fish water. Rinse again by repeating this procedure. Embryos can now be connected to the water system.

CONCLUSIONS

Forward genetics screens remain as an essential complement to genetic analysis, due to their ability to unbiasedly pinpoint factors involved in a biological process. We describe methodologies to identify and recover adult and parental-effect mutations through forward genetics screens, and to maintain and map these mutations towards their molecular cloning. Two main genetic screening strategies are described: an F_3/F_4 extended-family screen based solely on natural crosses, and F_2/F_3 gynogenesis-based screen. Each of these strategies has advantages and disadvantages. Extended-family screens based solely on natural crosses are technically relatively simple, and allow simultaneous mapping as well as the identification of maternal-zygotic mutations. However, such an approach requires larger amounts of space, generation time, and labor. On the other hand, gynogenesis screens require substantial selection of specialized lines amenable to the procedure but can be carried out in fewer generations and using less space. Both of these methods, however, have allowed the unbiased identification of many maternal-effect mutants (Dosch et al., 2004; Pelegri et al., 2004; Pelegri & Schulte-Merker, 1999; Wagner et al., 2004) and can be used for the identification of mutations in genes affecting juvenile and adult traits (Andreeva et al., 2011; Bauer & Goetz, 2001; Haffter et al., 1996; Harris et al., 2008; Johnson et al., 1995; Johnson & Weston, 1995; Lang et al., 2009; Parichy & Turner, 2003a; Rohner et al., 2009; Tschopp et al., 2010; Wagner et al., 2004). These efforts continue to pave the way towards the genetic analysis of early development in this vertebrate species and form the basis for the future genetic dissection of adult traits of biomedical relevance.

ACKNOWLEDGMENTS

F.P. is grateful to Stefan Schulte-Merker, Marcus Dekens, Hans-Martin Maischein and Catrin Weiler for their participation in the original EP screen, Christiane Nüsslein-Volhard for support and advice at the Max-Planck Institut für Entwicklungsbiologie, Tübingen, and Jonathon Heier and Kendra Takle for the development of HS2 at the University of Wisconsin–Madison. Support in the laboratory of F.P at the University of Wisconsin–Madison has been provided by an NIH grants R21 GM065303 and RO1 GM65303.

M.C.M. thanks: Daniel Wagner, Florence Marlow, Elliott Abrams, Hong Zhang, Tripti Gupta, Lee Kapp, Roland Dosch, and Keith Mintzer, and the fish facility staff for their contributions in performing the F_4 natural crosses screen described here and numerous discussions; Florence Marlow for contributing Fig 2A and B; M.C.M. is supported by NIH grants R01 R01GM117981 and R01HD069321.

REFERENCES

Ablain, J., Durand, E. M., Yang, S., Zhou, Y., & Zon, L. I. (2015). A CRISPR/Cas9 vector system for tissue-specific gene disruption in zebrafish. *Developmental Cell, 32*, 756–764.

Abrams, E. W., & Mullins, M. C. (2009). Early zebrafish development: it's in the maternal genes. *Current Opinion in Genetics and Development, 19*, 396–403.

Abrams, E. W., Zhang, H., Marlow, F. L., Kapp, L., Lu, S., & Mullins, M. C. (2012). Dynamic assembly of Brambleberry mediates nuclear envelope fusion during early development. *Cell, 150*, 521–532.

Andreeva, V., Connolly, M. H., Stewart-Swift, C., Fraher, D., Burt, J., Cardarelli, J., & Yelick, P. C. (2011). Identification of adult mineralized tissue zebrafish mutants. *Genesis, 49*, 360–366.

Bauer, M. P., & Goetz, F. W. (2001). Isolation of gonadal mutations in adult zebrafish from a chemical mutagenesis screen. *Biology of Reproduction, 64*, 548–554.

Bontems, F., Stein, A., Marlow, F., Lyautey, J., Gupta, T., Mullins, M. C., & Dosch, R. (2009). Bucky ball organizes germ plasm assembly in zebrafish. *Current Biology, 19*, 414–422.

Bowen, M. E., Henke, K., Siegfried, K. R., Warman, M. L., & Harris, M. P. (2012). Efficient mapping and cloning of mutations in zebrafish by low-coverage whole-genome sequencing. *Genetics, 190*, 1017–1024.

Cade, L., Reyon, D., Hwang, W. Y., Tsai, S. Q., Patel, S., Khayter, C., … Yeh, J. R. (2012). Highly efficient generation of heritable zebrafish gene mutations using homo- and heterodimeric TALENs. *Nucleic Acids Research, 40*, 8001–8010.

Campbell, P. D., Heim, A. E., Smith, M. Z., & Marlow, F. L. (2015). Kinesin-1 interacts with Bucky ball to form germ cells and is required to pattern the zebrafish body axis. *Development, 142*, 2996–3008.

Carneiro, M. C., Henriques, C. M., Nabais, J., Ferreira, T., Carvalho, T., & Ferreira, M. G. (2016). Short telomeres in key tissues initiate local and systemic aging in zebrafish. *PLoS Genetics, 12*, e1005798.

Chan, S. T. H., & Yeung, W. S. B. (1983). Sex control and sex reversal in fish under natural conditions. In W. S. Hoar, D. J. Randall, & E. M. Donaldson (Eds.), *Fish physiology* (Vol. IX, Part B, pp. 171–222). New York: Academic Press.

Corley-Smith, G. E., Brandhorst, B. P., Walker, C., & Postlethwait, J. H. (1999). Production of haploid and diploid androgenetic zebrafish (including methodology for delayed in vitro fertilization). In H. W. Detrich, III, L. I. Zon, & M. Westerfield (Eds.), *The zebrafish: Biology* (Vol. 60, pp. 45–60). San Diego, CA: Academic Press.

Dahlem, T. J., Hoshijima, K., Jurynec, M. J., Gunther, D., Starker, C. G., Locke, A. S., … Grunwald, D. J. (2012). Simple methods for generating and detecting locus-specific mutations induced with TALENS in the zebrafish genome. *PLoS Genetics, 8*, e1002861.

Dapprich, J., Ferriola, D., Magira, E. E., K, M., & Monos, D. (2008). SNP-specific extraction of haplotype-resolved targeted genomic regions. *Nucleic Acids Research, 36*, e94.

D'Ascenzo, M., Meacham, C., Kitzman, J., Middle, C., Knight, J., Winer, R., … Reinholdt, L. (2009). Mutation discovery in the mouse using genetically guided array capture and resequencing. *Mammalian Genome, 20*, 424–436.

Delattre, M., & Gönczy, P. (2004). The arithmetic of centrosome biogenesis. *Journal of Cell Science, 117*, 1619–1629.

Dosch, R., Wagner, D. S., Mintzer, K. A., Runke, G., Wiemelt, A. P., & Mullins, M. C. (2004). Maternal control of vertebrate development before the midblastula transition: mutants from the zebrafish I. *Developmental Cell, 6*, 771–780.

Doyon, Y., McGammon, J. M., Miller, J. C., Faraji, F., Ngo, C., Katibah, G. E., … Amacher, S. L. (2008). Heritable targeted gene disruption in zebrafish using designed zinc-finger nucleases. *Nature Biotechnology, 26*, 702–708.

Driever, W., Solnica-Krezel, L., Schier, A. F., Neuhauss, S. C. F., Malicki, J., Stemple, D. L., & Boggs, C. (1996). A genetic screen for mutations affecting embryogenesis in zebrafish. *Development, 123*, 37–46.

Eaton, R. C., & Farley, R. D. (1974). Spawning cycle and egg production of zebrafish, *Brachydanio rerio*, in the laboratory. *Copeia, 1*, 195–204.

van Eeden, F. J. M., Granato, M., Odenthal, J., & Haffter, P. (1999). Developmental mutant screens in the zebrafish. In H. W. Detrich, III, M. Westerfield, & L. I. Zon (Eds.), *The zebrafish: Genetics and genomics* (Vol. 60, pp. 21–41). San Diego: Academic Press.

Elkouby, Y. M., Jamieson-Lucy, A., & Mullins, M. C. (2016). Oocyte polarization is coupled to chromosomal bouquet, a conserved polarized nuclar configuration in meiosis. *PLoS Biol, 14*, e1002335.

Eno, C., Solanki, B., & Pelegri, F. (2016). *aura/mid1ip1L* regulates the cytoskeleton at the zebrafish egg-to-embryo transition. *Development* (in press).

Foley, J. E., Yeh, J. R., Maeder, M. L., Reyon, D., Sander, J. D., Peterson, R. T., & Joung, J. K. (2009). Rapid mutation of endogenous zebrafish genes using zinc finger nucleases made by Oligomerized Pool ENgineering (OPEN). *PLoS One, 4*, e4348.

Geisler, R. (2002). Mapping and cloning. In C. Nüslein-Volhard, & R. Dahm (Eds.), *Zebrafish* (Vol. 261, pp. 175–212). Oxford: Oxford University Press.

Gestl, E. E., Kauffman, E. J., Moore, J. L., & Cheng, K. C. (1997). New conditions for generation of gynogenetic half-tetrad embryos in the zebrafish (*Danio rerio*). *J Hered, 88*, 76–79.

Gupta, T., Marlow, F. L., Ferriola, D., Mackiewicz, K., Dapprich, J., Monos, D., & Mullins, M. C. (2010). Microtubule actin crosslinking factor 1 regulates the Balbiani body and animal-vegetal polarity of the zebrafish oocyte. *PLoS Genetics, 6*, e1001073.

Haffter, P., Granato, M., Brand, M., Mullins, M. C., Hamerschmidt, M., Kane, D. A., … Nüsslein-Volhard, C. (1996). The identification of genes with unique and essential functions in the development of the zebrafish, *Danio rerio*. *Development, 123*, 1–36.

Harris, M. P., Rohner, N., Schwarz, H., Perathoner, S., Konstantinidis, P., & Nüsslein-Volhard, C. (2008). Zebrafish *eda* and *edar* mutants reveal conserved and ancestral roles of ectodysplasin signaling in vertebrates. *PLoS Genetics, 4*, e1000206.

Harvey, S. A., Sealy, I., Kettleborough, R., Fenyes, F., White, R., Stemple, D., & Smith, J. C. (2013). Identification of the zebrafish maternal and paternal transcriptomes. *Development, 140*, 2703–2710.

Heier, J., Takle, K., Hasley, A., & Pelegri, F. (2015). Ploidy manipulation and induction of alternate cleavage patterns through inhibition of centrosome duplication in the early zebrafish embryo. *Developmental Dynamics, 244*, 1300–1312.

Henke, K., Bowen, M. E., & Harris, M. P. (2013a). Identification of mutations in zebrafish using next-generation sequencing. *Current Protocols in Molecular Biology, 104*. Unit 7.13.

Henke, K., Bowen, M. E., & Harris, M. P. (2013b). Perspectives for identification of mutations in the zebrafish: making use of next-generation sequencing technologies for forward genetic approaches. *Methods, 62*, 185–196.

Hill, J. T., Demarest, B. L., Bisgrove, B. W., Gorsi, B., Su, Y. C., & Yost, H. J. (2013). MMAPPR: mutation mapping analysis pipeline for pooled RNA-seq. *Genome Research, 23*, 687–697.

Holloway, B. A., Gomez de la Torre Canny, S., Ye, Y., Slusarski, D. C., Freisinger, C. M., Dosch, R., … Mullins, M. C. (2009). A novel role for MAPKAPK2 in morphogenesis during zebrafish development. *PLoS Genetics, 5*, e1000413.

Howe, K., Clark, A. G., Torroja, C. F., Torrance, C. J., Berthelot, C., Muffato, M., … Stemple, D. L. (2013). The zebrafish reference genome sequence and its relationship to the human genome. *Nature, 496*, 498–503.

Hruscha, A., Krawitz, P., Rechenberg, A., Heinrich, V., Hecht, J., Haas, C., & Schmid, B. (2013). Efficient CRISPR/Cas9 genome editing with low off-target effects in zebrafish. *Development, 140*, 4982–4987.

Huang, P., Xiao, A., Zhou, M., Zhu, Z.-Y., Lin, S., & Zhang, B. (2011). Heritable gene targeting in zebrafish using customized TALENs. *Nature Biotechnology, 29*, 699–700.

Hwang, W. Y., Fu, Y., Reyon, D., Maeder, M. L., Tsai, S. Q., Sander, J. D., … Joung, J. K. (2013). Efficient genome editing in zebrafish using a CRISPR-Cas system. *Nature Biotechnology, 31*, 227–229.

Irion, U., Krauss, J., & Nüsslein-Volhard, C. (2014). Precise and efficient genome editing in zebrafish using the CRISPR/Cas9 system. *Development, 141*, 4827–4830.

Jao, L.-E., Wente, S. R., & Chen, W. (2013). Efficient multiplex biallelic zebrafish genome editing using a CRISPR nuclease system. *Proceedings of the National Academy of Sciences of the United States of America, 110*, 13904–13909.

Jiang, L., Zhang, J., Wang, J.-J., Wang, L., Zhang, L., Li, G., … Liu, J. (2013). Sperm, but not oocyte, DNA methylome is inherited by zebrafish early embryos. *Cell, 153*, 773–784.

Johnson, S. L., & Weston, J. A. (1995). Temperature-sensitive mutations that cause stage-specific defects in zebrafish fin regeneration. *Genetics, 141*, 1583–1595.

Johnson, S. L., Africa, D., Walker, C., & Weston, J. A. (1995). Genetic control of adult pigment stripe development in zebrafish. *Developmental Biology, 167*, 27–33.

Kane, D. A., & Kimmel, C. B. (1993). The zebrafish midblastula transition. *Development, 119*, 447–456.

Kapp, L. D., Abrams, E. W., Marlow, F. L., & Mullins, M. C. (2013). The Integrator Complex subunit 6 (Ints6) confines the dorsal organizer in vertebrate embryogenesis. *PLoS Genetics, 9*, e1003822.

Kemphues, K. J., Kusch, M., & Wolf, N. (1988). Maternal-effect lethal mutations on linkage group II of *Caenorhabditis elegans*. *Genetics, 120*, 977–986.

Kemphues, K. J., & Strome, S. (1997). Fertilization and establishment of polarity in the embryo. In D. L. Riddle, T. Blumenthal, B. J. Meyer, & J. R. Priess (Eds.), C. elegans *II* (pp. 335–359). Cold Spring Harbor Laboratory Press.

Kettleborough, R. N. W., Busch-Nentwich, E. M., Harvey, S. A., Dooley, C. M., de Bruijn, E., van Eeden, F., … Stemple, D. L. (2013). A systematic genome-wide analysis of zebrafish protein-coding gene function. *Nature, 496*, 494–497.

Kimmel, C., Ballard, W. W., Kimmel, S. R., Ullmann, B., & Schilling, T. F. (1995). Stages of embryonic development in the zebrafish. *Developmental Dynamics, 203*, 253–310.

Kishimoto, Y., Koshida, S., Furutani-Seiki, M., & Kondoh, H. (2004). Zebrafish maternal-effect mutations causing cytokinesis defects without affecting mitosis or equatorial *vasa* deposition. *Mechanisms of Development, 121*, 79–89.

Lang, M. R., Patterson, L. B., Gordon, T. N., Johnson, S. L., & Parichy, D. M. (2009). *Basonuclin-2* requirements for zebrafish adult pigment pattern development and female fertility. *PLoS Genetics, 5*, e1000744.

Langdon, Y. G., Fuentes, R., Zhang, H., Abrams, E. W., Marlow, F. L., & Mullins, M. C. (2016). Split top: a maternal cathepsin B that regulates dorsoventral patterning and morphogenesis. *Development, 143*, 1016–1028.

Lee, M. T., Bonneau, A. R., Takacs, C. M., Bazzini, A. A., DiVito, K. R., Fleming, E. S., & Giraldez, A. J. (2013). Nanog, Pou5f1 and SoxB1 activate zygotic gene expression during the maternal-to-zygotic transition. *Nature, 503*, 360–364.

Lehmann, R., & Nüsslein-Volhard, C. (1986). Abdominal segmentation, pole cell formation, and embryonic polarity require the localized activity of *oskar*, a maternal gene in *Drosophila. Cell, 47*, 141–152.

Lindeman, R., & Pelegri, F. (2010). Vertebrate maternal-effect genes: insights into fertilization, early cleavage divisions, and germ cell determinant localization from studies in the zebrafish. *Molecular Reproduction and Development, 77*, 299–313.

Lindeman, R. E., & Pelegri, F. (2012). Localized products of *futile cycle/lrmp* promote centrosome-nucleus attachment in the zebrafish zygote. *Current Biology, 22*, 843–851.

Livak, K. J., Flood, S. J., Marmaro, J., Giusti, W., & Deetz, K. (1995). Oligonucleotides with fluorescent dyes at opposite ends provide a quenched probe system useful for detecting PCR product and nucleic acid hybridization. *PCR Methods and Applications, 4*, 357–362.

Lyman-Gingerich, J., Lindeman, R., Putiri, E., Stolzmann, K., & Pelegri, F. (2006). The analysis of axis induction mutant embryos reveals morphogenetic events associated with zebrafish yolk extension formation. *Developmental Dynamics, 235*, 2749–2760.

Lyman-Gingerich, J., Westfall, T. A., Slusarski, D. C., & Pelegri, F. (2005). *hecate*, a zebrafish maternal effect gene, affects dorsal organizer induction and intracellular calcium transient frequency. *Developmental Biology, 286*, 427–439.

Marlow, F. L., & Mullins, M. C. (2008). Bucky ball functions in Balbiani body assembly and animal-vegetal polarity in the oocyte and follicle cell layer in zebrafish. *Developmental Biology, 321*, 40–50.

Mathavan, S., Lee, S. G. P., Mak, A., Miller, L. D., Murthy, K. R. K., Govindarajan, K. R., … Lufkin, T. (2005). Transcriptome analysis of zebrafish embryogenesis using microarrays. *PLoS Genetics, 1*, 260–276.

Mei, W., Lee, K. W., Marlow, F. L., Miller, A. L., & Mullins, M. C. (2009). hnRNP I is required to generate the Ca^{2+} signal that causes egg activation in zebrafish. *Development, 136*, 3007–3017.

Meng, X., Noyes, M. B., Zhu, L. J., Lawson, N. D., & Wolfe, S. A. (2008). Targeted gen inactivation in zebrafish using engineered zinc-finger nucleases. *Nature Biotechnology, 26*, 695–701.

Miller, A. C., Obholzer, N. D., Shah, A. N., Megason, S. G., & Moens, C. B. (2013). RNA-seq-based mapping and candidate identification of mutations from forward genetics screens. *Genome Research, 23*, 679–686.

Moreno-Mateos, M. A., Vejnar, C. E., Beaudoin, J.-D., Fernandez, J. P., Mis, E. K., Khokha, M. K., & Giraldez, A. J. (2015). CRISPRscan: designing highly efficient sgRNAs for CRISPR/Cas9 targeting *in vivo*. *Nature Methods, 12*, 982–988.

Mullins, M. C., Hammerschmidt, M., Haffter, P., & Nüsslein-Volhard, C. (1994). Large-scale mutagenesis in the zebrafish: in search of genes controlling development in a vertebrate. *Current Biology, 4*, 189–202.

Neuhauss, S. (1996). *Craniofacial development in zebrafish (*Danio rerio*): Mutational analysis, genetic characterization, and genomic mapping*. Tübingen: Eberhard-Karl-Universität Tübingen.

Newport, J., & Kirschner, M. (1982a). A major developmental transition in early *Xenopus* embryos: II. Control of the onset of transcription. *Cell, 30*, 687–696.

Newport, J., & Kirschner, M. (1982b). A major developmental transition in early *Xenopus* embryos: I. Characterization and timing of cellular changes at the midblastula stage. *Cell, 30*, 675–686.

Nijman, I. J., Kuipers, S., Verheul, M., Guryev, V., & Cuppen, E. (2008). A genome-wide SNP panel for mapping and association studies in the rat. *BMC Genomics, 9*, 95.

Ota, S., Hisano, Y., Ikawa, Y., & Kawahara, A. (2014). Multiple genome modifications by the CRISPR/Cas9 system in zebrafish. *Genes to Cells, 19*, 555–564.

Parichy, D. M., Elizondo, M. R., Mills, M. G., Gordon, T. N., & Engeszer, R. E. (2009). Normal table of postembryonic zebrafish development: staging by externally visible anatomy of the living fish. *Developmental Dynamics, 238*, 2975–3015.

Parichy, D. M., & Turner, J. M. (2003a). Temporal and cellular requirements for Fms signaling during zebrafish adult pigment pattern development. *Development, 130*, 817–833.

Parichy, D. M., & Turner, J. M. (2003b). Zebrafish *puma* mutant decouples pigment pattern and somatic metamorphosis. *Developmental Biology, 256*, 242–257.

Pelegri, F., Dekens, M. P. S., Schulte-Merker, S., Maischein, H.-M., Weiler, C., & Nüsslein-Volhard, C. (2004). Identification of recessive maternal-effect mutations in the zebrafish using a gynogenesis-based method. *Developmental Dynamics, 231*, 325–336.

Pelegri, F., & Schulte-Merker, S. (1999). A gynogenesis-based screen for maternal-effect genes in the zebrafish, *Danio rerio*. In H. W. Detrich, III, L. I. Zon, & M. Westerfield (Eds.), *The zebrafish: Genetics and genomics* (Vol. 60, pp. 1–20). San Diego: Academic Press.

Poss, K. D., Nechiporuk, A., Stringer, K. F., Lee, C., & Keating, M. T. (2004). Germ cell aneuploidy in zebrafish with mutations in the mitotic checkpoint gene *mps1*. *Genes and Development, 18*, 1527–1532.

Postlethwait, J. H., & Talbot, W. S. (1997). Zebrafish genomics: from mutants to genes. *Trends in Genetics, 13*, 183–190.

Potok, M. E., Nix, D. A., Parnell, T. J., & Cairns, B. R. (2013). Reprogramming the maternal zebrafish genome after fertilization to match the paternal methylation pattern. *Cell, 153*, 759–772.

Ransom, D. G., & Zon, L. I. (1999). Collection, storage, and use of zebrafish sperm. In H. W. Detrich, III, M. Westerfield, & L. I. Zon (Eds.), *The zebrafish: Genetics and genomics* (Vol. 60, pp. 365–372). San Diego, CA: Academic Press.

Rehman, A. U., Morell, R. J., Belyantseva, I. A., Khan, S. Y., Boger, E. T., Shahzad, M., … Friedman, T. B. (2010). Targeted capture and next-generation sequencing identifies C9orf75, encoding taperin, as the mutated gene in nonsyndromic deafness DFNB79. *American Journal of Human Genetics, 86*, 378–388.

Rohner, N., Bercsényi, M., Orbaán, L., Kolanczyk, M. E., Linke, D., Brand, M., … Harris, M. P. (2009). Duplication of *fgfr1* permits Fgf signaling to serve as a target for selection during domestication. *Current Biology, 19*, 1642–1647.

Sakai, N., Burgess, S., & Hopkins, N. (1997). Delayed in vitro fertilization of zebrafish eggs in Hank's saline containing bovine serum albumin. *Molecular Marine Biology and Biotechnology, 6*, 84–87.

Schnabel, R., & Priess, J. R. (1997). Specification of cell fates in the early embryo. In D. L. Riddle, T. Blumenthal, B. J. Meyer, & J. R. Priess (Eds.), C. elegans *II* (pp. 361–382). Cold Spring Harbor Laboratory Press.

Selman, K., Petrino, T. R., & Wallace, R. A. (1994). Experimental conditions for oocyte maturation in the zebrafish, *Brachydanio rerio*. *Journal of Experimental Zoology, 269*, 538–550.

Signoret, J., & Lefresne, J. (1971). Contribution a l'etude de la segmentation de l'oeuf d'axolotl: I. Definition de la transition blastuleenne. *Annales d'Embryologie et de Morphogenese, 4*, 113.

Singer, A., Perlman, H., Yan, Y., Walker, C., Corley-Smith, G., Brandhorst, B., & Postlethwait, J. (2002). Sex-specific recombination rates in zebrafish (*Danio rerio*). *Genetics, 160*, 649–657.

Smith, S. M., & Maughan, P. J. (2015). SNP genotyping using KASPar assays. *Methods in Molecular Biology, 1245*, 243–256.

Solnica-Krezel, L., Schier, A. F., & Driever, W. (1994). Efficient recovery of ENU-induced mutations from the zebrafish germline. *Genetics, 136*, 1401–1420.

St. Johnston, D., & Nüsslein-Volhard, C. (1992). The origin of pattern and polarity in the *Drosophila* embryo. *Cell, 68*, 201–219.

Streisinger, G., Singer, F., Walker, C., Knauber, D., & Dower, N. (1986). Segregation analyses and gene-centromere distances in zebrafish. *Genetics, 112*, 311–319.

Streisinger, G., Walker, C., Dower, N., Knauber, D., & Singer, F. (1981). Production of clones of homozygous diploid zebra fish (*Brachydanio rerio*). *Nature, 291*, 293–296.

Talbot, S., & Schier, A. F. (1999). Positional cloning of mutated zebrafish genes. In H. W. Detrich, III, M. Westerfield, & L. I. Zon (Eds.), *The zebrafish: Genetics and genomics* (Vol. 60, pp. 259–286). San Diego: Academic Press.

Thorgard, G. H. (1983). Chromosome set manipulation and sex control in fish. In W. S. Hoar, D. J. Randall, & E. M. Donaldson (Eds.), *Fish physiology* (Vol. IX, Part B, pp. 405–434). New York: Academic Press.

Tschopp, M., Takamiya, M., Cerveny, K. L., Gestri, G., Biehlmaier, O., Wilson, S. W., … Neuhauss, S. C. (2010). Fundoscopy in adult zebrafish and its application to isolate mutant strains with ocular defects. *PLoS One, 5*, e15427.

Varshney, G. K., Pei, W., LaFave, M. C., Idol, J., Xu, L., Gallardo, V., … Burgess, S. M. (2015). High-throughput gene targeting and phenotyping in zebrafish using CRISPR/CAs9. *Genome Research, 25*, 1030–1042.

Vertii, A., Zimmerman, W., Ivshina, M., & Doxsey, S. (2015). Centrosome-intrinsic mechanisms modulate centrosome integrity during fever. *Molecular Biology of the Cell, 26*, 3451–3463.

Wagner, D. S., Dosch, R., Mintzer, K. A., Wiemelt, A. P., & Mullins, M. C. (2004). Maternal control of development at the midblastula transition and beyond: mutants from the zebrafish II. *Developmental Cell, 6*, 781–790.

Wakimoto, B. T., Lindsley, D. L., & Herrera, C. (2004). Toward a comprehensive genetic analysis of male fertility in *Drosophila melanogaster. Genetics, 167*, 207–216.

Wilson, C. A., High, S. K., McCluskey, B. M., Amores, A., Yan, Y.-L., Titus, T. A., … Postlethwait, J. H. (2014). Wild sex in zebrafish: loss of the natural sex determinant in domesticated strains. *Genetics, 198*, 1291–1308.

Yabe, T., Ge, X., Lindeman, R., Nair, S., Runke, G., Mullins, M., & Pelegri, F. (2009). The maternal-effect gene *cellular island* encodes Aurora B kinase and is essential for furrow formation in the early zebrafish embryo. *PLoS Genetics, 5*, e1000518.

Yabe, T., Ge, X., & Pelegri, F. (2007). The zebrafish maternal-effect gene *cellular atoll* encodes the centriolar component Sas-6 and defects in its paternal function promote whole genome duplication. *Developmental Biology, 312*, 44–60.

CHAPTER

Targeted candidate gene screens using CRISPR/Cas9 technology

4

A.N. Shah*[,1], C.B. Moens*[,1], A.C. Miller[§,1]

*Fred Hutchinson Cancer Research Center, Seattle, WA, United States

[§]University of Oregon, Eugene, OR, United States

[1]Corresponding authors: E-mail: arish.n.shah@gmail.com; cmoens@fhcrc.org; acmiller@uoregon.edu

CHAPTER OUTLINE

Abstract

In the postgenomic era, the ability to quickly, efficiently, and inexpensively assign function to the zebrafish proteome is critical. Clustered regularly interspaced short palindromic repeats (CRISPRs) have revolutionized the ability to perform reverse genetics because of its simplicity and broad applicability. The CRISPR system is composed of an engineered, gene-specific single guide RNA (sgRNA) and a Cas9 enzyme that causes double-stranded breaks in DNA at the targeted site. This simple, two-part system, when injected into one-cell stage zebrafish embryos, efficiently mutates target loci at a frequency such that injected embryos phenocopy known mutant phenotypes. This property allows for CRISPR-based F_0 screening in zebrafish, which provides a means to screen through a large number of

Methods in Cell Biology, Volume 135, ISSN 0091-679X, http://dx.doi.org/10.1016/bs.mcb.2016.01.008

candidate genes for their role in a phenotype of interest. While there are important considerations for any successful genetic screen, CRISPR screening has significant benefits over conventional methods and can be accomplished in any lab with modest molecular biology experience.

INTRODUCTION

Traditional forward genetic approaches have been a powerful means to identify genes involved in numerous zebrafish processes as the random nature of chemical mutagenesis allows for an unbiased screening method. However, due to the very large number of mutant fish required to screen the genome, as well as genetic redundancy and pleiotropy, many gene functions are not uncovered for a given screenable phenotype. *In vivo* reverse genetics methods such as Targeting Induced Local Lesions in Genomes (TILLING), Zinc-finger nucleases (ZFNs), Transcription Activator-Like Effector Nuclease (TALEN), and morpholinos provide a means to investigate the function of individual genes. While these reverse genetics approaches have complemented forward screens in zebrafish, they are laborious (Bedell et al., 2012; Foley et al., 2009; Pan et al., 2015) or are prone to nonspecific effects (Kok et al., 2015).

Recently, clustered regularly interspaced short palindromic repeats (CRISPRs) have revolutionized reverse genetics, with the most ambitious uses allowing for genome-wide screening of gene function in cell culture in a matter of days (Jinek et al., 2012; Shalem et al., 2014; Wang, Wei, Sabatini, & Lander, 2014). The CRISPR system targets the genome at specified loci and creates double-strand DNA breaks (DSBs). Cells repair the DSBs via either homology-directed repair, which usually preserves gene function, or the nonhomologous end joining (NHEJ) pathway, which can introduce insertion/deletion (indel) mutations at these targeted loci—potentially disrupting gene function. The CRISPR system is composed of two parts: the Cas9 protein containing enzymatic sites that cleave DNA and an engineered 100 nucleotide single guide RNA (sgRNA). The sgRNA contains a 20-nucleotide target sequence providing target specificity via Watson–Crick base pairing, and an 80-nucleotide loop sequence required for interaction with the Cas9 enzyme (Fig. 2) (Hsu, Lander, & Zhang, 2014; Jinek et al., 2012). CRISPR's ease of use and broad utility to modify seemingly any genome has quickly made it the tool of choice for reverse genetics.

CRISPR has been shown to efficiently mutate genes in zebrafish (Chang et al., 2013; Hwang et al., 2013; Peng et al., 2014). Injection of Cas9-encoding mRNA with sgRNA into the one-cell stage zebrafish efficiently induces indels at the target site resulting in a genetically mosaic embryo (see later). The frequency of biallelic disruption is high enough such that injected (F_0) embryos can phenocopy known mutant phenotypes (Jao, Wente, & Chen, 2013) and can reveal phenotypes for genes with previously unknown functions (Shah, Davey, Whitebirch, Miller, & Moens, 2015). Due to the cheap and efficient generation of engineered sgRNAs, a CRISPR-based reverse genetic screen can be conducted in a fraction of the time and cost of classical genetic screens (Fig. 1). The pilot screen we conducted took a total of 3 weeks, screened ~50 genes, and identified two genes previously unknown to be involved in synaptogenesis (Shah et al., 2015).

In this chapter we outline the considerations that will most impact a CRISPR F_0 screen's success. We examine the synthesis of Cas9-encoding mRNA, the engineering of sgRNAs, and consider the properties of each that are most important. Given the genetic mosaicism of F_0 embryos, we discuss several factors that need to be considered in order to successfully observe phenotypic changes in the screen. We also discuss the confirmation of results at the phenotypic and genotypic levels. While there are caveats given the mosaic nature of CRISPR-injected embryos, the screen hits are likely to

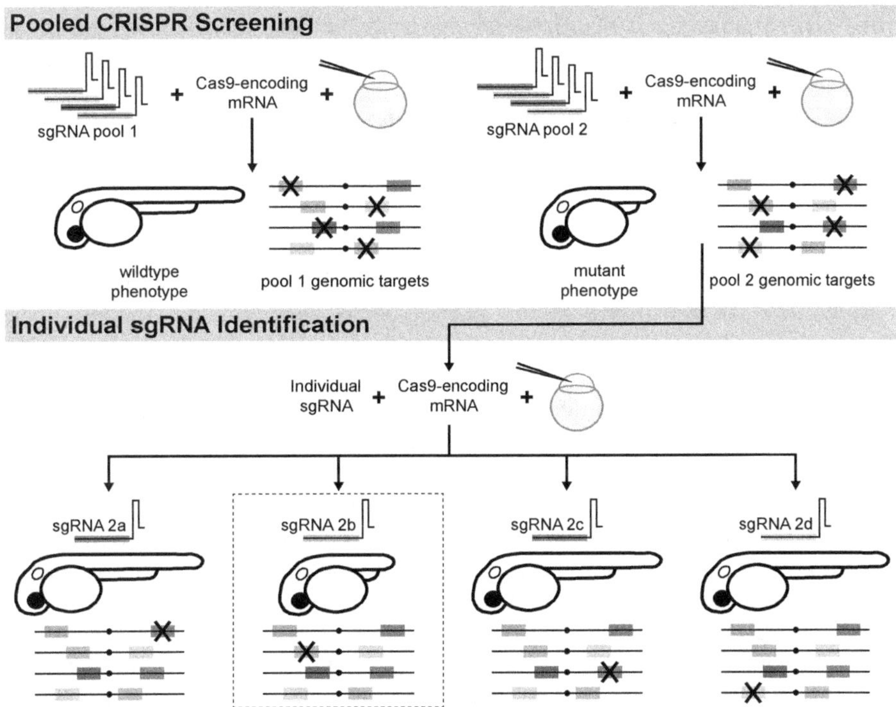

FIGURE 1

Multiplexed (clustered regularly interspaced short palindromic repeats) CRISPR-based F_0 screening in zebrafish. Pools of sgRNAs designed to target multiple candidate genes, along with Cas9-encoding mRNA, are injected into the single-cell stage embryo. CRISPR-induced indels (depicted as large X) occur at each of the pool targets (depicted as boxes on chromosomes) and mutant phenotypes (depicted as short embryos) can be scored in the injected (F_0) animals. Any sgRNA pool that caused a phenotype can be split (demultiplexed) with each sgRNA being reinjected individually to determine the sgRNA/gene responsible for the phenotype. (See color plate)

Modified with permission from Zebrafish, published by Mary Ann Liebert, Inc., New Rochelle, NY.

have important function for the process being investigated. CRISPR screening represents an innovative and effective first step in identifying new genes.

1. GENERAL CONSIDERATIONS BEFORE STARTING A CRISPR SCREEN

In order to use CRISPR to screen for phenotypes in F_0-injected fish, the main goal should be to maximize the efficiency of mutagenesis so that most cells in the embryo harbor biallelic mutations. Injection into the single-cell stage zebrafish embryo of either Cas9-encoding mRNA or Cas9 protein mixed with an sgRNA can induce high frequencies of indel mutations at the targeted gene (Gagnon et al., 2014; Jao et al., 2013; Shah et al., 2015). However, due to the fact that mutations do not occur immediately, the F_0 animals are mosaic for many different mutations, with each clonal population of cells containing a unique spectrum of no, mono-, or biallelic changes. Also, due to the random nature of NHEJ repair, in-frame indels will be created in 33% of affected alleles, and these may be tolerated depending on the location of the indel in the gene (Shah et al., 2015). Thus it is critical to achieve the highest mutagenesis efficiency possible to overcome the mosaicism and examine phenotypic changes.

When designing an F_0 CRISPR-based screen, it is very important to take into consideration the sensitivity of the phenotype to be examined. In particular, whether the mosaic knock out of gene function will allow for the identification of genes involved in the process. If the phenotype of interest is detectable in single cells and requires the cell-autonomous function of single genes, then it may be particularly well suited to screening. For example, we used multiplexed CRISPRs to identify genes required for the formation of a single synapse that forms between two identified neurons that are present bilaterally in each body segment (Shah et al., 2015). However, many processes are more complex and involve hundreds or thousands of cells working in concert. In such processes the behavior of mutant cells in mosaic embryos may be rescued by wild-type cells. For example, we injected an sgRNA against *vangl2* (a gene required broadly in the mesoderm for cell movements that elongate the body axis (Jessen et al., 2002)) and while we were able to see the mutant phenotype, a higher concentration of sgRNA was needed than for the cell-autonomous synapse phenotype (Shah et al., 2015). *A priori* it is difficult to predict whether a particular phenotype is accessible to CRISPR screening. We will discuss the use of positive control sgRNAs to determine this below.

2. CONSIDERATIONS FOR THE Cas9 ENZYME

The most commonly used Cas9 enzyme originates from *Streptococcus pyogenes*, which targets a 20-nucleotide (N_{20}) sequence directly 5′ of a 5′-NGG-3′ motif

(PAM) at the target site (Fig. 2) (Jinek et al., 2012). The 5′-NGG-3′ requirement of the PAM limits the number of sites in the genome that can be targeted by this Cas9. Despite this 5′- N_{20}NGG-3′ sequence constraint, most genes will contain many high-quality CRISPR targets that are likely to render the protein inactive (see Section 3). Alternatively, recent work has found that Cas9 can be mutated to recognize alternative PAM sites while retaining high mutagenic activity (Kleinstiver et al., 2015). This greatly increases the number of target sites in the genome; however, these mutated Cas9 enzymes have not yet been tested thoroughly *in vivo*.

The simplest method for Cas9 delivery into zebrafish is by injection of Cas9-encoding mRNA that has been transcribed *in vitro*, along with an sgRNA. This can lead to very high rates of mutagenesis and has been used to screen for genes involved in synapse formation (Shah et al., 2015). Alternatively, purified Cas9 protein can be hybridized *in vitro* to the sgRNA and the resulting complex can be injected (Gagnon et al., 2014). Theoretically this may increase mutagenesis efficiency by preassembling the components and bypassing the time required for translation of injected Cas9-encoding mRNA, but this has not been rigorously tested. Additionally, preparation of the protein requires some expertise in protein expression and purification.

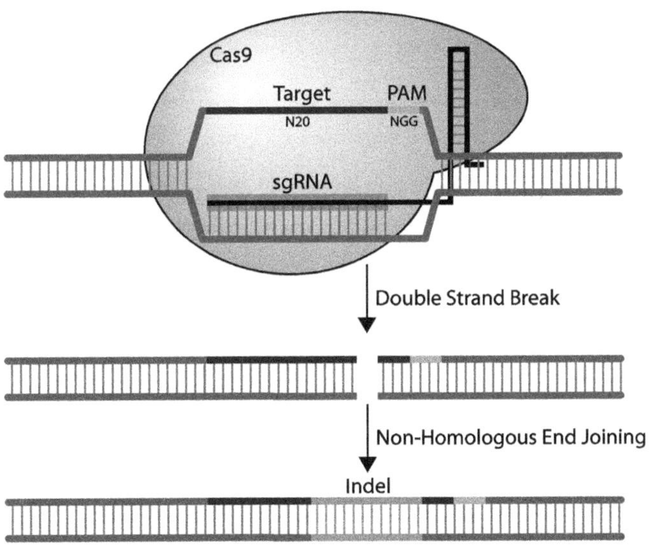

FIGURE 2

Cas9/sgRNA interaction with the target loci can result in insertion/deletion (indel) mutations. RNA-guided gene targeting involves Cas9 protein in complex with an sgRNA. Cas9 unwinds the DNA duplex and cleaves both strands upon recognition of a target sequence by the sgRNA, but only if the correct PAM (5′-"NGG"-3′) is present at the 3′ end. Double-strand DNA breaks are often erroneously repaired by nonhomologous end joining and can result in indel mutations of various sizes. (See color plate)

Given the ease of mRNA production via *in vitro* transcription (IVT), we recommend using the pT3Ts-nCas9n plasmid to generate zebrafish codon-optimized Cas9-encoding mRNA Addgene #46757 (Jao et al., 2013). Other enzymes have been found to work in zebrafish: Addgene #42251 (Hwang et al., 2013), and Addgene #51815, but the above Cas9 from pT3Ts-nCas9n has been shown by multiple groups to induce high levels of mutagenesis in zebrafish (Moreno-Mateos et al., 2015; Shah et al., 2015; Varshney et al., 2015). This plasmid encodes a Cas9 mRNA that includes two nuclear localization signal sequences and the SV40 polyadenylation signal, both of which might increase the mutagenesis efficiency (Jao et al., 2013).

3. Cas9 SYNTHESIS PROTOCOL

1. Using conventional molecular biology methods transform, grow, and prep plasmid Addgene #46757 pT3Ts-nCas9. We recommend a midi- or maxiprep for high yield (Zymo Research, D4202).
2. Linearize 1–2 μg of plasmid with XbaI (NEB, R0145S).

	Mix
1–2 μg plasmid DNA	X μL
CutSmart buffer (NEB)	5 μL
XbaI enzyme (NEB)	2 μL
Water	To 50 μL

3. Incubate at 37°C overnight.
4. Optional: run the linear product on an agarose gel. The plasmid is 4.18 kb.
5. Clean the linearized DNA using a DNA clean column kit, eluting in the smallest volume possible (6 μL for Zymo Research, D4013).
6. Use best practices for maintaining an RNAse free environment (ThermoFisher, AM9786M). Contamination by RNAses will degrade the sample. Use 1 μg of the linear DNA product as template for a T3 polymerase IVT reaction (Life, AM1348M).

	Mix
1 μg linear DNA	X μL
T3 10X buffer	2 μL
RNA 2X NTP	10 μL
T3 enzyme mix	2 μL
Water	To 20 μL

7. Incubate at 37°C for 3–4 h.
8. Stop the reaction by adding 1 μL DNase (included in kit). Incubate at 37°C for 15 min.

9. Bring the reaction volume to 50 μL with RNase-free water.
10. To clean the RNA, we highly recommend using an RNA clean column kit (Zymo Research, R1015).
    a. Refer to the manual for instructions.
    b. Elute the RNA in 15–20 μL RNase-free water.
11. Alternatively, a noncolumn method to clean the RNA is also effective.
    a. Add 10 μL 5M ammonium acetate (included) and 140 μL 100% ethanol to the reaction. Mix well.
    b. Incubate at −80°C for 20 min.
    c. Centrifuge at max speed at 4°C for 15 min.
    d. Remove supernatant without disturbing the pellet and wash with 1 mL 70% ethanol.
    e. Centrifuge at max speed at 4°C for 15 min.
    f. Remove supernatant and dry pellet at room temperature for 1–2 min.
    g. Resuspend the pellet in 15–20 μL RNase-free water.
12. Measure the RNA concentration. Bioanalyzer and Qubit are accurate methods of measuring high concentrations of RNA, but may not be accessible in a timely fashion. Nanodrop spectrophotometers are inaccurate at high concentrations of RNA. A diluted sample of the cleaned RNA product can be measured and the concentration can be calculated based on the dilution.
13. Dilute in RNase-free water (included) and aliquot the Cas9-encoding RNA in 1–2 μL aliquots. Store at −80°C.

4. CONSIDERATIONS FOR CHOOSING sgRNA TARGET SITES

There are two major factors that influence the efficiency of an sgRNA for making deleterious mutations via Cas9-mediated mutagenesis: the location of the target in the coding sequence of the gene and the nucleotide makeup of the target site.

When choosing a target site in a gene of interest, the goal is to create a frameshift indel mutation that will result in a premature stop codon. Ideally this frameshift mutation would occur early in the gene, thereby giving the highest likelihood of rendering the truncated protein nonfunctional. If there are no strong targets early in the gene, targeting a region of the protein that is conserved across species may cause a loss of function even when the mutation is in-frame (Shah et al., 2015). Therefore, the goal is to target functional domains and/or early in the coding sequence. However, it should be noted that the first exon of polyexonic genes should be avoided due to the prevalence of alternative first exon usage (Kimura et al., 2005). Additionally, using RNA-seq data and online databases such as Ensembl (http://www.ensembl.org/Danio_rerio/Info/Index) can help to avoid alternatively spliced exons.

While the position of the target site within the gene is important for eventual loss-of-function of the protein, the efficiency with which the sgRNA/Cas9 complex binds to DNA also has important effects on mutational efficiency. First, the sgRNA needs to be designed using the correct reference sequence, as single-nucleotide polymorphisms (SNPs) that cause a mismatch between the sgRNA and the genomic DNA will reduce Cas9 function (Cong et al., 2013; Hsu et al., 2013). If available, sequencing data from the strain to be used for screening will be helpful in designing around polymorphisms. Second, the sequence of the N_{20} targeting nucleotides of the sgRNA has been shown to influence the efficiency of DNA cleavage. Various *in vitro* assays and human cell culture studies have assessed sgRNAs at a genome-wide scale allowing for the determination of the sequence parameters that influence Cas9 enzymatic activity—namely that a higher GC content leads to a higher Cas9 activity (Chari, Mali, Moosburner, & Church, 2015; Doench et al., 2014; Wang et al., 2014). A recent study using a similar assay in zebrafish determined the sequence "rules" that affect the efficiency of sgRNAs *in vivo* and an algorithm was derived that effectively predicts this efficiency (Moreno-Mateos et al., 2015). This algorithm is incorporated into a design program that can be accessed at www.crisprscan.org (see Section 5).

It is important to note here that CRISPR is known to induce mutations at off-target sites (Cong et al., 2013; Hruscha et al., 2013; Hsu et al., 2013). For the purposes of a CRISPR-based screen, off-target indels that cause a phenotype of interest are unlikely to be problematic as they occur at much lower frequencies than on-target mutations; nevertheless, it is important to be aware of the possibility. While initially in F_0 animals the researcher will not recognize that it is an off-target mutation that is causing the phenotype, in the process of recovering stable mutant alleles they will note that the phenotype does not segregate with the genotype at the intended target locus. In this case if the phenotype persists in a stable mutant line, conventional genetic mapping approaches can be employed to identify this potentially important gene.

5. DESIGNING AN sgRNA

There are many sgRNA design algorithms available to the community, each having their own strengths. We recommend using CRISPRscan (www.criprscan.org) to design sgRNAs as it takes into consideration target sequence composition, potential off-target sites, and noncanonical 5′ guide sequences (Moreno-Mateos et al., 2015). However a number of other tools are available that offer their own strengths. ChopChop (chopchop.rc.fas.harvard.edu) has a very intuitive user interface with sgRNA targets across many genomes already calculated that are easy to visualize. It also has the added benefit of being able to export sgRNA targets in batch along with primers designed for genotyping (Montague, Cruz, Gagnon, Church, & Valen, 2014). A thoroughly tested off-target prediction algorithm is available at www.crispr.mit.edu which ranks sgRNA targets accordingly (Hsu et al., 2013). Additionally, performing CRISPR experiments in the NHGRI-1 (derived from TAB-5) strain can greatly reduce the possibility of a polymorphism existing at the target site as the line has been deep-sequenced and all polymorphism data has been made available at http://research.nhgri.nih.gov/manuscripts/Burgess/zebrafish/downloads/NHGRI-1/hub.txt (LaFave, Varshney, Vemulapalli, Mullikin, & Burgess, 2014). Furthermore, a track has been added to the UCSC genome browser that includes precomputed CRISPR sites for this strain, taking into account these polymorphisms.

6. sgRNA DESIGN AND SYNTHESIS

There are several methods that can be used to synthesize the sgRNA necessary for one-cell stage injections. The important factors to take into consideration with these methods are their sequence requirements, the ease/time of synthesis, and the reagent cost. All methods rely on first synthesizing a piece of double-stranded DNA (dsDNA) that encodes an RNA polymerase binding site (typically T7) followed by the sgRNA target site and the loop sequence (Fig. 3). This dsDNA is then used as a template for an IVT reaction that will generate the designed sgRNA. It should be noted that the first two nucleotides incorporated into the nascent RNA from the T7 polymerase are canonically 5′-"GG" (Milligan, Groebe, Witherell, & Uhlenbeck, 1987); this sequence requirement further constrains the number of potential target sequences in the genome.

The three major methods of generating this template DNA are ligation-based, enzyme-based, and PCR-based. The ligation-based method is simple and requires the synthesis of two overlapping oligonucleotides (oligos) that are ligated into a specially designed plasmid vector, which allows for the IVT of the desired sgRNA (Hwang et al., 2013). However, the protocol takes several days and requires sequencing of one to several plasmids for each designed sgRNA. The enzyme-based method is more involved, requiring first, the ligation into a plasmid similar to the previous one, then IVT, followed by enzymatic cleavage to release the sgRNA, then PAGE-purification of the resulting sgRNAs. The benefit comes in that many

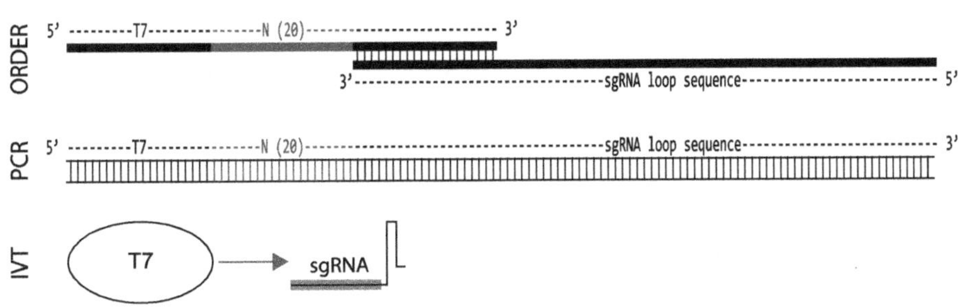

FIGURE 3

PCR-based sgRNA synthesis. One gene-specific oligo contains the T7 binding sequence, 20 targeting nucleotides, and homology to the loop-sequence oligo. The loop-sequence oligo contains a homology region and the loop sequence and is common to all sgRNA synthesis PCRs. A PCR is done using these two oligos to create a dsDNA template for a T7 *in vitro* transcription (IVT) reaction resulting in the synthesis of the sgRNA.

sgRNAs can be created simultaneously and it alleviates the 5′-“GG” T7-polymerase sequence requirement (Qin et al., 2015). Both of the above methods require cloning into plasmids with bacterial propagation, making them relatively costly and time-intensive.

The most cost-effective and least time-intensive protocol is the PCR-based method. Two overlapping oligos are synthesized: the first is the same for all sgRNAs and contains the full 80 nucleotide loop sequence and a short overlap region to anneal with the other oligo; the second is the designed, gene-specific oligo containing an IVT transcription start site, the 20 nucleotide target sequence, and the overlap region with the first oligo. These two oligos are then ligated and the sequence is “filled in” in a simple PCR reaction to create a dsDNA product that has the T7-sgRNA-loop sequence (Fig. 3) (Gagnon et al., 2014; Moreno-Mateos et al., 2015; Shah et al., 2015). This product can be used directly as the template of the IVT reaction. In our experience, T7 RNA polymerase transcribes sgRNA even in the context of 5′-“GA,” “GT,” “GC” scenarios with reduced, yet sufficient, yield for CRISPR mutagenesis in zebrafish embryos (Milligan and Uhlenbeck, 1989; Shah et al., 2015).

7. sgRNA SYNTHESIS PROTOCOL

1. After designing an sgRNA target sequence, the gene-specific oligo needs to be synthesized (Fig. 2B). The design software will generate a target sequence in the genome in the 5′-N_{20}NGG-3′ motif (eg, GGTCTCTCGCAGGATGTTGCTGG), of which only the variable N_{20} (GGTCTCTCGCAGGATGTTGC) will be included in the oligo. The T7 polymerase binding sequence (5′-aattaatacgactcactata-3′) is added 5′ to the target sequence. And the sequence overlap (5′-gttttagagctagaaatagc-3′) is added to the 3′. Therefore, the gene-specific oligo to be synthesized is 5′-aattaatacgactcactata[gene specific 20 nucleotides beginning with GN]gttttagagctagaaatagc-3′.
2. The loop-sequence oligo should be ordered as well and will be used for every sgRNA synthesis: 5′-gatccgcaccgactcggtgccactttttcaagttgataacggactagccttatttta-acttgctatttctagctctaaaac-3′.
3. Dilute a sample of each oligo to 10 μM. We recommend using Phusion Master Mix for the PCR step due to its high fidelity (NEB, M0531S); however, any PCR system should work.

	Mix
Gene-specific oligo (10 μM)	5 μL
Loop-sequence oligo (10 μM)	5 μL
2X Phusion Master Mix	12.5 μL
Water	2.5 μL

4. Run the sample in a thermocycler with the following program.
   a. 98°C for 30 s.
   b. [98°C for 10 s, 60°C for 10 s, 72°C for 10 s] X40.
   c. 72°C for 120 s.
5. Optional: run 1 μL of the PCR reaction on an agarose gel. The product size should be 120 bp.
6. Clean the PCR product using a DNA clean column kit, eluting in the smallest volume possible (6 μL for Zymo Research, D4013).
7. Use best practices for maintaining an RNAse free environment. Contamination by RNAses will degrade the sample (ThermoFisher, AM9786M). Use ~1 μg of the PCR product as template for a T7 polymerase IVT reaction (Life, AM1334M).

	Mix
1 μg PCR product	X μL
T7 10X buffer	1 μL
RNA 2.5X NTP	4 μL
T7 enzyme mix	1 μL
Water	To 10 μL

8. Incubate at 37°C for 3–4 h.
9. Stop the reaction by adding 1 μL DNase (included). Incubate at 37°C for 15 min.
10. Bring the reaction volume to 50 μL with RNase-free water.
11. To clean the RNA, we highly recommend using an RNA clean column kit (Zymo Research, R1015).
    a. Refer to the manual for instructions.
    b. Elute the RNA in 15–20 μL RNase-free water.
12. Alternatively, a noncolumn method to clean the RNA is also effective.
    a. Add 10 μL 5M ammonium acetate (included) and 140 μL 100% ethanol to the reaction. Mix well.
    b. Incubate at −80°C for 20 min.
    c. Centrifuge at maximum speed at 4°C for 15 min.
    d. Remove supernatant without disturbing the pellet and wash with 1 mL 70% ethanol.
    e. Centrifuge at max speed at 4°C for 15 min.
    f. Remove supernatant and dry pellet at room temperature for 1–2 min.
    g. Resuspend the pellet in 15–20 μL RNase-free water.
13. Measure the RNA concentration. Bioanalyzer and Qubit are accurate methods of measuring high concentrations of RNA, but may not be accessible in a timely fashion. Nanodrop spectrophotometers are inaccurate at high concentrations of RNA. A diluted sample of the cleaned RNA product can be measured and the concentration can be calculated based on the dilution.
14. Dilute the sgRNA in RNase-free water (included) and store at −80°C.

8. IDENTIFYING INJECTION CONDITIONS TO USE FOR THE SCREEN

To determine whether a phenotype is amenable to screening in mosaic CRISPR-injected F_0 animals it is advisable to perform positive controls to test for feasibility. A positive control targeting a gene known to be involved in the process of interest is likely the most informative and useful test to perform before designing a large-scale screen. By examining the CRISPR-injected embryos for a positive control target, a phenotypic threshold can be set for the screen. Knowing the range of phenotypes observed in CRISPR-injected animals provides critical information for what to expect when targeting new genes of unknown function. Negative controls targeting genes unlikely to be involved in the process can provide critical information about the robustness of the wild-type phenotype. For example, the pigmentation genes *slc24a5* and *tyr* are common controls for CRISPR injections in zebrafish. They provide a direct metric of how well the CRISPR injections are working given that pigmentation defects can be observed in injected embryos under a stereomicroscope (Dahlem et al., 2012; Jao et al., 2013; Shah et al., 2015). Moreover, the negative control will create a baseline measurement for the phenotype of interest indicating both the natural variability of the phenotype as well as the variability that can be induced by general stress brought on by CRISPR injection.

To find the maximal CRISPR-induced mutation efficiency, thereby providing the best chance to observe phenotypes of interest, it is useful to perform a number of titration controls. We suggest doing two separate titration experiments to determine the parameters of the screen: a titration of Cas9 with a fixed amount of sgRNA and a titration of sgRNA with a fixed amount of Cas9—ideally both with an sgRNA against a gene with known function in the process of interest. We have found that increasing the amounts of both Cas9-encoding mRNA and sgRNA (up to nanogram amounts of both) increases the mutagenesis rate, giving the best possible chance of observing a phenotype (Shah et al., 2015). However, with increasing concentrations we find increased levels of general toxicity, including edema, cellular necrosis, and other general developmental defects. While it is advisable to avoid screening for phenotypes that are associated with general toxicity, it is important to make sure that the toxicity does not create additional noise that would otherwise obfuscate measurement of the phenotype. When screening single genes we have found that injecting 1000 pg of Cas9-encoding RNA with 100–1000 pg of sgRNA is often effective. These titration controls will identify the concentrations of Cas9 and sgRNA to use for the screen that maximizes phenotypic identification while minimizing toxicity.

All of the above considerations will help to design the screening protocol and inform the strategy used to investigate a list of gene targets. If the target list is particularly large, multiplexing sgRNAs into a single injection can reduce the number of injections required (Fig. 1). We multiplexed up to eight sgRNAs into pools to reduce the number of injections for a 48-gene screen (Shah et al., 2015). Each pool which resulted in a positive phenotype was then demultiplexed (each sgRNA injected individually) to determine the unique sgRNA, and therefore gene, causative for the phenotype. In particular, pooling of sgRNAs targeting duplicated genes, and/or multiple pathway components could reveal gene/pathway redundancies in the injected embryos. This property is particularly important given the partial genome duplication present in the zebrafish genome (Howe et al., 2013). While pooling can increase the efficiency of screening a large gene list, there are a few drawbacks that should be taken into consideration. Injecting multiple sgRNAs reduces the efficiency at each target site, furthering the possibility of missing a positive phenotype (Shah et al., 2015). Additionally, multiplexing increases the toxicity observed in injected embryos. Therefore when pooling multiple sgRNAs, the concentration will likely need to be lowered in order to reduce toxicity, so it is advisable to perform titration experiments (as explained previously) to identify the appropriate Cas9 and pooled sgRNA concentrations to use for the screen. Also, while less frequent than NHEJ-induced indels, multiple DSBs on the same chromosome can lead to larger deletions and chromosomal rearrangements (Xiao et al., 2013). Therefore pooling of sgRNAs on the same chromosome should be approached with caution.

9. ASSAYING THE RATE OF MUTAGENESIS IN F_0-INJECTED EMBRYOS

When screening a list of genes using CRISPR, targets that produce phenotypic changes provide strong evidence that the gene is involved in the process of interest. However, many targeted genes do not cause a phenotype. It may be important to analyze whether such negative results mean that the targeted gene is not required for the process, or that the targeted gene is not being mutated. In order to do this there are a number of methods that measure Cas9-induced mutagenesis in the F_0-injected mosaic animal. Based on these data, candidate genes can be eliminated or instead new conditions or new sgRNAs can be designed for candidates that were not effectively mutated. Each method that assesses the efficacy of specific sgRNAs varies in accuracy. Some methods not only assay the level of mutagenesis but also the sequence characteristics of the induced indels. Assaying the level of mutagenesis in the F_0 is optional and should be done if the researcher needs to be confident that each gene was effectively screened.

PCR-based methods are fast, relatively inexpensive, and highly accessible. High resolution melt analysis is a post-PCR analysis method used to identify variations in DNA sequences. It provides only a rough estimate of the extent of mutagenesis and is not quantitative (Dahlem et al., 2012). Restriction enzyme or other endonuclease (CelI, Surveyor) digests of PCR amplicons surrounding the sgRNA target site offer a more accurate quantification of mutation efficiency by quantifying the relative amounts of digested and undigested bands on an agarose gel (Hwang et al., 2013; Jao et al., 2013). Digestion assays can be accurately designed as the *S. pyogenes*

Cas9 creates a dsDNA blunt-ended cut three nucleotides 5′ of the PAM sequence (Jinek et al., 2012), and most of the resulting indels disrupt sequence 5′ of the DSB site (Shah et al., 2015). Alternatively, a qPCR assay can be designed to quantify the relative amounts of wild-type and mutated sequence without the need for a restriction enzyme site to be disrupted (Shah et al., 2015; Yu, Zhang, Yao, & Wei, 2014).

While all of the previous methods assay the level of mutagenesis, they are not able to analyze the indel size induced by NHEJ. Instead, fragment analysis can generate both a size estimate of PCR amplicons as well as measure relative quantity of each amplicon. Fragment analysis involves fluorescence measurement of labeled PCR product run through capillary gel electrophoresis. The relative height difference between the intensity peaks from uninjected and injected PCR products is directly proportional to the mutagenesis efficiency, and measurements of amplicon size are done at the single-nucleotide resolution (Carrington, Varshney, Burgess, & Sood, 2015). For these reasons, we recommend using this method if a capillary gel electrophoresis machine is available.

Sequencing-based methods not only assay the level of mutagenesis, but also determine the mutation signature associated with the NHEJ event. PCR amplicons can be cloned into vectors, transformed, and sequencing from individual bacterial colonies provides information about a single allele—note that due to the mosaicism in F_0-injected embryos this provides a vast underrepresentation of the alleles generated (Gagnon et al., 2014; Shah et al., 2015). If higher accuracy or higher throughput is necessary, we recommend using next generation sequencing. Each amplicon is clonally amplified and sequenced allowing for an immense depth of coverage providing a much higher accuracy (Moreno-Mateos et al., 2015; Shah et al., 2015).

10. CONFIRMATION OF PHENOTYPES

Given a carefully setup CRISPR screen, a positive phenotypic hit suggests strongly that the gene being targeted is important for the phenotype of interest. We recommend as a best practice to confirm the hit with a different sgRNA targeting the same gene, with the expectation that a similar phenotype will be observed. Moreover, off targets between the two guides are likely to be different due to two designed targeting sequences, providing more confidence that the targeted gene is involved. While CRISPR represents a powerful reverse genetic approach to identify genes important for a process of interest, the phenotypic effects observed in mosaic animals should be viewed with caution. Mosaicism complicates phenotypic analysis given the variable and unknown genetic state of each cell and each animal examined. Conceivably, phenotypes could be due to the mosaicism itself, eg, due to aberrant cell sorting (Davy, Bush, & Soriano, 2006; Twigg et al., 2013). Therefore, affected F_0 embryos, or embryos injected with subphenotypic amounts of Cas9/sgRNA, should be raised to adulthood. In cases where mosaic animals display a lethal

Method	Nuclease Digest	High Resolution Melt (HRM)	qPCR	Fragment Analysis	Colony Sequencing	Next Generation Sequencing (NGS)
Specific allele data	None	None	None	Size only	Poor coverage	Very high
Embryo throughput	Low	Medium	Low	Medium	Very low	High
Quantitative (wt/mutant allele ratio)	Low	Low	Medium	High	Medium	Very high
Effort/time	Medium (PCR, digest, gel)	Low (PCR only)	Medium (nested PCR, optimization)	Low (PCR only)	High (PCR, cloning, PCR, sequencing)	High (library prep)
Cost	Low	Variable[a]	Medium	Low	High	High
Other	No special equipment	Requires qPCR machine with HRM module	Requires qPCR machine	Requires capillary-based sequencing machine	No special equipment	Requires NGS expertise

[a] *Cost depends on use of commercial or homemade PCR/dye mix.*

phenotype, use of Cas9 tagged with the *nanos* 3′UTR—which translationally represses mRNA in somatic cells (Köprunner, Thisse, Thisse, & Raz, 2001)—effectively restricts the mutagenesis event to the germ cells, allowing for a mutant germ line to be transmitted by a wild-type soma (Moreno-Mateos et al., 2015). For all hits in the screen, mutant alleles should be carried through the germ line and the phenotypic effects subsequently confirmed in stable Mendelian mutants.

11. SUMMARY

Phenotypic screening of CRIPSR-injected F_0 zebrafish allows for the fast, efficient, and inexpensive identification of genes involved in a phenotype of interest. While many considerations need to be made before embarking on a large-scale screen, this novel method demonstrates the power of reverse genetics in a vertebrate system. All of the considerations discussed in this chapter are based on information that has been rigorously tested *in vivo* over only the past couple of years. We predict that given the speed of innovation in CRISPR and other genome-engineering technologies, the scope, scale, and precision of screening will continue to open many new avenues, thus requiring constant adjustment to the latest practices. Nevertheless, the current technology fundamentally changes the approach to gene discovery in the postgenomic era and the speed at which it is possible to screen targets *in vivo* in zebrafish.

REFERENCES

Bedell, V. M., Wang, Y., Campbell, J. M., Poshusta, T. L., Starker, C. G., Krug, R. G.... Ekker, S. C. (2012). *In vivo* genome editing using a high-efficiency TALEN system. *Nature, 7422*, 114–118.

Carrington, B., Varshney, G. K., Burgess, S. M., & Sood, R. (2015). CRISPR-STAT: an easy and reliable PCR-based method to evaluate target-specific sgRNA activity. *Nucleic Acids Research, 43*(22).

Chang, N., Sun, C., Gao, L., Zhu, D., Xu, X., Zhu, X.... Xi, J. J. (2013). Genome editing with RNA-guided Cas9 nuclease in zebrafish embryos. *Cell Research, 4*, 465–472.

Chari, R., Mali, P., Moosburner, M., & Church, G. M. (2015). Unraveling CRISPR-Cas9 genome engineering parameters via a library-on-library approach. *Nature Methods, 9*, 823–826.

Cong, L., Ran, F. A., Cox, D., Lin, S., Barretto, R., Habib, N.... Zhang, F. (2013). Multiplex genome engineering using CRISPR/Cas systems. *Science (New York, NY), 6121*, 819–823.

Dahlem, T. J., Hoshijima, K., Jurynec, M. J., Gunther, D., Starker, C. G., Locke, A. S.... Grunwald, D. J. (2012). Simple methods for generating and detecting locus-specific mutations induced with TALENs in the zebrafish genome. *PLoS Genetics, 8*, e1002861.

Davy, A., Bush, J. O., & Soriano, P. (2006). Inhibition of gap junction communication at ectopic Eph/ephrin boundaries underlies craniofrontonasal syndrome. *PLoS Biology, 10*, e315.

Doench, J. G., Hartenian, E., Graham, D. B., Tothova, Z., Hegde, M., Smith, I.... Root, D. E. (2014). Rational design of highly active sgRNAs for CRISPR-Cas9-mediated gene inactivation. *Nature Biotechnology, 12*, 1262–1267.

Foley, J. E., Maeder, M. L., Pearlberg, J., Joung, J. K., Peterson, R. T., & Yeh, J. R. (2009). Targeted mutagenesis in zebrafish using customized zinc-finger nucleases. *Nature Protocols, 12*, 1855–1867.

Gagnon, J. A., Valen, E., Thyme, S. B., Huang, P., Akhmetova, L., Ahkmetova, L.… Schier, A. F. (2014). Efficient mutagenesis by Cas9 protein-mediated oligonucleotide insertion and large-scale assessment of single-guide RNAs. *PLoS One, 5*, e98186.

Howe, K., Clark, M. D., Torroja, C. F., Torrance, J., Berthelot, C., Muffato, M.… Stemple, D. L. (2013). The zebrafish reference genome sequence and its relationship to the human genome. *Nature, 7446*, 498–503.

Hruscha, A., Krawitz, P., Rechenberg, A., Heinrich, V., Hecht, J., Haass, C., & Schmid, B. (2013). Efficient CRISPR/Cas9 genome editing with low off-target effects in zebrafish. *Development (Cambridge, England), 24*, 4982–4987.

Hsu, P. D., Lander, E. S., & Zhang, F. (2014). Development and applications of CRISPR-Cas9 for genome engineering. *Cell, 6*, 1262–1278.

Hsu, P. D., Scott, D. A., Weinstein, J. A., Ran, F. A., Konermann, S., Agarwala, V.… Zhang, F. (2013). DNA targeting specificity of RNA-guided Cas9 nucleases. *Nature Biotechnology, 9*, 827–832.

Hwang, W. Y., Fu, Y., Reyon, D., Maeder, M. L., Tsai, S. Q., Sander, J. D.… Joung, J. K. (2013). Efficient genome editing in zebrafish using a CRISPR-Cas system. *Nature Biotechnology, 3*, 227–229.

Jao, L. E., Wente, S. R., & Chen, W. (2013). Efficient multiplex biallelic zebrafish genome editing using a CRISPR nuclease system. *Proceedings of the National Academy of Sciences of the United States of America, 34*, 13904–13909.

Jessen, J. R., Topczewski, J., Bingham, S., Sepich, D. S., Marlow, F., Chandrasekhar, A., & Solnica-Krezel, L. (2002). Zebrafish trilobite identifies new roles for Strabismus in gastrulation and neuronal movements. *Nature Cell Biology, 8*, 610–615.

Jinek, M., Chylinski, K., Fonfara, I., Hauer, M., Doudna, J. A., & Charpentier, E. (2012). A programmable dual-RNA-guided DNA endonuclease in adaptive bacterial immunity. *Science (New York, NY), 6096*, 816–821.

Kimura, K., Wakamatsu, A., Suzuki, Y., Ota, T., Nishikawa, T., Yamashita, R.… Sugano, S. (2005). Diversification of transcriptional modulation: large-scale identification and characterization of putative alternative promoters of human genes. *Genome Research, 1*, 55–65.

Kleinstiver, B. P., Prew, M. S., Tsai, S. Q., Topkar, V. V., Nguyen, N. T., Zheng, Z.… Joung, J. K. (2015). Engineered CRISPR-Cas9 nucleases with altered PAM specificities. *Nature, 7561*, 481–485.

Kok, F. O., Shin, M., Ni, C. W., Gupta, A., Grosse, A. S., van Impel, A.… Lawson, N. D. (2015). Reverse genetic screening reveals poor correlation between morpholino-induced and mutant phenotypes in zebrafish. *Developmental Cell*, (1), 97–108.

Köprunner, M., Thisse, C., Thisse, B., & Raz, E. (2001). A zebrafish nanos-related gene is essential for the development of primordial germ cells. *Genes & Development, 21*, 2877–2885.

LaFave, M. C., Varshney, G. K., Vemulapalli, M., Mullikin, J. C., & Burgess, S. M. (2014). A defined zebrafish line for high-throughput genetics and genomics: NHGRI-1. *Genetics, 1*, 167–170.

Milligan, J. F., Groebe, D. R., Witherell, G. W., & Uhlenbeck, O. C. (1987). Oligoribonucleotide synthesis using T7 RNA polymerase and synthetic DNA templates. *Nucleic Acids Research, 21*, 8783–8798.

Milligan, J. F., & Uhlenbeck, O. C. (1989). Synthesis of small RNAs using T7 RNA polymerase. *Methods in Enzymology*, 51–62.

Montague, T. G., Cruz, J. M., Gagnon, J. A., Church, G. M., & Valen, E. (2014). CHOPCHOP: a CRISPR/Cas9 and TALEN web tool for genome editing. *Nucleic Acids Research, Web Server Issue*, W401–W407.

Moreno-Mateos, M. A., Vejnar, C. E., Beaudoin, J. D., Fernandez, J. P., Mis, E. K., Khokha, M. K., & Giraldez, A. J. (2015). CRISPRscan: designing highly efficient sgRNAs for CRISPR-Cas9 targeting *in vivo*. *Nature Methods, 10*, 982–988.

Pan, L., Shah, A. N., Phelps, I. G., Doherty, D., Johnson, E. A., & Moens, C. B. (2015). Rapid identification and recovery of ENU-induced mutations with next-generation sequencing and Paired-End Low-Error analysis. *BMC Genomics, 83*.

Peng, Y., Clark, K. J., Campbell, J. M., Panetta, M. R., Guo, Y., & Ekker, S. C. (2014). Making designer mutants in model organisms. *Development (Cambridge, England), 21*, 4042–4054.

Qin, W., Liang, F., Feng, Y., Bai, H., Yan, R., Li, S., & Lin, S. (2015). Expansion of CRISPR/Cas9 genome targeting sites in zebrafish by Csy4-based RNA processing. *Cell Research, 9*, 1074–1077.

Shah, A. N., Davey, C. F., Whitebirch, A. C., Miller, A. C., & Moens, C. B. (2015). Rapid reverse genetic screening using CRISPR in zebrafish. *Nature Methods, 6*, 535–540.

Shalem, O., Sanjana, N. E., Hartenian, E., Shi, X., Scott, D. A., Mikkelsen, T. S. … Zhang, F. (2014). Genome-scale CRISPR-Cas9 knockout screening in human cells. *Science (New York, NY), 6166*, 84–87.

Twigg, S. R., Babbs, C., van den Elzen, M. E., Goriely, A., Taylor, S., McGowan, S. J. … Wilkie, A. O. (2013). Cellular interference in craniofrontonasal syndrome: males mosaic for mutations in the X-linked EFNB1 gene are more severely affected than true hemizygotes. *Human Molecular Genetics, 8*, 1654–1662.

Varshney, G. K., Pei, W., LaFave, M. C., Idol, J., Xu, L., Gallardo, V. … Burgess, S. M. (2015). High-throughput gene targeting and phenotyping in zebrafish using CRISPR/Cas9. *Genome Research, 7*, 1030–1042.

Wang, T., Wei, J. J., Sabatini, D. M., & Lander, E. S. (2014). Genetic screens in human cells using the CRISPR-Cas9 system. *Science (New York, NY), 6166*, 80–84.

Xiao, A., Wang, Z., Hu, Y., Wu, Y., Luo, Z., Yang, Z. … Zhang, B. (2013). Chromosomal deletions and inversions mediated by TALENs and CRISPR/Cas in zebrafish. *Nucleic Acids Research, 14*, e141.

Yu, C., Zhang, Y., Yao, S., & Wei, Y. (2014). A PCR based protocol for detecting indel mutations induced by TALENs and CRISPR/Cas9 in zebrafish. *PLoS One, 6*, e98282.

CHAPTER

5

TALEN- and CRISPR-enhanced DNA homologous recombination for gene editing in zebrafish

Y. Zhang[*,§], **H. Huang**[§], **B. Zhang**[¶], **S. Lin**[§,1]

**Peking University Shenzhen Graduate School, Shenzhen, China*

[§]University of California Los Angeles, Los Angeles, CA, United States

[¶]Peking University, Beijing, People's Republic of China

[1]Corresponding author: E-mail: shuolin@ucla.edu

CHAPTER OUTLINE

Abstract

The TALE nuclease and clustered regularly interspaced short palindromic repeats (CRISPR)/CRISPR-associated systems (Cas) have been developed as important tools for genome editing in zebrafish. Here we describe a CRISPR/Cas9-based approach for generating site-specific gene targeting in zebrafish using DNA double-strand breaks

Methods in Cell Biology, Volume 135, ISSN 0091-679X, http://dx.doi.org/10.1016/bs.mcb.2016.03.005

induced homologous recombination (HR)-dependent repair mechanism. Through comicroinjection of Cas9 mRNA, guide RNA targeting genomic DNA sequence corresponding to the twist2 gene and the corresponding double-strand long arm donor template with a point mutation identified in human, HR-mediated knock-in of the expected targeting sequence was obtained. To facilitate identification of germline transmission of targeted mutation, a method of screening sperms of male founder fish is designed.

INTRODUCTION

In last 20 years, the reverse genetics techniques of gene targeting have achieved great success. The engineered endonucleases, including zinc finger nuclease, TALE nuclease (TALEN), and the clustered regularly interspaced short palindromic repeats (CRISPR)/CRISPR-associated (Cas) system have been fully developed to manipulate the genome in various organisms, including zebrafish (Bedell et al., 2012; Chang et al., 2013; Cong et al., 2013; Huang et al., 2011; Hwang et al., 2013; Sander et al., 2011; Urnov, Rebar, Holmes, Zhang, & Gregory, 2010). DNA double-strand breaks (DSBs) introduced by TALEN and CRISPR at their target sites can be repaired through two pathways: error-prone nonhomologous end-joining (NHEJ) pathway and homologous recombination (HR) pathway (Fig. 1). The DSB repair by NHEJ could generate indel mutations, which can cause frame shift and then abolish gene function if the mutations occur in an exon. Thus, using TALENs or CRISPR to introduce frame-shift indel into genome would be the easiest way to disrupt a targeted gene, and allow us to study gene function by generating knockout mutants (Gupta et al., 2013; Huang, Zhu, Lin, & Zhang, 2012; Xiao et al., 2013). Meanwhile, the DSBs can also be repaired by the HR pathway in the presence of a homologous donor template, with higher HR efficiency than the natural status. Based on this mechanism, coinjection of a homologous donor template together with TALENs or CRISPR can theoretically achieve site-specific knock-in of any desired DNA sequence, including big reporter genes such as GFP, short DNA fragments like loxP, or other types of precise genome editing, like single nucleotide substitution (Hisano et al., 2015; Shin, Chen, & Solnica-Krezel, 2014; Zu et al., 2013). This HR-mediated gene editing can precisely modify genomes and provide a powerful tool for gene function study, for example, generating conditional gene knockout in a tissue or stage-specific manner. And it also offers a technology for disease modeling, especially for those dominant genetic disorders with gain-of-function mutations.

DSBs can be introduced by either TALEN or CRISPR, followed by HR-mediated DNA repair depending on a specific DNA template provided. In this chapter, we initially planned to provide a detailed protocol for generating and screening HR-targeted zebrafish enhanced by both TALEN and CRISPR. Since we have described the procedures of how to use TALEN to enhance HR (Huang, Xiao, Tong, Lin, & Zhang, 2016), here we focus on providing a guide of how to use the CRISPR system for this purpose. In terms of homologous donor template, it is

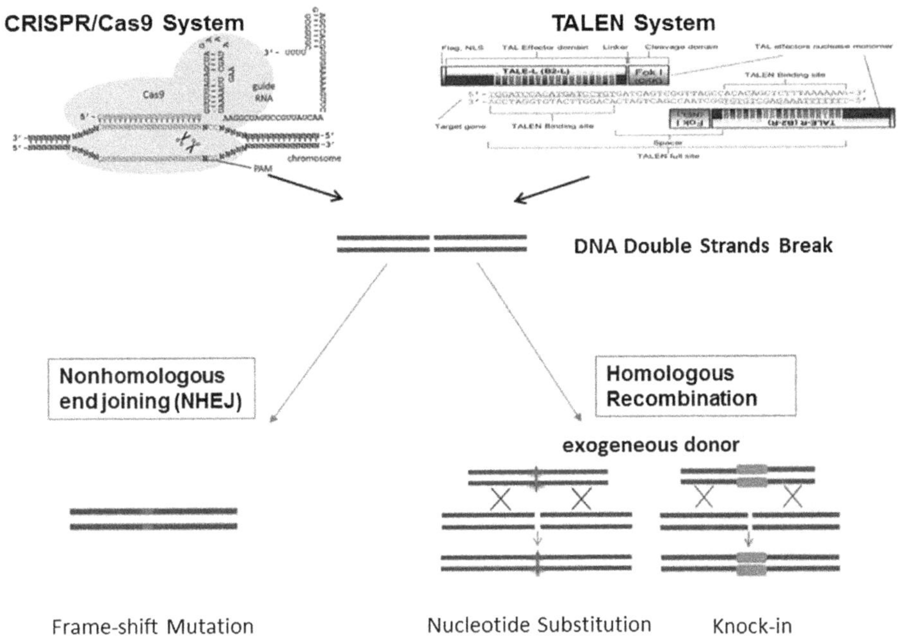

FIGURE 1

TALEN- and CRISPR-enhanced DNA homologous recombination for gene editing. After injection into zebrafish embryos, the TALEN/CRISPR cleaves the genome at the target sites and generates DSBs (Ma et al., 2012; Mali et al., 2013). Without homologous templates, the DSBs are repaired by the error-prone nonhomologous end-joining pathway, which could lead to indel mutations. In the presence of a homologous donor template, the chromosome can be repaired through the homologous recombination pathway, by which nucleotide substitution and knock-in of large or small DNA fragments could be achieved. (See color plate)

always composed of two homologous arms flanking the targeting sequence, but its length varies depending on the purpose of gene modification. For the insertion of large reporter genes like GFP, longer double-stranded DNA (dsDNA) fragments with two homologous arms, each >2kb, are recommended as the donor template (Hisano et al., 2015; Irion, Krauss, & Nusslein-Volhard, 2014; Shin et al., 2014; Zu et al., 2013). For the introduction of short sequences, both dsDNA and Single-Stranded Oligonucleotides (ssOligo) donors have been reported to work. An ssOligo donor of 50–200 nt can be commercially synthesized, and it works with higher efficiency but seems to have lower accuracy (Bedell et al., 2012; Chang et al., 2013; Chen et al., 2011; Hwang, Fu, Reyon, Maeder, Kaini, et al., 2013). To achieve precise modification, we prefer dsDNA donor template. In this chapter, we describe in detail how to perform HR-mediated nucleotide substitution by CRISPR and a dsDNA HR donor. In this case, a single base mutation c.223G>A [p.E75K] of Twist2 (Fig. 2A) found in human patients with Ablepharon

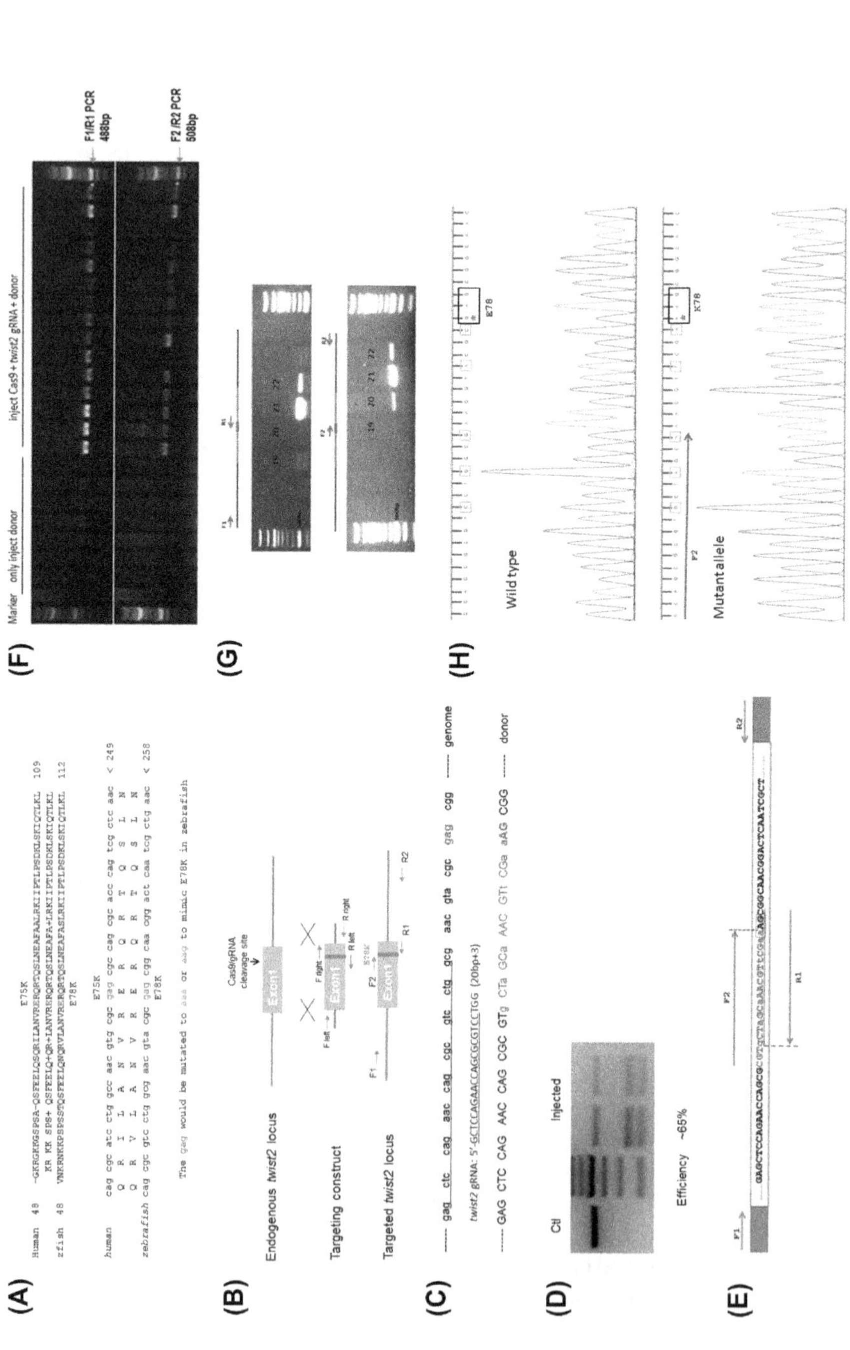
(A)
(B)
Endogenous twist2 locus
Targeting construct
Targeted twist2 locus
Cas9/gRNA cleavage site
Exon1
(C)
genome
donor
(D)
Ctl
Injected
Efficiency ~65%
(E)
(F)
Marker
only inject donor
inject Cas9 + twist2 gRNA + donor
F1/R1 PCR 488bp
F2/R2 PCR 508bp
(G)
(H)
Wild type
Mutant allele
E78
K78

FIGURE 2

Homologous recombination–mediated nucleotide substitution in zebrafish *twist2* gene to mimic the c.223G > A [p.E75K] mutation in human patients. (A) Protein sequence near the c.232G > A [p.E78K] mutation is conserved between human and zebrafish. (B) Schematic design of the targeted homologous recombination in zebrafish twist2 gene. Left arm is 468 bp and right arm is 452 bp. (C) Sequences of gRNA, targeting site, and donor template near the targeting site. (D) Analysis of the twist2 target site gRNA efficiency. (E) Design of the homologous recombination detection primers. (F) Detection of homologous recombination events in embryos injected with CRISPR and homologous recombination donor. (G) Detection of homologous recombination events in sperm DNA of F0 male founders. Individual #21 and #22 of F1 fish are positive for both primers detecting left and right arms of homologous recombination. (H) Sequencing validation of the integrated genome. *Yellow* (light gray in print versions) squares indicate altered bases. *Red* (gray in print versions) asterisk indicates G to A change, which causes amino acid alteration from E to K at position 78 in *twist2* allele of zebrafish chromosome.

macrostomia is introduced into the zebrafish genome, generating a zebrafish model for this dominant genetic disorder (Marchegiani et al., 2015). This protocol should enable similar types of precise genome modification for other genes of interest.

1. MATERIALS

1.1 REAGENTS

1. Appropriate restriction enzymes and reaction buffers for linearization of Cas9 expression vector to prepare the template for in vitro transcription
2. T7 endonuclease I (T7E1) and NEBuffer 2 (NEB, USA) for the evaluation of CRISPR efficiency
3. T7 (or other appropriate promoters) mMessage mMachine kit (Ambion, USA) for making Cas9 mRNA through in vitro transcription
4. T7/SP6 MAXIscript kit (Ambion, USA) for making CRISPR gRNA by in vitro transcription
5. DNA Clean & Concentrator-5 kit (Zymo Research, USA)
6. QIAquick Gel extraction kit (QIAGEN, USA)
7. RNeasy Mini Kit (QIAGEN, USA)
8. MicroRNA Isolation Kit (BioChain, USA)
9. Primers:
   a. To generate amplicons from targeted regions for indel analysis
   b. To amplify the gRNA template
   c. To amplify the donor template
   d. To detect the HR event
10. Phusion High-Fidelity PCR Master Mix (NEB, USA)
11. Taq enzyme
12. TOPO-TA Cloning Kit (Invitrogen, USA)
13. 50 mM NaOH
14. 1 M Tris−HCl (pH 8.0)
15. Lysis buffer: 10 mM Tris−HCl pH 8.2, 10 mM EDTA, 200 mM NaCl, 0.5% SDS, 200 μg/mL proteinase K
16. Agarose and Tris-Acetate-EDTA (TAE) buffer
17. RNase-free water, tips, and Eppendorf tubes.

1.2 ZEBRAFISH AND MICROINJECTION

1. Wild-type zebrafish or other desired zebrafish strains
2. Mating tanks
3. E3 embryo buffer: 5 mM NaCl, 0.17 mM KCl, 0.33 mM $CaCl_2$, 0.33 mM $MgSO_4$
4. Stereomicroscope
5. Glass capillaries (eg, outer diameter 1.0 mm, inner diameter 0.58 mm; Harvard Apparatus, USA), PN-30 Puller (Narishige, Japan), and Tweezer (Inox, 11 cm)

6. 1-μL disposable capillaries (R: 0.25%, CV: 0.6%) and a ruler: For quantification of injection volumes (CAMAG, Switzerland)
7. Nitrogen gas and PLI-90 Pico-Injector (Harvard Apparatus, USA)
8. Microinjection molds

2. METHODS

2.1 IN VITRO SYNTHESIS OF Cas9 mRNA

1. To synthesize Cas9 mRNA, the plasmid pGH-zCas9 is used, which contains the zebrafish-codon-optimized Cas9 (zCas9) cDNA with two flanking nuclear localization signals (Liu et al., 2014). The plasmid is linearized by a unique enzyme site XbaI, and transcribed to mRNA by T7 RNA polymerase. Other expression vectors can also be used if there are proper promoters available for in vitro transcription (ie, T3, T7, or SP6), and other unique restriction enzyme downstream of the NLS-Cas9-NLs coding sequence can be used for linearization. For linearization, 20 μg expression plasmids are digested by XbaI overnight at 37°C.
2. To monitor the completion of digestion, load 1 μL of the reaction mixture and 1 μL of undigested plasmids to 1% agarose gel and examine by electrophoresis.
3. When the digestion is complete, purify the linearized template by using DNA Clean & Concentrater kit and elute it with 20 μL nuclease-free water. Measure the concentration of the linearized template by a spectrophotometer.
4. Use the T7 mMessage mMachine kit to synthesize capped Cas9 mRNA. To set up an in vitro transcription reaction, mix 10 μL 2 × NTP/CAP with 2 μL 10 × reaction buffer, 1 μg linearized DNA template from the earlier step, and 2 μL T7 enzyme mix, and then supplement the volume to 20 μL with nuclease-free water, mix well by pipetting. Incubate the mixture at 37°C for 2 h. Add 1 μL TURBO DNase supplied by the kit and incubate at 37°C for 15 min to remove the DNA template.
5. To monitor the mRNA synthesis, load 0.5 μL reaction mixture to 1% agarose gel and examine by electrophoresis.
6. Purify the Cas9 mRNA by using RNeasy Mini Kit with an elution of 20 μL nuclease-free water. Determine the concentration of mRNA by a spectrophotometer and aliquot it into small volumes (eg, 5 μL). Store the aliquots at −80°C for later use. Since RNA degrades easily after thawing and freezing frequently, use a new aliquot of mRNA for each injection.

2.2 PREPARATION OF CRISPR gRNAs

1. The pMD19-gRNA scaffold plasmid (Chang et al., 2013) is used in the generation of the gRNA template by polymerase chain reaction (PCR),with a target-specific forward primer (T7-target-sfd/SP6-target-sfd) and a general reverse

primer (tracr rev). The PCR products are used as the template for in vitro transcription of the gRNA.
The forward primer (T7-target-sfd/SP6-target-sfd) contains the T7/SP6 promoter sequence, target site sequence, and part of the scaffold sequence. For instance, the primers below are designed to target zebrafish *twist2* locus:
T7-twist2-sfd: TAATACGACTCACTATAGCTCCAGAACCAGCGCGTCCGT TTTAGAGCTAGAAATAGC
The reverse primer (tracr rev) is of a short complementary sequence only targeting the end of the gRNA scaffold in the reverse direction: tracr rev: 5′-AAAAAAAGCACCGACTCGGTGCCAC-3′

2. When the PCR is complete, purify it by using DNA Clean & Concentrator kit and elute it with 20 μL nuclease-free water. Measure the concentration of the template by a spectrophotometer.
3. Use the T7/SP6 MAXIscript kit to synthesize CRIPSR gRNA. To set up an in vitro transcription reaction, mix 2 μL 10 × reaction buffer, 1 μL ATP, 1 μL GTP, 1 μL CTP, 1 μL UTP, 1 μg DNA template from the earlier step, 2 μL T7/SP6 enzyme mix, and then add nuclease-free water to 20 μL. Mix well by pipetting. Incubate the mixture at 37°C for 1 h.
4. Add 1 μL TURBO DNase supplied by the kit and incubate at 37°C for 15 min to remove the DNA template.
5. To monitor the gRNA synthesis, load 0.5 μL reaction mixture to 1.5% agarose gel and examine by electrophoresis.
6. Purify the gRNA by MicroRNA Isolation Kit and elute it with 20 μL nuclease-free water. Measure the concentration of gRNA by a spectrophotometer and aliquot it into small volumes (eg, 5 μL). Store the aliquots at −80°C for later use.

2.3 MICROINJECTION OF Cas9 mRNA/gRNA INTO ZEBRAFISH EMBRYOS

1. CRISPR gRNA is highly sequence specific, so mismatches in the target sites will dramatically decrease its activity. In zebrafish genomes, there are a great amount of sequence polymorphisms. Thus, to avoid the loss of targeting efficiency, we strongly recommend genotyping the adult zebrafish pairs intended to be used. gRNA is designed to perfectly match the sequence of target site.
2. Set up fish in the afternoon for the next day's injection.
3. Remove the divider from one tank each time in the morning of the injection day and collect embryos as soon as possible. Lay the embryos on the microinjection molds.
4 Cas9 mRNA and gRNA are mixed together with RNase-free water to a final concentration of 200 ng/μL and 25 ng/μL, respectively. The injection mixture sits on ice for use.
5. Break the needle with a tweezer and load the injection mixture into it. Use the 1-μL disposable capillaries and a ruler to quantify the injection volume. Adjust

the parameters of pressure and duration time of the microinjection machine to calibrate the volume to 1 or 2 nL per embryo.

6. The mixture (2 nL per embryo) will be injected into the cytoplasm of one-cell stage zebrafish embryos. When using different batches of CRISPR/Cas9 injection dosage should be optimized.
7. Incubate the injected embryos and uninjected sibling as a control in E3 embryo buffer at 28.5°C. Change the buffer every few hours and remove the dead and deformed embryos.

2.4 EVALUATION OF CRISPR EFFICIENCY IN INJECTED FOUNDER EMBRYOS

1. T7 endonuclease I enzyme assay is generally used to evaluate the efficiency of CRISPR. T7 endonuclease I assay is based on its mismatch recognition mechanism, so it is an effective and economical way for CRISPR efficiency measurement.
2. Genomic DNA is extracted from injected and control embryos using NaOH lysis method. Put five embryos into one PCR tube, remove extra E3 buffer, and add 50 μL 50 mM NaOH. Heat the embryos at 95°C in a PCR machine for 20 min and then cool down to 4°C. Briefly vortex the PCR tube to break the embryos, and then add 5 μL Tris–HCl (pH 8.0) to neutralize NaOH. Centrifuge for 5 min at 12,000 rpm (~13,500 × g), and the crude genomic DNA is in the supernatant (Meeker, Hutchinson, Ho, & Trede, 2007).
3. Mix 1 μL crude genomic DNA with a pair of primers, amplifying 300–500 bp fragment covering the target site, in a 50-μL PCR reaction. After PCR, take 2 μL PCR products to run a 1.5% agarose gel to confirm that there are expected PCR products.
4. Purify the PCR products with the DNA purification kit, and elute it into 20 μL ddH_2O. Determine the concentration of the purified PCR products by a spectrophotometer.
5. To obtain mismatched double-stranded PCR products, add 1 μg PCR products, 2 μl NEBuffer 2 and ddH_2O up to 20 μl in a PCR tube followed by running a denaturing/annealing program: (1) 95°C 2 min; (2) −2°C/s to 85°C; (3). −0.1°C/s to 25°C; (4) 16°C forever.
6. Add 0.5 μL T7E1 enzyme to the mixture and incubate at 37°C for 30 min.
7. Run the digested mixture and the undigested control on a 2% agarose gel. If the DNA from the injected embryos can be cleaved by T7E1, there will be several bands on the gel, while the control should only have a single band. The efficiency of the CRIPSR can be estimated by measuring the intensity of each band and calculate the percentage of the extra bands compared to the control band (Fig. 2D).
8. Precise genome editing at a specific site on DNA by HR would limit the usable CRISPR sites, as the DSBs must be introduced as close as possible to the targeted modification site. Efficiency of several CRIPSR sites around the targeted

modification site should be evaluated and the one with higher efficiency and the shortest distance would be the best choice for the HR.

2.5 PREPARATION OF THE HOMOLOGOUS DONOR TEMPLATE

1. The homologous donor template contains the targeted modification sequence of nucleotide substitution and two homologous arms flanking it. To introduce the modified sequence between the homologous arms, overlapping PCR is used.
   In the case of targeted nucleotide substitution, only one or few nucleotides need to be mutated, so the donor is composed of two homologous arms and a very short mutated sequence in the middle. Two pairs of primers should be carefully designed to individually amplify two homologous arms. Of each primer pairs, one covers the distal site on the genome, the other covers the target modification site. The targeted substituted sequence is included as an overhang at the 5′ end of this primer (Fig. 2B, shown by F left/right, R left/right). The two primers covering the modification site complementarily overlap with each other with ~20 bps, for the next step's overlapping PCR requirement. In addition, for the purpose of future PCR screening, few more nucleotides around the modification site should be substituted according to the amino acid codon degeneracy, so that the HR event can be easily detected by PCR with a primer specific to this altered sequence. Furthermore, if the modification site does not contain the CRISPR target site, some of the nucleotides in the CRISPR target site need to be substituted, so that the integrated DNA will not be cut again (Fig. 2C).
2. Two homologous arms of the donor template are amplified using DNA isolated from the same genotyped isogenic fish in 50 μL reaction solution containing Phusion High-Fidelity PCR Master Mix. After amplification, load all PCR products to run a 2% agarose gel. If the PCR products are correct, purify with gel extraction kit and elute DNA into 20 μL ddH2O.
3. Mix 2 μL each of these DNA fragments as templates with two outside primers described earlier to run overlapping PCR with a touchdown program, using Phusion PCR Master Mix.
4. After amplification, 2 μL PCR products are run on a 2% agarose gel, confirming that there are expected PCR products. To add adenines to the 3′ ends of the DNA fragments for TA cloning, 0.5 μL regular Taq enzyme is added to the PCR mixture for incubation at 72°C for another 20 min in the PCR machine.
5. Ligate the DNA fragment to a pCR2.1-TOPO vector, using the TOPO-TA Cloning Kit (Invitrogen, USA), and transform the ligation product into *E. coli* to obtain the donor plasmid. Confirm the construct by sequencing.
6. Digest out the donor DNA fragment containing the homologous arms and the modified sequence from the donor vector by proper restriction enzymes, purify by a gel extraction kit, and elute it into 20 μL RNase-free water.

2.6 MICROINJECTION OF Cas9 mRNA/gRNA AND DONOR TEMPLATE

1. The injection strategies are same as described in Section 2.3.
2. Inject 200–400 pg Cas9 mRNA, 25–50 pg CRISPR gRNA, and 50–100 pg linear donor DNA fragment into one-cell–stage zebrafish embryos. The embryos only injected with donor template are prepared as negative controls.

2.7 DETECTION OF HR EVENTS AND SCREENING OF POSITIVE FOUNDER FISH

1. To identify successful events of HR in the injected founder embryos, isolate the genomic DNA from groups of five embryos and run PCR using outside primers and modification-specific primers. The outside primers are specific to the sequence outside the donor homologous arms, and the modification-specific primers are specific to the targeted sequence. For each direction, one optimal pair of primers should be used (Fig. 2E, shown by F1/R1, F2/R2). If expected PCR products are obtained for both directions it is considered that successful HR events have been induced (Fig. 2F).
2. If the HR event can be detected, raise the remaining injected F0 embryos to adult. Outcross these mosaic F0 fish with wild-type fish and collect F1 embryos for extraction of genomic DNA and PCR detection as described earlier. Alternatively, we sacrifice F0 adult male fish and collect their sperms for cryo preservation and PCR on the sperm DNA. Once positive HR events are identified, F1 fish are generated through in vitro fertilization using frozen sperms. This approach saves a great deal of time and works well in our hand. Positive founder F0 are selected by positive PCR products (Fig. 2G) and sequencing data (Fig. 2H). In this case, we identified two positive F0 founders out of 60 fishes.
3. Grow up the F1 and genotype them using fin clip genomic DNA of each individual by the same method as in Section 2.7.1. Since 50% of F1 fish DNA should contain the targeted modification, HR events can be further confirmed by PCR amplifying the entire homologous region (including both homologous arms) by only using those two outside primers, followed by sequencing the cloned PCR product.

SUMMARY

HR-mediated genome editing allows precise manipulation of genes of interest, which enables gene function study as well as modeling human genetic diseases in zebrafish. To achieve genome editing in zebrafish by TALEN- and CRISPR-enhanced DNA HR, the following key points need to be kept in mind.

First, always take nucleotide polymorphism into consideration. Sequence mismatches between gRNA and the target site or between donor homologous arms and targeting genome both significantly reduce the efficiency of HR. Sequencing the entire homologous region in adult fish that will be used for HR is an essential step. It is desirable to have at least one pair of healthy adults with isogenic sequences in the target region. PCR products covering this region from tail clip DNA are used to generate the donor DNA construct. Progeny from this pair of fish will be injected with TALEN or CRISPR and a donor construct for HR purpose.

Second, choice of TALEN and CRISPR site is of great importance. Higher TALEN/CRISPR efficiency and shorter distance between DSBs and modification site lead to higher HR efficiency. Several target sites near the modification site should be evaluated to identify the one with the highest activity and nearest location.

Last but not the least, the donor template needs to be carefully designed. Immediately surrounding the targeting site within the modification-specific PCR primer region, altered sequence, as many as possible, should be introduced into the donor template by substituting some nucleotides according to amino acid codon degeneracy. This modification does not change nearby amino acid sequence but allows more specific PCR detection of HR events. In addition, the original TALEN/CRIPSR target site must be abolished from the donor template so that the integrated genome sequence is not targeted again by the TALEN/CRIPSR.

Compared to TALEN- or CRIPSR-induced indel mutations, the efficiency of HR events promoted by these two systems is far from satisfaction in zebrafish. We should initiate more mechanistic studies on this issue and further improvement of its efficiency is still of great significance.

ACKNOWLEDGMENTS

We thank members of our laboratory for technical assistance and discussion. This work was supported in part by a grant from the National Institutes of Health (NIH R21GM109908) to S.L.

REFERENCES

Bedell, V. M., Wang, Y., Campbell, J. M., Poshusta, T. L., Starker, C. G., Krug, R. G., 2nd, … Ekker, S. C. (2012). In vivo genome editing using a high-efficiency TALEN system. *Nature, 491*(7422), 114–118. http://dx.doi.org/10.1038/nature11537. pii:nature11537.

Chang, N., Sun, C., Gao, L., Zhu, D., Xu, X., Zhu, X., … Xi, J. J. (2013). Genome editing with RNA-guided Cas9 nuclease in zebrafish embryos. *Cell Research, 23*(4), 465–472. http://dx.doi.org/10.1038/cr.2013.45.

Chen, F., Pruett-Miller, S. M., Huang, Y., Gjoka, M., Duda, K., Taunton, J., … Davis, G. D. (2011). High-frequency genome editing using ssDNA oligonucleotides with zinc-finger nucleases. *Nature Methods, 8*(9), 753–755. http://dx.doi.org/10.1038/nmeth.1653.

Cong, L., Ran, F. A., Cox, D., Lin, S., Barretto, R., Habib, N., … Zhang, F. (2013). Multiplex genome engineering using CRISPR/Cas systems. *Science, 339*(6121), 819–823. http://dx.doi.org/10.1126/science.1231143. pii:science.1231143.

Gupta, A., Hall, V. L., Kok, F. O., Shin, M., McNulty, J. C., Lawson, N. D., & Wolfe, S. A. (2013). Targeted chromosomal deletions and inversions in zebrafish. *Genome Research, 23*(6), 1008–1017. http://dx.doi.org/10.1101/gr.154070.112. pii:gr.154070.112.

Hisano, Y., Sakuma, T., Nakade, S., Ohga, R., Ota, S., Okamoto, H., … Kawahara, A. (2015). Precise in-frame integration of exogenous DNA mediated by CRISPR/Cas9 system in zebrafish. *Scientific Reports, 5*, 8841. http://dx.doi.org/10.1038/srep08841. pii: srep08841.

Huang, P., Xiao, A., Tong, X., Lin, S., & Zhang, B. (2016). Targeted mutagenesis in zebrafish by TALENs. *Methods, 1338*, 191–206. http://dx.doi.org/10.1007/978-1-4939-2932-0_15.

Huang, P., Xiao, A., Zhou, M., Zhu, Z., Lin, S., & Zhang, B. (2011). Heritable gene targeting in zebrafish using customized TALENs. *Nature Biotechnology, 29*(8), 699–700. http://dx.doi.org/10.1038/nbt.1939.

Huang, P., Zhu, Z., Lin, S., & Zhang, B. (2012). Reverse genetic approaches in zebrafish. *Journal of Genetics and Genomics, 39*(9), 421–433. http://dx.doi.org/10.1016/j.jgg.2012.07.004. pii:S1673-8527(12)00145-2.

Hwang, W. Y., Fu, Y., Reyon, D., Maeder, M. L., Kaini, P., Sander, J. D., … Yeh, J. R. (2013). Heritable and precise zebrafish genome editing using a CRISPR-Cas system. *PLoS One, 8*(7), e68708. http://dx.doi.org/10.1371/journal.pone.0068708. pii:PONE-D-13-13968.

Hwang, W. Y., Fu, Y., Reyon, D., Maeder, M. L., Tsai, S. Q., Sander, J. D., … Joung, J. K. (2013). Efficient genome editing in zebrafish using a CRISPR-Cas system. *Nature Biotechnology, 31*(3), 227–229. http://dx.doi.org/10.1038/nbt.2501. pii:nbt.2501.

Irion, U., Krauss, J., & Nusslein-Volhard, C. (2014). Precise and efficient genome editing in zebrafish using the CRISPR/Cas9 system. *Development, 141*(24), 4827–4830. http://dx.doi.org/10.1242/dev.115584. pii:dev.115584.

Liu, D., Wang, Z., Xiao, A., Zhang, Y., Li, W., Zu, Y., … Zhang, B. (2014). Efficient gene targeting in zebrafish mediated by a zebrafish-codon-optimized cas9 and evaluation of off-targeting effect. *Journal of Genetics and Genomics, 41*(1), 43–46. http://dx.doi.org/10.1016/j.jgg.2013.11.004.

Ma, S., Zhang, S., Wang, F., Liu, Y., Xu, H., Liu, C., … Xia, Q. (2012). Highly efficient and specific genome editing in silkworm using custom TALENs. *PLoS One, 7*(9), e45035. http://dx.doi.org/10.1371/journal.pone.0045035. pii:PONE-D-12-15184.

Mali, P., Yang, L., Esvelt, K. M., Aach, J., Guell, M., DiCarlo, J. E., … Church, G. M. (2013). RNA-guided human genome engineering via Cas9. *Science, 339*(6121), 823–826. http://dx.doi.org/10.1126/science.1232033. pii:science.1232033.

Marchegiani, S., Davis, T., Tessadori, F., van Haaften, G., Brancati, F., Hoischen, A., … Markello, T. C. (2015). Recurrent mutations in the basic domain of TWIST2 cause Ablepharon macrostomia and Barber-Say syndromes. *American Journal of Human Genetics, 97*(1), 99–110. http://dx.doi.org/10.1016/j.ajhg.2015.05.017.

Meeker, N. D., Hutchinson, S. A., Ho, L., & Trede, N. S. (2007). Method for isolation of PCR-ready genomic DNA from zebrafish tissues. *Biotechniques, 43*(5), 610, 612, 614. pii:000112619.

Sander, J. D., Cade, L., Khayter, C., Reyon, D., Peterson, R. T., Joung, J. K., & Yeh, J. R. (2011). Targeted gene disruption in somatic zebrafish cells using engineered TALENs. *Nature Biotechnology, 29*(8), 697–698. http://dx.doi.org/10.1038/nbt.1934. pii:nbt.1934.

Shin, J., Chen, J., & Solnica-Krezel, L. (2014). Efficient homologous recombination-mediated genome engineering in zebrafish using TALE nucleases. *Development, 141*(19), 3807–3818. http://dx.doi.org/10.1242/dev.108019. pii:dev.108019.

Urnov, F. D., Rebar, E. J., Holmes, M. C., Zhang, H. S., & Gregory, P. D. (2010). Genome editing with engineered zinc finger nucleases. *Nature Reviews Genetics, 11*(9), 636–646. http://dx.doi.org/10.1038/nrg2842.

Xiao, A., Wang, Z., Hu, Y., Wu, Y., Luo, Z., Yang, Z., ... Zhang, B. (2013). Chromosomal deletions and inversions mediated by TALENs and CRISPR/Cas in zebrafish. *Nucleic Acids Research, 41*(14), e141. http://dx.doi.org/10.1093/nar/gkt464. pii:gkt464.

Zu, Y., Tong, X., Wang, Z., Liu, D., Pan, R., Li, Z., ... Lin, S. (2013). TALEN-mediated precise genome modification by homologous recombination in zebrafish. *Nature Methods, 10*(4), 329–331. http://dx.doi.org/10.1038/nmeth.2374. pii:nmeth.2374.

CHAPTER

6

Precise genome editing by homologous recombination

K. Hoshijima, M.J. Jurynec, D.J. Grunwald[1]

University of Utah, Salt Lake City, UT, United States

[1]*Corresponding author: E-mail: grunwald@genetics.utah.edu*

CHAPTER OUTLINE

Methods in Cell Biology, Volume 135, ISSN 0091-679X, http://dx.doi.org/10.1016/bs.mcb.2016.04.008

Abstract

Simple and efficient methods are presented for creating precise modifications of the zebrafish genome. Edited alleles are generated by homologous recombination between the host genome and double-stranded DNA (dsDNA) donor molecules, stimulated by the induction of double-strand breaks at targeted loci in the host genome. Because several kilobase-long tracts of sequence can be exchanged, multiple genome modifications can be generated simultaneously at a single locus. Methods are described for creating: (1) alleles with simple sequence changes or in-frame additions, (2) knockin/knockout alleles that express a reporter protein from an endogenous locus, and (3) conditional alleles in which exons are flanked by recombinogenic loxP sites. Significantly, our approach to genome editing allows the incorporation of a linked reporter gene into the donor sequences so that successfully edited alleles can be identified by virtue of expression of the reporter. Factors affecting the efficiency of genome editing are discussed, including the finding that dsDNA products of *I-SceI* meganuclease enzyme digestion are particularly effective as donor molecules for gene-editing events. Reagents and procedures are described for accomplishing efficient genome editing in the zebrafish.

INTRODUCTION

Studies in zebrafish have made substantial contributions to our understanding of gene function in vertebrates. The ability to conduct forward genetic screens combined with the ability often to detect phenotypic consequences that result immediately and directly from aberrant gene activity have contributed to use of the zebrafish as a platform for discovering gene functions and for obtaining novel insights into previously characterized genes. Implementation of antisense gene knockdown, targeted screening of randomly mutagenized genomes, and gene knockout methods that utilize targeted mutagenesis with programmable nucleases has expanded uses of the zebrafish by creating the opportunity to apply the power of phenotypic analyses in the zebrafish to selected genes of interest (Lawson & Wolfe, 2011). Hence genes initially discovered experimentally in other species (Giraldez et al., 2005), by virtue of their association with human diseases (Jurynec et al., 2008; Phillips & Westerfield, 2014), or predicted as a consequence of genome and expression analyses (Ulitsky, Shkumatava, Jan, Sive, & Bartel, 2011) have been studied through loss-of-function analysis in the zebrafish. Analysis of loss-of-function conditions is a powerful approach for uncovering the earliest acting functions of a gene and has particular usefulness for identifying genes that may contribute to a shared molecular or developmental pathway (Gritsman et al., 1999; Jurynec et al., 2008; Langdon & Mullins, 2011). Nevertheless, as discussed below, these methods of analysis do not give access to the full range of functions of a gene. New techniques for directed modification of the genome are being developed to expand the kinds of inquiry that can be conducted in the zebrafish.

Null alleles of essential genes are most useful for uncovering the first stage at which a gene is required. However, as many genes governing signal transduction pathways, tissue patterning, growth regulation, etc. are used in multiple contexts during development and tissue homeostasis, we need tools to control tissue-specific and temporal gene expression in order to study the context-specific functions of any gene. For example, as the pathways governing tissue turnover and regeneration often have essential roles also during embryogenesis (Beachy, Karhadkar, & Berman, 2004), the ability to generate and utilize conditional alleles will have dramatic impact on the ability to analyze genes governing tissue maintenance in adults.

Constitutive loss-of-function mutations also have limitations with respect to their ability to recapitulate disease states or provide insights into a gene's role in disease or its range of developmental activities. Many alleles associated with developmental, physiological, or behavioral disorders produce altered gene products or affect gene expression levels. Similarly, a main source of phenotypic variation among members of a species is likely due to sequence variants that do not eliminate gene function (Wray, 2007). Hence, tools for generating precise sequence modifications of the zebrafish genome are needed to more fully exploit this organism for purposes such as modeling disease states and understanding the roles of naturally occurring gene variants.

In addition, tools for modifying the zebrafish genome will allow study of the consequences of mutations that arise mosaically in the soma. Sporadically arising gain- or loss-of-function mutations are known to contribute to the origin of cancers and additional conditions including neurological disorders (Lupski, 2013; Poduri, Evrony, Cai, & Walsh, 2013). In sum, techniques to precisely modify the zebrafish genome will promote new approaches to the study of gene function that have the potential to render genuinely new insights into the phenotypic consequences of mutations.

Finally, the ability to precisely modify the genome may revolutionize developmental and cell biological studies in the zebrafish. Gene editing will allow us to generate modified proteins or introduce completely novel products that are expressed under conditions that precisely mimic an endogenous gene product. We anticipate these types of modifications will allow fate tracing or ablation of cells in developmental studies, visualization of the dynamic utilization of subcellular components, and identification of molecules that physically interact with the modified proteins of interest.

1. OVERVIEW OF CONTEMPORARY APPROACHES TO GENOME EDITING

The potential of genome-editing applications has spawned many efforts to develop flexible and reliable methods for modifying the zebrafish genome (Auer, Duroure,

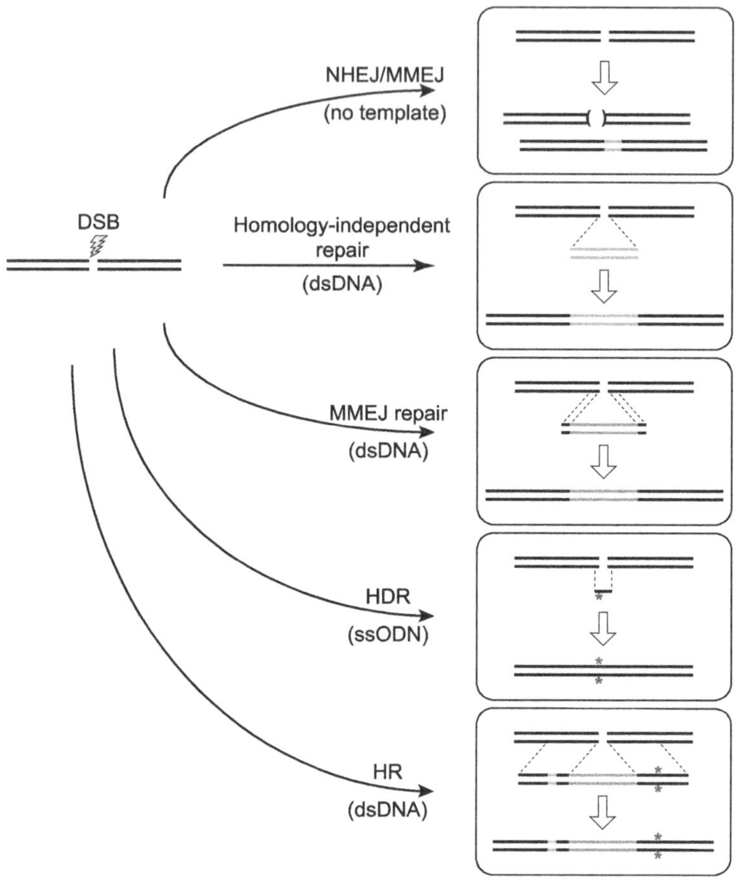

FIGURE 1 Approaches to genome editing.

Repair and recombination events are stimulated by double-strand breaks (DSBs) induced by targeted cleavage (lightning bolt) of the genome with programmable nucleases. In the absence of template DNA to guide repair, the nonhomologous end-joining (NHEJ) or microhomology-mediated end-joining (MMEJ) pathways heal broken ends of chromosomes in a process that often generates small deletions and/or insertions at the site of the lesion. These repair pathways can also facilitate integration of exogenously supplied DNA sequences at the lesion site. In the presence of linear double-stranded DNA (dsDNA) donor molecules that bear no homology to the targeted locus, the NHEJ pathway may join donor DNA sequences to the broken chromosome ends. End-joining occurs in a homology-independent, imprecise manner, and foreign sequences may be integrated in either orientation. If the borders of the donor DNA contain short sequences homologous to the regions immediately flanking the DSB site, end resection may uncover short stretches of complementarity that guide repair and result in incorporation of the donor sequences at the lesion. Two additional methods have been employed with the purpose of precisely modifying the host genome. When single-stranded oligodeoxynucleotides (ssODN) with close homology to the site of the

Concordet, et al., 2014; Auer, Duroure, De Cian, Concordet, & Del Bene, 2014; Bedell et al., 2012; Chang et al., 2013; Hisano et al., 2015; Hoshijima, Jurynec, & Grunwald, 2016; Hruscha et al., 2013; Irion, Krauss, & Nusslein-Volhard, 2014; Kimura, Hisano, Kawahara, & Higashijima, 2014; Li et al., 2015; Shin, Chen, & Solnica-Krezel, 2014; Zu et al., 2013). All current approaches to genome editing utilize the host cellular pathways that are normally responsible for repairing DNA damage (Fig. 1). The major advance that sets the groundwork for these methods was the development of sequence-specific programmable nucleases, which are used to initiate double-strand breaks (DSBs) at targeted loci and thus trigger repair (Carroll, 2014; Hsu, Lander, & Zhang, 2014; Kim & Kim, 2014). As the technology for efficiently inducing targeted DSBs is relatively young, the current approaches for modifying the genome are still at nascent stages.

Because chromosome breaks stimulate repair and recombination pathways at the site of the lesion, synthetic sequence-specific nucleases can be used to direct repair-dependent modifications of the zebrafish genome to a particular locus of interest (Fig. 1) (reviewed in Auer & Del Bene, 2014). Zinc-finger nucleases, TALENs, and RNA-guided nucleases of the clustered regularly interspaced short palindromic repeats (CRISPR)/Cas9 system have all been demonstrated capable of inducing DSBs at targeted sites in the zebrafish genome and triggering repair. In the absence of a DNA template to direct repair, DSB lesions can be healed by the nonhomologous end-joining (NHEJ) or the microhomology-mediated end-joining (MMEJ) pathway, both of which often result in small sequence modifications at the lesion site. This approach has become widely applied for the purposes of generating frame-shift or deficiency mutations at a targeted locus.

In addition, both repair pathways have been exploited as means for inserting exogenously provided donor sequences into a targeted zebrafish locus (Auer, Duroure, De Cian et al., 2014; Hisano et al., 2015; Kimura et al., 2014; Li et al., 2015). In the absence of homology between donor and targeted locus, donor sequences can be inserted imprecisely via NHEJ, with sequence alterations arising at the junction sites. When the ends of linear donor molecules are flanked by short sequences homologous to those flanking the DSB in the host, the homologies may facilitate MMEJ repair leading to intact incorporation of all novel sequences at the lesion site. These approaches have been used to insert protein-coding sequences in a manner that allows their expression to be regulated by the targeted locus.

DSB are available, the donor sequences may guide homology-directed repair (HDR) resulting in the modification of host sequences. Finally, when DSBs are induced in the presence of dsDNA molecules with extensive homology to the targeted region, true crossover events mediated by homologous recombination (HR) may occur, resulting in exchange of several kilobase-long regions between the host genome and the exogenously supplied DNA. This approach can be used to produce several closely linked genome modifications simultaneously. (See color plate)

In contrast to the editing methods described in the preceding paragraphs, two additional methods have been introduced with the express purpose of precisely revising the host genome. The exact molecular mechanisms mediating these repair/recombination events are not clear, and therefore, the factors that contribute to the efficiency of each type of editing event are not yet well defined.

In the first method, designated homology-directed repair (HDR), single-stranded oligodeoxynucleotides (ssODN) sharing homology with the lesion site are supplied as donor sequences to guide repair of the broken chromosome (Bedell et al., 2012; Chang et al., 2013; Hruscha et al., 2013). The repair process drives replacement of host sequences with donor sequences, allowing both the precise modification of host sequences and the insertion of short stretches of novel sequence into the genome. One strength of the approach is the ease with which donor templates can be designed and generated. However, as currently implemented, there are three substantial limitations to the use of HDR with ssODN donor molecules. First, because the donor templates are short, generally less than 100 nt, only small sequence modifications are produced by this method. Second, the recovery of edited alleles usually requires laborious DNA-based screening. Third, at least half the repair events are imprecise, accompanied by unintended changes to the genome. We expect the fidelity of this approach will improve with further study.

An alternative approach utilizes homologous recombination (HR) between genomic sequences flanking the DSB lesion site and double-stranded DNA (dsDNA) donor molecules, allowing the precise replacement of several kilobase-long stretches of host sequence with exogenously provided sequences (Hoshijima et al., 2016; Irion et al., 2014; Shin et al., 2014; Zu et al., 2013). Donor templates are supplied as dsDNA molecules that generally contain homology arms of 0.5–2 kbp, which may or may not border regions of nonhomology. Because this approach utilizes true crossover events involving the homology arms, it can be used simply to edit host sequences or to accomplish more complex genome modifications. As illustrated in Fig. 2A, one simple application is to produce the precise in-frame insertion of entire protein-coding sequences, leading to expression of a novel protein such as green fluorescent protein (GFP) from an endogenous locus. Significantly, the approach can also be used to produce multiple modifications distributed over several kilobases of a targeted locus (Hoshijima et al., 2016). As illustrated in Fig. 2B, in addition to critical sequence modifications of interest, we often include a reporter gene within the donor sequences. Recombination events that utilize both homology arms will produce the intended editing changes and also lead to incorporation of the linked reporter gene, which can be used initially to help identify an edited allele and subsequently to track its inheritance. Use of the linked reporter gene makes it easy to identify and recover edited alleles, such as loxP-flanked conditional mutations, that may not confer an overt phenotype on the carrier. In practice, inclusion of the reporter within the donor sequences greatly streamlines the recovery of edited alleles: we find 30–50% of the reporter-expressing genomes recovered from treated animals harbor precisely edited alleles that include all intended

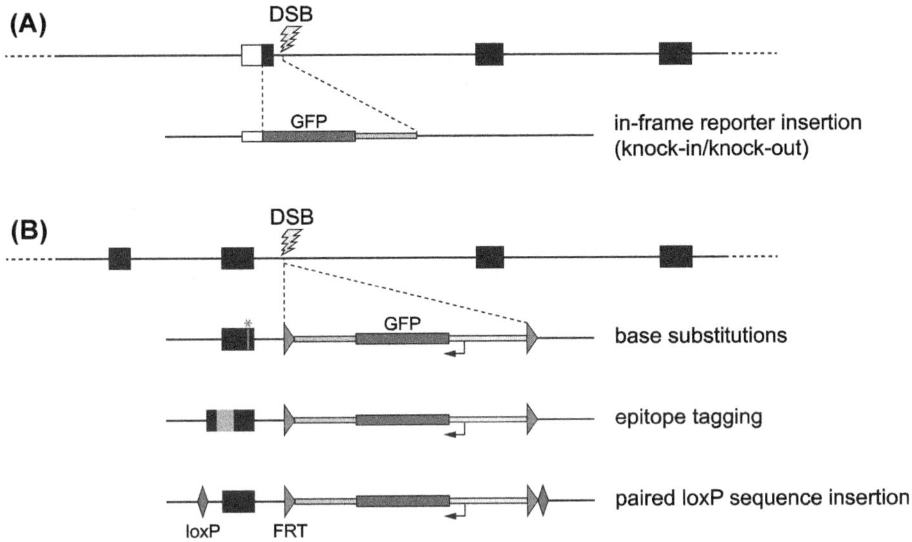

FIGURE 2 Examples of genome modifications produced by DSB-stimulated homologous recombination (HR) using dsDNA as donor template molecules.

Recombination events are stimulated by targeted cleavage (lightning bolts) of the genome with programmable nucleases. The top line of each panel represents the host genome, with *boxes* indicating exons (white, 5′ untranslated region; black, coding sequences). Below the genome cartoons are examples of dsDNA donor templates that have been used to generate the indicated types of edited alleles. Except as indicated, donor sequences are identical to those of the genome: *green boxes* indicate GFP coding sequences, *gray boxes* indicate transcription termination sequences, *yellow boxes* indicate *α-crystallin* promoter sequences, *orange arrowheads* are FRT sites, and pink diamonds are loxP sites. (A) Generation of a knockin/knockout allele. The DSB was induced in intron sequences just 3′ to the first exon. Recombination with donor sequences introduced GFP reporter sequences in-frame just downstream of the endogenous translation initiation codon. Animals inheriting this kind of edited allele can be recognized by virtue of expression of GFP under control of the endogenous promoter. (B) Use of a linked reporter gene to tag genomic modifications. In addition to the desired editing events, donor DNAs can be designed so they introduce into the host genome a small reporter gene (here *α-crystallin::GFP* flanked by directly repeated FRT sites). Edited alleles that incorporate precise alteration of coding sequences (blue *—base substitutions), in-frame introduction of epitope sequences (pale green—epitope tagging), or introduction of a pair of loxP recombination sites flanking an exon can be identified by expression of the linked tissue–specific reporter gene. Screening for GFP fluorescence allows for the efficient recovery of phenotypically silent modification events and obviates the need for laborious DNA-based screening of F1 individuals. Following identification of successful editing events, the reporter gene can be excised using FLP-mediated recombination. (See color plate) modifications. Moreover, when the reporter gene is bordered by FRT sites, it can be readily excised upon expression of FLP recombinase (Hoshijima et al., 2016).

Each of the approaches to genome editing described in the preceding paragraphs is sufficiently efficient to be practicable. Because of the range of types of genome modifications that can be produced, the precision with which edited alleles may be generated, and the ease with which reporter-marked alleles can be recovered, our focus has been on developing genome-editing methods that utilize DSB-stimulated HR with dsDNA donor molecules. Several groups using this approach have found between five percent and 15% of the animals that arise from zygotes injected with a combination of programmable nuclease and dsDNA donor molecules transmit precisely edited alleles to their progeny (Hoshijima et al., 2016; Irion et al., 2014; Shin et al., 2014). In our experience, about 6% (range 0.6–23) of the gametes of a founder with an edited germ line will transmit the modified allele of interest (Hoshijima et al., 2016). Thus marking donor sequences with a linked reporter gene can greatly facilitate the identification and recovery of edited alleles.

2. SUMMARY OF WORKFLOW

Our current focus is to edit genomes at early stages during the embryonic growth of a founder, with the expectation that a subset of the edited genomes will enter the germ line and become heritable. In experiments aimed at optimizing the induction of heritable modified alleles, we initially measured the production of edited alleles in the somatic tissues of 1–2-day-old founder embryos, assuming these events are representative of the germ-line genomes (Hoshijima et al., 2016). Our method of approach to gene editing has been informed by such experiments, some of which are described below.

Briefly, to stimulate HR events, just-fertilized zebrafish zygotes are injected with a mixture of synthetic nuclease targeting a unique locus and donor dsDNA consisting of novel sequences flanked by 1-kbp left and right homology arms (Fig. 3). At 2-day postfertilization (dpf), the genomes of 8–12 injected embryos are individually analyzed for the presence of correctly edited alleles. Polymerase chain reaction (PCR) analysis functions as a preliminary test to determine whether the targeted locus has acquired donor sequences in the expected configuration, using one primer specific to novel donor sequences and the second primer specific to host sequences distal to the homology regions. As shown in Fig. 3, nuclease activity has a profound effect stimulating HR. Under typical conditions, we cannot detect edited alleles in embryos injected only with donor DNA, but all embryos injected with nuclease and donor DNA have edited genomes. When injection leads to detectable genome editing in virtually all (>95%) founder embryos, sibling injected embryos are grown to adulthood and examined for the ability to transmit modified alleles to offspring. Depending on the type of genome modification that has been designed, transmission of edited alleles is detected in F1 offspring either by the expression of newly acquired coding sequences, such as a reporter protein, or by the analyses of genomic DNA.

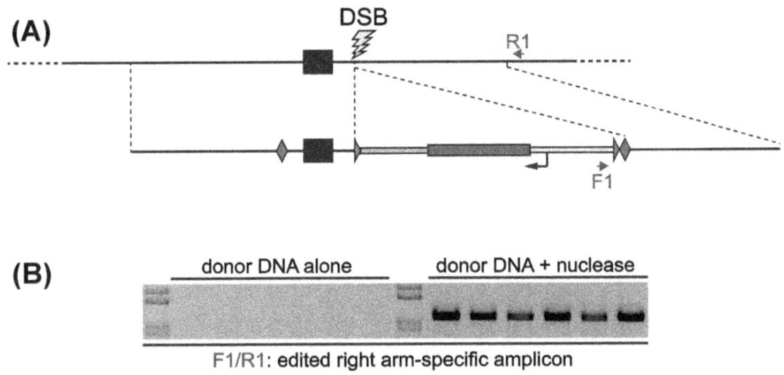

FIGURE 3 PCR-based screening to identify edited alleles in F0 embryos.

(A) Schematic representation (as in Fig. 2) of a genome-editing event in which HR with the dsDNA donor molecule introduces a pair of loxP recombination sites flanking an exon as well as the *α-crystallin::GFP* reporter gene. To detect the presence of correctly edited alleles in F0 embryos that had been injected with nuclease and donor molecules, genomic DNA is isolated from individual 1 dpf F0 embryos, and PCR analysis is performed using one primer (F1) specific to novel donor sequences and a second primer (R1) specific to endogenous host sequences distal to the homology region. (B) HR is dependent on nuclease activity. When donor DNA is injected without nuclease, edited alleles are not detected in F0 embryos. However, under typical efficient editing conditions, following injection of donor DNA with nuclease, every F0 embryo harbors precisely edited alleles.

3. FACTORS TO CONSIDER FOR GENOME EDITING VIA HOMOLOGOUS RECOMBINATION

Although the approach described here has produced precisely modified alleles of many loci (Hoshijima et al., 2016), we realize the methods are not yet perfected and many parameters that might affect the efficiency of editing have yet to be tested. In this section, we discuss factors that we consider in the design of a genome-editing experiment, highlighting aspects of the technology that need to be further investigated and optimized.

3.1 CHOICE OF NUCLEASE TARGET SITE

The induction of DSBs in the host genome is responsible for recruiting repair machinery and stimulating HR events between the affected locus and donor molecules harboring homologous sequences. However, DSBs also stimulate the NHEJ and MMEJ repair pathways, which simply rejoin the chromosome arms, often producing indel mutations at the DSB site (Fig. 1). Hence, there is potential for a conflict, for as the frequency of DSBs is increased to maximize the occurrence of HR, there will likely be a concomitant rise in the induction of indel mutations at targeted loci. For this reason, especially when we attempt to modify an essential gene, we prefer to design and utilize programmable nucleases that generate DSBs in intron regions devoid of conserved sequences, with the intent to minimize the production of mutant cells in the soma of injected embryos. In the future, the development of tools that stimulate gene editing specifically in the germ line may obviate this concern.

A second consideration is the distance between the DSB site and the closest position at which nonhomologous sequences are to be introduced. Indeed, there are no published empirical data addressing this issue in the zebrafish. Published studies have tended to induce DSBs at genomic sites very close to the sequences that are to be modified by the donor. Similarly, we do not know how far recombination events initiated at a DSB will travel, and thus how much donor sequence can be incorporated efficiently into the genome. Several reports indicate that tracts of about 2-kbp of sequence can be readily introduced into the host genome (Hoshijima et al., 2016; Shin et al., 2014; Zu et al., 2013).

Third, we found that DSBs with 5′ overhangs, produced by TALEN-mediated cleavage, and DSBs initiated by the blunt-cutting CRISPR/Cas9 system can each be used promote HR efficiently in the zebrafish. Either type of target site can be used to stimulate gene editing.

For several reasons, we attempt to minimize sequence heterogeneity at the target locus among the genomes that are subjected to editing. First, polymorphisms within the nuclease recognition site may reduce the induction of DSBs. Second, polymorphisms close to the targeted site may affect the ability to measure nuclease activity, because the induction of DSBs is often measured indirectly as the occurrence of repair-induced sequence heterogeneity close to the DSB site (Dahlem et al., 2012; Hwang et al., 2013; Jao, Wente, & Chen, 2013). Third, it is unclear how mismatches between the host locus and donor sequences affect recombination efficiency, although we note that donor sequences are generally designed so that their integration will destroy the nuclease recognition site and render the edited locus immune to further nuclease activity. For these reasons, we utilize a selected breeding population that lacks polymorphism near the DSB target site to produce the genomes to be edited.

3.2 INDUCTION OF TARGETED DSBs

To induce DSBs at a selected site in the host genome, we utilize TALENs or components of the CRISPR/Cas9 with equal abandon. As TALENs are delivered in the form of injected mRNA that needs to be translated prior to the onset of enzyme activity, our bias is to initiate DSBs by injecting a complex of sgRNA (single guide RNA) and Cas9 protein (CRISPR-associated protein 9) (Gagnon et al., 2014), components that should be capable of catalyzing DSBs soon after injection.

To achieve high rates of HR in the embryo, it is critical to identify nucleases that efficiently generate DSBs at the target site. To determine the relative cleavage activity of TALENs or sgRNAs in vivo, we inject candidate nucleases without donor DNA into zygotes and analyze the presence of mis-repaired alleles in the genomes of 8–12 individual 1 dpf embryos (Fig. 4) (Dahlem et al., 2012; Hoshijima et al., 2016). As the starting population of targeted genomes had been selected to lack polymorphisms in the region bordering the nuclease target site, the relative abundance of newly induced mutations reflects DSB activity. Newly arising sequence polymorphisms can be detected by any of a number of methods, including direct sequencing, high-resolution melt analysis (HRMA) (Dahlem et al., 2012), capillary electrophoresis–based fragment analysis (Carrington, Varshney, Burgess, & Sood, 2015), or the Surveyor or T7 endonuclease mismatch detection assays (Qiu et al., 2004; Reyon et al., 2012). We attempt genome editing only with programmable

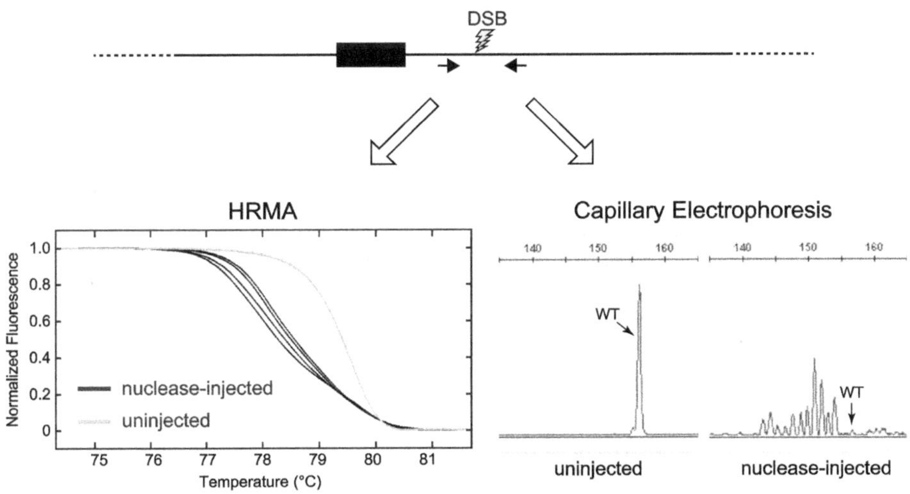

FIGURE 4 PCR-based methods to detect DSB repair-induced mutations.

Prior to conducting a genome-editing experiment, the in vivo activity of the programmable nuclease used to induce DSBs is assayed by the induction of small indels at the nuclease target site. Several methods are available for detecting repair-induced mutations. We use HRMA (Dahlem et al., 2012) or capillary electrophoresis (Carrington et al., 2015) for this purpose. Genomic DNA is isolated from individual 1 dpf embryos that had been injected at the one-cell stage with nuclease. PCR primers (*black arrows*) are used to amplify a 90–150-bp product that is centered on the nuclease target site. DNA from individual uninjected or nuclease-injected embryos is amplified in the presence of a dsDNA-binding dye (HRMA) or with fluorescently labeled primers (capillary electrophoresis). For HRMA, PCR products are denatured and renatured, and duplexes with mismatches are detected by their altered thermal denaturation profile (*red curves* (black in print versions)). For capillary electrophoresis analysis, products are resolved according to size, and the fraction of amplicons with altered size can be determined. Representative HRMA and capillary electrophoresis traces indicate a highly active nuclease.

nucleases that are shown to induce somatic mutations in every treated embryo. In experiments that have led to successful gene editing, we have used nucleases that induce mutations in 30–70% of the host genomes.

3.3 DESIGN OF DONOR SEQUENCES

Experiments have not yet been reported that determine requirements for homology between the donor and the sequences immediately surrounding the site of the induced lesion. For this reason, we usually generate donor molecules that carry sequences bridging the cleavage site. Furthermore, to prevent the programmed nuclease from cleaving a successfully edited allele, donor sequences must not contain an intact nuclease recognition site. If the nuclease target site lies within host coding sequences that are to be retained in the edited allele, silent mutation changes can be introduced into the donor that destroy the target site but maintain coding capacity.

Donor sequences sharing perfect homology with the targeted locus should flank the nonhomologous sequences to be introduced into the genome. We use 1-kbp homology arms because they are sufficient to achieve HR in zebrafish embryos with high efficiency, and yet they allow for the detection of precise integration events by PCR amplification (Fig. 3). However, we note the optimal extent of homology arms has yet to be determined, and some reports have found that longer homology arms in donor molecules enhance the frequency of HR (Shin et al., 2014). To create homology arms harboring sequences identical with the genomes to be targeted, we first identify a selected breeding population with absent/reduced polymorphism bordering the nuclease target site and then PCR amplify sequences from founder genomes or their offspring.

3.4 CONFIGURATION OF DONOR MOLECULES

As important as the induction of DSBs is for the stimulation of repair/recombination machinery, so too, is the availability of donor template for participation in HR with genomic sequences at the targeted site. We do not understand how injected donor DNA comes to be recruited to a recombination complex in zebrafish cleavage stage embryos. Three factors potentially affect the accessibility of donor DNA (Stuart, McMurray, & Westerfield, 1988; Udvadia & Linney, 2003). First, injected plasmid DNA does not readily diffuse or disperse throughout the cytoplasm of cleavage stage embryos. Second, injected circular or linear plasmid DNA is often rearranged into long molecules consisting of imperfectly repeated units. Third, exogenously supplied DNA may be subjected to degradation. In truth, there is little experimental evidence that informs us as to how best to deliver exogenously supplied donor DNA so as to maximize its ability to participate in HR events with the host genome.

As a result, different investigators have tested the effect of altering donor DNA configuration on the efficiency of gene editing. Although little consensus can be derived from these preliminary studies, we, and others, found injection of circular donor DNA is better than injection of linear molecules for producing edited alleles (Hoshijima et al., 2016; Irion et al., 2014).

In our efforts to identify factors that could improve the ability of donor DNA to participate in HR-mediated genome editing, we generated linear donor fragments produced by cleavage with *I-SceI* endonuclease, whose recognition sites were arranged in a head-to-head orientation bordering the donor sequences. *I-SceI* enzyme cleaves its 18-bp recognition site asymmetrically, producing a longer "head" portion to which the *I-SceI* protein remains associated (Perrin, Buckle, & Dujon, 1993). We reasoned that linear fragments associated with end-capped protein might provide good substrate for HR.

Cleavage of donor DNA molecules with *I-SceI* enzyme stimulates the ability of donor molecules to participate in HR and produce edited alleles. Fig. 5A illustrates an experiment that produced a knockin/knockout allele at the *kcnh6a* locus of zebrafish (Hoshijima et al., 2016). Donor sequences flanked by head-to-head *I-SceI* recognition sites were cloned into the pKHR4 vector (Fig. 6). Donor plasmid DNA was injected with or without targeting nuclease into one-cell embryos, and the generation of edited alleles in individual 2 dpf embryos was measured in quantitative PCR assays (Fig. 5B). Digestion of the circular plasmid donor DNA with *I-SceI* enzyme in vitro prior to being mixed with programmable nuclease and injected into zygotes stimulated the formation of edited *kcnh6a* alleles in a dose-dependent manner. We obtained similar results at additional loci. Hence, we developed a series of donor pKHR vectors (Fig. 6) that allow ready construction of donor sequences bordered by head-to-head-oriented *I-SceI* recognition sites.

4. METHOD OF APPROACH

4.1 DESIGN AND PREPARATION OF PROGRAMMED NUCLEASES

4.1.1 *Design of nucleases*

We utilize the TALEN or CRISPR/Cas9 nuclease systems to initiate targeted DSBs. Once a target region of interest has been selected, we retrieve 400–600 bp of genomic sequence covering the region from Ensembl and search for potential TALEN target sites using TALEN Targeter (https://tale-nt.cac.cornell.edu/) or potential sgRNA target sequences using web resources such as CRISPR Design (crispr.mit.edu) or CHOPCHOP (https://chopchop.rc.fas.harvard.edu/).

4.1.2 *Target sequence confirmation and selection of a breeding population*

Prior to the generation of nucleases, the actual fish genomes to be targeted should be analyzed to confirm they carry the target sequences as described in the reference sequence database. The purpose of this step is to establish a small breeding population that harbors the nuclease target and exhibits minimal sequence heterogeneity around the target site, a concern especially relevant when designing nucleases to

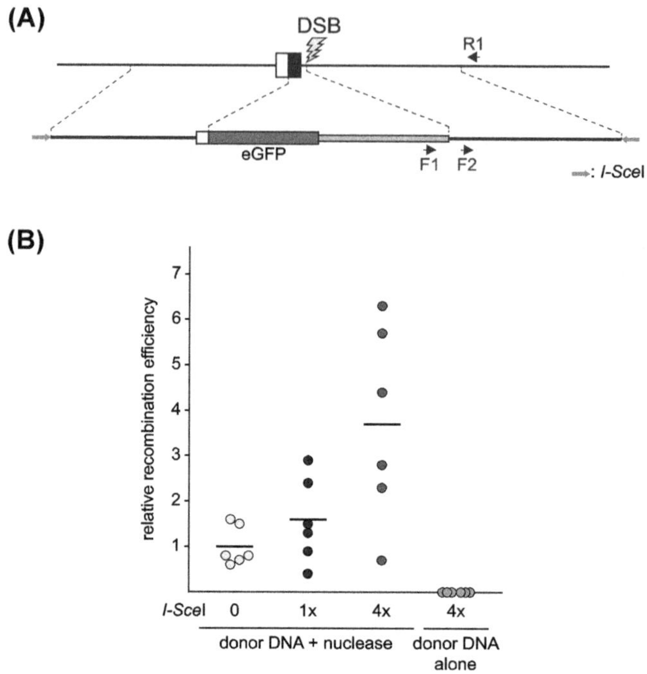

FIGURE 5 Enhancement of genome editing using donor molecules predigested with *I-SceI* meganuclease.

(A) Schematic representation of a genome-editing event to produce a reporter knockin/knockout allele, as in Fig. 2A. Donor DNA sequences were flanked by a pair of head-to-head oriented *I-SceI* sites (*red arrows*) present within the pKHR4 plasmid vector (Fig. 6). The relative abundance of edited alleles within the genomes of injected F0 embryos was determined following quantitative PCR (qPCR). Primers used for qPCR are depicted as *black arrows*: the F1/R1 pair specifically amplifies the edited allele, whereas the F2/R1 pair amplifies edited as well as unedited forms of the endogenous locus. (B) Genomic editing is enhanced by *I-SceI* digestion of a donor plasmid prior to injection. The donor plasmid was digested with increasing amounts of *I-SceI* enzyme in vitro and subsequently injected with programmable nuclease into zygotes. As a control, *I-SceI*-digested donor plasmid was injected alone, without added nuclease to target the host genome. The fraction of edited alleles (detected with the F1/R1 primer pair) relative to total targeted loci (detected with the F2/R1 primer pair) present in injected 2 dpf embryos was determined by qPCR. The relative recombination efficiency was determined by normalizing the mean fraction of edited alleles following injection of nuclease and undigested donor plasmid DNA to 1.0. For each condition, six individual embryos were analyzed (*circles*) and the mean relative recombination efficiency is indicated (*horizontal dash*). Unpaired *t*-test analysis indicated that digestion of donor DNA with 4× enzyme significantly stimulated the production of edited alleles as compared with untreated donor DNA ($p < 0.01$). Digestion with 1× enzyme did not yield a significant increase in recombination efficiency as compared with untreated DNA. Injection of *I-SceI*-digested DNA without programmable nuclease failed to produce a detectable level of edited target alleles. (See color plate)

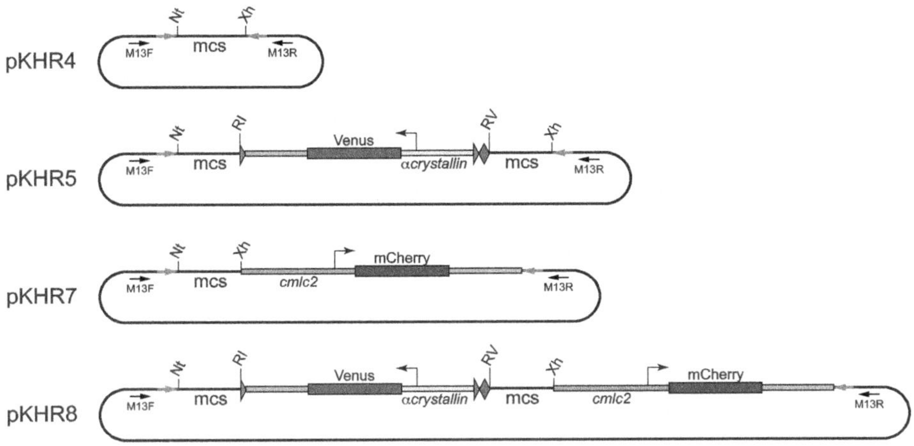

I-*Sce*I (18bp): TAGGGATAACAGGGTAAT

loxP (34bp): ATAACTTCGTATAGCATACATTATACGAAGTTAT FRT (34bp): GAAGTTCCTATTCTCTAGAAAGTATAGGAACTTC

FIGURE 6 pKHR plasmid vectors for genome editing.

pKHR4: pKHR4 is an *amp*^{R} plasmid built from Bluescript SK(+) with its multiple cloning site (mcs) flanked by inverted head-to-head oriented *I-Sce*I sites. *pKHR5*: pKHR5 is derived from pKHR4 by inserting the *CV* reporter gene cassette, FRT-*CV*-FRT-loxP, between the EcoRI and EcoRV sites. The cassette consists of the α-crystallin promoter driving expression of the Venus version of the GFP protein in the lens (Hesselson, Anderson, Beinat, & Stainier, 2009) flanked by FRT sites and bordered on the 3′ end with a single loxP site. pKHR5 contains two mcs for independently introducing left and right homology arms to flank the reporter cassette. *pKHR7*: pKHR7 contains a simple modification of pKHR4 that allows detection of imprecise or random insertion events. pKHR7 contains a *cmlc2::mCherry* (*red heart*) reporter gene that resides within the donor fragment that is produced by *I-Sce*I digestion, but outside any homology region. Incorporation of *cmlc2::mCherry* with donor sequences can only occur as a consequence of imprecise recombination or random integration but never as a result of precise HR. *pKHR8*: pKHR8 contains a simple modification of pKHR5 that allows detection of imprecise or random insertion events. pKHR8 carries the *CV* reporter gene cassette, as in pKHR5, but also has a *cmlc2::mCherry* (*red heart*) reporter gene, which resides within the donor fragment produced by *I-Sce*I digestion, but outside any homology region. Successful homologous recombination events of interest should acquire the *green lens* reporter but not the *red heart* reporter. Sequences for the donor vectors pKHR4, 5, 7, 8 have been deposited in GenBank (KU144822-KU144825). The plasmids are available through Addgene. (See color plate)

cleave within an intron. A 400–600-bp genomic region surrounding the nuclease recognition site should be amplified from genomic DNAs prepared from eight or more adult fish. Amplicons should be sequenced and the existence of target sequences verified. Simultaneously, sequence chromatographs are analyzed to determine the presence of polymorphic sequences. Adults that carry the exact target sequence and exhibit minimal additional polymorphisms should be selected as a breeding population. In our experience, wild-type strains differ greatly in the distributions of polymorphic sequences. While one strain (eg, AB) may be rich in sequence variation at a particular gene, another strain (eg, TU) may be completely devoid of such polymorphisms.

4.1.3 Synthesis of programmable nucleases

Programmable nucleases are prepared according to standard published methods. TALEN plasmids are constructed using the Golden Gate System using DDD/RRR FokI nuclease domains for obligate heterodimerization between left and right TALEN monomers (Dahlem et al., 2012). TALEN mRNAs are prepared from these templates by in vitro transcription. Target-specific sgRNAs for the CRISPR/Cas9 system are prepared by in vitro transcription from an oligonucleotide template DNA (Gagnon et al., 2014).

4.1.4 Determination of nuclease cleavage activity

Cleavage activity of the targeting nucleases must be analyzed in zebrafish embryos prior to performing gene editing with donor molecules. For TALENs, 50 pg left and 50 pg right TALEN mRNA are injected with 0.05% phenol red into the cytoplasm of one-cell stage of zygotes generated from the breeding population identified above. For the CRISPR/Cas9 system, 200–300 ng sgRNA is coinjected with 600 pg Cas9 protein (PNA BIO) and 0.05% phenol red. Under these conditions, most of the RNA-injected embryos should develop normally (80–90%). If a large fraction of injected embryos display developmental abnormalities, the amount of injected RNA should be reduced.

As described in **Section 3.2**, cleavage activity is actually measured as the induction of novel sequence changes at the nuclease target site in the somatic genomes of 1 dpf injected embryos. Genomic DNA is extracted from 8 to 12 individual embryos using the following protocol:

1. Place individual dechorionated embryos in 50 μL of 50 mM NaOH.
2. Incubate 95°C, 20 min.
3. Transfer onto ice, 4°C.
4. Add 5 μL 1 M Tris–HCl (pH 8.0).
5. Mix well and spin down.
   Store at −20°C.

To analyze the induction of polymorphisms, 90–150 bp bordering the targeted region is amplified from genomic and analyzed by HRMA, Surveyor/T7 Endonuclease activity, or capillary electrophoresis. Strong cleavage activity should be detected in every embryo before proceeding.

4.2 DESIGN AND PREPARATION OF DONOR MOLECULES USING pKHR VECTORS

A set of pKHR vectors (Fig. 6) was designed to allow ready construction of donor sequences bordered by head-to-head-oriented *I-SceI* recognition sites. pKHR4 simply has a multiple cloning site between *I-SceI* cleavage sites. pKHR5, pKHR7, and pKHR8 provide additional sequences including a reporter gene placed between the homology arms to track acquisition of donor sequences via HR and/or a reporter gene residing distal to the homology arms to track imperfect or random insertion events.

4.2.1 Preparation of homologous arms and novel donor sequences

Genomic DNA to be used as template to generate homology arms should be extracted from breeding population or F0 animals that are likely to have uniform sequence in the region of the target site. As illustrated in Figs. 7 and 8, typically three PCR-amplified fragments are initially generated and joined to vector sequences by classical ligation methods in the preparation of a donor plasmid. Each homology arm (about 1 kbp) is prepared by PCR amplification using high fidelity DNA polymerase with primers that introduce at the ends of the arms appropriate restriction enzyme recognition sequences. Nonhomologous sequences to be introduced, for example GFP coding sequences with 3′ UTR sequences, are also prepared by PCR amplification from source plasmids with primers that provide restriction enzyme sequences. As denoted in Fig. 7, a type-II restriction enzyme, such as BsaI, is convenient for generating amplicons with unique and complementary protruding ends.

4.2.2 Generating a loxP site within a homology arm

As illustrated in Fig. 8, to introduce the 34-bp *loxP* sequence (or other similar small sequence addition) into the host genome, an extended homology arm is created in the donor vector by overlapping PCR amplification. Using primers that introduce *loxP* sequences at the desired point of insertion, two overlapping amplicon fragments are initially generated from genomic template, one extending 5′ and the other extending 3′ from the point of insertion. Consequently, as shown in Fig. 8, each amplified fragment will have overlapping sequence that includes the *loxP* sequence. To create the extended homology arm, overlapping PCR amplification is carried out in two steps as follows:

1. Annealing and extension reaction to create a fused template:

25 μL	2× KAPA HiFi HotStart Ready Mix (KAPA)
4 μL	10 ng/μL gel-purified amplicon#1
4 μL	10 ng/μL gel-purified amplicon#2
17 μL	Nuclease-free water

Cycle conditions:
Denature: 95°C, 5 min
Ten cycles: [98°C, 20 s; 60°C, 15 s; 72°C, 30 s]
Final extension: 72°C, 1 min
Store: 10°C

Purify the fused template using QIAquick PCR purification Kit (QIAGEN).

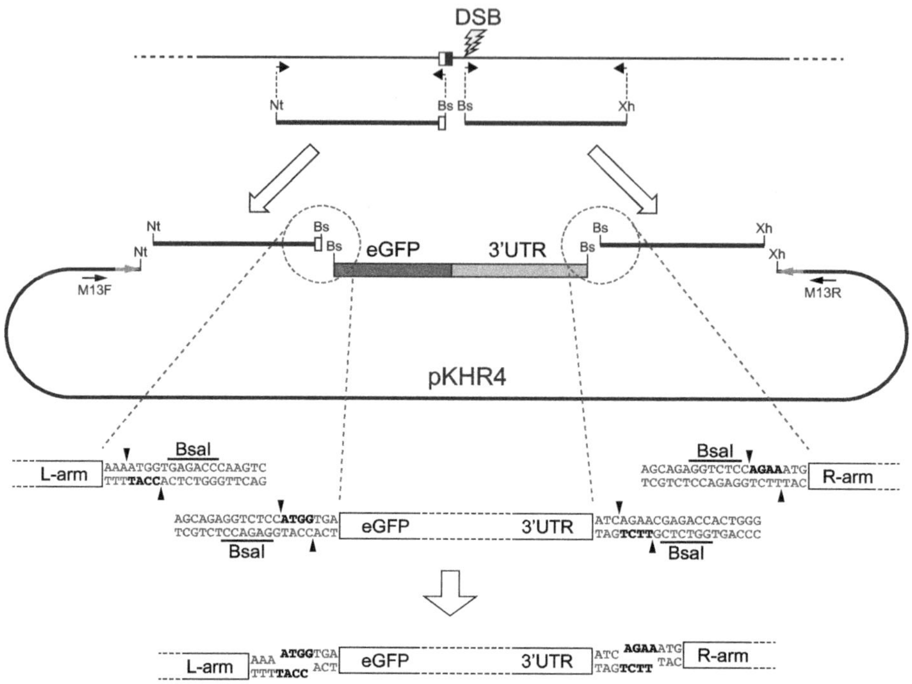

FIGURE 7 Preparation of a donor plasmid for targeted reporter integration.

To integrate a reporter gene consisting of eGFP coding sequences (green (dark gray in print versions)) and translation/transcription termination sequences (gray) at a specific target site, about 1 kbp of genomic sequences upstream and downstream of the reporter integration site are prepared by PCR amplification for left and right homology arms, respectively. The homology arms should include a mutated nuclease target sequence, so that integration of donor sequences will produce an edited allele that cannot be cleaved by the nuclease. Each amplified fragment is bordered by unique restriction enzyme recognition sequences derived from PCR primers. In this case, the left-arm fragment is bordered by *Not*I (Nt) and *Bsa*I (Bs) sites, and the right arm has *Bsa*I and *Xho*I (Xh) sites at its ends. *Bsa*I is a type-II restriction enzyme that produces a staggered cut next to the enzyme recognition site. As a result, it can be used to generate unique, complementary protruding ends (*red letters* (black in print versions)) independent of the enzyme recognition sequence (*underlined*). The reporter sequence is prepared as a middle fragment by PCR amplification with ends containing *Bsa*I recognition sites whose digestion would yield protruding ends complementary to the digested left- and right-arm fragments. Restriction enzyme-digested amplicons are individually purified. Because each digested end is complementary to a unique partner fragment, ligation of the digested fragments leads to ordered assembly, and a single cloning step is used to assemble the homology arms and the reporter middle fragment in correct sequence into the pKHR4 backbone, which had been predigested with *Not*I and *Xho*I.

To create the extended homology arm ready for cloning into a pKHR vector, the fused template is further amplified using forward and reverse primers that provide appropriate restriction enzyme sequences for subsequent cloning.

2. PCR amplification to generate extended homology arm:

25 μL	2× KAPA HiFi HotStart Ready Mix (KAPA)
7.5 μL	2 μM forward primer
7.5 μL	2 μM reverse primer
10 μL	10 ng/μL purified fused template

Cycle conditions (annealing temperature* is dependent on primer Tm):
Denature: 95°C, 5 min
Fifteen cycles: [98°C, 20 s; 66°C*, 15 s; 72°C, 30 s]
Final extension: 72°C, 1 min
Store: 10°C

Purify the amplicon using the QIAquick PCR purification Kit (QIAGEN).

4.2.3 Assembling donor sequences into pKHR vectors

As illustrated in Fig. 7, a single cloning step can be used to introduce donor sequences into the pKHR4 or pKHR7 vectors. These vectors do not themselves provide sequences to be incorporated into the genome. Hence, a single continuous assembly of donor sequences, generated from amplicons representing the left and right homology arms and a nonhomologous middle fragment is introduced into the vector. To ligate these fragments in correct orientation, each fragment should have unique protruding sequences that direct assembly in the desired order.

In contrast, the purpose of vectors pKHR5 and pKHR8 is to provide an *α-crystallin-Venus* reporter gene and associated recombination sites that will be incorporated into the host genome upon HR. The reporter gene can be used to identify inheritance of an edited allele, and subsequently it can be readily excised by FLP-mediated recombination (Hoshijima et al., 2016). Assembly of donor sequences in these vectors requires independent insertion of left and right homology arms so that they flank the reporter gene. As illustrated in Fig. 8, to generate a Cre-dependent, *loxP*-mediated conditional allele, left and right homology arm fragments are inserted sequentially into the pKHR vector.

4.2.4 Sequence verification of donor plasmids

The entire donor region of constructed plasmids should be sequenced to ensure that there have been no unintended amplification errors. Sequencing of several independent donor plasmids can indicate if there is extensive sequence polymorphism in the homology regions. If this occurs, one can identify a predominant haplotype and reselect breeder fish with the desired haplotype for subsequent donor injection.

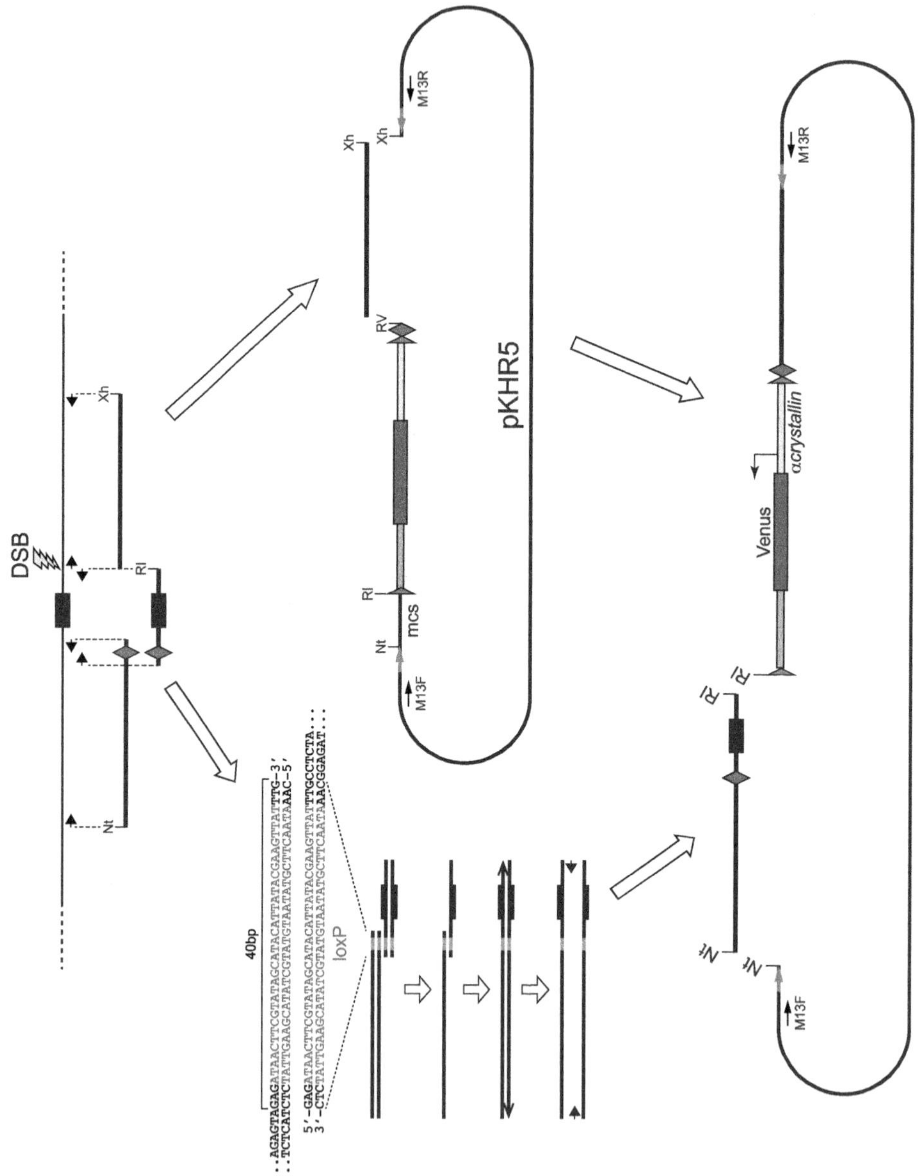
DSB
Xh
RI
Nt
pKHR5
M13R
M13F
RV
mcs
40bp
loxP
Venus
αcrystallin

FIGURE 8 Preparation of a donor plasmid to create a loxP-flanked conditional allele marked by a linked reporter gene.

A donor plasmid to create a conditional allele in which loxP sites flank an exon can be constructed using the pKHR5 or pKHR8 vectors (Fig. 6). These vectors provide the CV reporter gene cassette, which consists of the *α-crystallin::Venus* reporter gene flanked by FRT sites (*orange arrowheads*) with a single loxP site (*pink diamond*) located at the border between the reporter and the right homology arm. The right homology arm, consisting of about 1-kbp sequence extending 3′ from the lesion site, is amplified with primers to produce a blunt phosphorylated 5′ end and a 3′ end with an *XhoI* (Xh) site. The right arm is then cloned into an *EcoRV* + *XhoI*-digested vector. As illustrated, the vector-provided loxP site will be inserted into intron sequences downstream (3′) of the targeted exon, and so a second loxP site needs to be introduced into an extended left homology arm, 5′ to the exon. The left arm containing the loxP site is prepared by overlapping PCR amplification to create a fused template from two fragments initially generated from the targeted locus. One fragment contains the exon, extending from the DSB lesion site to the intended position of the second loxP site. Primers used to generate this fragment introduce novel end sequences, an *EcoRI* (RI) recognition sequence at the lesion site and a 40-bp sequence that includes the loxP site at the end terminating 5′ of the exon. The second fragment extends about 1 kbp further upstream from the position of the loxP site. Primers used to generate this amplicon produce a fragment bordered 5′ by a *NotI* (Nt) site and 3′ by the 40-bp sequence that includes the loxP site. Overlapping PCR is used to generate a single extended left-arm fragment that is bordered by *NotI* and *EcoRI* sites and extends from the lesion site into the upstream intron. The left arm is cloned into a *NotI* + *EcoRI* digested vector that already contains the right arm homology. (See color plate)

4.3 MICROINJECTION OF PROGRAMMED NUCLEASE AND *I-SceI*-DIGESTED DONOR DNA

Column-purified donor plasmid DNA is further purified to remove all traces of RNase activity by phenol/chloroform extraction, chloroform extraction, and ethanol precipitation with sodium acetate. Plasmid DNA is dissolved in nuclease-free water, quantified, diluted to 500 ng/μL, and stored at −20°C. Prior to injection into zebrafish embryos, 500 ng of donor plasmid DNA is digested with *I-SceI* enzyme in 5 μL 1× *I-SceI* buffer:

I-SceI Digestion:	
0.5 μL	10× *I-SceI* buffer (NEB)
1 μL	500 ng/μL donor plasmid DNA
2 μL	5 U/μL *I-SceI* (NEB)
1.5 μL	Nuclease-free water
Incubate 37°C, 1 h	
Store on ice, 4°C	

Meanwhile prepare sgRNA and Cas9 protein mixture or TALEN mRNA solution on ice.

sgRNA with Cas9 protein:	
2 μL	1 μg/μL sgRNA
1.2 μL	5 μg/μL Cas9 protein (PNA BIO)
1 μL	0.5% phenol red
0.8 μL	Nuclease-free water
TALEN mRNA solution:	
1 μL	500 ng/μL left TALEN mRNA
1 μL	500 ng/μL right TALEN mRNA
1 μL	0.5% phenol red
2 μL	Nuclease-free water

After *I-SceI* digestion, mix 5 μL donor DNA solution with 5 μL sgRNA/Cas9 protein mixture or 5 μL TALEN mRNA solution on ice and inject 1 nL of the cocktail into the cytoplasm of just-fertilized zebrafish zygotes. About 50–70% of embryos injected with cocktail mix should develop normally. The amount of donor DNA and nuclease should be adjusted if injection leads to excessive lethality.

4.4 DETECTION OF RECOMBINATION EVENTS

HR-mediated integration of donor sequences into the target site should be observed in almost every embryo injected with nuclease and donor DNA. It is important, and reassuring, to confirm the production of correctly edited alleles in treated embryos before growing them to adulthood. At 2 dpf, genomic DNA is extracted from 8 to 12 individual donor-injected embryos by the method described in **Section 4.1.4** and analyzed as illustrated in Fig. 3 by PCR amplification with a donor sequence–specific primer and a host genome–specific primer. As a control, confirm that the edited allele–specific fragment can be amplified from embryos injected with donor DNA and nuclease, but not from embryos injected only with donor DNA. We note that expression of the reporter genes present in pKHR vectors cannot be used as an indicator of integration, as these genes can be expressed in transient fashion without integration into the genome.

4.5 ISOLATION OF FOUNDER FISH THAT TRANSMIT PRECISELY EDITED GENOMES

Once it is confirmed that injected F0 embryos harbor correctly edited alleles, about 100 sibling injected F0 founders should be raised to adulthood. Analysis of F1 progeny is used to identify founders that have acquired and transmit donor plasmid-derived sequences.

4.5.1 Identification of F0 founders that transmit donor DNA sequences to progeny

In cases where acquisition of the *α-crystallin-Venus* reporter gene was used to indicate incorporation of donor sequences, about 100 F1 embryos should be generated from each F0 founder and analyzed at 2 dpf for expression of GFP in the lens. To expedite the screening process, we incross F0 adults. If GFP-positive embryos are identified among the progeny of an incross, then F0's can be outcrossed with a WT animal to identify the carrier. In cases where donor plasmids carried the *cmlc2::mCherry* reporter outside of the homology regions, expression of mCherry in the heart indicates imprecise integration of donor sequences. F1 embryos that express GFP but not mCherry are likely to harbor precisely edited alleles. The genomic DNA of a few of these should be characterized in detail (below) to confirm the presence of precise editing events, and then sibling embryos expressing appropriate reporter genes should be raised to adulthood.

In cases where integration of donor sequences does not lead to the expression of any visible marker, the genomic DNA of about 50 F1 embryos from each F0 founder should be analyzed for the inheritance of donor DNA. Typically, genomic DNA is prepared from 8 or more pools of 6 F1 embryos and analyzed by PCR amplification with primers that specifically detect donor sequences. Once transmission of donor DNA is detected, detailed analyses are performed to determine whether F1 individuals have inherited precisely modified alleles.

4.5.2 Identification of founder fish carrying precisely edited genomes

Integration of donor sequences into the host genome can arise as a consequence of: (1) precise integration via HR, (2) imprecise or partial integration of donor sequences at the target locus, or (3) nontargeted random insertion. In the case of precise target integration, novel donor sequences should be integrated as a single copy without additional indel mutations. As described in Fig. 9, preliminary analyses should be performed to detect (1) junction fragments that would be created upon precise integration (Fig. 3) and (2) donor—vector junction sequences whose presence would indicate imprecise integration events.

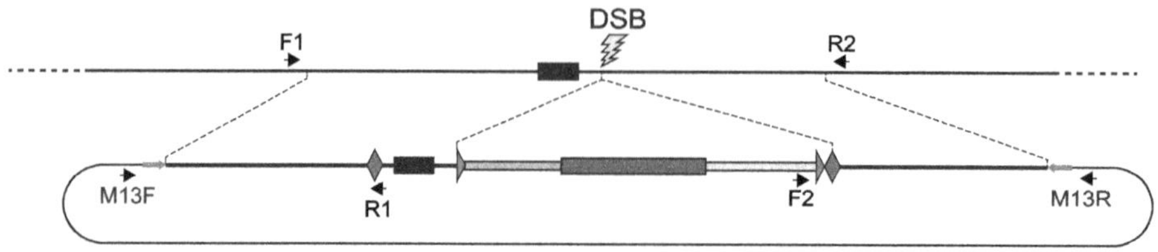

FIGURE 9 Diagnostic amplification to identify a precisely edited allele.

Once a candidate edited allele has been detected either by expression of the linked reporter gene or by preliminary genomic analysis, detailed PCR and sequence analysis of the targeted locus is performed to confirm that a precisely edited allele has been generated. Recombination events that correctly modify the targeted locus can be detected with primer pairs such as F1/R1 and F2/R2, which specifically amplify host–donor junction fragments and are expected to be present only in genomes harboring a precisely edited allele. In addition, to detect imprecise integration events, the edited genome is probed with primer pairs that amplify donor–vector backbone junction fragments, such as M13F/R1 and M13R/R2. Genomes that likely harbor only precisely edited alleles are finally analyzed by amplification of the entire edited region, using the F1 and R2 primers, which are complementary to host genomic sequences and not present in the donor DNA. The size of the F1/R2 amplicon should be consistent with that expected from a simple HR-mediated gene-editing event, and the accuracy of the editing event should be verified by sequencing the entire amplicon.

The entire sequence of an edited allele should be amplified from an F1 genome as a single intact amplicon and sequenced to confirm the accuracy of the edited allele. To accomplish this analysis, genomic DNA gently extracted from a pool of candidate embryos carrying precisely edited genomes should be amplified using primers that recognize genomic sequences outside of the homology regions. The size of the amplicon should be consistent with that of the expected edited allele, confirming single-copy integration. Once a founder carrying a precisely edited genome is isolated, carriers can be easily identified by expression of the GFP reporter or by DNA analysis from fin biopsies.

CONCLUSIONS

Although the methods presented here represent an early stage in the evolution of genome editing in the zebrafish, the approach described in this Chapter already enables precise modification of the zebrafish genome. Changes ranging from single base pair substitutions to additions and/or deletions of several kilobase-long stretches of sequence can be generated efficiently. Significantly, genome editing accomplished by DSB-stimulated HR can be used to produce several closely linked genome modifications simultaneously. Thus incorporation of a linked reporter gene can be used to tag edited alleles, and the expression of the linked reporter makes it very simple to identify and recover precisely edited alleles. The ease with which genome editing can be accomplished will revolutionize use of the zebrafish allowing new types of studies of cell biological processes, development and tissue homeostasis, and disease modeling.

ACKNOWLEDGMENTS

Work reported here utilized University of Utah Cores for sequencing, HRMA, and oligonucleotide synthesis. These studies were supported by grants to D.J.G. from the University of Utah and the National Institutes of Health (5R21HD073847, 1R21OD018323, and 1R01HD081950) and by a subaward from 5P30CA042014. The authors declare no conflicts of interest.

REFERENCES

Auer, T. O., & Del Bene, F. (2014). CRISPR/Cas9 and TALEN-mediated knock-in approaches in zebrafish. *Methods, 69*, 142–150.

Auer, T. O., Duroure, K., Concordet, J. P., & Del Bene, F. (2014). CRISPR/Cas9-mediated conversion of eGFP- into Gal4-transgenic lines in zebrafish. *Nature Protocols, 9*, 2823–2840.

Auer, T. O., Duroure, K., De Cian, A., Concordet, J. P., & Del Bene, F. (2014). Highly efficient CRISPR/Cas9-mediated knock-in in zebrafish by homology-independent DNA repair. *Genome Research, 24*, 142–153.

Beachy, P. A., Karhadkar, S. S., & Berman, D. M. (2004). Tissue repair and stem cell renewal in carcinogenesis. *Nature, 432*, 324–331.

Bedell, V. M., Wang, Y., Campbell, J. M., Poshusta, T. L., Starker, C. G., Krug, R. G., 2nd, … Ekker, S. C. (2012). In vivo genome editing using a high-efficiency TALEN system. *Nature, 491*, 114–118.

Carrington, B., Varshney, G. K., Burgess, S. M., & Sood, R. (2015). CRISPR-STAT: an easy and reliable PCR-based method to evaluate target-specific sgRNA activity. *Nucleic Acids Research, 43*, e157.

Carroll, D. (2014). Genome engineering with targetable nucleases. *Annual Review of Biochemistry, 83*, 409–439.

Chang, N., Sun, C., Gao, L., Zhu, D., Xu, X., Zhu, X., … Xi, J. J. (2013). Genome editing with RNA-guided Cas9 nuclease in zebrafish embryos. *Cell Research, 23*, 465–472.

Dahlem, T. J., Hoshijima, K., Jurynec, M. J., Gunther, D., Starker, C. G., Locke, A. S., … Grunwald, D. J. (2012). Simple methods for generating and detecting locus-specific mutations induced with TALENs in the zebrafish genome. *PLoS Genetics, 8*, e1002861.

Gagnon, J. A., Valen, E., Thyme, S. B., Huang, P., Akhmetova, L., Pauli, A., … Schier, A. F. (2014). Efficient mutagenesis by Cas9 protein-mediated oligonucleotide insertion and large-scale assessment of single-guide RNAs. *PLoS One, 9*, e98186.

Giraldez, A. J., Cinalli, R. M., Glasner, M. E., Enright, A. J., Thomson, J. M., Baskerville, S., … Schier, A. F. (2005). MicroRNAs regulate brain morphogenesis in zebrafish. *Science, 308*, 833–838.

Gritsman, K., Zhang, J., Cheng, S., Heckscher, E., Talbot, W. S., & Schier, A. F. (1999). The EGF-CFC protein one-eyed pinhead is essential for nodal signaling. *Cell, 97*, 121–132.

Hesselson, D., Anderson, R. M., Beinat, M., & Stainier, D. Y. (2009). Distinct populations of quiescent and proliferative pancreatic beta-cells identified by HOTcre mediated labeling. *Proceedings of the National Academy of Sciences of the United States of America, 106*, 14896–14901.

Hisano, Y., Sakuma, T., Nakade, S., Ohga, R., Ota, S., Okamoto, H., … Kawahara, A. (2015). Precise in-frame integration of exogenous DNA mediated by CRISPR/Cas9 system in zebrafish. *Scientific Reports, 5*, 8841.

Hoshijima, K., Jurynec, M. J., & Grunwald, D. J. (2016). Precise editing of the zebrafish genome made simple and efficient. *Developmental Cell, 36*, 654–667.

Hruscha, A., Krawitz, P., Rechenberg, A., Heinrich, V., Hecht, J., Haass, C., & Schmid, B. (2013). Efficient CRISPR/Cas9 genome editing with low off-target effects in zebrafish. *Development, 140*, 4982–4987.

Hsu, P. D., Lander, E. S., & Zhang, F. (2014). Development and applications of CRISPR-Cas9 for genome engineering. *Cell, 157*, 1262–1278.

Hwang, W. Y., Fu, Y., Reyon, D., Maeder, M. L., Tsai, S. Q., Sander, J. D., … Joung, J. K. (2013). Efficient genome editing in zebrafish using a CRISPR-Cas system. *Nature Biotechnology, 31*, 227–229.

Irion, U., Krauss, J., & Nusslein-Volhard, C. (2014). Precise and efficient genome editing in zebrafish using the CRISPR/Cas9 system. *Development, 141*, 4827–4830.

Jao, L. E., Wente, S. R., & Chen, W. (2013). Efficient multiplex biallelic zebrafish genome editing using a CRISPR nuclease system. *Proceedings of the National Academy of Sciences of the United States of America, 110*, 13904–13909.

Jurynec, M. J., Xia, R., Mackrill, J. J., Gunther, D., Crawford, T., Flanigan, K. M., … Grunwald, D. J. (2008). Selenoprotein N is required for ryanodine receptor calcium release channel activity in human and zebrafish muscle. *Proceedings of the National Academy of Sciences of the United States of America, 105*, 12485–12490.

Kim, H., & Kim, J. S. (2014). A guide to genome engineering with programmable nucleases. *Nature Reviews Genetics, 15*, 321–334.

Kimura, Y., Hisano, Y., Kawahara, A., & Higashijima, S. (2014). Efficient generation of knock-in transgenic zebrafish carrying reporter/driver genes by CRISPR/Cas9-mediated genome engineering. *Scientific Reports, 4*, 6545.

Langdon, Y. G., & Mullins, M. C. (2011). Maternal and zygotic control of zebrafish dorsoventral axial patterning. *Annual Review of Genetics, 45*, 357–377.

Lawson, N. D., & Wolfe, S. A. (2011). Forward and reverse genetic approaches for the analysis of vertebrate development in the zebrafish. *Developmental Cell, 21*, 48–64.

Li, J., Zhang, B. B., Ren, Y. G., Gu, S. Y., Xiang, Y. H., & Du, J. L. (2015). Intron targeting-mediated and endogenous gene integrity-maintaining knockin in zebrafish using the CRISPR/Cas9 system. *Cell Research, 25*, 634–637.

Lupski, J. R. (2013). Genetics. Genome mosaicism–one human, multiple genomes. *Science, 341*, 358–359.

Perrin, A., Buckle, M., & Dujon, B. (1993). Asymmetrical recognition and activity of the I-SceI endonuclease on its site and on intron-exon junctions. *EMBO Journal, 12*, 2939–2947.

Phillips, J. B., & Westerfield, M. (2014). Zebrafish models in translational research: tipping the scales toward advancements in human health. *Disease Models & Mechanisms, 7*, 739–743.

Poduri, A., Evrony, G. D., Cai, X., & Walsh, C. A. (2013). Somatic mutation, genomic variation, and neurological disease. *Science, 341*, 1237758.

Qiu, P., Shandilya, H., D'Alessio, J. M., O'Connor, K., Durocher, J., & Gerard, G. F. (2004). Mutation detection using surveyor nuclease. *Biotechniques, 36*, 702–707.

Reyon, D., Tsai, S. Q., Khayter, C., Foden, J. A., Sander, J. D., & Joung, J. K. (2012). FLASH assembly of TALENs for high-throughput genome editing. *Nature Biotechnology, 30*, 460–465.

Shin, J., Chen, J., & Solnica-Krezel, L. (2014). Efficient homologous recombination-mediated genome engineering in zebrafish using TALE nucleases. *Development, 141*, 3807–3818.

Stuart, G. W., McMurray, J. V., & Westerfield, M. (1988). Replication, integration and stable germ-line transmission of foreign sequences injected into early zebrafish embryos. *Development, 103*, 403–412.

Udvadia, A. J., & Linney, E. (2003). Windows into development: historic, current, and future perspectives on transgenic zebrafish. *Developmental Biology, 256*, 1–17.

Ulitsky, I., Shkumatava, A., Jan, C. H., Sive, H., & Bartel, D. P. (2011). Conserved function of lincRNAs in vertebrate embryonic development despite rapid sequence evolution. *Cell, 147*, 1537–1550.

Wray, G. A. (2007). The evolutionary significance of cis-regulatory mutations. *Nature Reviews Genetics, 8*, 206–216.

Zu, Y., Tong, X., Wang, Z., Liu, D., Pan, R., Li, Z., … Lin, S. (2013). TALEN-mediated precise genome modification by homologous recombination in zebrafish. *Nature Methods, 10*, 329–331.

CHAPTER

The zebrafish genome editing toolkit

7

H. Ata, K.J. Clark, S.C. Ekker[1]

Mayo Clinic, Rochester, MN, United States

[1]*Corresponding author: E-mail: ekker.stephen@mayo.edu*

CHAPTER OUTLINE

Abstract

Zebrafish (*Danio rerio*) is a unique model organism at the functional intersection between a high fecundity and conserved vertebrate physiology while being amenable to a multitude

Methods in Cell Biology, Volume 135, ISSN 0091-679X, http://dx.doi.org/10.1016/bs.mcb.2016.04.023

of genome editing techniques. The genome engineering field has experienced an unprecedented rate of growth in the recent years since the introduction of designer endonucleases, such as zinc finger nucleases, transcription activator-like effector nucleases, and clustered regularly interspaced short palindromic repeats—Cas9 systems. With the ever-evolving toolset available to the scientific community, the important question one should ask is not simply *how* to make a mutant line, but rather *how best* to do so. For this purpose, understanding the toolset is just one end of the equation; understanding how DNA is repaired once double-strand breaks are induced by designer endonucleases, as well as understanding proper fish handling and line maintenance techniques, are also essential to rapidly edit the zebrafish genome. This chapter is outlined to provide a bird's-eye view on each of these three components. The goal of this chapter is to facilitate the adoption of the zebrafish as a model to study human genetic disease and to rapidly analyze the function of the vertebrate genome.

INTRODUCTION

The first targeted gene knock-out was described in zebrafish in 2008 using zinc finger nucleases (ZFNs). Since that time, the tools for reverse genetics have seen an unprecedented rate of growth with the introductions of transcription activator-like effector nuclease (TALEN) and clustered regularly interspaced short palindromic repeats—Cas9 (CRISPR-Cas9) systems. These tools facilitate the generation of gene knock-out models by inducing targeted double-strand DNA breaks (DSBs) that are normally repaired by the robust but low-fidelity nonhomologous end-joining (NHEJ) repair pathway. Consequently, NHEJ often results in base insertions and/or deletions (indels) that give rise to frameshift or nonsense mutations, disrupting gene functions (Fig. 1). While gene knock-down technologies such as morpholinos (Nasevicius & Ekker, 2000) continue to exist as important toolsets for complementary functional validation science, gene knock-out technologies offer the advantage of traditional genetic approaches.

In addition to the ability to generate loss-of-function alleles, these designer endonucleases have enabled efficient targeted insertional mutagenesis. Most commonly, this is accomplished using homology directed repair (HDR) by providing exogenous donor DNA (Fig. 1). With an appropriate donor design, it is feasible to generate fish with gene knock-ins, epitope tags, or specific point mutations. However, the frequency of HDR events is much lower than that of NHEJ events, making it necessary to raise a large clutch of fish for subsequent screening by sequencing if positive screening markers such as ectopically expressed Green Fluorescent Protein (GFP) are not used. However, the zebrafish is an ideal system for even rare engineering events as it is feasible to generate and screen sufficient numbers thanks to the high fecundity and financial accessibility of this model organism.

In this chapter, we will cover the basic strategies for efficiently designing these modern gene editing tools and some points of consideration when choosing among the available options.

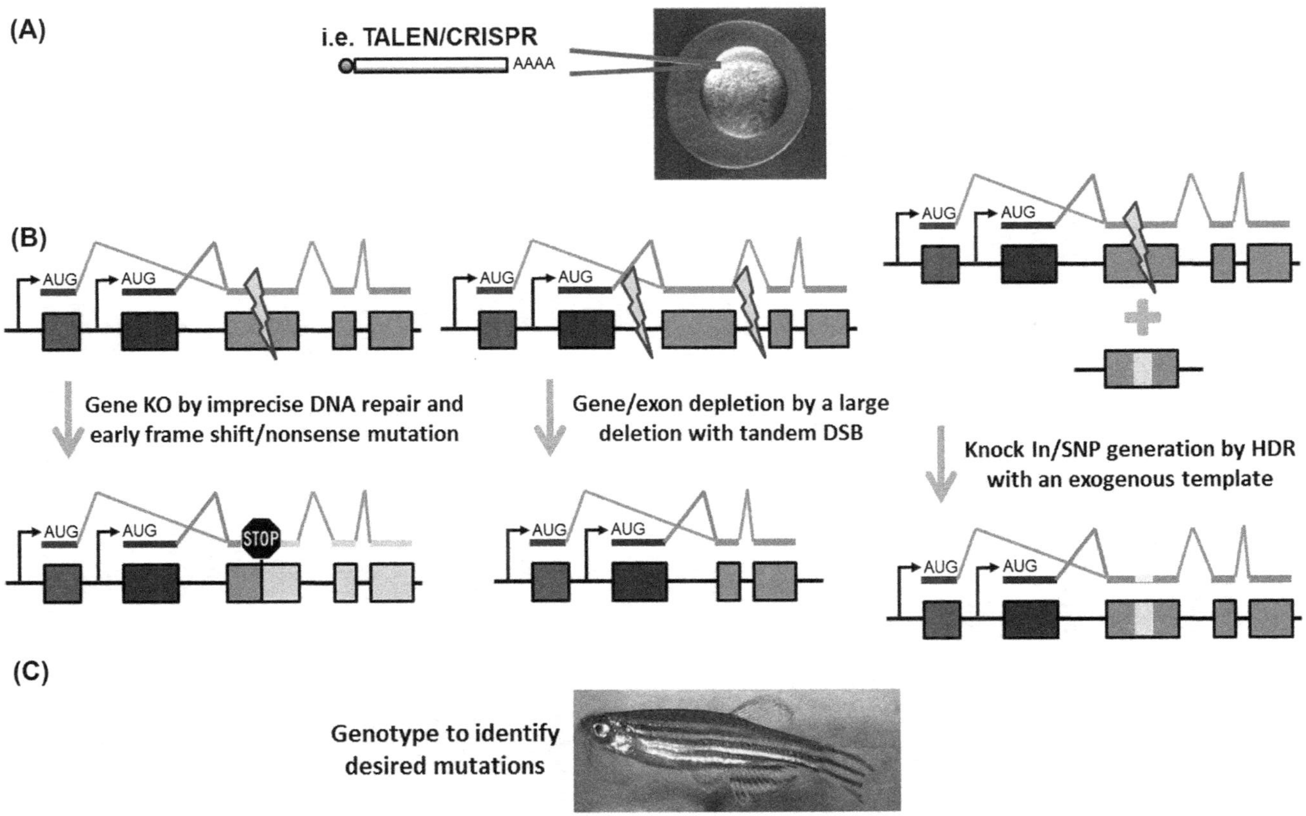

FIGURE 1 Mutant zebrafish generation using designer endonucleases.

(A) Designer endonucleases are injected into one-cell stage embryo as RNA. (B) Examples of mutations generated by designer endonucleases. A designer endonuclease is targeted to an early conserved exon to cause frameshift and/or nonsense mutation to generate a gene knock-out (KO) (left). A pair of designer endonucleases is targeted to both sides of an exon (or a gene) to cause an exon (or a gene) depletion (middle). An exogenous donor DNA harboring substantial homology arms is codelivered with a designer endonuclease to promote double-strand DNA break (DSB) repair by homology directed repair (HDR) (right). (C) F0 zebrafish harboring desired mutation is identified by genotyping and raised to sexual maturity for selective outcrossing. *CRISPR*, clustered regularly interspaced short palindromic repeats; *SNP*, single nucleotide polymorphism; *TALEN*, transcription activator-like effector nuclease. (See color plate)

1. REAGENTS

1.1 TALENs

Originally discovered in plant pathogens, native transcription activator-like effector (TALE) domains consist of clustered 33–35 amino acid repeats that make up modular units for recognizing DNA target sites. TALEs are a powerful tool for genome engineering as a repeat variable di-residue (RVD) contained within each repeat recognizes a single DNA base in a one-to-one manner and exhibits nearly no context dependency (Boch et al., 2009; Moscou & Bogdanove, 2009). Although DNA methylation has been reported to negatively affect TALE binding (Bultmann et al., 2012), the only common limitation in sequence targeting is the preferential binding to a 5′ T immediately preceding the target site. Even this limitation can be avoided, if necessary, by employing newer TALE scaffolds that do not have this restriction (Lamb, Mercer, & Barbas, 2013). TALEs can thus be targeted to virtually any sequence of the genome.

The history of TALEN development mirrored that of ZFN development; the catalytic domain of FokI monomers (a type IIS restriction enzyme) was fused at the C-termini of TALE for targeted endonuclease activity (Christian et al., 2010). As FokI functions as an obligate dimer, TALENs confer high sequence specificity by designating: (1) TALE-binding sites on both sides of the target DSB site and (2) the length of the spacer sequence that separates the two TALE-binding sites for optimal FokI dimerization. In addition, the FokI dimerization domain was engineered such that heterodimerization would be necessary for its catalytic activity (Cade et al., 2012), further reducing the off-target activity. Subsequently, TALE nickases were developed by coupling catalytically active and inactive halves (Wu et al., 2014). With appropriate selection of target sequence and TALEN scaffold, TALENs are generally regarded to have very few restrictions due to off-target activity compared with other modern designer endonucleases.

Various TALEN scaffolds have been described and are available for purchase through Addgene and the TALEN library resource (Table 1), along with kits for cloning in RVDs to generate unique DNA-binding domains (Bedell et al., 2012; Nakajima & Yaoita, 2013; Sun, Bao, Xiong, & Zhao, 2014). Among the available TALEN scaffolds, N- and C-termini truncated GoldyTALEN confers the highest activity and specificity, and thus is the scaffold of choice (Bedell et al., 2012; Nakajima & Yaoita, 2013). Each GoldyTALEN monomer specifies a 14–16 bp DNA-binding sequence separated by a 14–16 bp spacer sequence; DSB is regarded to occur at the halfway point of the spacer where each half of the FokI dimer has the highest chance of meeting the other.

Although generating TALEN constructs into plasmid DNA was historically difficult due to the highly repetitive nature of the TALE repeats, various methods have been described to circumvent this challenge. These methods include Golden Gate cloning (Cermak et al., 2011), serial ligation (Reyon et al., 2012), and ligation-independent cloning (Schmid-Burgk, Schmidt, Kaiser, Honing, & Hornung, 2013). In

Table 1

	Source	Resources for Help With Target Designing
TALEN	Addgene TALEN library resource (http://omictools.com/talenlibrary-resource-tool) Primary authors	Mojo Hand (www.talendegisn.org) ZiFit Targeter software ChopChop (https://chopchop.rc.fas.harvard.edu) E-TALEN (http://www.e-talen.org/E-TALEN/)
CRISPR-Cas9	Addgene Primary authors	Mojo Hand (www.talendegisn.org) ZiFit Targeter software ChopChop (https://chopchop.rc.fas.harvard.edu) E-CRISP (http://www.e-crisp.org/E-CRISP/) Zhang Lab's tool (http://crispr.mit.edu/) CrisprGE (for list of previously published gRNAs) (http://crdd.osdd.net/servers/crisprge/index.php)

addition, we have introduced a streamlined, Golden Gate cloning-based method, FusX: a single-tube, 3-day TALEN assembly process (Ma et al., 2016). As FusX assembly only requires standard molecular biology tools, TALEN synthesis can now be completed in a similar timeline as de novo single guide RNA (sgRNA) generation for the CRISPR-Cas9 system.

1.2 CRISPR-Cas9

The CRISPR-Cas array was first described in a bacterial genome in 1987. Since the discovery of its role as bacterial adaptive immunity against foreign DNA, Class II CRISPR-Cas systems, consisting only of one protein and two strands of RNA, have been rapidly adapted for the use of genome engineering (Cong et al., 2013; Gasiunas, Barrangou, Horvath, & Siksnys, 2012; Hwang et al., 2013; Jao, Wente, & Chen, 2013). Successful activation of native CRISPR-Cas9 depends on the identification of protospacer-adjacent motif (PAM) by Cas9 as well as target DNA sequence (protospacer) by CRISPR RNA (crRNA). In bacteria, Trans-activating crRNA (tracrRNA) and crRNA complex with the Cas9 protein to form ribonucleoprotein (RNP) complexes that scan for PAM sequences on DNA. Once a PAM sequence is identified, the RNP pauses and interrogates the target sequence in a 5′-to-3′ manner using crRNA by enabling it to form a heteroduplex with the target through Watson–Crick base pairing (Sternberg, Redding, Jinek, Greene, & Doudna, 2014). A sufficiently stable heteroduplex formation results in the activation of Cas9 for target DNA cleavage. Of the two catalytically active domains of Cas9, HNH cleaves the protospacer while RuvC cleaves the complimentary strand, both between the third and fourth bases upstream of the PAM (Gasiunas et al., 2012; Jinek et al., 2012).

In 2012, Charpentier's group first reported the successful use of chimeric sgRNA by fusing crRNA with tracrRNA via a tetraloop (Jinek et al., 2012). sgRNA was subsequently shown to have retained Cas9 targeting efficiency and has now become the guide RNA of choice (Cong et al., 2013; Hwang et al., 2013; Jao et al., 2013). Generation of sgRNAs can be accomplished using conventional cloning methods or in a clone-free manner from synthetic oligonucleotide DNA. Mature sgRNAs can be obtained in a matter of a day with the clone-free method (Varshney et al., 2015), whereas the more conventional cloning-based method takes a minimum of 3 days prior to performing in vitro transcription (Jao et al., 2013).

Currently, designing sgRNAs with high activity and specificity is a mix of empirical testing and design (Cho et al., 2014; Fu et al., 2013; Hsu et al., 2013). Although general rules that govern target recognition by CRISPR-Cas9 RNPs are being elucidated, deviations from these rules seem to be the norm rather than the exception (Cho et al., 2014; Lin et al., 2014; Pattanayak et al., 2013). Nonetheless, it is a good practice to keep them in mind as one designs an sgRNA against a new target. First, mismatches in the 12 nucleotides immediately upstream of the PAM are not well tolerated (Fu et al., 2013; Hsu et al., 2013), owing to the RNP interrogating the target sequence in a 5′-to-3′ manner as mentioned earlier. Most sgRNAs harboring mismatches to the target sequence in this region are thus quickly rejected by the RNP. Second, truncated sgRNAs carrying less than 20 nucleotides of targeting sequence provide a better targeting specificity (Fu, Sander, Reyon, Cascio, & Joung, 2014). Third, sgRNAs carrying an intentional sequence mismatch at the 5′ end seem to enhance targeting specificity (Cho et al., 2014). The idea behind the latter two points is to reduce DNA–RNA heteroduplex stability such that only a perfect match at the 3′ end of the sgRNA would be tolerated for Cas9-mediated DNA cleavage.

The biology of Cas9 is also an active area of research interest in zebrafish and other systems. In addition to zebrafish codon-optimized wild-type *Streptococcus pyogenes* Cas9 provided by Chen's group (Jao et al., 2013), many variants of Cas9 are publically available to enhance specificity and utility of the CRISPR-Cas system (Table 1). Among these are Cas9 variants capable of recognizing alternative PAM sequences (Kleinstiver et al., 2015). Although the canonical PAM sequence is NGG for *S. pyogenes* Cas9, new variants are capable of recognizing up to four bases at the PAM. Of note is the D1135E variant that has a much stricter restriction for NGG, thus reducing nonspecific activity at sites harboring NGA and NAG. In addition, Cas9 proteins from other bacterial species also provide alternative PAM recognition, often exceeding four bases in length (Deveau et al., 2008; Horvath et al., 2008; Ran et al., 2015). Finally, publications by Zhang's and Joung's groups attempted to improve CRISPR-Cas9 activity by reducing the affinity for protein–DNA interactions between Cas9 and the target DNA (Kleinstiver et al., 2016; Slaymaker et al., 2016). Although both seem to deliver enhanced specificity, to date, only the latter has been made available as a humanized form through Addgene.

As an alternative approach to reduce off-target activity, catalytically inactive Cas9 (dCas9 for dead Cas9) has been fused to FokI monomers for an improved sequence read length (Guilinger, Thompson, & Liu, 2014; Tsai et al., 2014). Furthermore, the paired nickase approach either with D10A or H840A has been successful in generating DSBs with specific overhangs (Cho et al., 2014; Ran et al., 2013). Note, however, that the "PAM-out" orientation of sgRNA with D10A mutant appears to work better than either with "PAM-in" orientation or with H840A (Ran et al., 2013).

1.3 OTHER REAGENTS

Cpf1 of *Francisella novicida* is a newly described Class II CRISPR-Cas system whose activity is not dependent on tracrRNA (Zetsche et al., 2015). Interestingly, the activity of Cpf1 uniquely depends on 5′ TTN PAM rather than on 3′ NGG PAM of *S. pyogenes* Cas9. In addition, Cas9 cleavage results in a blunt cut near the PAM, whereas Cpf1 cleavage occurs much distal to the PAM sequence and results in a four- or five-nucleotide 5′ overhang. Although its function in vivo needs to be validated, it may serve as an additional tool for enhanced targeting abilities and for applications in which having an overhang is preferred over a blunt end. Humanized Cpf1 from a few bacterial strains are deposited by Zhang's group to Addgene.

Bacterial artificial chromosomes (BACs) are an F-plasmid-based technology that enables integration of inserts up to a few hundred kilobases into the host genome. As BACs were originally generated for the purpose of whole genome sequencing, nearly every genomic segment of zebrafish is archived and is accessible to researchers. Once a particular BAC clone is obtained, a transgene can be modulated in *Escherichia coli* by taking advantage of their high-fidelity, high-efficiency homologous recombination (HR) (Yang et al., 2006). With appropriate donor DNA design, this critical step enables mutageneses that are difficult to achieve in vivo, such as introduction of a specific point mutation. Furthermore, its large cargo size facilitates the whole transgene to be inserted into the host genome with preserved regulatory elements; it is thus possible to conduct a rescue experiment within an appropriate genome context. Lastly, a *Tol2* transposase-mediated protocol was described in zebrafish, significantly improving the rate of BAC transgenesis (Fuentes, Reynolds, Lewellis, Venkiteswaran, & Knaut, 2016). Regardless of the use of *Tol2* transposase, insertion site context needs to be taken into consideration when using the BAC technique as the cargo is randomly integrated into the host genome, potentially disrupting functions of essential genes.

Argonautes are a family of nucleases whose role is best described in small RNA maturation. Intriguingly, the Argonaute proteins of *Thermus thermophilus* and *Pyrococcus furiosus* were found to be capable of functioning as DNA-guided-DNase against both single- and double-stranded DNA (Swarts et al., 2015, 2014) at supraphysiological temperatures. More recently, the *Natronobacterium gregoryi* Argonaute was demonstrated to be functional *in vitro* in human cells at physiological conditions (Gao et al., 2016). While its utility *in vivo* remains to be validated, the lack of PAM sequence requirement and its seemingly high sensitivity to base-pair mismatches are both attractive characteristics as an alternative to the CRISPR-Cas systems.

2. DNA REPAIR MECHANISMS—THE RISE OF MUTATIONS

2.1 NONHOMOLOGOUS END JOINING

Most DSBs in the cells are repaired by an error-prone NHEJ pathway (Fig. 2). Combined with its robust activity, NHEJ is the primary venue by which gene knock-outs are generated from DSBs induced by designer endonucleases. In classical NHEJ, DSB ends are directly ligated by the actions of ligase IV (Gu et al., 2007). Although the free DNA ends are largely protected by Ku 70/80 heterodimers from an extensive end resection (Sun, Lee, Davis, & Chen, 2012), some 5′ end processing is common, giving rise to heterogeneous population of mutant alleles. When DSBs are caused during the early phases of embryonic development, these mutations are readily transmitted through the germline. It is thus possible to select for a particular mutation of interest (eg, early frame-shift mutations and/or nonsense mutations) through selective outcrossing, provided that the allele does not cause dominant lethality during development.

Notably, the first example of tissue-specific somatic gene knock outs using designer nucleases in zebrafish was generated by deploying NHEJ; Zon's group coupled a tissue-specific expression of Cas9 with ubiquitous expression of sgRNAs for an erythrocyte-specific blood gene knock-out (Ablain, Durand, Yang, Zhou, & Zon, 2015). These fish recapitulated the predicted porphyria phenotype (red fluorescence in erythrocytes with inactivated UROD gene), and the sequence analyses demonstrated the lack of molecular signatures characteristic of other repair pathways. Although the limitation of NHEJ—namely the unpredictable, heterogeneous mutant allele generation that makes every different cell in each animal a potential different experiment—is not to be ignored, it can be a powerful tool for the purpose of reliably knocking out a gene.

2.2 HOMOLOGY-DIRECTED REPAIR

HDR is a mechanism of DSB repair whereby stretches of homologous sequences on both sides of a DSB are used to bring in a template for the repair (Fig. 2). For this reason, HDR is currently the pathway of choice in inducing targeted knock-ins (Bedell et al., 2012; Shin, Chen, & Solnica-Krezel, 2014; Zu et al., 2013). One well-characterized subset of HDR is HR. Unlike NHEJ that is active throughout the cell cycle, HR is active only during the S to G2 phases and requires an extensive end resection. The MRN (*Mre11-Rad50-Nbs1*) complex as well as CtIP participate in this process, exposing 3′ DNA ends (Takeda, Nakamura, Taniguchi, & Paull, 2007). The exposed ends are then used for homology search by *Rad51* (Qi et al., 2015), followed by strand invasion, DNA elongation, and subsequent resolution of the DSB. Recently, Grunwald et al. demonstrated HR to be an efficient and precise way to induce knock-ins and single nucleotide polymorphisms (SNPs) along with ectopical selection markers, such as GFP and Red Fluorescent Protein (RFP) (Hoshijima, Jurynec, & Grunwald, 2016). By selecting parental generation against intronic polymorphisms and by deriving long homology arms by polymerase chain

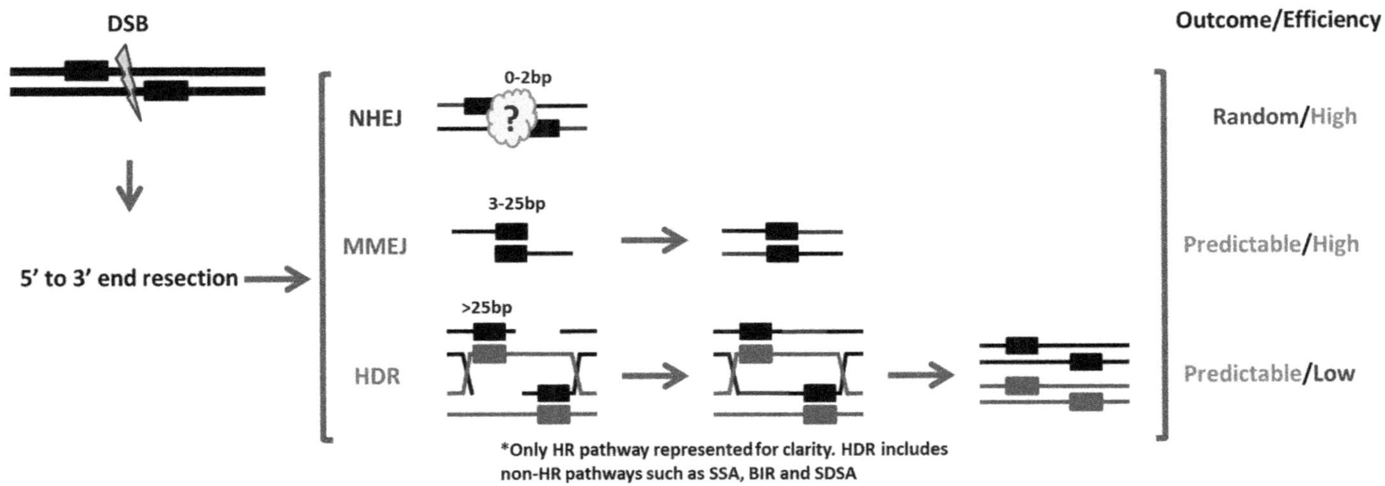

FIGURE 2 A schematic representation of nonhomologous end joining (NHEJ), microhomology-mediated end joining (MMEJ), and homologous recombination (HR) pathways following double-strand DNA breaks (DSB).

Each of the pathways begins with 5′ to 3′ end resection to a varying degree. Once the ends are processed, end joining in NHEJ often occurs between incompatible ends, producing unpredictable and heterogeneous mutant alleles. In contrast, MMEJ uses region of sequence microhomology flanking a DSB to temporarily appose the two strands. A polymerase will then elongate DNA from the homology arms in a templated fashion, resulting in predictable mutagenesis. Although both MMEJ and NHEJ can lead to high efficiency, biallelic mutagenesis of a target gene, HDR is usually a low-efficiency, monoallelic process. This is because DSB is repaired by recruiting homologous DNA as a template for repair. Of the HDR pathways, HR results in a high-fidelity repair, usually confined to late S ~ G2 phase because of the proximity requirement of the two chromosomes. Rectangular boxes represent homology arms of varying lengths indicated above them. *BIR*, break-induced replication; *SDSA*, synthesis-dependent strand annealing; *SSA*, single-stranded annealing. (See color plate) reaction (PCR) from these fish, they report a very high knock-in efficiency in F0 generation in all of the loci analyzed. Furthermore, germline transmissibility of these mutations was also robust with at least 10% of the mutants transmitted through the germline. Although HR requires long homology arms (~1 kbp), this protocol should make a multitude of gene-editing techniques (such as Cre-recombinase mediated, tissue-specific knock-outs) feasible and scalable.

In addition to HR, other characterized HDR pathways include single-stranded annealing (SSA) (Liu et al., 2012), break-induced replication (Kraus, Leung, & Haber, 2001), synthesis-dependent strand annealing (Nassif, Penney, Pal, Engels, & Gloor, 1994), and transcription-associated HDR (Davis & Maizels, 2014). These pathways are mediated by different protein machinery and require shorter homology arms compared to HR. Although HDR is a high-fidelity process, the junctions, especially on the 3′ end of the insert, may harbor erroneous mutations, thus care should be exercised when choosing a target using these approaches.

2.3 MICROHOMOLOGY-MEDIATED END JOINING

Microhomology-mediated end joining (MMEJ) is an alternative NHEJ pathway that repairs DSB in a manner similar to SSA but potentially at robustness comparable to NHEJ (Fig. 2). These characteristics have attracted several groups to investigate its utility in generating knock-in mutant zebrafish. In short, MMEJ repairs DSBs by apposing regions of local microhomology (3–25 bp in length) on both sides of the DSB, which are then used as templates for DNA elongation. Although much of the molecular mechanisms involved in the MMEJ pathway remains to be elucidated, the importance of ligase III for MMEJ activation was demonstrated in vivo using zebrafish (He et al., 2015). The most likely outcome of MMEJ is stereotyped deletion mutant generation in the absence of an exogenous donor. Additionally, insertion and chromosomal rearrangement events are common with MMEJ. In spite of these possible outcomes, MMEJ has been successfully deployed for precise integration of donor templates in zebrafish (Hisano et al., 2015).

3. MICROINJECTION

3.1 INJECTION SETUP

An example microinjection setup in zebrafish is fully described (Bill, Petzold, Clark, Schimmenti, & Ekker, 2009). Key components are the ability to visualize the procedure through the dissecting microscope, the micromanipulator control of the injection needle, and a volumetric regulator of the DNA and RNA reagents during the injection process.

3.2 PREPARATION OF REAGENTS

TALENs and CRISPR-Cas9 are most commonly delivered as RNA. Following appropriate in vitro transcription steps and RNA isolation, RNA should be aliquoted into single-use volumes to avoid freeze–thaw cycles. On the morning of the injection, make injection solutions containing either left and right arms of TALEN or CRISPR-Cas9. Typical starting injection doses for TALEN and CRISPR-Cas9 are 25 pg per arm and 25 pg gRNA plus 150 pg Cas9, respectively. To ensure the integrity of RNA used for injections, gloves and RNase-free reagent should always be used when handling these RNA; keep the RNA solution on ice for the duration of the injection session as well. When DNA is to be codelivered as a repair template, is it important that this too is prepared using all RNase-free reagents. Although the dosing is to be empirically determined, a typical starting dose is 10 pg.

3.3 INJECTION

1. Calibrate the injection volume (such as described (Bill et al., 2009) or other suitable method) to inject 1–2 nL of the injection solution. Between each different injection condition, the needle needs to be discarded and injection volume recalibrated.
2. Enhanced rates of mutagenesis are observed in embryos injected at the one-cell stage. Higher injection volumes are tolerated in the yolk. However, CRISPR-Cas9 injections and any injections containing DNA should be targeted to the cell/yolk interface or directly into the cell body for highest efficiency.
3. Injected embryos are transferred to petri dishes with embryo media and stored between 28 and 30°C.
4. At the end of the injection day, remove any dead or unhealthy embryos.

3.4 QUALITY CONTROL

After the round of injections, it is advisable to run both stock RNA solutions and final injection solution on agarose gel to ensure the quality (and quantity) of RNA used. Use a standard laboratory protocol to run denaturing RNA gel with precautions to avoid RNase contamination.

It is also important to remove (and euthanize, if necessary) any embryos that exhibit altered morphological and developmental defects. This is because (1) they are less likely to produce healthy, fertile adults and (2) these abnormal embryos could harbor off-target genomic changes.

4. SCREENING FOR F0 MUTANTS

4.1 KNOCK-INS

For knock-in applications, the most convenient way to screen for a successful integration is by adding a dominant marker to the donor construct. These may be ectopically expressed GFP proteins or other epitope tags. If the construct design does not allow for the inclusion of such markers, then the efficiencies of knock-in needs to be verified by performing PCR, bridging over both sides of the junctions or other suitable detection method. Even if a dominant marker is to be used, fidelity of the DNA sequences at both junctions should be verified by Sanger sequencing genomic DNA (gDNA) obtained by fin-clipping juvenile F0 fish.

4.2 KNOCK-OUTS

In generating gene knock-out mutants, having a dominant readout may not be feasible except when targeting well-defined genes with a known phenotype. Examples of such genes are *tyrosinase* and *golden* genes in which loss-of-function phenotypes result in lack of pigmentation. However, as DSB repair is a robust process with high rates of random mutagenesis, it is relatively fast and easy to score the efficiency of mutagenesis by various detection methods (discussed later).

To maximize the likelihood of obtaining desired, germ-line transmissible mutations, it is a good practice to assess the mutagenic efficiencies of particular injection conditions by extracting gDNA from a few injected embryos prior to raising an F0 population. The injection dose(s) of designer endonucleases should be optimized such that high rates of mutagenesis are attained while minimizing the population of injected embryos with aberrant morphological and developmental phenotypes. This can be achieved by using the lowest effective dose of designer endonuclease (Hsu et al., 2013). Subsequently, 50–100 embryos may be injected at the identified dose to be raised to sexual maturity. These fish may then be subjected to fin-clipping to maintain a clutch of fish harboring mutations. Note that these fish are most often genetically mosaic and may carry different germline mutations than that which are identified in somatic tissue. For this reason, any healthy fish harboring mutation(s) at the target site should be maintained for outcrossing even if a desired mutation was not initially identified by Sanger sequencing.

Most inexpensive and sequence nonspecific ways to assess for mutagenic efficiencies in F0 include Cel-I/Cel-II (Surveyor) (Vouillot, Thelie, & Pollet, 2015), T7E1 (Vouillot et al., 2015), and gel-mobility shift assays (Zhu et al., 2014) (Fig. 3A). To conduct these assays, a genomic region of interest flanking the DSB site is amplified with gene-specific primers. The amplicons are then subjected to melting and gradual reannealing for homo- and heteroduplex formations. Cel-I/Cel-II and T7E1 enzymes are single-stranded DNA-specific endonucleases that cleave heteroduplexed DNA at the site of base mismatch(es). Although the Surveyor assay is more sensitive at detecting SNPs, T7E1 is more efficient at cleaving longer base mismatches. Both assays are capable of detecting $<5\%$ mutant alleles in the pool of wild-type alleles (Vouillot et al., 2015). As DSB repair by NHEJ rarely results in SNP generation, T7E1 is the assay of choice among these two similar methods.

Unlike Surveyor and T7E1 assays, the gel-mobility shift assay does not depend on enzymatic cleavage of heteroduplexed DNA. Rather, this assay takes advantage of the slight change in mobility of duplexed DNA depending on both the sequence of the DNA and the presence of base pair mismatches (Zhu et al., 2014). Once the PCR amplicons are reannealed, they are run on nondenaturing, high-percentage polyacrylamide gel electrophoresis. Using 15% gel, it was demonstrated that this approach was sensitive enough to detect <1% mutation rate while the specificity was uncompromised when compared with T7E1.

High-resolution melt analysis (HRMA) is a more rapid alternative to gel-mobility shift assay if a quantitative Polymerase Chain Reaction (qPCR) machine is accessible (Fig. 3A) (Dahlem et al., 2012). The basic principle of HRMA is that heteroduplexed DNA is more thermoenergetically unstable than homoduplexed DNA and thus melts at higher temperature than do homoduplexes. When PCR is performed using double-strand DNA–specific dyes such as SYBR green, subsequent reannealing followed by HRMA will produce distinct curves depending on the DNA sequence as well as the nature of the base mismatches. Similarly to the gel-mobility shift assay, the detection limit for HRMA is <1% mutation rate.

The aforementioned methods generally provide qualitative assessment for the presence of mutant allele(s). In contrast, restriction fragment length polymorphism (RFLP) and Illumina-platform based next-generation sequencing offer more quantitative ways to analyze the mutagenic efficiencies. RFLP analysis requires that designer endonucleases be targeted to a sequence harboring a unique restriction enzyme target site on the PCR amplicon (Fig. 3A). Once NHEJ is triggered by DSB at this site, the repair outcome will likely result in destruction of the restriction site, thus producing a fragment resistant to restriction digest. One major limitation to this approach is that designer endonucleases need to be targeted to a unique restriction site in that genomic region, which may not always be possible. Furthermore, RFLP will not be able to detect mutations in rare events that DSB repair reconstitutes the restriction site. Nonetheless, the digest-resistant bands can be used for subcloning to eliminate the redundancy of having to sequence wild-type sequences. Thus RFLP can be a rapid and cost-effective way to characterize injected F0 fish.

Next-generation sequencing is a powerful tool that allows for a capture of a single PCR amplicon and sequencing in an unbiased manner (Varshney et al., 2015). Although this technology may become the gold standard for detecting designer endonuclease-mediated mutagenesis and for assessing precise mutagenic efficiencies, cost is prohibitive to most laboratories at present. When any other methods are used, mutant alleles need to be defined by follow-up subcloning into bacterial plasmid vector for subsequent Sanger sequencing.

5. SELECTING FOR DESIRABLE F1 MUTANTS

5.1 OUTCROSS F0 FISH

When F0 mutants become sexually mature, screen for germline-transmissible mutations by outcrossing them to a wild-type line. Outcrossing is necessary to avoid unintended homozygosity at off-target sites that may produce dominant phenotype. Depending on the skill levels of fish handlers, fish with easily distinguishable phenotype from F0 fish may be used to prevent cross-contamination. As an example,

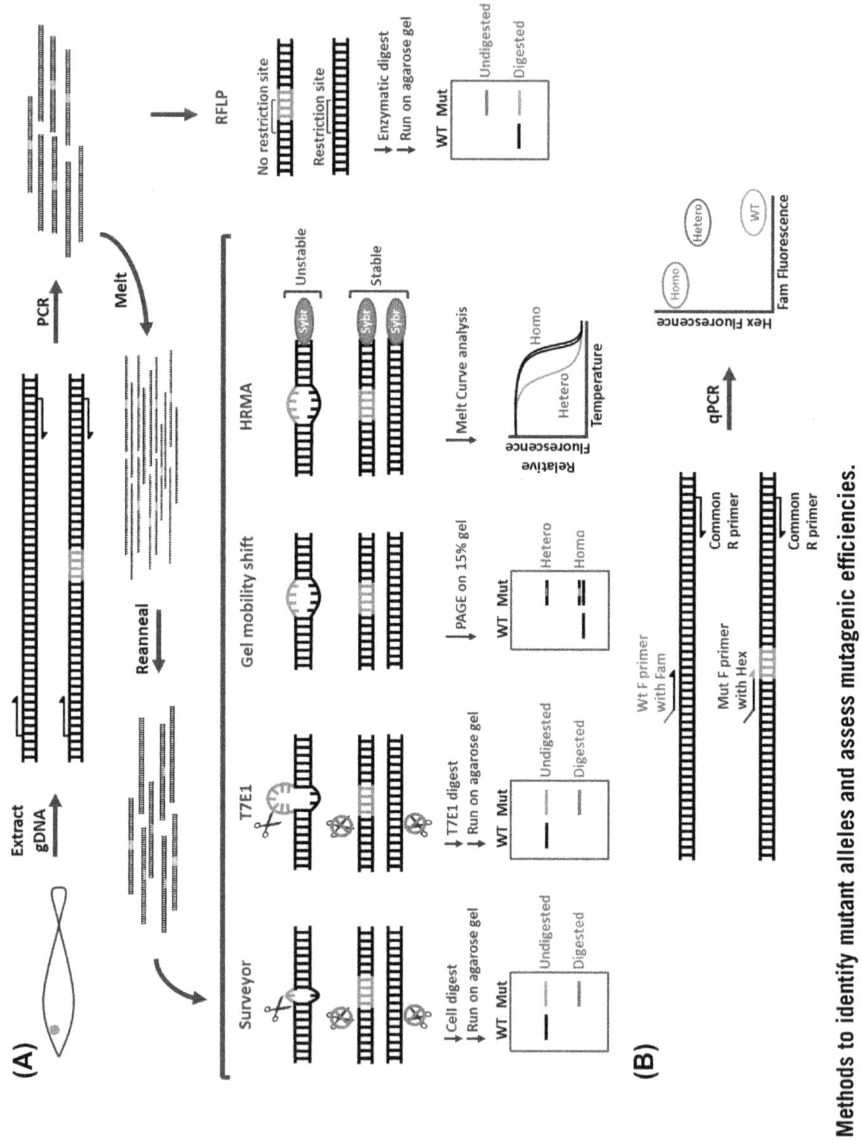

FIGURE 3 Methods to identify mutant alleles and assess mutagenic efficiencies.

(A) The process begins by extracting gDNA for subsequent polymerase chain reaction (PCR) on the segment of DNA harboring the desired mutations. The amplicon is melted by heating to 95°C followed by gradual cooling for reannealing. Should there be heterogeneous population of DNA present, this reannealing step would result in generation of both homo- and heteroduplexed DNA. Surveyor and T7E1 assays both depend on enzymatic cleavage of double-stranded DNA at the sites of mismatch. For this reason, homoduplexes are resistant to cleavage while heteroduplexes are not. Surveyor is more sensitive to detect single nucleotide polymorphisms and T7E1 is better at detecting longer indels. Once digestion is complete, DNA fragments can be resolved on agarose gel to determine the presence of digested bands in mutant fish. Gel-mobility shift assay depends on change in DNA duplex mobility based on both DNA sequence and presence/nature of the mismatches. Only one band is expected from a wild-type (WT) fish, whereas at least three bands are expected from a mutant fish. High-resolution melt assay is based on changes in DNA duplex stability; DNA base pair mismatches introduce substantial instability that can be detected by melt assay post qPCR. The curves presented here are representative of homogeneous population of either hetero- or homoduplexes. When DNA duplexes exist in a heterogeneous population, the curve will have multiple plateaus and shoulders. When double-strand DNA break is targeted to a unique restriction site on the amplicon, restriction fragment length polymorphism (RFLP) can be used to assess mutagenic efficiencies. Provided that the enzyme cleaves wild-type DNA at 100% efficiency, densitometric analyses of undigested and digested bands can be fully quantitative. (B) Once a desired mutation is identified and a line is established, primer sets can be designed for qPCR-based allele-specific quantitative genotyping. With appropriate primer designs, only a specific probe will be amplified from either wild-type or mutant DNA. Fluorescence can be later quantified to determine the genotype of a fish. Heterozygote fish will yield ample fluorescence signal for both probes, whereas wild-type or homozygous mutant fish will yield fluorescence signal for only one probe. *HRMA*, high-resolution melt analysis; *PAGE*, polyacrylamide gel electrophoresis. (See color plate)

$leo^{tq270/+}$ is one good choice as it produces dominant leopard patterns (Watanabe et al., 2006). Although germline transmissibility of the mutant allele(s) is being assessed, keep the F0 fish in individual housing with clear identifiers.

5.2 SCREEN FOR GERMLINE TRANSMISSIBILITY

Once F0 mutants reproduce, gDNA from a few F1 embryos should be used to assess germline transmissibility of the mutant allele(s). To this end, any of the screening methods described earlier can be used to check for heterozygosity. It is important to note that germline transmission is often mosaic and thus not Mendelian at this generation. F0s that produce 40 or more viable embryos that do not show any evidence of heterozygosity are considered negative for germline transmission and should be retired from the screen.

F1 offspring from germline positive F0 fish are raised to juvenile age until they are ready for fin-clipping. gDNA of confirmed F1 heterozygotes should be sequenced to establish a line with a defined gene mutation. In the case of a knock-in line, sequencing both 5′ and 3′ junctions may be necessary to ensure that inserts are integrated in the intended reading frame. In the case of a knock-out line, it is important to select for early frame-shift mutation and/or nonsense mutation to ensure the absence of functional protein. Fish harboring different mutations should be kept in separate tanks with mutation-specific identifiers.

5.3 F2 GENERATION AND ONWARD

F1 generation should again be outcrossed to an unrelated wild-type strain to dilute out potential off-target mutations. Subsequent generations can be maintained by alternating between in-crosses and outcrosses every generation. From the F2 onward, mutations will be inherited in a Mendelian manner, and thus screening gDNA from five embryos should give >95% confidence to ensure the heterozygosity of the F1 parent. It is good practice to maintain two independent lines with unique mutations for subsequent functional validations.

In addition to the aforementioned screening methods, once the F2 generation is established with a defined mutant allele, subsequent screening can be done more rapidly and inexpensively by a qRT-PCR-based allele-specific quantitative (ASQ) PCR method (Lee et al., 2016) (Fig. 3B). To do so, two forward primers with unique 5′ fluorescent probe sequences are designed for both wild-type and mutant alleles. These primers are mixed with a common reverse primer in a single tube for qRT-PCR for determining genotypes depending on the amount of fluorescence from each probe. Although care needs to be taken in designing and functionally validating the primer sets, ASQ PCR is a robust process allowing for genotypic determination in less than 2 h in a 96-well format. We have demonstrated that ASQ PCR is capable of discerning a SNP with 100% accuracy with an appropriate primer design (Lee et al., 2016).

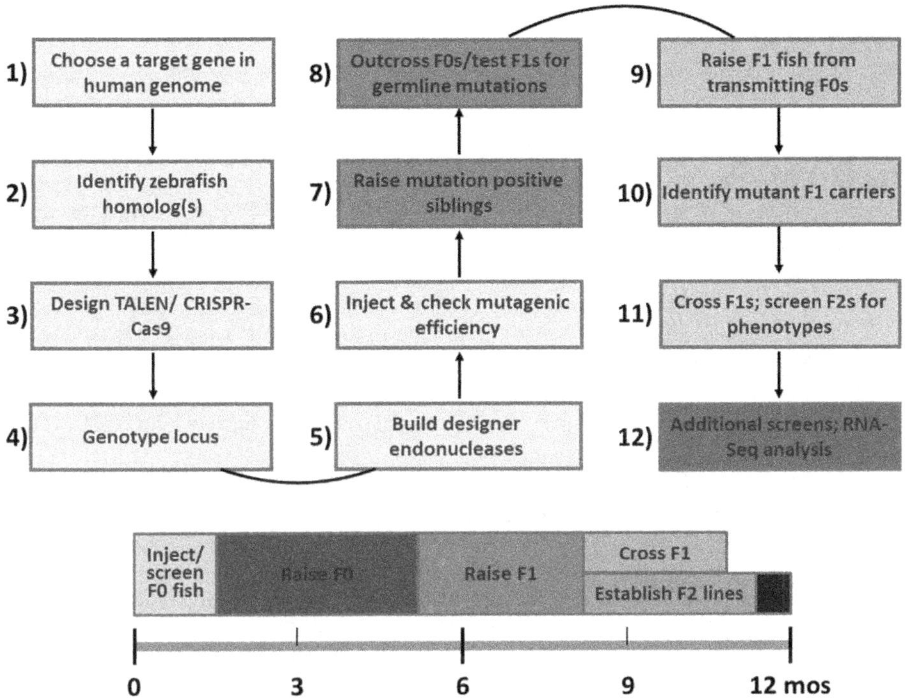

FIGURE 4 Sample timeline for establishing a mutant line.

For modeling human diseases, a target gene should be identified on a human genome first. Double-strand DNA break site should also be chosen based on transcript data from humans and not zebrafish. Once primer specificity is confirmed for genotyping, designer endonucleases are injected to assess mutagenic efficiency. With a satisfactory rate of mutagenesis and the lack of undesired developmental phenotypes, a batch of injected fish should be raised to the juvenile age. Check for somatic mutations by fin-clipping each injected fish and keep those with high rate of mutation. F1 and F2 generations should be made by two rounds of outcrossing. It is advisable to start using offspring of F2 generation in-crosses for the purpose of assessing functional/physiological consequences of the mutation. However, F1 generation fish can be in-crossed to generate preliminary results earlier. With the generation time of 3 months, sexually mature F2 fish can be obtained within a year of the start. *CRISPR*, clustered regularly interspaced short palindromic repeats; *TALEN*, transcription activator-like effector nuclease.

CONCLUSION

Zebrafish are remarkably compatible with a multitude of genome engineering technologies. With the rapid evolution of modern gene editing tools, it is now feasible to establish a mutant fish line of interest within a year (Fig. 4). Combined with high fecundity and highly conserved vertebrate physiology, zebrafish are an attractive animal model to study human genetic disorders and to explore the functional significance of the vertebrate genome.

REFERENCES

Ablain, J., Durand, E. M., Yang, S., Zhou, Y., & Zon, L. I. (2015). A CRISPR/Cas9 vector system for tissue-specific gene disruption in zebrafish. *Developmental Cell, 32*(6), 756–764. http://dx.doi.org/10.1016/j.devcel.2015.01.032.

Bedell, V. M., Wang, Y., Campbell, J. M., Poshusta, T. L., Starker, C. G., Krug, R. G., 2nd, ... Ekker, S. C. (2012). In vivo genome editing using a high-efficiency TALEN system. *Nature, 491*(7422), 114–118. http://dx.doi.org/10.1038/nature11537.

Bill, B. R., Petzold, A. M., Clark, K. J., Schimmenti, L. A., & Ekker, S. C. (2009). A primer for morpholino use in zebrafish. *Zebrafish, 6*(1), 69–77. http://dx.doi.org/10.1089/zeb.2008.0555.

Boch, J., Scholze, H., Schornack, S., Landgraf, A., Hahn, S., Kay, S., ... Bonas, U. (2009). Breaking the code of DNA binding specificity of TAL-type III effectors. *Science, 326*(5959), 1509–1512. http://dx.doi.org/10.1126/science.1178811.

Bultmann, S., Morbitzer, R., Schmidt, C. S., Thanisch, K., Spada, F., Elsaesser, J., ... Leonhardt, H. (2012). Targeted transcriptional activation of silent oct4 pluripotency gene by combining designer TALEs and inhibition of epigenetic modifiers. *Nucleic Acids Research, 40*(12), 5368–5377. http://dx.doi.org/10.1093/nar/gks199.

Cade, L., Reyon, D., Hwang, W. Y., Tsai, S. Q., Patel, S., Khayter, C., ... Yeh, J. R. (2012). Highly efficient generation of heritable zebrafish gene mutations using homo- and heterodimeric TALENs. *Nucleic Acids Research, 40*(16), 8001–8010. http://dx.doi.org/10.1093/nar/gks518.

Cermak, T., Doyle, E. L., Christian, M., Wang, L., Zhang, Y., Schmidt, C., ... Voytas, D. F. (2011). Efficient design and assembly of custom TALEN and other TAL effector-based constructs for DNA targeting. *Nucleic Acids Research, 39*(12), e82. http://dx.doi.org/10.1093/nar/gkr218.

Cho, S. W., Kim, S., Kim, Y., Kweon, J., Kim, H. S., Bae, S., & Kim, J. S. (2014). Analysis of off-target effects of CRISPR/Cas-derived RNA-guided endonucleases and nickases. *Genome Research, 24*(1), 132–141. http://dx.doi.org/10.1101/gr.162339.113.

Christian, M., Cermak, T., Doyle, E. L., Schmidt, C., Zhang, F., Hummel, A., ... Voytas, D. F. (2010). Targeting DNA double-strand breaks with TAL effector nucleases. *Genetics, 186*(2), 757–761. http://dx.doi.org/10.1534/genetics.110.120717.

Cong, L., Ran, F. A., Cox, D., Lin, S., Barretto, R., Habib, N., ... Zhang, F. (2013). Multiplex genome engineering using CRISPR/Cas systems. *Science, 339*(6121), 819–823. http://dx.doi.org/10.1126/science.1231143.

Dahlem, T. J., Hoshijima, K., Jurynec, M. J., Gunther, D., Starker, C. G., Locke, A. S., ... Grunwald, D. J. (2012). Simple methods for generating and detecting locus-specific mutations induced with TALENs in the zebrafish genome. *PLoS Genetics, 8*(8), e1002861. http://dx.doi.org/10.1371/journal.pgen.1002861.

Davis, L., & Maizels, N. (2014). Homology-directed repair of DNA nicks via pathways distinct from canonical double-strand break repair. *Proceedings of the National Academy of Sciences of the United States of America, 111*(10), E924–E932. http://dx.doi.org/10.1073/pnas.1400236111.

Deveau, H., Barrangou, R., Garneau, J. E., Labonte, J., Fremaux, C., Boyaval, P., ... Moineau, S. (2008). Phage response to CRISPR-encoded resistance in *Streptococcus thermophilus*. *Journal of Bacteriology, 190*(4), 1390–1400. http://dx.doi.org/10.1128/JB.01412-07.

Fu, Y., Foden, J. A., Khayter, C., Maeder, M. L., Reyon, D., Joung, J. K., & Sander, J. D. (2013). High-frequency off-target mutagenesis induced by CRISPR-Cas nucleases in human cells. *Nature biotechnology, 31*(9), 822–826. http://dx.doi.org/10.1038/nbt.2623.

Fu, Y., Sander, J. D., Reyon, D., Cascio, V. M., & Joung, J. K. (2014). Improving CRISPR-Cas nuclease specificity using truncated guide RNAs. *Nature Biotechnology, 32*(3), 279–284. http://dx.doi.org/10.1038/nbt.2808.

Fuentes, F., Reynolds, E., Lewellis, S. W., Venkiteswaran, G., & Knaut, H. (2016). A plasmid set for efficient bacterial artificial chromosome (BAC) transgenesis in zebrafish. *G3*. http://dx.doi.org/10.1534/g3.115.026344.

Gao, et al. (2016). http://dx.doi.org/10.1038/nbt.3547.

Gasiunas, G., Barrangou, R., Horvath, P., & Siksnys, V. (2012). Cas9-crRNA ribonucleoprotein complex mediates specific DNA cleavage for adaptive immunity in bacteria. *Proceedings of the National Academy of Sciences of the United States of America, 109*(39), E2579–E2586. http://dx.doi.org/10.1073/pnas.1208507109.

Gu, J., Lu, H., Tippin, B., Shimazaki, N., Goodman, M. F., & Lieber, M. R. (2007). XRCC4: DNA ligase IV can ligate incompatible DNA ends and can ligate across gaps. *EMBO Journal, 26*(4), 1010–1023. http://dx.doi.org/10.1038/sj.emboj.7601559.

Guilinger, J. P., Thompson, D. B., & Liu, D. R. (2014). Fusion of catalytically inactive Cas9 to FokI nuclease improves the specificity of genome modification. *Nature Biotechnology, 32*(6), 577–582. http://dx.doi.org/10.1038/nbt.2909.

He, M. D., Zhang, F. H., Wang, H. L., Wang, H. P., Zhu, Z. Y., & Sun, Y. H. (2015). Efficient ligase 3-dependent microhomology-mediated end joining repair of DNA double-strand breaks in zebrafish embryos. *Mutation Research, 780*, 86–96. http://dx.doi.org/10.1016/j.mrfmmm.2015.08.004.

Hisano, Y., Sakuma, T., Nakade, S., Ohga, R., Ota, S., Okamoto, H., ... Kawahara, A. (2015). Precise in-frame integration of exogenous DNA mediated by CRISPR/Cas9 system in zebrafish. *Scientific Reports, 5*, 8841. http://dx.doi.org/10.1038/srep08841.

Horvath, P., Romero, D. A., Coute-Monvoisin, A. C., Richards, M., Deveau, H., Moineau, S., ... Barrangou, R. (2008). Diversity, activity, and evolution of CRISPR loci in *Streptococcus thermophilus*. *Journal of Bacteriology, 190*(4), 1401–1412. http://dx.doi.org/10.1128/JB.01415-07.

Hoshijima, K., Jurynec, M. J., & Grunwald, D. J. (2016). Precise editing of the zebrafish genome made simple and efficient. *Developmental Cell, 36*(6), 654–667. http://dx.doi.org/10.1016/j.devcel.2016.02.015.

Hsu, P. D., Scott, D. A., Weinstein, J. A., Ran, F. A., Konermann, S., Agarwala, V., ... Zhang, F. (2013). DNA targeting specificity of RNA-guided Cas9 nucleases. *Nature Biotechnology, 31*(9), 827–832. http://dx.doi.org/10.1038/nbt.2647.

Hwang, W. Y., Fu, Y., Reyon, D., Maeder, M. L., Tsai, S. Q., Sander, J. D., ... Joung, J. K. (2013). Efficient genome editing in zebrafish using a CRISPR-Cas system. *Nature Biotechnology, 31*(3), 227–229. http://dx.doi.org/10.1038/nbt.2501.

Jao, L. E., Wente, S. R., & Chen, W. (2013). Efficient multiplex biallelic zebrafish genome editing using a CRISPR nuclease system. *Proceedings of the National Academy of Sciences of the United States of America, 110*(34), 13904–13909. http://dx.doi.org/10.1073/pnas.1308335110.

Jinek, M., Chylinski, K., Fonfara, I., Hauer, M., Doudna, J. A., & Charpentier, E. (2012). A programmable dual-RNA-guided DNA endonuclease in adaptive bacterial immunity. *Science, 337*(6096), 816–821. http://dx.doi.org/10.1126/science.1225829.

Kleinstiver, B. P., Pattanayak, V., Prew, M. S., Tsai, S. Q., Nguyen, N. T., Zheng, Z., & Joung, J. K. (2016). High-fidelity CRISPR-Cas9 nucleases with no detectable genome-wide off-target effects. *Nature, 529*(7587), 490–495. http://dx.doi.org/10.1038/nature16526.

Kleinstiver, B. P., Prew, M. S., Tsai, S. Q., Topkar, V. V., Nguyen, N. T., Zheng, Z., … Joung, J. K. (2015). Engineered CRISPR-Cas9 nucleases with altered PAM specificities. *Nature, 523*(7561), 481–485. http://dx.doi.org/10.1038/nature14592.

Kraus, E., Leung, W. Y., & Haber, J. E. (2001). Break-induced replication: a review and an example in budding yeast. *Proceedings of the National Academy of Sciences of the United States of America, 98*(15), 8255–8262. http://dx.doi.org/10.1073/pnas.151008198.

Lamb, B. M., Mercer, A. C., & Barbas, C. F., 3rd (2013). Directed evolution of the TALE N-terminal domain for recognition of all 5′ bases. *Nucleic Acids Research, 41*(21), 9779–9785. http://dx.doi.org/10.1093/nar/gkt754.

Lee, H. B., Schwab, T. L., Koleilat, A., Ata, H., Daby, C. L., Lopez Cervera, R., … Clark, K. J. (2016). Allele-specific quantitative PCR for accurate, rapid, and cost-effective genotyping. *Human Gene Therapy.* http://dx.doi.org/10.1089/hum.2016.011.

Lin, Y., Cradick, T. J., Brown, M. T., Deshmukh, H., Ranjan, P., Sarode, N., … Bao, G. (2014). CRISPR/Cas9 systems have off-target activity with insertions or deletions between target DNA and guide RNA sequences. *Nucleic Acids Research, 42*(11), 7473–7485. http://dx.doi.org/10.1093/nar/gku402.

Liu, J., Gong, L., Chang, C., Liu, C., Peng, J., & Chen, J. (2012). Development of novel visual-plus quantitative analysis systems for studying DNA double-strand break repairs in zebrafish. *Journal of Genetics and Genomics = Yi chuan xue bao, 39*(9), 489–502. http://dx.doi.org/10.1016/j.jgg.2012.07.009.

Ma, A. C., McNulty, M. S., Poshusta, T. L., Campbell, J. M., Martinez-Galvez, G., Argue, D. P., … Ekker, S. C. (2016). FusX: a rapid one-step transcription activator-like effector assembly system for genome science. *Human Gene Therapy.* http://dx.doi.org/10.1089/hum.2015.172.

Moscou, M. J., & Bogdanove, A. J. (2009). A simple cipher governs DNA recognition by TAL effectors. *Science, 326*(5959), 1501. http://dx.doi.org/10.1126/science.1178817.

Nakajima, K., & Yaoita, Y. (2013). Comparison of TALEN scaffolds in *Xenopus tropicalis. Biology Open, 2*(12), 1364–1370. http://dx.doi.org/10.1242/bio.20136676.

Nasevicius, A., & Ekker, S. C. (2000). Effective targeted gene 'knockdown' in zebrafish. *Nature Genetics, 26*(2), 216–220. http://dx.doi.org/10.1038/79951.

Nassif, N., Penney, J., Pal, S., Engels, W. R., & Gloor, G. B. (1994). Efficient copying of nonhomologous sequences from ectopic sites via P-element-induced gap repair. *Molecular and Cellular Biology, 14*(3), 1613–1625.

Pattanayak, V., Lin, S., Guilinger, J. P., Ma, E., Doudna, J. A., & Liu, D. R. (2013). High-throughput profiling of off-target DNA cleavage reveals RNA-programmed Cas9 nuclease specificity. *Nature Biotechnology, 31*(9), 839–843. http://dx.doi.org/10.1038/nbt.2673.

Qi, Z., Redding, S., Lee, J. Y., Gibb, B., Kwon, Y., Niu, H., … Greene, E. C. (2015). DNA sequence alignment by microhomology sampling during homologous recombination. *Cell, 160*(5), 856–869. http://dx.doi.org/10.1016/j.cell.2015.01.029.

Ran, F. A., Cong, L., Yan, W. X., Scott, D. A., Gootenberg, J. S., Kriz, A. J., … Zhang, F. (2015). In vivo genome editing using *Staphylococcus aureus* Cas9. *Nature, 520*(7546), 186–191. http://dx.doi.org/10.1038/nature14299.

Ran, F. A., Hsu, P. D., Lin, C. Y., Gootenberg, J. S., Konermann, S., Trevino, A. E., … Zhang, F. (2013). Double nicking by RNA-guided CRISPR Cas9 for enhanced genome editing specificity. *Cell, 154*(6), 1380–1389. http://dx.doi.org/10.1016/j.cell.2013.08.021.

Reyon, D., Tsai, S. Q., Khayter, C., Foden, J. A., Sander, J. D., & Joung, J. K. (2012). FLASH assembly of TALENs for high-throughput genome editing. *Nature Biotechnology, 30*(5), 460–465. http://dx.doi.org/10.1038/nbt.2170.

Schmid-Burgk, J. L., Schmidt, T., Kaiser, V., Honing, K., & Hornung, V. (2013). A ligation-independent cloning technique for high-throughput assembly of transcription activator-like effector genes. *Nature Biotechnology, 31*(1), 76–81. http://dx.doi.org/10.1038/nbt.2460.

Shin, J., Chen, J., & Solnica-Krezel, L. (2014). Efficient homologous recombination-mediated genome engineering in zebrafish using TALE nucleases. *Development, 141*(19), 3807–3818. http://dx.doi.org/10.1242/dev.108019.

Slaymaker, I. M., Gao, L., Zetsche, B., Scott, D. A., Yan, W. X., & Zhang, F. (2016). Rationally engineered Cas9 nucleases with improved specificity. *Science, 351*(6268), 84–88. http://dx.doi.org/10.1126/science.aad5227.

Sternberg, S. H., Redding, S., Jinek, M., Greene, E. C., & Doudna, J. A. (2014). DNA interrogation by the CRISPR RNA-guided endonuclease Cas9. *Nature, 507*(7490), 62–67. http://dx.doi.org/10.1038/nature13011.

Sun, N., Bao, Z., Xiong, X., & Zhao, H. (2014). SunnyTALEN: a second-generation TALEN system for human genome editing. *Biotechnology and Bioengineering, 111*(4), 683–691. http://dx.doi.org/10.1002/bit.25154.

Sun, J., Lee, K. J., Davis, A. J., & Chen, D. J. (2012). Human Ku70/80 protein blocks exonuclease 1-mediated DNA resection in the presence of human Mre11 or Mre11/Rad50 protein complex. *Journal of Biological Chemistry, 287*(7), 4936–4945. http://dx.doi.org/10.1074/jbc.M111.306167.

Swarts, D. C., Hegge, J. W., Hinojo, I., Shiimori, M., Ellis, M. A., Dumrongkulraksa, J., … van der Oost, J. (2015). Argonaute of the archaeon *Pyrococcus furiosus* is a DNA-guided nuclease that targets cognate DNA. *Nucleic Acids Research, 43*(10), 5120–5129. http://dx.doi.org/10.1093/nar/gkv415.

Swarts, D. C., Jore, M. M., Westra, E. R., Zhu, Y., Janssen, J. H., Snijders, A. P., … van der Oost, J. (2014). DNA-guided DNA interference by a prokaryotic Argonaute. *Nature, 507*(7491), 258–261. http://dx.doi.org/10.1038/nature12971.

Takeda, S., Nakamura, K., Taniguchi, Y., & Paull, T. T. (2007). Ctp1/CtIP and the MRN complex collaborate in the initial steps of homologous recombination. *Molecular Cell, 28*(3), 351–352. http://dx.doi.org/10.1016/j.molcel.2007.10.016.

Tsai, S. Q., Wyvekens, N., Khayter, C., Foden, J. A., Thapar, V., Reyon, D., … Joung, J. K. (2014). Dimeric CRISPR RNA-guided FokI nucleases for highly specific genome editing. *Nature Biotechnology, 32*(6), 569–576. http://dx.doi.org/10.1038/nbt.2908.

Varshney, G. K., Pei, W., LaFave, M. C., Idol, J., Xu, L., Gallardo, V., … Burgess, S. M. (2015). High-throughput gene targeting and phenotyping in zebrafish using CRISPR/Cas9. *Genome Research, 25*(7), 1030–1042. http://dx.doi.org/10.1101/gr.186379.114.

Vouillot, L., Thelie, A., & Pollet, N. (2015). Comparison of T7E1 and surveyor mismatch cleavage assays to detect mutations triggered by engineered nucleases. *G3, 5*(3), 407–415. http://dx.doi.org/10.1534/g3.114.015834.

Watanabe, M., Iwashita, M., Ishii, M., Kurachi, Y., Kawakami, A., Kondo, S., & Okada, N. (2006). Spot pattern of leopard *Danio* is caused by mutation in the zebrafish connexin41.8 gene. *EMBO Reports, 7*(9), 893–897. http://dx.doi.org/10.1038/sj.embor.7400757.

Wu, Y., Gao, T., Wang, X., Hu, Y., Hu, X., Hu, Z., … Liang, D. (2014). TALE nickase mediates high efficient targeted transgene integration at the human multi-copy ribosomal DNA locus. *Biochemical and Biophysical Research Communications, 446*(1), 261–266. http://dx.doi.org/10.1016/j.bbrc.2014.02.099.

Yang, Z., Jiang, H., Chaichanasakul, T., Gong, S., Yang, X. W., Heintz, N., & Lin, S. (2006). Modified bacterial artificial chromosomes for zebrafish transgenesis. *Methods, 39*(3), 183–188. http://dx.doi.org/10.1016/j.ymeth.2006.04.011.

Zetsche, B., Gootenberg, J. S., Abudayyeh, O. O., Slaymaker, I. M., Makarova, K. S., Essletzbichler, P., … Zhang, F. (2015). Cpf1 is a single RNA-guided endonuclease of a class 2 CRISPR-Cas system. *Cell, 163*(3), 759–771. http://dx.doi.org/10.1016/j.cell.2015.09.038.

Zhu, X., Xu, Y., Yu, S., Lu, L., Ding, M., Cheng, J., … Tian, Y. (2014). An efficient genotyping method for genome-modified animals and human cells generated with CRISPR/Cas9 system. *Scientific Reports, 4*, 6420. http://dx.doi.org/10.1038/srep06420.

Zu, Y., Tong, X., Wang, Z., Liu, D., Pan, R., Li, Z., … Lin, S. (2013). TALEN-mediated precise genome modification by homologous recombination in zebrafish. *Nature Methods, 10*(4), 329–331. http://dx.doi.org/10.1038/nmeth.2374.

CHAPTER

8

Clonal analysis of gene loss of function and tissue-specific gene deletion in zebrafish via CRISPR/Cas9 technology

F. De Santis[a], V. Di Donato[a], F. Del Bene[1]

PSL Research University, Paris, France

[1]*Corresponding author: E-mail: filippo.del-bene@curie.fr*

CHAPTER OUTLINE

[a]These two authors contributed equally to this work.

Methods in Cell Biology, Volume 135, ISSN 0091-679X, http://dx.doi.org/10.1016/bs.mcb.2016.03.006

Abstract

In the last few years the development of CRISPR/Cas 9-mediated genome editing techniques has allowed the efficient generation of loss-of-function alleles in several model organisms including zebrafish. However, these methods are mainly devoted to target-specific genomic loci leading to the creation of constitutive knock-out models. On the contrary, the analysis of gene function via tissue- or cell-specific mutagenesis remains challenging in zebrafish. To circumvent this limitation, we present here a simple and versatile protocol to achieve tissue-specific gene disruption based on the Cas9 expression under the control of the Gal4/upstream activating sequence binary system. In our method, we couple Cas9 with green fluorescent protein or Cre reporter gene expression. This strategy allows us to induce somatic mutations in genetically labeled cell clones or single cells, and to follow them *in vivo* via reporter gene expression. Importantly, because none of the tools that we present here are restricted to zebrafish, similar approaches are readily applicable in virtually any organism where transgenesis and DNA injection are feasible.

INTRODUCTION

TISSUE-SPECIFIC KNOCKOUT IN ANIMAL MODELS

The generation of transgenic animals carrying constitutive knockout alleles represents one of the most powerful approaches to investigate gene function. However, stable gene disruption can be linked to embryonic lethality, compensation mechanisms, and pleiotropic phenotypes that might limit the analysis of mutant animals. In these cases, the possibility of inducing targeted mutagenesis in a precisely controlled spatiotemporal manner represents a great advantage in reverse genetic studies. To date, different strategies have been developed to induce tissue-specific loss of function in diverse model organisms. The golden standard techniques for conditional mutagenesis in mouse and fruit fly (*Drosophila melanogaster*) rely on the Cre/loxP and flippase (Flp)/flippase recognition target (FRT) systems (Bouabe & Okkenhaug, 2013; Theodosiou & Xu, 1998). The first approach applies homologous recombination techniques to generate animals carrying a floxed target allele, in which loxP sites are inserted at the 5′ and 3′ ends of the selected locus. Cre recombinase catalyzes site-specific recombination events in genomic regions containing loxP sites and can be used to excise the floxed target allele, generating gene inactivation. By crossing animals carrying a floxed sequence to driver lines in which *Cre* expression is controlled by a tissue-specific promoter, it is possible to induce the excision of the target locus in the desired cell population, allowing for the generation of conditional gene disruption (Kuhn, Schwenk, Aguet, & Rajewsky, 1995). A modified version of the Cre enzyme, in which the recombinase is fused to the estrogen receptor (Cre-ER), allows tamoxifen-induced temporal control of Cre activity (Feil et al., 1996). The Flp/FRT technology is an analogous strategy based on the activity of the Flp enzyme, catalyzing site-directed recombination of DNA sequences flanked by FRT sites (Choi et al., 2009; Golic & Lindquist, 1989). Most recently, CRISPR (clustered regularly interspaced short palindromic repeats)/Cas (CRISPR-associated)−based approaches have been introduced to generate tissue-specific loss of function. The system is based on the combined activity of two components: a single guide RNA (sgRNA) containing 20 nucleotides complementary to the target genomic sequence and the Cas9 endonuclease catalyzing double-strand breaks (DSBs) at the targeted locus. In addition to the efficient generation of constitutive mutant alleles (produced by the imperfect DNA repair mechanisms occurring after Cas9-induced cleavage), the CRISPR/Cas9 technique has been used to insert loxP sites at the desired loci, thus generating floxed alleles for Cre-based conditional mutagenesis (Yang et al., 2013). In addition, it has been shown that it is possible to induce conditional knockout by coupling inducible expression of the *Cas9* enzyme with constitutive expression of sgRNAs (driven by U6 promoters recognized by type III RNA polymerase). In mouse, the generation of a floxed Cas9 allele (in which a *loxP-STOP-loxP* cassette is placed between a constitutive CAG promoter and the coding sequence (CDS) of the endonuclease) has been used to achieve Cre-dependent spatiotemporal regulation of *Cas9* expression (Platt et al., 2014). Similarly, control of Cas9 activity was obtained in *Drosophila* by using the Gal4/upstream activating sequence (UAS) binary system (Port, Chen, Lee, & Bullock, 2014). In this strategy, the use of a tissue-specific promoter to drive the transcription of the Gal4 transactivator enables conditional activation of the Cas9 enzyme, whose expression is controlled by the UAS recognized by Gal4.

CLONAL ANALYSIS OF GENE LOSS OF FUNCTION

To further improve the resolution of loss-of-function studies it is possible to manipulate gene activity in individual cells or single-cell clones and analyze the resulting phenotype within an otherwise wild-type tissue. Different strategies have been developed to generate chimeric animals in which the fate of clones derived from a mutant progenitor can be followed in a wild-type environment. In mouse, homozygous mutant embryonic stem cells, carrying the desired mutation together with an independent genetic marker, can be transplanted into host wild-type blastocysts to give rise to chimeric animals (Rossant & Spence, 1998). Similarly, transplantation strategies can be performed in *Drosophila* by introducing cells derived from a mutant donor embryo into wild-type hosts (Technau & Campos-Ortega, 1986). Nevertheless, transplantation-based approaches are often technically challenging and can only be applied to the analysis of genes acting early in development. Alternative methods are based on the Cre/loxP system and offer the possibility of inducing both targeted gene disruption and activation of a reporter for linage tracing of mutant cells (Nakazawa, Taniguchi, Okumura, Maeda, & Matsuno, 2012; Zong, Espinosa, Su, Muzumdar, & Luo, 2005). In *Drosophila* multiple strategies have been developed over the years to generate marked clones in genetic mosaics combining the flexibility of Gal4/UAS and Flp/FRT systems (Griffin, Binari, & Perrimon, 2014). Lately, a novel CRISPR-based strategy has been proposed in mouse. The technique takes advantage of in utero electroporation of a CRISPR-based plasmid to obtain gene inactivation together with a Piggybac transposase vector to express fluorescent reporters for lineage tracing of cells with loss-of-function mutations (Chen, Rosiene, Che, Becker, & LoTurco, 2015).

STATE OF THE ART IN ZEBRAFISH TISSUE-SPECIFIC GENE INACTIVATION

Despite the introduction in the last years of multiple approaches to induce targeted mutagenesis in the zebrafish genome, spatiotemporal control of gene inactivation remains challenging. Strategies based on the already described Cre/loxP and Flp/FRT systems have been also developed in zebrafish to achieve tissue-specific gene disruption (Ni et al., 2012). In this case, the generation of conditional alleles is induced by transgenic insertion of an inverted gene-trap cassette in an intronic region of a gene. Because the mutagenicity of the transgene is orientation dependent, Cre- and Flp-mediated recombination enables a switch from a state of conditional rescue in one orientation to a state of conditional knockout in the other. Nevertheless, the targeting of a precise locus is not possible with this method that relies on random transgene integration. More recently to circumvent this limitation, several techniques based on the TALEN or CRISPR/Cas9 systems have been established to introduce DNA fragments of variable size at predefined genomic locations (Auer & Del Bene, 2014). Among the different applications of these genome editing approaches, the efficient insertion of loxP sites in a targeted fashion has been achieved via TALEN-mediated DSB and subsequent homology-directed repair (Bedell et al., 2012). Although such a tool allows the creation of conditional alleles, the long generation time of zebrafish transgenic lines carrying stable genomic integrations of floxed alleles represents an important constraint. An alternative way to inactivate target genes in a tissue-specific manner has been developed by combining the CRISPR/Cas9 and Tol2 technologies (Ablain, Durand, Yang, Zhou, & Zon, 2015). In this report, a modular construct provides cell-type-specific expression of the Cas9 endonuclease (*Tissue-specific promoter:cas9*) and ubiquitous transcription of the sgRNA by a U6 promoter (*U6:sgRNA*). While mosaic phenotypes can be induced by transient expression of the plasmid, its inclusion into the genome by Tol2 transposition results in higher levels of tissue-restricted gene inactivation. The versatility of the vector system relies on the possibility to readily change the promoter driving *Cas9* transcription and the sgRNA to target a specific open reading frame (ORF) to obtain spatially controlled gene loss of function. A similar approach has been proposed lately to achieve also a tight temporal regulation of Cas9 expression (Yin et al., 2015). This methodology is based on the generation of two transgenic lines: one carries a transgene allowing heat-shock induction of Cas9 expression after Cre mRNA injection (*hp70:loxP-mCherry-STOP-loxP-cas9*); the other harbors U6 cassettes for the transcription of sgRNAs (*U6:sgRNA*). Therefore, double carriers would show tissue-specific phenotypes subsequent to heat-shock and Cre activation.

CLONAL ANALYSIS OF GENE INACTIVATION IN ZEBRAFISH

CRISPRs-based methods for conditional mutagenesis represent a valid alternative to Cre- and Flp-mediated techniques. However, to date, their application has been limited to the induction and the analysis of tissue-specific gene disruption resulting in visually detectable phenotypes. In fact, genetic labeling of mutant cells, required for the examination of phenotypes that are not readily detectable, is not addressed by these systems.

In this chapter, we present two approaches that combine the Gal4/UAS (Asakawa & Kawakami, 2008) and CRISPR/Cas9 systems in order to induce simultaneously cell-type-specific locus inactivation and fluorescent marking of the targeted cells (Di Donato et al. *Genome Research Accepted*). The first strategy that we describe permits the identification of *Cas9*-expressing cells by concomitant expression of the green fluorescent protein (GFP) and Cas9, both driven by the activation of a UAS sequence (*UAS:Cas9T2AGFP*). Thus, GFP-positive cells can be detected in the spatiotemporal domain of Gal4 expression, which is defined by a chosen tissue-specific promoter. The second strategy provides a tool that enables both the permanent labeling of *Cas9*-expressing cells and the visualization of their clonal progeny, through UAS-dependent coexpression of the Cas9 endonuclease and Cre recombinase (*UAS:Cas9T2ACre*).

For both methods, the vector system used contains two U6 promoters allowing the simultaneous targeting of two sequences at any genomic locus of interest, to increase the percentage of out-of-frame mutations and large deletions.

1. METHODS FOR TISSUE-SPECIFIC GENE INACTIVATION AND CLONAL ANALYSIS OF MUTANT CELLS

1.1 RATIONALE FOR THE DESIGN OF A VECTOR SYSTEM FOR CRISPR/Cas9-Induced Conditional Gene Disruption via Gal4/UAS

CRISPR/Cas9-mediated tissue-specific gene inactivation and concomitant visualization of mutant cells requires: (1) spatiotemporal regulation of the expression of the Cas9 endonuclease; (2) expression of sgRNAs; (3) a fluorescent transgene to label cells where Cas9/sgRNA complex is present. The cell-type-specific expression of the Cas9 enzyme is achieved by using the yeast Gal4/UAS system in which the Gal4 transcriptional activator drives the expression of transgenes linked to the UAS. In our vector design, the CDS of the Cas9, flanked by two SV40 nuclear localization signals, is placed downstream of a 5xUAS cassette. To monitor *Cas9* expression, we use a viral T2A self-cleaving peptide followed by the GFP CDS enabling the synthesis of the fluorescent reporter from the same transcript (*UAS:Cas9T2AGFP*). Two U6 promoters are subcloned in the same construct to drive ubiquitous transcription of the sgRNAs. BsmbI and BsaI restriction sites allow the cloning of the 20-bp target sequence at the transcription start site (TSS = +1) of the U6 promoters (Figs. 1A and 2). The vector contains Tol2 sites for efficient transposition of the transgenic cassette in the fish genome.

1.2 EXPERIMENTAL WORKFLOW

1.2.1 Overview of the method

The induction of conditional gene inactivation via the Gal4/UAS system consists of three steps. Step 1: design of sgRNAs targeting the genomic locus of interest and assessment of their mutagenesis efficiency. Step 2: cloning of the target sequences of the sgRNAs displaying the highest mutation rate at the TSS of the U6 promoters in the *pUAS:Cas9T2AGFP* plasmid (Fig. 1C and D). Step 3: microinjection of p*UAS:CAS9T2AGFP;U6:sgRNA1;U6:sgRNA2* plasmid for transient expression or generation of a transgenic line carrying stable genomic integration of the cassette (Fig. 1E).

1.2.2 Technical procedure

Step 1: Design of sgRNAs targeting the genomic locus of interest and assessment of their mutagenesis efficiency

sgRNA design

- Select a target site for sgRNA synthesis by using one of the available online tools (ie, http://crispor.tefor.net/crispor.cgi; http://crispr.mit.edu/; http://zifit.partners.org/ZiFiT/; http://www.broadinstitute.org/rnai/public/analysis-tools/sgrna-design; https://chopchop.rc.fas.harvard.edu/; http://www.crisprscan.org/; http://crispr.cos.uni-heidelberg.de/help.html).
- Filter the research for oligonucleotides suitable for cloning at the TSS of the T7 in the pDR274 vector (Addgene ref 42250).

Oligonucleotides cloning in the pDR274 vector

- Digest 1–5 μg of the pDR274 plasmid with BsaI overnight at 37°C.
- Run the digestion product on a 0.8% agarose gel and extract the linear plasmid with a gel and polymerase chain reaction (PCR) extraction kit.
- Mix 3 μL of 100 μM forward and reverse oligonucleotides containing the target sequence and add 14 μL of TE buffer (10 mM Tris-Cl, pH 7.5. 1 mM EDTA).
- Heat the mixture 10 min at 95°C and cool down for 30 min at room temperature to allow the annealing of the oligonucleotides.
- Ligate 2 μL of annealed oligonucleotides and 100 ng of the digested pDR274 plasmid.
- Transform the reaction in DH5α competent cells.

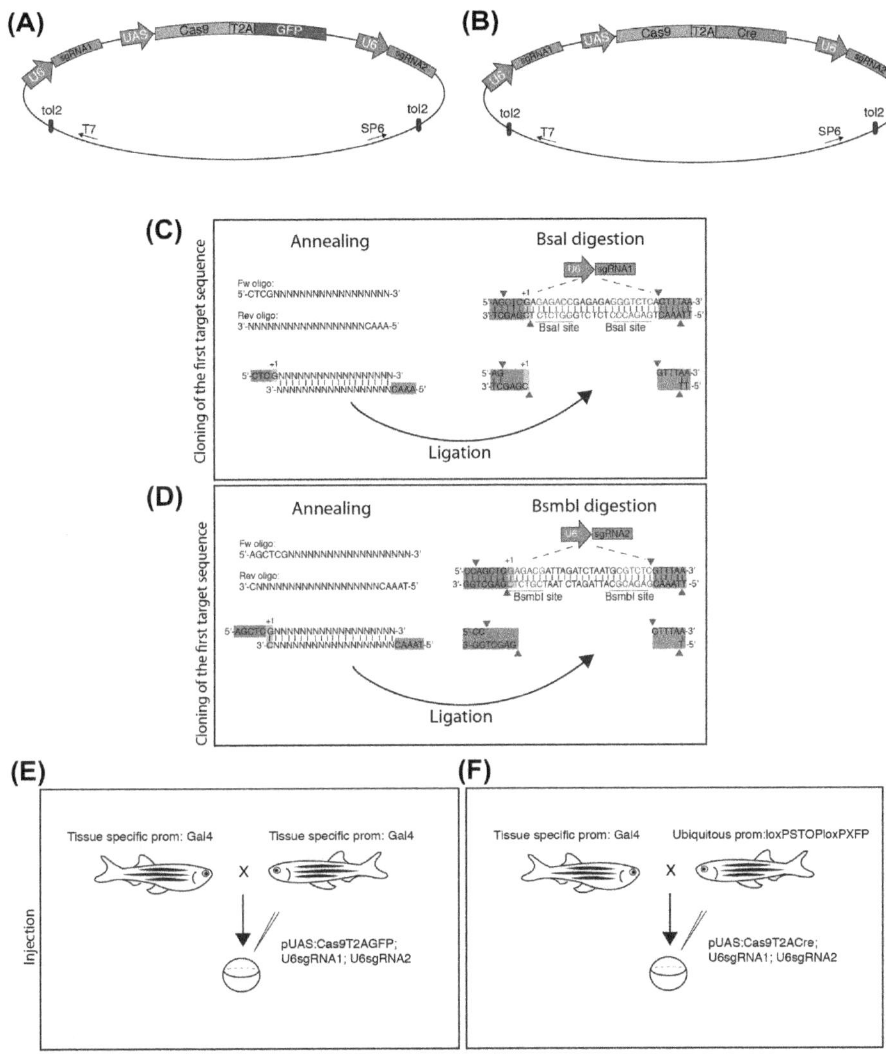

FIGURE 1

(A) Schematic of the *pUAS:Cas9T2AGFP;U6sgRNA1;U6sgRNA2* plasmid; (B) schematic of the *pUAS:Cas9T2ACre;U6sgRNA1;U6sgRNA2* plasmid; (C) insertion of the first target sequence into the transgenesis vector at the BsaI cloning sites; (D) insertion of the second target sequence into the transgenesis vector at the BsmbI cloning sites; (E) schematic view of cross of zebrafish Gal4 transgenic driver lines and injection of the *pUAS:Cas9T2AGFP; U6sgRNA1;U6sgRNA2* plasmid; (F) schematic representation of cross of zebrafish Gal4 transgenic driver to line carrying a floxed reporter allele and injection of the *pUAS: Cas9T2ACre;U6sgRNA1;U6sgRNA2* plasmid. (See color plate)

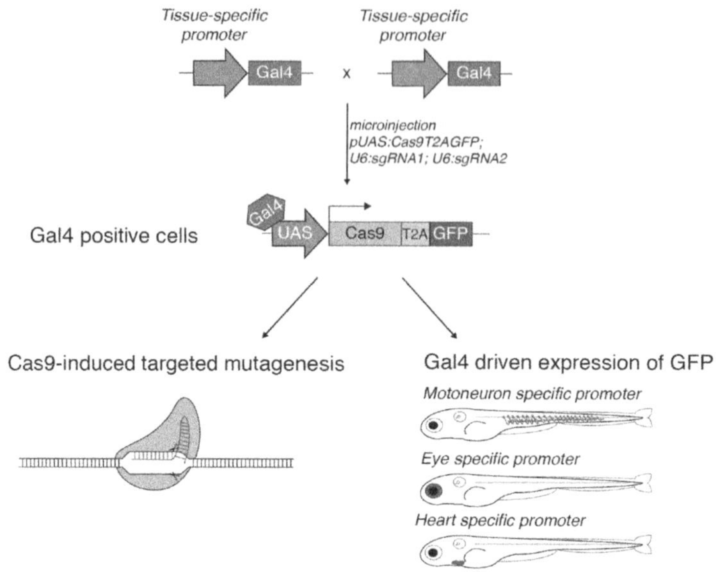

FIGURE 2

Schematic of conditional gene inactivation and genetic labeling of *Cas9*-expressing cells after injection of the *pUAS:Cas9T2AGFP;U6sgRNA1;U6sgRNA2* plasmid in specific Gal4 driver lines.

- Plate on LB (Lysogeny broth)-kanamycin plates.
- The next day, pick two colonies from the transformation plate. Inoculate each colony into 2 mL of LB-kanamycin medium and incubate at 37°C overnight with agitation.
- Isolate DNA using a Miniprep kit.
- Sequence the plasmid with M13 forward primer to verify the presence of the target sequence.

  sgRNA in vitro synthesis
- Use the pDR274 plasmid containing the target sequence as template for a PCR reaction with the primers:

  Fw oligo: 5′-GCTCGTCCTGAATGATATGCGACC-3′

  Rev oligo: 5′-CCAGTAGTGATCGACACTGCTCG-3′
- Set up the PCR reaction as follows:

  pDR274 template: 10 ng

  5X buffer HF: 10 μL

  dNTP (10 mM): 1 μL

  Primers (10 μM): 2.5 μL each

  DMSO: 1.5 μL

  Phusion polymerase: 0.5 μL

  H_2O up to 50 μL

Use the program:

98° 30 s

–

98° 30 s
58° 20 s } ×35
72° 20 s

–

72° 5 min

The expected size of the band is 253 bp.

The PCR product will contain a T7 promoter sequence and the sgRNA sequence.

- Gel purify the amplicon and elute in 20 μL. The PCR product will be used as template for the sgRNA synthesis.
- Proceed with sgRNA synthesis using the MegaScript T7 kit (AMBION—AM1334) assembling the reaction in a final volume of 20 μL as described below:

  8 μL of PCR product
  +2 μL 10X transcription buffer
  +2 μL of each NTP
  +2 μL of T7 RNA polymerase
- Transcribe for 4 h at 37°C.
- Add 1 μL of TurboDNase: 15 min 37°C.
- Recover RNA with RNA purification with RNeasy Mini Kit (QIAGEN—74104) according to the manufacturer's instructions and perform a double elution with 30 μL of water (Rnase/Dnase free).
- Check the quality of the sgRNA on a 2% agarose gel and with a spectrophotometer.

Microinjection of sgRNA in fish embryos

- Prepare injection solution as follows:

  300 ng/μL sgRNA
  25 μM Cas9 protein (NEB-M0386M or an equivalent homemade protein)
  1X Cas9 10X buffer
  H_2O
- Inject 1 nL of solution into one-cell–stage wild-type embryos.

Estimation of the mutagenesis rate

- Perform genomic extraction on a pool of 25 injected embryos.
- Amplify by PCR the genomic locus containing the sequence targeted by sgRNAs used.
- Clone PCR amplicons into the pCRII-TOPO (Zero Blunt TOPO PCR Cloning Kit, ThermoFisher Scientific—K2800-02) vector and sequence DNA recovered from 20 colonies.
- The frequency of insertion and deletion (indel) mutations at the targeted locus is estimated by comparison with the wild-type allele.

Step 2: Generation of pUAS:CAS9T2AGFP;U6:sgRNA1;U6:sgRNA2 construct

Design of the oligonucleotides for cloning at the TSS of U6 promoters

Two U6 promoters flank the *UAS:CAS9T2AGFP* cassette. They contain at the +1 position restriction sites for BsaI and BsmBI suitable for the cloning of desired target sequences. U6 promoters require transcription to start with a guanine nucleotide (G). Design the oligonucleotides as follows:

BsaI site (Fig. 1C):

For a target sequence starting with a G add the overhang CTC- to the 5′ of the 20-nt target and the AAAC- overhang to the 5′ of the reverse complement of the last 19 nt of the target (removing the final C).

Example:

Target: 5′-GCAGATACACATACAGATAC-3′
Fw oligo: 5′-CTCGCAGATACACATACAGATAC -3′
Rev oligo: 5′-AAACGTATCAGTATGTGTATCTG- 3′

For a target sequence not starting with a G substitute the first nucleotide with a G and proceed as previously described.

Example:

Target: 5′-CAGATACACATACAGATACA-3′
Fw oligo: 5′-CTCGAGATACACATACAGATACA -3′
Rev oligo: 5′-AAACTGTATCAGTATGTGTATCT- 3′

BsmbI site (Fig. 1D):

For a target sequence starting with a G add the overhang AGCTC- to the 5′ of the 20-nt target and the TAAAC- overhang to the 5′ of the reverse complement of the 20-nt target.

Example:

Target: 5′-GCAGATACACATACAGATAC-3′
Fw oligo: 5′-AGCTCGCAGATACACATACAGATAC -3′
Rev oligo: 5′-TAAACGTATCAGTATGTGTATCTGC- 3′

For a target sequence not starting with a G substitute the first nucleotide with a G and proceed as previously described.

Example:

Target: 5′-CAGATACACATACAGATACA-3′
Fw oligo: 5′-AGCTCGAGATACACATACAGATACA -3′
Rev oligo: 5′-TAAACTGTATCAGTATGTGTATCTC- 3′

Oligonucleotides cloning in the pUAS:CAS9T2AGFP;U6:sgRNA1;U6:sgRNA2 vector (Fig. 1C and D)

- Digest 1–2 μg of the p*UAS:CAS9T2AGFP;U6:sgRNA1;U6:sgRNA2* vector with BsaI overnight at 37°C.
- Run the digestion product on a 0.8% agarose gel and extract the linear plasmid with a gel extraction kit.
- Mix 3 μL of 100 μM forward and reverse oligonucleotides containing the target sequence and overhangs for BsaI cloning and add 14 μL of TE buffer.
- Heat the mixture 10 min at 95°C and cool down for 30 min at room temperature to allow the annealing of the oligonucleotides.
- Ligate 2 μL of annealed oligonucleotides and 100 ng of the digested p*UAS:CAS9T2AGFP;U6:sgRNA1;U6:sgRNA2* plasmid (Fig. 1C).

- Transform the reaction in TOP10 competent cells (ThermoFisher Scientific—K2800-02).
- Plate on LB-ampicillin plates
- The next day, pick two colonies from the transformation plate. Inoculate each colony into 2 ml of LB-ampicillin medium and incubate at 37°C overnight with agitation.
- Purify DNA using a miniprep kit
- Sequence the plasmid with T7 specific primer to verify the presence of the target sequence
- Repeat the entire process with BsmBI enzyme using oligonucleotides containing overhangs for BsmBI cloning (Fig. 1D), digesting the plasmid where the first target sequence at the BsaI site has already been inserted.
- Sequence the plasmid for correct insertion of the Sp6 specific primer.

Step 3: Establishing tissue-specific gene inactivation (Fig. 1E and 2)

Microinjection of pUAS:CAS9T2AGFP;U6:sgRNA1;U6:sgRNA2 plasmid

- Choose a specific Gal4 driver line based on the population of cells where gene function needs to be analyzed.
- Set up mating couples of the chosen Gal4 transgenic line.
- Prepare the injection mix as described:
  30 ng/μL of pUAS:CAS9T2AGFP;U6:sgRNA1;U6:sgRNA2 construct
  +50 ng/μL of Tol2 mRNA
  +RNAse, DNAse-free water up to 5 μL
- Inject with p*UAS:CAS9T2AGFP;U6:sgRNA1;U6:sgRNA2* into one-cell–stage embryos from an incross of a *Tg(Tissue-specific:Gal4)* line.
- Screen embryos with GFP-positive cells in the expected Gal4 expression pattern.
- If a stable *UAS:CAS9T2AGFP;U6:sgRNA1;U6:sgRNA2* is needed, raise the embryos to adulthood after injection in wild-type embryos.
- Screen for Fo adults carrying the transgene in the germline and cross them with the chosen Gal4 transgenic line.

1.3 RATIONALE FOR THE DESIGN OF A VECTOR SYSTEM FOR CLONAL ANALYSIS OF MUTANT CELLS BY COMBINING THE Gal4/UAS WITH THE CRE/LOXP SYSTEM

The use of a GFP reporter to mark *Cas9*-expressing cells allows for the phenotypic analysis of potentially mutant cells during the time span of activity of the promoter driving Gal4 transcription. Long-term tracking of these cells needs a strategy to permanently mark the cells displaying a loss-of-function phenotype. To stably label the population of *Cas9*-expressing cells, we replaced the GFP CDS in the p*UAS:CAS9T2AGFP;U6:sgRNA1;U6:sgRNA2* with the ORF of the Cre recombinase (Fig. 1B). Thus, potentially mutant cells can be visualized by using fish lines carrying a transgene where a ubiquitous promoter drives the expression of a fluorescent reporter upon Cre-mediated excision of a floxed STOP sequence. The tissue-specific promoter regulating Gal4 transcription guarantees synchronized translation of Cas9 and Cre enzymes from the same mRNA, ensuring targeted mutagenesis at the chosen genomic locus and expression of the floxed reporter within the same cell. The fluorescence of the reporter will be stable throughout the entire life of the fish and, most importantly, will be inherited in all the cells deriving from the same *Cas9*-expressing progenitor (Fig. 3). Our strategy, named 2C-Cas9 (**C**re-mediated recombination for **C**lonal analysis of **Cas9** mutant cells) offers the possibility to induce site-specific mutagenesis in a selected cell population and, at the same time, enables genetic labeling of the clones derived from the targeted cells (Di Donato et al. *Genome Research accepted*).

1.4 EXPERIMENTAL WORKFLOW

The procedures for the cloning of the target sequences in the plasmid and for the microinjection of the resulting transgenesis vector have been previously described and are summarized in Fig. 1C,D and F. The injection of the p*UAS:CAS9T2ACre; U6:sgRNA1;U6:sgRNA2* plasmid is done in a cross between the selected Gal4 driver line and a transgenic line carrying a floxed allele. Different floxed lines can be used (Di Donato et al. *Genome Research accepted*; Bertrand et al., 2010; Hans et al., 2011; Mosimann et al., 2011).

1.5 STRATEGIES FOR THE DETECTION OF GENE LOSS OF FUNCTION

By using the 2C-Cas9 approach, all the *Cas9*-expressing cells can be identified by the expression of a fluorescent reporter. Nevertheless, the marked cells will represent a heterogeneous population in which each cell (or clone) will have independently induced mutations (some of them might not be out-of-frame mutations). To simplify the statistical analysis of the phenotypes arising from 2C-Cas9-mediated targeted gene disruption it is necessary to evaluate the proportion of mutant cells within the labeled population.

1.5.1 Detection of protein loss in Cas9-expressing cells

The most direct approach to identify the cells carrying *null* mutations is to perform immunohistochemistry (IHC) experiments by using an antibody recognizing the protein encoded by the targeted gene. This strategy requires a double staining to detect simultaneously the fluorescent reporter linked to *Cas9* expression and the protein coded by the targeted locus. The absence of colocalization will reveal efficient gene disruption.

1.5.1.1 Protocol for IHC on whole mount embryos

- Fix embryos in 4% paraformaldehyde in 1X PBS (pH 7.4) for 2 h at room temperature or overnight at 4°C.
- Wash three times in 1X PBS/0.1% Tween-20 (PBS-Tw), for 5 min each time.
- Incubate for 20 min in ice-cold acetone.

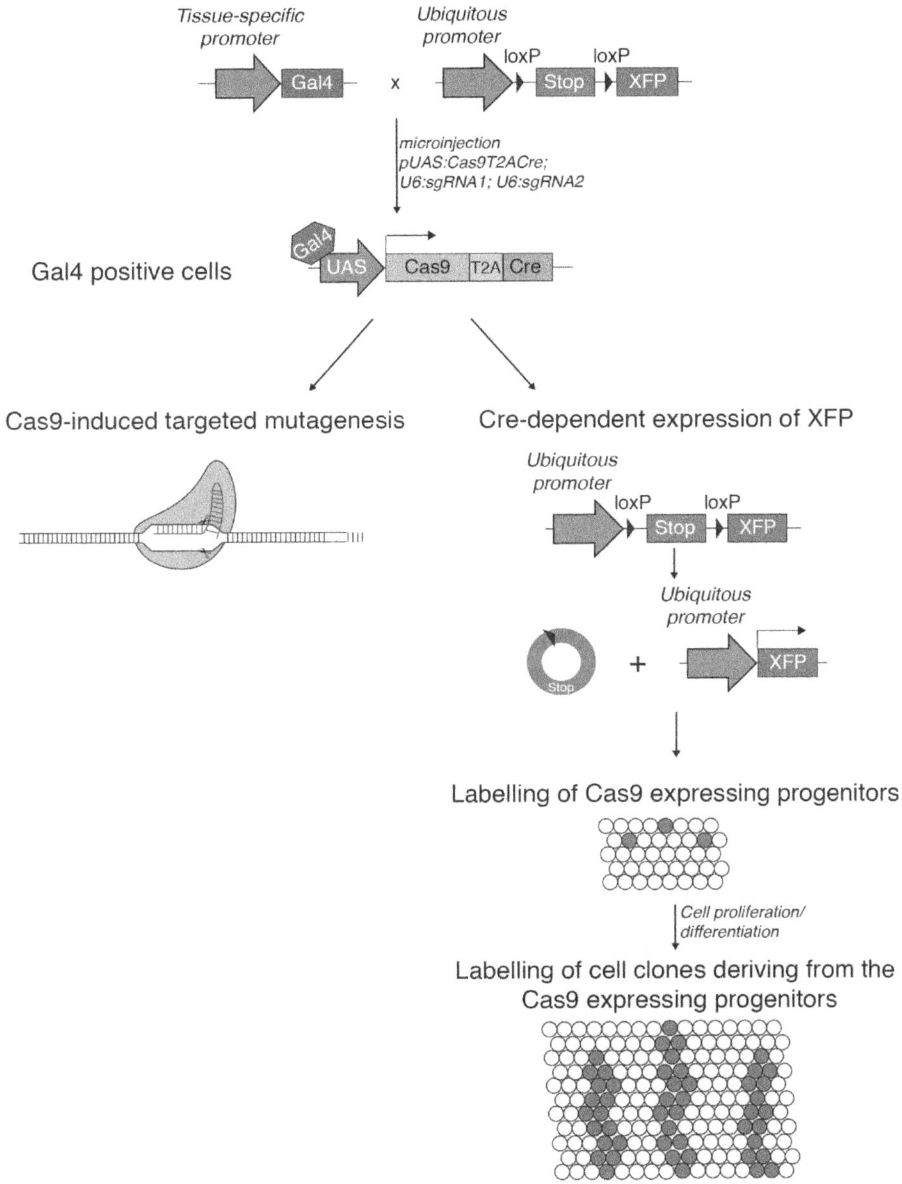

FIGURE 3

Schematic of tissue-specific gene disruption and permanent clonal labeling of *Cas9*-expressing cells after injection of the *pUAS:Cas9T2ACre;U6sgRNA1;U6sgRNA2* plasmid in a cross of a specific Gal4 driver to a floxed reporter line (2C-Cas9). (See color plate)

- Rehydrate in three steps (75%; 50%; 25% acetone in PBS).
- Wash twice in PBS-Tw, 5 min each.
- Incubate with 1 mg/mL collagenase diluted in PBS according to the stage (1dpf, day post fertilization: 10 min; 2 dpf: 20 min; 3 dpf: 75 min; 4 and 5 dpf: 120 min).
- Wash three times in PBS-Tw, 5 min each.
- Incubate for 1 h at room temperature in blocking solution (2% BSA; 0.5% Triton-X100 in PBS).
- Wash three times in PBS-Tw, 5 min each.
- Incubate overnight at 4°C with primary antibodies diluted in blocking solution at the appropriate concentrations.
- Wash three times in PBS-Tw, 5 min each.
- Incubate overnight at 4°C with secondary antibodies diluted in blocking solution at the appropriate concentrations.
- Wash three times in PBS-Tw, 5 min each.
- Mount for imaging in 1% low-melting agarose.

1.5.1.2 Protocol for IHC on cryosections

- Fix embryos in 4% paraformaldehyde in 1X PBS (pH 7.4) for 2 h at room temperature or overnight at 4°C.
- Incubate overnight in 30% sucrose; 0.02% sodium azide; PBS solution.
- Embed the embryos in Tissue-Tek O.C.T. compound after removal of the sucrose and place the resulting blocks on dry ice before sectioning.
- Section the blocks with a thickness of 14 μm and mount the sections on Fisherbrand Superfrost plus slides (No 12-550-15).
- Wash twice in PBS-Tw, 5 min each time.
- Incubate for 1 h at room temperature in blocking solution (10% normal goat serum in PBS-Tw).
- Incubate overnight at 4°C with primary antibodies diluted in blocking solution.
- Wash three times in PBS-Tw, 5 min each.
- Incubate for 2 h with secondary antibodies diluted in blocking solution.
- Wash five times in PBS-Tw, 5 min each.
- Add Vectashield drops on the sections and place coverslips on the slides.

1.5.2 Molecular assessment of the mutagenesis efficiency via fluorescence-activated cell sorting and genome locus sequencing

The molecular assessment of the mutagenesis efficiency of the 2C-Cas9 system in the selected Gal4 line requires the isolation of the fluorescent *Cas9*-expressing cells. Fluorescence-activated cell sorting (FACS) might be used to isolate the marked *Cas9*-expressing cells from the wild-type nonfluorescent cell population. After separation, total DNA can be extracted from the two pools of cells and the targeted locus can be analyzed to screen for the presence and the frequency of out-of-frame mutations.

1.5.2.1 Protocol for FACS

- Dissociate the embryos ($n > 200$) as previously described by Manoli & Driever (2012).
- Perform cell sorting and collect cells in lysis buffer (E.G. NucleoSpin Tissue kit; Macherey–Nagel—740952.10).
- Extract genomic DNA following kit instructions.
- PCR amplify the targeted genomic loci.
- Clone PCR amplicons into the pCR-bluntII-TOPO vector.
- Isolate plasmid DNA from single colonies.
- Sequence the plasmids and identify mutant alleles by comparison with the wild-type sequence.

2. DISCUSSION

In this chapter we describe a new tool allowing, with a single plasmid injection, both conditional mutagenesis and lineage tracing of potentially mutant cells. The flexibility of the approach is based on the use of the Gal4/UAS system to drive *Cas9* expression. Many Gal4 driver transgenic lines showing diverse cell-type–specific expression patterns are available to the zebrafish community. The transient or stable UAS-dependent expression of *Cas9,* coupled with constitutive transcription of gene-specific sgRNAs, allows the inactivation of any gene of interest in any Gal4 line. Therefore, gene disruption can be achieved in multiple tissues by generating a unique transgenesis vector, containing gene-specific sgRNA, to be injected into different Gal4 driver lines. Unambiguous identification of cells displaying loss-of-function phenotypes is provided by the expression of a reporter gene (GFP or Cre) together with Cas9 endonuclease. While GFP fluorescence enables visualization of *Cas9*-expressing cells during the activity time window of the chosen Gal4 line, the use of Cre allows permanent labeling and clonal analysis of potentially mutated cells even after *Gal4* expression has terminated. Both methods represent a valuable alternative to laborious transplantation experiments that represent, to date, the only method to analyze cell-autonomous behavior of mutant cells in a wild-type genetic background. Furthermore, the analysis of genetic chimeras resulting from transplantations is restricted to early phenotypes and does not allow the investigation of the role of genes and pathways that are used repeatedly during development. If well established, Cre/loxP-based approaches for conditional gene inactivation in zebrafish would overcome these issues. Nevertheless they would need long generation time and are currently limited by the low efficiency of homologous recombination-based genomic manipulations.

Although the system described here is useful to score conditional loss-of-function phenotypes by using genetic fluorescent labeling, it does not mark exclusively populations of cells carrying *null* alleles but all *Cas9*-expressing cells. Because this is a mixed population of cells, only a percentage will harbor out-of-frame mutations.

We propose to discriminate mutant from wild-type cells by performing immunofluorescence experiments to detect loss of protein by the targeted gene at a single-cell level. Alternatively, molecular analysis of gene disruption can be performed, after DNA extraction from FACS-sorted fluorescent cells, to estimate the proportion of truly mutant cells within the labeled population.

Several parameters need to be considered to obtain highly penetrant tissue-specific phenotypes by using the 2C-Cas9 system. The selection of sgRNA displaying efficient mutagenesis rates is crucial to successfully induce DSBs at the target genomic locus. In addition, to achieve loss of function in the cases where indel mutations do not disrupt the ORF of the targeted gene, designing of sgRNAs targeting essential functional domains of a protein can be advantageous. The injection of our vector system into a heterozygous mutant background, where one allele carries a *null* mutation, represents another way to enhance the efficiency of gene disruption. Furthermore, the use of a strong Gal4 driver line, leading to a high intracellular expression of Cas9 endonuclease levels, is recommended. In the future, additional improvements in the *Cas9* expression cassette may be needed. A possibility could be the incorporation into our vector system of noncoding elements of the zebrafish genome that have been shown to boost expression of transgenes in UAS-based plasmids (Horstick et al., 2015).

The strategy presented in this chapter can be widely applied for the analysis of gene function in single cells or in clonal populations, from larval stage to adulthood. The genetic labeling of potentially mutant cells generated by the 2C-Cas9 method provides a versatile technique to correlate phenotype and genotype, a challenge that cannot be addressed with other currently available genetic tools in zebrafish.

REFERENCES

Ablain, J., Durand, E. M., Yang, S., Zhou, Y., & Zon, L. I. (2015). A CRISPR/Cas9 vector system for tissue-specific gene disruption in zebrafish. *Developmental Cell, 32*(6), 756–764. http://dx.doi.org/10.1016/j.devcel.2015.01.032.

Asakawa, K., & Kawakami, K. (2008). Targeted gene expression by the Gal4-UAS system in zebrafish. *Development, Growth & Differentiation, 50*(6), 391–399. http://dx.doi.org/10.1111/j.1440-169X.2008.01044.x. pii:DGD1044.

Auer, T. O., & Del Bene, F. (2014). CRISPR/Cas9 and TALEN-mediated knock-in approaches in zebrafish. *Methods*. http://dx.doi.org/10.1016/j.ymeth.2014.03.027.

Bedell, V. M., Wang, Y., Campbell, J. M., Poshusta, T. L., Starker, C. G., Krug, R. G., 2nd, … Ekker, S. C. (2012). In vivo genome editing using a high-efficiency TALEN system. *Nature, 491*(7422), 114–118. http://dx.doi.org/10.1038/nature11537.

Bertrand, J. Y., Chi, N. C., Santoso, B., Teng, S., Stainier, D. Y., & Traver, D. (2010). Haematopoietic stem cells derive directly from aortic endothelium during development. *Nature, 464*(7285), 108–111. http://dx.doi.org/10.1038/nature08738.

Bouabe, H., & Okkenhaug, K. (2013). Gene targeting in mice: a review. *Methods in Molecular Biology, 1064*, 315–336. http://dx.doi.org/10.1007/978-1-62703-601-6_23.

Chen, F., Rosiene, J., Che, A., Becker, A., & LoTurco, J. (2015). Tracking and transforming neocortical progenitors by CRISPR/Cas9 gene targeting and piggyBac transposase lineage labeling. *Development, 142*(20), 3601–3611. http://dx.doi.org/10.1242/dev.118836.

Choi, C. M., Vilain, S., Langen, M., Van Kelst, S., De Geest, N., Yan, J., … Hassan, B. A. (2009). Conditional mutagenesis in *Drosophila. Science, 324*(5923), 54. http://dx.doi.org/10.1126/science.1168275.

Feil, R., Brocard, J., Mascrez, B., LeMeur, M., Metzger, D., & Chambon, P. (1996). Ligand-activated site-specific recombination in mice. *Proceedings of the National Academy of Sciences of the United States of America, 93*(20), 10887–10890. Retrieved from http://www.ncbi.nlm.nih.gov/pubmed/8855277.

Golic, K. G., & Lindquist, S. (1989). The FLP recombinase of yeast catalyzes site-specific recombination in the *Drosophila* genome. *Cell, 59*(3), 499–509. Retrieved from http://www.ncbi.nlm.nih.gov/pubmed/2509077.

Griffin, R., Binari, R., & Perrimon, N. (2014). Genetic odyssey to generate marked clones in *Drosophila* mosaics. *Proceedings of the National Academy of Sciences of the United States of America, 111*(13), 4756–4763. http://dx.doi.org/10.1073/pnas.1403218111.

Hans, S., Freudenreich, D., Geffarth, M., Kaslin, J., Machate, A., & Brand, M. (2011). Generation of a non-leaky heat shock-inducible Cre line for conditional Cre/lox strategies in zebrafish. *Developmental Dynamics, 240*(1), 108–115. http://dx.doi.org/10.1002/dvdy.22497.

Horstick, E. J., Jordan, D. C., Bergeron, S. A., Tabor, K. M., Serpe, M., Feldman, B., & Burgess, H. A. (2015). Increased functional protein expression using nucleotide sequence features enriched in highly expressed genes in zebrafish. *Nucleic Acids Research, 43*(7), e48. http://dx.doi.org/10.1093/nar/gkv035.

Kuhn, R., Schwenk, F., Aguet, M., & Rajewsky, K. (1995). Inducible gene targeting in mice. *Science, 269*(5229), 1427–1429. Retrieved from http://www.ncbi.nlm.nih.gov/pubmed/7660125.

Manoli, M., & Driever, W. (2012). Fluorescence-activated cell sorting (FACS) of fluorescently tagged cells from zebrafish larvae for RNA isolation. *Cold Spring Harbor Protocols, 2012*(8). http://dx.doi.org/10.1101/pdb.prot069633.

Mosimann, C., Kaufman, C. K., Li, P., Pugach, E. K., Tamplin, O. J., & Zon, L. I. (2011). Ubiquitous transgene expression and Cre-based recombination driven by the ubiquitin promoter in zebrafish. *Development, 138*(1), 169–177. http://dx.doi.org/10.1242/dev.059345.

Nakazawa, N., Taniguchi, K., Okumura, T., Maeda, R., & Matsuno, K. (2012). A novel Cre/loxP system for mosaic gene expression in the *Drosophila* embryo. *Developmental Dynamics, 241*(5), 965–974. http://dx.doi.org/10.1002/dvdy.23784.

Ni, T. T., Lu, J., Zhu, M., Maddison, L. A., Boyd, K. L., Huskey, L., … Chen, W. (2012). Conditional control of gene function by an invertible gene trap in zebrafish. *Proceedings of the National Academy of Sciences of the United States of America, 109*(38), 15389–15394. http://dx.doi.org/10.1073/pnas.1206131109.

Platt, R. J., Chen, S., Zhou, Y., Yim, M. J., Swiech, L., Kempton, H. R., … Zhang, F. (2014). CRISPR-Cas9 knockin mice for genome editing and cancer modeling. *Cell, 159*(2), 440–455. http://dx.doi.org/10.1016/j.cell.2014.09.014.

Port, F., Chen, H. M., Lee, T., & Bullock, S. L. (2014). Optimized CRISPR/Cas tools for efficient germline and somatic genome engineering in *Drosophila. Proceedings of the National Academy of Sciences of the United States of America, 111*(29), E2967–E2976. http://dx.doi.org/10.1073/pnas.1405500111.

Rossant, J., & Spence, A. (1998). Chimeras and mosaics in mouse mutant analysis. *Trends in Genetics, 14*(9), 358–363. Retrieved from http://www.ncbi.nlm.nih.gov/pubmed/9769731.

Technau, G., & Campos-Ortega, J. (1986). Lineage analysis of transplanted individual cells in embryos of *Drosophila melanogaster. Roux's Archives of Developmental Biology, 195*(8), 489–498. http://dx.doi.org/10.1007/BF00375889.

Theodosiou, N. A., & Xu, T. (1998). Use of FLP/FRT system to study *Drosophila* development. *Methods, 14*(4), 355–365. http://dx.doi.org/10.1006/meth.1998.0591.

Yang, H., Wang, H., Shivalila, C. S., Cheng, A. W., Shi, L., & Jaenisch, R. (2013). One-step generation of mice carrying reporter and conditional alleles by CRISPR/Cas-mediated genome engineering. *Cell, 154*(6), 1370–1379. http://dx.doi.org/10.1016/j.cell.2013.08.022.

Yin, L., Maddison, L. A., Li, M., Kara, N., LaFave, M. C., Varshney, G. K., … Chen, W. (2015). Multiplex conditional mutagenesis using transgenic expression of Cas9 and sgRNAs. *Genetics, 200*(2), 431–441. http://dx.doi.org/10.1534/genetics.115.176917.

Zong, H., Espinosa, J. S., Su, H. H., Muzumdar, M. D., & Luo, L. (2005). Mosaic analysis with double markers in mice. *Cell, 121*(3), 479–492. http://dx.doi.org/10.1016/j.cell.2005.02.012.

CHAPTER

Tissue-specific gene targeting using CRISPR/Cas9

9

J. Ablain[1], L.I. Zon

Howard Hughes Medical Institute and Harvard Medical School, Boston, MA, United States
[1]Corresponding author: E-mail: jablain@enders.tch.harvard.edu

CHAPTER OUTLINE

Abstract

The zebrafish has been a powerful model in forward genetic screens to identify genes essential for organogenesis and embryonic development. Conversely, using reverse genetics to investigate specific gene function requires phenotypic analysis of complete gene inactivation. Despite the availability and efficacy of morpholinos, the lack of tractable and efficient knockout technologies has impeded reverse genetic studies in the zebrafish, particularly in adult animals. The recent development of genome-editing technologies such as CRISPR/Cas9 greatly widened the scope of loss-of-function studies in the zebrafish, allowing for the rapid phenotypic assessment of gene silencing in embryos, the generation of knockout lines, and large-scale reverse genetic screens. Tissue-specific gene inactivation would be ideal for these studies given the caveats of whole-embryo gene silencing, yet spatial control of gene targeting remains a challenge. In this chapter,

Methods in Cell Biology, Volume 135, ISSN 0091-679X, http://dx.doi.org/10.1016/bs.mcb.2016.03.004

we focus on tissue-specific gene inactivation using the CRISPR/Cas9 technology. We first explain the rationale for this technique, including some of its potential applications to tackle important biological issues and the inability of current technologies to address these issues. We then present a method to target genes in a tissue-specific manner in the zebrafish. Finally, we discuss technical difficulties and limitations of this method as well as possible future developments.

INTRODUCTION

Zebrafish can generate large numbers of embryos that develop quickly and externally, and are therefore easy to analyze phenotypically. The zebrafish has thus been a vertebrate model of choice to perform forward genetic screens that have led to the identification of numerous genes regulating organogenesis (Driever et al., 1996; Haffter et al., 1996; Mullins, Hammerschmidt, Haffter, & Nusslein-Volhard, 1994). Reverse genetic screens as a complementary approach have been hindered by the lack of high-throughput gene-silencing technology in the zebrafish. Although morpholino antisense oligomers (morpholinos) have been used extensively to analyze loss-of-function phenotypes during early stages of embryogenesis, they present important limitations and have rarely been used for large scale screens.

The CRISPR/Cas9 technology of genome editing has opened new vistas for loss-of-function studies in zebrafish. CRISPR/Cas9 is an adaptive immune response developed by bacteria and archae to fight phage invasion through RNA-based recognition and subsequent cleavage of foreign DNA sequences (Barrangou et al., 2007; Brouns et al., 2008; Garneau et al., 2010). Deciphering this molecular mechanism rapidly led to the realization that it could be applied to genome editing (Jinek et al., 2012). Engineered short guide RNAs (gRNA) are comprised of a 20-nucleotide 5′ component (referred to as the seed) complementary to the target DNA sequence, and of a 3′ motif that forms a secondary structure capable of interacting with CRISPR-associated 9 (Cas9) endonuclease. Cas9 loads the gRNA and scans the genome in search of target sequences complementary to the seed (Sternberg, Redding, Jinek, Greene, & Doudna, 2014). When recognized sequences are followed by a proto-spacer adjacent motif (PAM), Cas9 cleaves DNA within the seed. The resulting double-strand breaks are resolved by the cell, most likely through alternative nonhomologous end joining, which leads to mutations, insertions, or deletions (indels). When occurring in coding sequences, these indels frequently disrupt the targeted gene by introducing frameshifts and early stop codons. Note however that statistically, a third of indels leave the open reading frame unaltered and may therefore not affect gene function. While the CRISPR/Cas9 technology was initially used to generate knockout cell or animal lines (Cho, Kim, Kim, & Kim, 2013; Cong et al., 2013; Hwang et al., 2013; Jiang, Bikard, Cox, Zhang, & Marraffini, 2013; Mali et al., 2013), its applications now range from the fine modulation of gene expression (Gilbert et al., 2013) to genome-wide genetic screens in vitro (Koike-Yusa, Li, Tan, Velasco-Herrera Mdel, & Yusa, 2014; Wang, Wei,

Sabatini, & Lander, 2014; Zhou et al., 2014) and in vivo (Chen et al., 2015). The CRISPR technology has also joined the fight against viruses with strategies to target essential viral genes or cell surface receptors involved in viral entry, and against genetic disorders to correct disease-causing mutations.

In the zebrafish, the CRISPR/Cas9 technology not only allows us to generate mutant lines (Hwang et al., 2013) but also to observe phenotypes rapidly in vivo by direct injection of a gRNA and *Cas9* mRNA into one-cell stage embryos (Jao, Wente, & Chen, 2013). Several reports have also established knockin zebrafish lines with this technology (Auer, Duroure, De Cian, Concordet, & Del Bene, 2014; Hisano et al., 2015; Kimura, Hisano, Kawahara, & Higashijima, 2014; Li et al., 2015). Its ease of use, low cost, and the possibility of multiplexing make it a prime tool for large-scale reverse genetic screens in the zebrafish (Shah, Davey, Whitebirch, Miller, & Moens, 2015). In many cases, the investigation of gene function in vivo requires the spatiotemporal control of gene silencing. Here, we present the rationale for tissue-specific gene inactivation in zebrafish using the CRISPR/Cas9 technology and detail a method to achieve it. We also discuss some limitations and potential future developments of this technique.

1. RATIONALE

1.1 NEED FOR TISSUE-SPECIFIC TUNING OF GENE EXPRESSION

Embryonic lethality represents a major hurdle when investigating gene function in vivo. Indeed, many genes play essential roles in the proper development of an organ, and animals with mutations in these genes are not viable. In zebrafish it has been estimated that about 1400 genes are embryonic lethal (Amsterdam et al., 2004). In these cases, gene knockout in the whole organism may not be informative, requiring the development of alternative experimental approaches.

Tissue-specific gene targeting can readily address cell autonomy questions, whereas other methods are generally complex and/or indirect. For example, to prove that a gene or a pathway affects a function (eg, proliferation, self-renewal, and survival) directly in a specific cell type (eg, hematopoietic stem cell) and not via an effect on neighboring cells (eg, stromal cells), it is necessary to target the gene or pathway of interest specifically in that cell type. Transplantation between different transgenic animals or parabiosis are methods to address cell autonomy, but they present a number of technical challenges and biological caveats. Specifically inactivating a gene in different cell types allows identification of the precise cell type responsible for a phenotype observed in a knockout line or induced by systemic drug treatment. Additionally, to look for modulators of a specific trait in vivo, it may be more informative to inactivate genes specifically in the corresponding tissue.

Precise disease modeling may also require the tissue-specific inactivation of a gene. While some conditions such as muscular dystrophies are generally caused by inherited mutations and can therefore be appropriately modeled by full gene knockouts, most cancers arise from somatic mutations only present in the neoplastic tissue. The inactivation of tumor-suppressor genes can alter cell homeostasis in many different cell types. For instance, p53 mutant zebrafish predominantly develops malignant peripheral nerve sheath tumors over time (Berghmans et al., 2005). It may therefore be preferable to target tumor-suppressor genes in a tissue-restricted manner to accurately model malignant transformation in a particular organ.

1.2 LIMITATIONS OF EXISTING GENE-SILENCING METHODS

In the absence of an established knockout line for a specific gene, or if the knockout is embryonic lethal at a late stage of development, it is possible to transiently knock-down a gene to assess loss-of-function in embryos. Equivalent to the mammalian shRNA technology, morpholinos are injected into zebrafish embryos to silence a gene of interest and study its function during early development (Nasevicius & Ekker, 2000). Morpholinos are nucleotide analogs that recognize and bind short sequences (about 25 nucleotides) at the transcription start site or at splice sites of pre-mRNAs, and thus block the translation or proper splicing of the mRNA (Summerton & Weller, 1997). They can be injected into zebrafish zygotes to inhibit gene expression. Due to its ease of use and remarkable efficacy, this technology has played an instrumental role in the development of the zebrafish as a model for cell biology studies. Morpholinos have also been useful in reverse genetic screens (Huang et al., 2013). Nonspecific toxicity and off-target effects limited the utility of morpholinos and forced the zebrafish community to adopt a series of guidelines to adequately control morpholino experiments (Eisen & Smith, 2008). In addition, morpholinos function in embryos for about 3 days, which limits loss-of-function analyses to the early stages of development and precludes studies in adults. Finally, the cost of morpholino synthesis practically prevents systematic, large-scale screens.

Zinc-finger nucleases (ZFN) and transcription activator-like effector nucleases (TALEN) are enzymes engineered to recognize and cleave specific DNA sequences (Miller et al., 2011; Urnov et al., 2005). Once expressed in zebrafish embryos, they induce double-strand breaks at their target sites in the genome, which are resolved by nonhomologous end joining (Bedell et al., 2012; Doyon et al., 2008; Meng, Noyes, Zhu, Lawson, & Wolfe, 2008; Sander et al., 2011). Resulting indels often disrupt gene function. The mutations are heritable, which allows for the generation of stable knockout lines. These genome-editing technologies enable loss-of-function studies both in embryos and adults, thus complementing the transient knockdown approaches through morpholinos. Note that the phenotypes observed by acute gene down-regulation with morpholinos or permanent gene knockout with genome-editing technologies can be discordant (Kok et al., 2015), partially due to compensation mechanisms (Rossi et al., 2015), raising questions about the respective significance of either approach. Targeting a new gene with ZFN or TALEN requires the design of a new nuclease, and the mutation efficiency is generally low, which renders the intensive use of these technologies challenging.

In contrast, the CRISPR/Cas9 technology uses a single endonuclease, Cas9, to target specific genomic sequences in an RNA-dependent manner (Jinek et al., 2012). The co-injection of *Cas9* mRNA or protein and a short guide RNA (gRNA) recognizing a 20-bp genomic sequence into one-cell stage zebrafish embryos yields frequent indels at the target locus (Hwang et al., 2013). The relative ease of use and efficiency of this technology have made it a prime tool to assess loss-of-function phenotypes in injected embryos and to establish mutant lines quickly. It does not allow, however, for the spatiotemporal control of gene inactivation.

Of note, alternatives to direct gene editing have been developed to perform tissue-specific loss-of-function studies. The conditional Cre/lox technology has been adapted to the zebrafish for lineage tracing experiments (Mosimann et al., 2011) and could serve the same purpose as in mice (ie, spatiotemporal control of gene knockout) but the generation of floxed alleles is complex and time consuming. Interestingly, tissue-specific overexpression of dominant-negative mutants has also been used to circumvent the lack of spatial control of gene inactivation. However, gene silencing remains the method of choice to study gene function because it most faithfully recapitulates physiologic expression levels.

2. METHODS

Here we detail a method based on the CRISPR/Cas9 technology to target genes in a tissue-specific fashion in the zebrafish (Ablain, Durand, Yang, Zhou, & Zon, 2015). Developments and alternatives to this method are discussed in the next section.

2.1 IDENTIFICATION OF CRISPR TARGET SITES

The first step consists of identifying gRNAs that efficiently mutate the gene of interest. This can be done following published protocols.

Prediction algorithms such as CHOPCHOP (https://chopchop.rc.fas.harvard.edu/) are based on rules inferred from in vivo measurement of gRNA efficiency and can be used to design gRNAs (Montague, Cruz, Gagnon, Church, & Valen, 2014; Moreno-Mateos et al., 2015). The presence of a G as the last base of the seed sequence and a high GC content seem to be the two most important rules. The number of predicted off-target sites is also taken into account in the ranking of potential target sequences. The choice of the first two bases will determine the type of in vitro transcription enzyme needed in the next step. We recommend picking target sequences that start with a G since the U6 promoter used in the tissue-specific vector (see Section 2.2) requires a G as a transcription start site. If the proposed target sequence does not start with GG or GA, it is possible to replace the second base by a G or an A since the presence of mismatches at the 5′ end of the gRNA is generally well tolerated by Cas9 for the recognition of the cleavage site (Fu et al., 2013). Depending on the structure of the gene of interest, it may or may not be suitable to target early exons preferentially. In general, if a gene contains known functional domains (eg, DNA-binding or catalytic or ligand-binding domains), we recommend targeting these in priority. In our experience, about a third of gRNAs designed following these guidelines show a good targeting efficiency, although we and others have observed significant gene-dependent variability. Designing at least three gRNAs per gene may therefore be warranted to find an efficient seed sequence.

Once potential CRISPR target sequences have been chosen, a DNA template for in vitro transcription can be prepared either by cloning the seed sequence into a vector that contains both a T7 or SP6 promoter and the constant 3′ part of the gRNA (Hwang et al., 2013), or by T4 DNA polymerase (or PCR) synthesis following annealing of two long oligonucleotides (Gagnon et al., 2014) (Fig. 1). The purified DNA template can then be transcribed in vitro into a gRNA using either T7 (for target sequences starting with GG) or SP6 (for sequences starting with GA) RNA polymerases.

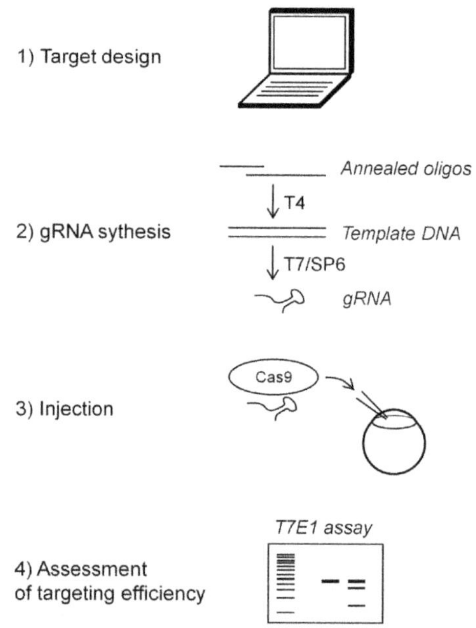

FIGURE 1

CRISPR sequences against any gene of interest can be identified in silico through published algorithms. gRNAs can then be prepared by in vitro transcription (using T7 or SP6 RNA polymerases) from DNA templates comprising an RNA polymerase promoter, the gene-specific seed sequence, and the sequence encoding the constant 3′ part of the gRNA. Purified gRNAs can be injected into single-cell zebrafish embryos along with Cas9 mRNA or protein. Finally, genomic DNA can be extracted from injected embryos and analyzed for CRISPR-induced mutations, for example, by direct sequencing, HRM, or T7E1 mutagenesis assay.

After the gRNAs have been purified (on column or by ammonium acetate precipitation), their quality can be controlled by gel electrophoresis or using a fragment analyzer. The presence of shorter transcripts is expected due to impurity of the oligonucleotides used to synthesize the DNA template and to incomplete RNA elongation, but the major band on the gel has to correspond to the full-size transcript. gRNAs can then be co-injected into one-cell stage zebrafish embryos with Cas9 mRNA or protein. The reported gRNA concentration in the injection mix greatly varies in the literature (from 12.5 ng/μL to almost 1 μg/μL), but it seems that even very high doses show limited toxicity to the embryos while the impact of concentration on targeting efficiency is still debated. *Cas9* mRNA can be easily synthesized in vitro from plasmid DNA (Jao et al., 2013) and injected at 300 ng/μL, while Cas9 protein can be produced (Gagnon et al., 2014) or purchased and used at 500 ng/μL.

Genomic DNA can be extracted from embryos 24–48 h post-injection, and several methods can be used to assess mutagenesis efficiency including: enzymatic methods (restriction, T7E1, or Surveyor assays (Kim, Lee, Kim, Cho, & Kim, 2009; Qiu et al., 2004)), PCR-based methods (High Resolution Melt (HRM) (Parant, George, Pryor, Wittwer, & Yost, 2009), direct PCR in the case of expected deletions between two CRISPR target sites), or sequencing of the CRISPR target site. Only the latter is truly quantitative and gives access to the full range of mutations. Enzymatic methods are arguably cheaper and easier to perform, but also more susceptible to technical variability. In our experience, about half of gRNAs display poor targeting efficiency (less than 10% of mutated alleles, which is generally under the detection threshold of enzymatic methods), and a third of gRNAs target 20% or more of alleles. Only the latter category should be further considered for loss-of-function studies.

2.2 CONSTRUCTION OF VECTORS

Here we describe vectors based on the Tol2 transposase technology (Kawakami et al., 2004) that can be injected into zebrafish embryos along with *Tol2* mRNA for integration into the genome. Other vectors (eg, viral vectors) can be constructed on the same model. Gateway technology (Hartley, Temple, & Brasch, 2000) or regular cloning methods can easily insert a *Cas9* sequence codon-optimized for zebrafish (Jao et al., 2013) under the control of a tissue-specific promoter (Fig. 2). It is possible to adapt a vector already containing a transgenesis marker (eg, *cmlc2:GFP* that yields green fluorescent hearts) or to add GFP linked to Cas9 via a 2A peptide in order to visualize the cells that express the vector in vivo. Once efficient CRISPR seed sequences have been identified, they can be cloned under the control of a zebrafish U6 promoter at the 5′ end of a gRNA scaffold (Ablain et al., 2015). A protocol for gRNA and *Cas9* cloning into Tol2 vectors is detailed in (Ablain et al., 2015).

Note: In some cases, it may be possible to skip the identification of efficient gRNAs (Section 2.1) and either identify them after injection of the vectors (this requires mutation detection in the targeted tissue) or pool several vectors directed against the same gene to statistically increase targeting.

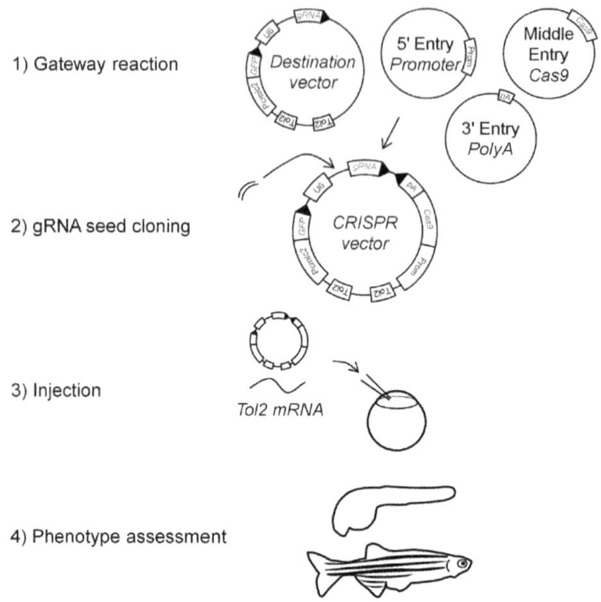

FIGURE 2

A Gateway reaction allows to assemble a *Cas9* sequence under the control of a tissue-specific promoter and followed by polyA into a backbone containing a U6:gRNA cassette and possibly a transgenesis marker. One can then clone any gene-specific seed sequence into the U6:gRNA cassette of the resulting vector. This tissue-specific CRISPR vector can be injected with Tol2 mRNA into one cell—stage embryos that will later be submitted to phenotype evaluation. (See color plate)

2.3 INJECTION, EVALUATION OF MOSAICISM, AND PHENOTYPE ASSESSMENT

The generated CRISPR vectors are then integrated into the zebrafish genome by direct co-injection with *Tol2* mRNA (20 ng/μL) into one-cell stage embryos. The DNA concentration in the injection mix varies according to the size of the plasmid and is typically between 10 and 25 ng/μL. Depending on the transgenesis marker used, mosaicism can be approximated by the ratio of fluorescent cells to the total number of cells in the tissue of interest. For example, in the case of the vector expressing GFP under the control of the heart-specific promoter cmlc2, the proportion of the heart that displays green fluorescence at 48 hpf provides an estimate of mosaicism. When using a *Cas9-T2A-GFP* construct, injection into a transgenic line expressing a different fluorescent protein in the tissue of interest, or comparison of injected embryos with embryos of such a transgenic line, may allow a better assessment of mosaicism.

We showed that the zebrafish U6-3 promoter drives early, continuous, and ubiquitous expression of gRNAs (Ablain et al., 2015). Thus, the regulation and specificity of gene targeting stem from *Cas9* expression. Theoretically, the consequences of gene inactivation can be observed shortly after the tissue-specific promoter has turned on. Indeed, *Cas9* transcription and translation; subsequent gene targeting; and the decrease in mRNA and protein levels of the targeted gene presumably all occur within a few hours. Some phenotypes may be noticed at the level of the whole embryo (eg, survival and defects in organ development). Phenotypes at the cellular (eg, proliferation and migration) or molecular levels (eg, gene expression and pathway activation) may be more difficult to detect because of the mosaicism of vector expression in F0-injected embryos. To overcome this issue, it may be possible to image or analyze the cells that express the *Cas9-T2A-GFP* construct in situ, or sort them by fluorescence-activated cell sorting (FACS) and proceed with molecular assays.

3. DISCUSSION

3.1 TECHNICAL CAVEATS AND TROUBLESHOOTING

The method for tissue-specific gene targeting described earlier presents several caveats that may result in poor gene targeting.

One of the main potential limitations resides in the tissue-specific promoter used to drive *Cas9* expression. Weak promoters may not produce enough Cas9 in order to recognize and cleave the CRISPR target. Measures of *Cas9* mRNA levels by whole mount in situ hybridization or qPCR in the tissue of interest may provide insights into promoter activity.

It remains unclear whether the intrinsic quality of gRNAs (ie, their efficiency at inducing mutations at the CRISPR site after co-injection with Cas9 mRNA or protein into one-cell embryos) impacts targeting efficiency by the vector. Yet, it is conceivable that the continuous production of *Cas9* mRNA and gRNA by the vector will force targeting over time, normalizing the efficiency of different gRNAs. Along the same line, the vector may favor homozygous over heterozygous targeting in each cell, thus circumventing one of the issues associated with transient Cas9 injection. Single-cell sequencing of the CRISPR site would be required to evaluate whether one or both alleles have been hit.

If the CRISPR vector fails to induce mutations at the target locus, injection of the vector with either the gRNA itself or *Cas9* mRNA, and subsequent assessment of targeting may help determine whether the problem stems from the gRNA or from the promoter driving *Cas9* expression. Injection of a vector producing the same gRNA and expressing *Cas9* under a ubiquitous promoter such as ubi (Mosimann et al., 2011) may also prove informative in that case.

A consequence of continual vector targeting is the higher likelihood of off-target effects. While the issue of off-target effects by the CRISPR technology is still debated (Fu et al., 2013; Hsu et al., 2013; Kuscu, Arslan, Singh, Thorpe, & Adli, 2014; Wang et al., 2015), a clear assessment of the number and frequency of off-targets may be warranted in the case of the tissue-specific CRISPR vectors.

3.2 PHENOTYPES IN F0 ANIMALS VERSUS STABLE LINES

The mosaicism associated with Tol2-mediated insertion of the vectors may also be an issue. Indeed, some phenotypes may require a high frequency of gene inactivation to be detected. For example, deleterious phenotypes that lead to the death or proliferation arrest of the targeted cells may be masked by the growth of cells in the same tissue that do not express the vector. This problem is exacerbated by the semirandom nature of DNA double-strand break resolution by alternative nonhomologous end-joining mechanisms, which induce insertions and deletions at the CRISPR target site that can either result in frameshift or retain the reading frame of the gene. In some cells, CRISPR targeting may thus lead to "silent" mutations that leave a functional copy of the gene (eg, point mutations or short inframe indels). These cells, although expressing the vector, may escape the deleterious consequences of gene inactivation and obscure the phenotype. One way to minimize this risk is to design gRNAs against domains essential for the activity of the corresponding protein, so that even in-frame mutations may disrupt gene function. Additionally, it is possible to generate large deletions of entire exons by combining several gRNAs directed against the same gene.

To circumvent the impact of mosaicism, it may be beneficial to grow injected F0 embryos up to adulthood and cross them to establish a line of zebrafish stably expressing the vector. This would allow virtually every cell in the tissue of interest to target the gene. In addition, this would eliminate the ectopic expression occasionally observed with Tol2 vectors in F0. The main drawbacks of this method are the time required to generate stable lines, which makes it unsuitable for a screen (particularly if the phenotype is to be observed in adults), and the possible silencing of Tol2 vectors in the F1 generation, which complicates the selection of founders. In addition, due to the relatively large size of the CRISPR vectors, germline transmission rates may be low. Taken together, these considerations point to the importance of adapting the method according to the expected phenotype.

3.3 DEVELOPMENTS AND ALTERNATIVES

The CRISPR vector system described earlier provides a modular platform for a range of developments and improvements. It can easily be adapted for use in species other than the zebrafish by modifying the codon-optimized *Cas9* and U6 promoter sequences. Note that systems for tissue-specific gene inactivation have also been developed in fly and worm (Shen et al., 2014; Xue et al., 2014). We have also added a second U6:gRNA cassette in the vector backbone to allow for multiplexing. This is particularly useful to perform reverse genetic studies and analyze the loss-of-function phenotypes of human genes in the zebrafish since many human genes have more than one zebrafish orthologue (due to the whole-genome duplication that occurred in the ray-finned fish lineage after its separation from the tetrapods (Taylor, Braasch, Frickey, Meyer, & Van de Peer, 2003)).

The power of the Gal4-UAS system (Brand & Perrimon, 1993) may also be harnessed for tissue-specific gene targeting. Indeed, a vector containing a U6:gRNA cassette and *Cas9* under the control of a UAS element can be injected into a stable zebrafish line expressing Gal4 in a specific tissue. Alternatively, a stable line can be generated with the Cas9/gRNA vector and crossed with the Gal4 line. This method may circumvent problems linked to the strength of the tissue-specific promoters used to drive *Cas9* expression.

Finally, the applications of the vector system would be tremendously extended if it also enabled the temporal control of gene silencing. Several options could be explored including coupling the CRISPR/Cas9 technique with the ERT2 inducible system to control the activity of Cre, Gal4, or Cas9 itself.

Owing to the high-throughput nature of the CRISPR/Cas9 technology, the tissue-specific CRISPR system is a tractable tool for precise in vivo reverse genetic screens. It may make it possible to uncover new genes that regulate the biology of a particular tissue, without affecting the rest of the organism. It may also allow researchers to revisit or clarify the role of known genes by inactivating them in different cell types, measuring their respective contributions to the global phenotype.

ACKNOWLEDGMENTS

We thank Jonathan E. Henninger for critical reading of the manuscript. This work was supported by NIH grants R01 CA103846, R01 HL04880, R01 DK53298, PO1 HL32262, P30 DK49216, U01 HL10001, and R24 DK092760. In addition, L. I. Z. is a Howard Hughes Medical Institute Investigator. L.I.Z. is a founder and stock holder of Fate, Inc., Scholar Rock, and a scientific advisor for Stemgent.

REFERENCES

Ablain, J., Durand, E. M., Yang, S., Zhou, Y., & Zon, L. I. (2015). A CRISPR/Cas9 vector system for tissue-specific gene disruption in zebrafish. *Developmental Cell, 32*(6), 756–764.

Amsterdam, A., Nissen, R. M., Sun, Z., Swindell, E. C., Farrington, S., & Hopkins, N. (2004). Identification of 315 genes essential for early zebrafish development. *Proceedings of the National Academy of Sciences of the United States of America, 101*(35), 12792–12797.

Auer, T. O., Duroure, K., De Cian, A., Concordet, J. P., & Del Bene, F. (2014). Highly efficient CRISPR/Cas9-mediated knock-in in zebrafish by homology-independent DNA repair. *Genome Research, 24*(1), 142–153.

Barrangou, R., Fremaux, C., Deveau, H., Richards, M., Boyaval, P., Moineau, S., … Horvath, P. (2007). CRISPR provides acquired resistance against viruses in prokaryotes. *Science, 315*(5819), 1709–1712.

Bedell, V. M., Wang, Y., Campbell, J. M., Poshusta, T. L., Starker, C. G., Krug, R. G., 2nd, … Ekker, S. C. (2012). In vivo genome editing using a high-efficiency TALEN system. *Nature, 491*(7422), 114–118.

Berghmans, S., Murphey, R. D., Wienholds, E., Neuberg, D., Kutok, J. L., Fletcher, C. D., … Look, A. T. (2005). tp53 mutant zebrafish develop malignant peripheral nerve sheath tumors. *Proceedings of the National Academy of Sciences of the United States of America, 102*(2), 407–412.

Brand, A. H., & Perrimon, N. (1993). Targeted gene expression as a means of altering cell fates and generating dominant phenotypes. *Development, 118*(2), 401–415.

Brouns, S. J., Jore, M. M., Lundgren, M., Westra, E. R., Slijkhuis, R. J., Snijders, A. P., … van der Oost, J. (2008). Small CRISPR RNAs guide antiviral defense in prokaryotes. *Science, 321*(5891), 960–964.

Chen, S., Sanjana, N. E., Zheng, K., Shalem, O., Lee, K., Shi, X., … Sharp, P. A. (2015). Genome-wide CRISPR screen in a mouse model of tumor growth and metastasis. *Cell, 160*(6), 1246–1260.

Cho, S. W., Kim, S., Kim, J. M., & Kim, J. S. (2013). Targeted genome engineering in human cells with the Cas9 RNA-guided endonuclease. *Nature Biotechnology, 31*(3), 230–232.

Cong, L., Ran, F. A., Cox, D., Lin, S., Barretto, R., Habib, N., … Zhang, F. (2013). Multiplex genome engineering using CRISPR/Cas systems. *Science, 339*(6121), 819–823.

Doyon, Y., McCammon, J. M., Miller, J. C., Faraji, F., Ngo, C., Katibah, G. E., … Amache, S. L. (2008). Heritable targeted gene disruption in zebrafish using designed zinc-finger nucleases. *Nature Biotechnology, 26*(6), 702–708.

Driever, W., Solnica-Krezel, L., Schier, A. F., Neuhauss, S. C., Malicki, J., Stemple, D. L., … Boggs, C. (1996). A genetic screen for mutations affecting embryogenesis in zebrafish. *Development, 123*, 37–46.

Eisen, J. S., & Smith, J. C. (2008). Controlling morpholino experiments: don't stop making antisense. *Development, 135*(10), 1735–1743.

Fu, Y., Foden, J. A., Khayter, C., Maeder, M. L., Reyon, D., Joung, J. K., & Sander, J. D. (2013). High-frequency off-target mutagenesis induced by CRISPR-Cas nucleases in human cells. *Nature Biotechnology, 31*(9), 822–826.

Gagnon, J. A., Valen, E., Thyme, S. B., Huang, P., Ahkmetova, L., Pauli, A., … Schier, A. F. (2014). Efficient mutagenesis by Cas9 protein-mediated oligonucleotide insertion and large-scale assessment of single-guide RNAs. *PLoS One, 9*(5), e98186.

Garneau, J. E., Dupuis, M. E., Villion, M., Romero, D. A., Barrangou, R., Boyaval, P., … Moineau, S. (2010). The CRISPR/Cas bacterial immune system cleaves bacteriophage and plasmid DNA. *Nature, 468*(7320), 67–71.

Gilbert, L. A., Larson, M. H., Morsut, L., Liu, Z., Brar, G. A., Torres, S. E., … Qi, L. S. (2013). CRISPR-mediated modular RNA-guided regulation of transcription in eukaryotes. *Cell, 154*(2), 442–451.

Haffter, P., Granato, M., Brand, M., Mullins, M. C., Hammerschmidt, M., Kane, D. A., … Nüsslein-Volhard, C. (1996). The identification of genes with unique and essential functions in the development of the zebrafish, *Danio rerio*. *Development, 123*, 1–36.

Hartley, J. L., Temple, G. F., & Brasch, M. A. (2000). DNA cloning using in vitro site-specific recombination. *Genome Research, 10*(11), 1788–1795.

Hisano, Y., Sakuma, T., Nakade, S., Ohga, R., Ota, S., Okamoto, H., … Kawahara, A. (2015). Precise in-frame integration of exogenous DNA mediated by CRISPR/Cas9 system in zebrafish. *Scientific Reports, 5*, 8841.

Hsu, P. D., Scott, D. A., Weinstein, J. A., Ran, F. A., Konermann, S., Agarwala, V., … Zhang, F. (2013). DNA targeting specificity of RNA-guided Cas9 nucleases. *Nature Biotechnology, 31*(9), 827–832.

Huang, H. T., Kathrein, K. L., Barton, A., Gitlin, Z., Huang, Y. H., Ward, T. P., … Zon, L. I. (2013). A network of epigenetic regulators guides developmental haematopoiesis in vivo. *Nature Cell Biology, 15*(12), 1516–1525.

Hwang, W. Y., Fu, Y., Reyon, D., Maeder, M. L., Tsai, S. Q., Sander, J. D., … Joung, J. K. (2013). Efficient genome editing in zebrafish using a CRISPR-Cas system. *Nature Biotechnology, 31*(3), 227–229.

Jao, L. E., Wente, S. R., & Chen, W. (2013). Efficient multiplex biallelic zebrafish genome editing using a CRISPR nuclease system. *Proceedings of the National Academy of Sciences of the United States of America, 110*(34), 13904–13909.

Jiang, W., Bikard, D., Cox, D., Zhang, F., & Marraffini, L. A. (2013). RNA-guided editing of bacterial genomes using CRISPR-Cas systems. *Nature Biotechnology, 31*(3), 233–239.

Jinek, M., Chylinski, K., Fonfara, I., Hauer, M., Doudna, J. A., & Charpentier, E. (2012). A programmable dual-RNA-guided DNA endonuclease in adaptive bacterial immunity. *Science, 337*(6096), 816–821.

Kawakami, K., Takeda, H., Kawakami, N., Kobayashi, M., Matsuda, N., & Mishina, M. (2004). A transposon-mediated gene trap approach identifies developmentally regulated genes in zebrafish. *Developmental Cell, 7*(1), 133–144.

Kim, H. J., Lee, H. J., Kim, H., Cho, S. W., & Kim, J. S. (2009). Targeted genome editing in human cells with zinc finger nucleases constructed via modular assembly. *Genome Research, 19*(7), 1279–1288.

Kimura, Y., Hisano, Y., Kawahara, A., & Higashijima, S. (2014). Efficient generation of knock-in transgenic zebrafish carrying reporter/driver genes by CRISPR/Cas9-mediated genome engineering. *Scientific Reports, 4*, 6545.

Koike-Yusa, H., Li, Y., Tan, E. P., Velasco-Herrera Mdel, C., & Yusa, K. (2014). Genome-wide recessive genetic screening in mammalian cells with a lentiviral CRISPR-guide RNA library. *Nature Biotechnology, 32*(3), 267–273.

Kok, F. O., Shin, M., Ni, C. W., Gupta, A., Grosse, A. S., van Impel, A., … Lawson, N. D. (2015). Reverse genetic screening reveals poor correlation between morpholino-induced and mutant phenotypes in zebrafish. *Developmental Cell, 32*(1), 97–108.

Kuscu, C., Arslan, S., Singh, R., Thorpe, J., & Adli, M. (2014). Genome-wide analysis reveals characteristics of off-target sites bound by the Cas9 endonuclease. *Nature Biotechnology, 32*(7), 677–683.

Li, J., Zhang, B. B., Ren, Y. G., Gu, S. Y., Xiang, Y. H., & Du, J. L. (2015). Intron targeting-mediated and endogenous gene integrity-maintaining knockin in zebrafish using the CRISPR/Cas9 system. *Cell Research, 25*(5), 634–637.

Mali, P., Yang, L., Esvelt, K. M., Aach, J., Guell, M., DiCarlo, J. E., … Church, G. M. (2013). RNA-guided human genome engineering via Cas9. *Science, 339*(6121), 823–826.

Meng, X., Noyes, M. B., Zhu, L. J., Lawson, N. D., & Wolfe, S. A. (2008). Targeted gene inactivation in zebrafish using engineered zinc-finger nucleases. *Nature Biotechnology, 26*(6), 695–701.

Miller, J. C., Tan, S., Qiao, G., Barlow, K. A., Wang, J., Xia, D. F., … Rebar, E. J. (2011). A TALE nuclease architecture for efficient genome editing. *Nature Biotechnology, 29*(2), 143–148.

Montague, T. G., Cruz, J. M., Gagnon, J. A., Church, G. M., & Valen, E. (2014). CHOPCHOP: a CRISPR/Cas9 and TALEN web tool for genome editing. *Nucleic Acids Research, 42*(Web Server issue), W401–W407.

Moreno-Mateos, M. A., Vejnar, C. E., Beaudoin, J. D., Fernandez, J. P., Mis, E. K., Khokha, M. K., & Giraldez, A. J. (2015). CRISPRscan: designing highly efficient sgRNAs for CRISPR-Cas9 targeting in vivo. *Nature Methods, 12*(10), 982–988.

Mosimann, C., Kaufman, C. K., Li, P., Pugach, E. K., Tamplin, O. J., & Zon, L. I. (2011). Ubiquitous transgene expression and Cre-based recombination driven by the ubiquitin promoter in zebrafish. *Development, 138*(1), 169–177.

Mullins, M. C., Hammerschmidt, M., Haffter, P., & Nusslein-Volhard, C. (1994). Large-scale mutagenesis in the zebrafish: in search of genes controlling development in a vertebrate. *Current Biology, 4*(3), 189–202.

Nasevicius, A., & Ekker, S. C. (2000). Effective targeted gene 'knockdown' in zebrafish. *Nature Genetics, 26*(2), 216–220.

Parant, J. M., George, S. A., Pryor, R., Wittwer, C. T., & Yost, H. J. (2009). A rapid and efficient method of genotyping zebrafish mutants. *Developmental Dynamics, 238*(12), 3168–3174.

Qiu, P., Shandilya, H., D'Alessio, J. M., O'Connor, K., Durocher, J., & Gerard, G. F. (2004). Mutation detection using Surveyor nuclease. *Biotechniques, 36*(4), 702–707.

Rossi, A., Kontarakis, Z., Gerri, C., Nolte, H., Holper, S., Kruger, M., & Stainier, D. Y. R. (2015). Genetic compensation induced by deleterious mutations but not gene knockdowns. *Nature, 524*(7564), 230–233.

Sander, J. D., Cade, L., Khayter, C., Reyon, D., Peterson, R. T., Joung, J. K., & Yeh, J. R. (2011). Targeted gene disruption in somatic zebrafish cells using engineered TALENs. *Nature Biotechnology, 29*(8), 697–698.

Shah, A. N., Davey, C. F., Whitebirch, A. C., Miller, A. C., & Moens, C. B. (2015). Rapid reverse genetic screening using CRISPR in zebrafish. *Nature Methods, 12*(6), 535–540.

Shen, Z., Zhang, X., Chai, Y., Zhu, Z., Yi, P., Feng, G., … Ou, G. (2014). Conditional knockouts generated by engineered CRISPR-Cas9 endonuclease reveal the roles of coronin in *C. elegans* neural development. *Developmental Cell, 30*(5), 625–636.

Sternberg, S. H., Redding, S., Jinek, M., Greene, E. C., & Doudna, J. A. (2014). DNA interrogation by the CRISPR RNA-guided endonuclease Cas9. *Nature, 507*(7490), 62–67.

Summerton, J., & Weller, D. (1997). Morpholino antisense oligomers: design, preparation, and properties. *Antisense and Nucleic Acid Drug Development, 7*(3), 187–195.

Taylor, J. S., Braasch, I., Frickey, T., Meyer, A., & Van de Peer, Y. (2003). Genome duplication, a trait shared by 22,000 species of ray-finned fish. *Genome Research, 13*(3), 382–390.

Urnov, F. D., Miller, J. C., Lee, Y. L., Beausejour, C. M., Rock, J. M., Augustus, S., … Holmes, M. C. (2005). Highly efficient endogenous human gene correction using designed zinc-finger nucleases. *Nature, 435*(7042), 646–651.

Wang, T., Wei, J. J., Sabatini, D. M., & Lander, E. S. (2014). Genetic screens in human cells using the CRISPR-Cas9 system. *Science, 343*(6166), 80–84.

Wang, X., Wang, Y., Wu, X., Wang, J., Qiu, Z., Chang, T., … Yee, J. K. (2015). Unbiased detection of off-target cleavage by CRISPR-Cas9 and TALENs using integrase-defective lentiviral vectors. *Nature Biotechnology, 33*(2), 175–178.

Xue, Z., Wu, M., Wen, K., Ren, M., Long, L., Zhang, X., & Gao, G. (2014). CRISPR/Cas9 mediates efficient conditional mutagenesis in Drosophila. *G3 (Bethesda), 4*(11), 2167–2173.

Zhou, Y., Zhu, S., Cai, C., Yuan, P., Li, C., Huang, Y., & Wei, W. (2014). High-throughput screening of a CRISPR/Cas9 library for functional genomics in human cells. *Nature, 509*(7501), 487–491.

PART

Transgenesis and Functional Genomics

2

CHAPTER

Transcriptional regulation using the Q system in transgenic zebrafish

10

A. Ghosh[a], M.E. Halpern[1]

Carnegie Institution for Science, Baltimore, MD, United States

[1]Corresponding author: E-mail: halpern@ciwemb.edu

CHAPTER OUTLINE

Abstract

Methods to label cell populations selectively or to modify their gene expression are critical tools in the study of developmental or physiological processes in vivo. A variety of approaches have been applied to the zebrafish model, capitalizing on Tol2 transposition to generate transgenic lines with high efficiency. Here we describe the adoption of the Q system of *Neurospora crassa*, which includes the QF transcription factor and the upstream activating sequence (QUAS) to which it binds. These components function as a bipartite regulatory system similar to that of yeast Gal4/UAS, producing robust expression in transient assays of zebrafish embryos injected with plasmids and in stable transgenic lines. An important advantage, however, is that QUAS-regulated transgenes appear far less susceptible to transcriptional silencing even after seven generations. This chapter describes some of the Q system reagents that have been developed for zebrafish, as well as the use of the QF transcription factor for isolation of tissue-specific driver lines from gene/enhancer trap screens. Additional strategies successfully implemented in invertebrate models, such as a truncated QF transcription factor (QF2) or the reassembly of a split QF,

[a]Current address: Tata Institute of Fundamental Research, Mumbai, India

Methods in Cell Biology, Volume 135, ISSN 0091-679X, http://dx.doi.org/10.1016/bs.mcb.2016.05.001

are also discussed. The provided information, and available Gateway-based vectors, should enable those working with the zebrafish model to implement the Q system with minimal effort or to use it in combination with Gal4, Cre, or other regulatory systems for further refinement of transcriptional control.

1. BACKGROUND

Transgenic approaches to manipulate gene expression in spatially or temporally restricted ways have been instrumental in the study of biological processes. The transparency of zebrafish embryos and larvae make them particularly well suited for selective labeling of cell populations and observing their behavior over time, for assessing the effects of genetic perturbations, for optogenetic methods to alter neural activity, and for monitoring of genetically encoded biosensors. In these and other strategies to explore cellular function in vivo, versatility is an important property of the transgenic tools being generated. One way to increase flexibility and maximize utility is with a binary system, in which the expression of a collection of reporter or effector genes is activated by a transcription factor that is itself placed under the control of different cell type–specific promoters/enhancers.

The use of bipartite transcriptional regulatory systems in zebrafish was first accomplished in the laboratory of Jose Campos-Ortega with promoters driving near ubiquitous expression of the gene encoding the yeast Gal4 transcription factor. Five copies of an optimized upstream activating sequence (UAS) where Gal4 binds, placed upstream of the minimal *E1b* promoter of adenovirus and a synthetic start site from the carp β-actin promoter, controlled transcription of sequences encoding the intracellular domain of the zebrafish Notch1 protein fused with six myc tags (Scheer & Campos-Ortega, 1999). Expression was visualized in embryos derived from matings between the driver and responder transgenic lines by anti-myc antibody labeling. This early demonstration showed that the Gal4/UAS system could function in zebrafish embryos, albeit at low levels of expression as detection of the UAS-regulated gene product required signal amplification by immunolabeling.

Expression levels were significantly increased through the adoption of methods to overexpress randomly targeted genes in a *Drosophila* insertional mutagenesis screen (Rorth, 1996). Widespread and intense labeling from the green fluorescent protein (GFP) was achieved in zebrafish embryos using a self-reporting vector that contained the transcriptional activation domain (AD) of the VP16 protein of Herpes simplex virus fused to the Gal4 DNA binding domain and 14 UAS (14X) binding sites in a tandem array upstream of the *GFP* gene (Koster & Fraser, 2001). Injection of circular or linearized plasmids in transient assays yielded robust GFP labeling in a variety of zebrafish embryonic tissues depending on the promoter driving Gal4-VP16. Transgenic lines were not established on account of lethality, presumably due to "squelching" of factors required for the transcription of endogenous genes (Koster & Fraser, 2001). Moreover, the injected DNA likely integrated into the genome in high copy number as complex concatemers that would be targets for transcriptional silencing.

The discovery and application of Tol2 transposition (Kawakami, Shima, & Kawakami, 2000; Urasaki, Morvan, & Kawakami, 2006) circumvented the problem of high copy number and lethality due to overexpression. Several groups modified Gal4-VP16 constructs with the addition of Tol2 arms (eg, Asakawa & Kawakami, 2008; Davison et al., 2007; Scott et al., 2007). When injected with Tol2 transposase into 1-cell stage embryos, integration into the zebrafish genome by transposition results in single copy insertions, survival of the resultant embryos and germ-line propagation of transgene. Construction of gene/enhancer trap vectors for Tol2 transposition has yielded numerous Gal4 driver lines from screens for tissue-specific patterns of expression (eg, Distel, Wullimann, & Koster, 2009; Kawakami et al., 2010; Marquart et al., 2015; Otsuna et al., 2015; Takeuchi et al., 2015) as well as generated new insertional mutations (Balciuniene & Balciunas, 2013).

Owing to its widespread use in the *Drosophila* model, many zebrafish researchers have used the Gal4/UAS system to manipulate gene expression in cells or tissues of interest, but with varying results. Both the Gal4-VP16 and UAS components have been revised to maximize expression levels while reducing toxicity (Akitake, Macurak, Halpern, & Goll, 2011; Asakawa & Kawakami, 2008; Distel et al., 2009). However, a persistent problem has been the loss of expression from transgenes in which transcription is under the control of a multicopy UAS. The UAS contains essential CpG dinucleotides for Gal4 binding, which also makes it a preferred site for DNA methylation and, hence, transcriptional silencing (Goll, Anderson, Stainier, Spradling, & Halpern, 2009). Although UAS-regulated transgenes may be robustly expressed initially, in subsequent generations, expression often becomes variable in individuals from the same clutch, and, in extreme cases, fully extinguished (Akitake et al., 2011; Goll et al., 2009; Pang, Wang, Zhu, & Sun, 2015). Transgenic lines have sometimes been maintained by selecting embryos or larvae that show more intense or complete labeling patterns, but this only is possible when a transgene contains fluorescent or visible markers that can be readily scored. Alternatively, constructs have been reinjected and lines rederived from new transgenic founders.

To potentially combat transgene silencing and expand the repertoire of transgenic tools, another binary system derived from *Neurospora crassa* was adapted for Tol2 transposition in zebrafish (Subedi et al., 2014). The *qa* gene cluster was originally identified from the discovery of *Neurospora* mutants that failed to metabolize quinic acid as a carbon source, and its regulatory genes were later determined (refer to Giles et al., 1985). Components of this cluster (named the Q system) were shown to activate transcription in *Drosophila* embryonic and adult stages at a higher level than the Gal4/UAS system and to operate in cultured mammalian cells (Potter, Tasic, Russler, Liang, & Luo, 2010) and, subsequently, in nematodes (Wei, Potter, Luo, & Shen, 2012). This chapter reviews how the Q system has been applied to zebrafish transgenesis, as well as for gene/enhancer trap screens to recover new tissue-specific driver lines. The wild-type AB laboratory strain (Walker, 1999) was used for all of the described experiments and to generate new transgenic lines.

2. COMPONENTS OF THE Q TRANSCRIPTIONAL REGULATORY SYSTEM

The key features of the Q system are depicted in Fig. 1. The QF transcription factor is an 816 amino acid protein comprised of structurally distinct regions: a DNA binding domain (DBD), a middle region of unknown function (DM), and a transcriptional AD. QF binds to QUAS sites to induce transcription of adjacent genes. A repressor protein, QS, functions to inhibit QF binding, while exposure to quinic acid relieves this inhibition and restores QF-mediated transcriptional activation.

2.1 QF AND QF DERIVATIVES

To assess whether QF would be active in zebrafish embryos, the entire QF coding sequence was initially placed under the control of either the *Xenopus laevis elongation factor 1-alpha* (*EF1*α) or zebrafish *ubiquitin B* (*ubb*) promoter, promoters that are known to drive high levels of expression throughout the zebrafish embryo (Amsterdam, Lin, & Hopkins, 1995; Mosimann et al., 2011). Constructs were produced in Tol2 vectors as previously described (Subedi et al., 2014) and coinjected in one to two cell stage embryos along with a *QUAS:GFP* Tol2 plasmid (see Section 2.2) and RNA encoding the Tol2 transposase (25 ng/μL). Robust fluorescent labeling was observed in diverse tissues of the injected embryos after 1 day post fertilization (dpf), validating the use of the Q system in transient expression assays. Lethality was measured and the optimal concentration for the driver and reporter plasmids was

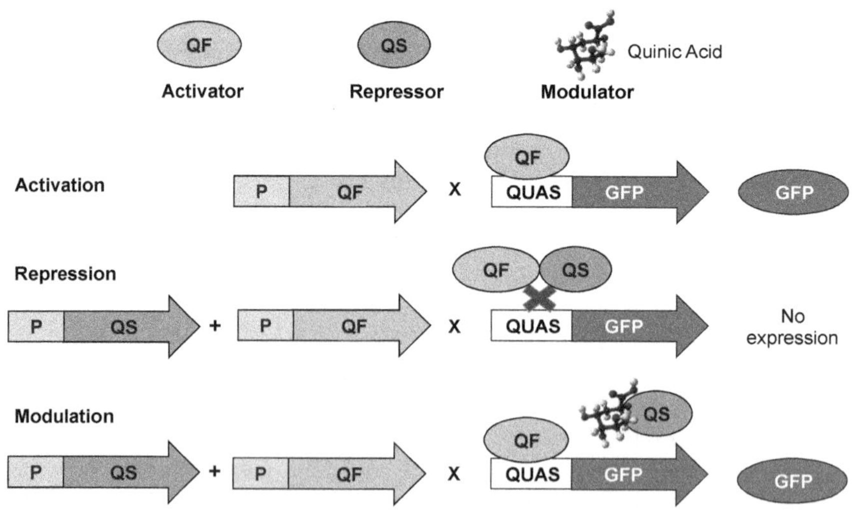

FIGURE 1

Schematic of the Q system. *GFP*, green fluorescent protein; *QUAS*, Q upstream activating sequence. (See color plate)

determined (1−10 ng yielded a survival rate of ranging from 88% to 95%). Following these encouraging findings, Tol2 vectors with tissue-specific promoters (eg, *mnx1*, *fabp1a*, and *ins*) driving QF expression were assembled for the generation of stable transgenic lines (Subedi et al., 2014). Transgenic QF reporters have also been produced to label cells activated by Notch or retinoic acid (RA) signaling using tp1 (Parsons et al., 2009; Subedi et al., 2014) and RA response elements (Huang et al., 2014), respectively. To facilitate the inclusion of promoters or regulatory sequences of interest, a PCR fragment containing the QF gene and the SV40 termination sequence was inserted as a middle entry (ME) clone in the Gateway Recombination Cloning system (Invitrogen, Thermo Fisher Scientific Inc.). In this modular system, any promoter in a 5' entry clone can be recombined along with the pME-QF clone and a 3' reporter gene into a Gateway destination vector reengineered for Tol2 transposition (refer to Kwan et al., 2007).

A drawback to QF is that, as with ectopic Gal4-VP16 activity, widespread overexpression of this transcriptional activator appears to be toxic to developing embryos. It has therefore been challenging to recover zebrafish transgenic founders with broadly expressed promoters driving QF. To circumvent this problem in *Drosophila*, Riabinina et al. (2015) identified the region of the QF protein responsible for toxicity and devised other versions of the protein that would significantly reduce it. By testing a number of QF variants, they determined that lethality corresponded with the presence of the middle region that, fortunately, was dispensable for QF function. One of the resultant truncated and codon-optimized products, QF2 (Fig. 2), fulfilled the criteria of promoting high levels of gene expression, being

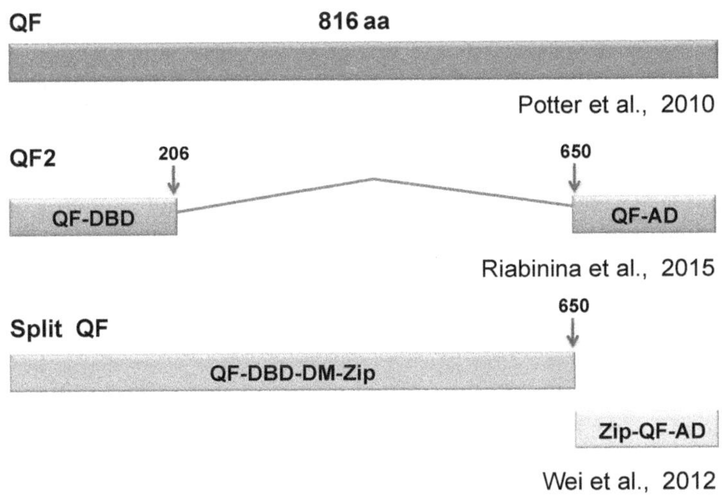

FIGURE 2

The QF transcription factor and modifications.

suppressible by QS and nontoxic when activated throughout the tissues of larval or adult *Drosophila* (Riabinina et al., 2015).

In preliminary experiments to test the efficacy of QF2 in zebrafish, a 1061 bp fragment (including the terminal 58 bp of the *synaptobrevin* (*syb*) promoter and a translation initiation sequence 5′ and 28 bp of the Hsp70 termination site 3′) was amplified from the *pattB-synaptobrevin-7-QFBDAD-hsp70* plasmid developed for *Drosophila* (Riabinina et al., 2015), and introduced as a Gateway ME clone. This clone was recombined with a Gateway 5′ entry clone bearing the *ubb* promoter (Mosimann et al., 2011) into a Tol2 destination vector, and the resulting plasmid coinjected into one to two cell embryos together with *QUAS:GFP* Tol2 plasmid and RNA encoding Tol2 transposase. By 1 dpf, robust labeling was observed. Moreover, QF2 appeared to be less toxic than in parallel tests with a comparable range of QF concentrations (Monge and Halpern, unpublished results). Thus, QF2 is functional in zebrafish, and its smaller size may be a significant advantage for gene/enhancer trap vectors or for integration into genes of interest by Crispr/Cas9 genome editing.

A further modification of QF is the division of the coding sequence into two parts that can each be expressed under the control of a different promoter for the purpose of intersectional gene expression (Fig. 2). The split QF system is based on split Gal4, whereby sequences containing the DNA binding and transcriptional ADs are separated and fused with heterodimeric leucine zipper sequences (Luan, Peabody, Vinson, & White, 2006). With transcription of each module regulated by different promoters, the two peptides are synthesized only in those cells in which both promoters are active and, upon binding through the leucine zipper, assemble into a functional transcription factor. This method was recently applied to zebrafish by splitting an optimized version of Gal4 and demonstrating its reassembly by activation of UAS reporters (Almeida & Lyons, 2015). In a similar fashion, split QF components were designed through fusion of sequences for leucine zipper heterodimers to those encoding the QF DNA binding and ADs. In transgenic worms, fluorescent labeling was observed in a single neuron, in which expression from a promoter driving the DBD construct overlapped with that of another promoter driving the AD construct (Wei et al., 2012). Although a powerful technique to restrict gene expression, a downside to the split QF approach is that the activity of the reconstituted protein, as measured by fluorescence intensity, was only 42% of that of the intact transcription factor and not all of expected cells were labeled (Wei et al., 2012).

The functionality of split QF in zebrafish has, so far, been verified in transient expression assays. Sequences corresponding to the CZipper(−)::AD(QF) (678 bp) and the NLS-BD-DM(QF)::NZipper(+) (2118 bp) regions were amplified, respectively, from plasmids XW52 and XW54 of Wei et al. (2012) to produce ME Gateway clones. The ME clones were individually recombined with the *ubb* promoter 5’entry clone and an SV40 polyA tail 3’ entry clone into a Tol2 destination vector (Kwan et al., 2007). The two destination vectors were co-injected with *QUAS:GFP* plasmid and Tol2 RNA into one to two cell stage zebrafish embryos or, alternatively, co-injected into *QUAS:GFP* transgenic embryos (see Section 2.2). All groups were scored for the presence of fluorescently labeled cells a day later. Relative to sibling embryos injected with a *ubb:QF* construct that produces widespread GFP labeling, those expressing the split QF components displayed far fewer, brightly labeled cells, scattered throughout a variety of tissues (Monge and Halpern, unpublished results). GFP labeled cells were not detected when the AD or BD construct was injected alone. The split QF ME clones were also assembled into Gateway destination vectors with a 5′ entry clone containing the promoter of the *Xenopus neural beta tubulin* gene (*XlTubb*) (gift of Paul Krieg, University of Arizona). In *QUAS:GFP* transgenic embryos, coinjected with the two vectors, GFP positive cells were only recovered in the central nervous system (CNS) where the *XlTubb* promoter selectively functions (Ghosh and Halpern, unpublished results). Thus, reassembly of split QF components can be successfully accomplished in zebrafish tissues, however, in stable lines, optimal activation of QUAS-regulated genes will likely require strong promoters driving expression of the separate DBD and AD modules.

2.2 Q UPSTREAM ACTIVATING SEQUENCE

The QUAS is a conserved 16 bp sequence that is present at multiple sites and in varying numbers of tandem copies within the qa gene cluster (Giles et al., 1985). In the first plasmid produced (Subedi et al., 2014), a restriction fragment containing the five copy QUAS array was obtained from *pQUAST* (Potter et al., 2010) and cloned upstream of the carp *β-actin* minimal promoter and GFP coding sequence in a Tol2 plasmid in reverse orientation relative to the original construct used in *Drosophila*. As described earlier, coinjection into one to two cell embryos of the reversed *QUAS:GFP* Tol2 vector and plasmids containing ubiquitous promoters controlling QF results in robust mosaic labeling in many tissues. A stable transgenic line was established, $Tg(QUAS^R{:}GFP)^{c403}$, that gives consistently high levels of GFP labeling. Fluorescence from this transgenic reporter does not vary when tested with the same QF driver over multiple generations and high levels of activation have persisted for over seven generations (Fig. 3). These results contradict the assertion that QUAS-regulated transgenes will be transcriptionally silenced similar to those regulated by a multicopy UAS (Suli, Guler, Raible, & Kimelman, 2014). $Tg(QUAS^R{:}GFP)^{c403}$ can also be maintained as a homozygous line.

An unexpected finding was that the embryonic progeny of $Tg(QUAS^R{:}GFP)^{c403}$ mothers show faint GFP labeling from the 1-cell stage to several days, even in the absence of QF activity. Because early embryos are weakly fluorescent, irrespective of their genotype, it is most likely due to activation of the QUAS in the maternal germ line and deposition of GFP. Accordingly, when the $Tg(QUAS^R{:}GFP)^{c403}$ reporter is transmitted from the paternal genome, basal expression from the transgene is not observed (Subedi et al., 2014). To determine whether reversal of the 5X QUAS relative to the GFP transcription start site was the basis for the maternal expression, a Gateway 5′ entry clone was constructed with the QUAS sequences in the original orientation (Potter et al., 2010) and recombined into a Tol2 destination vector. To expedite the identification of transgenic founders and long-term maintenance of

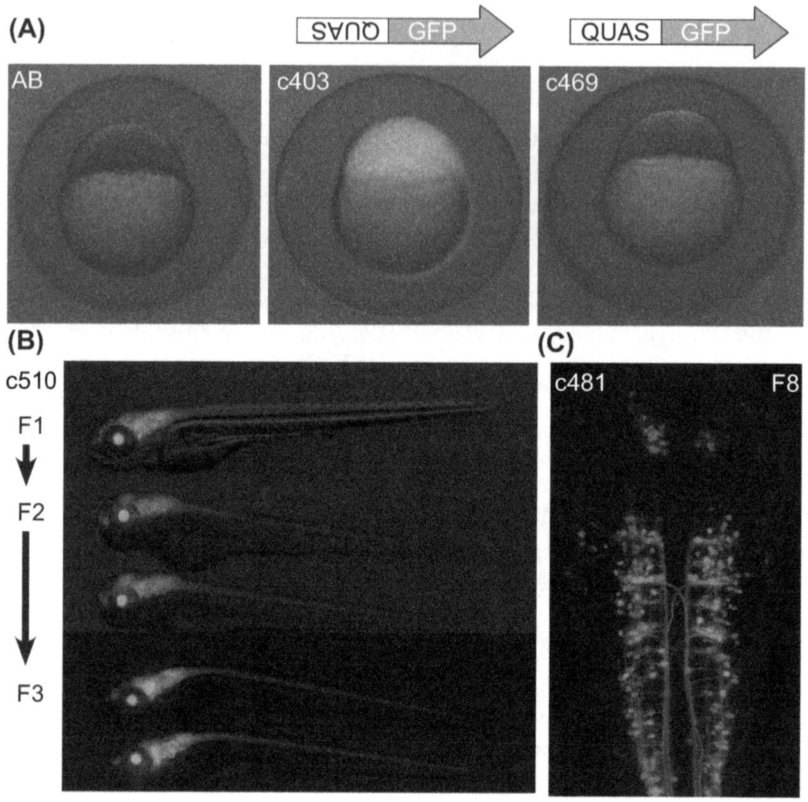

FIGURE 3 QUAS:GFP transgenic reporters.

(A) In the absence of QF, fluorescence in blastula stage embryos derived from Tg(QUAS^R:GFP) mothers correlates with the orientation of the 5 copy Q upstream activating sequence (QUAS) array relative to the GFP transcriptional start site. Maternal expression is abolished in the c469 line. (*Image courtesy of Estela Monge*). (B) Labeling from the Tg(QUAS^R:GFP)^{c403} transgene regulated by the ET(QF)^{c510} driver is undiminished in 5-day-old larvae over three generations. (C) In the eighth generation of Tg(QUAS^R:GFP)^{c403} progeny, green fluorescent protein (GFP) is robustly expressed at 48 hpf in a subset of neurons in which the ET(QF)c481 driver is active. (See color plate)

Image courtesy of Jean Michael Chanchu. The chromosomal locations of the c510 and c481 transgenic insertions have not yet been determined.

the QUAS reporter line in the absence of QF, an additional marker for the lens was supplied with a 3′ entry clone consisting of the zebrafish *β-crystallin b1* (*crybb1*) promoter driving an *enhanced cyan fluorescent protein* (*CFP*) gene (a gift of M. Parsons, Johns Hopkins University). The *Tg(QUAS:GFP)*^{c469} line, established from this vector, has been bred for over five generations. Embryos derived from either heterozygous or homozygous c469 mothers are brightly labeled upon QF activation but do not show any evidence of basal fluorescence in the absence of a QF driver (Fig. 3), indicating that the QUAS transgene is not constitutively expressed. Therefore, all QUAS reporter and effector lines have been subsequently produced with the QUAS array in the same orientation as in *Tg(QUAS:GFP)*^{c469}. Screening for CFP labeling of the lens also permits identification of larvae bearing QUAS-regulated transgenes or propagation of QUAS responder lines independent of QF.

2.3 QS AND QUINIC ACID

QS, the QF repressor, was assessed for its ability to reduce or eliminate GFP labeling in *Tg(QUAS*^{R}*:GFP)*^{c403} embryos coinjected with Tol2 plasmids containing the *EF1α* or *ubb* ubiquitous promoters driving either QF or QF2 expression (Monge and Halpern, unpublished results; Subedi et al., 2014). The QS gene was also placed under the control of the same two strong promoters in Tol2 vectors and those plasmids were injected at a higher concentration (50 ng/μL) relative to the QF plasmids (3.2 ng/μL). In all experiments, QF and QF2-dependent GFP labeling was significantly reduced in the presence of either QS construct and unaffected in the absence of QS. Fluorescence intensity measurements recorded from individual *Tg(QUAS*^{R}*:GFP)*^{c403} embryos at 1 dpf corroborated that, on average, the level of GFP labeling decreased by approximately 50% in those that had received the QS plasmid. Repression of QF activity by QS proteins synthesized from transcription of integrated transgenes has not yet been accomplished and will require careful calibration of the relative concentrations of both proteins. Until this is achieved, it will be difficult to assay the ability of quinic acid to alleviate QS-mediated repression of QF or QF2. Since a saturated quinic acid solution (~300 mg/mL) is added as a supplement to fly food vials (Potter & Luo, 2011) or to nematode growth medium (Wei et al., 2012), it is not possible to extrapolate the equivalent amount to administer to zebrafish embryos in system water. In a dose-response test, embryonic and larval zebrafish were found to survive concentrations of quinic acid equal to or lower than 0.3 mg/mL for up to 5 days of exposure. Higher doses produced morphological abnormalities and lethality. Whether this baseline of quinic acid treatment will be adequate to disrupt the QS—QF interaction remains to be determined.

3. GENE/ENHANCER TRAPPING USING THE Q SYSTEM

In a pilot screen, the QF transcription factor was used effectively for gene/enhancer trapping with a focus on isolating driver lines specific for the CNS. A gene trap vector, *p(GT-QF)*, similar to the Gal4 gene trap of Davison et al. (2007), was generated through recombination of a rabbit β-globin splice acceptor, the QF transcription factor, and SV40 polyA termination sequences into a Tol2 destination vector (Kwan et al., 2007). A related plasmid, p(GT-QF2), was produced with the QF2 ME clone. In contrast to the Gal4 gene trap vector, a QUAS-regulated reporter gene was provided *in trans*, rather than included within the same construct. To test the effectiveness of p(GT-QF) and p(GT-QF2), the Tol2 vectors were separately injected into 1-cell stage Tg(QUAS:GFP) embryos, which were screened for GFP expression from 1 to 5 days later and raised to adulthood. Transgenic founders were identified by screening for patterns of GFP labeling in their F1 progeny, which were raised and outcrossed to establish QF driver lines. GFP labeled cells were confined to the CNS (Fig. 3B and C) in approximately 20% of the QF driver lines derived from 33 identified founders. Cloning and sequencing of the Tol2 QF insertion sites is necessary to confirm that they correspond to gene trap events, as prior work indicated that some Gal4 drivers map outside of coding sequences and are likely regulated by nearby enhancers (Davison et al., 2007). In contrast to p(GT-QF), the QF2 Tol2 gene trap vector induced basal GFP labeling in the midline of the brain and spinal cord, which confounded the identification of founders. We are currently testing whether this is due to residual *syb* promoter sequences amplified along with the QF2 gene. However, the preliminary screens support gene/enhancer trapping as a means of obtaining QF driver lines in cell types of interest at a frequency similar to that of Gal4-based vectors.

4. FUTURE PROSPECTS

For any transcriptional regulatory system to be adopted by the zebrafish research community, reagents and transgenic lines must be available, easy to apply, and validated in vivo. We are in the process of generating new QUAS lines that will be useful tools for a variety of applications, including fluorescent reporters to label subcellular structures, effectors for cell-autonomous ablation, and optogenetic modulators and genetically encoded calcium indicators of neural activity. A limitation to any bipartite system in zebrafish is the paucity of identified tissue or cell type–specific promoters to drive expression of QF or other regulatory genes. This can, to some extent, be overcome by expanding the collection of tissue-specific driver lines isolated from gene/enhancer trap screens. Exciting advances in Crispr/Cas9 technology will enable the creation of driver lines in a more targeted manner. In pilot studies, QF2 was integrated into specific gene loci following the strategy of Kimura, Hisano, Kawahara, and Higashijima (2014), and transgenic embryos were recovered that activate transcription of QUAS reporters in patterns recapitulating endogenous gene expression (Choi and Halpern, unpublished results). Even so, to perform experiments on selective populations of cells, particularly for functional studies of subsets of neurons in specific regions of the zebrafish brain, will require additional refinements.

The split QF approach, designed for intersectional gene expression, is a promising direction but whether it will provide sufficient levels of activity in single copy transgenic lines remains to be demonstrated. Alternative strategies involve combining the Q system with other binary regulatory systems. QF/QUAS and Gal4/UAS have been shown to operate independently in the fly (Potter et al., 2010) and zebrafish (Subedi et al., 2014). However, to achieve the sophistication of methods commonly used in *Drosophila*, such as mosaic analysis with a repressible cell marker (MARCM) or coupled MARCM that capitalizes on both Gal4 and QF (refer to Potter & Luo, 2011), will necessitate optimization of their respective repressors to confine the spatial or temporal extent of their activity. Suppression of Gal4 by transgene supplied Gal80 has been demonstrated in zebrafish embryos (Faucherre & Lopez-Schier, 2011; Fujimoto, Gaynes, Brimley, Chien, & Bonkowsky, 2011), but does not work for modified versions of the protein that have the native transcriptional AD replaced by another. While the QS repressor inhibits QF and QF2 activation of QUAS-regulated transgenes when provided from injected plasmids, appropriate levels still must be attained in transgenic lines. This should be possible using strong promoters to drive QS expression or by developing lines that have multiple copies of QS transgenes.

Cre recombinase-based technology has also been successfully used in zebrafish in concert with Gal4 to restrict gene expression to more limited numbers of cells. For example, stochastic Cre-mediated activation of Gal4-VP16 permitted labeling of single tectal neurons and their dendritic and axonal processes in the brains of larval zebrafish (Sato, Hamaoka, Aizawa, Hosoya, & Okamoto, 2007). Techniques to induce Cre activity conditionally in zebrafish (eg, Hans, Kaslin, Freudenreich, & Brand, 2009; Hesselson, Anderson, Beinat, & Stainier, 2009) provide further ways of regulating QF alleles disrupted by loxP-stop-loxP cassettes, in subsets of cells at desired developmental stages.

The initial investigation into the Q system indicates that it will be a valuable addition to the increasing genetic toolkit for transgenic approaches in zebrafish. As more QF drivers and QUAS responders are developed and shared, this alternative regulatory system, and the potential for coupling it with other methods, is certain to become widely exploited.

ACKNOWLEDGMENTS

We thank Jean Michael Chanchu, Tyler Harvey, Ayse Keskus, Michelle Macurak, and Estela Monge for their expert technical assistance in developing reagents and providing results described in this chapter. Abigyna Subedi and Mary Goll were instrumental in initiating this project and Chris Potter has generously provided plasmids and advice. We are grateful to Shannon Fisher, Koichi Kawakami, Kristen Kwan, Michael Parsons, and Kang Shen for sharing plasmids. The zebrafish Q project is supported by the National Institute of Child Health and Human Development (4R01HD078220).

REFERENCES

Akitake, C. M., Macurak, M., Halpern, M. E., & Goll, M. G. (2011). Transgenerational analysis of transcriptional silencing in zebrafish. *Developmental Biology, 352*(2), 191–201. http://dx.doi.org/10.1016/j.ydbio.2011.01.002.

Almeida, R. G., & Lyons, D. A. (2015). Intersectional gene expression in zebrafish using the split KalTA4 system. *Zebrafish, 12*(6), 377–386. http://dx.doi.org/10.1089/zeb.2015.1086.

Amsterdam, A., Lin, S., & Hopkins, N. (1995). The Aequorea victoria green fluorescent protein can be used as a reporter in live zebrafish embryos. *Developmental Biology, 171*(1), 123–129. http://dx.doi.org/10.1006/dbio.1995.1265.

Asakawa, K., & Kawakami, K. (2008). Targeted gene expression by the Gal4-UAS system in zebrafish. *Development, Growth & Differentiation, 50*(6), 391–399. http://dx.doi.org/10.1111/j.1440-169X.2008.01044.x.

Balciuniene, J., & Balciunas, D. (2013). Gene trapping using gal4 in zebrafish. *Journal of Visualized Experiments: JoVE [Electronic Resource], 79*, e50113. http://dx.doi.org/10.3791/50113.

Davison, J. M., Akitake, C. M., Goll, M. G., Rhee, J. M., Gosse, N., Baier, H., … Parsons, M. J. (2007). Transactivation from Gal4-VP16 transgenic insertions for tissue-specific cell labeling and ablation in zebrafish. *Developmental Biology, 304*(2), 811–824. http://dx.doi.org/10.1016/j.ydbio.2007.01.033.

Distel, M., Wullimann, M. F., & Koster, R. W. (2009). Optimized Gal4 genetics for permanent gene expression mapping in zebrafish. *Proceedings of the National Academy of Sciences of the United States of America, 106*(32), 13365–13370. http://dx.doi.org/10.1073/pnas.0903060106.

Faucherre, A., & Lopez-Schier, H. (2011). Delaying Gal4-driven gene expression in the zebrafish with morpholinos and Gal80. *PLoS One, 6*(1), e16587. http://dx.doi.org/10.1371/journal.pone.0016587.

Fujimoto, E., Gaynes, B., Brimley, C. J., Chien, C. B., & Bonkowsky, J. L. (2011). Gal80 intersectional regulation of cell-type specific expression in vertebrates. *Developmental Dynamics: an Official Publication of the American Association of Anatomists, 240*(10), 2324–2334. http://dx.doi.org/10.1002/dvdy.22734.

Giles, N. H., Case, M. E., Baum, J., Geever, R., Huiet, L., Patel, V., & Tyler, B. (1985). Gene organization and regulation in the qa (quinic acid) gene cluster of *Neurospora crassa. Microbiological Reviews, 49*(3), 338–358.

Goll, M. G., Anderson, R., Stainier, D. Y., Spradling, A. C., & Halpern, M. E. (2009). Transcriptional silencing and reactivation in transgenic zebrafish. *Genetics, 182*(3), 747–755. http://dx.doi.org/10.1534/genetics.109.102079.

Hans, S., Kaslin, J., Freudenreich, D., & Brand, M. (2009). Temporally-controlled site-specific recombination in zebrafish. *PLoS One, 4*(2), e4640. http://dx.doi.org/10.1371/journal.pone.0004640.

Hesselson, D., Anderson, R. M., Beinat, M., & Stainier, D. Y. (2009). Distinct populations of quiescent and proliferative pancreatic beta-cells identified by HOTcre mediated labeling. *Proceedings of the National Academy of Sciences of the United States of America, 106*(35), 14896–14901. http://dx.doi.org/10.1073/pnas.0906348106.

Huang, W., Wang, G., Delaspre, F., Vitery Mdel, C., Beer, R. L., & Parsons, M. J. (2014). Retinoic acid plays an evolutionarily conserved and biphasic role in pancreas development. *Developmental Biology, 394*(1), 83–93. http://dx.doi.org/10.1016/j.ydbio.2014.07.021.

Kawakami, K., Abe, G., Asada, T., Asakawa, K., Fukuda, R., Ito, A., … Yoshida, M. (2010). zTrap: zebrafish gene trap and enhancer trap database. *BMC Developmental Biology, 10*, 105. http://dx.doi.org/10.1186/1471-213X-10-105.

Kawakami, K., Shima, A., & Kawakami, N. (2000). Identification of a functional transposase of the Tol2 element, an Ac-like element from the Japanese medaka fish, and its transposition in the zebrafish germ lineage. *Proceedings of the National Academy of Sciences of the United States of America, 97*(21), 11403–11408. http://dx.doi.org/10.1073/pnas.97.21.11403.

Kimura, Y., Hisano, Y., Kawahara, A., & Higashijima, S. (2014). Efficient generation of knock-in transgenic zebrafish carrying reporter/driver genes by CRISPR/Cas9-mediated genome engineering. *Scientific Reports, 4*, 6545. http://dx.doi.org/10.1038/srep06545.

Koster, R. W., & Fraser, S. E. (2001). Tracing transgene expression in living zebrafish embryos. *Developmental Biology, 233*(2), 329–346. http://dx.doi.org/10.1006/dbio.2001.0242.

Kwan, K. M., Fujimoto, E., Grabher, C., Mangum, B. D., Hardy, M. E., Campbell, D. S., … Chien, C. B. (2007). The Tol2kit: a multisite gateway-based construction kit for Tol2 transposon transgenesis constructs. *Developmental Dynamics: an Official Publication of the American Association of Anatomists, 236*(11), 3088–3099. http://dx.doi.org/10.1002/dvdy.21343.

Luan, H., Peabody, N. C., Vinson, C. R., & White, B. H. (2006). Refined spatial manipulation of neuronal function by combinatorial restriction of transgene expression. *Neuron, 52*(3), 425–436. http://dx.doi.org/10.1016/j.neuron.2006.08.028.

Marquart, G. D., Tabor, K. M., Brown, M., Strykowski, J. L., Varshney, G. K., LaFave, M. C., … Burgess, H. A. (2015). A 3D searchable database of transgenic zebrafish Gal4 and Cre lines for functional neuroanatomy studies. *Frontiers in Neural Circuits, 9*, 78. http://dx.doi.org/10.3389/fncir.2015.00078.

Mosimann, C., Kaufman, C. K., Li, P., Pugach, E. K., Tamplin, O. J., & Zon, L. I. (2011). Ubiquitous transgene expression and Cre-based recombination driven by the ubiquitin promoter in zebrafish. *Development, 138*(1), 169–177. http://dx.doi.org/10.1242/dev.059345.

Otsuna, H., Hutcheson, D. A., Duncan, R. N., McPherson, A. D., Scoresby, A. N., Gaynes, B. F., … Dorsky, R. I. (2015). High-resolution analysis of central nervous system expression patterns in zebrafish Gal4 enhancer-trap lines. *Developmental Dynamics: an Official Publication of the American Association of Anatomists, 244*(6), 785–796. http://dx.doi.org/10.1002/dvdy.24260.

Pang, S. C., Wang, H. P., Zhu, Z. Y., & Sun, Y. H. (2015). Transcriptional activity and DNA methylation dynamics of the gal4/UAS system in zebrafish. *Marine Biotechnology (New York, N.Y.), 17*(5), 593–603. http://dx.doi.org/10.1007/s10126-015-9641-0.

Parsons, M. J., Pisharath, H., Yusuff, S., Moore, J. C., Siekmann, A. F., Lawson, N., & Leach, S. D. (2009). Notch-responsive cells initiate the secondary transition in larval zebrafish pancreas. *Mechanisms of Development, 126*(10), 898–912. http://dx.doi.org/10.1016/j.mod.2009.07.002.

Potter, C. J., & Luo, L. (2011). Using the Q system in *Drosophila melanogaster. Nature Protocols Other Titles: Protocols, 6*(8), 1105–1120. http://dx.doi.org/10.1038/nprot.2011.347.

Potter, C. J., Tasic, B., Russler, E. V., Liang, L., & Luo, L. (2010). The Q system: a repressible binary system for transgene expression, lineage tracing, and mosaic analysis. *Cell, 141*(3), 536–548. http://dx.doi.org/10.1016/j.cell.2010.02.025.

Riabinina, O., Luginbuhl, D., Marr, E., Liu, S., Wu, M. N., Luo, L., & Potter, C. J. (2015). Improved and expanded Q-system reagents for genetic manipulations. *Nature Methods, 12*(3), 219–222. http://dx.doi.org/10.1038/nmeth.3250, 215 p. following 222.

Rorth, P. (1996). A modular misexpression screen in *Drosophila* detecting tissue-specific phenotypes. *Proceedings of the National Academy of Sciences of the United States of America, 93*(22), 12418–12422.

Sato, T., Hamaoka, T., Aizawa, H., Hosoya, T., & Okamoto, H. (2007). Genetic single-cell mosaic analysis implicates ephrinB2 reverse signaling in projections from the posterior tectum to the hindbrain in zebrafish. *Journal of Neuroscience: the Official Journal of the Society for Neuroscience, 27*(20), 5271–5279. http://dx.doi.org/10.1523/JNEUROSCI.0883-07.2007.

Scheer, N., & Campos-Ortega, J. A. (1999). Use of the Gal4-UAS technique for targeted gene expression in the zebrafish. *Mechanisms of Development, 80*(2), 153–158.

Scott, E. K., Mason, L., Arrenberg, A. B., Ziv, L., Gosse, N. J., Xiao, T., … Baier, H. (2007). Targeting neural circuitry in zebrafish using GAL4 enhancer trapping. *Nature Methods, 4*(4), 323–326. http://dx.doi.org/10.1038/nmeth1033.

Subedi, A., Macurak, M., Gee, S. T., Monge, E., Goll, M. G., Potter, C. J., … Halpern, M. E. (2014). Adoption of the Q transcriptional regulatory system for zebrafish transgenesis. *Methods, 66*(3), 433–440. http://dx.doi.org/10.1016/j.ymeth.2013.06.012.

Suli, A., Guler, A. D., Raible, D. W., & Kimelman, D. (2014). A targeted gene expression system using the tryptophan repressor in zebrafish shows no silencing in subsequent generations. *Development, 141*(5), 1167–1174. http://dx.doi.org/10.1242/dev.100057.

Takeuchi, M., Matsuda, K., Yamaguchi, S., Asakawa, K., Miyasaka, N., Lal, P., … Hibi, M. (2015). Establishment of Gal4 transgenic zebrafish lines for analysis of development of cerebellar neural circuitry. *Developmental Biology, 397*(1), 1–17. http://dx.doi.org/10.1016/j.ydbio.2014.09.030.

Urasaki, A., Morvan, G., & Kawakami, K. (2006). Functional dissection of the Tol2 transposable element identified the minimal cis-sequence and a highly repetitive sequence in the subterminal region essential for transposition. *Genetics, 174*(2), 639–649. http://dx.doi.org/10.1534/genetics.106.060244.

Walker, C. (1999). Haploid screens and gamma-ray mutagenesis. *Methods in Cell Biology, 60*, 43–70.

Wei, X., Potter, C. J., Luo, L., & Shen, K. (2012). Controlling gene expression with the Q repressible binary expression system in *Caenorhabditis elegans*. *Nature Methods, 9*(4), 391–395. http://dx.doi.org/10.1038/nmeth.1929.

CHAPTER

11

Contemporary zebrafish transgenesis with *Tol2* and application for Cre/*lox* recombination experiments

A. Felker, C. Mosimann[1]

University of Zürich, Zürich, Switzerland

[1]*Corresponding author: E-mail: christian.mosimann@imls.uzh.ch*

CHAPTER OUTLINE

Methods in Cell Biology, Volume 135, ISSN 0091-679X, http://dx.doi.org/10.1016/bs.mcb.2016.01.009

Abstract

Spatiotemporal transgene regulation by transgenic DNA recombinases is a central tool for reverse genetics in multicellular organisms, with excellent applications for misexpression and lineage tracing experiments. One of the most widespread technologies for this purpose is Cre recombinase-controlled *lox* site recombination that is attracting increasing interest in the zebrafish field. *Tol2*-mediated zebrafish transgenesis provides a stable platform to integrate *lox* cassette transgenes, while the amenability of the zebrafish embryo to drug treatments makes the model an ideal candidate for tamoxifen-inducible CreERT2 experiments. In addition, advanced transgenesis technologies such as *phiC31* or CRISPR-Cas9-based knock-ins are even further promoting zebrafish transgenesis for Cre/*lox* applications. In this chapter, we will first introduce the basics of Cre/*lox* methodology, CreERT2 regulation by tamoxifen, as well as the utility of *Tol2* and other contemporary transgenesis techniques for Cre/*lox* experiments. We will then outline in detail practical experimental steps for efficient transgenesis toward the creation of single-insertion transgenes and will introduce protocols for 4-hydroxytamoxifen-mediated CreERT2 induction to perform spatiotemporal *lox* transgene regulation experiments in zebrafish embryos. Last, we will discuss advanced experimental applications of Cre/*lox* beyond traditional lineage tracing approaches.

1. INTRODUCTION

1.1 TRANSGENESIS

A transgene is an experimentally constructed piece of DNA integrated into the genome of a recipient organism. Once integrated into germ cells, subsequent generations genetically inherit the transgene, referred to as stable transgene transmission. Engineered transgenes allow testing of molecular mechanisms by expressing a desired gene at different developmental times and in different cell types. Transgenes can further express an experimentally detectable marker in particular cell types or cellular structures for detailed, high-resolution observations.

Experimentally controlling spatiotemporal transgene expression by regulated DNA recombination has become a central method in genetics. A transgenic DNA recombinase enzyme recognizes two genomic DNA target sequences in *cis* or in *trans* and recombines the target DNA sequences. Consequently, the recombinase removes or inverts a piece of transgenic DNA flanked by target sequences, resulting in a recombined DNA product based on the initial relative orientation of the target sequences. In mice and *Drosophila*, cassette excision is a classic application wherein a recombinase excises a target sequence-flanked transgene cassette. The zebrafish field is increasingly embracing this powerful genetic tool since the introduction of efficient single-insertion transgenesis.

1.2 THE Cre/*lox* SYSTEM AND INDUCIBLE CreERT2

The current workhorse for spatiotemporally controlled recombination genetics is the bacteriophage P1-derived cyclic recombinase Cre, which catalyzes the site-specific recombination between 34-base pair (bp) inverted repeat *lox* sites (from *locus of X-ing over*) (Sauer, 1987; Sauer & Henderson, 1988). Cre-mediated recombination of two *lox* sites in head-to-head orientation inverts the intervening DNA cassette, whereas two *lox* sites in tandem orientation cause circularization, excision, and subsequent loss of the cassette, leaving one functional *lox* site at the transgenic locus. Several variations of the original *loxP* sites have been created by mutagenesis (such as *lox2272* and *loxN*) (Livet et al., 2007), which can be combined in the same transgene to create mutually exclusive, predictable cassette recombination patterns. Of note, the term *loxP* is frequently used to describe *lox* sites in general, but is also used to specifically refer to the original *loxP* sequence (Fig. 1). In our text, we call general recombination sites *lox* and use *loxP* in individual cases (and to the best of our knowledge) to refer to the original sequence.

Cre/*lox* recombination finds widespread application as a genetic two-component transgene system (Sauer, 1998): in the default state, transgene "*A*" drives expression of a *lox*-flanked cassette, usually containing an *open reading frame* (*ORF*) or *STOP* cassette (Fig. 1). Cre expressed from transgene "*B*" excises the *lox* cassette and brings a second cassette coding for a gene of interest, here referred to as *cargo,* under direct control of the promoter in transgene "*A*". Thus expression is switched from the cassette template or a stopped condition to *cargo* transcription. Such modular transgenes, which we refer to here as *lox* switch constructs, are key tools for lineage tracing and conditional loss- and gain-of-function experiments in the mouse (Bailey,

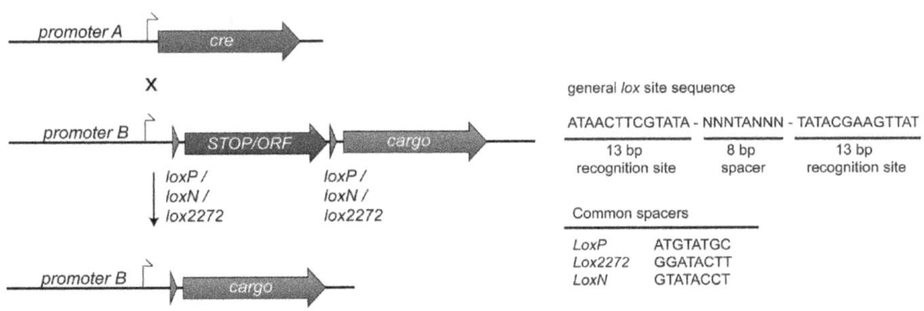

FIGURE 1 Schematics of the Cre/*lox* system using a two-component transgenic switching approach.

Cre is driven by *promoter A*, most commonly restricting its expression to a specific cell type or lineage and providing spatial control. The *lox* switch expresses an *ORF* (or a transcriptional *STOP* cassette) framed by *lox* recombination sites (36 bp) under the control of *promoter B* in the default state. In switch constructs, this cassette is followed by a second *ORF*, named *cargo* in this review. In cells where *promoter A* is active, Cre recombinase recombines the *lox* sites and excises the flanked cassette, putting the *cargo* under the control of *promoter B*.

Creamer, & Hollingsworth, 2009; Sauer, 1998). Importantly, the Cre/*lox* system uncouples the Cre-driving promoter from the effector transgene that is mostly driven by a constitutively active *cis*-regulatory element to achieve stable transgene activation in the progeny of all genetically recombined cells.

To achieve not only spatial but also tightly regulated temporal control over recombination, Cre is fused to the human ligand-binding domain of the estrogen receptor (ER), resulting in posttranslational estrogen-dependent activation of CreER (Feil et al., 1996; Metzger, Clifford, Chiba, & Chambon, 1995). In the default state, heat shock protein 90 (Hsp90) binds the incompletely folded ER and traps the receptor in the cytoplasm. Binding of estrogen or structurally related steroids induces a conformational change in the ER ligand-binding domain, which subsequently sheds Hsp90 and translocates CreER to the nucleus. By this mechanism, a CreER fusion protein cannot reach nuclear *lox* sites until it binds estrogen. High expression of the CreER protein can, however, overwhelm endogenous Hsp90 levels and cause "leaky" *lox* cassette recombination. To circumvent potential CreER activation by endogenous estrogen, mutant versions of the ER domain have been developed that have high affinity for estrogen derivatives not found in metazoans. The most commonly used version is CreERT2, which is insensitive to natural estrogen but binds tamoxifen (TAM) or its derivate 4-hydroxytamoxifen (4-OHT, Fig. 2) (Feil et al., 1996; Feil, Wagner, Metzger, & Chambon, 1997). In mice, body size poses an obstacle to TAM or 4-OHT administration and distribution; in contrast, zebrafish embryos are small and allow direct 4-OHT addition to the embryo medium, where 4-OHT is rapidly taken up and triggers fast CreERT2 responses (Hans, Kaslin, Freudenreich, & Brand, 2009; Mosimann et al., 2011).

1.3 Cre/*lox* AND ZEBRAFISH TRANSGENESIS: CURRENT STATUS

The first efforts to establish Cre/*lox* transgenes in zebrafish relied on rare integration of concatemers of linearized plasmid DNA (Iyengar, Muller, & Maclean, 1996; Stuart, McMurray, & Westerfield, 1988). Concatemeric insertions result in tandem repeats of *lox* cassette transgenes. As Cre can act on any two *lox* sites in the multicopy concatemer, individual embryonic cells may retain different copy numbers of the recombined transgene or experience unproductive recombination. Consequently, reported recombination efficiencies were low (Feng et al., 2007; Hans et al., 2009; Le et al., 2007). Furthermore, high-copy-number concatemeric transgenes are prone to silencing and to DNA repair-mediated recombination, leading to variable transgene expression or loss over successive generations.

Single-insertion transgenesis via *Tol2* transposons resolves these issues (Hans et al., 2009; Kawakami, 2007) and has triggered a strong increase in Cre/*lox* applications in zebrafish. The first *Tol2*-integrated Cre/*lox* transgenes in zebrafish contained *hsp70*-controlled *cre* (*hsp70:cre* and *hsp70:EGFP-cre*) offering temporal control through the heat shock-inducible activity of the promoter (Le et al., 2007; Thummel et al., 2005). However, this strategy has been found to be problematic

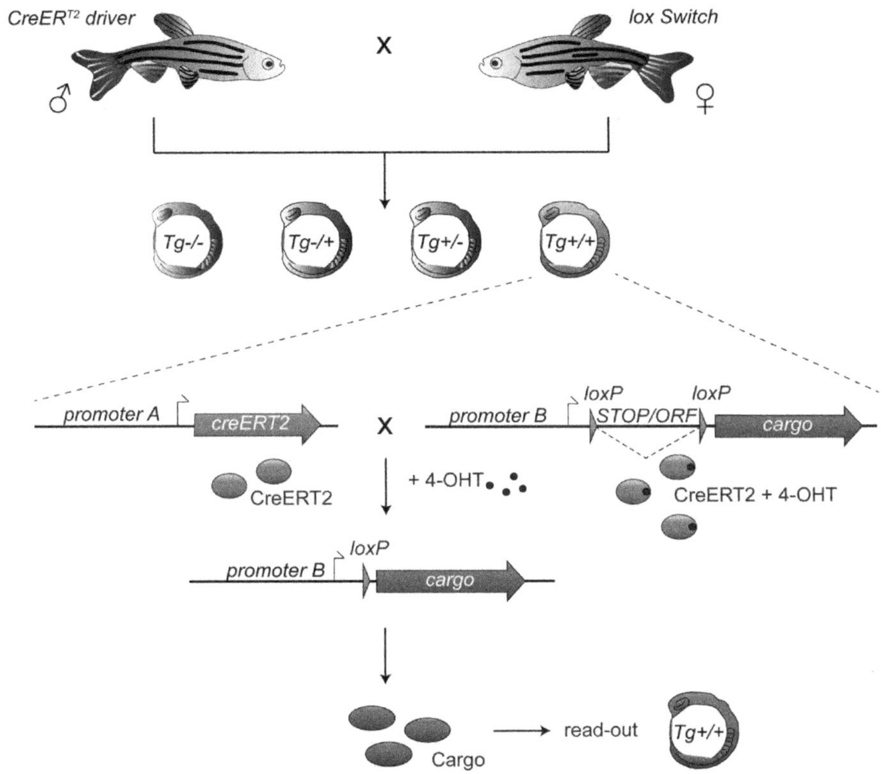

FIGURE 2 Chemically inducible CreERT2-mediated recombination.

Male *creERT2* transgenics are crossed to female *lox* switch transgenics to avoid maternal contribution of CreERT2. CreERT2 is sequestered in the cytoplasm in the default state and CreERT2-mediated *lox* excision is only activated upon addition of 4-OHT, providing tight temporal control over recombination. (See color plate)

due to the "leakiness" of the promoter with potential constitutive expression of the transgene at low levels presumably resulting from *hsp70* sensitivity to various stresses and transgene locus-dependent effects (Hans et al., 2010). Furthermore, *hsp70* transgenes express during oogenesis, and *hsp70:cre* features maternally contributed constitutive Cre activity in the embryo (Hans et al., 2010, 2009). Other pioneering efforts employed *rag2* or *β-actin* promoter-controlled *loxP* cassettes followed by *c-MYC* or *kRAS*^{V12} oncogenes; switching to oncogene expression by injecting Cre-encoding messenger RNA (mRNA) into transgenic embryos or by heat-induced expression of *hsp70:cre* established first inducible tumor models in zebrafish (Feng et al., 2007; Langenau et al., 2005; Le et al., 2007). More tightly controlled and specific recombination could be achieved establishing ubiquitously expressed inducible *creERT2* transgenics such as *ubi:creERT2* (Mosimann et al., 2011), and a growing number of reported lines with tissue-specific CreERT2 expression including *myl7:creERT2*, *gata4:creERT2*, *hmcn2:creERT2*, *krttlc19e:CreERt2*, *sox10:creERT2*, or *drl:creERT2* (Boniface, Lu, Victoroff, Zhu, & Chen, 2009; Hans et al., 2010; Jopling et al., 2010; Kikuchi et al., 2010; Lee, Asharani, & Carney, 2014, Lee, Knapik, Thiery, Carney, 2013; Mongera et al., 2013; Mosimann et al., 2015).

Overall, the zebrafish field has seen a steady increase in established tissue-specific *cre* or *creERT2* transgenic strains; nonetheless, reliable *lox* reporter transgenes are still limited in number. Both components are required for synergistic development of the technology. Several reports have described *lox* switch lines with *Xef1α* or *β-actin* promoter-controlled *lox* cassettes containing fluorescent protein genes (Bertrand et al., 2010; Boniface et al., 2009; Hans et al., 2010, 2009; Langenau et al., 2005; Yoshikawa, Kawakami, & Zhao, 2008). These promoters, however, do not drive ubiquitous transgene expression in embryonic or adult tissues, comparable to the gold standard *Rosa26* transgenes in mice (Soriano, 1999; Zambrowicz et al., 1997). We have previously isolated the *ubiquitin* (*ubi*, also referred to as *ubiquitin B*, *ubb*) promoter, which is highly active at all stages of zebrafish development and in adulthood, and have created the Cre-sensitive *ubi:loxP-EGFP-loxP-mCherry* (*ubi:Switch*) transgene (Mosimann et al., 2011). Besides serving as a sensitive lineage tracing tool, *ubi:Switch* facilitates screening of new tissue-specific *creERT2* drivers and assessment of their temporal 4-OHT responsiveness. *ubi:switch* has been successfully used for several lineage tracing approaches (Ghaye et al., 2015; Lee et al., 2014; Mosimann et al., 2015; Zhao et al., 2014) and the ubiquitous expression of *ubi*-driven *lox* cassettes has stimulated the generation of several new transgenes for this purpose and lineage-specific genetic manipulation (Huang et al., 2014; Park & Leach, 2013; Zhou et al., 2011). Based on the *ubi* 5′ elements, we further engineered *ubi:creERT2* as a zebrafish line for ubiquitous expression of 4-OHT-inducible CreERT2 to simplify screening for functional *lox* switch transgenes (Centanin et al., 2014; Huang et al., 2014; Mosimann et al., 2011).

The demand for Cre/*lox* transgenic technology applicable to the zebrafish is high and likely increasing with the advent of introducing *lox* sites or even bigger cassettes into the genome using matured editing technologies such as *phiC31* or CRISPR/Cas9-mediated genome integration. In the following sections we discuss key points concerning transgenesis, such as basic transgene design, *Tol2* transposon handling, and postinjection genetics for creation of high-quality single-insertion transgenic lines. Subsequently, we outline the key principles and protocols for Cre/*lox* experimentation in zebrafish including CreERT2 induction by 4-OHT and the application of transgenic lines for lineage tracing and general transgene manipulation.

2. TRANSGENE DESIGN, CLONING, AND TRANSGENESIS

2.1 BASIC CONSIDERATIONS OF TRANSGENE DESIGN

Transgene-driven recombination experiments in vivo critically depend on efficient gene expression from the involved transgenes. Transcription of mRNA from a transgene and final protein translation is a multistep process including several co- and posttranscriptional processes such as 5′-capping, intron excision, and 3′-polyadenylation. Here, we consider the basic principles of eukaryotic RNA polymerase II-based gene expression and their systematic implementation to create effective transgenes.

The minimal features of a protein-coding, mRNA-expressing transgene are the following:

1. A *cis*-regulatory element that contains, or is paired with, a minimal promoter with transcription start site. Enhancer elements missing a minimal promoter, as when working with intronic or downstream *cis*-regulatory elements of a gene, require coupling with a minimal promoter. Several minimal promoter elements have been reported in the literature to successfully control transgenes in zebrafish, including upstream elements from the zebrafish *gata2* gene or of the mouse *β-globinE1* gene (Bessa et al., 2009; Tamplin et al., 2015).
2. A 5′-untranslated region (5′-*UTR*) sequence recognized by the ribosome to initiate efficient translation at the start codon of the desired protein-coding region. In practical terms, the *5′-UTR* sequence is often defined by the used (minimal) promoter, which provides the transcription start site and 5′ sequence of the resulting mRNA.
   In 1987, Marilyn Kozak described a common consensus sequence *5′-cgccRcc<u>ATG</u>G-3′*, in which the key residue is a purine (abbreviated *R*) three bases upstream (position *-3*) of the protein's start codon (underlined), consequently named the Kozak sequence (Kozak, 1987, 1999). Common cloning vectors contain some variant of the consensus, such as the minimal *5′-caccATG-3′*. Kozak sequences show many gene-specific variations, which reflects both the recognition tolerance of the ribosome and gene-specific regulatory features that modulate translation efficiency. Recent studies revealed that the most common variation of the canonical Kozak sequence found in zebrafish is *5′-aaacATGGC-3′* and experimentally enhances protein expression, suggesting that small variations in Kozak sequence can tweak translation efficiency (Grzegorski, Chiari, Robbins, Kish, & Kahana, 2014). Nevertheless, transgenes using the *5′-cgccacc<u>ATG</u>-3′* consensus or even *5′-cacc<u>ATG</u>-3′* are efficiently translated in zebrafish. 5′-UTRs with in-frame and out-of-frame *ATG* triplets upstream of the main *ATG* may severely impair efficient translation of the desired transgene protein (Kozak, 1996) and should be avoided when designing control elements.
3. A protein-coding *ORF*, starting with an *ATG* in a Kozak consensus context and ending in a stop codon (*TAG*, *TAA*, *TGA*).
4. A *3′-UTR* with a polyadenylation signal sequence (i.e. *5′-AAUAAA-3′*). As for *5′-UTRs*, little information is available on transgene-tested *3′-UTRs* and their impact on RNA stability or protein expression. The minimal *3′-UTRs* of *SV40* and the *Bovine Growth Hormone* (*BGH*) gene are used in the vast majority of zebrafish transgenes reported to date. Both are relatively short (~200 bp) and can be included as a module after an ORF to ensure efficient polyadenylation.

Of note, the SV40-derived polyadenylation signal is bidirectional (Carswell & Alwine, 1989), and although both directions are used for transgenesis, only little is known about their impact on different transgenes (Fink, Flekna, Ludwig, Heimbucher, & Czerny, 2006).

The minimal modules to drive protein production therefore contain a *5′-UTR-Kozak-[ORF]-3′-UTR-polyA* transgene plus a suitable promoter (Fig. 3). Assembly of the modules required in a transgene is easily accomplished via contemporary recombinant DNA methods, such as the MultiSite Gateway® in vitro recombination cloning system (Invitrogen Corp., Carlsbad, CA, United States) (Kwan et al., 2007). Transgenes intended to express noncoding RNAs require different structural features, which are generally poorly defined. *lox* sites are small (36 bp) and can be flexibly placed at any position of a transgene, most often flanking the first ORF of a switching transgene (Fig. 1).

2.2 TRANSGENESIS FOR SINGLE INSERTIONS

True single-insertion transgenesis in zebrafish became possible when the Medaka-derived *Tol2* transposon was engineered (Kawakami, 2007; Kawakami et al., 2004), minimally containing a gene cassette and flanking sequences for transposase-catalyzed genomic integration (200 bp at the *5′*-end and 150 bp at the *3′*-end).

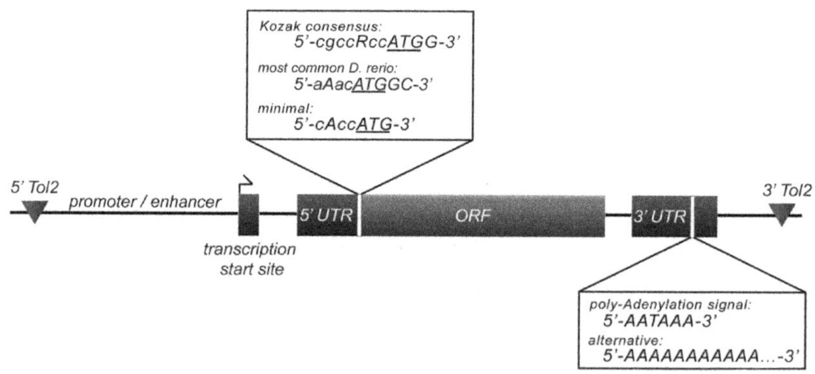

FIGURE 3 Overview of individual components required for efficient expression of a transgene *ORF* from a given promoter/enhancer.

The *5′*-regulatory element (promoter or enhancer) contains a transcription start site as a docking station for the polymerase. When not inherently included in a given enhancer, a minimal promoter is placed between enhancer and *ORF*. The translation start site is preceded with the Kozak sequence which is variable and can lead to differential expression levels. A polyadenylation signal in the *3′-UTR* is important for genuine mRNA processing. Of note, *UTRs* on both sites of the *ORF* are reduced to a minimum (Kozak in *5′* and SV40 polyA signal in *3′*) in some transgenes such as the commonly used *EGFP* or *creERT2* plasmids.

The *5′*- and *3′*- *Tol2* arms include 12-bp terminal inverted repeats and subterminal regions that are recognized by the *Tol2* transposase and are joined to the transgene cassette, flanking it in a plasmid vector backbone. For transgenesis, one-cell-stage zebrafish embryos are injected with a mix of *Tol2* transposase-encoding mRNA and *Tol2* transposon transgene vector. *Tol2* transgenes do not seem to be size-restricted; indeed, *Tol2* transposons have successfully integrated bacterial artificial chromosome transgenes in the 70-kb range into the zebrafish genome (Suster, Sumiyama, & Kawakami, 2009). Nonetheless, data regarding genomic integration locus preferences are limited (Kondrychyn, Garcia-Lecea, Emelyanov, Parinov, & Korzh, 2009). *Tol2* transposons efficiently transmit through the germline, and we routinely observe *Tol2* transposon germline transmission frequencies of 20–100% of injected F0 animals, with 3–100% of F1 embryos per clutch positive for the transgene, similar to the published range (Kawakami, 2007). *Tol2* transposons are therefore ideally suited for complex transgenes, such as complex *lox* switch vectors.

While *Tol2* enables efficient transgenesis, particular drawbacks call for alternatives: multiple integrations at random genomic loci impose complicated crossing schemes and screening through multiple generations to generate stable single-insertion transgenics. *phiC31*-mediated recombination, the predominant transgenesis technique in *Drosophila* (Bischof, Maeda, Hediger, Karch, & Basler, 2007; Venken & Bellen, 2005), is now also implemented in zebrafish (Hu, Goll, & Fisher, 2011; Kirchmaier et al., 2013; Mosimann et al., 2013; Roberts et al., 2014). In the *phiC31* system, transgene recombination of heterotypic binding sites termed *attB* (*attachment site Bacterium*) and *attP* (*attachment site Phage*) is catalyzed by the serine family SSR *phiC31* integrase from the *phiC31* phage (Groth, Fish, Nusse, & Calos, 2004). Recombination at *attP/attB* sites creates *phiC31* integrase-incompatible hybrid *attL* and *attR* (Left and Right) sites, leading to irreversible transgene integration (Groth, Olivares, Thyagarajan, & Calos, 2000). To achieve transgenesis, *attP* sites previously introduced into the zebrafish genome are recombined with *attB* site-containing transgenesis vectors, which are coinjected with *phiC31* integrase-encoding mRNA (Fig. 4) (Hu et al., 2011; Kirchmaier et al., 2013; Lu, Maddison, & Chen, 2011; Mosimann et al., 2013; Roberts et al., 2014). This approach depends on zebrafish lines carrying single insertions of *attP* sites, so far generated through *Tol2* transgenesis and validated through genomic mapping of integration loci and thorough follow-up of transgene stability throughout several generations. *attB*-containing vectors that carry a flexible MultiSite Gateway-compatible transgenesis cassette can be integrated at one stable locus, leading to single-insertion transgenics with comparable transgene expression levels within one generation (Mosimann et al., 2013; Roberts et al., 2014). Targeted knock-in of *attP* sites through CRISPR-Cas9-mediated genome editing provides a novel approach to integrate transgenes into a desired genomic location of choice, as DNA oligomers have been successfully knocked-in using this method (Gagnon et al., 2014). Mapping of previously characterized transgenes with faithful expression over several generations would be greatly desirable to identify suitable transgene integration spots in the zebrafish genome. In our hands, integration of short

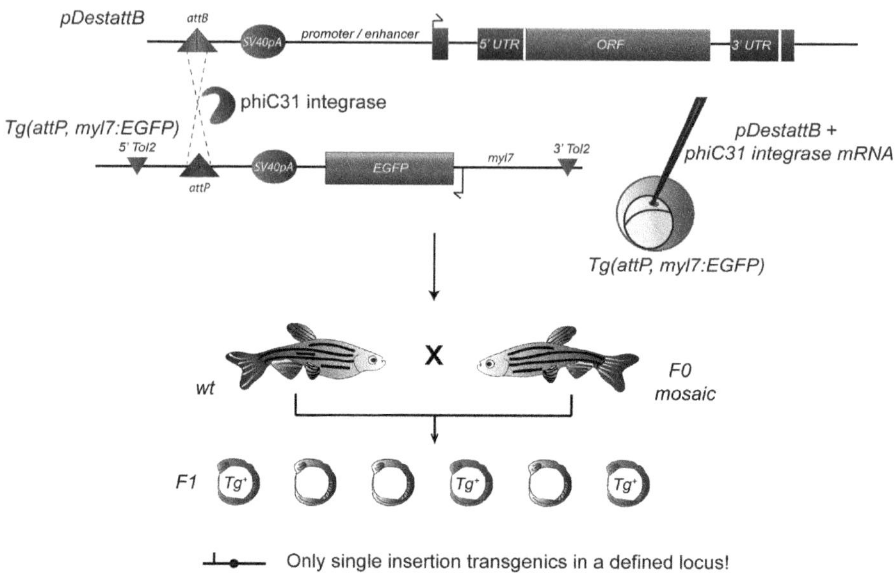

FIGURE 4 The phiC31 integrase technology for locus-defined single-insertion transgenesis.

See text for details. Note that *attP* site transgenics have to be established via *Tol2* transgenesis. *attP* recombination is achieved through simultaneous injection of a *pDestattB* vector (containing the transgene of choice) and *phiC31* mRNA. (See color plate)

oligomers up to 200 bp into a CRISPR-Cas9-mediated double-strand break is highly efficient and reproducibly results in high germline transmission. On the contrary, the knock-in of long cassettes encoding for genes such as fluorescent reporters are feasible but more challenging and variable for both homology-independent double-strand break repair (Auer, Duroure, De Cian, Concordet, & Del Bene, 2014) and homology recombination (HR)-based approaches (Hisano et al., 2015; Irion, Krauss, & Nusslein-Volhard, 2014; Kimura, Hisano, Kawahara, & Higashijima, 2014).

2.3 REAGENT CONSIDERATIONS FOR TRANSGENESIS

Tol2 mRNA and the *Tol2* transposon transgene vector are easily generated and purified using standard laboratory techniques and commercial kits. We emphasize that optimal transgenesis critically depends on proper preparation and quality. To generate *Tol2* mRNA, we linearize the *pCS3FA-transposase* vector (Tol2kit *#396*) (Kwan et al., 2007) by digestion with NotI, purify the plasmid template by agarose gel electrophoresis, and in vitro transcribe *5′*-capped mRNA using Sp6 polymerase (mMESSAGE mMACHINE kit, Applied Biosystems/Ambion, Austin, TX, USA). We store *Tol2* mRNA at −80°C in single-use aliquots of 2.5–3 μL at 100 ng/μL. Our standard 10 μL injection mix to generate stable transgene integrations contains

25 ng/μL transgene vector plus 15–25 ng/μL *Tol2* mRNA. The same principles hold true for *phiC31*-mediated transgenesis, with the *phiC31* mRNA and suitable transgenesis vectors processed akin to Tol2 reagents (Mosimann et al., 2013).

Routine laboratory practice should include quality-control of *Tol2* or *phiC31* mRNA. mRNA quality can be reliably assessed by following integration efficiency of an easily monitored transgene, such as *ubi:EGFP* (Mosimann et al., 2013, 2011). When coinjected with functional *Tol2* mRNA, *ubi:EGFP* yields extensive mosaic expression of Enhanced Green Fluorescent Protein (EGFP), which is still detectable 3–4 weeks postinjection. In comparison, injection of *ubi:EGFP* plasmid DNA alone leads to a markedly weaker EGFP mosaicism. The fluorescence fades over the course of a few weeks, as the vector rarely integrates into the genome and the plasmid is degraded. Observation of adult EGFP expression is facilitated by ultraviolet (UV) headsets ("goggles") allowing nonintrusive fluorescence observation (Mosimann et al., 2011) (ie, BlueStar flashlight package or Dual Fluorescent Protein Flashlight from Nightsea, Lexington, MA, United States). To avoid inefficient transgenesis rates, *Tol2* mRNA batches that do not result in extensive EGFP mosaicism over several weeks should be discarded.

The purity of the *Tol2* transposon and *phiC31* transgene DNA is critical to successful transgenesis. Contaminants, such as ethanol, chaotropic salts, bacterial DNA/RNA, or RNase, severely impair embryo viability, and likely, transposase/integrase mRNA stability. Injection-grade DNA preparations (ie, minipreps) should feature a 260/230 nm ratio of 1.8–1.9 (lower values indicate salt or ethanol contamination), and a 260/280 nm ratio of 2.0–2.3 (values below 2.0 indicate protein contamination). Altogether, a thorough assessment of injection reagents is key for successful transgenesis, but may not overcome difficulties in transgene expression resulting from the integration locus or negative selection of transgenic cells due to biological effects.

3. *Tol2* TRANSGENE GENETICS

3.1 "EVERY FISH IS UNIQUE": BASIC GENETIC CONSIDERATIONS FOR *Tol2* TRANSGENESIS

Tol2-mediated transgenesis is the most commonly used and established technique within the zebrafish community to date. Therefore, understanding the underlying genetic principles of *Tol2* transgenesis is indispensable for the successful generation of new transgenic lines. Upon injection of the *Tol2* transgenesis mix into one-cell-stage embryos, some, but not all, of the cells in cleavage will successfully catalyze *Tol2* integration. The resulting embryo is thus mosaic for the transgene, with each cell potentially carrying 0–10 or more integrations spread randomly across the 50 zebrafish chromosomes (Kawakami, 2007). Consequently, the germ cells of a mosaic F0 animal will differ in transgene composition and will transmit transgenic chromosomes at an unpredictable ratio to the next generation. The non-Mendelian nature of transgene inheritance typically yields 3–100% positive embryos, and each embryo will harbor an unpredictable number of independent insertions. Consequently, achieving predictable Mendelian inheritance of functional transgenes (that is, reliable single insertions) frequently requires several generations of outcrosses and each fish of the first generation is considered a unique transgene carrier. In the following paragraphs, we discuss and elaborate on distinct points during the process:

1. Microinjected zebrafish (F0 generation) are screened for transgene transmission by outcrossing to wild-type zebrafish (Fig. 5). Each *Tol2* transposon-injected F0 animal will give rise to a distinctive collection of F1 animals. F0 parents with transgene transmission to the F1 are subsequently called F0 founders. A possible workflow starts by injecting *AB* or *Tu* wild-type strains as the F0 and screening for F0 founders by mating to the *TL* line, which contains the mutations *tuepfel* (*tup*, also called leo^{t1}, recessive mutation causing spotted pigment instead of stripes) and *longfin* (lof^{dt2}, dominant homozygous viable mutation causing overgrowing fins) (ZFIN, http://zfin.org). Mating of injected *AB* or *Tu* F0 to *TL* enables simple separation of the parents after the cross and introduces the leo^{t1} and lof^{dt2}markers into transgenic lines. When the transgene is coupled to a transgenic marker in *cis* (see Steps and Challenges generating Novel *creERT2* and *lox* Lines), we prioritize F0 animals with visible transgene marker expression as it indicates successful injection (but does not predict successful transgene transmission).
2. Individual F0 animals that provide offspring are consecutively numbered and kept in isolation. Their F1 embryos are subsequently screened for transgene integration by fluorescence microscopy or by polymerase chain reaction (PCR). If a reasonable clutch size (50–100 embryos) is transgene-negative, the corresponding F0 animal is culled. If the clutch is positive, the positive F1 embryos are raised, and the F0 parent is kept for further rounds of outcrosses.
3. Each F1 offspring is still potentially unique due to unpredictable transgene segregation in the F0 germline. For example, fluorescent reporter transgenes may reveal mixed transgene numbers in F1 clutches by variable fluorophore intensity between siblings. Transgenes can also independently insert onto the same chromosome and cause complex inheritance due to chromosome crossover (Fig. 5). We label the F1 offspring of a given F0 animal by adding a consecutive letter to the F0 designation (ie, 1_A, 1_B, 1_C). Between two and five transgene-positive F1 animals from an F0 founder are again individually outcrossed to wild type to assess transgene segregation in the F2 generation. If the F2 embryos are 50% transgene-positive, then it is likely that the F1 parent contains a single transgenic insertion (or may have two or more closely linked insertions on the same chromosome; see also Fig. 5). Higher F2 percentages reveal multiple transgene insertions. In the latter case, if required, several individual F2 animals from a single F1 parent are further outcrossed to wild type

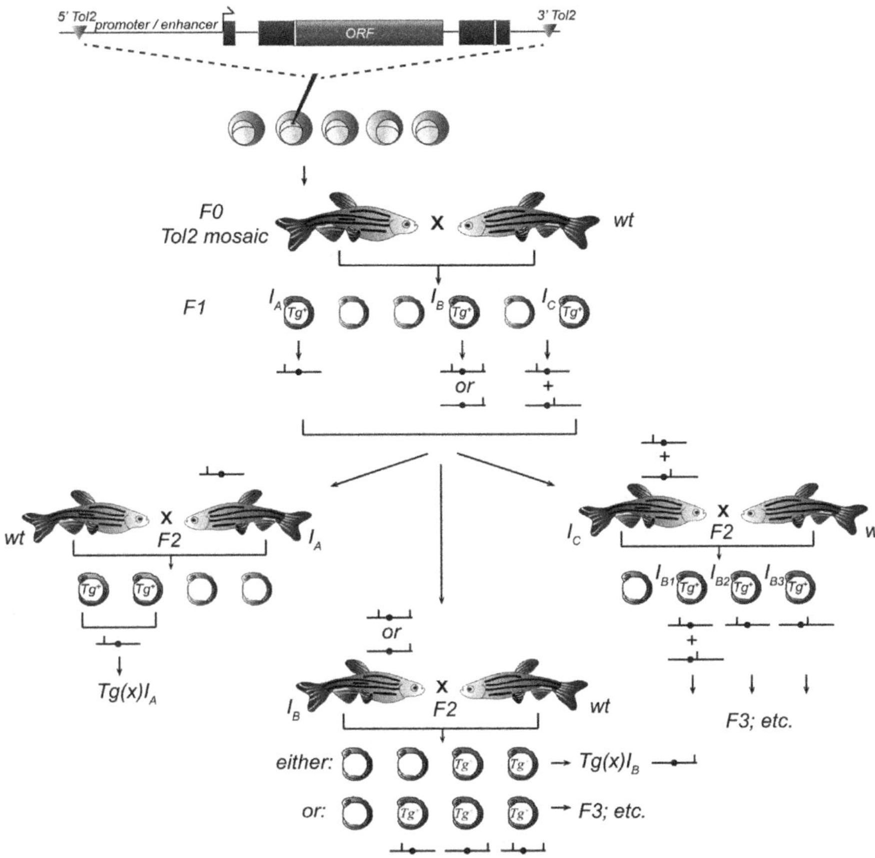

FIGURE 5 Transgenic crossing scheme to screen single-insertion *Tol2* transgenesis.

See text for details. Note the three different possible combinations of transgene insertions in the F1, labeled I_A, I_B, and I_C, and the resulting downstream crossing scheme required to obtain single-insertion transgenic strains. Chromosome schematics contain either one (I_A, I_C) or two (I_B) chromosomal insertions. In the case of two insertions on the same chromosome (I_B), the insertion number cannot be properly deduced by Mendelian genetics in the F1 but can be revealed in the F2 generation (or subsequent generations, depending on the centiMorgan distance of the two insertions). (See color plate)

until 50% segregation is observed (Fig. 5). Southern blot analysis for the integrated vector is a viable method to molecularly confirm single insertions (Kikuta & Kawakami, 2009).

4. *Tol2* transposons can also integrate onto the same chromosome and segregate by cross-over, resulting in >50% transgene transmission in subsequent generations. In contrast, we rarely observe <50% transgene segregation, which would

indicate transgene silencing after the F1; if observed, such lines are not pursued further. We regularly detect silenced transgene copies in transgene-negative embryos from F2 clutches by PCR (thus, undetectable by fluorescence microscopy), which stems from heterochromatin-mediated silencing or defective transposition in the F0.

As the field currently lacks a simple method to permanently label individual zebrafish, this labor-, time-, and space-consuming crossing scheme results in rapidly expanding numbers of individual zebrafish in isolation tanks. Therefore, we prioritize F1 animals with 50% segregation in the F2 generation since they present the greatest probability of already harboring single-copy insertions. *phiC31*-mediated transgenesis overcomes the problem of multiple insertions enabling the generation of single-insertion transgenics within one generation (Fig. 4) (Mosimann et al., 2013; Roberts et al., 2014).

3.2 TRANSGENE QUALITY ASSESSMENT

Single-insertion zebrafish transgenics feature predictable inheritance patterns and expression, yet long-term transgene stability is critical to sustainable experimentation. Transgene silencing can occur randomly and eliminates the utility of any established line. It is good practice to establish at least three (or more) independent transgenic lines for a given transgene. Several independent lines may further help to uncover individual position effects: eg, nonspecific transgene expression in inappropriate cell types. Complementary analyses, such as mRNA in situ hybridization (ISH) to detect *cre*/*creERT2* transgenes, also help to ensure faithful transgene expression.

When using fluorescent reporters, the commonly observed decrease in insertion copy number after the F1 generation will cause progressively dimmer, but more consistent, reporter fluorescence in subsequent generations until single-copy transgenes are established. Thus, F1 animals can lead to erroneous overestimations of reporter potency. In the interest of subsequent reproducibility, drawing experimental conclusions from F1 animals should be discouraged.

3.3 STEPS AND CHALLENGES GENERATING NOVEL *creERT2* AND *lox* LINES

The inducible Cre/*lox* system has revolutionized biological research by facilitating genetic lineage tracing, conditional alleles, and spatiotemporally controlled genetic modifications in living organisms. In mice, *creERT2* or *lox* transgenes/*lox* sites are commonly integrated into the genome through homologous HR, incorporating the transgene in the endogenous locus of the gene of choice and under the control of an endogenous promoter. The zebrafish has only recently started to embrace HR for transgene insertions, and widespread use is currently still restricted by variable efficiency. Thus, the most common generation of new *cre* driver lines requires the extraction and characterization of isolated regulatory elements from the zebrafish genome which are first characterized as fluorescent reporter lines and are then reintroduced into the genome as *creERT2*-coupled transgenes. This imposes the generation of two different sets of transgenic lines with double efforts of screening and validating transgene insertions. Moreover, in our hands, the generation of *lox* lines usually requires extensive screening through well above 10 founders to find transgenic lines with functional Cre-responsive recombination. Any transgene integration locus contributes chromatin effects that can influence Cre access, reactivity at *lox* sites, and resulting recombination efficiency due to unfavorable histone modifications or protein binding to DNA in the proximity (Hans et al., 2010, 2009; Mosimann et al., 2011). While position effect influence on recombination efficiency is a likely obstacle, we have currently no experimental evidence to concisely explain the variable *lox* efficiency.

When not inherently coupled with fluorescent reporters, screening and maintenance of Cre/*lox* lines are facilitated by fluorescent reporters and transgenes engineered with a fluorescent marker in *cis*. Zebrafish transgene markers express a fluorescent protein in a particular cell type or confined developmental structure and are cloned into the *Tol2* transgene vector backbone. Currently, the choice for zebrafish transgene markers is rather limited and will greatly benefit from future additions. The cardiomyocyte-specific *cmlc2:EGFP/myl7:EGFP* cassette is common in *Tol2* cloning vectors (Kwan et al., 2007). Other options are the eye lens-labeling *α-crystallin*-driven cassettes (Hesselson, Anderson, Beinat, & Stainier, 2009; Villefranc, Amigo, & Lawson, 2007), as well as the skin-specific *krt3:RFP* (Liu, Li, Emelyanov, Parinov, & Gong, 2008). Various *lox* transgenes contain a *lox*-flanked fluorescent cassette, such as *FlEx (Flip Excision)* or *ubi:Switch* (Boniface et al., 2009; Mosimann et al., 2011), inherently marking transgenic zebrafish before the switch.

When screening for positive founders, not all independent F1 lines carrying the transgenesis marker will express CreERT at sufficient levels, or have integrated a *lox* transgene that is readily accessible for Cre-mediated recombination. We routinely screen new Cre/*lox* lines by crossing to ubiquitously expressing CreERT or *lox* transgenes, namely, *ubi:creERT2* and *ubi:switch*. When screening for tissue-specific *creERT2* lines, crossing to *ubi:switch* reveals efficient switching from Green Fluorescent Protein (GFP) to mCherry expression in the desired tissues. Alternatively, when screening high numbers of founders, a preassessment by performing ISH for *cre*/*creERT2* expression can assist in ruling out lines that express CreERT2 nonspecifically or at low levels. When screening for new *lox* transgenic lines, we found *cre* mRNA injections (delivering 7.5 ng/μL as injection mix concentration, our unpublished results) to be a reliable technique to assess functional switching. Nonetheless, in our hands, crossing to *ubi:creERT2* followed by TAM induction produces more reliable results, as *cre* mRNA injections generally overestimate switching efficiencies.

4. CreERT2-CONTROLLED *lox* RECOMBINATION USING 4-OHT

Zebrafish embryos are ideal for lineage tracing studies as they develop rapidly, are nearly transparent, can be staged precisely, and are highly amenable to drug treatments. TAM and 4-OHT are lipid-soluble steroids that can pass through the chorion and cell membranes. While 4-OHT has a reported 30- to 100-fold higher affinity to ERT2 (Katzenellenbogen, Norman, Eckert, Peltz, & Mangel, 1984); the relative efficacies of TAM and 4-OHT for CreERT2 induction are difficult to assess due to their instability and markedly different water solubility (0.3 mg/L for TAM, 4-OHT is described as insoluble in water). Nonetheless, CreERT2 is sensitive to micromolar concentrations of both drugs in E3 embryo medium (Hans et al., 2010, 2009; Mosimann et al., 2011). Use of TAM, a prodrug that requires metabolic transformation to produce the higher affinity 4-OHT, introduces a potential lag in CreERT2 response. We predominantly use 4-OHT for our experiments due to its direct activity without need for prior metabolic conversion. As ERT2 preferentially binds the *trans* isomer (Katzenellenbogen et al., 1984), we advise the use of *trans*-4-OHT for induction experiments.

4.1 4-OHT CHEMISTRY AND HANDLING

Trans-4-OHT is commercially available from various sources (such as H7904 from Sigma–Aldrich, St. Louis, MO, United States). We routinely create 10 mM working stock solutions in dimethyl sulfoxide (DMSO) or ethanol and store 15 μL single-use aliquots in the dark at −20°C. Published studies also report 4-OHT stocks of 25 mM in ethanol and 50 mM TAM in DMSO (Hans et al., 2009). We and several others have found that prolonged storage decreases the potency of dissolved 4-OHT over time, with a drop in activity of more than 50% within 12 weeks (our unpublished observations). This occurs even under strict storage conditions at −20°C in the dark. This so far chemically uninvestigated decline of 4-OHT activity accounts for a lower reproducibility of switching efficiencies when using the same stock over time and thus can heavily influence biological results. Use of fresh or recently dissolved 4-OHT stocks is therefore advised. Undissolved 4-OHT powder kept in the dark at 2–8°C is stable for years.

CreERT2-mediated *lox* recombination is triggered at concentrations as low as 0.5 μM TAM or 4-OHT (Hans et al., 2009; Mosimann et al., 2011). Zebrafish embryos tolerate 10 μM 4-OHT treatment without obvious phenotypic consequences, and we routinely apply 10–20 μM to achieve maximal and saturated CreERT2 responses to test new transgenics or induction time points. As in comparable mouse experiments, the optimal TAM/4-OHT dose for a specific CreERT2 driver paired with a *lox* switch requires individual elucidation (Hans et al., 2009).

4.2 THE "GOLDEN RULE" FOR ZEBRAFISH LINEAGE TRACING: PATERNAL Cre, MATERNAL *lox* SWITCH

Successful Cre/*lox* experiments rely on faithful Cre activity in the investigated cell lineage. To avoid unwanted maternal recombinase contribution, we advise to use males as the source of the Cre driver (Fig. 2). Genuine examples of recombinase transgenes that contribute maternally are *ubi:creERT2* (Mosimann et al., 2011) and *hsp70:creERT2* lines (Hans et al., 2010, 2009). Tissue-specific *cre/creERT2* drivers may also unexpectedly contribute maternally due to chromatin position effects of individual transgene insertions or maternal activity of the promoter, as we frequently find while screening seemingly tissue-specific lines. Consequently, we recommend to always cross Cre transgene males to *lox* switch females for tissue-specific developmental lineage tracing. Cre drivers and *lox* transgenes should also not be propagated together in double-transgenic animals as leaky *cre/creERT2* transgenes would lead to irreversible recombination of the *lox* transgene.

Maternally contributed Cre may further lead to an overestimation of *lox* excision efficiency of new *lox* switch transgenes, such as when testing with the commonly available *hsp70:cre*. Nonetheless, maternal CreERT2 from *hsp70:creERT2* or *ubi:creERT2* offers a potential tool to create discrete clones of cells with recombined *lox* switches by 4-OHT induction before the onset of zygotic transcription.

4.3 4-OHT-MEDIATED CreERT2 INDUCTION

A simple format for embryo CreERT2 experiments is to array embryos prior to induction in fresh E3 into multiwell plates, such as 6- or 12-well format (with up to 35 embryos per well in 6-well plates). The multiwell format facilitates testing of several transgenic lines as well as subsequent handling and microscopy. It also reduces the amount of 4-OHT consumed and allows induction of embryos in different wells at different time points. Dechorionating the embryos does not significantly increase 4-OHT potency in our hands, yet it may be advisable with less sensitive *creERT2* drivers.

The potential instability of 4-OHT in water dictates fresh preparation of 4-OHT-containing E3 induction medium immediately before treatment. In a 50-μL Falcon tube, the desired final volume of E3 medium is prepared and the required 4-OHT added with a pipette from a single-use aliquot, taken from −20°C storage immediately prior to use. With stocks of >1 month age, heating to 65°C or sonicating the stock for 10 min prior to use can improve recombination efficiency (our unpublished results). When adding 4-OHT to E3, the drug will temporarily go out of solution when the DMSO or ethanol mixes with water and will be visible as a cloudy precipitate trail at the pipette tip. The precipitate can flake and drift to the medium surface. Brief vortexing effectively dissolves the precipitate.

The E3 medium on the arrayed embryos is removed using a vacuum pipette with a fine tip while tilting the plate toward the experimenter to facilitate careful yet rapid aspiration starting from the top of the well. 4-OHT induction medium is then immediately applied after aspiration to prevent the embryos from drying out. We found 3–5 mL of medium to be sufficient for undisturbed embryo growth in 6-well plates. We routinely perform experiments combining 4-OHT with bioactive chemicals or 1-phenyl 2-thiourea (PTU) (to inhibit melanization for enhanced microscopy) without detrimental effects on 4-OHT or CreERT2 activity.

Experimental observations indicate a short half-life of waterborne 4-OHT (Mosimann et al., 2011), requiring repeated addition of fresh 4-OHT-containing E3 to maintain CreERT2 activity for extended induction time windows. Nonetheless, 15-min 4-OHT treatment followed by thorough washing has been shown to trigger a strong response in the majority of cells when tested with *ubi:creERT2;ubi:Switch* transgenics (Mosimann et al., 2011). In our experience, 4-OHT handling does not require working in dimmed rooms or shielding of embryo-containing plates while performing experiments on the laboratory bench. Nevertheless, to minimize undesired decomposition of the drug as observed upon UV exposure, embryos should be kept in the dark over extended periods of time and handling on bench tops should be minimized during induction.

4.4 INDUCTION TIME POINT

The timing of 4-OHT induction needs to be coordinated with (1) anticipated transgenic CreERT2 expression, (2) the 4-OHT activity window in embryo medium, and (3) the optimal 4-OHT concentration to activate sufficient CreERT2 to recombine a given *lox* switch at a given time point. Adding 4-OHT prior to CreERT2 expression might not leave sufficient active drug to activate newly synthesized recombinase. The dynamics of a given *lox* switch's expression is a further factor to consider when determining the optimal induction time point. Although successful *lox* recombination can be easily and rapidly detectable by suitable reporters (Hans et al., 2009) or within 1–3 h by PCR (our unpublished observations), a considerable lag time passes between 4-OHT administration and appearance of the actual *lox* switch transgene readout: cytoplasmic CreERT2 needs to bind 4-OHT, translocate to the nucleus, and engage two *lox* sites in recombination; the newly formed reporter requires mRNA transcription and subsequent protein translation, followed by proper folding of the protein and sufficient accumulation. The rapid development of the zebrafish embryo pushes the limit of how fast transcription- and protein translation-dependent readouts can be.

5. BEYOND TRADITIONAL Cre/*lox* LINEAGE TRACING

5.1 SINGLE-CELL INDUCTION

Understanding lineage relationships of cells is key in developmental biology. Cre/*lox* tools enable following distinct lineages over time, but distinguishing and tracking adjacent cells to understand complex cell dynamics such as divisions, morphological changes, and migration remains complex. Approaches using photo-convertible proteins such as Kaede or Dendra allow following single cells and their progeny over limited periods, but highly depend on the half-life of the protein and proliferation dynamics of the labeled cells (Dempsey, Fraser, & Pantazis, 2012; Hatta, Tsujii, & Omura, 2006). To overcome these limitations, a new approach using laser-inducible caged cyclofen (abbreviated as cInd for caged inducer), a synthetic ERT2 ligand, was introduced for CreERT2 activation and has been established in zebrafish, combining permanent Cre/*lox*-based marking with single cell-labeling techniques (Sinha, Neveu, Gagey, Aujard, Benbrahim-Bouzidi, et al., 2010; Sinha, Neveu, Gagey, Aujard, Le Saux, et al., 2010).

To achieve single-cell switching, embryos expressing CreERT2 as well as a *lox* transgene are incubated in 3 μM cInd in embryo medium from epiboly stages. Adding the compound at later stages has been observed to result in lower penetration of the drug into the embryos (Sinha, Neveu, Gagey, Aujard, Le Saux, et al., 2010). cInd stocks are prepared just as 4-OHT stocks described above. Of note, no switching is observed without UV illumination, suggesting low to absent leakiness of the system (Sinha, Neveu, Gagey, Aujard, Benbrahim-Bouzidi, et al., 2010). At the developmental stage of choice, cInd-incubated embryos are moved to an illumination chamber. Two-photon illumination with a focused infrared laser beam uncages cInd specifically in illuminated cells and releases the ligand, activating CreERT2-mediated DNA recombination. Caged ERT2 ligands enable targeting cells in the desired location to detect their progeny in transgenic animals expressing ubiquitous CreERT2 adding a lineage tracing approach without the generation of tissue-specific *creERT2* lines. Importantly, the system reacts within 10 min from uncaging to recombination activity, providing high temporal precision (Sinha, Neveu, Gagey, Aujard, Le Saux, et al., 2010).

5.2 MULTICOLOR SWITCHING FOR CLONAL ANALYSIS

Labeling single cells in a population enables cell tracking of individual cells, yet imposes limitations due to the lack of information such as cell–cell interactions and migration behaviors of distinct cells in relation to each other. In contrast, multicolor staining allows distinguishing many individual cells within a single sample due to large number of cell-specific labels. The so-called *Brainbow* approach mixes at least three fluorophore cassettes flanked by *lox* site variants in a single transgene to generate up to 90 distinguishable colors resulting from irreversible recombination events (Livet et al., 2007). Originally implemented to distinguish single neurons in the brain, multispectral *Brainbow* transgenes can be applied in any tissue of choice, also in zebrafish referred to as *Zebrabow* (Pan et al., 2013).

In the original transgene, dTomato, Cerulean, and Yellow Fluorescent Protein (YFP) are arranged in a tandem array, with two distinct pairs of *lox* sites (*lox2272* and *loxP*) flanking the first two fluorescent proteins. By default, this transgene expresses dTomato; random Cre-mediated recombination results in one of the colors in a mosaic fashion. Due to stochastic recombination and combined expression of fluorescent proteins, a multiplication of possible colors is achieved through the insertion of several copies of the *Brainbow* cassette. Unique combinations of the recombined fluorescent cassettes label each cell in a different color, distinguishing the cell boundaries of adjacent cells (Livet et al., 2007; Pan et al., 2013).

Few stable multispectral lineage tracing zebrafish lines have been published to date including *β-actin-2:Brainbow* (Gupta & Poss, 2012), *ubi:Zebrabow* (Pan et al., 2013), and *Multibow* (Xiong, Obholzer, Noche, & Megason, 2015). Additionally, the *UAS:Zebrabow* line enables combining the *Zebrabow* transgene with existing Gal4 transgenics driven by a promoter of choice (Pan et al., 2013). Of note, since single-insertion lines are not desired in the *Brainbow* concepts, the lines should never be propagated as pure outcrosses.

To determine the clonal origins of distinct cells within organs or tumors pointing to common precursors, *Brainbow* approaches rely on DNA rearrangements that are stably propagated to the progeny of recombined cells, resulting in clones of the same color. To meet this assumption, fluorescent protein combinations need to reach a steady state through transient Cre activity restricting recombination events that would otherwise continuously excise the first fluorophore cassettes until no recombination-competent *lox* sites are left. To analyze color inheritance and stability in individual experiments, the hue—saturation plot has been proposed as a reliable method. On the hue—saturation graph, measurements from different time points form a cluster when colors are stably rearranged, while nonstable rearrangements result in dispersed data points (Pan et al., 2013). Furthermore, the following points need to be considered when analyzing a *Brainbow* experiment which can heavily influence the color mixture: the color diversity resulting from differential Cre activity, the maturation time for the nondefault fluorophores versus the degradation time of the default protein, and the imaging conditions and photo—stability profiles of the fluorophores. When adjusting conditions to individual experiments, *Brainbow* approaches offer the potential for clonal analysis in a wide variety of tissues.

5.3 SPATIOTEMPORALLY CONTROLLED SIGNALING PERTURBATIONS

A key challenge in developmental biology remains the dissection of specific signaling cues and their cooperation with lineage-specific transcription factors to instruct cell fate decisions. The experimental challenge today is to separate organ-specific influences of molecular pathways such as the Wnt, Hh, Fibroblast Growth Factor (*FGF*), or Nodal pathways from their broader roles in embryo patterning, which is challenging with the traditional methods of morpholino-mediated knockdown, chemical genetics, and mutants. Advanced transgenic methods to activate or perturb molecular mechanisms and to test the function of novel candidate genes under spatiotemporal control provide powerful means to address this complex biology.

Using the Cre/*lox* technology, signaling pathways can be activated or downregulated in a spatiotemporal manner as has been shown by Huang et al. (2014): by coupling CreERT2 to a pancreas-specific regulatory element (*Tp1glob:creERT2)* and combining it with a ubiquitous transgene carrying a dominant-negative retinoic acid receptor as *cargo* after a *lox*-flanked *eCFP* cassette (*ubi:lox-eCFP-lox-dnRAR-GFP)*, retinoic acid signaling was inhibited specifically in islet cell precursors, demonstrating its cell-autonomous role in pancreas development (Huang et al., 2014).

6. DISCUSSION

Tol2-mediated single-integration transgenes have greatly invigorated the zebrafish community and triggered a new boom in creating transgenic zebrafish strains. Moreover, the *phiC31*-mediated transgenesis and the emerging CRISPR-Cas9 insertion techniques offer new opportunities to establish novel transgenic lines in a faster and more reproducible manner. As the zebrafish field ventures into more complex molecular genetics applications, transgenesis and Cre/*lox* technologies are constantly refined and improved leading to more elaborate approaches such as single-cell labeling, *Brainbow*, or spatiotemporal signaling modulation transgenes.

Despite great advances in the field, the following basic challenges lie ahead: (1) the lack of reliable new transgene markers to complement the few currently existing to facilitate complex genetic experiments combining multiple transgenes; (2) the lack of *attP* landing sites to advance *phiC31*-mediated transgenesis; and (3) community involvement to organize reagent sharing, maintenance of transgenic lines, and publication of detailed protocols upon a rising number of transgenic zebrafish lines and increasingly sophisticated experimental strategies. We did not discuss the application of TAM/4-OHT transgene induction to adult zebrafish here, and such experiments pose different challenges compared to embryo experiments (Jopling et al., 2010; Kikuchi et al., 2010). Efforts to overcome these challenges are on the way and warrant a fast advancement in the described technologies, as exemplified by community projects such as the *Tol2 kit*, the *CreZoo*, and *ZCre*. Nonetheless, we envision that the zebrafish model system will, in the near future, feature an array of transgenic experimental protocols that rivals those of *Drosophila* and the mouse.

ACKNOWLEDGMENTS

We thank Christopher Hess and Karin Prummel for their ideas and critical reading of the manuscript, Dr. Alexa Burger and Dr. Elena Chiavacci for their valuable input, and all other members of the Mosimann lab for constructive support.

REFERENCES

Auer, T. O., Duroure, K., De Cian, A., Concordet, J.-P., & Del Bene, F. (2014). Highly efficient CRISPR/Cas9-mediated knock-in in zebrafish by homology-independent DNA repair. *Genome Research, 24*, 142–153. http://dx.doi.org/10.1101/gr.161638.113.

Bailey, J. M., Creamer, B. A., & Hollingsworth, M. A. (2009). What a fish can learn from a mouse: principles and strategies for modeling human cancer in mice. *Zebrafish, 6*, 329–337.

Bertrand, J. Y., Chi, N. C., Santoso, B., Teng, S., Stainier, D. Y., & Traver, D. (2010). Haematopoietic stem cells derive directly from aortic endothelium during development. *Nature, 464*, 108–111.

Bessa, J., Tena, J. J., de la Calle-Mustienes, E., Fernández-Miñán, A., Naranjo, S., Fernández, A. … Gómez-Skarmeta, J. L. (2009). Zebrafish enhancer detection (ZED) vector: a new tool to facilitate transgenesis and the functional analysis of cis-regulatory regions in zebrafish. *Developmental Dynamics, 238*, 2409–2417. http://dx.doi.org/10.1002/dvdy.22051.

Bischof, J., Maeda, R. K., Hediger, M., Karch, F., & Basler, K. (2007). An optimized transgenesis system for *Drosophila* using germ-line-specific phiC31 integrases. *Proceedings of the National Academy of Sciences of the United States of America, 104*, 3312–3317.

Boniface, E. J., Lu, J., Victoroff, T., Zhu, M., & Chen, W. (2009). FlEx-based transgenic reporter lines for visualization of Cre and Flp activity in live zebrafish. *Genesis, 47*, 484–491.

Carswell, S., & Alwine, J. C. (1989). Efficiency of utilization of the simian virus 40 late polyadenylation site: effects of upstream sequences. *Molecular and Cellular BiologyMolecular and Cellular Biology, 9*, 4248–4258. http://dx.doi.org/10.1128/MCB.9.10.4248.

Centanin, L., Ander, J.-J., Hoeckendorf, B., Lust, K., Kellner, T., Kraemer, I. … Wittbrodt, J. (2014). Exclusive multipotency and preferential asymmetric divisions in post-embryonic neural stem cells of the fish retina. *Development, 141*, 3472–3482. http://dx.doi.org/10.1242/dev.109892.

Dempsey, W. P., Fraser, S. E., & Pantazis, P. (2012). PhOTO zebrafish: a transgenic resource for in vivo lineage tracing during development and regeneration. *PLoS One, 7*, e32888. http://dx.doi.org/10.1371/journal.pone.0032888.

Feil, R., Brocard, J., Mascrez, B., LeMeur, M., Metzger, D., & Chambon, P. (1996). Ligand-activated site-specific recombination in mice. *Proceedings of the National Academy of Sciences of the United States of America, 93*, 10887–10890.

Feil, R., Wagner, J., Metzger, D., & Chambon, P. (1997). Regulation of Cre recombinase activity by mutated estrogen receptor ligand-binding domains. *Biochemical and Biophysical Research, 237*, 752–757.

Feng, H., Langenau, D. M., Madge, J. A., Quinkertz, A., Gutierrez, A., Neuberg, D. S. … Look, A. T. (2007). Heat-shock induction of T-cell lymphoma/leukaemia in conditional Cre/lox-regulated transgenic zebrafish. *British Journal of Haematology, 138*, 169–175.

Fink, M., Flekna, G., Ludwig, A., Heimbucher, T., & Czerny, T. (2006). Improved translation efficiency of injected mRNA during early embryonic development. *Developmental Dynamics, 235*, 3370–3378. http://dx.doi.org/10.1002/dvdy.20995.

Gagnon, J. A., Valen, E., Thyme, S. B., Huang, P., Ahkmetova, L., Pauli, A. … Schier, A. F. (2014). Efficient mutagenesis by Cas9 protein-mediated oligonucleotide insertion and large-scale assessment of single-guide RNAs. *PLoS One, 9*, e98186. http://dx.doi.org/10.1371/journal.pone.0098186.

Ghaye, A. P., Bergemann, D., Tarifeño-Saldivia, E., Flasse, L. C., Von Berg, V., Peers, B. … Manfroid, I. (2015). Progenitor potential of nkx6.1-expressing cells throughout zebrafish life and during beta cell regeneration. *BMC Biology, 13*, 70. http://dx.doi.org/10.1186/s12915-015-0179-4.

Groth, A. C., Fish, M., Nusse, R., & Calos, M. P. (2004). Construction of transgenic *Drosophila* by using the site-specific integrase from phage phiC31. *Genetics, 166*, 1775–1782. pii:166/4/1775.

Groth, A. C., Olivares, E. C., Thyagarajan, B., & Calos, M. P. (2000). A phage integrase directs efficient site-specific integration in human cells. *Proceedings of the National Academy of Sciences of the United States of America, 97*, 5995–6000. http://dx.doi.org/10.1073/pnas.090527097. pii:090527097.

Grzegorski, S. J., Chiari, E. F., Robbins, A., Kish, P. E., & Kahana, A. (2014). Natural variability of Kozak sequences correlates with function in a zebrafish model. *PLoS One, 9*, e108475. http://dx.doi.org/10.1371/journal.pone.0108475.

Gupta, V., & Poss, K. D. (2012). Clonally dominant cardiomyocytes direct heart morphogenesis. *Nature, 484*, 479–484. http://dx.doi.org/10.1038/nature11045.

Hans, S., Freudenreich, D., Geffarth, M., Kaslin, J., Machate, A., & Brand, M. (2010). Generation of a non-leaky heat shock-inducible Cre line for conditional Cre/lox strategies in zebrafish. *Developmental Dynamics, 240*, 108–115.

Hans, S., Kaslin, J., Freudenreich, D., & Brand, M. (2009). Temporally-controlled site-specific recombination in zebrafish. *PLoS One, 4*, e4640.

Hatta, K., Tsujii, H., & Omura, T. (2006). Cell tracking using a photoconvertible fluorescent protein. *Nature Protocols, 1*, 960–967. http://dx.doi.org/10.1038/nprot.2006.96.

Hesselson, D., Anderson, R. M., Beinat, M., & Stainier, D. Y. (2009). Distinct populations of quiescent and proliferative pancreatic beta-cells identified by HOTcre mediated labeling. *Proceedings of the National Academy of Sciences of the United States of America, 106*, 14896–14901. http://dx.doi.org/10.1073/pnas.0906348106.

Hisano, Y., Sakuma, T., Nakade, S., Ohga, R., Ota, S., Okamoto, H. … Kawahara, A. (2015). Precise in-frame integration of exogenous DNA mediated by CRISPR/Cas9 system in zebrafish. *Scientific Reports, 5*, 8841. http://dx.doi.org/10.1038/srep08841.

Huang, W., Wang, G., Delaspre, F., Vitery, M. D. C., Beer, R. L., & Parsons, M. J. (2014). Retinoic acid plays an evolutionarily conserved and biphasic role in pancreas development. *Developmental Biology, 394*, 83–93. http://dx.doi.org/10.1016/j.ydbio.2014.07.021.

Hu, G., Goll, M. G., & Fisher, S. (2011). PhiC31 integrase mediates efficient cassette exchange in the zebrafish germline. *Developmental Dynamics, 240*, 2101–2107. http://dx.doi.org/10.1002/dvdy.22699.

Irion, U., Krauss, J., & Nusslein-Volhard, C. (2014). Precise and efficient genome editing in zebrafish using the CRISPR/Cas9 system. *Development*, 115584. http://dx.doi.org/10.1242/dev.115584.

Iyengar, A., Muller, F., & Maclean, N. (1996). Regulation and expression of transgenes in fish – a review. *Transgenic Research, 5*, 147–166.

Jopling, C., Sleep, E., Raya, M., Marti, M., Raya, A., & Belmonte, J. C. (2010). Zebrafish heart regeneration occurs by cardiomyocyte dedifferentiation and proliferation. *Nature, 464*, 606–609.

Katzenellenbogen, B. S., Norman, M. J., Eckert, R. L., Peltz, S. W., & Mangel, W. F. (1984). Bioactivities, estrogen receptor interactions, and plasminogen activator-inducing activities of tamoxifen and hydroxy-tamoxifen isomers in MCF-7 human breast cancer cells. *Cancer Research, 44*, 112–119.

Kawakami, K. (2007). Tol2: a versatile gene transfer vector in vertebrates. *Genome Biology, 8*(Suppl. 1), S7.

Kawakami, K., Takeda, H., Kawakami, N., Kobayashi, M., Matsuda, N., & Mishina, M. (2004). A transposon-mediated gene trap approach identifies developmentally regulated genes in zebrafish. *Developmental Cell, 7*, 133–144.

Kikuchi, K., Holdway, J. E., Werdich, A. A., Anderson, R. M., Fang, Y., Egnaczyk, G. F. … Poss, K. D. (2010). Primary contribution to zebrafish heart regeneration by gata4(+) cardiomyocytes. *Nature, 464*, 601–605.

Kikuta, H., & Kawakami, K. (2009). Transient and stable transgenesis using tol2 transposon vectors. *Methods in Molecular Biology, 546*, 69–84.

Kimura, Y., Hisano, Y., Kawahara, A., & Higashijima, S. (2014). Efficient generation of knock-in transgenic zebrafish carrying reporter/driver genes by CRISPR/Cas9-mediated genome engineering. *Scientific Reports, 4*, 6545. http://dx.doi.org/10.1038/srep06545.

Kirchmaier, S., Höckendorf, B., Möller, E. K., Bornhorst, D., Spitz, F., & Wittbrodt, J. (2013). Efficient site-specific transgenesis and enhancer activity tests in medaka using PhiC31 integrase. *Development, 140*, 4287–4295. http://dx.doi.org/10.1242/dev.096081.

Kondrychyn, I., Garcia-Lecea, M., Emelyanov, A., Parinov, S., & Korzh, V. (2009). Genome-wide analysis of Tol2 transposon reintegration in zebrafish. *BMC Genomics, 10*, 418.

Kozak, M. (1987). An analysis of 5'-noncoding sequences from 699 vertebrate messenger RNAs. *Nucleic Acids Research, 15*, 8125–8148.

Kozak, M. (1996). Interpreting cDNA sequences: some insights from studies on translation. *Mammalian Genome, 7*, 563–574.

Kozak, M. (1999). Initiation of translation in prokaryotes and eukaryotes. *Gene, 234*, 187–208.

Kwan, K. M., Fujimoto, E., Grabher, C., Mangum, B. D., Hardy, M. E., Campbell, D. S.… Chien, C. B. (2007). The Tol2kit: a multisite gateway-based construction kit for Tol2 transposon transgenesis constructs. *Developmental Dynamics, 236*, 3088–3099.

Langenau, D. M., Feng, H., Berghmans, S., Kanki, J. P., Kutok, J. L., & Look, A. T. (2005). Cre/lox-regulated transgenic zebrafish model with conditional myc-induced T cell acute lymphoblastic leukemia. *Proceedings of the National Academy of Sciences of the United States of America, 102*, 6068–6073.

Le, X., Langenau, D. M., Keefe, M. D., Kutok, J. L., Neuberg, D. S., & Zon, L. I. (2007). Heat shock-inducible Cre/Lox approaches to induce diverse types of tumors and hyperplasia in transgenic zebrafish. *Proceedings of the National Academy of Sciences of the United States of America, 104*, 9410–9415.

Lee, R. T. H., Asharani, P. V., & Carney, T. J. (2014). Basal keratinocytes contribute to all strata of the adult zebrafish epidermis. *PLoS One, 9*, e84858. http://dx.doi.org/10.1371/journal.pone.0084858.

Lee, R. T. H., Knapik, E. W., Thiery, J. P., & Carney, T. J. (2013). An exclusively mesodermal origin of fin mesenchyme demonstrates that zebrafish trunk neural crest does not generate ectomesenchyme. *Development, 140*, 2923–2932. http://dx.doi.org/10.1242/dev.093534.

Liu, X., Li, Z., Emelyanov, A., Parinov, S., & Gong, Z. (2008). Generation of oocyte-specifically expressed cre transgenic zebrafish for female germline excision of loxP-flanked transgene. *Developmental Dynamics, 237*, 2955–2962.

Livet, J., Weissman, T. A., Kang, H., Draft, R. W., Lu, J., Bennis, R. A.… Lichtman, J. W. (2007). Transgenic strategies for combinatorial expression of fluorescent proteins in the nervous system. *Nature, 450*, 56–62.

Lu, J., Maddison, L. A., & Chen, W. (2011). PhiC31 integrase induces efficient site-specific excision in zebrafish. *Transgenic Research, 20*, 183–189. http://dx.doi.org/10.1007/s11248-010-9394-5.

Metzger, D., Clifford, J., Chiba, H., & Chambon, P. (1995). Conditional site-specific recombination in mammalian cells using a ligand-dependent chimeric Cre recombinase. *Proceedings of the National Academy of Sciences of the United States of America, 92*, 6991–6995. http://dx.doi.org/10.1073/pnas.92.15.6991.

Mongera, A., Singh, A. P., Levesque, M. P., Chen, Y.-Y., Konstantinidis, P., & Nüsslein-Volhard, C. (2013). Genetic lineage labeling in zebrafish uncovers novel neural crest contributions to the head, including gill pillar cells. *Development, 140*, 916–925. http://dx.doi.org/10.1242/dev.091066.

Mosimann, C., Kaufman, C. K., Li, P., Pugach, E. K., Tamplin, O. J., & Zon, L. I. (2011). Ubiquitous transgene expression and Cre-based recombination driven by the ubiquitin promoter in zebrafish. *Development, 138*, 169–177. http://dx.doi.org/10.1242/dev.059345.

Mosimann, C., Panáková, D., Werdich, A. A., Musso, G., Burger, A., Lawson, K. L.… Zon, L. I. (2015). Chamber identity programs drive early functional partitioning of the heart. *Nature Communications, 6*, 8146. http://dx.doi.org/10.1038/ncomms9146.

Mosimann, C., Puller, A.-C., Lawson, K. L., Tschopp, P., Amsterdam, A., & Zon, L. I. (2013). Site-directed zebrafish transgenesis into single landing sites with the phiC31 integrase system. *Developmental Dynamics, 242*, 949–963. http://dx.doi.org/10.1002/dvdy.23989.

Pan, Y. A., Freundlich, T., Weissman, T. A., Schoppik, D., Wang, X. C., Zimmerman, S.… Schier, A. F. (2013). Zebrabow: multispectral cell labeling for cell tracing and lineage analysis in zebrafish. *Development, 140*, 2835–2846. http://dx.doi.org/10.1242/dev.094631.

Park, J. T., & Leach, S. D. (2013). TAILOR: transgene activation and inactivation using lox and rox in zebrafish. *PLoS One, 8*, e85218. http://dx.doi.org/10.1371/journal.pone.0085218.

Roberts, J. A., Miguel-Escalada, I., Slovik, K. J., Walsh, K. T., Hadzhiev, Y., Sanges, R.… Müller, F. (2014). Targeted transgene integration overcomes variability of position effects in zebrafish. *Development, 141*, 715–724. http://dx.doi.org/10.1242/dev.100347.

Sauer, B. (1987). Functional expression of the cre-lox site-specific recombination system in the yeast *Saccharomyces cerevisiae*. *Molecular and Cellular Biology, 7*, 2087–2096.

Sauer, B. (1998). Inducible gene targeting in mice using the Cre/lox system. *Methods, 14*, 381–392.

Sauer, B., & Henderson, N. (1988). Site-specific DNA recombination in mammalian cells by the Cre recombinase of bacteriophage P1. *Proceedings of the National Academy of Sciences of the United States of America, 85*, 5166–5170.

Sinha, D. K., Neveu, P., Gagey, N., Aujard, I., Benbrahim-Bouzidi, C., Le Saux, T.… Jullien, L. (2010). Photocontrol of protein activity in cultured cells and zebrafish with one- and two-photon illumination. *ChemBioChem, 11*, 653–663. http://dx.doi.org/10.1002/cbic.201000008.

Sinha, D. K., Neveu, P., Gagey, N., Aujard, I., Le Saux, T., Rampon, C.… Vriz, S. (2010). Photoactivation of the CreER T2 recombinase for conditional site-specific recombination with high spatiotemporal resolution. *Zebrafish, 7*, 199–204. http://dx.doi.org/10.1089/zeb.2009.0632.

Soriano, P. (1999). Generalized lacZ expression with the ROSA26 Cre reporter strain. *Nature Genetics, 21*, 70–71.

Stuart, G. W., McMurray, J. V., & Westerfield, M. (1988). Replication, integration and stable germ-line transmission of foreign sequences injected into early zebrafish embryos. *Development, 103*, 403–412.

Suster, M. L., Sumiyama, K., & Kawakami, K. (2009). Transposon-mediated BAC transgenesis in zebrafish and mice. *BMC Genomics, 10*, 477.

Tamplin, O. J., Durand, E. M., Carr, L. A., Childs, S. J., Hagedorn, E. J., Li, P.… Zon, L. I. (2015). Hematopoietic stem cell arrival triggers dynamic remodeling of the perivascular niche. *Cell, 160*, 241–252. http://dx.doi.org/10.1016/j.cell.2014.12.032.

Thummel, R., Burket, C. T., Brewer, J. L., Sarras, M. P., Jr., Li, L., Perry, M.… Godwin, A. R. (2005). Cre-mediated site-specific recombination in zebrafish embryos. *Developmental Dynamics, 233*, 1366–1377.

Venken, K. J. T., & Bellen, H. J. (2005). Emerging technologies for gene manipulation in *Drosophila* melanogaster. *Nature Reviews Genetics, 6*, 167–178.

Villefranc, J. A., Amigo, J., & Lawson, N. D. (2007). Gateway compatible vectors for analysis of gene function in the zebrafish. *Developmental Dynamics, 236*, 3077–3087.

Xiong, F., Obholzer, N. D., Noche, R. R., & Megason, S. G. (2015). Multibow: digital spectral barcodes for cell tracing. *PLoS One, 10*, e0127822. http://dx.doi.org/10.1371/journal.pone.0127822.

Yoshikawa, S., Kawakami, K., & Zhao, X. C. (2008). G2R Cre reporter transgenic zebrafish. *Developmental Dynamics, 237*, 2460–2465.

Zambrowicz, B. P., Imamoto, A., Fiering, S., Herzenberg, L. A., Kerr, W. G., & Soriano, P. (1997). Disruption of overlapping transcripts in the ROSA beta geo 26 gene trap strain leads to widespread expression of beta-galactosidase in mouse embryos and hematopoietic cells. *Proceedings of the National Academy of Sciences of the United States of America, 94*, 3789–3794.

Zhao, L., Borikova, A. L., Ben-Yair, R., Guner-Ataman, B., MacRae, C. A., Lee, R. T.… Burns, C. E. (2014). Notch signaling regulates cardiomyocyte proliferation during zebrafish heart regeneration. *Proceedings of the National Academy of Sciences of the United States of America, 111*, 1403–1408. http://dx.doi.org/10.1073/pnas.1311705111.

Zhou, Y., Cashman, T. J., Nevis, K. R., Obregon, P., Carney, S. A., Liu, Y.… Burns, C. G. (2011). Latent TGF-beta binding protein 3 identifies a second heart field in zebrafish. *Nature, 474*, 645–648. http://dx.doi.org/10.1038/nature10094.

CHAPTER

Method for somatic cell nuclear transfer in zebrafish

12

K. Siripattarapravat*[,¶], S. Prukudom*[,¶], J. Cibelli[||,#,1]

*Kasetsart University, Bangkok, Thailand

[¶]Center of Excellence on Agricultural Biotechnology: (AG-BIO/PERDO-CHE), Bangkok, Thailand

[||]Michigan State University, East Lansing, MI, United States

[#]BIONAND, Andalucía, Spain

[1]Corresponding author: E-mail: cibelli@anr.msu.edu

CHAPTER OUTLINE

Abstract

This chapter presents a detailed methodology for somatic cell nuclear transfer—cloning of zebrafish. We aim to place the reader in a virtual lab experience to assist acquisition of the technical skills required for reproducing the published protocol. All materials, including catalog numbers for reagents and techniques for their preparation, are provided. Our protocols describe laser inactivation of egg chromosomes, the transfer of a cell through the oocyte micropyle, and spontaneous activation of the reconstructed embryo. High-quality eggs are the key to cloning success, and Chinook salmon ovarian fluid is

Methods in Cell Biology, Volume 135, ISSN 0091-679X, http://dx.doi.org/10.1016/bs.mcb.2016.04.022

indispensable for keeping eggs arrested at the metaphase of meiosis II. This protocol continues to be refined by our laboratory. However, naive investigators should be able to apply it in its present form to generate cloned zebrafish.

INTRODUCTION

Since our first publication on somatic cell nuclear transfer (SCNT) in zebrafish (*Danio rerio*) in 2009, we have been able to refine the technique as reported in 2010 (Siripattarapravat et al., 2010; Siripattarapravat, Pinmee, Venta, Chang, & Cibelli, 2009) and later in two book chapters that followed (Siripattarapravat & Cibelli, 2011; Siripattarapravat, Pinmee, & Cibelli, 2013). In this article, we attempt to describe SCNT as if the reader is not familiar with the methodological steps, and we are adding the latest refinements as of December 2014. However, the major steps of zebrafish SCNT (also called cloning throughout this chapter) remain unchanged and we recommend the reader to review our 2009 publication.

The technique of cloning itself is an acquired skill that requires as much practice as patience. We usually remind students that one of the most important features of a successful "cloner" is the capacity to rebound quickly from failure, since there is not a single cloning session in which even the most skilled technician will not experience it. The work must be performed in a semi-dark room, and all the motor skills required (both hands and both feet) must be practiced to the point that muscle memory will free the technician's mind to make determinations as to which somatic cell is the best to use, which is the best oocyte to enucleate, when is it necessary to change needles, and even when is the time to take a break. Sessions last from 2 to 4 h depending on the number of reconstructed embryos. We recommend having two people working in tandem, with one person enucleating the oocytes and the other transferring the cell. It is fair to say that we are still far from knowing the best approaches to obtaining the largest number of viable clones. Nonetheless, if applied as here described, clones can be successfully generated, thereby enabling the experimenter to begin to answer some of the most fascinating questions regarding embryonic development, cellular differentiation, nuclear reprogramming, and aging.

1. MATERIALS

1. Hank's balance salt solution (HBSS).
   - **1.1** HBSS

     HBSS is prepared from Hank's balanced salt, with Ca^{2+} and Mg^{2+} (Cat no. B2261, Sigma), according to the manufacturer's instructions and kept at 4°C.
   - **1.2** HBSS with 0.5% (w/v) bovine serum albumin (H-BSA, Cat no. A8806, Sigma).

     H-BSA is prepared, filtered, aliquoted into 5-mL tubes, and kept at −80°C until use.

2. Chinook salmon ovarian fluid (CSOF).
We have described extensively the process of CSOF collection in our previous publication (Siripattarapravat & Cibelli, 2011). We have access to Chinook salmon through the Michigan Department of Natural Resources at the Little Manistee River weir. Ovarian fluid surrounds matured oocytes in oviducts of gravid salmonids. Any technique for collection of CSOF free of contaminants (ie, water or fecal material) should work. We generally squeeze healthy females to obtain eggs and the fluid simultaneously. CSOF is separated from eggs using a strainer and kept on ice until processing. We recommend using the same batch of CSOF for all CSOF-based solutions to be used in one cloning session.

2.1 Filtered CSOF.
The largest bottle top-filter unit with a 0.22 μm filter (Cat no. 431098, Corning) is generally used. Only small volumes of CSOF are filtered at a time. Filtered CSOF must be kept on ice throughout this step. Aliquots of filtered CSOF, 10 mL, are stored at −80°C until use.

2.2 CSOF with 5% (w/v) polyvinylpyrrolidone (PVP, Cat no. P0930, Sigma).
PVP is difficult to dissolve in CSOF. We recommend adding the PVP on top of CSOF and placing it overnight at 4°C. By the next morning the PVP should be dissolved. Homogenize the solution using a vortex at low speed, make 1-mL aliquots, and store them at −20°C until use.

2.3 CSOF with 50 μg/mL bis-benzimide Hoechst 33342 trihydrochloride (H33342, Cat no. B2261, Sigma).
H33342 is prepared as a 1000X stock solution (50 mg/mL of H33342 in HBSS). When needed for staining, add 1 μL of stock solution to 1 mL of CSOF, and vortex. Be aware that there might be a precipitate in the final solution that could affect the final concentration of H33342.

3. Reagents for preparation of donor cells.

3.1 Laboratory of Human Carcinogenesis (LHC) basal media (Cat no. 12677, Gibco), kept at 4°C.

3.2 TrypLE express (Cat no. 12604, Gibco), kept at 4°C.

3.3 1%(w/v) PVP in serum-free growth medium, kept at 4°C.
Growth medium is prepared using a base solution of Dulbecco's Modified Eagle Medium (DMEM) high glucose (Cat no. 11995-065, Gibco), supplemented with 2 mM N-acetyl-L-cysteine (Cat no. A9165, Sigma), 1 μM L-ascorbic acid 2-phosphate sesquimagnesium salt hydrate (Cat no. A8960, Sigma), 10 ng/mL bovine insulin (Cat no. 128-100, Cell Applications), 5% fetal bovine serum, and 1% Trout serum (SeaGrow, Cat no. JJ80, EastCoast Bio).

4. Embryo medium.
Embryo medium is prepared as described by the Zebrafish book (Westerfield, 2000) using full-strength Hank's balanced salt solution (Cat no. H1387, Sigma) as a 10X stock. Filter the stock with bottle top-filter unit 0.22 μm (Cat no. 431098, Corning) and keep at 4°C. When needed, dilute 10X stock solution with distilled water and keep it at room temperature. Adjust the pH to 7.2 with a few drops of 1 M NaOH.

5. Setup for micromanipulation.

5.1 Inverted microscope.

We use an ultraviolet (UV) phase contrast inverted microscope equipped with 4× and 20× objective lenses, and a 40× objective lens equipped with a laser-drilling system. Our experience is limited to the Xyclone system (RED-i class 1 laser product, Hamilton Thorne), but other systems that have been used to drill the zona pellucida for in vitro fertilization (IVF) should work. For practical purposes, we recommend footswitches to operate both the UV shutter and the laser-drilling system. The diaphragm of the UV light source must be set to the smallest diameter to limit exposure of the egg to UV light.

5.2 Micromanipulators.

Any micromanipulator set that allows fine control of the glass pipettes at 40× magnification, such as those used for cloning in other species or for human intracytoplasmic sperm injection (ICSI) can be used. In this case, Narishige hydraulic (Cat no. 202ND) or Eppendorf electronic manipulators work well for zebrafish SCNT.

5.3 Microinjectors.

Our preference is to use CellTram Air (Cat no. 5176000017, Eppendorf) for the egg holder and CellTram vario (Cat no. 5176000033, Eppendorf) using oil for the cell transfer needle. The "supporting needle" does not require a connection to a microinjector.

6. Holder, supporting, and injection needles.

We custom-make our holder and supporting needles in our laboratory. For "home made" needles we use borosilicate glass pipettes with an outside diameter of 1 mm and an inside diameter at approximately 0.58 mm (Cat no. B100-58-10, Sutter Instrument Company). For the sake of consistency, we purchase injection needles from Sutter Instrument Company.

6.1 Preparation of holder needle (Fig. 1).

Use a glass pipette puller (Cat no. P-2000, Sutter Instrument Company) to pull the glass microtube and then cut the tip to a diameter of approximately 300 μm with a diamond cutter. Dip the cut edge in distilled water and sand it with fine (No. 600) and extrafine (No. 2000) silicon carbide sandpaper. Observe the polished edge under a microscope and continue sanding until a straight and smooth cut with approximately 500 μm outside diameter is obtained. Wash the needle tip thoroughly with distilled water and absolute ethanol to remove any debris. Dry the needle and fire-polish its tip to smooth sharp edges. Bend the needle approximately 8 mm from the tip using a flame. The desired angle is 25–30 degrees.

6.2 Supporting needle preparation (Fig. 2).

Pull a glass pipette as described earlier and use a Microforge (Cat no. MF-900, Narishige) to cut the tip at an outside diameter of 20 μm. Fuse the tip of the needle by heating.

6.3 Injection–cell transfer–needle.

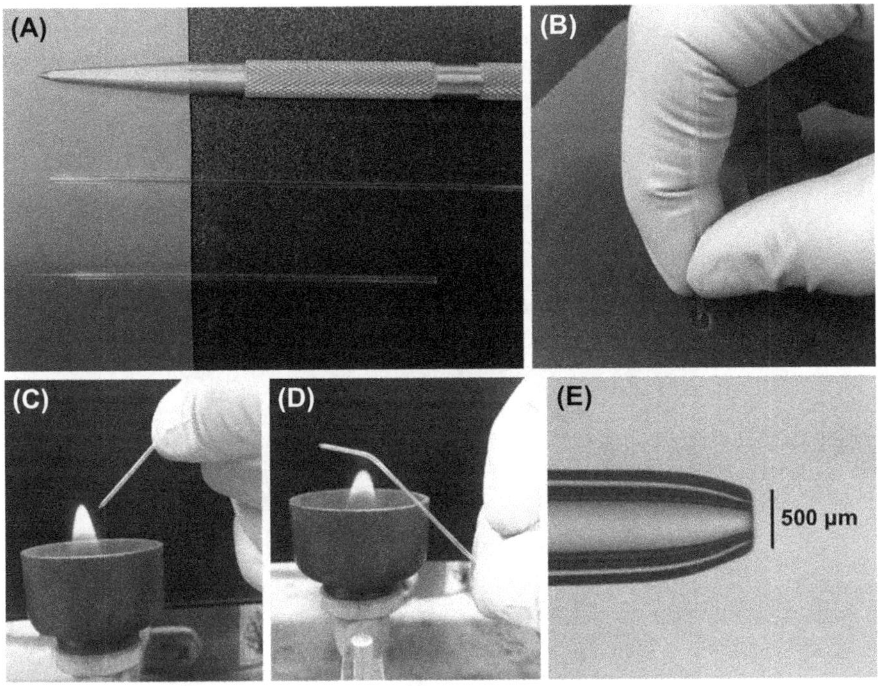

FIGURE 1

Glass holder preparation steps. (A) Fine sandpaper (black), extrafine sandpaper (gray), diamond cutter, glass microtube, and pulled glass needle; (B) polishing edges of the glass needle with sandpapers; (C) fire-polishing edges; (D) bending to get 25–30 degrees angle; and (E) close-up of the 500 μm outside diameter holder needle tip.

We recommend purchasing ICSI needles used with human sperm injections. Depending on the size range of the donor cells, ICSI needles with an inner diameter of 6–10 μm and bent at 20 degrees are most useful. We have successfully used needles from European (Cat no. BM100T-15, Biomedical instrument), Australian (Ref no. LISR, The pipette company), and US companies (Cat no. MIC-50-20, MIC-8-20, and MIC-9-20, Origio).

7. Glass pipettes.
   - 7.1 Microdispenser.
     We use the microdispenser to transfer MII eggs (arrested in the metaphase of meiosis II) in and out of a manipulation drop, ie, the place where eggs are enucleated and cells are transferred. The Drummond microdispenser of 20 μL fixed volume (Cat no. 3-000-320) works perfectly for zebrafish MII eggs. The glass capillary of the microdispenser must be fire-polished to smooth any sharp edges prior to use.
   - 7.2 Glass Pasteur pipettes with rubber bulbs.

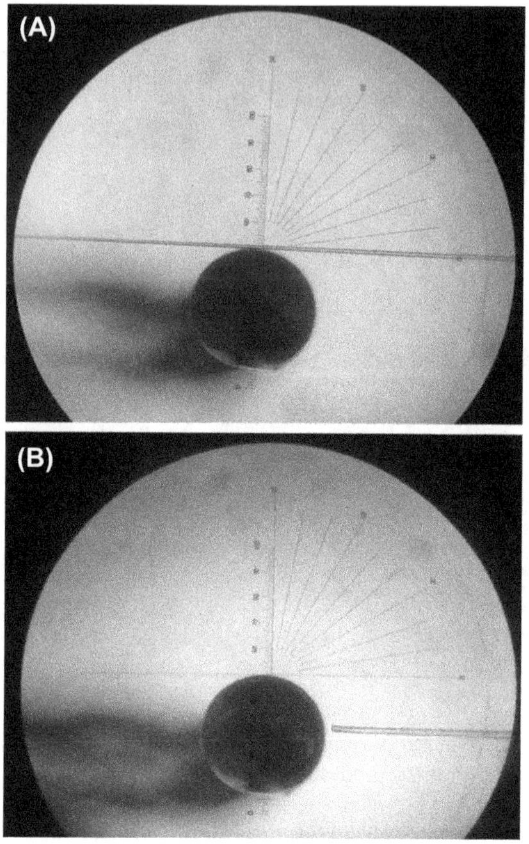

FIGURE 2

Supporting needle preparation steps. (A) Cutting pulled glass needle with heat at the point of 20 μm outside diameter and (B) heating the tip until smoothly closed.

Nine-inch glass Pasteur pipettes (Cat no. 14672-380, VWR) are cut and fire-polished to reduce their internal tip diameters to approximately 2 mm. These pipettes are used for transferring eggs and embryos between different solutions at low power under a stereomicroscope.

8. Miscellaneous items.
- **8.1** Stereomicroscope.
- **8.2** Warm plate and incubator, set at 28.5°C.
- **8.3** Mineral oil (Cat no. M8410, Sigma).
- **8.4** Fluorinert (Cat no. F9755, Sigma).
- **8.5** Glass rod for squeezing female and male fish.
- **8.6** Kimwipes (Cat no. 06-666A).
- **8.7** Dark-moist chamber.

8.8 Falcon plastic dishes, noncoated, 35 × 10 mm (Cat no. 351,008), 60 × 15 mm (Cat no. 351007), and 100 × 15 mm (Falcon, Cat no. 351,029).
8.9 Cosmetic sponge with slit and glass microcapillary for milt collection.
8.10 MS222 (Cat no. A5040, Sigma) (Westerfield, 2000).
8.11 Fish breeding box. Any breeding apparatus that allows a single mating pair setting can be used.

Remarks: Unless otherwise indicated, all procedures are carried out at room temperature (23°C).

2. METHOD

We summarize the process below in chronological order.

2.1 FISH BREEDING

We have found that the single most important factor determining the success of zebrafish SCNT is egg quality. Assuming that you are working with the Association for Assessment and Accreditation of Laboratory Animal Care International (AAALAC)-approved animal facility and are working under the animal husbandry conditions described by the zebrafish book (Westerfield, 2000), we also recommend the following animal care steps. Breeder fish should be fed three times daily, at least twice with live hatched brine shrimp (atremia) and once with supplements such as cyclops, worms, and spirulina. Use eggs from 6- to 15-month-old females of an outbred line such as TAB (F1 cross between Tübingen and AB). Select male/female pairs that breed well in a breeding apparatus and monitor the quality of eggs (as measured by percentage of fertilized embryos). Other characteristics of good breeders include bright skin coloration—golden/black stripes for males and silver/black stripes for females. Females should be allowed to breed regularly every 5–7 days and should ideally yield fertilization rates >90%.

Late afternoon on the day before cloning, prepare approximately 10 breeding pairs, which should be sufficient to obtain, at minimum, 60 to 70 high-quality but unfertilized, eggs for each individual who will perform SCNT the following day. The male and female must be kept in a breeding tank but separated by a screen to avoid premature mating.

On the morning of the cloning procedure, immediately after the room light comes on, remove the screen separating the male and female in each breeding tank, working with two or three tanks at a time. Continuously observe the mating behavior of each pair. Typically, the male will chase a gravid female and, if the courting is successful, she will lay eggs multiple times. After the female releases eggs a couple of times, remove the male from the tank. Prepare the inverted microscope and the micromanipulation tools. Then proceed with the collection of eggs not yet released by each female (squeezing).

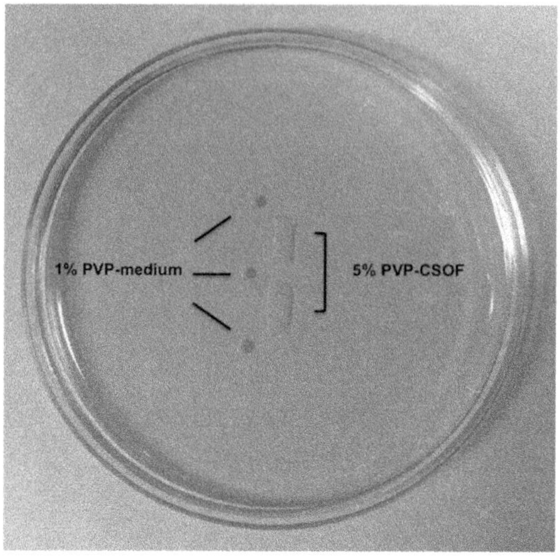

FIGURE 3

Typical setting of the manipulation drops under mineral oil.

2.2 PREPARATION OF REAGENTS

Once you know that you have at least one female able to provide matured eggs, thaw your stock solutions and CSOF. Most of reagents can be thawed in a water-bath at room temperature including filtered CSOF, 5% PVP in CSOF, H-BSA, and MS222. The egg staining solution is freshly prepared at this point by adding 1 μL of stock H33342 to 1 mL of CSOF, vortexing, and storing in the dark until use. Keep the remaining CSOF for egg collection and washing solution. Prepare the manipulation drops—two middle drops of 5% PVP-CSOF, and three left-side drops of 1% PVP media—then overlay these drops with mineral oil (Fig. 3).

2.3 SET UP THE MICROMANIPULATOR

Test the laser beam and adjust its target alignment prior to setting the glass needles. To test the laser, mark a glass coverslip with a black, erasable-ink marker and place it on a 100 × 15 mm petri dish, thereby emulating the distance between the laser and an egg's chromosomes. When the laser is properly adjusted, the laser pulse will produce a small hole in the black ink.

Connect the egg-holder needle to the CellTram pneumatic microinjector and position the needle within the field of view of the microscope using the 4× objective. The holder needle should be positioned on the left side at 9 o'clock. Fill the injection-cell transfer needle with fluorinert (Cat no. F9755, Sigma) and connect it to the oil-filled CellTram vario. The injection needle should be on the right side of the field of view, at 3 o'clock. The supporting needle must be set almost parallel with the injection needle, somewhere between 2 and 3 o'clock within the field of view (Fig. 4).

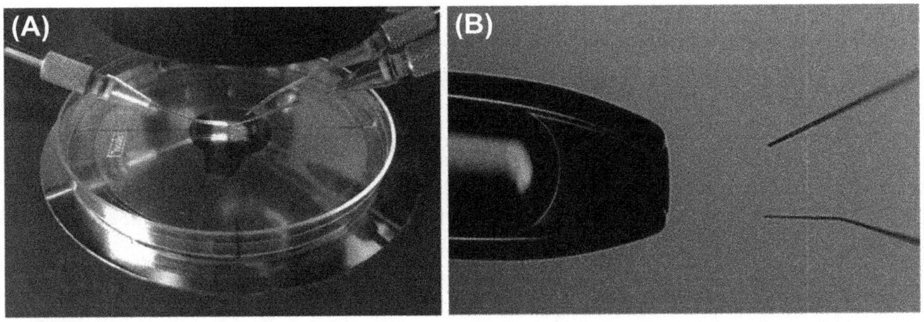

FIGURE 4

Setting of glass needles used for micromanipulation. (A) Stage view showing holder needle on the left and injection and supporting needles on the right and (B) microscopic view of the needles described in (A) at 4×.

2.4 EGG COLLECTION

Return to the breeding tanks to determine whether the eggs that were released by each female are fertilized and developing. At this point in time, they should be at the two-cell stage. The best quality eggs should have a fertilization rate close to 100%. These observations determine the female(s) that you are going to squeeze.

Prepare MS222 according to the Zebrafish book (Westerfield, 2000). Place 1 mL of CSOF in a 35 × 10 mm petri dish and set it aside for egg collection. Place female fish in the MS222 bath until fully anesthetized. As soon as slow movement of gills and fins is observed, rinse each female briefly in fish water and place her on a Kimwipe towelette. Hold the fish gently and belly-up in the folded Kimwipe with the mid-body between your thumb and middle finger. Dry the posterior part of the body (including pelvic fin, anal fin, dorsal fin, and caudal fin) with another Kimwipe to remove residual water that will prematurely activate the eggs. Position the female over the edge of the 35 mm petri dish in such a way that the genital opening is almost touching the CSOF and squeeze the female's abdomen gently with a glass rod to release eggs into the CSOF. Make a small caudal fin biopsy while the female is under anesthesia and then place her in fish water until she fully recovers. The section of fin that was removed should be kept frozen at −20°C for DNA analysis. Place breeders that have shed milt or eggs together in an isolation tank and leave them undisturbed for at least four weeks. After this period they can again be used as donors.

The unfertilized eggs that were collected by squeezing females should be at the MII stage of meiosis. High-quality eggs will appear yellowish and uniformly agranular, and their chorions will be intact and in direct contact with the cell membrane of the egg. If you observe eggs that have granular cytoplasm and/or have detached chorions, we recommended that you discard the batch, even if good quality eggs are present. Assuming that you obtain good eggs, carefully move them to the H33342 staining solution using a glass Pasteur pipette and rubber bulb. Avoid carryover of excess CSOF when moving the eggs since it will dilute the staining solution. Let the eggs settle in the H33342 solution by gravity. Place the 35-mm dish with the eggs in a dark-moist chamber at room temperature for 20 min. Finally, transfer eggs to 2.5 mL of CSOF in a 35-mm petri dish and place them in the humidified, dark chamber until use.

Note: It is important to use the H33342 product described in the Section 1 (Cat no. B2261, Sigma). If you have difficulty obtaining a strong DNA signal, consider remaking the stock solution according to the solubility reported in the H33342 specification sheet.

2.5 DONOR CELL PREPARATION

Prepare the donor somatic cells while eggs are in the H33342 solution (20 min). Donor cells can be obtained both from embryonic tissue and from cultured cells. Mechanical dissociation is best suited for embryos at early stages (<24 h postfertilization; hpf). Use TrypLE express to dissociate cultured cells or to release cells from embryos beyond 24 hpf. Use LHC medium to inactivate the enzymes. Wash cells prior to centrifugation at 3500 rpm for 3 min using growth medium (DMEM) and resuspend the cells in a solution of 1% PVP in serum-depleted growth medium.

2.6 MICROMANIPULATION

The zebrafish egg is approximately 800 μm in diameter, much bigger than a human or bovine egg (~120 μm), and manipulating it under an inverted microscope requires some modifications from the conventional setup for mammalian species. In addition to a larger egg holder needle in the left manipulator, two parallel needles—the supporting needle and the injection-cell transfer needle—are needed in the right manipulator. The two needles in the left manipulator resemble a pair of chopsticks at a fixed angle. The holder needle, injection-transfer needle, and supporting needle enable you to control the rotation of the egg in all directions.

Transfer stained eggs to the manipulation drop containing 5% PVP in CSOF using the microdispenser. The goal is to maneuver the egg with the "chopsticks" to position the micropyle (sperm entry site) at the bottom of the petri dish. The micropyle also indicates the region where the egg's chromosomes are located, usually within 10–20 μm in any direction. Once the desired position is achieved, the egg is secured in place using suction from the holder needle on the left. The PVP in the manipulation drop slows movement of the egg.

To perform the laser-assisted enucleation, the operator first positions the micropyle at the bottom of the petri dish using the 4× objective and UV illumination. In this position, the stained metaphase plate is obvious, and the laser beam is more effective. After aligning the micropyle, switch to the 40× objective equipped with the laser. Identify the metaphase plate (ie, the oocytes chromosomes) using only a few seconds of UV exposure, then trigger the laser. This step requires some practice because the left joystick, which controls the egg holder and thus the movements of the micropyle, must be moved very slowly to re-align the metaphase plate under the laser's target zone without losing grip on the egg and without exposing the egg to excessive UV light. At maximum power, the laser beam will raise the temperature at the center of the target to 400°C. With the Hamilton Thorne laser system, we recommend using 100% power and delivering two consecutive pulses of 500 μs each aimed directly at the metaphase plate. Depending upon the dexterity of the operator, one can perform 5 to 10 enucleations at a time (1 min per egg) before moving to the next step, cell injection.

To inject the somatic cell, the micropyle must be positioned in the equatorial region at 3 o'clock in the field of view in direct alignment with the transfer needle. The donor cell is delivered through the micropyle to minimize an injury to the recipient egg and to limit the injection volume. At the sperm entry site, the micropyle coincides with the region where the animal pole will develop. The micropyle is shaped as a funnel with the larger opening facing the outside of the egg. At its smallest, the micropylar pit accommodates only a single sperm. Since the pit is small, one should break the donor cell membrane using the injection-cell transfer needle. Chose a single donor cell, break its plasma membrane, and transfer the broken cell to the enucleated oocyte. Using a broken cell rather than a naked nucleus helps prevent clogging the transfer needle.

The alignment of the egg for cell injection is important. The micropyle should be positioned using the 4× objective in the equatorial region at 3 o'clock facing the injection needle. Secure the egg using gentle suction on the holder needle. Switching to the 20× objective, confirm that you can see the whole shape of the micropyle as if you are looking at a longitudinal cut of the funnel, from the largest opening to the smallest. Lock your focus point and bring into the same focal plane the injection needle, with the broken cell already loaded. Use the z-axis joystick control to move the needle into position near the micropyle. Then, and only then, attempt to go through the micropyle. Do not use excessive force; too much resistance indicates that you have not aligned the injection needle with the longitudinal axis of the funnel. If you encounter this problem, release the suction on the egg holder needle, return to the 4× objective, and reposition the micropyle to establish better longitudinal alignment of the micropyle with respect to the injection needle. Returning to the 20× objective, re-attempt to insert the injection needle into the micropyle. Successful insertion is recognized by the absence of resistance as the needle penetrates the egg cytosol. Inject the smallest amount of medium that guarantees delivery of the cell into the egg. Due to the lack of transparency of zebrafish eggs, the release of the somatic cell into the cytosol cannot be seen. You must rely upon visual observation of the distance traveled within the injection needle by the meniscus formed at the fluorinert/manipulation medium interface. Once you have determined that the donor cell has been transferred, gently remove the injection needle.

2.7 ACTIVATION OF RECONSTRUCTED EGGS

Rinse the reconstructed eggs twice using CSOF and incubate them for 15 min at room temperature in a dark-moist chamber. Immediately prior to activation of the egg, rinse the reconstructed eggs twice with H-BSA, then transfer them into embryo medium, and place them in a 28°C incubator. Monitoring and recording of the development is done daily throughout the development phase until they are moved to the conventional fish tanks. It is important to remove all embryos that show signs of abnormal or no development. Alternatively, one may move healthy embryos to fresh embryo medium each time development is monitored.

2.8 IN VITRO FERTILIZATION AT THE END OF THE SOMATIC CELL NUCLEAR TRANSFER PROCEDURE

We described previously the method for determination of egg quality via natural fertilization while eggs were squeezed from the donor female for SCNT. An equally important control is to perform IVF on unused eggs from the same SCNT batch. This should be done at the end of the SCNT procedure to ascertain whether the squeezed eggs decayed in any fashion during the SCNT session. Depending on the level of expertise of the operator, SCNT can last between 2 and 4 h, during which the eggs can lose potency. A fertilization rate of >70% after completion of SCNT indicates that the eggs retained their developmental potential while they were kept in CSOF awaiting use.

2.9 CONFIRMATION THAT A "CLONE" IS A REAL CLONE

Cloned fish and the somatic cell donor must have identical phenotypes and genotypes. Phenotypic screening can provide evidence that the clone originated from the somatic cell donor. However, it is restricted to instances in which the phenotypes of the egg donor and the somatic cell donor are quite obvious; for example, when the egg donor is wild-type TAB and the cell donor is either a *golden* or *casper* mutant.

To evaluate the stability of the cloned genome, one should obtain a karyotype. Replication banding, which has been recommended for its accuracy in identifying zebrafish chromosomes (Amores & Postlethwait, 1999), can be applied to cultured fibroblasts obtained from a fin clip of the cloned fish. The most stringent confirmation that a clone has been generated from a somatic cell is obtained by Single nucleotide polymorphism (SNP) analysis. We have described polymerase chain reaction - restriction fragment length polymorphism analysis of 11 genomic SNP markers for genotyping in our 2009 publication (Siripattarapravat et al., 2009). SNPs from the donor cells and the cloned animals should match 100% and be different than those from the egg donor.

ACKNOWLEDGMENTS

This work has been supported by the Thailand research fund (MRG5480228 – KS); by the Center for Advanced Studies for Agriculture and Food—Kasetsart University Institute for Advanced Studies, by the Center of Excellence on Agricultural Biotechnology, Science and

Technology—Postgraduate Education and Research Development Office, by the Commission on Higher Education, Ministry of Education (AG-BIO/PERDO-CHE), by the Center for Advanced Studies for Agriculture and Food, Institute for Advanced Studies, Kasetsart University Under the Higher Education Research Promotion and National Research University Project of Thailand, Office of the Higher Education Commission, Ministry of Education, Thailand (KS and SP); and by Michigan AgBio Research (JC).

REFERENCES

Amores, A., & Postlethwait, J. H. (1999). Banded chromosomes and the zebrafish karyotype. *Methods in Cell Biology, 60*, 323–338.

Siripattarapravat, K., & Cibelli, J. B. (2011). *Methods in Cell Biology, 104*, 209–217. http://dx.doi.org/10.1016/B978-0-12-374814-0.00012-4. Elsevier.

Siripattarapravat, K., Pinmee, B., Chang, E.-A., Munoz, J. D., Kawakami, K., & Cibelli, J. B. (2010). The influence of donor nucleus source on the outcome of zebrafish somatic cell nuclear transfer. *International Journal of Developmental Biology, 54*(11–12), 1679–1683. http://dx.doi.org/10.1387/ijdb.103189ks.

Siripattarapravat, K., Pinmee, B., & Cibelli, J. (2013). *Chapter 16. Somatic cell nuclear transfer in zebrafish. Principles of cloning* (pp. 1–15). Elsevier Inc. http://dx.doi.org/10.1016/B978-0-12-386541-0.00016-3.

Siripattarapravat, K., Pinmee, B., Venta, P. J., Chang, C.-C., & Cibelli, J. B. (2009). Somatic cell nuclear transfer in zebrafish. *Nature Publishing Group, 6*(10), 733–735. http://dx.doi.org/10.1038/nmeth.1369.

Westerfield, M. (2000). *The zebrafish book* (4th ed.). Eugene: University of Oregon Press.

CHAPTER

13 Experimental approaches to studying the nature and impact of splicing variation in zebrafish

M.C. Keightley[*,¶¶], **S. Markmiller**[§], **C.G. Love**[¶,||], **J.E.J. Rasko**[#,**,§§], **G.J. Lieschke**[*,¶¶,1,a], **J.K. Heath**[¶,1,a]

*Australian Regenerative Medicine Institute, Clayton, VIC, Australia
§University of California San Diego, La Jolla, CA, United States
¶Walter and Eliza Hall Institute of Medical Research, Parkville, VIC, Australia
||University of Melbourne, Parkville, VIC, Australia
#Centenary Institute, Camperdown, NSW, Australia
**University of Sydney, Sydney, NSW, Australia
§§Royal Prince Alfred Hospital, Newtown, NSW, Australia
¶¶Monash University, Clayton, VIC, Australia
[1]Corresponding authors: E-mail: graham.lieschke@monash.edu; joan.heath@wehi.edu.au

CHAPTER OUTLINE

[a]These two authors contributed equally.

Methods in Cell Biology, Volume 135, ISSN 0091-679X, http://dx.doi.org/10.1016/bs.mcb.2016.02.006

Abstract

From a fixed number of genes carried in all cells, organisms create considerable diversity in cellular phenotype through differential regulation of gene expression. One prevalent source of transcriptome diversity is alternative pre-mRNA splicing, which is manifested in many different forms. Zebrafish models of splicing dysfunction due to mutated spliceosome components provide opportunity to link biochemical analyses of spliceosome structure and function with whole organism phenotypic outcomes. Drawing from experience with two zebrafish mutants: *cephalophŏnus* (a *prpf8* mutant, isolated for defects in granulopoiesis) and *caliban* (a *rnpc3* mutant, isolated for defects in digestive organ development), we describe the use of glycerol gradient sedimentation and native gel electrophoresis to resolve components of aberrant splicing complexes. We also describe how RNAseq can be employed to examine relatively rare alternative splicing events including intron retention. Such experimental approaches in zebrafish can promote understanding of how splicing variation and dysfunction contribute to phenotypic diversity and disease pathogenesis.

1. OVERVIEW OF RNA SPLICING

A huge component of the proteomic diversity that exists between cells of different lineages is generating by splicing. This process is required for proper gene expression since almost all eukaryotic genes are expressed as precursor mRNAs (pre-mRNAs) comprising coding sequences (exons) interspersed with noncoding sequences (introns). Splicing, which occurs concurrently with transcription, is required for the production of mature mRNA molecules that are ready to be translated into proteins. In *constitutive splicing*, exons are generally spliced together in the order in which they occur in the primary transcript. In *alternative splicing*, different mRNA isoforms are generated from the same pre-mRNA, providing a mechanism for producing greater transcriptome and proteome diversity from genetically economical genomes. It is thought that >95% of all multi-exon genes are subject to alternative splicing (Pan, Shai, Lee, Frey, & Blencowe, 2008). Different forms of alternative splicing identified thus far include exon skipping (inclusion of alternative exons), use of alternative (cryptic) 5′ or 3′ splice sites, mutually exclusive exons, alternative initial or terminal exons, and intron retention (Fig. 1). Additionally, there is increasing evidence that alternative splicing can produce qualitative and quantitative changes in gene expression that are pivotally involved in homeostatic regulation. Furthermore, splicing dysfunction is increasingly being recognized as a contributor to disease pathogenesis (Kalsotra & Cooper, 2011; Padgett, 2012; Singh & Cooper, 2012). This is particularly true for cancer, where dysregulation of pre-mRNA splicing is emerging as a hallmark of cancer (Jung et al., 2015; Sveen, Kilpinen, Ruusulehto, Lothe, & Skotheim, 2015).

The integrity of splicing depends on short, conserved sequences in introns. These are found at the 5′ splice site, 3′ splice site, and at the branch point, a motif that is typically located 18–40 nucleotides upstream of the 3′ splice site. These features guide the precise cleavage of pre-RNA molecules at the 5′ termini of introns

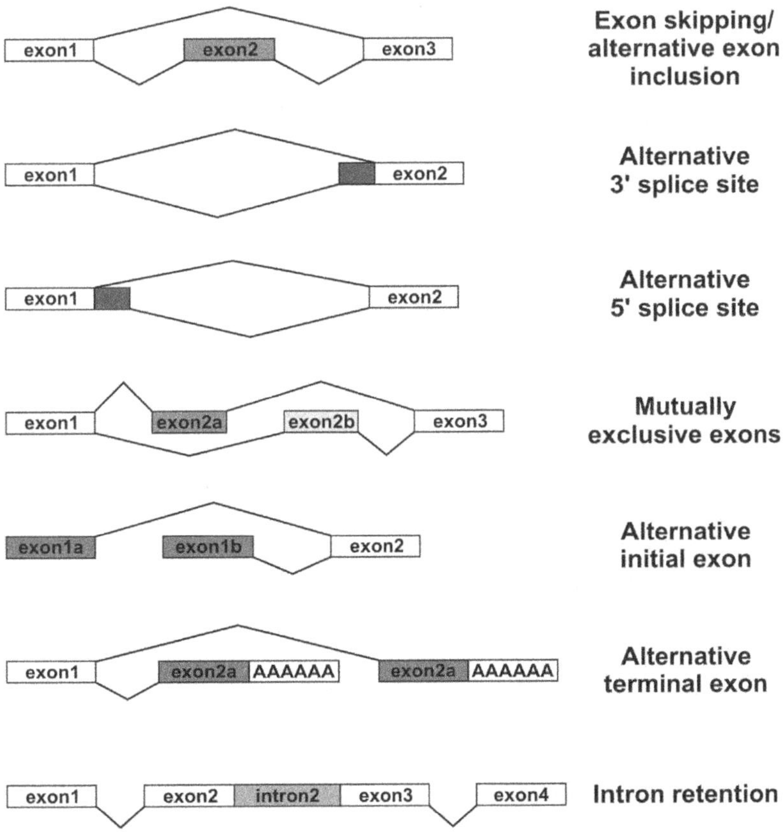

FIGURE 1 Schematic representation of the most common forms of alternative splicing.

Virtually all human multi-exon genes express more than one spliced isoform. The different forms shown here can be combined to generate more complex alternative splicing events and further increase transcriptome complexity. (See color plate)

followed by ligation of adjacent exons to produce uninterrupted coding sequence. These steps are achieved by two successive transesterification reactions (Fig. 2). First, an adenosine in the branch point carries out a nucleophilic attack on the 5′ splice site, which cleaves the RNA and ligates the 5′ end of the intron to the branch point. Second, the 3′ splice site is attacked by the 3′-OH of the 5′ exon. This releases

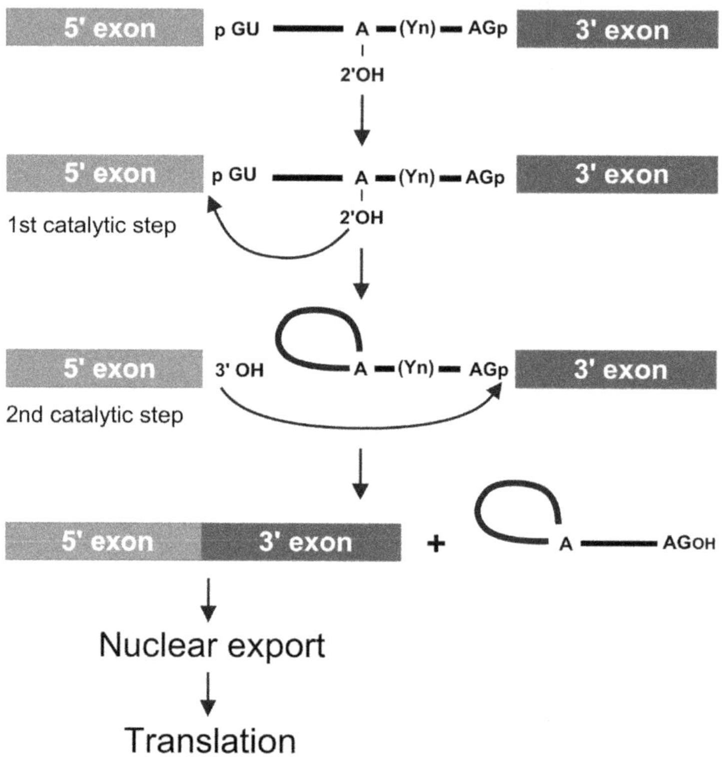

FIGURE 2 Overview of splicing.

The splicing of pre-mRNA is a multistep process beginning with the recognition of specific sequences in the intron by spliceosomal snRNPs. The spliceosome is then assembled on the pre-mRNA and undergoes a series of conformational rearrangements, resulting in a splicing-competent spliceosome. The 5′ splice donor site of each intron is recognized by an almost invariant GT while the 3′ splice acceptor site terminates with an AG. Upstream from the 3′ AG is the polypyrimidine tract (Yn), a region rich in C and U (pyrimidine) nucleotides. Further upstream from this is the branch-point sequence, which includes an adenosine nucleoside. In the first step of splicing, the 2′-OH of the branch-point adenosine nucleophilically attacks the 5′ splice site, resulting in cleavage of the pre-mRNA to release the 5′ exon and a looped form of the intron known as the lariat. In the second transesterification step of the reaction, the 3′-OH of the 5′ exon nucleophilically attacks the 3′ splice site, resulting in the ligation of the exons and release of the lariat. Once the chemical reactions are completed, the spliceosome disassembles, and its components are recycled for another round of splicing.

the intron in a looped or lariat structure, and the adjacent exons are ligated together. Reviews by Fu and Ares (2014) and Jangi and Sharp (2014) provide comprehensive overviews of the molecular processes of RNA splicing.

1.1 SPLICEOSOME: COMPOSITION AND FUNCTION

Splicing is a tightly coordinated sequential process carried out by spliceosomes, massive macromolecular machines composed of >300 proteins, and five small nuclear RNAs (snRNAs). The structure of the yeast spliceosome was described by Yan et al. (2015). Most introns (>200,000 in the human genome (Hoskins & Moore, 2012)) are excised by the major class or U2-type spliceosome and contain snRNA components known as U1, U2, U4, U5 and U6. A tiny fraction of introns (~800 in human) require splicing by low abundance minor class or U12-type spliceosomes, which contain U11, U12, U4atac, U5, and U6atac snRNAs (Hoskins & Moore, 2012; Patel & Steitz, 2003; Wahl, Will, & Lührmann, 2009). Spliceosomal proteins, which include structural proteins, RNA-binding complexes such as heterogeneous nuclear and small nuclear ribonucleoproteins (hnRNPs and snRNPs, respectively), ATPases, and helicases, are collectively known as splicing factors (SFs). The function of SFs is to recognize and select authentic paired splice sites from among the nonauthentic sites commonly present in pre-mRNA, and avoid unintended intron retention or exon skipping that could occur from splice site mispairing. Considering this, the consensus 5′ and 3′ splice sites of major class introns in higher eukaryotic genes are surprisingly degenerate. The removal of introns is therefore not only dependent on the presence of the correct consensus splice site sequences at the exon–intron boundaries and the branch-point sequence. A number of *trans*-acting RNA-binding proteins and *cis*-regulatory sequence elements act in concert to determine whether an alternative exon is included or skipped (Black, 2003). Thus, splice site recognition may depend on the relative strengths of competing splice sites, neighboring splicing enhancers and splicing silencers, secondary structure of the pre-mRNA and the levels and configuration of SFs in a given spliceosome. It is thought that this variability can confer different splicing outcomes specific to a given cell-type or tissue by modulating splice site usage in a particular gene depending on its cellular context.

1.2 REGULATION OF SPLICE SITE SELECTION

The majority of known SFs belong to one of two classes, the serine/arginine (SR) proteins (Anko, 2014; Howard & Sanford, 2015) or the hnRNPs (Jean-Philippe, Paz, & Caputi, 2013). The first canonical SR proteins were discovered in the late 1980s in *Drosophila* (Amrein, Gorman, & Nothiger, 1988; Boggs, Gregor, Idriss, Belote, & McKeown, 1987; Chou, Zachar, & Bingham, 1987) and, using SF2/ASF (SFRS1) as the prototypical family member, were soon shown to be required for constitutive splicing (Krainer, Conway, & Kozak, 1990) as well as the regulation of alternative splicing (Ge & Manley, 1990). Their common mode of action is to bind to *cis*-elements known as exonic splicing enhancers (ESEs) (Blencowe, 2000; Chen & Manley, 2009) where they facilitate the recruitment of spliceosomal snRNPs to the 5′ and 3′ splice sites (Chen & Manley, 2009; Long & Caceres, 2009; Shepard & Hertel, 2009). Positive regulation through SR proteins and other SFs has been shown to also occur from intronic splicing enhancers (ISEs) although these are less common and less well characterized (Black, 2003; Chen & Manley, 2009).

The recruitment of spliceosome components is mediated by either protein–protein interactions of an ESE-bound SR protein with core spliceosomal components through the Arg/Ser (RS) domain (Adams et al., 2001; Kohtz et al., 1994; Wu & Maniatis, 1993) or by direct contact of the RS domain with the branch-point sequence in the pre-mRNA (Shen & Green, 2004; Shen, Kan, & Green, 2004). Post-translational modifications, in particular phosphorylation of serines in the RS domain, are critical for the activity of SR proteins (Xiao & Manley, 1997). In the case of SRSF10 (also known as SRp38), this can even determine whether the protein functions as a splicing repressor or activator (Feng, Chen, & Manley, 2008). The mostly splice-enhancing effect of SR proteins is offset by the activity of negative regulators, most notably the hnRNPs, which bind to intronic (ISS) and exonic (ESS) splicing silencer sequences (Chen & Manley, 2009; Dreyfuss, Kim, & Kataoka, 2002). Their simplest mode of action includes sterically inhibiting the recruitment of spliceosome components. This is achieved either directly by interfering with sequences or pre-mRNA secondary structure necessary for recruitment of snRNPs, or indirectly by blocking the binding of splicing enhancers to adjacent enhancer sequences as was shown for hnRNP A1 and hnRNP 1 (also known as polypyrimidine tract–binding , PTB, protein) (Black, 2003; Jean-Philippe et al., 2013).

1.3 INTRON RETENTION

The demonstration that intron retention, previously thought to be a form of missplicing, plays an obligate role in the normal development of several cell lineages (Boutz, Bhutkar, & Sharp, 2015; Braunschweig et al., 2014; Cho et al., 2014; Pimentel et al., 2015; Wong et al., 2013) supports the notion that "master" splicing regulators are required for lineage specification and the maintenance of homeostasis (Jangi & Sharp, 2014). Intron retention was recognized in nearly 15% of human transcripts more than a decade ago (Galante, Sakabe, Kirschbaum-Slager, & de Souza, 2004). The authors noted that such a number was likely "not spurious and might reflect biological significance." Subsequent studies have proven them correct. Current estimates indicate that nearly 75% of multi-exon transcripts are affected by intron retention (Braunschweig et al., 2014). Intron retention is now recognized as a regulatory mechanism for down-regulating gene expression in which transcripts containing retained introns are degraded as a result of incorporation of premature termination codons (PTCs) and nonsense-mediated decay (NMD) (Wong, Au, Ritchie, & Rasko, 2015). It is deployed in the normal development of particular cell lineages such as granulocytes, lymphocytes, erythrocytes, and neurons (Cho et al., 2014; Pimentel et al., 2015; Wong et al., 2013; Yap, Lim, Khandelia,

Friedman, & Makeyev, 2012) and we have found that intron retention occurs in zebrafish granulocytes (unpublished).

Intron retention is also a conspicuous feature of most cancer transcriptomes, even in the absence of mutations in SFs (Dvinge & Bradley, 2015). In a study by Jung et al. (2015), RNA sequencing and exome analysis of >1800 primary cancer specimens across six cancer types profiled in The Cancer Genome Atlas (TCGA) project identified ~900 exonic single nucleotide variants (SNVs) that disrupt splicing. Of these, 163 SNVs caused intron retention or exon skipping in an allele-specific manner. Of particular note, SNVs causing intron retention were enriched in tumor suppressor genes and, in 97% of these cases, generated a PTC, leading to loss of function through NMD or expression of truncated protein. This led the authors to conclude that intron retention is a common mechanism of tumor suppressor gene inactivation (Jung et al., 2015).

1.4 SPLICING DYSFUNCTION AND DISEASE

An overarching theme in the regulation of alternative splicing is one of combinatorial control by both activating and inhibiting factors that together determine whether an alternative exon is included or skipped. This regulation can be modulated and fine-tuned in response to the requirements of cell type, developmental stage, cellular signaling or stress. The complexity of splicing regulation provides large scope for vulnerabilities leading to disease.

About one-third of all disease alleles are thought to affect splicing (Lim, Ferraris, Filloux, Raphael, & Fairbrother, 2011). Congenital mutations in SFs (including PRPF8, PRPF6, PRPF31, and PRPF3) are associated with retinitis pigmentosa (RP), which can lead to blindness (Chen et al., 2014; Daiger, Bowne, & Sullivan, 2007; Martinez-Gimeno, 2003; Maubaret et al., 2011; McKie et al., 2001; Sullivan, 2006; Ziviello, 2005). Somatic mutations in multiple SFs (including SF3B1, SRSF2, U2AF1, ZRSR2, PRPF8, PRPF40B, U2AF2, and SF1) are associated with acquired myelodysplastic syndromes (MDS) and hematopoietic malignancies (Abdel-Wahab & Levine, 2011; Bejar et al., 2011; Cazzola, Della Porta, & Malcovati, 2013; Graubert et al., 2012; Haferlach et al., 2014; Kurtovic-Kozaric et al., 2015; Papaemmanuil et al., 2011; Thol et al., 2012). Well-characterized spliceopathies affecting TDP-43, NOVA, RBFOX1, and SMN1/2 result in neurological diseases (amyotrophic lateral sclerosis (ALS), paraneoplastic neurological disorders, autism, and spinal muscular atrophy (SMA), respectively) (Brzustowicz et al., 1990; Buckanovich, Posner, & Darnell, 1993; Dredge, Polydorides, & Darnell, 2001; Lefebvre et al., 1995; Martin et al., 2007). In 2015, cancer-specific dysregulation of SFs was shown to contribute to disease progression with >30 SF genes identified as oncogenes (Sveen et al., 2015).

The role of splicing variation in normal development and disease requires methodologies and models for identifying alternative splicing events and quantifying their prevalence and contribution to normal cell biology and disease pathogenesis. Advances in next-generation sequencing technologies have made it possible to move from studying the complexity of RNA regulation for individual transcripts to examining global transcriptomes of tissues and even single cells (Tang et al., 2009). In addition, methods such as HITS-CLIP (High-Throughput Sequencing of Cross-Linked and ImmunoPrecipitated cDNA library) (Chi, Zang, Mele, & Darnell, 2009; Licatalosi et al., 2008), and its variations, iCLIP (individual-nucleotide resolution CLIP) (Hauer et al., 2015; Konig et al., 2010; Sutandy, Hildebrandt, & Konig, 2016), and PAR-CLIP (PhotoActivatable-Ribonucleoside-enhanced CLIP) (Hafner et al., 2010; Kishore et al., 2011) combine biochemical methods with high-throughput sequencing technology and bioinformatics approaches to determine single base–resolution mapping of protein RNA–binding sites *in vivo*. These sophisticated approaches provide technologies for elucidating regulatory networks and determining alternative splicing outcomes.

In this chapter, we survey methods for the unbiased evaluation of the biological consequences of splicing variation, with a particular focus on zebrafish, an attractive *in vivo* model for studying splicing regulation. We describe methods for the global analysis of spliceosome function using RNAseq, the subsequent validation of individual candidates, and the biochemical characterization of snRNP assembly. We conclude by suggesting opportunities that exist in zebrafish to expand the number of splicing modulators that may ultimately prove to be useful for therapeutic purposes.

2. EMPLOYING ZEBRAFISH MODELS FOR STUDYING SPLICING VARIATION

2.1 ZEBRAFISH GENOMICS AND REVERSE GENETICS

Zebrafish provide an established biomedical research model that particularly excels in generating insights into dynamic biological questions that are best answered *in vivo*. Their fecundity coupled with tractable genetics and rapid development make zebrafish a flexible model, particularly for developmental genetics (Lieschke & Currie, 2007). The optical transparency of zebrafish embryos facilitates observational studies of *in vivo* biology.

About 70% of human genes have zebrafish orthologs (Howe et al., 2013), allowing the effective modeling of human disease in zebrafish. Moreover, the potential relevance of mutations in, or adjacent to, genes that are significantly associated with human disease is increasingly being elucidated in zebrafish. The many advantages of the model have been enhanced by the remarkable impact of CRISPR/Cas9 technology (Jinek et al., 2012), which has overcome the erstwhile lack of a stable and effective reverse genetics approach in zebrafish (Hwang et al., 2013). In particular, CRISPR/Cas9 has been used efficiently for targeted gene modification. The basic methodology relies on the error-prone nonhomologous end-joining DNA repair mechanism to introduce indels at the target site. Meanwhile, increasingly sophisticated modifications to the technology are permitting even greater precision of targeting, including tissue-specific knockout, knock-ins of reporter genes, or precise, disease-mimicking mutations (Auer, Duroure, De Cian, Concordet, & Del Bene, 2014; Hwang et al., 2013; Maruyama et al., 2015; Sander & Joung, 2014). The ease and efficiency of CRISPR/Cas9-targeted gene editing has transformed our ability to undertake large-scale functional analyses of sets of candidate disease genes (Varshney et al., 2015), including large sets of SF genes. The feasibility of these investigations has also been enhanced by the availability of a much-improved zebrafish genome assembly (GRCz10), now in its 10th reiteration, which is well annotated with few gaps, making informative RNA-splicing analysis viable in this model.

2.2 BIOLOGICAL INSIGHTS FROM ZEBRAFISH SF MUTANTS

Our groups have characterized mutants with defects in major and minor class splicing (Keightley et al., 2013; Markmiller et al., 2014). The *cephalophŏnus* (*cph*) mutant harbors a mutation in *prpf8*, encoding a component common to both major and minor class spliceosomes. Although *pre-mRNA processing factor 8 (prpf8)* is widely expressed and *cph* has a phenotype affecting multiple tissues, *cph* was identified for its role in myelopoiesis (Keightley et al., 2013). Phenotypic specificity, despite the ubiquity of SF expression and splicing in all cells is a surprising feature of some germline human splicing mutations (see Section 2.3). The hematopoietic defects in *cph* support the hypothesis that PRPF8 mutations prevalent in about 5% of MDS and secondary acute myeloid leukemia (Haferlach et al., 2014; Kurtovic-Kozaric et al., 2015; Makishima et al., 2012) may make a functional contribution to the disease phenotype.

Minor class or U12-type splicing is catalyzed by a spliceosome that recognizes distinct, noncanonical consensus sequences in U12-type introns (Patel & Steitz, 2003). The zebrafish intestinal mutant, *caliban* (*clbn*), was the first vertebrate model to undergo genome-wide analysis of minor class splicing defects *in vivo* (Markmiller et al., 2014). *clbn* harbors a mutation in *RNA-binding region (RNP1, RRM) containing 3* (*rnpc3*), encoding one of only seven proteins specific to the U12-type spliceosome (Will et al., 2004). The *rnpc3* deficiency of *clbn* resulted in aberrant U11−U12 di-snRNP assembly and U12-type intron retention (Markmiller et al., 2014). U12-type introns, although comprising only 0.34% of all human introns, are often found in genes with essential roles in DNA replication and repair, transcription, RNA processing, and translation (Burge, Padgett, & Sharp, 1998). In *clbn*, the retention of U12-type introns in U12-type intron-containing genes resulted in frame-shifted transcripts that generated PTCs. This in turn triggered NMD leading to severely impaired expression of mutant alleles. A large number of genes lacking minor class introns are also dysregulated in *clbn* compared to wild type (WT), most likely due to the fact that adequate rates of transcription, RNA processing, and translation are required for the expression of all genes, especially in metabolically active tissues.

Other zebrafish SF mutants include *earl grey* (*egy*), which carries a mutation in the general SF gene *p110* that is required for snRNP recycling (Trede et al., 2007). Characterization of *egy* revealed a network of >30 cofactors involved in the spliceosome cycle and snRNP biogenesis that are co-regulated by p110. In the zebrafish U1C mutant (*hi1371*), a large set of specific target genes with altered splicing as a consequence of U1C deficiency, was identified demonstrating U1C to be a regulator of alternative splicing (Rosel et al., 2011). U1C (*snrnpc*) encodes a protein component of the U1 snRNP important for 5′ splice site recognition and subsequent assembly of the spliceosome. The zebrafish *mummy* (*mmy*) mutant carries a truncation of *dhx8*, the ortholog of yeast *prp22* encoding an RNA helicase that is involved in releasing fully processed RNA from the spliceosome for nuclear export. This mutant has a complex phenotype that includes disruption of early hematopoiesis. Retention of introns in 11/16 selected genes was demonstrated (English et al., 2012).

2.3 ZEBRAFISH MODELS OF HUMAN DISEASES DUE TO SF DYSFUNCTION

Several spliceopathies (human diseases caused by defects in SFs) are modeled by zebrafish carrying mutated SF genes or by zebrafish with transient SF deficiency due to knockdown approaches, most often with antisense morpholino oligonucleotides (AMOs). While studies by Kok et al. (2015) have drawn attention to some technical limitations of this widely employed reverse genetics approach in zebrafish, well-controlled transient knockdown experiments retain usefulness for studying gene function in zebrafish (Eisen & Smith, 2008; Pauli, Montague, Lennox, Behlke, & Schier, 2015).

2.3.1 *Retinitis pigmentosa*

RP is a degenerative disease with progressive loss of rod photoreceptor cells in the retina and subsequent disruption of the adjacent retinal pigment epithelium, leading to vision loss. Mutations in several genes give rise to RP. Of particular relevance here are mutations in five pre-mRNA processing factors PRPF3, PRPF4, PRPF8, PRPF31, and Pim1-associated protein (PAP1) that cause autosomal dominant RP. The remarkable tissue specificity of the disease, given the ubiquitous expression and function of the contributing SFs, remains an enigma, although it is thought to indicate that photoreceptor cells have a higher requirement for mRNA processing than any other cells, or that SF mutations disrupt an as yet unidentified tissue-specific pathway. Zebrafish have been used to functionally evaluate two pathogenic *prpf31* mutations that individually generate a similar RP phenotype through different mechanisms. While one mutation results in instability of the encoded protein and its mis-localization to the cytoplasm, the other results in aberrant splicing of retinal transcripts with no general defects in gene expression and increased apoptosis of photoreceptor cells (Linder et al., 2011; Yin, Brocher, Fischer, & Winkler, 2011). A zebrafish *prpf4* model based on AMO gene knockdown demonstrated tissue-specific retinal defects in gene expression in advance of the subsequent association of this gene with RP in humans (Chen et al., 2014; Linder et al., 2011). Together these studies in zebrafish pointed to tri-snRNP assembly as a limiting regulatory checkpoint in RP, and indeed biochemical studies in zebrafish have shown that deficiency of Prpf8, another SF associated with RP (Martinez-Gimeno, 2003) leads to impaired U5 and U4 snRNP assembly (Keightley et al., 2013).

2.3.2 Spinal muscular atrophy

Another spliceopathy successfully modeled in zebrafish is SMA. SMA is a degenerative motor neuron disease resulting from mutations in the *survival motor neuron* (*SMN*) gene. SMN is known to have a role in snRNP biogenesis, specifically of the Sm class of snRNPs. Studies in zebrafish have complemented those in mouse with the initial zebrafish model identifying the role of Smn in motor axon development *in vivo* (McWhorter, Monani, Burghes, & Beattie, 2003). Subsequent identification of *stasimon* in *Drosophila* as a U12-type intron-containing SMN target that is required for motor neuron function was validated in *Drosophila* and zebrafish (Lotti et al., 2012). A zebrafish SMA model expressing human SMN2 to more closely mimic human disease has shown that splicing of human SMN is conserved in zebrafish and exhibits the same splicing pattern as in human and mouse (Hao le, Burghes, & Beattie, 2011). Although this model postdates a number of elegant mouse models for SMA, it provides the additional advantage of economically feasible high-throughput drug screening (Lieschke & Currie, 2007).

2.3.3 Other diseases

Recurrent mutations in SF3B1, a core component of the U2 snRNP involved in branch-point recognition, are highly prevalent in refractory anemia with ring sideroblasts, a subtype of myelodysplasia (Haferlach et al., 2014; Papaemmanuil et al., 2011; Yoshida et al., 2011). A zebrafish mutant in *sf3b1* has been identified and characterized for defects in neural crest development but not for its hematopoietic development (An & Henion, 2012).

AMO knockdown of the *c16orf57* gene encoding Usb1 (U6 biogenesis 1) (Mroczek et al., 2012) models the rare inherited disorder poikiloderma with neutropenia (Patil, Uechi, & Kenmochi, 2015). In this zebrafish model, aberrant splicing of neutrophil-specific genes, but not of genes encoding erythroid or hemopoietic stem cell markers, was thought to contribute to the observed neutropenia.

In addition to the use of AMOs to treat or create transient models of spliceopathies, transcription activator-like effector nucleases (TALENs) have been used to engineer a stable mutation into a splice donor site of the *col6a1* gene, thereby creating a mutant line for studying collagen VI-related diseases in adult zebrafish as well as larvae (Radev et al., 2015). In the future, these genomic-editing approaches will be greatly facilitated by the availability of CRISPR/Cas technology (see Section 2.1).

3. TOOLS FOR GLOBAL SPLICEOME ANALYSIS IN ZEBRAFISH

Next-generation sequencing methods and a well-annotated zebrafish genome allow for global analysis of the zebrafish spliceome, defined as the set of all possible alternatively spliced mRNA transcripts. Combined with zebrafish reverse genetics, this enables a comprehensive, systematic approach to the study of splicing, for example, allowing concurrent evaluation and comparison of splicing defects in multiple SFs using panels of zebrafish SF mutants. Although a one-by-one examination of each SF has its role, given the complexity of the multicomponent spliceosome, a systems biology approach to functional analysis of the whole complex is likely to be particularly revealing. A global approach will also facilitate the identification of fine distinctions between cell- and/or tissue-specific transcriptomes, which warrants attention given the remarkable differential tissue specificity exhibited by particular mutations in general SFs.

3.1 RNAseq

Intron retention can be assessed in cells or tissues from mutant or overexpressing transgenic lines and compared to corresponding samples from WT age-matched controls. In our previous analysis of intron retention in zebrafish mutants, intron retention was assessed by aligning RNAseq tags to the annotated zebrafish (Zv9) genome assembly that contains 606 manually curated U12-type introns. To normalize for the more common occurrence of nonunique sequences, such as repetitive elements in introns compared with exons, we generated a zebrafish uniqueome (Koehler, Issac, Cloonan, & Grimmond, 2011) and calculated all expression levels as reads per unique kilobase per million reads (RPuKM). Plotting expression levels of all introns that satisfy minimum expression requirements (RPuKM $>$ 0.05) in both WT and *clbn* revealed a clear bias toward retention of U12-type introns in *clbn* at 72hpf (Fig. 3). We found that a significantly higher fraction of U12-type introns also showed some degree of intron retention in WT larvae, compared to U2-type introns, consistent with U12-type introns being processed more slowly than their U2-type counterparts (Patel, McCarthy, & Steitz, 2002; Singh & Padgett, 2009)

Current approaches to assessment of intron retention and other modes of alternative splicing generally require at least 2×10^7 cells to generate enough RNA for each sample, especially if examining terminally differentiated cells with low RNA content. Poly(A)-enriched mRNA/cDNA libraries can be prepared using a commercial library preparation kit such as TruSeq (Illumina) or, for enrichment of noncoding RNAs, a ribosomal RNA depletion protocol such as RiboZero/ScriptSeq is more appropriate, followed by sequencing using Illumina Hi-Seq at a minimum of 60 M paired-end reads (120 M total). This high sequencing depth is required for efficient detection of comparatively rare intron retention transcripts. In addition to the rarity of these alternatively spliced transcripts there are several concerns that need to be addressed when designing the analysis pipeline. First, certain mapping algorithms may bias toward splitting reads across two exons rather than mapping to the exon and through a retained intron. Secondly, intronic sequences are frequently low complexity and contain repeats, which means in standard RNAseq pipelines these sequences may be filtered out and therefore lost. New algorithms are continually being developed in order to maximize signal to noise in the pursuit of a specific

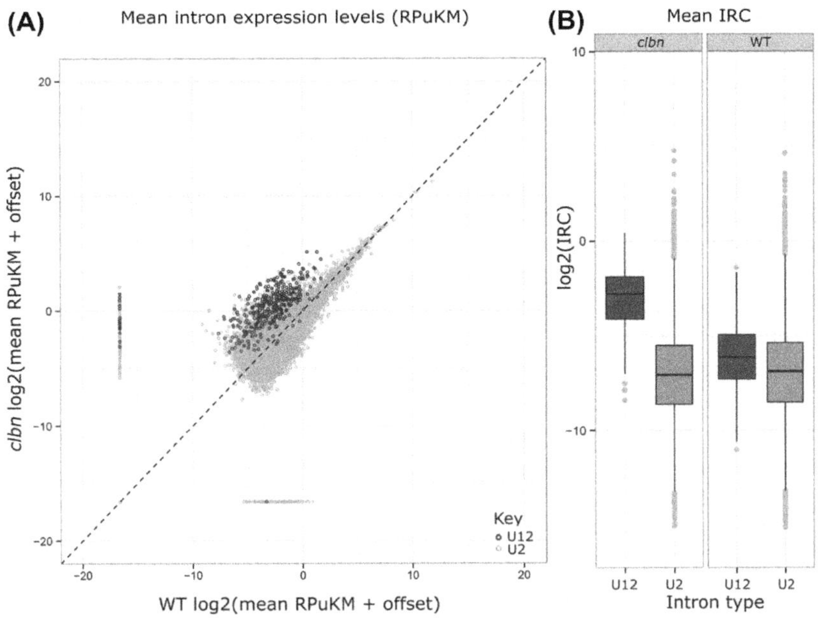

FIGURE 3 U12-type introns are selectively retained in the zebrafish mutant, *caliban* (*clbn*).

(A) Scatterplot of normalized intron expression levels derived by RNAseq analysis of *clbn* and WT zebrafish larvae at 72hpf ($n = 3$). All (Refseq) introns are shown as gray dots and U12-type introns as black dots. Calculation of RPuKM (reads per unique kilobase per million reads) was as described (Markmiller et al., 2014). RPuKM values are plotted with an offset of 1E−05 to account for 0 values. U12 introns highlighted in black lie almost entirely above the midline, where the midline represents equal expression, indicating enhanced U12-type intron retention in *clbn*. (B) Boxplot displaying the higher level of U12-type intron retention in *clbn* mutants compared to WT larvae by mean intron retention coefficient (IRC) ($n = 3$). The IRC normalizes intron expression levels according to the level of expression of the corresponding gene (Markmiller et al., 2014). The splicing defect in *clbn* is specific to U12-type introns.

biological question that is being addressed by the data. A new algorithm, IRFinder, was developed to detect intron retention with greater accuracy and precision. It has been used successfully for mouse and human transcriptomes (Wong et al., 2013) and has recently been applied to zebrafish (our unpublished work).

Although the above-mentioned work flows are specifically for intron retention analysis, they can be adapted for investigation of other splicing variants of interest by altering the data analysis pipeline. In general, most forms of alternative splicing result in differential exon–exon junctions (Fig. 1) and so accurate detection of exon–exon, rather than exon–intron, junctions becomes paramount. This kind of analysis can generally be carried out at a more standard sequencing depth of 20 M reads. The current trade-offs between different aligners are more stringent mapping relying on splice annotations giving good positive predictive value but potentially neglecting allocation of a considerable number of reads (eg, STAR) versus a less conservative aligner (eg, TopHat2), which provides greater sensitivity and mapping accuracy leading to better quantification but at the expense of false positives (eg, reads spanning multiple splice sites). A 2014 paper performed side-by-side comparisons of the common aligners and found TopHat2 to be superior in terms of alignment precision and uniquely mapped reads (Gatto et al., 2014). To take advantage of this strength and simultaneously address its poorer positive predictive value, another layer of additional processing termed FineSplice including logistic regression and posterior probability estimates can be applied post-alignment to almost eliminate spurious gapped alignments and generate 99% precision in detecting expressed features (Gatto et al., 2014). Algorithms will continue to evolve, and it is important to be clear on the biological questions at the outset and to consult with bioinformaticians at the experimental design stage to ensure the data are generated and analyzed appropriately in order to answer the question at hand.

3.2 RT-PCR VALIDATION

Once candidate genes have been identified from a next-generation sequencing pipeline, splicing variations require validation by an independent method such as RT-PCR. Primers are designed to span the exon–exon or exon–intron junctions of interest and transcripts analyzed from control and test samples using random primed cDNA. Typically, primers are designed to ensure the expected differences will be resolvable by gel electrophoresis. Intron size is a major consideration for these experiments, as introns may be many kilobases in length and not readily amplifiable. Hence, small "model" introns are generally selected for analysis of intron retention (Fig. 4). Similarly, analysis of exon skipping needs to take into account the size of skipped exons. Controls for these experiments include: (1) reverse transcriptase negative samples which should not generate a signal and control for contaminating genomic or cDNA in the sample; (2) genomic DNA, positive control. In the case of intron retention, samples may be treated with and without NMD inhibitors such as caffeine and cycloheximide (and/or siRNA inhibition of the regulator of nonsense transcripts homolog, Upf1/Upf2) to enhance low intron retention signals. PCR products can then be sequence verified.

3.3 IN VIVO MINIGENE ASSAY

In vivo splicing assays can be conducted in zebrafish by transient expression of minigenes (Barboric et al., 2009; Markmiller et al., 2014). Either commercially available or customized minigenes can be used. For custom assays, a minigene encompassing exons and intervening introns of interest is amplified. Restriction enzyme site containing 23-bp tails of customized sequence that does not map to the zebrafish genome by BLAST analysis may be added at the 5′ end of each primer to allow cloning into the vector of choice, such as pCS2+. At the same time this

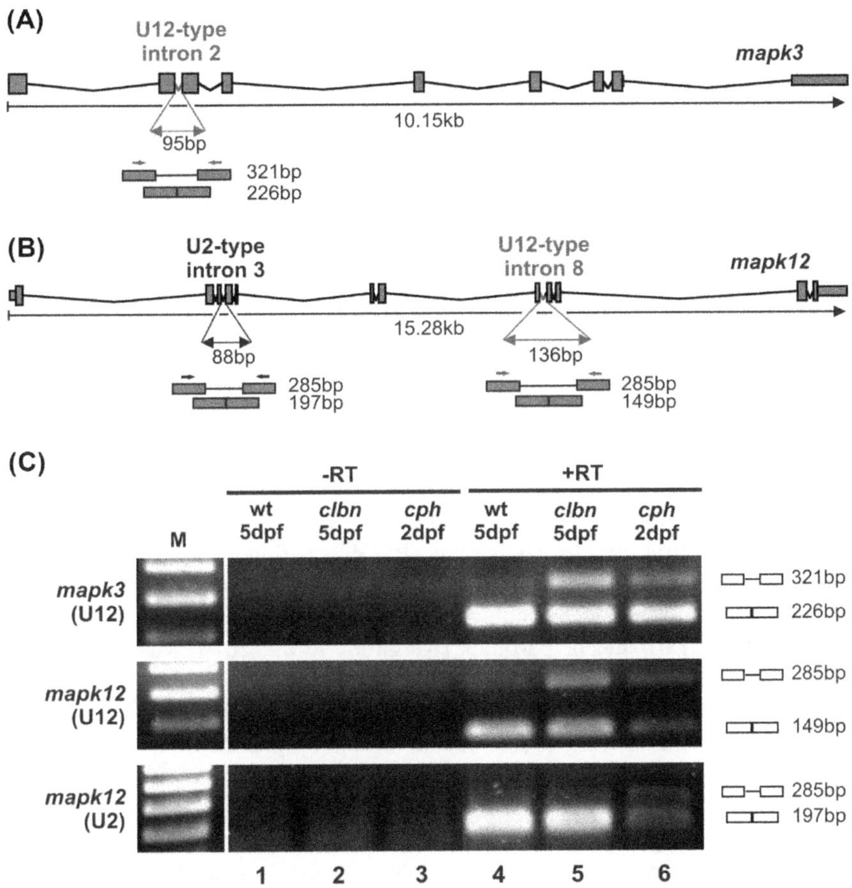

FIGURE 4 Splicing efficiency is generally impaired in *cph* embryos whereas *clbn* larvae are specifically impaired in U12-type splicing.

(A) The zebrafish *mitogen-activated kinase 3* (*mapk3*) gene contains eight introns, including a single U12-type intron (intron 2) of 95bp. (B) The *mapk12a* gene contains 11 introns, including a single U12-type intron (136bp) between exons 8 and 9 and a U2-type intron of similar size (intron 3; 88bp). These introns were selected for RT-PCR analysis to assess the efficiency of U2-type splicing and U12-type splicing by intron retention. (C) RT-PCR shows that splicing of the U12-type introns is inhibited in *clbn* larvae at 5 dpf compared to wild type (top two panels; lanes 4 and 5), while splicing of U2-type introns is not affected (lower panel, lanes 4 and 5). Meanwhile *cph* embryos at 2 dpf show accumulation of incompletely spliced precursor mRNAs that is independent of intron type (lane 6, all panels). No signals are detected in RT-negative control samples (lanes 1–3). M, markers. Primer sequences for this analysis are provided in Table 1.

permits the specific amplification of the exogenous minigene and distinction from host genomic DNA. Microinjection of linearized minigene DNA into single cell–stage embryos achieves transient mosaic expression. One-step RT-PCR analysis using the minigene-specific primers can then be performed (Section 3.2) on total RNA extracted from embryos at the desired age (for instance, when mutant and WT embryos are readily distinguishable by phenotype). This assay provides a powerful approach to assessing whether specific mutations in a gene of interest, recapitulated in the minigene, affect splicing.

4. BIOCHEMICAL ANALYSIS OF SPLICEOSOME COMPONENT ASSEMBLY

4.1 GLYCEROL GRADIENT SEDIMENTATION

The integrity of spliceosome assembly in zebrafish larvae can be readily evaluated by analyzing the composition of snRNP complexes resolved by glycerol gradients or native polyacrylamide gel electrophoresis (PAGE). Following transfer of resolved snRNA species to nylon membranes, northern analysis with probes designed to hybridize to the different snRNA components is used to detect the presence/absence of mono-, di-, and tri-snRNP particles (Keightley et al., 2013; Markmiller et al., 2014).

To examine the composition of snRNPs by comparative sedimentation on glycerol gradients, SF mutant and WT larvae are sorted by phenotype and whole larval extracts prepared (about 100 larvae per group) and layered onto 10–30% glycerol gradients (Fig. 5). Following centrifugation, 20 samples of equal volume are collected sequentially from the top of the gradient prior to RNA extraction. The snRNAs extracted from the fractionated snRNP samples and 10% input samples are then separated on high-resolution 10% polyacrylamide/8 M urea TBE gels and transferred to Hybond-N nylon membranes (GE Healthcare), for northern analysis. Alternatively, native snRNP size and conformation can be compared in whole larval extracts from WT and mutant SF larvae on native (nondenaturing) polyacrylamide gels. Larval lysates are prepared and incubated with or without heparin (to reduce potential nonspecific interactions and facilitate characterization of stable complexes) prior to loading on 4% native polyacrylamide gels. Electrophoresis is conducted at 4°C with constant recirculation of cold buffer. snRNPs are then transferred to a membrane such as Hybond-N (GE Healthcare) for northern analysis.

4.2 NORTHERN ANALYSIS

To detect the presence of immobilized snRNPs on northern blots, membranes are incubated with ^{32}P-radiolabeled probes designed to hybridize to zebrafish major and/or minor class snRNAs (Table 1). Background is minimized using Ultrahyb hybridization buffer (Ambion) for the pre-hybridization and hybridization steps.

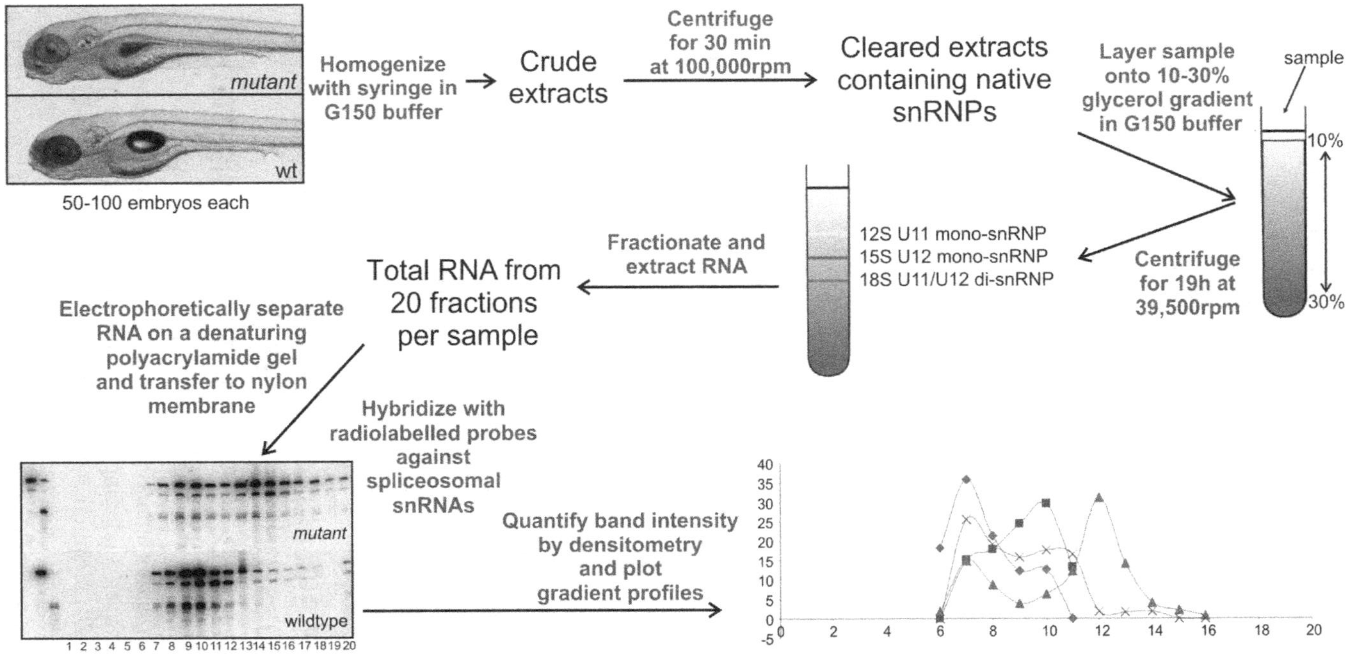

FIGURE 5 Analysis of snRNP size and conformation by glycerol gradient and northern blot analysis

At 6 dpf, zebrafish larvae are separated into pools of either homozygous mutant and wild type siblings by visual inspection. Larvae are homogenized on ice in G150 buffer and lysates cleared of debris by ultracentrifugation (eg, for 30 min at 100,000 rpm/435,000*g* in a TLA100.2 rotor in a Beckman Coulter ultracentrifuge). Cleared lysates are separated on 10–30% glycerol gradients (eg, for 19 h at 39,500 rpm/277,000*g* in a Beckman SW40 swinging bucket rotor in a Beckman Coulter Optima L-90K ultracentrifuge. Upon fractionation of gradients into 20 fractions of equal volume, RNA is extracted with phenol–chloroform, separated by denaturing PAGE and transferred to a nylon membrane. Immobilized RNA molecules are hybridized with ^{32}P-radiolabeled cRNA (U11, U6atac, and U5) or cDNA (U12) probes to zebrafish snRNAs and signals quantified to calculate gradient profiles. (See color plate)

Table 1 Oligonucleotide Primers

Amplicon	Forward Primer (5′−3′)	Reverse Primer (5′−3′)
RT-PCR splicing assays		
mapk3 U12 (intron 2)	AGACCTACTG CCAGCGCACCCTG	AGCCCTCGCA GGATCTGATACAG
mapk12 U12 (intron 8)	GACATCTGGTC AGTCGGGTGCATC	TCTTCAGACT GTAGCTTGGCTGTG
mapk12 U2 (intron 3)	GTTATCGGACT TGTGGATGTGTTC	CATCTGATAGAC CAGATACTGCAC
Templates for northern probes		
U11	GCATCTGCTG TGAATAGCGTA	GAGGCACCAA GATAACAGATCA
U12	TGCCTTAAACTGA TGAGTAAGGAAAA	CGCGGCATC TCGCTAAAGTA
U5	TGTTTCTCTTC ATATCGAATAAGTC	AAAATTAGTAA ATACTCAAGGTGTTCC
U6atac	CTGTTGTTTGAGA GGAGAGAAGGT	AAACCACCC CGATCATGG

Sensitivity of signal detection is maximized using full-length radiolabeled cRNA probes generated from linearized plasmid DNA template. Primers for cloning zebrafish snRNAs are provided in Table 1. Washed membranes are exposed to Phosphor Storage screens (GE Healthcare) before image acquisition and analysis using ImageQuant TL software (GE Healthcare) to quantify full-length signals for snRNPs of interest. Background is subtracted manually for each lane, and signals in lanes with >2× signal to background ratio are expressed as a percentage of the lane with the greatest signal intensity. Normalized intensities of bands versus fraction number are computed for each fraction and plotted as a single gradient profile.

5. ASSESSING THE BIOLOGICAL IMPACT OF SPLICING VARIATION OR DYSFUNCTION IN ZEBRAFISH

5.1 ZEBRAFISH AS A MODEL TO DEVELOP THERAPIES DIRECTED AT DISEASE CORRECTION BY TARGETING SPLICING

Modulation of splicing is gaining attention as a therapeutic strategy for treatment of splicing dysfunction diseases including various myotonic dystrophies (Barrie, Smith, Sanford, & Sadee, 2012; Spitali & Aartsma-Rus, 2012; Tse, 2012) and cancer (Dehm, 2013). Among the most common and effective agents for this modulation are antisense oligonucleotides (ASOs). ASOs are synthesized complementary to a splice site sequence in the target mRNA and are generally chemically modified by either phosphodiamidate or phosphorothioate, or the introduction of 2'-methoxyethoxyoligoribonucleotide gapmers (Hong et al., 2011) to avoid digestion by endogenous nucleases prior to binding target mRNA and increase stability of the mRNA–ASO hybrid. Target binding by ASOs can execute a range of splicing modulations including exon skipping, exon inclusion, restoration of cryptic splicing, favorable skewing of alternative splicing, and indirect normalization of splicing—for example, by targeting abnormal expansion of CUG repeats in the *DMPK* gene of patients with myotonic dystrophy type 1 (DM1), an incurable, multisystemic neuromuscular disease. In DM1, the expanded CUG transcripts sequester important proteins, including the alternative splicing regulator muscleblind–like protein (MBNL), leading to splicing defects and disease symptoms. De-sequestering of MBNL from the expanded CUG repeats by competition with ASOs leads to correction of splicing dysfunction (Spitali & Aartsma-Rus, 2012).

The success of this therapeutic approach depends on effective delivery as well as target specificity. Although these therapeutics have been in development for over 25 years, and have been plagued with problems in the transition from bench to clinical trial (Stein & Goel, 2011), there are some successes (Koo & Wood, 2013; Sharma, Sharma, & Singh, 2014; Tse, 2013). One of the main stumbling blocks is that while these agents are very effective when delivered at high concentration via liposomes into cell lines, they can be mired in the far more complex in vivo environment (Spitali & Aartsma-Rus, 2012; Stein & Goel, 2011). It is therefore of paramount importance to develop robust disease models that can enable expeditious screening for therapeutic efficacy. Zebrafish are a model organism that may provide such utility.

Duchenne muscular dystrophy (DMD) is a debilitating muscle-wasting syndrome caused by mutations in the large *DMD* gene encoding dystrophin. Following the discovery that deletion of large regions of DMD results in a milder syndrome (Becker's dystrophy), it was hypothesized that removal of deleterious exons in DMD could result in gains in protein function (Harper et al., 2002; Klein et al., 1992), but the required efficacy of exon skipping for therapeutic benefit was not known. While DMD zebrafish models that closely resemble the human disease have been available for some time (Bassett et al., 2003; Harper et al., 2002; Klein et al., 1992), the nature of the underlying mutation means that they are not useful for examining exon skipping as a therapy. In 2011, the development of a new dystrophin-deficient zebrafish mutant has enabled evaluation of exon skipping. It was found that 30–40% exon skipping efficiency/restoration of dystrophin was required to produce robust phenotypic amelioration but improvements were seen with as little as 10% exon skipping efficiency (Berger, Berger, Jacoby, Wilton, & Currie, 2011). This model is an example of the utility of zebrafish for preclinical assessment and refinement of therapeutic ASO strategies.

In another example, zebrafish has been used to model Menkes disease and examines ASOs as a potentially effective therapy (Madsen, Morcos, Mendelsohn, & Gitlin, 2008). Menkes disease is characterized by impaired extracellular matrix formation, neurodegeneration, and infant mortality and results from loss-of-function mutations in the *ATP7A* gene that encodes a P-type ATPase required for copper absorption and homeostasis (Kaler, 1998). A zebrafish model of this disease was developed (Mendelsohn et al., 2006) and used to screen for effective ASOs, demonstrating that even a small increase in WT Atp7a significantly improved the phenotype and that modest augmentation of ATP7A within a specific developmental window might be useful for *in utero* treatment of the disease (Madsen et al., 2008).

5.2 USING ZEBRAFISH TO SEARCH FOR DRUGS THAT MODULATE SPLICING

Relatively few reagents have been described that modulate pre-mRNA splicing (Effenberger et al., 2014; Kaida et al., 2007; Kotake et al., 2007; Younis et al., 2010). In principle, transgenic zebrafish larvae harboring splicing reporter genes could provide a platform to greatly expand the number of splicing modulators available for testing in preclinical models.

Chemical genetic screens in zebrafish are in widespread use to identify compounds that target specific signaling pathways and regulate developmental processes and disease phenotypes (Lieschke & Currie, 2007). In a small number of cases the identified compounds have progressed to clinical trials (Cutler et al., 2013; White et al., 2011). Zebrafish embryos and larvae provide particularly attractive *in vivo* models for chemical screens because they are small in size and available in large numbers. In a typical screen, up to 20 zebrafish embryos are allocated to individual wells of a 96-well plate and exposed to hundreds, if not thousands, of compounds, depending on the robotic capability of the laboratory (Jing, Durand, Ezzio, Pagliuca, & Zon, 2012; Tat, Liu, & Wen, 2013; White, Rose, & Zon, 2013).

Previous efforts to identify splicing modulators have been carried out largely *in vitro* or in mammalian cell culture systems (Effenberger et al., 2014; Patel et al., 2002; Younis et al., 2010). In an approach that could potentially be extrapolated to zebrafish, Younis et al. (2010) described a high-throughput cell-based assay to accurately measure levels of pre-mRNA splicing. In their quantitative luciferase-based reporter system, a luciferase cDNA is interrupted with a small U2-type intron and juxtaposed with a PEST sequence, which upon transcription generates a short-lived mRNA. Upon optimizing analogous reporter constructs for use in zebrafish, it should be feasible to generate lines stably expressing the splicing reporter transgenes for high-throughput screening. In general, libraries of known bioactive compounds with annotated mechanisms of action, including drugs already achieving FDA approval, will provide the best starting point for screens designed to identify drugs capable of correcting splicing dysfunction that may ultimately be of use in the clinic.

6. SUMMARY

Alternative splicing is a major determinant of gene expression in physiological and pathological processes. Studies of splicing in zebrafish have been instrumental in showing how mutations in specific components of the splicing machinery are capable of generating tissue-specific, rather than global, defects in morphology and function, and how intron retention in a small number of susceptible genes can bring about widespread changes in gene expression.

ACKNOWLEDGMENTS

The authors are supported by the National Health and Medical Research Council of Australia (NHMRC, 1024878, 1044754, 1070687, 1061906, 1080530), Cancer Council Victoria (1047660), Cancer Council of NSW (RG1405 and RG1409), Cure the Future, an anonymous foundation (JEJR) and Ludwig Cancer Research. The Australian Regenerative Medicine Institute and The Walter and Eliza Hall Institute of Medical Research are supported by funds from the State Government of Victoria and the Australian Federal Government.

REFERENCES

Abdel-Wahab, O., & Levine, R. (2011). The spliceosome as an indicted conspirator in myeloid malignancies. *Cancer Cell, 20*(4), 420–423. http://dx.doi.org/10.1016/j.ccr.2011.10.004. pii:S1535-6108(11)00363-1.

Adams, D. J., van der Weyden, L., Mayeda, A., Stamm, S., Morris, B. J., & Rasko, J. E. (2001). ZNF265–a novel spliceosomal protein able to induce alternative splicing. *Journal of Cell Biology, 154*(1), 25–32.

Amrein, H., Gorman, M., & Nothiger, R. (1988). The sex-determining gene tra-2 of *Drosophila* encodes a putative RNA binding protein. *Cell, 55*(6), 1025–1035.

An, M., & Henion, P. D. (2012). The zebrafish sf3b1b460 mutant reveals differential requirements for the sf3b1 pre-mRNA processing gene during neural crest development. *International Journal of Developmental Biology, 56*(4), 223–237. http://dx.doi.org/10.1387/ijdb.113383ma.

Anko, M. L. (2014). Regulation of gene expression programmes by serine-arginine rich splicing factors. *Seminars in Cell & Developmental Biology, 32*, 11–21. http://dx.doi.org/10.1016/j.semcdb.2014.03.011.

Auer, T. O., Duroure, K., De Cian, A., Concordet, J. P., & Del Bene, F. (2014). Highly efficient CRISPR/Cas9-mediated knock-in in zebrafish by homology-independent DNA repair. *Genome Research, 24*(1), 142–153. http://dx.doi.org/10.1101/gr.161638.113.

Barboric, M., Lenasi, T., Chen, H., Johansen, E. B., Guo, S., & Peterlin, B. M. (2009). 7SK snRNP/P-TEFb couples transcription elongation with alternative splicing and is essential for vertebrate development. *Proceedings of the National Academy of Sciences of the United States of America, 106*(19), 7798–7803. http://dx.doi.org/10.1073/pnas.0903188106.

Barrie, E. S., Smith, R. M., Sanford, J. C., & Sadee, W. (2012). mRNA transcript diversity creates new opportunities for pharmacological intervention. *Molecular Pharmacology, 81*(5), 620–630. http://dx.doi.org/10.1124/mol.111.076604.

Bassett, D. I., Bryson-Richardson, R. J., Daggett, D. F., Gautier, P., Keenan, D. G., & Currie, P. D. (2003). Dystrophin is required for the formation of stable muscle attachments in the zebrafish embryo. *Development, 130*(23), 5851–5860. http://dx.doi.org/10.1242/dev.00799.

Bejar, R., Stevenson, K., Abdel-Wahab, O., Galili, N., Nilsson, B., Garcia-Manero, G. ... Ebert, B. L. (2011). Clinical effect of point mutations in myelodysplastic syndromes. *New England Journal of Medicine, 364*(26), 2496–2506. http://dx.doi.org/10.1056/NEJMoa1013343.

Berger, J., Berger, S., Jacoby, A. S., Wilton, S. D., & Currie, P. D. (2011). Evaluation of exon-skipping strategies for Duchenne muscular dystrophy utilizing dystrophin-deficient zebrafish. *Journal of Cellular and Molecular Medicine, 15*(12), 2643–2651. http://dx.doi.org/10.1111/j.1582-4934.2011.01260.x.

Black, D. L. (2003). Mechanisms of alternative pre-messenger RNA splicing. *Annual Review of Biochemistry, 72*, 291–336. http://dx.doi.org/10.1146/annurev.biochem.72.121801.161720.

Blencowe, B. J. (2000). Exonic splicing enhancers: mechanism of action, diversity and role in human genetic diseases. *Trends in Biochemical Sciences, 25*(3), 106–110.

Boggs, R. T., Gregor, P., Idriss, S., Belote, J. M., & McKeown, M. (1987). Regulation of sexual differentiation in *D. melanogaster* via alternative splicing of RNA from the transformer gene. *Cell, 50*(5), 739–747.

Boutz, P. L., Bhutkar, A., & Sharp, P. A. (2015). Detained introns are a novel, widespread class of post-transcriptionally spliced introns. *Genes & Development, 29*(1), 63–80. http://dx.doi.org/10.1101/gad.247361.114.

Braunschweig, U., Barbosa-Morais, N. L., Pan, Q., Nachman, E. N., Alipanahi, B., Gonatopoulos-Pournatzis, T. ... Blencowe, B. J. (2014). Widespread intron retention in mammals functionally tunes transcriptomes. *Genome Research, 24*(11), 1774–1786. http://dx.doi.org/10.1101/gr.177790.114.

Brzustowicz, L. M., Lehner, T., Castilla, L. H., Penchaszadeh, G. K., Wilhelmsen, K. C., Daniels, R. ... Wood, D. (1990). Genetic mapping of chronic childhood-onset spinal muscular atrophy to chromosome 5q11.2-13.3. *Nature, 344*(6266), 540–541. http://dx.doi.org/10.1038/344540a0.

Buckanovich, R. J., Posner, J. B., & Darnell, R. B. (1993). Nova, the paraneoplastic Ri antigen, is homologous to an RNA-binding protein and is specifically expressed in the developing motor system. *Neuron, 11*(4), 657–672.

Burge, C. B., Padgett, R. A., & Sharp, P. A. (1998). Evolutionary fates and origins of U12-type introns. *Molecular Cell, 2*(6), 773–785.

Cazzola, M., Della Porta, M. G., & Malcovati, L. (2013). The genetic basis of myelodysplasia and its clinical relevance. *Blood.* http://dx.doi.org/10.1182/blood-2013-09-381665. pii: blood-2013-09-381665.

Chen, X., Liu, Y., Sheng, X., Tam, P. O., Zhao, K., Chen, X. ... Zhao, C. (2014). PRPF4 mutations cause autosomal dominant retinitis pigmentosa. *Human Molecular Genetics, 23*(11), 2926–2939. http://dx.doi.org/10.1093/hmg/ddu005.

Chen, M., & Manley, J. L. (2009). Mechanisms of alternative splicing regulation: insights from molecular and genomics approaches. *Nature Reviews Molecular Cell Biology, 10*(11), 741–754. http://dx.doi.org/10.1038/nrm2777.

Chi, S. W., Zang, J. B., Mele, A., & Darnell, R. B. (2009). Argonaute HITS-CLIP decodes microRNA-mRNA interaction maps. *Nature, 460*(7254), 479–486. http://dx.doi.org/10.1038/nature08170.

Cho, V., Mei, Y., Sanny, A., Chan, S., Enders, A., Bertram, E. M. ... Andrews, T. D. (2014). The RNA-binding protein hnRNPLL induces a T cell alternative splicing program delineated by differential intron retention in polyadenylated RNA. *Genome Biology, 15*(1), R26. http://dx.doi.org/10.1186/gb-2014-15-1-r26.

Chou, T. B., Zachar, Z., & Bingham, P. M. (1987). Developmental expression of a regulatory gene is programmed at the level of splicing. *EMBO Journal, 6*(13), 4095–4104.

Cutler, C., Multani, P., Robbins, D., Kim, H. T., Le, T., Hoggatt, J.… Shoemaker, D. D. (2013). Prostaglandin-modulated umbilical cord blood hematopoietic stem cell transplantation. *Blood, 122*(17), 3074–3081. http://dx.doi.org/10.1182/blood-2013-05-503177.

Daiger, S. P., Bowne, S. J., & Sullivan, L. S. (2007). Perspective on genes and mutations causing retinitis pigmentosa. *Archives of Ophthalmology, 125*(2), 151–158. http://dx.doi.org/10.1001/archopht.125.2.151. pii:125/2/151.

Dehm, S. M. (2013). Test-firing ammunition for spliceosome inhibition in cancer. *Clinical Cancer Research, 19*(22), 6064–6066. http://dx.doi.org/10.1158/1078-0432.CCR-13-2461.

Dredge, B. K., Polydorides, A. D., & Darnell, R. B. (2001). The splice of life: alternative splicing and neurological disease. *Nature Reviews Neuroscience, 2*(1), 43–50. http://dx.doi.org/10.1038/35049061.

Dreyfuss, G., Kim, V. N., & Kataoka, N. (2002). Messenger-RNA-binding proteins and the messages they carry. *Nature Reviews Molecular Cell Biology, 3*(3), 195–205. http://dx.doi.org/10.1038/nrm760.

Dvinge, H., & Bradley, R. K. (2015). Widespread intron retention diversifies most cancer transcriptomes. *Genome Medicine, 7*(1), 45. http://dx.doi.org/10.1186/s13073-015-0168-9.

Effenberger, K. A., Anderson, D. D., Bray, W. M., Prichard, B. E., Ma, N., Adams, M. S.… Jurica, M. S. (2014). Coherence between cellular responses and in vitro splicing inhibition for the anti-tumor drug pladienolide B and its analogs. *Journal of Biological Chemistry, 289*(4), 1938–1947. http://dx.doi.org/10.1074/jbc.M113.515536.

Eisen, J. S., & Smith, J. C. (2008). Controlling morpholino experiments: don't stop making antisense. *Development, 135*(10), 1735–1743. http://dx.doi.org/10.1242/dev.001115.

English, M. A., Lei, L., Blake, T., Wincovitch, S. M., Sr., Sood, R., Azuma, M.… Liu, P. P. (2012). Incomplete splicing, cell division defects, and hematopoietic blockage in dhx8 mutant zebrafish. *Developmental Dynamics, 241*(5), 879–889. http://dx.doi.org/10.1002/dvdy.23774.

Feng, Y., Chen, M., & Manley, J. L. (2008). Phosphorylation switches the general splicing repressor SRp38 to a sequence-specific activator. *Nature Structural & Molecular Biology, 15*(10), 1040–1048. http://dx.doi.org/10.1038/nsmb.1485.

Fu, X. D., & Ares, M., Jr. (2014). Context-dependent control of alternative splicing by RNA-binding proteins. *Nature Reviews Genetics, 15*(10), 689–701. http://dx.doi.org/10.1038/nrg3778.

Galante, P. A., Sakabe, N. J., Kirschbaum-Slager, N., & de Souza, S. J. (2004). Detection and evaluation of intron retention events in the human transcriptome. *RNA, 10*(5), 757–765.

Gatto, A., Torroja-Fungairino, C., Mazzarotto, F., Cook, S. A., Barton, P. J., Sanchez-Cabo, F., & Lara-Pezzi, E. (2014). FineSplice, enhanced splice junction detection and quantification: a novel pipeline based on the assessment of diverse RNA-Seq alignment solutions. *Nucleic Acids Research, 42*(8), e71. http://dx.doi.org/10.1093/nar/gku166.

Ge, H., & Manley, J. L. (1990). A protein factor, ASF, controls cell-specific alternative splicing of SV40 early pre-mRNA in vitro. *Cell, 62*(1), 25–34.

Graubert, T. A., Shen, D., Ding, L., Okeyo-Owuor, T., Lunn, C. L., Shao, J. … Walter, M. J. (2012). Recurrent mutations in the U2AF1 splicing factor in myelodysplastic syndromes. *Nature Genetics, 44*(1), 53–57. http://dx.doi.org/10.1038/ng.1031. pii:ng.1031.

Haferlach, T., Nagata, Y., Grossmann, V., Okuno, Y., Bacher, U., Nagae, G. … Ogawa, S. (2014). Landscape of genetic lesions in 944 patients with myelodysplastic syndromes. *Leukemia, 28*(2), 241–247. http://dx.doi.org/10.1038/leu.2013.336.

Hafner, M., Landthaler, M., Burger, L., Khorshid, M., Hausser, J., Berninger, P. … Tuschl, T. (2010). Transcriptome-wide identification of RNA-binding protein and microRNA target sites by PAR-CLIP. *Cell, 141*(1), 129–141. http://dx.doi.org/10.1016/j.cell.2010.03.009.

Hao le, T., Burghes, A. H., & Beattie, C. E. (2011). Generation and characterization of a genetic zebrafish model of SMA carrying the human SMN2 gene. *Molecular Neurodegeneration, 6*(1), 24. http://dx.doi.org/10.1186/1750-1326-6-24. pii:1750-1326-6-24.

Harper, S. Q., Hauser, M. A., DelloRusso, C., Duan, D., Crawford, R. W., Phelps, S. F. … Chamberlain, J. S. (2002). Modular flexibility of dystrophin: implications for gene therapy of Duchenne muscular dystrophy. *Nature Medicine, 8*(3), 253–261. http://dx.doi.org/10.1038/nm0302-253.

Hauer, C., Curk, T., Anders, S., Schwarzl, T., Alleaume, A. M., Sieber, J. … Kulozik, A. E. (2015). Improved binding site assignment by high-resolution mapping of RNA-protein interactions using iCLIP. *Nature Communications, 6*, 7921. http://dx.doi.org/10.1038/ncomms8921.

Hong, D. S., Kurzrock, R., Oh, Y., Wheler, J., Naing, A., Brail, L. … Simon, G. (2011). A phase 1 dose escalation, pharmacokinetic, and pharmacodynamic evaluation of eIF-4E antisense oligonucleotide LY2275796 in patients with advanced cancer. *Clinical Cancer Research, 17*(20), 6582–6591. http://dx.doi.org/10.1158/1078-0432.CCR-11-0430.

Hoskins, A. A., & Moore, M. J. (2012). The spliceosome: a flexible, reversible macromolecular machine. *Trends in Biochemical Sciences, 37*(5), 179–188. http://dx.doi.org/10.1016/j.tibs.2012.02.009.

Howard, J. M., & Sanford, J. R. (2015). The RNAissance family: SR proteins as multifaceted regulators of gene expression. *Wiley Interdisciplinary Reviews RNA, 6*(1), 93–110. http://dx.doi.org/10.1002/wrna.1260.

Howe, K., Clark, M. D., Torroja, C. F., Torrance, J., Berthelot, C., Muffato, M. … Stemple, D. L. (2013). The zebrafish reference genome sequence and its relationship to the human genome. *Nature, 496*(7446), 498–503. http://dx.doi.org/10.1038/nature12111.

Hwang, W. Y., Fu, Y., Reyon, D., Maeder, M. L., Kaini, P., Sander, J. D. … Yeh, J. R. (2013). Heritable and precise zebrafish genome editing using a CRISPR-Cas system. *PLoS One, 8*(7), e68708. http://dx.doi.org/10.1371/journal.pone.0068708.

Hwang, W. Y., Fu, Y., Reyon, D., Maeder, M. L., Tsai, S. Q., Sander, J. D. … Joung, J. K. (2013). Efficient genome editing in zebrafish using a CRISPR-Cas system. *Nature Biotechnology, 31*(3), 227–229. http://dx.doi.org/10.1038/nbt.2501.

Jangi, M., & Sharp, P. A. (2014). Building robust transcriptomes with master splicing factors. *Cell, 159*(3), 487–498. http://dx.doi.org/10.1016/j.cell.2014.09.054.

Jean-Philippe, J., Paz, S., & Caputi, M. (2013). hnRNP A1: the Swiss army knife of gene expression. *International Journal of Molecular Sciences, 14*(9), 18999–19024. http://dx.doi.org/10.3390/ijms140918999.

Jinek, M., Chylinski, K., Fonfara, I., Hauer, M., Doudna, J. A., & Charpentier, E. (2012). A programmable dual-RNA-guided DNA endonuclease in adaptive bacterial immunity. *Science, 337*(6096), 816–821. http://dx.doi.org/10.1126/science.1225829.

Jing, L., Durand, E. M., Ezzio, C., Pagliuca, S. M., & Zon, L. I. (2012). In situ hybridization assay-based small molecule screening in zebrafish. *Current Protocols in Chemical Biology, 4*(2). http://dx.doi.org/10.1002/9780470559277.ch110236.

Jung, H., Lee, D., Lee, J., Park, D., Kim, Y. J., Park, W. Y. … Lee, E. (2015). Intron retention is a widespread mechanism of tumor-suppressor inactivation. *Nature Genetics, 47*(11), 1242–1248. http://dx.doi.org/10.1038/ng.3414.

Kaida, D., Motoyoshi, H., Tashiro, E., Nojima, T., Hagiwara, M., Ishigami, K. … Yoshida, M. (2007). Spliceostatin A targets SF3b and inhibits both splicing and nuclear retention of pre-mRNA. *Nature Chemical Biology, 3*(9), 576–583. http://dx.doi.org/10.1038/nchembio.2007.18.

Kaler, S. G. (1998). Metabolic and molecular bases of Menkes disease and occipital horn syndrome. *Pediatric and Developmental Pathology, 1*(1), 85–98.

Kalsotra, A., & Cooper, T. A. (2011). Functional consequences of developmentally regulated alternative splicing. *Nature Reviews Genetics, 12*(10), 715–729. http://dx.doi.org/10.1038/nrg3052.

Keightley, M. C., Crowhurst, M. O., Layton, J. E., Beilharz, T., Markmiller, S., Varma, S. … Lieschke, G. J. (2013). In vivo mutation of pre-mRNA processing factor 8 (Prpf8) affects transcript splicing, cell survival and myeloid differentiation. *FEBS Letters, 587*(14), 2150–2157. http://dx.doi.org/10.1016/j.febslet.2013.05.030.

Kishore, S., Jaskiewicz, L., Burger, L., Hausser, J., Khorshid, M., & Zavolan, M. (2011). A quantitative analysis of CLIP methods for identifying binding sites of RNA-binding proteins. *Nature Methods, 8*(7), 559–564. http://dx.doi.org/10.1038/nmeth.1608.

Klein, C. J., Coovert, D. D., Bulman, D. E., Ray, P. N., Mendell, J. R., & Burghes, A. H. (1992). Somatic reversion/suppression in Duchenne muscular dystrophy (DMD): evidence supporting a frame-restoring mechanism in rare dystrophin-positive fibers. *American Journal of Human Genetics, 50*(5), 950–959.

Koehler, R., Issac, H., Cloonan, N., & Grimmond, S. M. (2011). The uniqueome: a mappability resource for short-tag sequencing. *Bioinformatics, 27*(2), 272–274. http://dx.doi.org/10.1093/bioinformatics/btq640.

Kohtz, J. D., Jamison, S. F., Will, C. L., Zuo, P., Luhrmann, R., Garcia-Blanco, M. A., & Manley, J. L. (1994). Protein-protein interactions and 5'-splice-site recognition in mammalian mRNA precursors. *Nature, 368*(6467), 119–124. http://dx.doi.org/10.1038/368119a0.

Kok, F. O., Shin, M., Ni, C. W., Gupta, A., Grosse, A. S., van Impel, A. … Lawson, N. D. (2015). Reverse genetic screening reveals poor correlation between morpholino-induced and mutant phenotypes in zebrafish. *Developmental Cell, 32*(1), 97–108. http://dx.doi.org/10.1016/j.devcel.2014.11.018.

Konig, J., Zarnack, K., Rot, G., Curk, T., Kayikci, M., Zupan, B. … Ule, J. (2010). iCLIP reveals the function of hnRNP particles in splicing at individual nucleotide resolution. *Nature Structural & Molecular Biology, 17*(7), 909–915. http://dx.doi.org/10.1038/nsmb.1838.

Koo, T., & Wood, M. J. (2013). Clinical trials using antisense oligonucleotides in duchenne muscular dystrophy. *Human Gene Therapy, 24*(5), 479–488. http://dx.doi.org/10.1089/hum.2012.234.

Kotake, Y., Sagane, K., Owa, T., Mimori-Kiyosue, Y., Shimizu, H., Uesugi, M. … Mizui, Y. (2007). Splicing factor SF3b as a target of the antitumor natural product pladienolide. *Nature Chemical Biology, 3*(9), 570–575. http://dx.doi.org/10.1038/nchembio.2007.16.

Krainer, A. R., Conway, G. C., & Kozak, D. (1990). Purification and characterization of pre-mRNA splicing factor SF2 from HeLa cells. *Genes & Development, 4*(7), 1158–1171.

Kurtovic-Kozaric, A., Przychodzen, B., Singh, J., Konarska, M. M., Clemente, M. J., Otrock, Z. K. ... Padgett, R. A. (2015). PRPF8 defects cause missplicing in myeloid malignancies. *Leukemia, 29*(1), 126–136. http://dx.doi.org/10.1038/leu.2014.144.

Lefebvre, S., Burglen, L., Reboullet, S., Clermont, O., Burlet, P., Viollet, L. ... Melki, J. (1995). Identification and characterization of a spinal muscular atrophy-determining gene. *Cell, 80*(1), 155–165.

Licatalosi, D. D., Mele, A., Fak, J. J., Ule, J., Kayikci, M., Chi, S. W. ... Darnell, R. B. (2008). HITS-CLIP yields genome-wide insights into brain alternative RNA processing. *Nature, 456*(7221), 464–469. http://dx.doi.org/10.1038/nature07488.

Lieschke, G. J., & Currie, P. D. (2007). Animal models of human disease: zebrafish swim into view. *Nature Reviews Genetics, 8*(5), 353–367. http://dx.doi.org/10.1038/nrg2091.

Lim, K. H., Ferraris, L., Filloux, M. E., Raphael, B. J., & Fairbrother, W. G. (2011). Using positional distribution to identify splicing elements and predict pre-mRNA processing defects in human genes. *Proceedings of the National Academy of Sciences of the United States of America, 108*(27), 11093–11098. http://dx.doi.org/10.1073/pnas.1101135108.

Linder, B., Dill, H., Hirmer, A., Brocher, J., Lee, G. P., Mathavan, S. ... Fischer, U. (2011). Systemic splicing factor deficiency causes tissue-specific defects: a zebrafish model for retinitis pigmentosa. *Human Molecular Genetics, 20*(2), 368–377. http://dx.doi.org/10.1093/hmg/ddq473. pii:ddq473.

Long, J. C., & Caceres, J. F. (2009). The SR protein family of splicing factors: master regulators of gene expression. *Biochemical Journal, 417*(1), 15–27. http://dx.doi.org/10.1042/BJ20081501.

Lotti, F., Imlach, W. L., Saieva, L., Beck, E. S., Hao le, T., Li, D. K. ... Pellizzoni, L. (2012). An SMN-dependent U12 splicing event essential for motor circuit function. *Cell, 151*(2), 440–454. http://dx.doi.org/10.1016/j.cell.2012.09.012. pii:S0092-8674(12)01115-4.

Madsen, E. C., Morcos, P. A., Mendelsohn, B. A., & Gitlin, J. D. (2008). In vivo correction of a Menkes disease model using antisense oligonucleotides. *Proceedings of the National Academy of Sciences of the United States of America, 105*(10), 3909–3914. http://dx.doi.org/10.1073/pnas.0710865105.

Makishima, H., Visconte, V., Sakaguchi, H., Jankowska, A. M., Abu Kar, S., Jerez, A. ... Maciejewski, J. P. (2012). Mutations in the spliceosome machinery, a novel and ubiquitous pathway in leukemogenesis. *Blood, 119*(14), 3203–3210. http://dx.doi.org/10.1182/blood-2011-12-399774.

Markmiller, S., Cloonan, N., Lardelli, R. M., Doggett, K., Keightley, M. C., Boglev, Y. ... Heath, J. K. (2014). Minor class splicing shapes the zebrafish transcriptome during development. *Proceedings of the National Academy of Sciences of the United States of America*. http://dx.doi.org/10.1073/pnas.1305536111.

Martin, C. L., Duvall, J. A., Ilkin, Y., Simon, J. S., Arreaza, M. G., Wilkes, K. ... Geschwind, D. H. (2007). Cytogenetic and molecular characterization of A2BP1/FOX1 as a candidate gene for autism. *American Journal of Medical Genetics Part B Neuropsychiatric Genetics, 144B*(7), 869–876. http://dx.doi.org/10.1002/ajmg.b.30530.

Martinez-Gimeno, M. (2003). Mutations in the Pre-mRNA splicing-factor genes PRPF3, PRPF8, and PRPF31 in Spanish families with autosomal dominant retinitis pigmentosa. *Investigative Ophthalmology & Visual Science, 44*(5), 2171–2177. http://dx.doi.org/10.1167/iovs.02-0871.

Maruyama, T., Dougan, S. K., Truttmann, M. C., Bilate, A. M., Ingram, J. R., & Ploegh, H. L. (2015). Increasing the efficiency of precise genome editing with CRISPR-Cas9 by inhibition of nonhomologous end joining. *Nature Biotechnology, 33*(5), 538–542. http://dx.doi.org/10.1038/nbt.3190.

Maubaret, C. G., Vaclavik, V., Mukhopadhyay, R., Waseem, N. H., Churchill, A., Holder, G. E.… Webster, A. R. (2011). Autosomal dominant retinitis pigmentosa with intrafamilial variability and incomplete penetrance in two families carrying mutations in PRPF8. *Investigative Ophthalmology & Visual Science, 52*(13), 9304–9309. http://dx.doi.org/10.1167/iovs.11-8372. pii:iovs.11-8372.

McKie, A. B., McHale, J. C., Keen, T. J., Tarttelin, E. E., Goliath, R., van Lith-Verhoeven, J. J.… Inglehearn, C. F. (2001). Mutations in the pre-mRNA splicing factor gene PRPC8 in autosomal dominant retinitis pigmentosa (RP13). *Human Molecular Genetics, 10*(15), 1555–1562.

McWhorter, M. L., Monani, U. R., Burghes, A. H., & Beattie, C. E. (2003). Knockdown of the survival motor neuron (Smn) protein in zebrafish causes defects in motor axon outgrowth and pathfinding. *Journal of Cell Biology, 162*(5), 919–931. http://dx.doi.org/10.1083/jcb.200303168.

Mendelsohn, B. A., Yin, C., Johnson, S. L., Wilm, T. P., Solnica-Krezel, L., & Gitlin, J. D. (2006). Atp7a determines a hierarchy of copper metabolism essential for notochord development. *Cell Metabolism, 4*(2), 155–162. http://dx.doi.org/10.1016/j.cmet.2006.05.001.

Mroczek, S., Krwawicz, J., Kutner, J., Lazniewski, M., Kucinski, I., Ginalski, K., & Dziembowski, A. (2012). C16orf57, a gene mutated in poikiloderma with neutropenia, encodes a putative phosphodiesterase responsible for the U6 snRNA 3' end modification. *Genes & Development, 26*(17), 1911–1925. http://dx.doi.org/10.1101/gad.193169.112.

Padgett, R. A. (2012). New connections between splicing and human disease. *Trends in Genetics, 28*(4), 147–154. http://dx.doi.org/10.1016/j.tig.2012.01.001.

Pan, Q., Shai, O., Lee, L. J., Frey, B. J., & Blencowe, B. J. (2008). Deep surveying of alternative splicing complexity in the human transcriptome by high-throughput sequencing. *Nature Genetics, 40*(12), 1413–1415. http://dx.doi.org/10.1038/ng.259.

Papaemmanuil, E., Cazzola, M., Boultwood, J., Malcovati, L., Vyas, P., Bowen, D.… Campbell, P. J. (2011). Somatic SF3B1 mutation in myelodysplasia with ring sideroblasts. *New England Journal of Medicine, 365*(15), 1384–1395. http://dx.doi.org/10.1056/NEJMoa1103283.

Patel, A. A., McCarthy, M., & Steitz, J. A. (2002). The splicing of U12-type introns can be a rate-limiting step in gene expression. *EMBO Journal, 21*(14), 3804–3815. http://dx.doi.org/10.1093/emboj/cdf297.

Patel, A. A., & Steitz, J. A. (2003). Splicing double: insights from the second spliceosome. *Nature Reviews Molecular Cell Biology, 4*(12), 960–970. http://dx.doi.org/10.1038/nrm1259. pii:nrm1259.

Patil, P., Uechi, T., & Kenmochi, N. (2015). Incomplete splicing of neutrophil-specific genes affects neutrophil development in a zebrafish model of poikiloderma with neutropenia. *RNA Biology, 12*(4), 426–434. http://dx.doi.org/10.1080/15476286.2015.1017240.

Pauli, A., Montague, T. G., Lennox, K. A., Behlke, M. A., & Schier, A. F. (2015). Antisense oligonucleotide-mediated transcript knockdown in zebrafish. *PLoS One, 10*(10), e0139504. http://dx.doi.org/10.1371/journal.pone.0139504.

Pimentel, H., Parra, M., Gee, S. L., Mohandas, N., Pachter, L., & Conboy, J. G. (2015). A dynamic intron retention program enriched in RNA processing genes regulates gene expression during terminal erythropoiesis. *Nucleic Acids Research*. http://dx.doi.org/10.1093/nar/gkv1168.

Radev, Z., Hermel, J. M., Elipot, Y., Bretaud, S., Arnould, S., Duchateau, P. … Sohm, F. (2015). A TALEN-Exon skipping design for a Bethlem Myopathy model in zebrafish. *PLoS One, 10*(7), e0133986. http://dx.doi.org/10.1371/journal.pone.0133986.

Rosel, T. D., Hung, L. H., Medenbach, J., Donde, K., Starke, S., Benes, V. … Bindereif, A. (2011). RNA-Seq analysis in mutant zebrafish reveals role of U1C protein in alternative splicing regulation. *EMBO Journal, 30*(10), 1965–1976. http://dx.doi.org/10.1038/emboj.2011.106. pii:emboj2011106.

Sander, J. D., & Joung, J. K. (2014). CRISPR-Cas systems for editing, regulating and targeting genomes. *Nature Biotechnology, 32*(4), 347–355. http://dx.doi.org/10.1038/nbt.2842.

Sharma, V. K., Sharma, R. K., & Singh, S. K. (2014). Antisense oligonucleotides: modifications and clinical trials. *MedChemComm, 5*(10), 1454–1471. http://dx.doi.org/10.1039/C4MD00184B.

Shen, H., & Green, M. R. (2004). A pathway of sequential arginine-serine-rich domain-splicing signal interactions during mammalian spliceosome assembly. *Molecular Cell, 16*(3), 363–373. http://dx.doi.org/10.1016/j.molcel.2004.10.021.

Shen, H., Kan, J. L., & Green, M. R. (2004). Arginine-serine-rich domains bound at splicing enhancers contact the branchpoint to promote prespliceosome assembly. *Molecular Cell, 13*(3), 367–376.

Shepard, P. J., & Hertel, K. J. (2009). The SR protein family. *Genome Biology, 10*(10), 242. http://dx.doi.org/10.1186/gb-2009-10-10-242.

Singh, R. K., & Cooper, T. A. (2012). Pre-mRNA splicing in disease and therapeutics. *Trends in Molecular Medicine, 18*(8), 472–482. http://dx.doi.org/10.1016/j.molmed.2012.06.006.

Singh, J., & Padgett, R. A. (2009). Rates of in situ transcription and splicing in large human genes. *Nature Structural & Molecular Biology, 16*(11), 1128–1133. http://dx.doi.org/10.1038/nsmb.1666.

Spitali, P., & Aartsma-Rus, A. (2012). Splice modulating therapies for human disease. *Cell, 148*(6), 1085–1088. http://dx.doi.org/10.1016/j.cell.2012.02.014.

Stein, C. A., & Goel, S. (2011). Therapeutic oligonucleotides: the road not taken. *Clinical Cancer Research, 17*(20), 6369–6372. http://dx.doi.org/10.1158/1078-0432.CCR-11-2013.

Sullivan, L. S. (2006). Prevalence of disease-causing mutations in families with autosomal dominant retinitis pigmentosa: a screen of known genes in 200 families. *Investigative Ophthalmology & Visual Science, 47*(7), 3052–3064. http://dx.doi.org/10.1167/iovs.05-1443.

Sutandy, F. X., Hildebrandt, A., & Konig, J. (2016). Profiling the binding sites of RNA-binding proteins with nucleotide resolution using iCLIP. *Methods in Molecular Biology, 1358*, 175–195. http://dx.doi.org/10.1007/978-1-4939-3067-8_11.

Sveen, A., Kilpinen, S., Ruusulehto, A., Lothe, R. A., & Skotheim, R. I. (2015). Aberrant RNA splicing in cancer; expression changes and driver mutations of splicing factor genes. *Oncogene*. http://dx.doi.org/10.1038/onc.2015.318.

Tang, F., Barbacioru, C., Wang, Y., Nordman, E., Lee, C., Xu, N. … Surani, M. A. (2009). mRNA-Seq whole-transcriptome analysis of a single cell. *Nature Methods, 6*(5), 377–382. http://dx.doi.org/10.1038/nmeth.1315.

Tat, J., Liu, M., & Wen, X. Y. (2013). Zebrafish cancer and metastasis models for in vivo drug discovery. *Drug Discovery Today Technologies, 10*(1), e83–89. http://dx.doi.org/10.1016/j.ddtec.2012.04.006.

Thol, F., Kade, S., Schlarmann, C., Loffeld, P., Morgan, M., Krauter, J. … Heuser, M. (2012). Frequency and prognostic impact of mutations in SRSF2, U2AF1, and ZRSR2 in patients with myelodysplastic syndromes. *Blood*. http://dx.doi.org/10.1182/blood-2011-12-399337. pii:blood-2011-12-399337.

Trede, N. S., Medenbach, J., Damianov, A., Hung, L. H., Weber, G. J., Paw, B. H. … Bindereif, A. (2007). Network of coregulated spliceosome components revealed by zebrafish mutant in recycling factor p110. *Proceedings of the National Academy of Sciences of the United States of America, 104*(16), 6608–6613. http://dx.doi.org/10.1073/pnas.0701919104. pii:0701919104.

Tse, M. T. (2012). Antisense therapeutics: nuclear RNA more susceptible to knockdown. *Nature Reviews Drug Discovery, 11*(9), 674. http://dx.doi.org/10.1038/nrd3825.

Tse, M. T. (2013). Regulatory watch: antisense approval provides boost to the field. *Nature Reviews Drug Discovery, 12*(3), 179.

Varshney, G. K., Pei, W., LaFave, M. C., Idol, J., Xu, L., Gallardo, V. … Burgess, S. M. (2015). High-throughput gene targeting and phenotyping in zebrafish using CRISPR/Cas9. *Genome Research, 25*(7), 1030–1042. http://dx.doi.org/10.1101/gr.186379.114.

Wahl, M. C., Will, C. L., & Lührmann, R. (2009). The spliceosome: design principles of a dynamic RNP machine. *Cell, 136*(4), 701–718. http://dx.doi.org/10.1016/j.cell.2009.02.009.

White, R. M., Cech, J., Ratanasirintrawoot, S., Lin, C. Y., Rahl, P. B., Burke, C. J. … Zon, L. I. (2011). DHODH modulates transcriptional elongation in the neural crest and melanoma. *Nature, 471*(7339), 518–522. http://dx.doi.org/10.1038/nature09882.

White, R., Rose, K., & Zon, L. (2013). Zebrafish cancer: the state of the art and the path forward. *Nature Reviews Cancer, 13*(9), 624–636. http://dx.doi.org/10.1038/nrc3589.

Will, C. L., Schneider, C., Hossbach, M., Urlaub, H., Rauhut, R., Elbashir, S. … Luhrmann, R. (2004). The human 18S U11/U12 snRNP contains a set of novel proteins not found in the U2-dependent spliceosome. *RNA, 10*(6), 929–941.

Wong, J. J., Au, A. Y., Ritchie, W., & Rasko, J. E. (2016 Jan). Intron retention in mRNA: no longer nonsense: known and putative roles of intron retention in normal and disease biology. *Bioessays, 38*(1), 41–49. http://dx.doi.org/10.1002/bies.201500117. Epub 2015 Nov 27.

Wong, J. J., Ritchie, W., Ebner, O. A., Selbach, M., Wong, J. W., Huang, Y. … Rasko, J. E. (2013). Orchestrated intron retention regulates normal granulocyte differentiation. *Cell, 154*(3), 583–595. http://dx.doi.org/10.1016/j.cell.2013.06.052.

Wu, J. Y., & Maniatis, T. (1993). Specific interactions between proteins implicated in splice site selection and regulated alternative splicing. *Cell, 75*(6), 1061–1070.

Xiao, S. H., & Manley, J. L. (1997). Phosphorylation of the ASF/SF2 RS domain affects both protein-protein and protein-RNA interactions and is necessary for splicing. *Genes & Development, 11*(3), 334–344.

Yan, C., Hang, J., Wan, R., Huang, M., Wong, C. C., & Shi, Y. (2015). Structure of a yeast spliceosome at 3.6-angstrom resolution. *Science, 349*(6253), 1182–1191. http://dx.doi.org/10.1126/science.aac7629.

Yap, K., Lim, Z. Q., Khandelia, P., Friedman, B., & Makeyev, E. V. (2012). Coordinated regulation of neuronal mRNA steady-state levels through developmentally controlled intron retention. *Genes & Development, 26*(11), 1209–1223. http://dx.doi.org/10.1101/gad.188037.112.

Yin, J., Brocher, J., Fischer, U., & Winkler, C. (2011). Mutant Prpf31 causes pre-mRNA splicing defects and rod photoreceptor cell degeneration in a zebrafish model for Retinitis pigmentosa. *Molecular Neurodegeneration, 6*, 56. http://dx.doi.org/10.1186/1750-1326-6-56. pii:1750-1326-6-56.

Yoshida, K., Sanada, M., Shiraishi, Y., Nowak, D., Nagata, Y., Yamamoto, R. … Ogawa, S. (2011). Frequent pathway mutations of splicing machinery in myelodysplasia. *Nature, 478*(7367), 64–69. http://dx.doi.org/10.1038/nature10496. pii:nature10496.

Younis, I., Berg, M., Kaida, D., Dittmar, K., Wang, C., & Dreyfuss, G. (2010). Rapid-response splicing reporter screens identify differential regulators of constitutive and alternative splicing. *Molecular and Cellular Biology, 30*(7), 1718–1728. http://dx.doi.org/10.1128/MCB.01301-09.

Ziviello, C. (2005). Molecular genetics of autosomal dominant retinitis pigmentosa (ADRP): a comprehensive study of 43 Italian families. *Journal of Medical Genetics, 42*(7), e47. http://dx.doi.org/10.1136/jmg.2005.031682.

CHAPTER

PICCORO: A technique for manipulating the activity of transcription factors with blue light

14

S. Masuda[1], M. Tanaka[1]

Tokyo Institute of Technology, Yokohama, Japan

[1]*Corresponding authors: E-mail: shmasuda@bio.titech.ac.jp; mitanaka@bio.titech.ac.jp*

CHAPTER OUTLINE

Abstract

The ability to modulate transcription factor activity is critically important for deciphering many cellular and developmental processes. Light is an attractive signal for artificially controlling transcription factor activity because it has high spatial and temporal precision. Several light-switchable technologies have been developed to control gene expression negatively or positively. Although these systems allow spatiotemporal control of transcription, the target gene must be placed downstream of a specific promoter, which is time consuming and applicable to a limited number of genes at the same time. On the other hand, we recently developed a technique for manipulation of transcription factor activity with blue light, termed PixD complex-dependent control (PICCORO) using the bacterial blue-light photoreceptor PixD. This method has the advantage of controlling transcription factor activity at the posttranslational level. Using this method, we were able to control activity of the chimeric dominant-negative transcription factor No tail (Ntl) in zebrafish with blue light. Here we describe the mechanism of PICCORO action and a detailed protocol to apply the technique for controlling other transcription factors.

Methods in Cell Biology, Volume 135, ISSN 0091-679X, http://dx.doi.org/10.1016/bs.mcb.2016.03.009

INTRODUCTION

Recent progress in the identification of various types of photoreceptor proteins and their mechanisms of light-signal transduction offer opportunities to genetically engineer biological applications in the field of optogenetics, in which the photoreceptor proteins are exploited as tools for artificial control of biological events by light. The photoreceptors identified to date are classified into seven groups: rhodopsins, xanthopsins, phytochromes, cryptochromes, light-oxygen-voltage (LOV) domain-containing proteins, ultraviolet-resistance-locus 8 (UVR8) family proteins, and sensor of blue-light-using-flavin (BLUF) domain-containing proteins (Heijde & Ulm, 2012; Möglich, Yang, Ayers, & Moffat, 2010). In 2002, Shimizu-Sato, Huq, Tepperman, & Quail (2002) first demonstrated light-dependent switching of gene expression by coupling a plant phytochrome and the yeast GAL4 transcription factor. Several light-dependent gene expression systems have since been developed by use of LOV-, cryptochrome-, and UVR8-family proteins (Crefcoeur, Yin, Ulm, & Halazonetis, 2013; Kennedy, et al., 2010; Wang, Chen, & Yang, 2012; Yazawa, Sadaghiani, Hsueh, & Dolmetsch, 2009). These systems have advantages of spatiotemporal control of gene expression over previous chemical-dependent induction methods. However, these light-dependent induction systems are still not widely used for the control of gene expression in multicellular organisms, perhaps due to the low dynamic range of regulation or low functional performance of the recombinant photoreceptors in heterologous organisms.

The BLUF-type photoreceptor was identified in 2002 as a member of a distinct flavin-based blue-light photoreceptor family (Gomelsky & Klug, 2002). We characterized the cyanobacterial BLUF photoreceptor PixD as a model to understand how BLUF senses light and induces light-dependent signal transduction in cells (Masuda, 2013). Based on the results obtained and reports from other laboratories, we recently established a new light-controllable gene expression system termed PixD complex-dependent control (PICCORO) (Masuda, Nakatani, Ren, & Tanaka, 2013). Here, we summarize the method of PICCORO and its application to light-dependent control of transcription factor activity in zebrafish.

1. MECHANISM OF PICCORO

PixD is a small ($\sim$15 kDa) flavin-binding protein that controls phototaxis (movement toward a light source) in the photosynthetic cyanobacterium *Synechocystis* sp. PCC6803 (Masuda, 2013). By use of yeast two-hybrid analysis, a PixD-interacting protein, PixE, was found (Sato et al., 2007). Biochemical analyses revealed that PixD forms a decamer in the dark, and that light excitation of the bound flavin results in a conformational change of PixD, leading to formation of a dimer (Fig. 1) (Ren, Sato, Hasegawa, Ohta, & Masuda, 2013). PixE interacts with the PixD decamer in the dark to form a large $PixD_{10}$–$PixE_4$ complex. Under these conditions, PixE is in

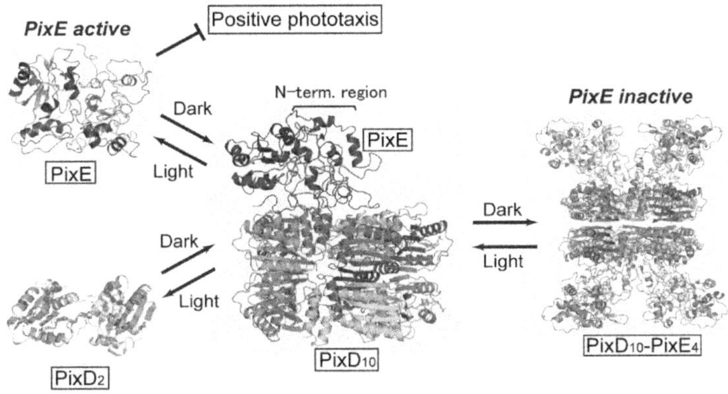

FIGURE 1

Schematic model of the actions of PixD and PixE in control of phototaxis of the cyanobacterium *Synechocystis* sp. PCC6803. In the dark, PixD forms a decamer and interacts with PixE to form a $PixD_{10}$–$PixE_4$ complex. Upon activation by light, PixD dissociates to form homodimers and PixE is released from the complex. The released monomeric PixE suppresses positive phototaxis by an unknown mechanism. (See color plate)

The figure is reproduced with permission with a slight modification Ren, S., Sugimoto, Y., Kobayashi, T., & Masuda, S. (2015). Cross-linking analysis reveals the putative dimer structure of the cyanobacterial photoreceptor PixD. FEBS Letters, 589, *1879–1882.*

an inactive form and cannot send a light signal downstream to control phototaxis. Upon irradiation with light, PixD is converted to a dimer and PixE is released from the large complex, leading to the functional monomer (Fig. 1) (Ren et al., 2013). Deletion analysis indicated that the N-terminal ~250 amino acids of PixE (PixE_N) interact with the decameric PixD (Masuda et al., 2013). We utilized the formation of the light-dependent PixD–PixE large oligomeric complex to control the activity of transcription factors of interest.

PICCORO can control chimeric transcription factors by blue light. We established this method with a dominant negative version of the transcription factor No tail (Ntl), fusing the DNA-binding domain of Ntl with the repressor domain of the transcription factor Engrailed (En^R) (Masuda et al., 2013). The resulting chimeric transcription factor (Ntl-En^R) inhibits Ntl activity in vivo, leading to suppression of tail formation in wild-type zebrafish (Goering et al., 2003). PICCORO was designed to control Ntl-En^R function by use of light-dependent aggregation between PixD and PixE, which modulates the transcriptional repressor activity of Ntl-En^R (Fig. 2). Specifically, the PixE N-terminal domain that interacts with PixD is fused to the N-terminus of Ntl-En^R; the resulting construct was named Ntl^{PixE} (Fig. 3A). For in vivo analysis, we constructed a transgenic zebrafish, Tg(EF1α:PixD), that expresses PixD from a constitutive EF1α promoter (Fig. 3B) (Masuda et al., 2013). The PixD-dependent inhibition of Ntl^{PixE} activity is controllable in a blue light-dependent

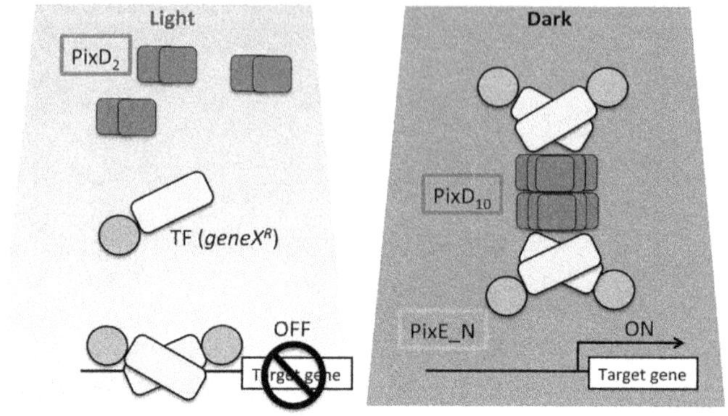

FIGURE 2

Schematic model of the PICCORO technique. A decamer of PixD ($PixD_{10}$) interacts with a chimeric transcription repressor (such as Ntl^{PixE}) under dark conditions. Upon blue-light illumination, the interaction is negated and the chimeric transcription repressor suppresses transcription of target genes. (See color plate)

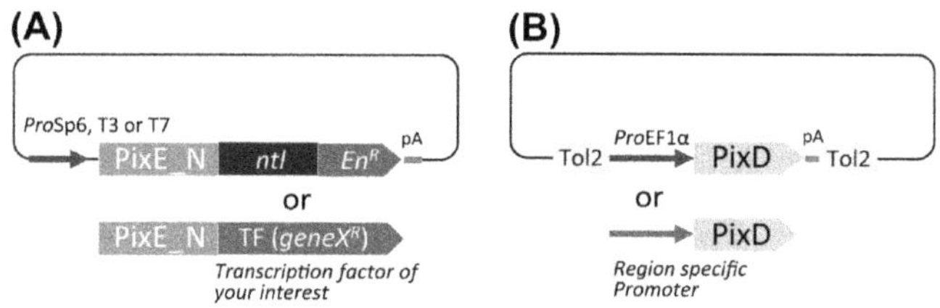

FIGURE 3

Schematic view of PICCORO constructs. (A) Plasmid construct for synthesis of mRNA coding for the chimeric transcription repressor used for PICCORO. Although we used Ntl-En^R as a model (Masuda et al., 2013), the method could be applicable to any transcription factor. (B) We constructed a transgenic zebrafish, Tg(EF1α:PixD), that expresses PixD by a constitutive EF1α promoter. The Tol2 system (Kawakami et al., 2004) was employed to integrate the DNA construct into the zebrafish genome. If a tissue-specific promoter is used in the construct, chimeric transcription factor activity could be controlled locally.

manner in the transgenic zebrafish through light-dependent modulation of PixD-Ntl^{PixE} complex formation (Fig. 2) (Masuda et al., 2013). This method may be applicable to other transcription factors.

2. METHOD FOR CONSTRUCTING A PICCORO PLASMID

2.1 CONSTRUCTION OF A CHIMERIC TRANSCRIPTION FACTOR

1. Fuse the repressor En^R domain of the *Drosophila* Engrailed (amino acids 2–299) (GenBank: M10017.1) to the DNA-binding domain of the transcription factor of

interest (*geneX*); the resulting chimeric transcription factor is designated *geneX*^{R} (Fig. 3). Standard double polymerase chain reaction technology could be applied for the construction.

2. Fuse the PixD-interacting domain of PixE (amino acids 1–256) to the N-terminus of *geneX*^{R}; the resulting chimeric transcription factor is designated *PixE-geneX*^{R} (Fig. 3). Nucleotide and amino acid sequences of PixE (annotated as slr1693) are available at CyanoBase (http://genome.microbedb.jp/CyanoBase).
3. Insert the *PixE-geneX*^{R} into any expression plasmid (eg, pCS2) to allow in vitro transcription by a highly inducible SP6, T3, or T7 promoter (Fig. 3). The authors' laboratories can provide plasmids encoding *pixE*, *pixD*, and/or *pixE-Ntl-En*^{R}.

2.2 EVALUATING THE APPROPRIATE AMOUNT OF mRNA FOR TRANSFORMATION

1. Prepare capped mRNA coding for the chimeric transcription factor (*PixE-geneX*^{R}) using a commercially available RNA transcription kit.
2. Dissolve mRNA for *PixE-geneX*^{R} in sterile H_2O, at final concentrations ranging from 0.10 to 0.50 μg/μL.
3. Inject 1 nL of each concentration of *PixE-geneX*^{R} mRNA solution into one-cell–stage wild-type zebrafish eggs using a microinjector (IM30; Narishige).
4. Transfer embryos to a dish with warm fish water (Westerfield, 2006), maintain at 28°C, and examine the phenotype at appropriate stages.

2.3 INJECTING PixE-geneX^R mRNA INTO EGGS OF Tg(EF1α:PixD)

1. Prepare *PixE-geneX*^{R} mRNA solution at the concentration that provided 50–70% of the expected phenotype when injected into the eggs of wild-type zebrafish.
2. Inject 1 nL of the appropriate concentration of *PixE-geneX*^{R} mRNA solution into one-cell–stage Tg(EF1α:PixD) eggs using a microinjector. Tg(EF1α:PixD) is available at the National BioResource Project (http://www.shigen.nig.ac.jp/zebra/).
3. For dark conditions: Transfer the embryos into a dish with warm fish water immediately after injection, cover the dish with aluminum foil, and maintain at 28°C.
4. For illuminated conditions: Transfer the embryos into a dish with warm fish water, and then illuminate with blue light provided by a light-emitting diode at λ_{max} 472 nm, 250 μmol/m^2 s (Fig. 4; MIL-B18, Sanyo) at 28°C. Place dishes containing distilled water on the top and bottom sides of the dish containing embryos for insulation (Fig. 4).

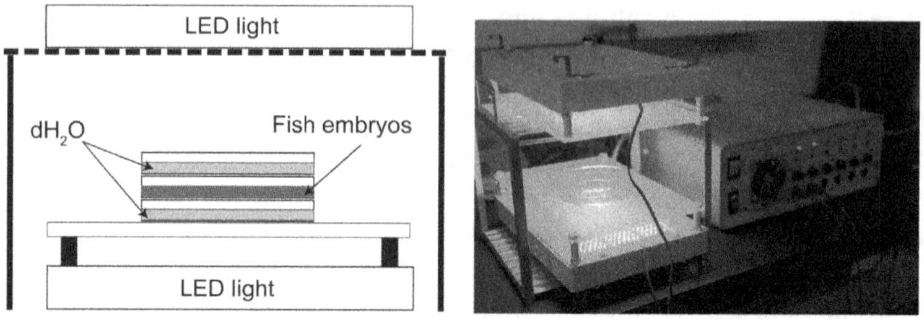

FIGURE 4

Schematic view of how zebrafish embryos are illuminated.

CONCLUSION

Although light-controllable gene expression systems other than PICCORO have been reported (Crefcoeur et al., 2013; Kennedy, et al., 2010; Wang et al., 2012; Yazawa et al., 2009), most of these systems utilize the yeast GAL4 transcription factor and its *cis*-acting regulatory sequence (UAS), and the target gene must be placed downstream of the specific UAS promoter sequence. This means that the genes that can be regulated by these systems are limited. On the other hand, PICCORO is designed to directly control transcription factor activity at the posttranslational level. This advantage of PICCORO would be useful for characterizing transcription factors involved in developmental processes.

PICCORO has the potential to control transcription factor activity in a more spatiotemporal manner. Once PixD-expressing transgenic zebrafish were constructed with a tissue-specific promoter, chimeric transcription factor activity could be controlled locally (Fig. 3). In addition, because PICCORO is designed to control protein aggregation in a light-dependent manner, it could also be applied to controlling other cellular processes such as translation, splicing, and histone modification. Further characterization of the PixD photoreceptor as well as further optimization of PICCORO will be important for engineering such methodologies.

REFERENCES

Crefcoeur, R. P., Yin, R., Ulm, R., & Halazonetis, T. D. (2013). Ultraviolet-B-mediated induction of protein–protein interaction in mammalian cells. *Nature Communications, 4*, 1779.

Goering, L. M., Hoshijima, K., Hyg, B., Bisgrove, B., Kispert, A., & Grunwald, D. J. (2003). An interacting network of T-box genes directs gene expression and fate in the zebrafish mesoderm. *Proceedings of the National Academy of Sciences of the United States of America, 100*, 9410–9415.

Gomelsky, M., & Klug, G. (2002). BLUF: a novel FAD-binding domain involved in sensory transduction in microorganisms. *Trends in Biochemical Sciences, 27*, 497–500.

Heijde, M., & Ulm, R. (2012). UV-B photoreceptor-mediated signaling in plants. *Trands in Plant Science, 17*, 230–237.

Kawakami, K., Takeda, H., Kawakami, N., Kobayashi, M., Matsuda, N., & Mishima, M. (2004). A transposon-mediated gene trap approach identifies developmentally regulated genes in zebrafish. *Developmental Cell, 7*, 133–144.

Kennedy, M. J., Hughes, R. M., Peteya, L. A., Schwartz, J. W., Ehllers, M. D., & Tucker, C. L. (2010). *Nature Methods, 7*, 973–975.

Masuda, S. (2013). Light detection and signal transduction in the BLUF photoreceptors. *Plant and Cell Physiology, 54*, 171–179.

Masuda, S., Nakatani, Y., Ren, S., & Tanaka, M. (2013). Blue light-mediated manipulation of transcription factor activity in vivo. *ACS Chemical Biology, 8*, 2649–2653.

Möglich, A., Yang, X., Ayers, R. A., & Moffat, K. (2010). Structure and function of plant photoreceptors. *Annual Review of Plant Biology, 61*, 21–47.

Ren, S., Sato, R., Hasegawa, K., Ohta, H., & Masuda, S. (2013). A predicted structure for the PixD-PixE complex determined by homology modeling, docking simulations, and a mutagenesis study. *Biochemistry, 52*, 1272–1279.

Ren, S., Sugimoto, Y., Kobayashi, T., & Masuda, S. (2015). Cross-linking analysis reveals the putative dimer structure of the cyanobacterial photoreceptor PixD. *FEBS Letters, 589*, 1879–1882.

Sato, S., Shimada, Y., Muraki, A., Kohara, M., Nakamura, Y., & Tabata, S. (2007). A large-scale protein–protein interaction analysis in *Synechocystis* sp. PCC6803. *DNA Research, 14*, 207–216.

Shimizu-Sato, S., Huq, E., Tepperman, J. M., & Quail, P. H. (2002). A light-switchable gene promoter system. *Nature Biotechnology, 20*, 1041–1044.

Wang, X., Chen, X., & Yang, Y. (2012). Spatiotemporal control of gene expression by a light-switchable transgene system. *Nature Methods, 9*, 266–269.

Westerfield, M. (2006). *The zebrafish book: A guide for the laboratory use of zebrafish (*Danio rerio*)*. Eugene: University of Oregon Press.

Yazawa, Y., Sadaghiani, A. R., Hsueh, B., & Dolmetsch, R. E. (2009). Induction of protein–protein interaction in live cells using light. *Nature Biotechnology, 27*, 941–945.

PART

Transcriptomics

3

CHAPTER

Tomo-seq: a method to obtain genome-wide expression data with spatial resolution

15

F. Kruse, J.P. Junker, A. van Oudenaarden, J. Bakkers[1]

Hubrecht Institute and University Medical Centre Utrecht, Utrecht, The Netherlands

[1]*Corresponding author: E-mail: j.bakkers@hubrecht.eu*

CHAPTER OUTLINE

Abstract

To improve our understanding of pattern formation during development and disease we heavily rely on the identification of novel regulators and pathways. While RNA sequencing yields genome-wide expression data that suit this purpose, it lacks spatial resolution. Such spatial resolution can be obtained by microscopy-based methods like in situ hybridization, but these fail to provide information on more than a few genes at a time. Here, we describe *tomo*-seq, a technique that combines the advantages of the above-mentioned approaches and provides genome-wide expression data with spatial information. The *tomo*-seq technique is based on cryosectioning of an embryo or tissue of interest and performing RNA-seq on individual sections. Using this method, we have generated genome-wide transcriptomics with high spatial resolution of the whole zebrafish embryo at various stages of development (Junker et al., 2014) and of adult zebrafish hearts after injury (Wu et al., 2016).

Methods in Cell Biology, Volume 135, ISSN 0091-679X, http://dx.doi.org/10.1016/bs.mcb.2016.01.006

INTRODUCTION

Embryogenesis could not proceed if specific regulators and pathways were not restricted to well-defined regions to ensure asymmetry, directionality, and growth into the proper shape. As such, it is fundamental for development that certain genes are expressed in a temporally and spatially restricted domain. Early dorsoventral patterning, for example, requires expression and function of Bone Morphogenetic Protein (BMP) inhibitors such as *chordin* in the dorsal organizer (reviewed in Langdon & Mullins, 2011). Thus far, the study of such factors relied on immunohistochemistry or mRNA in situ hybridization, which limits the focus to only a few genes per sample. RNA sequencing, on the contrary, provides the power to study the whole transcriptome. While the requirement for high amounts of input material has limited its use for small samples in the past, recent development of strategies to amplify RNA for single-cell sequencing approaches has attenuated this restriction (Hashimshony, Wagner, Sher, & Yanai, 2012; Islam et al., 2011; Picelli et al., 2013; Ramskold et al., 2012; Tang et al., 2009). While these techniques have certainly been proven to be very useful, they take cells out of their natural context. Computational strategies are required to infer cellular localization by integrating in situ hybridization patterns with single-cell RNA-sequencing data (Satija, Farrell, Gennert, Schier, & Regev, 2015).

By adopting protocols for single-cell RNA-sequencing, we have established a method to detect spatially resolved genome-wide transcriptomics experimentally, called *tomo*-seq (Junker et al., 2014). This method is based on cryosectioning of frozen samples, followed by RNA extraction from individual sections and cDNA synthesis with bar-coded primers. In vitro transcription is then used to linearly amplify the cDNA. The addition of bar codes allows identifying which transcript originated from which section. As a result, expression profiles of all detected genes can be generated. Moreover, bioinformatics analysis can generate lists of genes that have a similar expression pattern to a gene of interest. As such, this analysis can identify new factors that are differentially expressed in a region of interest. Moreover, samples taken from multiple embryos or tissues can be pooled in a single sequencing run, reducing the costs for these experiments. In summary, *tomo*-seq provides genome-wide sensitivity with high spatial resolution and can be used to identify genes that are upregulated in a region of interest.

1. TISSUE EXTRACTION AND EMBEDDING

The following protocol is very versatile in its applicability. It can be used for a whole (zebrafish) embryo or different organs of various organisms, given a reasonably well-annotated genome. To start of with, euthanize the individual animal and directly extract the organ of interest (if applicable). Without any delay, place the tissue in a Tissue-Tek cryomold (Sakura) containing tissue freezing medium (Jung). The tissue can be oriented in the desired position (in relation to the sectioning plane) under a dissection microscope. Affi-Gel Blue Gel beads (100–200 mesh BioRad) can be positioned right in front and behind the tissue. This is helpful during the sectioning process to know when to start and when to stop collecting sections. After proper orientation of the tissue, the mold containing the tissue is placed on dry ice until completely frozen and stored at −80°C until further use.

2. CRYOSECTIONING AND TRIZOL EXTRACTION

Before sectioning, labeled LoBind 1.5-mL tubes (Eppendorf) should be prechilled on dry ice for >30 min. Cryosections are made at the desired thickness and intervals; however, it is not recommended to section at <10 μm due to inaccuracy of the cryotome. The diameter of the tissue that is sectioned may be as small as a few cells, since the protocol was adapted from single-cell RNA-seq methods. Importantly, one should try to cut away large areas of empty tissue freezing medium prior to sectioning. Ideally, the tissue is surrounded by less than 3 mm of tissue freezing medium. Each section is then placed in the bottom of a different tube and kept on dry ice. The maximum number of sections per sample is 96, due to limitation in bar codes of the primers. Prior to Trizol extraction, the tubes containing the sections should be "warmed up" at −20°C for 10 min. In the meantime a master mix containing Trizol (Ambion, for the RNA-extraction), GlycoBlue (Ambion, for visualization of the pellet), and ERCC spike-in RNA (Ambion, to keep track of technical variation) should be prepared: 0.5 mL Trizol, 0.5 μL GlycoBlue, 2 μL ERCC spike-in (stock 1:50,000) per tube. The concentration of the spike-in RNA varies depending on the size of the tissue per section, the above concentration is used for 12-μm sections of zebrafish embryos, where one section contains ~100 cells (Junker et al., 2014). Ideally, one would like to end up with ~200 reads spike-in per section. The RNA extraction should not be performed on more than 48 sections at once to ensure sufficient RNA quality. Next, the tubes are taken out of the freezer and RNA extraction is performed as follows:

1. Add 0.5 mL of the Trizol mix to each Eppendorf tube and incubate for 3′ at RT.
2. Add 0.1 mL chloroform, vigorously shake all tubes together by hand for 15″, incubate 2–3′ at RT.
3. Centrifuge at 12,000*g* for 15′at 4°C, transfer upper aqueous phase to a fresh tube (note: avoid transferring any other phase, as it will interfere with the following steps).
4. Add 250 μL isopropanol and incubate at −20°C for 1 h or o/N.
5. Centrifuge at 12,000*g* for 10′at 4°C, remove supernatant (the pellet is visible in blue).
6. Add 0.5 mL of ice-cold 75% EtOH and vortex briefly.
7. Centrifuge at 7500*g* for 5′at 4°C, remove supernatant, air-dry 5–10′.

For all steps, low-bind tubes and pipette-tips should be used. While working with RNA, perform the steps without delay in order to prevent RNA degradation.

3. cDNA SYNTHESIS

Now, the RNA pellets should quickly be resuspended by pipetting in 1.2 μL of the bar-coded primer mix (at 5 ng/μL). cDNA synthesis and cleanup using the MessageAmpII kit (Ambion) is performed as below. The bar-coded RT primer has several distinct regions: an anchored polyT, a unique bar code (8 nucleotides, differs for every tube/section), a unique molecule identifier (4 nucleotide random bar code, helps identify amplification biases), the 5′ Illumina adapter and a T7 promoter (sequences available in Junker et al., 2014; Wu et al., 2016).

1. Incubate primer-RNA-mix 5′ at 70°C.
2. Add 0.2 μL first strand buffer + 0.4 μL dNTP + 0.1 μL RNase inhibitor + 0.1 μL ArrayScript and incubate for 2 h at 42°C.
3. Add 6.3 μL of water + 1 μL second strand buffer + 0.4 μL dNTP + 0.2 μL DNA pol + 0.1 μL RNase H.
4. Incubate for 2 h at 16°C (thermal cycler with unheated lid).
5. Pool all tubes of the same sample in a 1.5-mL-tube (samples can now be stored at −20°C) and add 250 μL cDNA binding buffer per 100 μL (volume of one section is ~10 μL). In case the volume is lower (ie, less than 10 sections are combined), adjust the volume to 100 μL. Load the sample mix onto a cDNA cleanup column.
6. Centrifuge 1′ at 10,000*g*, discard flow-through (if necessary, repeat this step until whole sample is loaded).
7. Add 500 μL wash buffer, centrifuge 1′ at 10,000*g*, discard flow-through, spin once more to dry the column.
8. Transfer spin-column to fresh round-bottom 2-mL tube.
9. Add 15 μL preheated (55°C) nuclease-free water (Ambion), incubate for 2′ and spin 1.5′ at 10,000*g*.
10. Repeat elution (samples can now be stored at −20°C).

In case of samples with low amounts of input (<20 cells per section) it is advisable to dry the sample in a speedvac and dissolve the pellet in 6.4 μL nuclease-free water.

4. IN VITRO TRANSCRIPTION

In order to minimize amplification biases, in vitro transcription is performed for linear amplification of cDNA (Hashimshony et al., 2012). To 6.4 μL of sample from the previous step add 1.6 μL of each nucleotide (dATP, dGTP, dCTP, and dUTP), 1.6 μL of 10× T7 buffer and 1.6 μL of T7 enzyme (MessageAmpII kit, Ambion), and incubate for 13 h at 37°C (thermal cycler lid at 70°C). Next, the amplified RNA (aRNA) is fragmented by adding 4 μL Fragmentation buffer (200 mM Tris-acetate, pH 8.1, 500 mM KOAc, 150 mM MgOAc) and incubated for exactly 1′40″ at 94°C before quenching on ice and immediately adding 2 μL

Fragmentation STOP buffer (0.5 M EDTA). The sample volume is adjusted by adding 8 μL nuclease-free water prior to cleanup (MessageAmpII kit, Ambion). The following steps (loading and binding of aRNA to column) should not be performed for more than one sample at a time. Before loading the sample onto an aRNA spin column, 105 μL aRNA binding buffer is added to the sample, followed by 75 μL EtOH. The sample is immediately mixed by pipetting 3–4 times, loaded onto the column, and centrifuged for 1′ at 10,000*g*. Discard the flow-through, add 0.5 mL wash buffer to the column and proceed with the next sample. When all samples are loaded to the columns, continue with all samples together to perform the wash-step by spinning for 1′ at 10,000*g*. Discard the flow-through and centrifuge again for 1′ at 10,000*g* before transfer of the spin-column to a fresh tube. Elute twice by adding 15 μL preheated (55°C) nuclease-free water, incubating for 2′ and spinning for 1.5′ at 10,000*g*. Samples can now be stored at −80°C. Check the aRNA amount and quality by loading 1 μL onto Bioanalyzer RNA pico chip (Agilent). The size distribution should peak at 500 bp. If not, the time of the fragmentation should be adjusted accordingly. When starting the in vitro transcription with 0.5 ng total RNA the expected yield is 500–1000 pg/μL.

5. LIBRARY PREPARATION

For a successful library preparation (Illumina TruSeq small RNA sample prep kit, unless otherwise specified), an aRNA concentration of >1 μg/μL is recommended. If the concentration is lower, it is advisable to concentrate the sample in a speedvac until a sufficient concentration is reached.

First, ligate the 3′adapter to the aRNA from the previous step. To 5 μL of aRNA add 1 μL of 5× diluted RA3 and incubate for 2′ at 70°C before quenching on ice. Add 2 μL HML + 1 μL RNase inhibitor + 1 μL truncated T4 RNA ligase (NEB) and incubate for 1 h at 28°C. Add 4 μL nuclease-free water to adjust the volume.

Now, the reverse transcription reaction can be set up. To 6 μL of adapter-ligated RNA from previous step (rest can be stored at −80°C) add 1 μL RTP and incubate for 2′ at 70°C before quenching on ice. Add 2 μL 5× first strand buffer (from superscript kit, Invitrogen), 0.5 μL of 2× diluted dNTPs, 1 μL 100 mM DTT (from superscript kit), 1 μL RNase inhibitor, 1 μL Superscript III reverse transcriptase, and incubate for 1 h at 50°C.

For the following PCR amplification add 8.5 μL ultra pure water, 25 μL PLM, and 2 μL RP1. To each library add 2 μL of uniquely indexed RNA PCR primer (RPIX). Run the following PCR program: 98°C for 30″, 14× (98°C for 10″, 60°C for 30″, 72°C for 30″), 72°C for 10′, 4°C final hold. Samples can now be stored at −20°C.

Cleanup of the libraries is performed using AMPure XP beads (Beckman Coulter). Beads should be prewarmed at room temperature. Vortex beads until well dispersed before adding 45 μL of beads to 50 μL PCR reaction. Mix well by pipetting and incubate for 15 min.

On a magnetic stand that holds 1.5 μL tubes:

1. Place tubes on stand until liquid appears clear (>5′) and then remove supernatant.
2. Add 200 μL fresh 80% EtOH (room temperature) and incubate >30″ and then remove supernatant.
3. Add 200 μL fresh 80% EtOH and incubate >30″ and then remove supernatant.
4. Air-dry for 15′ (or until completely dry).

Resuspend with 32.5 μL EBT (10 mM Tris pH8.5, 0.02% Tween-20) and mix by pipetting. Incubate for 2′ to allow the DNA to dissociate from the beads.

On a magnetic stand:

1. Place tubes on stand until liquid appears clear (>5′).
2. Transfer 30 μL of supernatant to new tube.

Repeat the bead cleanup, but this time use 35-μL beads and resuspend with 12.5-μL EBT. Check the DNA concentration in 1 μL on the Qubit Fluorometer (Invitrogen) using high-sensitivity reagents (dsDNA HS Assay, Invitrogen). The expected concentration should be above 1 ng/μL. Check the size distribution by running 1 μL on a Bioanalyzer (high sensitivity DNA kit, Agilent): expected peak at 300–400 bp. The libraries are now ready to be sequenced by paired-end Illumina sequencing at 50-bp read length.

6. DATA ANALYSIS

The obtained paired-end reads can be aligned to the transcriptome using Burrows-Wheeler Aligner's Smith-Waterman Alignment (BWA-SW) (Li & Durbin, 2010). The 5′ mate of each pair should be mapped to the reference gene model and reads that map equally well to multiple loci should be discarded. The 3′ mate is used for bar code information. Mappability, total number of reads and the number of mapped genes depend on several factors like the origin and amount of input material and the quality of the reference genome. In any case, these parameters should be tracked across different samples to ensure reliable and persistent quality.

The following data analysis can best be performed using the freeware R or MATLAB (MathWorks). Read counts can be normalized either to total counts per section or by linear fit to spike-in RNA reads. The data should afterward be normalized to the median of total reads across sections to ensure rough correspondence between count numbers and number of mapped reads. Subsequent analyses should be performed on data normalized in such a way.

There are many different ways to analyze the *tomo*-seq data, which will depend on the biological question to be addressed. A good start is a correlation analysis. Here, pairs of sections are compared to each other across all expressed genes. The resulting heatmap plots will show clusters of sections that are similar (or different) to each other, thereby recapitulating the architecture of the tissue that was sectioned and sequenced. For example, the correlation plot that was made with *tomo*-seq data of the adult zebrafish heart after cryoinjury showed a distinction between injury area, border zone, and remote/uninjured myocardium (Wu et al., 2016). To identify genes with localized expression, clustering of expression traces can be very useful. This analysis can be performed after feature selection (eg, after filtering for genes with a peak of expression along the sectioning coordinate). The filtered genes are then clustered (eg by hierarchical clustering) by similarity of their expression profiles, thereby revealing groups of genes that are possibly coregulated.

Together, the cluster and correlation analysis provide a good overview of the different zones that are found in the tissue at hand. In most cases it will be very helpful to validate the *tomo*-seq data by in situ hybridization on selected candidate genes that are predicted to be upregulated in a specific region. As an example we like to point at the *tomo*-seq analysis on the cryoinjured zebrafish heart. The *tomo*-seq analysis predicted the presence of two distinct border zones in the injured heart which was confirmed by in situ hybridizations (Wu et al., 2016).

SUMMARY AND CONCLUSIONS

Here, we describe *tomo*-seq, a method to gain whole-genome transcriptomics data in a spatially resolved manner. The protocol described above is easy to implement because it relies on standard laboratory equipment and commercially available reagents. It is very broadly applicable since it can be used on whole organisms (eg, the zebrafish embryo), but also on isolated organs or tissues (eg, zebrafish adult heart). We have also successfully used *tomo*-seq on isolated embryonic zebrafish hearts indicating that the method can be downscaled to very small samples. Only few adjustments are necessary to adopt the protocol. The section-size can be varied depending on the research question and the input of spike-in RNA should be adjusted to the amount of input tissue per section (see Section 2). For very low input of RNA it might be necessary to concentrate the sample before continuation of the in vitro transcription (Section 4) and again before the library preparation (Section 5).

In most cases, *tomo*-seq will only be possible in one direction, thus providing 1D data. Since pattern formation often happens along one body axis, this is sufficient in most cases. However, in the case of the zebrafish embryo, which develops in a highly synchronized and staged manner, we had the unique opportunity to perform 3D RNA tomography (Junker et al., 2014). The spatial information provided by 3D RNA tomography recapitulates the actual zone of expression in the embryo.

Moreover, *tomo*-seq provides several advantages over computational analyses that infer cellular localization of RNA-seq data (Satija et al., 2015). This method heavily relies on reference genes for which in situ patterns are available. As such, it will have a big disadvantage working with mutants, for example. In mutants, the expression of some reference genes is likely disturbed and computational analyses will have difficulty reconstructing a cellular context. On the contrary, *tomo*-seq will not encounter this problem since it does not rely on in situ patterns.

One limitation of *tomo*-seq is that it does not provide the spatial resolution of microscopy-based techniques. We recommend always validating candidate genes by immunohistochemistry or in situ hybridization, also because these methods provide additional information like cell-type specificity.

Moreover, *tomo*-seq data can complement single cell RNA-seq results. While single cell RNA-sequencing has the power to identify expression programs in defined cell populations, *tomo*-seq will be able to put these results in the context of the tissue that is under investigation. Vice versa, *tomo*-seq can identify genes that are upregulated in a specific region, which subsequently can be used to genetically mark these cells for more detailed analysis by Fluorescence-Activated Cell Sorting (FACS) sorting and single cell RNA-sequencing.

Openly accessible *tomo*-seq data sets are of tremendous value to the zebrafish community, since searching for a gene-of-interest can be done in a simple manner. Moreover, genes with similar expression traces to a gene-of-interest can easily be identified. Our analysis revealed that over 10% of the differentially expressed genes in the regenerating heart are uncharacterized, highlighting the potential of *tomo*-seq to identify novel players during development and disease.

REFERENCES

Hashimshony, T., Wagner, F., Sher, N., & Yanai, I. (2012). CEL-Seq: single-cell RNA-Seq by multiplexed linear amplification. *Cell Reports, 2*(3), 666–673. http://dx.doi.org/10.1016/j.celrep.2012.08.003.

Islam, S., Kjallquist, U., Moliner, A., Zajac, P., Fan, J. B., Lonnerberg, P., & Linnarsson, S. (2011). Characterization of the single-cell transcriptional landscape by highly multiplex RNA-seq. *Genome Research, 21*(7), 1160–1167. http://dx.doi.org/10.1101/gr.110882.110.

Junker, J. P., Noel, E. S., Guryev, V., Peterson, K. A., Shah, G., Huisken, J. … van Oudenaarden, A. (2014). Genome-wide RNA Tomography in the zebrafish embryo. *Cell, 159*(3), 662–675. http://dx.doi.org/10.1016/j.cell.2014.09.038.

Langdon, Y. G., & Mullins, M. C. (2011). Maternal and zygotic control of zebrafish dorsoventral axial patterning. *Annual Review of Genetics, 45*, 357–377. http://dx.doi.org/10.1146/annurev-genet-110410-132517.

Li, H., & Durbin, R. (2010). Fast and accurate long-read alignment with Burrows-Wheeler transform. *Bioinformatics, 26*(5), 589–595. http://dx.doi.org/10.1093/bioinformatics/btp698.

Picelli, S., Bjorklund, A. K., Faridani, O. R., Sagasser, S., Winberg, G., & Sandberg, R. (2013). Smart-seq2 for sensitive full-length transcriptome profiling in single cells. *Nature Methods, 10*(11), 1096–1098. http://dx.doi.org/10.1038/nmeth.2639.

Ramskold, D., Luo, S., Wang, Y. C., Li, R., Deng, Q., Faridani, O. R. … Sandberg, R. (2012). Full-length mRNA-Seq from single-cell levels of RNA and individual circulating tumor cells. *Nature Biotechnology, 30*(8), 777–782. http://dx.doi.org/10.1038/nbt.2282.

Satija, R., Farrell, J. A., Gennert, D., Schier, A. F., & Regev, A. (2015). Spatial reconstruction of single-cell gene expression data. *Nature Biotechnology, 33*(5), 495–502. http://dx.doi.org/10.1038/nbt.3192.

Tang, F., Barbacioru, C., Wang, Y., Nordman, E., Lee, C., Xu, N. ... Surani, M. A. (2009). mRNA-Seq whole-transcriptome analysis of a single cell. *Nature Methods, 6*(5), 377–382. http://dx.doi.org/10.1038/nmeth.1315.

Wu, C.-C., Kruse, F., Vasudevaro, M. D., Junker, J. P., Zebrowski, D. C., Fischer, K. ... Bakkers, J. (2016). Spatially-resolved genome-wide transcriptional profiling identifies BMP signaling as essential regulator of zebrafish cardiomyocyte regeneration. *Developmental Cell, 36*(1), 36–49. http://dx.doi.org/10.1016/j.devcel.2015.12.010.

CHAPTER

16

Cell type-specific transcriptomic analysis by thiouracil tagging in zebrafish

T. Erickson[1], T. Nicolson

Oregon Health & Science University, Portland, OR, United States

[1]Corresponding author: E-mail: ericksti@ohsu.edu

CHAPTER OUTLINE

Methods in Cell Biology, Volume 135, ISSN 0091-679X, http://dx.doi.org/10.1016/bs.mcb.2016.04.009

Abstract

Transcriptomic studies are important tools for understanding the development and function of the different cell types that make up complex tissues. Zebrafish (*Danio rerio*) is a valuable organism for modeling key aspects of vertebrate development, cell biology, and human disease. However, the small size of individual larvae and relative scarcity of certain cell types in zebrafish can hamper efforts to collect enough pure material for cell type-specific transcriptomic studies. Thus, there is a need in the zebrafish field for spatially and temporally resolved gene expression assays. This chapter will discuss the general principles behind the TU-tagging method to isolate cell type-specific RNAs and provide guidance in designing and executing TU-tagging experiments in zebrafish.

INTRODUCTION

A cell's transcriptome—the identity and relative quantities of RNAs transcribed from the genome—is a key measure of cellular form and function. By knowing the spatial and temporal coordinates of gene expression within a complex population of cells, one can make some inferences as to the identity and function of a particular cell type, and the genes expressed in them.

In addition to intrinsic mechanisms, factors such as cell adhesion, mechanical stress, and other extracellular signals can also influence gene expression (Balda, 2003; Hill & Treisman, 1995; Mammoto, Mammoto, & Ingber, 2012). Thus, the challenge for transcriptomic experiments has been to optimize the spatial and/or temporal resolution of analysis without disturbing a cell's endogenous program of transcription. These considerations are especially important when using a small, rapidly developing model system such as the embryonic zebrafish. Analyzing whole tissues or individuals may faithfully preserve the native cellular environment, but the results tend to lack in spatial information and sensitivity. Conversely, spatial fidelity (down to single cells) can be improved by sorting dissociated cells, but these invasive techniques risk disrupting a cell's native transcriptional itinerary. Laser-capture microdissection can enrich for cells of interest without first removing them from their native environment. However, the technique is labor intensive, requires expensive equipment, and may not be suitable for all tissues or organisms. For these reasons, laser-capture microdissection has not been widely used on zebrafish larvae. With regards to temporal resolution, all of these techniques fail to distinguish between preexisting and newly transcribed RNA. As such, dynamic changes in gene transcription may be lost among the bulk of steady-state RNAs.

Several recently developed techniques promise to improve both the spatial and temporal resolution of transcriptome analyses while circumventing the possible biases introduced by cell dissociation and sorting. What makes these techniques special is that they all take an in situ approach to distinguishing the RNA in a cell type of interest from that of all other cells. This is achieved by the cell type-restricted expression of a transgene that allows the researcher to later purify a subset of RNAs originating from these cells. These methods and their variants can be grouped into three classes: (1) INTACT (isolation of nuclei tagged in specific cell types) nuclei purification (Amin et al., 2014; Deal & Henikoff, 2010, 2011; Henry, Davis, Picard, & Eddy, 2012; Steiner, Talbert, Kasinathan, Deal, & Henikoff, 2012), (2) Protein-RNA complex purification such as RiboTag or TRAP (translating ribosome affinity purification) for isolating mRNA (Heiman et al., 2008; Housley et al., 2014; Sanz et al., 2009; Tryon, Pisat, Johnson, & Dougherty, 2013) or miRAP for purifying miRNA (He et al., 2012), and (3) UPRT-mediated thiouracil tagging of RNA (TU-tagging) (Cleary, 2008; Erickson & Nicolson, 2015; Gay et al., 2013; Miller, Robinson, Cleary, & Doe, 2009). The details of these techniques can be found in the original papers, and their relative merits and pitfalls have been thoroughly compared by others (McClure & Southall, 2015; Tallafuss, Washbourne, & Postlethwait, 2014), so they need not be fully reiterated here. However, prior to undertaking the nontrivial task of cell type-specific RNA purification, it is important to consider some of the major characteristics of each method and decide if they address the needs of one's particular experimental question.

INTACT and TRAP capture specific subsets of cellular RNA representing a "snapshot" of the transcriptome at a given time. The INTACT procedure purifies nuclei, potentially allowing for analysis of any molecule held within this organelle. For transcriptomic experiments, INTACT provides nuclear RNA that represents transcripts that are being actively transcribed and spliced at the time of isolation. In this way, INTACT should distinguish between preexisting RNA in the cytoplasm and actively transcribed nuclear RNA. If one is interested in capturing dynamic changes in gene transcription, INTACT can provide excellent spatial and temporal resolution, as long as the relatively low yield and possible transcript bias of nuclear RNA is not an issue. TRAP captures RNA that is associated with ribosomes, ie, the translatome. As such, this technique does not directly measure transcriptional dynamics, but rather changes in RNA translation. This technique is suitable if one is interested in a cell's translational state or response within a narrow window of time.

In contrast to INTACT and TRAP, TU-tagging offers more flexibility, both with regards to the types of RNA that can be analyzed and to the temporal window of active transcription that can be captured. Instead of purifying an RNA-associated organelle or molecular complex, TU-tagging directly labels all RNA molecules that are synthesized during a defined period. As such, TU-tagging is well suited to capture the active transcriptome of a particular cell type. The technique hinges on a special property of the *Toxoplasma gondii* uracil phospho-ribosyltransferase (UPRT) enzyme to efficiently use 4-thiouracil (4TU) as a substrate to make 4-thiouridine monophosphate (4-TUMP), a thiol-substituted uridine that can be readily incorporated into the RNA biosynthetic pathway. While the 4TU substrate is applied globally, spatial specificity for the experiment is achieved by expressing the UPRT enzyme only in the cells of interest. UPRT-positive cells will have enriched 4TU incorporation into nascent RNA relative to UPRT-negative cells. The temporal coordinates for the experiment are defined by removing the 4TU substrate and immediately processing the tissue for RNA isolation. By taking advantage of the fact that thiol groups do not normally exist in ribonucleic acids, thiol-labeled RNA can be biotinylated in vitro and selectively purified from the unlabeled RNA pool. Like INTACT and TRAP, TU-tagging alleviates concerns about disrupting endogenous patterns of gene expression by invasive cell isolation techniques.

The advantages of TU-tagging do not come without their limitations however. Most important is the compromise in spatial fidelity and sensitivity caused by thiouracil RNA labeling in nontarget cell types (Erickson & Nicolson, 2015; Tallafuss et al., 2014). In theory, only cells expressing the *T. gondii* UPRT should be able to use the 4TU substrate; but because 4TU is applied globally, even low levels of endogenous UPRT activity can cause background RNA labeling. Indeed, *Drosophila* possess a UPRT enzyme (CG5537) whose activity can lead to the incorporation of 4TU into RNA (Ghosh, Shimell, Leof, Haley, & O'Connor, 2015). Zebrafish also have a broadly expressed *uprt* ortholog (ENSDARG00000035549) (Thisse & Thisse, 2004) that may account for background 4TU incorporation in vivo, though this activity has not been directly demonstrated for the *Danio* Uprt. Additionally, it is possible that 4TUMP from UPRT-positive cells is transported or leaked into adjacent cells, contributing to background noise (Tallafuss et al., 2014). This potential background problem is not an issue with INTACT and TRAP, which rely solely on the cell type-specific expression of a transgene and do not require an exogenous substrate.

Thus, like all methods to analyze gene expression, TU-tagging has its own set of pros and cons, the weight of each will differ depending on the goals of a particular experiment. Before settling on a technique, one should determine whether cell dissociation is a problem, and define the level of temporal and spatial resolution required to answer the experimental question. For some questions, TU-tagging, INTACT or TRAP simply may not be necessary. Also, one should consider the type of RNA that each cell type-specific technique provides. In general, TU-tagging promises the most flexibility in terms of temporal resolution, and in the variety of RNA that can be analyzed. As such, TU-tagging should definitely be considered when examining transcriptional dynamics in zebrafish cells that are difficult or impossible to purify by cell sorting.

A typical TU-tagging experiment consists of four parts: (1) Establishing and characterizing a cell type-specific UPRT transgenic line (*Tg(UPRT)* for short); (2) Experimental design and 4TU treatments; (3) RNA manipulations, biotinylation, and purification; and (4) RNA sequencing, data analysis, and validation. For each of these sections, we will discuss the general principles of the workflow (Fig. 1A–D) and point out potential pitfalls that should be considered when initiating a TU-tagging experiment in larval zebrafish.

1. PART 1. ESTABLISHING AND CHARACTERIZING A CELL TYPE-RESTRICTED UPRT TRANSGENIC LINE OF FISH

1.1 CREATING A CELL TYPE-SPECIFIC UPRT TRANSGENIC

For TU-tagging to be successful, one *must* be able to establish a stable line of transgenic fish that express high levels of UPRT in the cell type of interest during the appropriate developmental stage. Cell type specificity is critical; less restricted

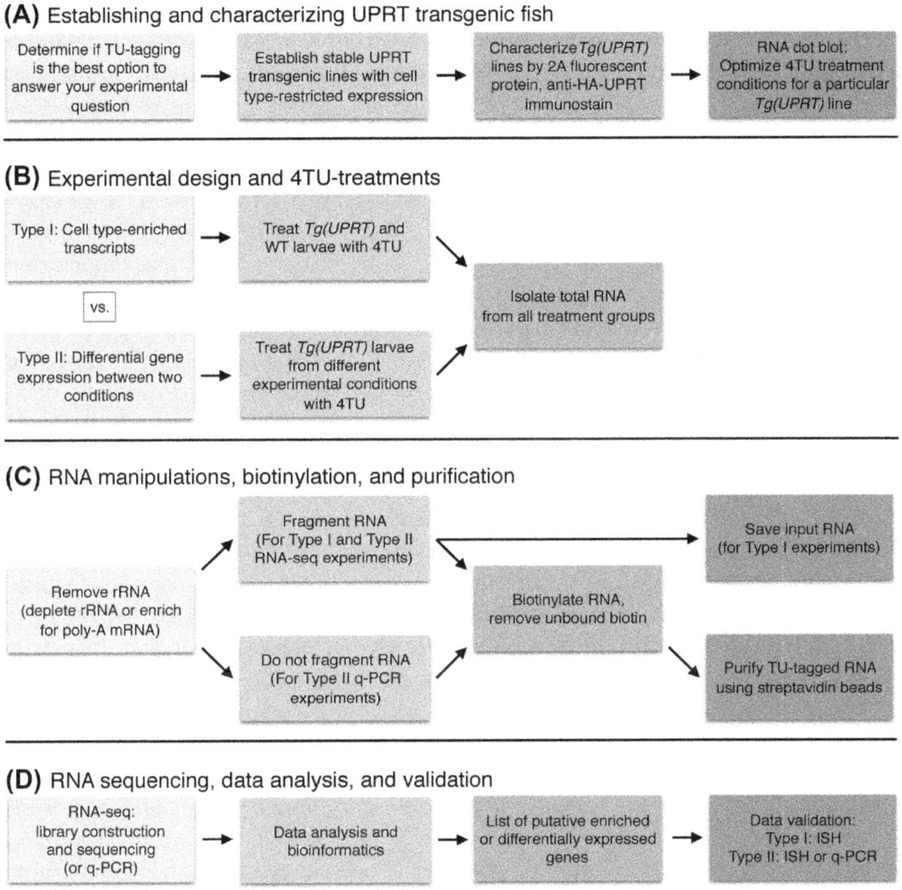

FIGURE 1

Workflow diagrams for a typical TU-tagging experiment. (A) Establishing and characterizing a cell type-specific UPRT transgenic—*Tg(UPRT)*. (B) Experimental design and 4TU treatments. (C) RNA manipulations, biotinylation, and purification. (D) RNA sequencing, data analysis, and validation.

UPRT expression means that the researcher will have to tolerate noisier data, or rely on invasive techniques to enrich for the cells of interest. The most common approaches to achieve cell type-specific transgenic expression in zebrafish are: (1) use a fragment of DNA that is sufficient to drive transgene expression in the cells of interest; (2) use a cell type-restricted Gal4 driver in combination with a UAS promoter line.

For our TU-tagging studies in sensory hair cells (Erickson & Nicolson, 2015), we used a well-characterized 6.5 kb fragment of the *myo6b* promoter whose activity is restricted to sensory hair cells of the ear and lateral line system from ~24 h postfertilization (hpf) into adulthood (Obholzer et al., 2008). Standard Tol2-Gateway techniques were used to assemble a *-6.5myo6b:HA-UPRT-P2A-mCherry* vector and create a stable transgenic line of fish (Kwan et al., 2007). We tagged UPRT with an N-terminal HA-epitope for immunostaining, and used a viral P2A peptide for bicistronic expression of mCherry. The advantage of using a 2A peptide to coexpress a fluorescent marker is that 2A does not add a bulky tag to the UPRT enzyme itself, yet the fluorescent protein is expressed in a 1:1 ratio with UPRT (Kim et al., 2011; Szymczak et al., 2004). However, an HA-UPRT-mCherry fusion protein has also been successfully used for TU-tagging (Tallafuss et al., 2015). Fluorescent protein expression allows one to screen live fish for transgenics, and permits a direct comparison of different lines to pick the one with the highest levels of cell type-specific UPRT expression. Additionally, to reduce the chance of UPRT expression in nontarget cell types, we recommend avoiding use of the standard "heart marker" Tol2 transgenesis vector (*pDestTol2CG*) when making a UPRT construct. We have observed some incidences where the *cmlc2* promoter can drive heart expression of transgenes intended for hair cells.

It is also possible to create a *UAS:UPRT* line and use the wide array of zebrafish gene/enhancer trap Gal4 driver lines to express UPRT in specific cell types (Kawakami et al., 2010), or to create one's own Gal4 driver using a favorite promoter. We explored this possibility by establishing a *14xUAS:HA-UPRT-T2A-eGFP* line and pairing it with the *s1001t* driver that expresses Gal4 in hair cells (Scott & Baier, 2009). We observed very robust eGFP fluorescence in cells where Gal4 was active, suggesting that this approach could work well in some circumstances. However, many of the enhancer trap Gal4 lines exhibit fairly broad or nonspecific patterns of activity (Scott, 2009), and may not be restricted enough to proceed with TU-tagging without first manually enriching for the cells of interest. Our experience with the hair cell Gal4 driver line shared these drawbacks, and this option was not explored further. However, if one's experimental design permits it, using a *UAS:UPRT* line may be a good option for expressing high levels of UPRT in the cells of interest (Tallafuss et al., 2015).

1.2 USING RNA DOT BLOTS TO CHARACTERIZE UPRT TRANSGENICS AND OPTIMIZE 4TU TREATMENT CONDITIONS

In TU-tagging experiments, the total pool of thiol-labeled RNA originates from two sources (1) the signal: UPRT-dependent labeling in the cell type of interest and (2) the noise: background labeling in all cells. Among the various factors that can influence the rate at which 4TU is converted to 4-TUMP, the researcher has control over only two: the concentrations of enzyme and substrate. Since it is not possible to preferentially increase 4TU availability to target cells only, the main way to maximize the signal-to-noise ratio is to increase the concentration of UPRT in the cells of interest.

Thus, to be suitable for TU-tagging, a *Tg(UPRT)* line must exhibit high levels of cell type-specific expression. Even an exquisitely specific promoter may not be suitable if it does not drive high levels of UPRT expression. If using a transposon-based transgenesis technique (Kawakami, 2005), keep in mind that the genomic location of insertion may affect the level of UPRT expression (ie, the position effect). It may be possible to address variability in expression using site-directed transgenesis (Mosimann et al., 2013; Roberts et al., 2014). In our experience with hair cells, although the *myo6b* promoter is capable of driving high levels of transgene expression, in practice we recovered some transgenic lines where the level of UPRT expression was quite poor and the level of RNA labeling was not sufficiently above background to warrant further experiments.

Comparing the relative levels of fluorescence from the 2A peptide construct is a good way to choose among the transgenic lines for the best level of UPRT expression, but will not indicate if a particular line expresses *enough* UPRT to be effective for TU-tagging. Thus, it is essential to directly test the signal-to-noise ratio in each *Tg(UPRT)* line prior to undertaking the full experiment. One could attempt to quantify the amount of 4TU incorporation into nascent RNA using a spectrophotometric assay (Zeiner et al., 2008). However, this approach was not sensitive enough to detect a small proportion of thiol-tagged RNA among a larger untagged pool in our experience. The RNA dot blot is a sensitive method to find a suitable *Tg(UPRT)* insertion and to optimize the 4TU treatment conditions for that particular transgenic line. By using total RNA, the dot blots do not require the time-consuming and expensive steps of rRNA depletion and TU-tagged transcript purification. As an added benefit, the dot blot procedure provides an opportunity to optimize many of the steps—total RNA isolation, RNA biotinylation and cleanup—that will be critical for the full TU-tagging experiment.

1.3 DOT BLOT MATERIALS

- 4-Thiouracil (4TU) (Sigma—440736)
- 4-Thiouridine (4sU) (Sigma—T4509)
- Dimethyl sulfoxide (DMSO) (Sigma—D8418)
- E3 embryo media (5 mM NaCl, 0.17 KCl, 0.33 mM $CaCl_2$, and 0.33 mM $MgSO_4$)
- Thermo Scientific Nalgene bottle top vacuum filter 0.2 μm (VWR—73520-984)
- Dumont Tweezers #5, 0.1 × 0.06 mm (World Precision Instruments—14098)
- Pronase E (Sigma—P5147)
- Falcon Cell Strainer 70 μm (Fisher Scientific—08-771-2)
- Falcon 6-well tissue culture plates 353224 (Fisher Scientific—08-772-1G)
- TRIzol reagent (Ambion—15596–026)
- Turbo DNA-free (Ambion—AM1907)
- DEPC-treated water (Bioexpress—G-3223-125)
- EZ-Link HPDP-Biotin (Thermo Scientific—21341)
    - Alternatively: MTSEA biotin-XX (Biotium—90066)
- N,N-dimethylformamide (DMF) (Sigma—D4551)
- Qiagen RNeasy mini kit (Qiagen—74104)

- Alternatively: Chloroform-isoamyl alcohol 24:1 (Sigma—25666); Phase Lock Gel—Heavy—1.5 mL tubes (5 Prime—2302810)
- Dot blot blocking solution (1× PBS/1 mM EDTA/1% SDS)
- Nylon membrane (Bio-Rad—162-0165)
- Streptavidin-HRP (Thermo Scientific—21126)
- Chemiluminescent substrate—Pierce SuperSignal West Pico (Thermo Scientific—34077)
- Autorad Film—GeneMate Blue Ultra (BioExpress—F-9029-8x10)

1.4 DOT BLOT PROCEDURE (ESTIMATED TIME ~2 DAYS)

1. Collect embryos for the dot blot experiment. There should be four treatment groups for each concentration of 4TU and duration of labeling: (1) Wild type (WT) *non-Tg* with 4sU (positive control), (2)] WT *non-Tg* with 4TU (background noise), (3) *Tg(UPRT)* with 4TU (signal), and (4) *Tg(UPRT)* with DMSO (negative control). If needed, dechorionate embryos either manually (using Dumont #5 tweezers) or enzymatically (with 2 mg/mL Pronase E). Because the dot blot is done with total RNA, treating 10–20 individuals per condition should provide more than enough material. This number can be adjusted depending on development stage or other experimental parameters.
2. Make stock solutions of 0.5 M 4TU and 4sU in DMSO and store at −80°C. When the embryos are at the appropriate stage, prepare fresh 5 mL working solutions in filter-sterilized embryo media (E3), keeping the DMSO concentration constant with that of the highest 4TU concentration.
   a. The optimal concentration and duration of 4TU treatment will vary depending upon the particular experiment. However, when testing a line for the first time, 4TU concentrations of 1, 5, and 10 mM are good starting points, with labeling times ranging between 1 and 4 h.
   b. Longer labeling times increase the proportion of TU-tagged RNA relative to the total pool of cellular RNAs.
   c. Generally, background labeling increases with time and 4TU concentration.
   d. We advise the use of filter-sterilized embryo media prior to and during the 4TU-labeling period in order to lessen the chance that the 4TU will be incorporated into the RNA of microorganisms that are cohabitant with the fish (Cleary, 2008; Tallafuss et al., 2014). Extra antibiotics or other precautions may be used as necessary.
3. Transfer the embryos to 6-well plates, one well per 5 mL of 4TU or DMSO solution, and one plate per labeling duration. Incubate in the dark at 29°C.
   a. Minimize the amount of liquid carry over during the transfer so as to not distort the chosen concentrations of 4TU. Using Falcon 70 μm cell strainers during the transfer may aid in this aspect.
   b. Best practice is to perform each treatment condition and RNA purification in duplicate, and dot these duplicates on the same blot.

4. At the end of the labeling period, transfer the embryos to new dishes containing fresh E3, anesthetize with Tricaine, and transfer to 1.5 mL microfuge tubes. Remove as much liquid as possible and homogenize in TRIzol. Isolate total RNA according to the manufacturer's protocol and remove DNA contamination by DNase-treatment.
   a. Other total RNA isolation procedures may be substituted for TRIzol.
   b. Unlike the full TU-tagging procedure, there is no need for rRNA depletion or RNA fragmentation before proceeding to the biotinylation step.
5. Biotinylate the total RNA. We and others have used the HPDP-biotin reagent to biotinylate TU-tagged RNA, but recent work has shown that MTS-biotin significantly improves the biotinylation reaction (Duffy et al., 2015). Therefore, we recommend that readers refer to the detailed RNA biotinylation and purification protocols supplied with the Duffy et al. paper for their TU-tagging experiments. For the dot blot, the relevant steps include the following:
   a. RNA biotinylation: incubate total RNA with the activated biotin reagent.
   b. RNA cleanup to remove excess biotin. Success at this step is critical; unbound biotin will cause high background on the blot, and will compromise the streptavidin purification of TU-tagged RNA when performing the full experiment. Using the Qiagen RNeasy mini kit is acceptable for the dot blot when working with large quantities of intact RNA. However, when cleaning up small quantities of fragmented RNA during the full TU-tagging procedure, we found that a chloroform-isoamyl alcohol (24:1) extraction using Phase Lock Gel tubes followed by ethanol precipitation helped to minimize the loss of precious material (Dölken et al., 2008; Erickson & Nicolson, 2015).
6. Following the biotinylation and RNA cleanup steps, determine the quantity of RNA for each treatment group by spectrophotometry and make 50 and 250 ng/μL aliquots for each. It is recommended to dot at least two quantities of biotinylated RNA for each replicate condition to assure that the exposure is within the linear range of the film for analysis. The suggested 50 and 250 ng quantities may be adjusted depending on the particular conditions of the experiment.
7. Construct the dot blot by pipetting RNA directly onto the membrane. A suggested layout and expected results are diagrammed in Fig. 2.
8. Immobilize the RNA on the membrane by UV cross-linking.
9. Block the membrane 15 min in 1x PBS/1 mM EDTA/1% SDS at room temperature on an orbital shaker.
10. Dilute 1 mg/mL Streptavidin-HRP 1:5000 in fresh block solution and incubate for 10 min on an orbital shaker.
11. Wash the membrane 1 × 10 min in block, 3 × 5 min in 1x PBS/0.1% SDS, and 1 × 5 min in 1x PBS.
12. Incubate the membrane in chemiluminescence substrate according to the manufacturer's protocol.

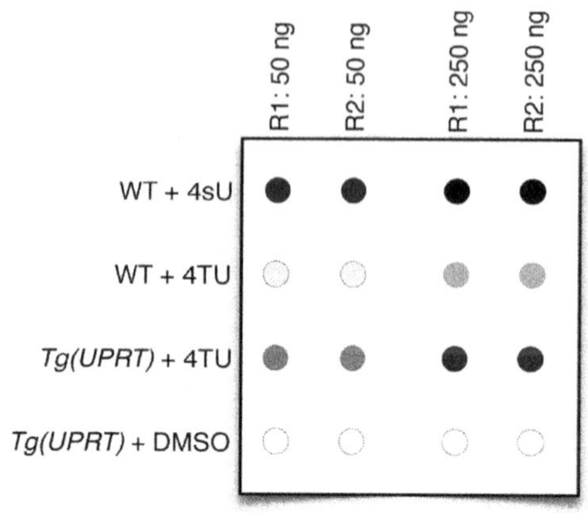

FIGURE 2

Using an RNA dot blot to characterize a *Tg(UPRT)* line and optimize 4TU treatment conditions. A suggested layout and expected results are diagrammed. For each set of parameters (4TU concentration and length of exposure), treatments and purifications should be done in duplicate (R1 and R2), and two different quantities from each replicate set should be arrayed on the blot.

13. Expose and develop film as one would for a Western blot. Typically, exposure times of less than 1 min are required.
14. Quantitatively (measure and compare dot intensities) or qualitatively analyze the results for all of the labeling periods. As diagrammed in Fig. 2, one expects dot intensities to be very strong for the WT + 4sU positive control, low or undetectable for WT + 4TU (background), medium-to-high for the *Tg(UPRT)* + 4TU (signal), and completely absent for *Tg(UPRT)* + DMSO (negative control). Dot intensities should increase as the quantity of RNA increases.
    a. If one fails to detect biotinylated RNA for any of the treatments, including the WT + 4sU positive control, one may include a 5′-biotinylated oligo on the blot as a positive control for the blotting procedure. If this can be detected, then review the 4TU treatment, biotinylation, and purification steps for possible problems.
    b. If one detects a signal in the *Tg(UPRT)* + DMSO negative control, this indicates a problem with the postbiotinylation RNA cleanup step. When performing the full TU-tagging experiment, it is essential to remove all unbound biotin prior to performing the streptavidin purification of TU-tagged RNA to assure maximum recovery.
    c. The acceptable level of enrichment between the WT + 4TU and the *Tg(UPRT)* + 4TU treatments will vary depending on which cell type is being studied and its relative abundance in the embryo, the developmental stage, and the particular characteristics of the *Tg(UPRT)* line being used in

the experiments. However, for a particular transgenic line, if there is consistently less than twofold enrichment in the *Tg(UPRT)* sample for any labeling protocol, isolating a stronger line might be beneficial.

2. PART 2. EXPERIMENTAL DESIGN AND 4TU TREATMENTS

Each TU-tagging experiment will be characterized by unique parameters such as the specific experimental question, the cell type being examined, the developmental stage, and the particular UPRT transgenic line. As such, many details will have to be worked out on an experiment-to-experiment basis. These include the experimental design, the number of experimental replicates, the appropriate developmental time frame for analysis, and the number of animals required to purify enough TU-tagged RNA.

2.1 EXPERIMENTAL DESIGN: TYPE I VERSUS TYPE II EXPERIMENTS

As outlined by Gay, Karfilis, Miller, Doe, and Stankunas (2014) there are two types of TU-tagging experimental designs: Type I for cell type-enriched transcripts, and Type II for differential gene expression between two conditions. Since the workflow differs between the experimental designs, it is important to consider how the data will be analyzed prior to starting the 4TU-labeling procedure so that the correct treatments are completed at the same time (Fig. 1B).

Type I experiments are appropriate for discovering nascent RNAs that are enriched in the cell type of interest. These Type I experiments come in two flavors. The first of these can be called "Type Ia"; this is where transcript abundance in the TU-tagged RNA sample is compared to that of the whole organism or tissue (input RNA). For Type Ia experiments, one should treat *Tg(UPRT)* fish with 4TU and save an aliquot of "input" RNA—a sample of RNA before the purification of TU-tagged RNA. However, this comparison has two potential shortcomings, both related to the strong bias of TU-tagging towards actively transcribed genes. First, because there is 4TU incorporation in nontarget cells, genes for which the bulk of their transcription has occurred during the 4TU labeling period may appear as "enriched" in the TU-tagged RNA sample, even though they are not expressed in the cell type of interest. This background incorporation would produce a false positive. Secondly, since TU-tagging labels actively transcribed RNA only, it is possible for a gene product to be enriched in the cell type of interest, but not be enriched in the TU-tagged sample if its rate of transcription during the labeling period is very low. This low rate of transcription would produce a false negative.

To control for this bias, one may perform a "Type Ib" experiment where TU-tagged RNA from *Tg(UPRT)* fish is compared to TU-tagged RNA from nontransgenic WT fish. As a result, active transcription from nontarget cell types will be equally represented in both pools, and abundant preexisting transcripts will not mask nascent cell type-specific RNAs with low rates of transcription. To perform a Type Ib experiment, in addition to treating *Tg(UPRT)* fish with 4TU, one will have to treat stage-matched, nontransgenic fish with either 4TU (to mimic the background labeling observed in *Tg(UPRT)* larvae), or with 4sU directly to efficiently label actively transcribed genes throughout the fish.

Realistically, when using TU-tagging as a tool for transcript discovery, it may be beneficial to perform both Type Ia and Ib comparisons in order to compile a list of candidate transcripts. This approach entails collecting and sequencing at least three RNA samples: (1) TU-tagged RNA from *Tg(UPRT)*; (2) input RNA from *Tg(UPRT)*; and (3) TU-tagged RNA from WT. Whichever comparison is performed, the cell type-specific or differential expression of candidate genes will have to be confirmed by an independent method such as mRNA in situ hybridization (ISH) or q-PCR.

Type II experiments compare TU-tagged RNA from two or more conditions (ie, WT vs mutant or WT vs drug-treated). In this case, *Tg(UPRT)* larvae from condition A and condition B will be exposed to 4TU for the same amount of time, and then analyzed for differential gene expression between the two conditions. If one intends to perform Type II experiments comparing WT versus mutant, it will save time to establish and characterize a *Tg(UPRT)* line in the genetic background of interest. If you have already optimized 4TU conditions for a particular *Tg(UPRT)* line in a different genetic background, that transgenic insertion should be crossed into the mutant background to avoid having to re-optimize 4TU conditions for a new transgenic line.

2.2 BIOLOGICAL REPLICATES

Whether you are conducting a Type I (enrichment) or Type II (differential expression) experiment, all treatments and purifications should be carried out in duplicate. This strengthens the statistical analysis, and bolsters confidence in the results. Replicates are essential for Type II experiments, where biological and technical variation may confound the comparison of two conditions, especially if the differences are subtle. However, when working with zebrafish, the small size of individual embryos means that one is never working with individual biological samples; for practical purposes, each experimental replicate is actually a large pool of individual biological replicates. As such, for Type I enrichment experiments, it may be acceptable to analyze a large number of biological replicates (individuals) as a single pool, as long as the results are confirmed using an independent technique such as ISH or q-PCR from a unique sample of RNA.

2.3 DEVELOPMENTAL STAGING

TU-tagging excels at allowing the researcher to define the temporal window of active gene expression for analysis. However, since zebrafish development is so rapid, this window can be quite narrow for some cell types. Thus, extra care must be taken to minimize staging differences between individuals so as to not bias the outcome of a TU-tagging experiment. Using retinal photoreceptor cell differentiation as an example, the difference between one group of larvae exhibiting low levels of opsin expression and another group expressing high levels could be only a matter of a few hours difference in age (Raymond, Barthel, & Curran, 1995); this can be the difference between a clutch born at 9 am and one born at noon. Pooling these clutches could increase the false negative rate, while comparing these groups could produce false positives. Thus, regardless of what cell type is being analyzed, it is imperative that all control, experimental, and replicate groups are stage matched as closely as possible.

2.4 NUMBER OF LARVAE REQUIRED FOR TU-TAGGING

The number of larvae required for each experiment will depend on the details of the particular experiment, such as the *Tg(UPRT)* line, the cell type of interest, and the duration of 4TU labeling. For example, an analysis of trunk muscle (an abundant cell type) gene expression at 5 days postfertilization (dpf) using a long labeling period will require relatively fewer larvae than an experiment where the cell type is rare and the labeling period is short. For our Type I TU-tagging experiment on hair cells, we used a long 15-h labeling period and approximately 250 larvae analyzed at 4 dpf (Erickson & Nicolson, 2015). Hair cells are a rare cell type, but most hair cells are located at or near the surface of the larva. As such, it is possible that they have more access to the 4TU substrate than, for example, a group of neurons buried deep in the brain. For these reasons and other factors that will be unique to one's own experiment, it is difficult to make a general recommendation as to the minimum number of required larvae. Another consideration is that a new, high efficiency biotinylation chemistry for TU-tagged RNA (Duffy et al., 2015) (see *Purification of TU-tagged RNA*) may require less starting material for TU-tagging experiments. Regardless, except for the high cost of rRNA depletion, there is no reason not to use as many individual specimens as possible.

3. PART 3. RNA MANIPULATIONS, BIOTINYLATION, AND PURIFICATION

3.1 REMOVING RIBOSOMAL RNA

Once total RNA has been isolated from all of the treatment groups and checked for quality, the next step is to prepare the RNA for biotinylation and purification (Fig. 1C). For typical transcriptomic experiments, this will first involve either an rRNA depletion step, or a poly-A mRNA enrichment. Although it is expensive, rRNA depletion is the preferable method because it retains non-poly-A RNA that may be of interest, such as miRNAs and unspliced pre-mRNA. Retaining pre-mRNA is particularly important when performing short 4TU-labeling periods that are expected to capture a narrow window of active gene expression (Windhager et al., 2012). We have used both the Ribo-Zero Gold rRNA removal kit (Illumina—MRZG126) and poly-A enrichment to successfully deplete rRNA from zebrafish samples.

3.2 RNA FRAGMENTATION

The ability to analyze the active transcriptome in an unbiased way using RNA-seq is one of the major benefits of TU-tagging. If RNA-seq is the end point assay, fragmenting the rRNA-depleted sample prior to biotinylation and streptavidin purification can reduce background noise by omitting weakly labeled transcripts from the TU-tagged pool (Gay et al., 2013). However, in some cases, one may want to forego a full scale RNA-seq experiment in favor of testing only a few genes of interest by q-PCR. Because fragmented or degraded RNA can lower the quality of q-PCR data (Fleige & Pfaffl, 2006), it may be beneficial to eliminate the fragmentation step to purify RNA of the highest possible quality for q-PCR experiments.

3.3 BIOTINYLATION AND PURIFICATION OF TU-TAGGED RNA

The basis of TU-tagging is that the thiol-substituted uracil can be selectively biotinylated in vitro via a chemical reaction that will not occur with regular nucleotides. This selective chemistry allows for the purification of TU-tagged RNA from a larger pool of untagged transcripts. Previously published experiments have used the reaction between the sulfhydryl group in 4-thiouridine and a pyridylthio-activated biotin (HPDP-biotin) to biotinylate the tagged RNA. However, this reaction has recently been found to be inefficient, leading to a suboptimal yield of TU-tagged RNA. Duffy et al. (2015) have demonstrated that the reaction between 4-thiouridine and methylthiosulfonate-activated biotin (MTS-biotin) is nearly 100% efficient, representing a fivefold increase over the HPDP-biotin reaction. This higher efficiency should translate into a number of advantages: (1) Fewer embryos will be required for the experiment. If using the rRNA depletion method to remove ribosomal RNA, this step translates into cost savings. If performing Type II experiments comparing autosomal recessive mutants and siblings, this step means less time spent collecting specimens for the experiment. (2) Increased sensitivity to analyze rare cell types. (3) Improved sensitivity for the detection of rare transcripts expressed by the cell type of interest. (4) Decreased bias towards long transcripts, thereby increasing the sensitivity for short transcripts, especially miRNA.

For these reasons, it is recommended that everyone take advantage of this new biotinylation chemistry. The detailed protocol that accompanies the Duffy et al. paper is an excellent resource, although some of the reactions may have to be scaled back to account for lower amounts of thiol-labeled RNA resulting from a cell type-specific TU-tagging experiment. Additionally, some of the general recommendations in this article may have to be adjusted in light of this more efficient thiol–biotin coupling. These include a reduction in the amount of biotinylated total RNA used in the dot blot assays, and a reduction in the number of larvae required to perform the TU-tagging experiment.

4. PART 4: RNA SEQUENCING, DATA ANALYSIS, AND VALIDATION

Once all of the final RNA samples have been collected, many researchers may seek outside expertise for the final steps of library preparation, RNA sequencing, and data analysis. Indeed, even though the details of these procedures are outside the scope of this review, there are a few general considerations that one should be aware of as the TU-tagging experiment enters its final stages (Fig. 1D).

4.1 LIBRARY PREPARATION AND RNA SEQUENCING

The details of the RNA-seq library preparation will depend on the sequencing platform of choice. If a sequencing core or another third-party service is making the library, it is recommended to consult closely with them since your sample (a small quantity of rRNA-depleted, fragmented RNA) may differ from what they are normally used to handling. In general, the quantity of purified, TU-tagged RNA will be very small, and will require a library preparation kit geared towards low input samples. Keep in mind that these low input kits include amplification steps that may introduce additional biases into the data (van Dijk, Jaszczyszyn, & Thermes, 2014; Head et al., 2014).

4.2 DATA ANALYSIS

The raw result of an RNA-seq experiment is tens of millions of short sequence reads (amounting to gigabytes of data) that need to be processed, mapped, counted, and statistically analyzed in order to produce meaningful data. However, unless one is already set up with the equipment and expertise to analyze next generation sequencing data, performing the analysis oneself may require more time and computing power than the average researcher has access to. Therefore it is recommended that the nonpower computing biologist seek assistance from a friendly bioinformatician for the mapping and statistical analysis of RNA-seq data. Regardless of whether the analysis is performed by oneself or others, the TU-tagging biologist is advised to become familiar with the data analysis workflow. Besides the primary literature (Anders & Huber, 2010; Kim et al., 2013; Langmead & Salzberg, 2012; Love, Huber, & Anders, 2014), there are good community-based web resources available to help the neo-bioinformatician with RNA-seq data analysis, such as SeqAnswers or Biostar (Li et al., 2012; Parnell et al., 2011).

4.3 VALIDATION OF THE RNA-seq RESULTS

If the experiment was successful, the end result will be a list of significantly enriched genes (Type I experiment), or differentially expressed genes (Type II experiment). For the genes that warrant further investigation, it is essential to validate the results of the RNA-seq experiment by an independent method such as ISH or q-PCR. Of the two methods, ISH is best suited for detailing the spatial pattern of gene expression (Thisse & Thisse, 2014), and should be used for validating the results of a Type I enrichment experiment whenever possible. And although standard colorimetric ISH is not quantitative, one can use it to qualitatively compare relative levels of gene expression between two conditions in order to validate the results of a Type II experiment (Hensley et al., 2011). However, if the gene expression changes are subtle, standard ISH may not sensitive enough. To quantitatively verify the results of a Type II experiment, one should perform q-PCR on an independent sample of RNA. A crude dissection to enrich for the target cell type may be an acceptable way to gather starting material for q-PCR validation, provided that the particular experimental question allows for it, and the spatial restriction of a candidate transcript has already been established by ISH. However, in some cases, validation may require a repeat of the TU-tagging experiment, this time with q-PCR as the end point instead of RNA-seq.

CONCLUSIONS

TU-tagging allows for the analysis of active gene transcription in specific cell types of an intact organism. Compared to other noninvasive, spatially resolved transcriptomic techniques (INTACT, TRAP), TU-tagging affords the most control over the temporal coordinates, and offers the broadest range of RNA types that can be purified. However, with the finding that metazoans may possess their own active UPRT enzymes (Ghosh et al., 2015), the actual sensitivity and spatial fidelity of TU-tagging may be less than ideal. Future efforts to optimize TU-tagging may involve knocking down the endogenous UPRT activity, or engineering a *UPRT* transgene with unique substrate requirements such that endogenous UPRT activity no longer limits the sensitivity of the technique.

If spatial resolution is of the utmost importance, the temporal control afforded by TU-tagging may be combined with crude tissue dissection or cell-sorting procedures. TU-tagging is blind to RNA produced both before and after the 4TU-labeling window. Thus, one could label the active transcriptome of a specific cell type during the 4TU exposure period, then dissociate the cells and enhance spatial resolution by cell sorting. If cell sorting induces undesirable gene expression changes, the data should be largely ignorant of these effects because they occurred outside of the labeling period. Alternatively, provided that the cells of interest express a fluorescent marker and are amenable to FACS, it may be possible to perform a UPRT-free thiouracil-labeling experiment by simply exposing larvae to 4sU and using FACS to provide cell type specificity. In this way, thiouracil tagging can provide the temporal dimension while dissection or cell sorting can enhance or provide the spatial resolution.

REFERENCES

Amin, N. M., Greco, T. M., Kuchenbrod, L. M., Rigney, M. M., Chung, M.-I., Wallingford, J. B., ... Conlon, F. L. (2014). Proteomic profiling of cardiac tissue by isolation of nuclei tagged in specific cell types (INTACT). *Development (Cambridge, England), 141*(4), 962–973. http://dx.doi.org/10.1242/dev.098327.

Anders, S., & Huber, W. (2010). Differential expression analysis for sequence count data. *Genome Biology, 11*(10), R106. http://dx.doi.org/10.1186/gb-2010-11-10-r106.

Balda, M. (2003). Epithelial cell adhesion and the regulation of gene expression. *Trends in Cell Biology, 13*(6), 310–318. http://dx.doi.org/10.1016/S0962-8924(03)00105-3.

Cleary, M. D. (2008). Cell type-specific analysis of mRNA synthesis and decay in vivo with uracil phosphoribosyltransferase and 4-thiouracil. *Methods in Enzymology, 448*, 379–406. http://dx.doi.org/10.1016/S0076-6879(08)02619-0.

Deal, R. B., & Henikoff, S. (2010). A simple method for gene expression and chromatin profiling of individual cell types within a tissue. *Developmental Cell, 18*(6), 1030–1040. http://dx.doi.org/10.1016/j.devcel.2010.05.013.

Deal, R. B., & Henikoff, S. (2011). The INTACT method for cell type-specific gene expression and chromatin profiling in *Arabidopsis thaliana*. *Nature Protocols, 6*(1), 56–68. http://dx.doi.org/10.1038/nprot.2010.175.

van Dijk, E. L., Jaszczyszyn, Y., & Thermes, C. (2014). Library preparation methods for next-generation sequencing: tone down the bias. *Experimental Cell Research, 322*(1), 12–20. http://dx.doi.org/10.1016/j.yexcr.2014.01.008.

Dölken, L., Ruzsics, Z., Rädle, B., Friedel, C. C., Zimmer, R., Mages, J., ... Koszinowski, U. H. (2008). High-resolution gene expression profiling for simultaneous kinetic parameter analysis of RNA synthesis and decay. *RNA (New York, NY), 14*(9), 1959–1972. http://dx.doi.org/10.1261/rna.1136108.

Duffy, E. E., Rutenberg-Schoenberg, M., Stark, C. D., Kitchen, R. R., Gerstein, M. B., & Simon, M. D. (2015). Tracking distinct RNA populations using efficient and reversible covalent chemistry. *Molecular Cell, 59*(5), 858–866. http://dx.doi.org/10.1016/j.molcel.2015.07.023.

Erickson, T., & Nicolson, T. (2015). Identification of sensory hair-cell transcripts by thiouracil-tagging in zebrafish. *BMC Genomics, 16*(1), 842. http://dx.doi.org/10.1186/s12864-015-2072-5.

Fleige, S., & Pfaffl, M. W. (2006). RNA integrity and the effect on the real-time qRT-PCR performance. *Molecular Aspects of Medicine, 27*(2–3), 126–139. http://dx.doi.org/10.1016/j.mam.2005.12.003.

Gay, L., Karfilis, K. V., Miller, M. R., Doe, C. Q., & Stankunas, K. (2014). Applying thiouracil tagging to mouse transcriptome analysis. *Nature Protocols, 9*(2), 410–420. http://dx.doi.org/10.1038/nprot.2014.023.

Gay, L., Miller, M. R., Ventura, P. B., Devasthali, V., Vue, Z., Thompson, H. L., ... Doe, C. Q. (2013). Mouse TU tagging: a chemical/genetic intersectional method for purifying cell type-specific nascent RNA. *Genes & Development, 27*(1), 98–115. http://dx.doi.org/10.1101/gad.205278.112.

Ghosh, A. C., Shimell, M., Leof, E. R., Haley, M. J., & O'Connor, M. B. (2015). UPRT, a suicide-gene therapy candidate in higher eukaryotes, is required for *Drosophila* larval growth and normal adult lifespan. *Scientific Reports, 5*, 13176. http://dx.doi.org/10.1038/srep13176.

He, M., Liu, Y., Wang, X., Zhang, M. Q., Hannon, G. J., & Huang, Z. J. (2012). Cell-type-based analysis of microRNA profiles in the mouse brain. *Neuron, 73*(1), 35–48. http://dx.doi.org/10.1016/j.neuron.2011.11.010.

Head, S. R., Komori, H. K., LaMere, S. A., Whisenant, T., Van Nieuwerburgh, F., Salomon, D. R., & Ordoukhanian, P. (2014). Library construction for next-generation sequencing: overviews and challenges. *BioTechniques, 56*(2). http://dx.doi.org/10.2144/000114133, 61–64, 66, 68, passim.

Heiman, M., Schaefer, A., Gong, S., Peterson, J. D., Day, M., Ramsey, K. E., … Heintz, N. (2008). A translational profiling approach for the molecular characterization of CNS cell types. *Cell, 135*(4), 738–748. http://dx.doi.org/10.1016/j.cell.2008.10.028.

Henry, G. L., Davis, F. P., Picard, S., & Eddy, S. R. (2012). Cell type-specific genomics of Drosophila neurons. *Nucleic Acids Research, 40*(19), 9691–9704. http://dx.doi.org/10.1093/nar/gks671.

Hensley, M. R., Emran, F., Bonilla, S., Zhang, L., Zhong, W., Grosu, P., … Leung, Y. F. (2011). Cellular expression of Smarca4 (Brg1)-regulated genes in zebrafish retinas. *BMC Developmental Biology, 11*, 45. http://dx.doi.org/10.1186/1471-213X-11-45.

Hill, C. S., & Treisman, R. (1995). Transcriptional regulation by extracellular signals: mechanisms and specificity. *Cell, 80*(2), 199–211. http://dx.doi.org/10.1016/0092-8674(95)90403-4.

Housley, M. P., Reischauer, S., Dieu, M., Raes, M., Stainier, D. Y. R., & Vanhollebeke, B. (2014). Translational profiling through biotinylation of tagged ribosomes in zebrafish. *Development (Cambridge, England), 141*(20), 3988–3993. http://dx.doi.org/10.1242/dev.111849.

Kawakami, K. (2005). Transposon tools and methods in zebrafish. *Developmental Dynamics : An Official Publication of the American Association of Anatomists, 234*(2), 244–254. http://dx.doi.org/10.1002/dvdy.20516.

Kawakami, K., Abe, G., Asada, T., Asakawa, K., Fukuda, R., Ito, A., … Yoshida, M. (2010). zTrap: zebrafish gene trap and enhancer trap database. *BMC Developmental Biology, 10*(1), 105. http://dx.doi.org/10.1186/1471-213X-10-105.

Kim, D., Pertea, G., Trapnell, C., Pimentel, H., Kelley, R., & Salzberg, S. L. (2013). TopHat2: accurate alignment of transcriptomes in the presence of insertions, deletions and gene fusions. *Genome Biology, 14*(4), R36. http://dx.doi.org/10.1186/gb-2013-14-4-r36.

Kim, J. H., Lee, S.-R., Li, L.-H., Park, H.-J., Park, J.-H., Lee, K. Y., … Choi, S.-Y. (2011). High cleavage efficiency of a 2A peptide derived from porcine teschovirus-1 in human cell lines, zebrafish and mice. *PLoS One, 6*(4), e18556. http://dx.doi.org/10.1371/journal.pone.0018556.

Kwan, K. M., Fujimoto, E., Grabher, C., Mangum, B. D., Hardy, M. E., Campbell, D. S., … Chien, C.-B. (2007). The Tol2kit: a multisite gateway-based construction kit for Tol2 transposon transgenesis constructs. *Developmental Dynamics : An Official Publication of the American Association of Anatomists, 236*(11), 3088–3099. http://dx.doi.org/10.1002/dvdy.21343.

Langmead, B., & Salzberg, S. L. (2012). Fast gapped-read alignment with Bowtie 2. *Nature Methods, 9*(4), 357–359. http://dx.doi.org/10.1038/nmeth.1923.

Li, J.-W., Schmieder, R., Ward, R. M., Delenick, J., Olivares, E. C., & Mittelman, D. (2012). SEQanswers: an open access community for collaboratively decoding genomes. *Bioinformatics (Oxford, England), 28*(9), 1272–1273. http://dx.doi.org/10.1093/bioinformatics/bts128.

Love, M. I., Huber, W., & Anders, S. (2014). Moderated estimation of fold change and dispersion for RNA-seq data with DESeq2. *Genome Biology, 15*(12), 550. http://dx.doi.org/10.1186/PREACCEPT-8897612761307401.

Mammoto, A., Mammoto, T., & Ingber, D. E. (2012). Mechanosensitive mechanisms in transcriptional regulation. *Journal of Cell Science, 125*(13), 3061–3073. http://dx.doi.org/10.1242/jcs.093005.

McClure, C. D., & Southall, T. D. (2015). Getting down to specifics: profiling gene expression and protein-DNA interactions in a cell type-specific manner. *Advances in Genetics, 91*, 103–151. http://dx.doi.org/10.1016/bs.adgen.2015.06.003.

Miller, M. R., Robinson, K. J., Cleary, M. D., & Doe, C. Q. (2009). TU-tagging: cell type-specific RNA isolation from intact complex tissues. *Nature Methods, 6*(6), 439–441. http://dx.doi.org/10.1038/nmeth.1329.

Mosimann, C., Puller, A.-C., Lawson, K. L., Tschopp, P., Amsterdam, A., & Zon, L. I. (2013). Site-directed zebrafish transgenesis into single landing sites with the phiC31 integrase system. *Developmental Dynamics : An Official Publication of the American Association of Anatomists, 242*(8), 949–963. http://dx.doi.org/10.1002/dvdy.23989.

Obholzer, N., Wolfson, S., Trapani, J. G., Mo, W., Nechiporuk, A., Busch-Nentwich, E., … Nicolson, T. (2008). Vesicular glutamate transporter 3 is required for synaptic transmission in zebrafish hair cells. *The Journal of Neuroscience : The Official Journal of the Society for Neuroscience, 28*, 2110–2118. http://dx.doi.org/10.1523/JNEUROSCI.5230-07.2008.

Parnell, L. D., Lindenbaum, P., Shameer, K., Dall'Olio, G. M., Swan, D. C., Jensen, L. J., … Albert, I. (2011). BioStar: an online question & answer resource for the bioinformatics community. *PLoS Computational Biology, 7*(10), e1002216. http://dx.doi.org/10.1371/journal.pcbi.1002216.

Raymond, P. A., Barthel, L. K., & Curran, G. A. (1995). Developmental patterning of rod and cone photoreceptors in embryonic zebrafish. *The Journal of Comparative Neurology, 359*(4), 537–550. http://dx.doi.org/10.1002/cne.903590403.

Roberts, J. A., Miguel-Escalada, I., Slovik, K. J., Walsh, K. T., Hadzhiev, Y., Sanges, R., … Müller, F. (2014). Targeted transgene integration overcomes variability of position effects in zebrafish. *Development (Cambridge, England), 141*(3), 715–724. http://dx.doi.org/10.1242/dev.100347.

Sanz, E., Yang, L., Su, T., Morris, D. R., McKnight, G. S., & Amieux, P. S. (2009). Cell-type-specific isolation of ribosome-associated mRNA from complex tissues. *Proceedings of the National Academy of Sciences of the United States of America, 106*(33), 13939–13944. http://dx.doi.org/10.1073/pnas.0907143106.

Scott, E. K. (2009). The Gal4/UAS toolbox in zebrafish: new approaches for defining behavioral circuits. *Journal of Neurochemistry, 110*(2), 441–456. http://dx.doi.org/10.1111/j.1471-4159.2009.06161.x.

Scott, E. K., & Baier, H. (2009). The cellular architecture of the larval zebrafish tectum, as revealed by gal4 enhancer trap lines. *Frontiers in Neural Circuits, 3*, 13. http://dx.doi.org/10.3389/neuro.04.013.2009.

Steiner, F. A., Talbert, P. B., Kasinathan, S., Deal, R. B., & Henikoff, S. (2012). Cell type-specific nuclei purification from whole animals for genome-wide expression and chromatin profiling. *Genome Research, 22*(206), 766–777. http://dx.doi.org/10.1101/gr.131748.111.

Szymczak, A. L., Workman, C. J., Wang, Y., Vignali, K. M., Dilioglou, S., Vanin, E. F., & Vignali, D. A. A. (2004). Correction of multi-gene deficiency in vivo using a single "self-cleaving" 2A peptide-based retroviral vector. *Nature Biotechnology, 22*(5), 589–594. http://dx.doi.org/10.1038/nbt957.

Tallafuss, A., Kelly, M., Gay, L., Gibson, D., Batzel, P., Karfilis, K. V., … Washbourne, P. (2015). Transcriptomes of post-mitotic neurons identify the usage of alternative pathways during adult and embryonic neuronal differentiation. *BMC Genomics, 16*(1), 1100. http://dx.doi.org/10.1186/s12864-015-2215-8.

Tallafuss, A., Washbourne, P., & Postlethwait, J. (2014). Temporally and spatially restricted gene expression profiling. *Current Genomics, 15*(4), 278–292. http://dx.doi.org/10.2174/1389202915666140602230106.

Thisse, B., & Thisse, C. (2004). *Fast release clones: A high throughput expression analysis.* ZFIN Direct Data Submission http://zfin.org.

Thisse, B., & Thisse, C. (2014). In situ hybridization on whole-mount zebrafish embryos and young larvae. *Methods in Molecular Biology (Clifton, NJ), 1211*, 53–67. http://dx.doi.org/10.1007/978-1-4939-1459-3_5.

Tryon, R. C., Pisat, N., Johnson, S. L., & Dougherty, J. D. (2013). Development of translating ribosome affinity purification for zebrafish. *Genesis (New York, NY: 2000), 51*(3), 187–192. http://dx.doi.org/10.1002/dvg.22363.

Windhager, L., Bonfert, T., Burger, K., Ruzsics, Z., Krebs, S., Kaufmann, S., … Dölken, L. (2012). Ultra short and progressive 4sU-tagging reveals key characteristics of RNA processing at nucleotide resolution. *Genome Research.* http://dx.doi.org/10.1101/gr.131847.111.

Zeiner, G. M., Cleary, M. D., Fouts, A. E., Meiring, C. D., Mocarski, E. S., & Boothroyd, J. C. (2008). RNA analysis by biosynthetic tagging using 4-thiouracil and uracil phosphoribosyltransferase. *Methods in Molecular Biology (Clifton, NJ), 419*, 135–146. http://dx.doi.org/10.1007/978-1-59745-033-1_9.

CHAPTER

RT-qPCR gene expression analysis in zebrafish: preanalytical precautions and use of expressed repetitive elements for normalization

17

S. Vanhauwaert, S. Lefever, P. Coucke, F. Speleman, A. De Paepe, J. Vandesompele, A. Willaert[1]

Ghent University, Ghent, Belgium

[1]*Corresponding author: E-mail: Andy.Willaert@UGent.be*

CHAPTER OUTLINE

Methods in Cell Biology, Volume 135, ISSN 0091-679X, http://dx.doi.org/10.1016/bs.mcb.2016.02.002

Abstract

Gene expression analysis is increasingly important in many fields of biological research. Understanding patterns of expressed genes is assumed to provide insight into complex regulatory networks and can lead to the identification of genes relevant to specific biological processes, including disease. Among different techniques, reverse transcription quantitative polymerase chain reaction (RT-qPCR) is currently regarded as the gold standard for targeted quantification of RNA gene expression, especially because of its high sensitivity, specificity, accuracy, and precision, and also because of its practical simplicity and processing speed. However, different critical factors can influence the outcome of RT-qPCR studies, including isolation of RNA, reverse transcription to cDNA, and data analysis. These factors need to be addressed in order to obtain biologically meaningful results. In this chapter, we describe how RT-qPCR can be used in a reliable way to successfully study differential gene expression in zebrafish. Hereby, we especially focus on how expressed repetitive elements can be employed as reference targets in zebrafish RT-qPCR studies and how they can further improve the quality of the data.

INTRODUCTION

Reverse transcription quantitative polymerase chain reaction (RT-qPCR) is currently considered the gold standard for accurate, sensitive, and fast measurement of gene expression. Due to its apparent simplicity and processing speed, inexperienced users can also rapidly produce results. However, numerous critical issues in the preanalytical steps for RT-qPCR analysis, including variable yields of RNA extraction and reverse transcription and also variable amplification efficiencies can greatly affect RT-qPCR results (Bustin et al., 2009; Derveaux, Vandesompele, & Hellemans, 2010). Therefore, in order to measure true biological variation, it is important to control and also correct for this technically induced variation. In the first section of this chapter, zebrafish sample collection, sample homogenization, RNA extraction, and cDNA synthesis are described and it is shown how these steps can influence RT-qPCR results and how their variability can be controlled. In the second section, PCR assay efficiency and normalization of zebrafish RT-qPCR data is discussed. To measure only the biological variation in samples, it is important to correct for technical variation by applying a good normalization strategy. The use of multiple reference genes as internal controls is the most frequently applied and recommended procedure for normalizing RT-qPCR data (Dheda et al., 2005; Goossens et al., 2005; Kim et al., 2004; Tricarico et al., 2002; Vandesompele et al., 2002). In this respect, specific attention should be given to the correct selection and validation of reference genes for normalization, as stated in the MIQE (Minimum Information for Publication of Quantitative Real-Time PCR Experiments guidelines (Bustin et al., 2009). The selected reference genes should be stably expressed in the studied samples and should thus show a strong correlation with the total amount of mRNA present in the samples. Importantly, many commonly used reference genes in zebrafish expression studies are not generally applicable as their expression stability greatly varies under different experimental conditions, introducing bias and noise (Casadei et al., 2011; McCurley & Callard, 2008; Tang et al., 2007). Therefore, in this chapter, we propose a novel accurate method of normalization using expressed repetitive elements (EREs) (Vanhauwaert et al., 2014). EREs generally outperform classically used reference genes as reference targets for normalization in a wide range of zebrafish RT-qPCR studies. Collectively, the described methods and approaches in this chapter show that standardization and quality control is essential throughout the preanalytical steps of the RT-qPCR workflow in zebrafish expression studies.

1. SAMPLE COLLECTION, RNA EXTRACTION AND QUALITY CONTROL, AND REVERSE TRANSCRIPTION

Sample collection is one of the most important steps in RT-qPCR since it is the first potential source of experimental variability. Correct sample collection is essential to prevent RNA degradation that negatively impacts the outcome of an RT-qPCR experiment (Bustin et al., 2009). In many cases it is useful to collect all samples before proceeding to RNA extraction. In this way, bias introduced by different handling of the samples can be prevented. We have observed that for zebrafish embryos and early larvae (up till 14 days post fertilization (dpf)) storage in RNAlater (Sigma-Aldrich, St. Louis, United States) and for adult zebrafish snap freezing of tissue and organs in liquid nitrogen provides the best preservation and quality of RNA. When samples are collected in RNAlater it is vital to store them properly. At room temperature, they can only be stored for 1–2 weeks, while at 4°C they can be stored for months. After snap freezing, it is of major importance that the samples are never thawed before the actual cell lysis preceding RNA extraction, since this substantially decreases the RNA quality.

Prior to nucleic acid extraction it is important to properly homogenize the samples. For zebrafish embryos and early larvae (up till 14 dpf) it is sufficient to submerge the embryos in TRIzol reagent and homogenize them immediately with a pestle or pass them through an 18G syringe. For adult zebrafish tissues and organs, two approaches can be used that, in our experience, both result in good RNA yield and quality. A first procedure involves the transfer of tissues to TRIzol or Qiazol reagent followed by homogenization using a TissueLyser (Qiagen) or similar instrument. According to a second procedure, tissues are imbedded in optimal cutting temperature medium, cut into 50-μm cryosections, and then transferred to TRIzol reagent. In the latter protocol, it is very important that all handling is performed at −20°C to prevent RNA degradation. Nucleic acid extraction is the next crucial preparation step in the RT-qPCR workflow, since extraction efficiency and RNA quality depends on the type and physiological state of the samples and on the amount of biomaterial to be processed (Morrogh et al., 2007). Therefore, it is recommended to process all the samples that are intended to be used in a specific expression study, in a similar way prior to RNA extraction and to perform the RNA extraction itself with the same method/kit. One method that provides good-quality RNA from zebrafish samples is the TRIzol/RNeasy hybrid RNA extraction protocol, where TRIzol-based RNA extraction is followed by Qiagen RNeasy purification of the isolated RNA. Of note, there is no RNA isolation method that generates RNA that is completely free of genomic DNA contamination. Subsequently, it is advisable to treat the RNA samples with a DNase for removal of contaminating DNA. The efficiency of this procedure can be tested by a careful check for absence of DNA through qPCR analysis of a DNA target on the extracted RNA. In addition, the amplification of potential contaminating genomic DNA can also be excluded by designing assays that span large introns (>700 nucleotides) such that residual DNA is not coamplified in the RT-qPCR analysis. This works only if no processed pseudogenes (retropseudogenes) exist for the gene of interest.

When limited RNA is available, one could consider preamplification of the samples (Derveaux et al., 2010). RNA preamplification methods address this issue by producing micrograms of cDNA starting from a few nanograms of total RNA. Previously, *Vermeulen* et al. could clearly demonstrate that differential expression is preserved after whole transcriptome preamplification and that no substantial bias is introduced, which further proves the usefulness of this approach (Vermeulen et al., 2009).

It is essential to assess the quality of the RNA before proceeding to cDNA synthesis and RT-qPCR. Measurement of the A_{260}/A_{280} absorbance ratio provides a rough indication of the RNA purity since the presence of DNA or phenol alters the ratio (ratio of ~2 for pure RNA in pH neutral buffer). For assessment of RNA integrity, we recommend an electrophoresis-based or a microfluidics-based rRNA analysis method using an Experion (Bio-Rad), Bioanalyzer (Agilent), Fragment Analyzer (Advanced Analytics), or other device (Fleige & Pfaffl, 2006). Compromised RNA quality is known to lead to unreliable results in gene expression studies as it has an impact on the variability of reference genes and significance of differential expression (De Preter et al., 2011).

Synthesis of cDNA can introduce substantial variation to RT-qPCR results (Stahlberg et al., 2004; Stahlberg, Kubista, & Pfaffl, 2004). It can introduce errors produced by effects of RNA secondary and tertiary structures, by variation in efficiency of the priming and by the variable activity of the reverse transcriptase. Reverse transcription yields may vary up to 100-fold depending on the choice of reverse transcriptase and on the target gene (Stahlberg et al., 2004). As a result, it is suggested that all samples are processed with the same reverse transcriptase kit and that the total RNA input is identical in every sample. It may also be advisable to perform RT replicates instead of or on top of PCR replicates.

Here, we describe detailed protocols for sample collection and sample homogenization for RT-qPCR in zebrafish samples. Procedures for cDNA synthesis, RNA extraction, and RNA quality measurements are not described here, since comprehensive manuals are delivered with the kits and devices that are used for this purpose.

1.1 METHOD A: SAMPLE COLLECTION ZEBRAFISH EMBRYOS AND TISSUES

1.1.1 Reagents

- Ethyl 3-aminobenzoate methanesulfonate salt (Tricaine, MS-222) (A5040-25G, Sigma-Aldrich)
- Embryos or adult zebrafish
- RNAlater (R0901-100ML, Sigma-Aldrich)
- Qiazol lysis reagent (79306, Qiagen, Hilden, Germany) or TRIzol (Invitrogen, 15596-025)
- Water ice (0°C)
- Liquid nitrogen

1.1.2 Protocol

1. For embryo collection, euthanize embryos (10–15) by adding an overdose of tricaine (0.4% or 400 mg/100 mL) and transfer the embryos to a new eppendorf tube. Remove as much medium as possible and add dropwise a sufficient amount (≥50 μL) of RNAlater. Store at 4°C (RNA remains stable up to 1 month). Prior to RNA extraction, remove RNAlater using a glass pipet and add 700 μL Qiazol or TRIzol reagent.
2. For collection of adult zebrafish tissues and organs, euthanize the adult fish by adding an overdose of tricaine (0.4% or 400 mg/100 mL). Transfer the euthanized fish immediately to ice to prevent RNA degradation. Dissect the desired tissue/organ, transfer the tissue/organ to an eppendorf tube, and put the sample immediately in liquid nitrogen (snap freezing). Store at −80°C.

1.2 METHOD B: TISSUE HOMOGENIZATION USING THE TISSUELYSER

1.2.1 Reagents/devices

- TissueLyser (85300, Qiagen)
- Stainless-steel beads 5 mm (69989, Qiagen)
- Qiazol lysis reagent (79306, Qiagen)

1.2.2 Protocol

1. Remove tubes with zebrafish tissues/organs from −80°C and keep on dry ice until further processing. Place at room temperature (15–25°C) and immediately add 700 μL of Qiazol or TRIazol reagent to each tube. Add one stainless-steel bead per tube.

Note: Do not use Qiazol lysis reagents with tungsten carbide beads, as these buffers can react with the bead and damage the bead (and contaminate the samples).

2. Place the tubes in the TissueLyser adapter set 2 × 24 (if using 2-mL tubes) or the TissueLyser adapter set 2 × 96 (if using 1.2-mL tubes). Ensure equal division of the samples between both sides of the TissueLyser. Operate the TissueLyser for

2 min at 20–30 Hz. Disassemble the adaptor set and rotate the rack of the tubes so that the tubes nearest to the TissueLyser are now outermost. Operate the TissueLyser for another 2 min at 20–30 Hz. The duration of disruption and homogenization depends on the tissue being processed and can be extended until no tissue debris are visible.

3. Proceed with RNA extraction as described by the manufacturer. Do not reuse the stainless-steel beads. To keep the quality and quantity of the RNA extraction as high as possible, ignore the presence of the bead in the sample and proceed with the extraction protocol immediately.

1.3 METHOD C: TISSUE HOMOGENIZATION USING CRYOTOME

1.3.1 Reagents

- OCT (LAMB-OCT-USA, ThermoScientific, Waltham, United States)
- Cryotome blades (20750003 feather microtome blade C35, pfm medical, Cologne, Germany)
- Liquid nitrogen

1.3.2 Protocol

1. Switch on and cool cryostat to −20°C. Remove the tubes with zebrafish tissues/organs samples from −80°C and transfer the samples to liquid nitrogen. Put a drop of OCT on the metal grid that fits the cryostat. Put the metal grid in the cryostat and let the OCT freeze.
2. Remove a sample from the liquid nitrogen, take the metal grid with OCT, and add the sample. Add OCT to the sample till the sample is completely covered. Place the metal grid with the sample on the cryobar and cut ±15 50 μm sections. Transfer the sections to 700 μL Qiazol or TRIazol and snap freeze the sample or immediately proceed with RNA isolation.

2. ASSAY DESIGN AND QUALITY CONTROL

The design and validation of an RT-qPCR assay is another crucial preanalytical step to obtain robust and accurate results. First of all, the assay should be highly efficient and also specific for the target sequence. This strongly depends on primer design. A number of free and commercially available software programs are available for primer design. A popular primer design program that is free of charge is Primer3 (http://bioinfo.ut.ee/primer3/). A more elaborate pipeline, which can also be used for zebrafish RT-qPCR assay design, is the primerXL tool (www.primerxl.org). PrimerXL is a Primer3-based high-throughput primer design pipeline with in silico assay evaluation. It performs some essential downstream primer pair evaluations such as specificity and secondary structure and single nucleotide polymorphism assessment in primer annealing sites.

It is strongly recommended to validate the specificity of the assay by doing an actual RT-qPCR experiment and inspecting the length and the sequence of the amplicon, using gel electrophoresis or melting curve analysis and by performing sequencing of the amplicon. In addition, the PCR efficiency of the primers needs to be determined. An optimal primer efficiency of 100% means that the product of interest is doubling with each cycle; in reality, this is often not the case. A simple way to determine PCR efficiencies is to run serial dilutions of a template and use the results to generate a standard curve. This standard curve is constructed by plotting the log of the starting quantity against the Cq value obtained during amplification of each dilution. The equation of the linear regression line, along with the Pearson correlation coefficient (r) of the coefficient of determination (R^2), can then be used to evaluate PCR efficiencies. The more dilution points and the wider the range (dilution factor), the more precisely can the PCR efficiency be determined. It is best to use a mixture of representative samples as input material for the dilution series. For zebrafish, a standard dilution series can be made using a mixture of RNA derived from zebrafish embryos from different developmental time points and/or from several zebrafish organs and tissues. Although standard dilution series are frequently used to test primer efficiencies, PCR efficiency values can also be directly obtained from the observed data. Throughout the years several algorithms have been developed (e.g. LinRegPCR and PCR-miner) for this purpose. LinRegPCR calculates PCR efficiency values per sample by fitting a regression line to a subset of data points in the log-linear phase (Ruijter et al., 2009). It was shown that the most reliable efficiency value is obtained when PCR efficiencies of different samples for the same amplicon were averaged.

Finally, an important aspect of assay design is to select an appropriate target sequence, which in many cases is not easy to do, especially for genes that have alternatively spliced isoforms. If no prior knowledge is available on the function of these transcripts, one may decide to design an assay for a part of the transcript that is expressed in most or all the isoforms.

Here, we describe a detailed protocol for the determination of PCR efficiencies in zebrafish expression studies using standard curve analysis.

2.1 METHOD D: ASSESSMENT OF PCR EFFICIENCY USING STANDARD CURVE ANALYSIS

2.1.1 Reagents

- Template: zebrafish RNA from different time points (2 hours post fertilization (hpf), 8 hpf, 24 hpf, 48 hpf, 72 hpf, 96 hpf, 120 hpf)
- Carrier nucleic acid: tRNA from brewer's yeast (Roche #10109517001, 100 mg, lyophilized, dissolve to 10 mg/mL)
- Nuclease-free water (Sigma #W-4502)
- iScript cDNA synthesis kit (Bio-Rad #170-8891)

2.1.2 Protocol

1. To perform cDNA synthesis, mix the same amounts of DNA-free zebrafish RNA from different developmental time points in an RNAse-free 0.2-mL tube to obtain 10 μg of pooled zebrafish RNA. Add 20 μL of iScript reaction mix and 5 μL of iScript reverse transcriptase and fill to 100 μL with nuclease-free water. Incubate 5 min at 25°C followed by 30 min at 42°C and 5 min at 85°C.
2. Prepare 50X and 1X carrier tRNA solutions as follows: Dilute 0.9 μL carrier with 35 μL nuclease-free water to obtain a 50X carrier solution (250 ng/μL) and vortex thoroughly. Dilute 29 μL 50X solution with 1421 μL nuclease-free water to obtain a 1X carrier solution (5 ng/μL) and vortex thoroughly.
3. Make a cDNA dilution series (Table 1):
   a. Dilute 100 μL of cDNA with 6.25 μL 50X carrier and 206.25 μL H20 (point 1: 32 ng/μL cDNA and 5 ng/μL carrier), vortex thoroughly
   b. Dilute 78 μL of the point 1 solution with 234 μL 1X carrier (point 2: 8 ng/μL cDNA and 5 ng/μL carrier), vortex thoroughly
   c. Repeat the previous steps four times to obtain the dilution points 3 to 6 (see Table 1)
   d. Add 240 μL of 1X carrier to a tube marked as negative control
4. Cq values (dependent variable, Y-axis) of the cDNA dilution series are plotted against the cDNA quantity (independent variable, X axis). The slope of the standard curve can be calculated by linear regression and, subsequently, from the value of the slope, PCR efficiencies can be calculated according to the following formula: PCR efficiency = $10^{-1/\text{slope}}-1$(Bustin et al., 2009; Hellemans et al., 2007). In general, efficiency values between 90% and 110% are considered as sufficient for reliable quantification. PCR efficiencies can also be calculated automatically using software such as qbase+, a tool for automated data analysis of real-time PCR experiments, including error propagation, determination of reference gene stability, multiple reference gene normalization, interrun calibration, and basic statistical analysis (http://www.qbaseplus.com) (Hellemans et al., 2007).

Table 1 cDNA Standard Dilution Series

Final cDNA Concentration	Starting Volume	Diluted With
Point 1: 32 ng/μL	100 μL cDNA	6.25 μL 50X carrier + 206.25 μL nuclease-free water
Point 2: 8 ng/μL	78 μL point 1	234 μL 1X carrier
Point 3: 2 ng/μL	78 μL point 2	234 μL 1X carrier
Point 4: 0.5 ng/μL	78 μL point 3	234 μL 1X carrier
Point 5: 0.125 ng/μL	78 μL point 4	234 μL 1X carrier
Point 6: 0.03125 ng/μL	78 μL point 5	234 μL 1X carrier
Negative control	240 μL 1X carrier	-

2.1.3 Notes

- Prevent repeated freeze–thaw cycles of cDNA dilution series aliquots (maximum 10 times).
- Handle RNA on ice.
- Vortex and spin down cDNA upon thawing.

3. REFERENCE TARGET VALIDATION

The final step before an RT-qPCR experiment can be performed is the identification and selection of reference genes for data normalization. The purpose of normalization is to remove or reduce the technically or experimentally induced variation, ending up with only the true biological variation. Technical variation can be introduced because of variable input quantity (RNA quantity, cDNA synthesis efficiency, see Section 2) and input quality (RNA integrity, RNA purity, etc., see Section 2). While the use of endogenous reference genes is the most universal and most appropriate method for normalization, it is important to identify and select (at least two) reference genes that are stably expressed throughout the studied samples and thus show a strong correlation with the total amount of mRNA present in the samples (Fig. 1) (Dheda et al., 2005; Goossens et al., 2005; Kim et al., 2004; Tricarico et al., 2002; Vandesompele et al., 2002). Importantly, many commonly used reference genes are not generally applicable as their expression stability greatly varies under different experimental conditions (Dhorne-Pollet, Thelie, & Pollet, 2013; Jacob et al., 2013; Ledderose et al., 2011; McCurley & Callard, 2008). Therefore, it is essential to determine the optimal number and choice of reference genes for the specific experimental conditions in every study. A number of studies have measured and compared the expression stability of a set of commonly used reference genes in samples derived from different zebrafish organs, developmental stages, and treatments, using one of the available tools that automatically calculate expression stability values (geNorm (available in qbase+), BestKeeper, Normfinder) (Casadei et al., 2011; Dhorne-Pollet et al., 2013; Jacob et al., 2013; Ledderose et al., 2011; McCurley & Callard, 2008; Tang et al., 2007). These studies propose the set of most stable reference genes as being the most suitable for normalizing gene expression data. However, the determination of stable reference genes only occurs in a comparative fashion and the detection of the "most stably" expressed genes does not necessarily mean they are stably expressed in other conditions. In the zebrafish, studying developmental time series and comparing different tissues are particularly challenging to normalize, most likely because of dramatic changes in expression patterns during zebrafish development and major differences in expression between different matured organs (Abramsson et al., 2010; Dhorne-Pollet et al., 2013; Hellemans et al., 2007; McCurley & Callard, 2008; Yang et al., 2013). Therefore, the ideal situation of using only one set of reference genes to cover all experimental conditions in the zebrafish has not been feasible when using the traditional reference genes. This is due to variations in expression levels of these genes across different experimental

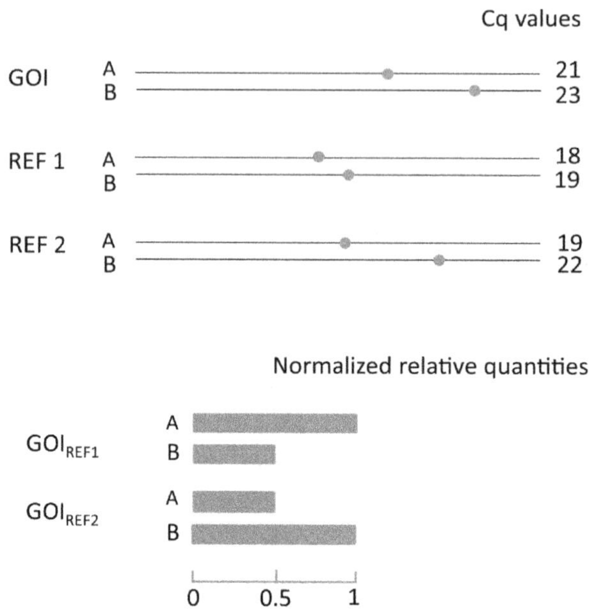

FIGURE 1

Schematic overview of RT-qPCR normalization. A gene of interest (GOI) in two different conditions (A and B) is depicted together with two unvalidated reference genes (REF1 and REF2). Depending on which reference gene is used for normalization, a four fold difference in expression can be observed. This shows that the usage of multiple validated reference genes is critical to achieve more reliable expression data.

conditions, developmental stages, or across different tissues. Inappropriate use of these reference genes may lead to false interpretation of expression data and unreliable conclusions. To tackle the aforementioned issues, we developed a novel normalization method for zebrafish expression studies (Vanhauwaert et al., 2014). This normalization method uses expressed repetitive elements (ERE) as reference targets instead of protein coding mRNAs.

4. EXPRESSED REPEAT RT-qPCR NORMALIZATION IN ZEBRAFISH

The underlying strategy for the development of a new RT-qPCR normalization approach in zebrafish expression studies was based on the identification of specific types of repetitive elements that have spread throughout the zebrafish genome during evolution and that are also present in genomic sequences that are transcribed to RNA (Vanhauwaert et al., 2014). With a single pair of RT-qPCR primers, one specific ERE can be amplified, thereby simultaneously detecting numerous different transcripts in which the specific ERE is present (Fig. 2). When mapping

FIGURE 2

Schematic representation of ERE amplification (hypothetical example). The full-length consensus repeat element (dark gray line, top) and a number of aligned ERE sequence-containing fragments present in different transcripts (each line represents a different transcript) are depicted. Sequences that are commonly present in most of the fragments are used as a template for qPCR amplification. It was shown that the amplified ERE fragments are exclusively located in untranslated gene regions.

the primers on the zebrafish transcriptome (Ensembl release 82, *Danio rerio* build GRCz10) using Bowtie (version 0.12.7), a median number of 20–30 target genes are predicted to be amplified using a single ERE assay, considering a maximum of two or three mismatches per primer, respectively (Table 2). The underlying assumption is that by measuring many transcripts at the same time, differential expression of a few of them will not drastically alter the total level of ERE expression. Therefore, expression of this set of repeats is expected to be highly stable throughout different experimental situations, as it serves as an estimation of the general mRNA fraction abundance. When the expression stability of a number of EREs (Table 2) is compared to that of commonly used zebrafish reference genes such as *gapdh*, *elfa,* and *bactin2* in a diverse set of experimental conditions including a developmental time series, a set of different organs from adult fish, and different treatments of zebrafish embryos including morpholino injections and administration of chemicals, it was demonstrated that EREs indeed have a higher overall expression stability. Moreover, a limited set of ERE reference targets (*hatn10*, *dna15ta1,* and loopern4) that show stable expression throughout the wide range of experiments in this study can be recommended as strong candidates for evaluation of their suitability as normalization targets for qPCR normalization in zebrafish expression studies. Nevertheless, it remains important to assess the expression stability of the selected reference genes in the actual experiment.

Table 2 Overview of EREs and Commonly Used Reference Genes in Zebrafish Expression Studies and Their Corresponding Primer Sequence

Reference Target	Forward Primer	Reverse Primer	Predicted Number of Targets (a)	Predicted Number of Targets (b)
Tc1n1	TGTCTGGGTTGGTGTTGTAT	GCTCTGTCGACTTTTGATGT	31	17
Dna11ta1	GGGACAACATGAAGGAATTGT	AAAAATGCAGGGTTCCACACA	53	21
Tdr7	GCAGCATAATTGAGTACACCC	TTGCCTATATTCACTGAGAAATGGA	27	8
Dna15ta1	TACTGTGCTCAAATTGCTTCA	AATGAGTACTGTGAACTTAATCCAT	30	19
Cr1-1	GCTCTTCAGTGTTTGAACTCTCAGT	CAATGTAGATTGTGTCAAAGCAG	26	19
Hatn8	CAATGACGGTTGGGGTTAGG	TTTAAAAAGGAGGCGTGGCA	93	53
Hatn10	TGAAGACAGCAGAAGTCAATG	CAGTAAACATGTCAGGCTAAATAA	31	24
Hatn4	ACCCTGATCAAACACACCTG	TCAAGTGCTGTTCAGGTCCTA	216	114
Loopern4	TGAGCTGAAACTTTACAGACACAT	AGACTTTGGTGTCTCCAGAATG	16	11
Sine3	GGAGACCACATGGGAAAACT	AGAGTCAGGACCTCGGTTTA	20	13
Tuba1	TCATCTTCTCCTTCCACACT	GTACGTGGGTGAGGGTAT	ND[a]	ND[a]
Tbp	AAGTTTACGGTGGACACAAT	CAGGCAACACACCACTTTAT	ND[a]	ND[a]
B2m	ACGCTGCAGGTATATTCATC	TCTCCATTGAACTGCTGAAG	ND[a]	ND[a]
Elfa	GGAGACTGGTGTCCTCAA	GGTGCATCTCAACAGACTT	ND[a]	ND[a]
Cyp19a1b	AAGGCCATCCTAGTAACCAT	GGTTGTTGGTCTGTCTGATG	ND[a]	ND[a]
Bactin2	ACGATGGATGGGAAGACA	AAATTGCCGCACTGGTT	ND[a]	ND[a]
Rpl13a	AGGCTGAAGGTGTTTGATG	TTTCAGACGCACAATCTTGA	ND[a]	ND[a]
Hprt1	GAGGAGCGTTGGATACAGA	CTCGTTGTAGTCAAGTGCAT	ND[a]	ND[a]
Rps18	AGTTCTCCAGCCCTCTTATT	TCAACACGAACATTGATGGA	ND[a]	ND[a]
Gapdh	GTGGAGTCTACTGGTGTCTTC	GTGCAGGAGGCATTGCTTACA	ND[a]	ND[a]

The predicted number of targets is given for all EREs with (a) maximum three mismatches per primer pair and (b) maximum two mismatches per primer pair.
[a] The predicted number of targets is not determined for common reference genes, as primers were designed with primerXL software, which generates designs of which the number of targets is only 1.

This can be easily done in the qbase + software by calculating the expression stability measure (geNorm) M.

In case the proposed set of ERE reference targets (*hatn10*, *dna15ta1*, and *loop-ern4*) does not show stable expression in a specific zebrafish expression study, it is strongly recommended to set up a pilot experiment to select another set of reference targets. A typical pilot experiment for evaluation of reference gene expression stability measures around 10 candidate reference targets (Table 2 lists primer sequences for ERE and classic reference genes) in a dozen samples that are representative for the specific experiment. Identification of stably expressed reference targets across this set of samples can be done using geNorm or similar algorithms. The geNorm algorithm is available as a module in qbase+, as of version 2.0 (http://www.qbaseplus.com). As input for geNorm analysis, either Cq values or efficiency-corrected Cq values from LinRegPCR can be used. GeNorm calculates the gene expression stability measure M for a reference gene as the average pairwise variation V for that gene with all other tested reference genes. Stepwise exclusion of the gene with the highest M value allows accurate ranking of the reference genes according to their expression stability in the samples. Reference targets with M values below 0.5 and 0.2 are considered having a "high" and "very high" expression stability, respectively (Hellemans et al., 2007). Moreover, geNorm will also determine the optimal number of reference genes that should be included in the experiment, to remove most of the technical variation. The geNorm algorithm determines the pairwise variation Vn/*n*+1, between two sequential normalization factors containing an increasing number of reference targets. A large variation means that the added reference target has a significant effect and should preferably be included for calculation of a reliable normalization factor. Vandesompele et al. (2002) used 0.15 as the cutoff value, below which the inclusion of an additional reference gene is not required. Typically, between two and five reference targets are required for accurate normalization, depending on the heterogeneity in the sample set. It has been shown that the use of a single reference gene can lead to incorrect data with an error up to threefold in 25% of the cases (Vandesompele et al., 2002).

REFERENCES

Abramsson, A., et al. (2010). Proteomics profiling of single organs from individual adult zebrafish. *Zebrafish, 7*(2), 161–168.

Bustin, S. A., et al. (2009). The MIQE guidelines: minimum information for publication of quantitative real-time PCR experiments. *Clinical Chemistry, 55*(4), 611–622.

Casadei, R., et al. (2011). Identification of housekeeping genes suitable for gene expression analysis in the zebrafish. *Gene Expression Patterns, 11*(3–4), 271–276.

De Preter, K., et al. (2011). miRNA expression profiling enables risk stratification in archived and fresh neuroblastoma tumor samples. *Clinical Cancer Research, 17*(24), 7684–7692.

Derveaux, S., Vandesompele, J., & Hellemans, J. (2010). How to do successful gene expression analysis using real-time PCR. *Methods, 50*(4), 227–230.

Dheda, K., et al. (2005). The implications of using an inappropriate reference gene for real-time reverse transcription PCR data normalization. *Analytical Biochemistry, 344*(1), 141–143.

Dhorne-Pollet, S., Thelie, A., & Pollet, N. (2013). Validation of novel reference genes for RT-qPCR studies of gene expression in *Xenopus tropicalis* during embryonic and post-embryonic development. *Developmental Dynamics, 242*(6), 709–717.

Fleige, S., & Pfaffl, M. W. (2006). RNA integrity and the effect on the real-time qRT-PCR performance. *Molecular Aspects of Medicine, 27*(2–3), 126–139.

Goossens, K., et al. (2005). Selection of reference genes for quantitative real-time PCR in bovine preimplantation embryos. *BMC Developmental Biology, 5*, 27.

Hellemans, J., et al. (2007). qBase relative quantification framework and software for management and automated analysis of real-time quantitative PCR data. *Genome Biology, 8*(2), R19.

Jacob, F., et al. (2013). Careful selection of reference genes is required for reliable performance of RT-qPCR in human normal and cancer cell lines. *PLoS One, 8*(3), e59180.

Kim, B. S., et al. (2004). Spearman's footrule as a measure of cDNA microarray reproducibility. *Genomics, 84*(2), 441–448.

Ledderose, C., et al. (2011). Selection of reliable reference genes for quantitative real-time PCR in human T cells and neutrophils. *BMC Research Notes, 4*, 427.

McCurley, A. T., & Callard, G. V. (2008). Characterization of housekeeping genes in zebrafish: male-female differences and effects of tissue type, developmental stage and chemical treatment. *BMC Molecular Biology, 9*, 102.

Morrogh, M., et al. (2007). Tissue preparation for laser capture microdissection and RNA extraction from fresh frozen breast tissue. *Biotechniques, 43*(1), 41–42, 44, 46 passim.

Ruijter, J. M., et al. (2009). Amplification efficiency: linking baseline and bias in the analysis of quantitative PCR data. *Nucleic Acids Research, 37*(6), e45.

Stahlberg, A., et al. (2004). Properties of the reverse transcription reaction in mRNA quantification. *Clinical Chemistry, 50*(3), 509–515.

Stahlberg, A., Kubista, M., & Pfaffl, M. (2004). Comparison of reverse transcriptases in gene expression analysis. *Clinical Chemistry, 50*(9), 1678–1680.

Tang, R., et al. (2007). Validation of zebrafish (*Danio rerio*) reference genes for quantitative real-time RT-PCR normalization. *Acta Biochimca et Biophysica Sinica (Shanghai), 39*(5), 384–390.

Tricarico, C., et al. (2002). Quantitative real-time reverse transcription polymerase chain reaction: normalization to rRNA or single housekeeping genes is inappropriate for human tissue biopsies. *Analytical Biochemistry, 309*(2), 293–300.

Vandesompele, J., et al. (2002). Accurate normalization of real-time quantitative RT-PCR data by geometric averaging of multiple internal control genes. *Genome Biology, 3*(7). RESEARCH0034.

Vanhauwaert, S., et al. (2014). Expressed repeat elements improve RT-qPCR normalization across a wide range of zebrafish gene expression studies. *PLoS One, 9*(10), e109091.

Vermeulen, J., et al. (2009). RNA pre-amplification enables large-scale RT-qPCR gene-expression studies on limiting sample amounts. *BMC Research Notes, 2*, 235.

Yang, H., et al. (2013). Deep mRNA sequencing analysis to capture the transcriptome landscape of zebrafish embryos and larvae. *PLoS One, 8*(5), e64058.

PART

4

Epigenetics

CHAPTER

Genome-wide DNA methylation profiling in zebrafish

18

P.J. Murphy, B.R. Cairns[1]

University of Utah School of Medicine, Salt Lake City, UT, United States

[1]*Corresponding author: E-mail: Brad.Cairns@hci.utah.edu*

CHAPTER OUTLINE

Abstract

Genomic DNA methylation functions to repress gene expression by interfering with transcription factor binding and/or recruiting repressive chromatin machinery. Recent data support contribution of regulated DNA methylation to embryonic pluripotency, development, and tissue differentiation; this important epigenetic mark is chemically stable yet enzymatically reversible—and heritable through the germline. Importantly, all the major components involved in dynamic DNA methylation are conserved in zebrafish, including the factors that "write, read, and erase" this mark. Therefore, the zebrafish has become an excellent model for studying most biological processes associated with DNA methylation in mammals. Here we briefly review the zebrafish model for studying DNA methylation and describe a series of methods for performing genome-wide DNA methylation analysis. We address and provide methods for

Methods in Cell Biology, Volume 135, ISSN 0091-679X, http://dx.doi.org/10.1016/bs.mcb.2016.05.002

methylated DNA immunoprecipitation followed by sequencing (MeDIP-Seq), bisulfite sequencing (BS-Seq), and reduced representation bisulfite sequencing (RRBS-Seq).

INTRODUCTION

METHODS FOR STUDYING DNA METHYLATION IN ZEBRAFISH

The precise spatial and temporal regulation of transcription is necessary for both basic and higher-order biological processes, especially in complex multicellular organisms. One of the modes utilized to achieve dynamic transcriptional control involves direct chemical methylation of DNA (termed hereafter "DNAme"). In vertebrates, DNAme occurs almost exclusively at the fifth carbon of cytosine in the context of CpG dinucleotides (Adalsteinsson & Ferguson-Smith, 2014; Andersen et al., 2013; Marcho, Cui, & Mager, 2015; Wu et al., 2011). DNAme regulates transcription by two main modes: first, DNAme directly blocks the binding to DNA of certain site-specific transcriptional activators; and second, in an indirect manner, DNAme can promote the binding of methyl-CpG binding proteins, which associate with repressive histone deacetylase enzymes (Fig. 1). Therefore, by knowing the patterns of DNAme in a genome, one can identify candidate genes that are repressed by this epigenetic mode. Importantly, DNAme is both stably heritable and fully reversible. For example, DNAme at the promoter of a particular gene can be added to actively repress transcription and then subsequently removed—often as a response to an external stimulus—to allow transcription. These types of dynamic changes to DNAme tend to occur during embryonic development as cells differentiate or become reprogrammed.

Notably, there is evolutionary conservation in zebrafish for nearly all enzymes and factors involved in the deposition, binding, and removal of DNAme—establishing zebrafish as a powerful and effective model system for studying DNAme. Here, we briefly discuss the importance of DNAme for gene regulation in humans and the zebrafish and describe methods to quantify DNAme: (1) for the whole genome at moderate resolution using methyl-DNA immunoprecipitation sequencing (MeDIP-Seq); (2) for the whole genome at base-pair resolution using bisulfite sequencing (BS-Seq); and (3) for the CpG dense portion of the genome at base-pair resolution using reduced representation bisulfite sequencing (RRBS-Seq).

SIGNIFICANCE—DNAme IN HUMAN DEVELOPMENT AND DISEASE

The CpG dinucleotide pair occurs in mammalian genomes at only 20–25% the expected frequency. This depletion is thought to result from evolutionary selection

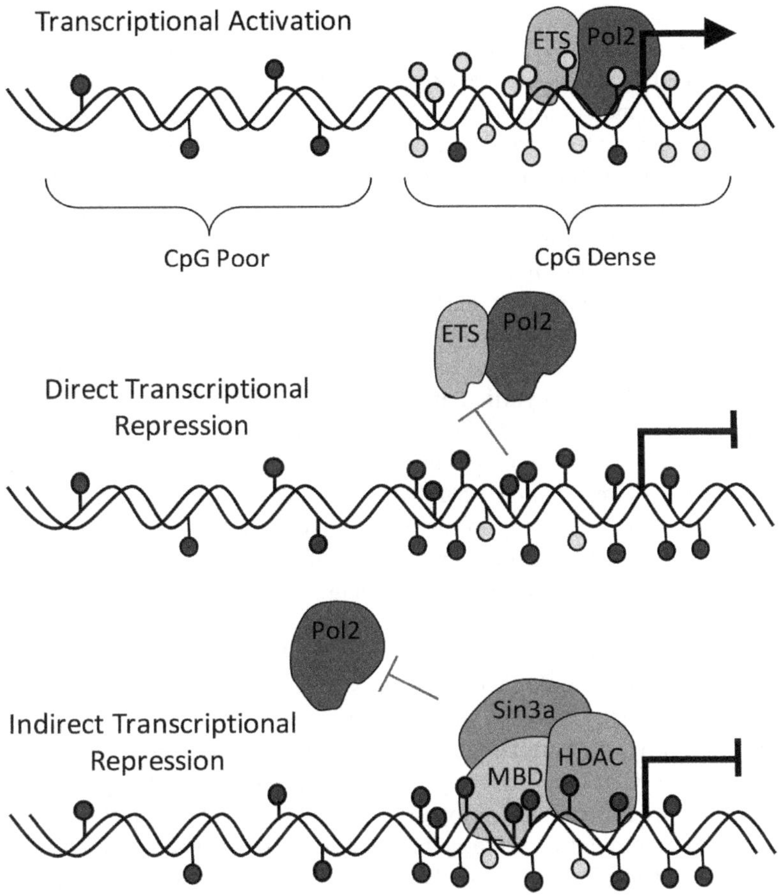

FIGURE 1

DNAme regulation of transcription. Filled in *lollipop circles* represent methylated CpGs and *open circles* are unmethylated. Constitutively high DNAme levels occur at low CpG content regions (CpG Poor) and dynamically regulated DNAme levels occur within high CpG content (CpG dense) regions. Top—Typical mode of gene expression where by a transcription factor (TF-green) enables polymerase (Pol2-red) binding and activation when DNAme levels are low. Middle—Direct repulsion of transcription factors occurs when DNAme levels are high to prevent gene expression. Bottom—Indirect repulsion of polymerase occurs via chromatin modifiers that recognize high DNAme levels. (See color plate)

against deleterious mutations, as methylated cytosine can be spontaneously deaminated and cause cytosine to thymine transitions. CpGs tend to occur clustered in high-density regions known as CpG islands (Antequera & Bird, 1993). These islands often reside in regulatory regions including the enhancers and promoters of housekeeping and developmentally regulated genes. Interestingly, the density of CpG dinucleotides is not consistent across vertebrates (Fig. 2) (Potok, Nix, Parnell, & Cairns, 2013). As compared to zebrafish and *Xenopus*, there are significantly fewer CpGs in mammals than would be expected based on random occurrence. Along these lines, CpG islands in mammals are less CpG rich than either zebrafish or *Xenopus* (Fig. 2) (Potok et al., 2013). What accounts for this difference in CpG density is unknown, but one possibility is that zebrafish and *Xenopus* harbor protective mechanisms to combat mutagenesis by more efficiently repairing C to T transitions.

At CpG islands, DNAme can negatively regulate transcription either directly—by preventing the binding of methylation sensitive transcription factors, or indirectly—by recruiting repressive machinery like histone deacetylases or histone methyltransferases and inhibiting polymerase recruitment (Fig. 1). Notably, cancer is strongly associated with abnormal genomic DNAme patterns, which often emerge as focal hypermethylation at particular gene promoters (eg, tumor suppressors) and/or hypomethylation at the genome-wide level, which can promote the expression of factors that affect growth, development, or apoptosis (Feinberg, 2007; Robertson, 2005). However, it is important to note that the causal role and molecular mechanism that underlie DNAme patterning in many human diseases including cancer are unknown. Thus, results from directed research studies on DNAme programming at its most fundamental level are exceedingly valuable for clinical and basic research fields.

In humans the higher-order functional role of DNAme has been well studied in a number of prominent biological processes and disease states, including X-chromosome inactivation (XCI), genomic imprinting, and cancer. During human development, shortly after implantation, female embryos undergo a process referred to as XCI. During XCI, one randomly selected (with respect to the parental origin) X-chromosome becomes transcriptionally silenced (Lyon, 1961; Robertson, 2005), and proper DNA methylation patterns have been shown to be essential for this process (Nesterova et al., 2008). Similarly, genomic imprinting occurs when a single allelic copy for a given gene is transcribed in a parent-of-origin–based manner, and in nearly every known example, imprinted expression is regulated by allele-specific DNAme patterns (Adalsteinsson & Ferguson-Smith, 2014). Notably, failed XCI has been linked to a host of diseases, including hemophilia, manic depression, and bipolar disorder (Dobyns et al., 2004; Dorus, 1983; Valleix et al., 2002), and failed imprinting causes a number of developmental diseases, including Beckwith–Weidemann, Prader–Willi, and Angelman syndromes (Robertson, 2005). Although XCI and genomic imprinting do not occur in zebrafish, the zebrafish model is quite powerful for studying the basic machinery (including DNAme) that regulates these complex disease states. Mutant phenotypes for depleted or catalytically inactive DNA methyltransferases in zebrafish will be addressed later.

Aside from its utility at the basic mechanistic level, the zebrafish system has emerged as a model for direct study of diseases that result from improper DNAme patterning, including Triplet Repeat Expansion Diseases (TRED) (Das & Rajanikant, 2014; Kelley, Chang, & Lin, 2012; Todd et al., 2014), which occur when a simple three nucleotide repeat region expands within a given gene and causes

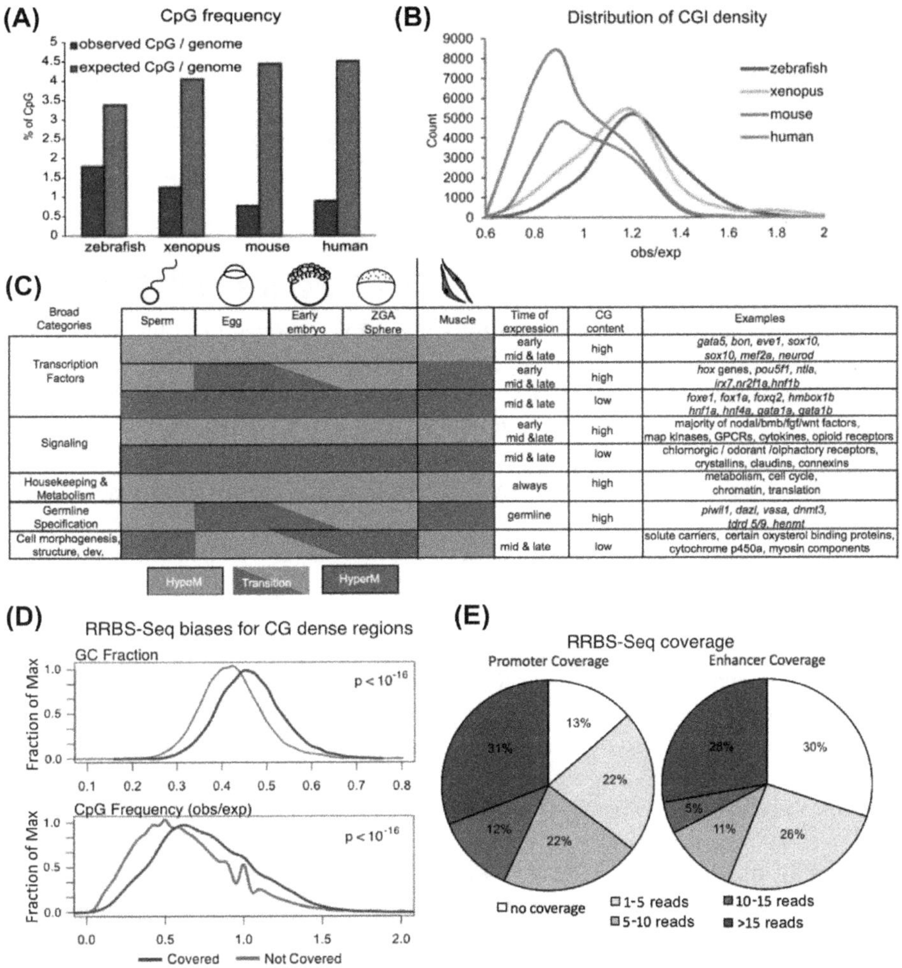

FIGURE 2

(A) Zebrafish display modest CpG depletion relative to mammals. Observed and expected CpG fractions are displayed. (B) Zebrafish CpG islands (zCGIs, UCSC) have extremely high obs/exp CpG frequencies. (C) DNAme dynamics (colored red for high DNAme and green for low DNAme) during zebrafish development (columns 2–5), as a function of category (first column), expression pattern (seventh column), and CpG density (eighth column). Expression timing obtained from public sources (http://ZFIN.org) where "early" indicates germ layer specification/gastrulation, "mid" stages refer to progenitor cell differentiation, and "late" refers to terminal differentiation. (D–E) Pilot of RRBS-Seq studies (30,000,000 mapped reads) in zebrafish. (D) Reduced representation bisulfite sequencing (RRBS) coverage (blue) is biased for detection of region with high GC content (top) and CpG density (bottom) as compared to the undetectable portion of the genome (red). (E) A breakdown of read coverage based on two genomic features, promoters (left) and enhancer (right). (See color plate)

Panels A–C are from Potok, M. E., Nix, D. A., Parnell, T. J., & Cairns, B. R. (2013). Reprogramming the maternal zebrafish genome after fertilization to match the paternal methylation pattern. Cell, 153, *759–772.*

inappropriate gene expression. Examples of these diseases in humans include: Huntington's disease; Fragile-X syndrome; Myotonic dystrophy; and Facioscapulohumeral muscular dystrophy. Although the cause of TRED is unknown, DNAme has been shown to antagonizing the effects of repeat expansion, either directly or indirectly via histone modifications and the insulator protein CTCF (Robertson, 2005).

WRITERS, ERASERS, AND READERS OF DNAme IN ZEBRAFISH

Zebrafish have become a major model for studying vertebrate development because they are small, hardy, relatively inexpensive, have a short generation time—and are easily visualized (transparent during embryogenesis), and manipulable for functional studies. In part for these reasons, they have recently emerged as a powerful tool for the study of DNAme regulation, as almost all of the enzymes involved in DNAme and gene silencing in mammals are likewise found in zebrafish (with notable exceptions).

Writing DNAme: The enzymes that account for genomic DNA methylation can be classified into two families: maintenance DNA methyltransferases, which act during replication to reliably preserve already present methyl-cytosine—and de novo DNA methyltransferases, which act at previously unmethylated genomic loci. While the mammalian and zebrafish genomes both harbor a single gene within the maintenance methyltransferase family, DNA methyltransferase 1 (DNMT1), each vertebrate also has numerous genes that are part of the de novo methyltransferase family. In mammals there are three genes of this type, *DNMT3a*, *DNMT3b*, and the catalytically inactive *DNMT3L*. In zebrafish the de novo methyltransferase family includes six members, *Dnmt3aa, Dnmt3ab, Dnmt3ba, Dnmt3bb.1, Dnmt3bb.2*, and *Dnmt3bb.3*. These zebrafish de novo DNMTs do not clearly partition into sets that are similar to mammalian *DNMT3a* or *DNMT3b*. Notably, *DNMT2* falls into neither family, as it functions as an RNA methyltransferase (Khoddami & Cairns, 2013). In zebrafish, there is no evidence for an ortholog of *DNMT3L*—the DNMT that regulates maternal imprinted DNAme in mammals (Hata, Okano, Lei, & Li, 2002; Wu et al., 2011), and likewise, there is no evidence for imprinting either.

Functional studies in zebrafish have measured the phenotypic consequence of DNMT depletion, either by genetic mutation or by morpholino injection. *Dnmt1* genetic nulls largely exhibited defects associated with terminal differentiation and maintenance of somatic cells. For example, catalytically inactive Dnmt1 mutations caused apoptotic oblation of ankyrin cells and degradation of both the endocrine panaceas and liver (Anderson et al., 2009), and *Dnmt1* morpholino injections lead to defects in intestinal and exocrine pancreas development (Rai et al., 2010, 2006). Additional studies confirmed these findings and also demonstrated that Dnmt1 is required for proper embryonic retinal development (Tittle et al., 2011). Relative to the maintenance methyltransferase, far fewer studies have been performed on the de novo DNA methyltransferases in zebrafish. Morpholino-based studies aimed to determine the function of zebrafish Dnmt3 indicate its necessity for both brain and retina development, likely working in conjunction with the histone methyltransferase G9a, whereas other organs appeared grossly unaffected (Rai et al., 2010), suggesting broad tissue-specific function for the zebrafish de novo methyltransferases. Furthermore, a series of morpholinos were used to systemically characterize the effects of depleting each of the six zebrafish *DNMT3* orthologs individually (Shimoda, Yamakoshi, Miyake, & Takeda, 2005). Somewhat surprisingly, strong morphological defects were not observed, suggesting either mechanistic overlap function for the de novo methyltransferases, or a more subtle impact, requiring further studies.

Erasing DNAme: The zebrafish genome also harbors all orthologs involved in DNA demethylation, including Aicda, Mbd4(AB918737), Tdg, and the Tet1/2/3 family (note: the complex phylogenetic details regarding orthologs will not be addressed here). However, the role of these factors during developmental reprogramming of DNAme differs from mammals at a few key stages. For example, in mammals, active DNA demethylation mediated by TET3 is a fundamental feature of cleavage phase development (Marcho et al., 2015), but there is no evidence for Tet or Aicda/Mbd4 function at this developmental stage in zebrafish (Jiang et al., 2013; Li et al., 2015; Potok et al., 2013; Shimoda et al., 2014), presumably due to the timing of zygotic genome transcriptional activation—discussed later. Furthermore, zebrafish Tet null mutants are embryonic viable, including the Tet1/2/3 triple knockout, with phenotypes excluded to blood cell development (Li et al., 2015).

Reading DNAme: Much of the machinery for reading DNAme, which include a host of protein factors that recognize CpG methylation and function indirectly to mediate transcription, are also conserved from zebrafish to mammals. These include methyl-binding domain (MBD) containing and Zinc-Finger CxxC (ZF-CxxC) domain containing proteins, which recognize methylated and unmethylated DNA, respectively. All mammalian MDB-containing proteins are conserved in zebrafish, including orthologs of Mecp2, Mbd1, Mbd2, Mbd3, and Mbd4. Mbd5 and Mbd6 are also conserved, and like mammals, lack full MBD-domain conservation, suggesting an alternative function to binding methylated DNA. Interestingly, unlike its mammalian ortholog, the zebrafish Mbd3 has a high degree of MBD-family conservation within it the MBD-domain, which may enable binding to DNAme (Wade, 2001). Many Zinc-finger—containing proteins, which are conserved in zebrafish, have been shown to recognize DNAme, including Ctcf, Kaiso, and a host of ZF-CxxC domain containing factors (Gómez-Marín et al., 2015; Ma et al., 2011; Ruzov et al., 2004). Notably, many ZF-CxxC domain-containing proteins have histone modifying capabilities, including KDM1/2 and MLL1/2 (Blackledge, Thomson, & Skene, 2013). This may enable cross talk between DNAme and histone modifications, which may facilitate high-order epigenetic mark placement and indirectly affect DNAme and DNAme-mediated transcriptional regulation.

DNAme REPROGRAMMING DURING VERTEBRATE DEVELOPMENT—A COMPARISON OF MAMMALS AND ZEBRAFISH

One finds both commonalities and differences when comparing DNAme reprogramming in zebrafish and mammalians. Regarding gametic DNAme patterns, in both mammals and zebrafish, the DNAme patterns of sperm and egg are quite different from one another, as the egg genome is undermethylated relative to the sperm genome (Potok et al., 2013; Smith et al., 2012). However, opposite patterns can be found at particular regulatory regions, such as certain promoters and CpG islands; high DNAme levels in the egg and low levels in sperm. Indeed, in zebrafish, CpG islands located at promoters in sperm are almost uniformly demethylated, whereas a considerable fraction are methylated in eggs, especially at genes for developmental transcription factors (which must be demethylated during cleavage, see later discussion) (Potok et al., 2013; Smith et al., 2012). In this manner, the egg is similar to a somatic cell, and the sperm is similar to a pluripotent stem cell—and this aspect holds true for both zebrafish and mammals.

Notable differences between zebrafish and mammals occur during cleavage stages and during zygotic genome activation (Potok et al., 2013; Smith et al., 2012). In mammals the paternal inherited genome is largely hypermethylated and undergoes an active phase of DNA demethylation that occurs within the first three cell cycles—prior to zygotic genome activation. The mammalian maternally inherited genome, on the other hand, is demethylated in a gradual manner over six to eight cell cycles and may rely on more on passive mechanisms (Marcho et al., 2015; Smith et al., 2012). The zebrafish DNAme cleavage phase reprogramming process is similar in many respects but also quite different. Here, the paternally inherited genome maintains its DNAme patterns throughout cleavage, and the maternally inherited genome conforms to reflect the paternal (Fig. 2). For example, genes which function during germline specification are inherited as hypo-DNAme from sperm, and hyper-DNAme from the oocyte. During cleavage, DNAme appears to be passively lost at these loci on the maternal allele (as TET proteins are not present at this stage) such that both alleles resemble the paternal state. Indeed, a major current question involves how these maternal loci lose methylation—which may involve the implementation of chromatin structures/modifications that deter DNAme. A number of genes involved in differentiation and morphogenesis are inherited as hyper-DNAme from sperm and hypo-DNAme from oocyte (Potok et al., 2013). These genes are then reprogrammed such that the maternally inherited loci gain DNAme. Notably, in both mammals and zebrafish, paternally inherited DNAme patterns are stabilized prior to zygotic genome activation, and maternally inherited genomes undergo passive DNAme reprogramming throughout the pluripotent conversion process. This is true for both genomes as a whole, and for the promoters of developmentally regulated genes, which are paternally inherited as hypo-methylated and require reprogramming only on the maternal allele—in both mammals and zebrafish (Potok et al., 2013; Smith et al., 2012).

As previously alluded to, the differences in DNAme reprogramming between zebrafish and mammals likely reflect differences in developmental requirements and dissimilarity in timing of zygotic transcriptional initiation. In mammals the genome becomes transcriptionally active between the 1-cell and 4-cell stages, and pluripotency (ie, presence of NANOG and POU5F1 in the nucleus) is not achieved until 5 days later after cell cycle 8 (Marcho et al., 2015). Alternatively, in zebrafish, pluripotency is achieved during zygotic genome activation, taking only 4 h and encompassing over 10 cell cycles (Andersen et al., 2013). Put simply, the mouse genome is transcriptionally activated quickly, and a pluripotent inner cell mass is established more gradually. In zebrafish, unlike mammals, genome activation coincides with pluripotent conversion at a much later stage of cleavage phase, but this occurs more rapidly with respect to time. Thus, the greater number of cell cycles associated with zebrafish cleavage embryogenesis may enable the genome to utilize a passive mode of DNAme reprogramming (ie, deter new DNAme in regions and let replication dilute DNAme) as compared to the active targeting of DNA demethylation utilized by mammals. In other words, zebrafish DNAme reprogramming does not need to be rapid, as there are 10 cell cycles before zygotic genome activation to passively reprogram the genomes via default DNA methylation and passive DNA demethylation. Because mammalian transcriptional initiation occurs after only 1−3 cell cycles, reprogramming must rely on active targeted processes.

1. RATIONALE

Clearly much associated with genomic DNAme is conserved from zebrafish to mammals, including: (1) the passive DNAme reprogramming that occurs on the maternal allele during the cleavage phase of embryonic development; (2) localization of hypo-DNAme at important developmentally regulated promoters in pluripotent embryos; (3) the DNA methyltransferase machinery; (4) the DNA demethylation machinery; and (5) the proteins that recognize DNAme and facilitate transcriptional regulation. Importantly, the ability to isolate large numbers of embryos provides a level of source material that is difficult to replicate in mice. Thus, the zebrafish model is an excellent system for studying DNAme and all its accompanying factors during development. Here we provide the community with an overview of methods for performing genome-wide DNA methylation analysis in zebrafish.

2. METHODS

2.1 EXTRACTION AND FRAGMENTATION OF GENOMIC DNA

In our hands, 50,000 cells are required for both whole genome bisulfite sequencing (WG-BS-Seq) and reduced representation bisulfite sequencing (RRBS-Seq), and methyl-DNA immunoprecipitation sequencing (MeDIP-Seq) requires 500,000 cells as input. This may vary depending on the efficiency of DNA purification and tissue type but is a good general rule of thumb. To extract DNA from zebrafish adult tissue, we utilized the Gentra Puregene Tissue Kit (Qiagen Inc., Valencia, CA) with the following modifications. Rather than using a mortar and pestle to grind frozen tissues, fresh chunks of tissues (eg, muscle or liver) are minced quickly on ice. Zebrafish embryos and activated oocytes have a high level of lipid content within the yolk, which we find inhibits DNA extraction. Thus, we performed nuclei purification prior to lysis. Here 1 mL of a cold cell lysis buffer (10 mM Tris-Cl, 10 mM NaCl, 0.5% NP-40, pH 8.0) was added to each sample and with a 3 mL syringe and 20 gauge needle, chorions were removed. Next we used a manual pipette to discard chorions, as they float to the top of lysis solution. To collect nuclei, samples were centrifuged for 5 min at 4°C and 3,500 g. All samples, including tissue lysate, embryo/oocyte nuclei, and pelleted sperm, were lysed using RIPA buffer (150 mM sodium chloride, 1.0% NP-40, 0.5% sodium deoxycholate, 0.1% SDS (sodium dodecyl sulfate), 50 mM Tris, pH 8.0) at 65°C for 4 h with gentle agitation and then incubated with RNaseA (4 mg/mL) at 37 C for 2 h. Lysates were then phenol extracted and alcohol precipitated to purify DNA. Ten millimolar Tris (pH 8.0) was used to rehydrate. We then fragmented DNA in the 200–600 bp range by either sonication or by enzymatic digestion (dependent on the source of DNA and the measured degree of overfragmentation). Oocyte samples were fragmented using a cocktail of restriction enzymes as overfragmentation occurred even with minimal sonication (more detailed methods pertaining to this can be found in Potok et al.). These samples were then purified and reconcentrated using standard PCR column purification methods (Qiagen kit). Fragmented DNA was then used as input for either the WG-BS-Seq or MeDIP-Seq method. DNA fragmentation for RRBS was nearly identical to these methods, except DNA which was fragmented exclusively by MspI digestion.

2.2 METHYLATED DNA IMMUNOPRECIPITATION AND SEQUENCING

Fragmented DNA (4 μg in 50 μL −10 mM Tris buffer) was denatured at 94°C for 10 min and then transferred immediately to ice to prevent reannealing. Ten micrograms of antibody raised against 5-methylcytidine (Eurogentec BI-MECY-0100) in ice-cold IP buffer (20 mM Tris, 140 mM NaCl, 0.05% Triton X-100, pH 7.5) was then added to the sheared genomic DNA to reach a final volume of 500 μL and incubated with rotation at 4°C for 2 h. Antimouse IgG magnetic Dynabeads (Invitrogen 11202D) were first washed in PBS containing 5 mg/mL BSA, then added to the DNA–antibody mixture and incubated for an additional 2 h at 4°C. After incubation, to remove unbound proteins and DNA, we performed three sequential washes of the Dynabeads in IP buffer. DNA was then eluted from the bead-antibody complexes by treating samples with 2 mg/mL Proteinase K in IP buffer for 3 h at 50°C. Eluted DNA was purified by PCR-column purification (Qiagen kit) and reconstituted in 10 mM Tris pH 8.0. This DNA was then used to generate Illumina-HiSeq compatible libraries by standard methods (NEBNext Kit). Briefly, fragmented DNA is end-repaired and adenine-tailed (using Klenow exo-), and adapters are ligated to the ends of each fragment. PCR amplification was then performed to generate libraries. For a more detailed library preparation protocol, see the NEBNext DNA Library Prep manual. Six to eight MeDIP-Seq samples can be multiplexed during sequencing, dependent on desired read depth. Sequencing was performed on an Illumina HiSeq2500 machine using 50-base single end format (Note: this method can be easily adapted for measuring 5 hmC abundance by using a different antibody (Eurogentec BI-HMCY-0100)). In our hands, 40,000,000 reads is sufficient for relative quantification genome-wide DNAme. Bioinformatic analysis for DNAme quantification was performed using the USeq pipeline (Nix et al., 2010), which we find provides results nearly identical to other analysis methods, including MACS2 (Zhang et al., 2008).

2.3 WHOLE GENOME SHOTGUN BISULFITE SEQUENCING

Sequencing of bisulfite treated and amplified DNA provides a base-pair resolution approach for measuring DNAme levels at individual cytosines. Notably, this method has been adapted to measure cytosine methylation in a genome-wide manner (Cokus et al., 2008; Lister et al., 2009; Meissner et al., 2008). Notably, there is a remarkably high degree of concordance comparing the zebrafish DNAme levels reported using a variation on MeDIP-Seq called MethylCRF (Li et al., 2015) and WG-BS-Seq (Potok et al., 2013). WG-BS-Seq involves bisulfite treatment, where unmethylated cytosines are converted to uracils, and methylated cytosines remain unaltered. Uracils are then converted to thymines during amplification at the library preparation phase, and next generation sequencing methods enable measurement of DNA methylation at base-pair resolution for the whole genome. To begin we added unmethylated lambda DNA to the previously noted fragmented DNA at 1% total quantity. This serves as a spike-in control for bisulfite conversion efficiency, such that samples where lambda specific C to T conversion are below 99%, are disregarded—this is extremely rare in our hands. After lambda is added, we performed end-repair and added an adenine to the ends of each fragment (according to manufacturer's instructions: TruSeq DNA Methylation Sample Prep Kit—Illumina Inc.). Next, we ligated methylated bar-coded adaptors provided by Illumina, Inc. The DNA is then cleaned up using Agencourt AMPure Beads and eluted in 30 μL water. Adapter ligated DNA was then bisulfite converted using the EpiTect Bisulfite Kit (QIAGEN) with one additional incubation step at 60°C for 2 h to increase conversion rate. We find this modification ensures a high degree of C to T conversion without compromising DNA quality or integrity. Following bisulfite treatment, DNA was PCR amplified with 16 cycles, and then cleaned up using Agencourt AMPure Beads. Purity and size of amplified libraries were confirmed using Agilent DNA 1000 Bioanalyzer. High-quality libraries were then sequenced on an Illumina HiSeq2500 machine using 101-base paired end format. We find 300—500 million reads are sufficient for DNAme quantification as mid-to high-level resolution (5—1 CpG per scoreable region respectively).

Bioinformatics analysis of DNAme levels was performed using the USeq pipeline (Nix et al., 2010), which we find to provide nearly identical results to the Bismark pipeline (Krueger & Andrews, 2011).

2.4 REDUCED REPRESENTATION BISULFITE SEQUENCING

RRBS-Seq is an alternative to WG-BS-Seq and MeDIP that enables genome-wide base-pair resolution DNAme quantification but at a significantly lower cost than WG-BS-Seq. Comparisons of RRBS-Seq with WG-BS-Seq and MeDIP-Seq indicate a high level of correlation between methods (Harris et al., 2010). There are some noteworthy trade-offs that come with this approach, however. To test these, we performed pilot studies (1/10th Illumina High-Seq Library, approximately 30,000,000 reads) and measured genome coverage from RRBS-Seq performed on a zebrafish sperm sample. For low-resolution DNAme quantification (5 CpGs per analyzed per region), which we performed here, 5–10 reads per region is sufficient. For higher resolution (single CpG dinucleotide level), we recommend 50–100 reads per CpG dinucleotide. While the RRBS method can be used to measure DNAme at the majority of zebrafish promoters, and enhancers (Fig. 2), because there is less CpG depletion in zebrafish than in mammals (Fig. 2a + b), RRBS assays in zebrafish offer nearly twice the genome coverage than assays in mammals (~2.5% as compared to ~1.5%) (unpublished zebrafish RRBS data from our lab). For those aiming to measure DNAme levels at low-density CpG regions, RRBS-Seq is not suitable, and for those wishing to assess putative genomic regulator regions, which typically reside in CpG islands, RRBS is quite appropriate. Furthermore, on a per read basis, RRBS-Seq offers deeper coverage over these putative regulatory regions than any other method (Gu et al., 2011) at a cost comparable to MeDIP-Seq, which does not offer base-pair resolution. Additionally, because RRBS-Seq regions cover only 2.2% of the zebrafish genome, multiplexing libraries can enable sequencing of numerous samples at a fraction of the cost of WG-BS-Seq. While much of the protocol for preparing RRBS-Seq libraries is similar to WG-BS-Seq, there are some key differences. Notably, DNA is fragmented with MspI rather than by sonication. Thus, fragmented DNA will appear on an agarose gel as a smear with the majority of fragment at high molecular weight, rather than the narrow normal distribution 200–400 bp smear that results from randomly fragmented DNA. Because only a small portion of fragments (those 50–300 bp) are compatible with RRBS library prep, most fragments are lost during preparation, and input amounts must be higher than typical WG-BS-Seq methods (250 ng as compared to 5 ng in our hands). In an identical fashion to the WG-BS-Seq method described previously, MspI-digested fragmented DNA is blunted, adenylated, adapter ligated, bisulfite treated, PCR amplified, and cleaned up during library preparation. However, because we typically perform analysis on four to eight RRBS samples in multiplex, methylated bar-coded adapters are used (dependent on desired read depth). Like WG-BS-Seq, samples were run on an Illumina HiSeq2500 machine using 101-base paired end format. We find 30,000,000 reads are sufficient for DNAme quantification as mid-to low-level resolution (3–5 CpGs per scoreable region). Bioinformatics analysis of DNAme levels was performed using the USeq pipeline (Nix et al., 2010), which we find to provide nearly identical results to the Bismark pipeline (Krueger & Andrews, 2011).

DISCUSSION AND CONCLUSIONS

Understanding DNAme patterning and reprogramming, during normal developmental transitions as well as transition to disease state, is of great importance for human health. The zebrafish system presents an excellent model for studying these processes and offers a number of advantages over other models. Here we provided an overview of DNAme reprogramming in zebrafish as compared to mammals and highlight the evolutionary conservation for nearly all factors involved in reading, writing, and erasing DNAme.

We also detail three methods to perform genome-wide DNAme analysis and denote advantages and disadvantages of each method. Briefly: WG-BS-Seq is the most comprehensive but most costly; MeDIP-Seq is the least expensive and most straightforward but provides the poorest sequence resolution; and RRBS is both cost-effective and high resolution but does not enable whole genome coverage. We hope the methods described here will enable a wide range of researchers to perform DNAme analysis and will stimulate the study of epigenetics in zebrafish, in other model and nonmodel organisms, and in the clinical setting.

REFERENCES

Adalsteinsson, B. T., & Ferguson-Smith, A. C. (2014). Epigenetic control of the Genome—Lessons from genomic imprinting. *Genes, 5*, 635–655.

Andersen, I. S., Lindeman, L. C., Reiner, A. H., Østrup, O., Aanes, H., Aleström, P., & Collas, P. (2013). Epigenetic marking of the zebrafish developmental program. *Current Topics in Developmental Biology, 104*, 85–112.

Anderson, R. M., Bosch, J. A., Goll, M. G., Hesselson, D., Dong, P. D. S., Shin, D., … Stainer, D. Y. (2009). Loss of Dnmt1 catalytic activity reveals multiple roles for DNA methylation during pancreas development and regeneration. *Developmental Biology, 334*, 213–223.

Antequera, F., & Bird, A. (1993). Number of CpG islands and genes in human and mouse. *Proceedings of the National Academy of Sciences of the United States of America, 90*, 11995–11999.

Blackledge, N. P., Thomson, J. P., & Skene, P. J. (2013). CpG island chromatin is shaped by recruitment of ZF-CxxC proteins. *Cold Spring Harbor Perspectives in Biology, 5*, a018648.

Cokus, S. J., Feng, S., Zhang, X., Chen, Z., Merriman, B., Haudenschild, C. D., … Jacobsen, S. E. (2008). Shotgun bisulphite sequencing of the *Arabidopsis* genome reveals DNA methylation patterning. *Nature, 452*, 215–219.

Das, S., & Rajanikant, G. K. (2014). Huntington disease: can a zebrafish trail leave more than a ripple? *Neuroscience & Biobehavioral Reviews, 45*, 258–261.

Dobyns, W. B., Filauro, A., Tomson, B. N., Chan, A. S., Ho, A. W., Ting, N. T., … Ober, C. (2004). Inheritance of most X-linked traits is not dominant or recessive, just X-linked. *American Journal of Medical Genetics, Part A, 129A*, 136–143.

Dorus, E. (1983). X-chromosome inactivation and the study of X-linked dominant transmission of bipolar illness. *Archives of General Psychiatry, 40*, 698–699.

Feinberg, A. P. (2007). Phenotypic plasticity and the epigenetics of human disease. *Nature, 447*, 433–440.

Gómez-Marín, C., Tena, J. J., Acemel, R. D., López-Mayorga, M., Naranjo, S., de la Calle-Mustienes, E., … Gómez-Skarmeta, J. (2015). Evolutionary comparison reveals that diverging CTCF sites are signatures of ancestral topological associating domains borders. *Proceedings of the National Academy of Sciences of the United States of America, 112*, 7542–7547.

Gu, H., Smith, Z. D., Bock, C., Boyle, P., Gnirke, A., & Meissner, A. (2011). Preparation of reduced representation bisulfite sequencing libraries for genome-scale DNA methylation profiling Nature Protocols. *6*, 468–481.

Harris, R. A., Wang, T., Coarfa, C., Nagarajan, R. P., Hong, C., Downey, S. L., … Costello, J. F. (2010). Comparison of sequencing-based methods to profile DNA methylation and identification of monoallelic epigenetic modifications. *Nature Biotechnology, 28*, 1097–1105.

Hata, K., Okano, M., Lei, H., & Li, E. (2002). Dnmt3L cooperates with the Dnmt3 family of de novo DNA methyltransferases to establish maternal imprints in mice. *Development (Cambridge, England), 129*, 1983–1993.

Jiang, L., Zhang, J., Wang, J.-J., Wang, L., Zhang, L., Li, G., … Jiang, L. (2013). Sperm, but not oocyte, DNA methylome is inherited by zebrafish early embryos. *Cell, 153*, 773–784.

Kelley, K., Chang, S.-J. E., & Lin, S.-L. (2012). Mechanism of repeat-associated MicroRNAs in fragile X syndrome. *Neural Plasticity, 2012*.

Khoddami, V., & Cairns, B. R. (2013). Identification of direct targets and modified bases of RNA cytosine methyltransferases. *Nature Biotechnology, 31*, 458–464.

Krueger, F., & Andrews, S. R. (2011). Bismark: a flexible aligner and methylation caller for Bisulfite-Seq applications. *Bioinformatics (Oxford, England), 27*, 1571–1572.

Li, C., Lan, Y., Schwartz-Orbach, L., Korol, E., Tahiliani, M., Evans, T., & Goll, M. G. (2015). Overlapping requirements for Tet2 and Tet3 in normal development and hematopoietic stem cell emergence. *Cell Reports (Electronic Resource), 12*, 1133–1143.

Li, D., Zhang, B., Xing, X., & Wang, T. (2015). Combining MeDIP-seq and MRE-seq to investigate genome-wide CpG methylation. *Methods, 72*, 29–40.

Lister, R., Pelizzola, M., Dowen, R. H., Hawkins, R. D., Hon, G., Tonti-Filippini, J., … Ecker, J. R. (2009). Human DNA methylomes at base resolution show widespread epigenomic differences. *Nature, 462*, 315–322.

Lyon, M. F. (1961). Gene action in the X-chromosome of the mouse (*Mus musculus* L.). *Nature, 190*, 372–373.

Ma, S., Huang, W., Zhang, L., Zhao, S., Tong, Y., Liu, Z., … Luo, C. (2011). Germ cell-specific DNA methylation and genome diploidization in primitive vertebrates. *Epigenetics, 6*, 1471–1480.

Marcho, C., Cui, W., & Mager, J. (2015). Epigenetic dynamics during preimplantation development. *Reproduction: The Official Journal of the Society for the Study of Fertility, 150*, R109–R120.

Meissner, A., Mikkelsen, T. S., Gu, H., Wernig, M., Hanna, J., Sivachenko, A., ... Lander, E. S. (2008). Genome-scale DNA methylation maps of pluripotent and differentiated cells. *Nature, 454*, 766–770.

Nesterova, T., Popova, B., Cobb, B., Norton, S., Senner, C., Tang, Y. A., ... Brockdorff, N. (2008). Dicer regulates Xist promoter methylation in ES cells indirectly through transcriptional control of Dnmt3a. *Epigenetics Chromatin, 1*, 2.

Nix, D. A., Di Sera, T. L., Dalley, B. K., Milash, B. A., Cundick, R. M., Quinn, K. S., & Courdy, S. J. (2010). Next generation tools for genomic data generation, distribution, and visualization. *BMC Bioinformatics, 11*, 455.

Potok, M. E., Nix, D. A., Parnell, T. J., & Cairns, B. R. (2013). Reprogramming the maternal zebrafish genome after fertilization to match the paternal methylation pattern. *Cell, 153*, 759–772.

Rai, K., Jafri, I. F., Chidester, S., James, S. R., Karpf, A. R., Cairns, B. R., & Jones, D. A. (2010). Dnmt3 and G9a cooperate for tissue-specific development in zebrafish. *Journal of Biological Chemistry, 285*, 4110–4121.

Rai, K., Nadauld, L. D., Chidester, S., Manos, E. J., James, S. R., Karpf, A. R., ... Jones, D. A. (2006). Zebra fish Dnmt1 and Suv39h1 regulate organ-specific terminal differentiation during development. *Molecular and Cellular Biology, 26*, 7077–7085.

Robertson, K. D. (2005). DNA methylation and human disease. *Nature Reviews Genetics, 6*, 597–610.

Ruzov, A., Dunican, D. S., Prokhortchouk, A., Pennings, S., Stancheva, I., Prokhortchouk, E., & Meehan, R. R. (2004). Kaiso is a genome-wide repressor of transcription that is essential for amphibian development. *Development (Cambridge, England), 131*, 6185–6194.

Shimoda, N., Hirose, K., Kaneto, R., Izawa, T., Yokoi, H., Hashimoto, N., & Kikuchi, Y. (2014). No evidence for AID/MBD4-coupled DNA demethylation in zebrafish embryos. *PLoS One, 9*, e114816.

Shimoda, N., Yamakoshi, K., Miyake, A., & Takeda, H. (2005). Identification of a gene required for de novo DNA methylation of the zebrafish no tail gene. *Developmental Dynamics: An Official Publication of American Association of Anatomists, 233*, 1509–1516.

Smith, Z. D., Chan, M. M., Mikkelsen, T. S., Gu, H., Gnirke, A., Regev, A., & Meissner, A. (2012). A unique regulatory phase of DNA methylation in the early mammalian embryo. *Nature, 484*, 339–344.

Tittle, R. K., Sze, R., Ng, A., Nuckels, R. J., Swartz, M. E., Anderson, R. M., ... Gross, J. M. (2011). Uhrf1 and Dnmt1 are required for development and maintenance of the zebrafish lens. *Developmental Biology, 350*, 50–63.

Todd, P. K., Ackall, F. Y., Hur, J., Sharma, K., Paulson, H. L., & Dowling, J. J. (2014). Transcriptional changes and developmental abnormalities in a zebrafish model of myotonic dystrophy type 1. *Disease Models & Mechanisms, 7*, 143–155.

Valleix, S., Vinciguerra, C., Lavergne, J.-M., Leuer, M., Delpech, M., & Negrier, C. (2002). Skewed X-chromosome inactivation in monochorionic diamniotic twin sisters results in severe and mild hemophilia A. *Blood, 100*, 3034–3036.

Wade, P. A. (2001). Methyl CpG binding proteins: coupling chromatin architecture to gene regulation. *Oncogene, 20*, 3166–3173.

Wu, S.-F., Zhang, H., Hammoud, S. S., Potok, M., Nix, D. A., Jones, D. A., & Cairns, B. R. (2011). DNA methylation profiling in zebrafish. *Methods in Cell Biology, 104*, 327–339.

Zhang, Y., Liu, T., Meyer, C. A., Eeckhoute, J., Johnson, D. S., Bernstein, B. E., ... Liu, X. S. (2008). Model-based analysis of ChIP-Seq (MACS). *Genome Biology, 9*, R137.

CHAPTER

Histone modifications in zebrafish development

19

V.T. Cunliffe

University of Sheffield, Sheffield, United Kingdom

E-mail: v.t.cunliffe@sheffield.ac.uk

CHAPTER OUTLINE

Methods in Cell Biology, Volume 135, ISSN 0091-679X, http://dx.doi.org/10.1016/bs.mcb.2016.05.005

Abstract

Reversible covalent histone modifications are known to influence spatiotemporal patterns of gene transcription during development. Here I review recent advances in the development and use of methods to analyze the distribution and functions of histone modifications in zebrafish chromatin. I discuss the roles of dynamic histone modification patterns at the promoters and enhancers of genes during the process of zygotic gene activation at blastula stages and the interplay between the molecular machinery responsible for histone modifications, chromatin remodeling and DNA methylation. Interactions are also described between developmentally regulated enhancer sequences and modified histones. A detailed method for chromatin immunoprecipitation using antibodies is provided, and I describe the use of high-throughput whole genome sequencing technology to generate DNA sequence data from chromatin immunoprecipitates. I also discuss computational approaches to integrating DNA sequence data obtained from chromatin immunoprecipitates with annotated reference genome sequences, transcriptome and methylome sequence data, transcription factor binding motif databases, and gene ontologies and describe the types of software tools currently available for visualizing the results.

List of Abbreviations

BWA	Burrows-Wheeler Aligner
$CaCl_2$	Calcium chloride
CAGE	Cap analysis of gene expression
CDK	Cyclin-dependent kinase
ChIP-Seq	Chromatin immunoprecipitation sequencing
DMR	Differentially methylated region
DMSO	Dimethyl sulfoxide
EDTA	Ethylenediaminetetraacetate
EGS	Ethylene glycol bis(succinic acid N-hydroxysuccinimide ester)
EGTA	Ethyleneglycoltetraacetate
ESC	Embryonic stem cell
GO	Gene Ontology
GRB	Genomic regulatory block
H3K27ac	Histone H3 Lysine 27 acetylation
H3K27me3	Histone H3 Lysine 27 trimethylation
H3K36me3	Histone H3 Lysine 36 trimethylation
H3K4me1	Histone H3 Lysine 4 monomethylation
H3K4me3	Histone H3 Lysine 4 trimethylation
H3K9me3	Histone H3 Lysine 9 trimethylation
HAT	Histone Acetyltransferase

HDAC	Histone Deacetylase
HMT	Histone Methyltransferase
hpf	hours postfertilization
IgG	Immunoglobulin
KAT	Lysine Acetyltransferase
KCl	Potassium chloride
KDM	Lysine Demethylase
KEGG	Kyoto Encyclopedia of Genes and Genomes
LiCl	Lithium chloride
MACS	Model-based analysis of ChIP-Seq
MeDIP-Seq	Methylated DNA immunoprecipitation sequencing
MSK1	Mitogen and stress-activated Protein Kinase-1
NaAc	Sodium acetate
NaCl	Sodium chloride
$NaHCO_3$	Sodium hydrogen carbonate
PBS	Phosphate buffered saline
PCIA	Phenyl-chloroform-isoamyl alcohol
PHD	Plant homeo domain
PMSF	Phenylmethylsulfonyl fluoride
RIPA Buffer	Radioimmunoprecipitation Assay Buffer
RNA-Seq	RNA sequencing
SDS	Sodium dodecyl sulfate
SOAP2	Short Oligonucleotide Analysis Package 2
TSS	Transcription start site
ZGA	Zygotic genome activation

INTRODUCTION

ROLES FOR CHROMATIN MODIFICATIONS IN THE EPIGENETIC REGULATION OF GENE EXPRESSION

Through their interactions with specific genomic DNA sequences, constitutive and signal-regulated DNA-binding transcription factors direct and coordinate the gene transcription processes that generate form and function during development. In adults, proper regulation of gene transcription is also fundamental to the maintenance of physiological homeostasis, generating the phenotypic plasticity that enables adaptive responses to environmental challenges and facilitating healthy aging across the life course. Moreover, dysregulated gene transcription is a common characteristic of many chronic diseases. The accuracy, reliability, and plasticity of the cellular and organismal phenotypes that are determined by regulated gene transcription are influenced by a dynamic "epigenetic" landscape of patterned covalent modifications within chromatin, which is known as the epigenome, involving changes to both DNA and histones (Kouzarides, 2007). Recent research has shed much new light on the functions of the enzymes responsible for generating and modifying the complex patterns of epigenetic modifications that comprise the epigenome and has begun to reveal the biological functions of these modifications (Chen & Dent, 2014; Henikoff & Shilatifard, 2011). The epigenetic machinery participates in many of the molecular transactions that take place within the nucleus, such as DNA repair, recombination, replication, and transcription. Nevertheless, most of the insights obtained to date into the biological functions of epigenomic modifications relate to their roles in regulating the access of transcription factors and RNA polymerase to their DNA targets. Accordingly, in this chapter I will consider how the analysis of histone modifications in zebrafish informs our understanding of the molecular mechanisms that regulate gene transcription. I will also consider the prospects for further applying these techniques in future research with zebrafish to elucidate how the epigenome regulates developmental, physiological, and behavioral processes more broadly across the life course, both in health and disease.

Gene transcription is regulated by interactions between sequence-specific DNA binding factors, their target sites in DNA, RNA polymerases and a complex array of chromatin components. These components include core and linker histones, enzymes that can add or remove covalent modifications to histones, other chromatin proteins or DNA, and chromatin remodeling complexes that can disrupt or promote noncovalent interactions between histones and the DNA templates with which they interact, making DNA sequences either more or less accessible to other proteins. While the focus of this review is on histone modifications, it must be noted that there are many functional interactions between histone modification and DNA methylation processes, which are themselves further influenced by, and in turn influence, patterns of transcription factor binding, recruitment of chromatin remodeling machinery, as well as interactions with chromatin-associated, noncoding regulatory RNAs (Bartholomew, 2014).

TYPES OF HISTONE MODIFICATIONS: PRODUCTION, RECOGNITION, AND REMOVAL

Many different types of histone modification have been described, and extensive research has identified a wide variety of roles for these modifications in gene regulation. An emerging consensus indicates that the vast majority of known histone modifications are associated with either transcriptional activation or repression of the genes with which they are physically associated and that they act in combinatorial ways to create a molecular framework within which DNA-binding transcription factors, RNA polymerase, and their auxiliary complex components can interact, both with one another and with physically associated DNA sequences. The dynamic nature of the interactions between DNA-binding transcription factors, RNA polymerase and the enzymes that modify histones and DNA, further enables the patterns of histone modifications to change over time, as circumstances dictate, which thus facilitates plasticity in gene expression.

A wide variety of different covalent histone modifications have been described. The most extensively studied modifications of core histones include methylation, acetylation, and ubiquitylation of lysine residues, as well as phosphorylation of serines and methylation of arginines (Kouzarides, 2007). All of the four core histones H2A, H2B, H3, and H4 are substrates for lysine methylation and acetylation, of which the predominant target for modification that appears to be most closely involved in regulation of transcription is the core histone H3 subunit (Table 1). Histone H3 is subject to mono-, di-, and trimethylation of lysines 4, 9, 29, and 36 by residue-specific histone methyltransferases (HMTs). Lysines 4, 9, 14, 18, 23, and 27 are also targeted for acetylation by Histone Lysine (K) Acetyltransferases (HATs or KATs).

A growing body of evidence indicates that the activities and specificities of many HMTs are regulated by signaling pathway-dependent, cyclin-dependent kinase (CDK)-mediated phosphorylation of serine and threonine residues, which affects the catalytic activities of these enzymes (Badeaux & Shi, 2013). Downstream of HMT and HAT/KAT activities, proteins that contain chromodomain and plant homeo domain (PHD) finger motifs are able to recognize core histones bearing methylated lysines, whereas acetylated lysines are recognized by bromodomain-containing proteins (Patel & Wang, 2013).

Modifications of specific histone residues are dynamic and may be readily reversed by recruitment of Histone Lysine Demethylase (KDM) or Histone Deacetylases (HDAC) to the chromatin domain within which the modified residues lie. These enzymes catalytically cleave methyl or acetyl modifications from lysines and arginines of core histones, restoring the protein to its unmodified state. Similarly, protein phosphatase PP1α reverses the phosphorylation of Histone H3 on Serine 10 catalyzed by the MSK1 protein kinase.

THE IMPACTS OF HISTONE MODIFICATIONS ON GENE TRANSCRIPTION

Despite the progress that has been made in understanding the biological roles of proteins that generate, recognize, or remove histone modifications, the precise mechanistic functions of histone modifications themselves, in the regulation of gene transcription, remain unresolved. An extensive and growing body of data has documented the biochemical properties for many individual histone modifications and the wide varieties of proteins that create, bind to, or excise them. Moreover, mutual interdependencies through cross talk between histone-modifying enzymes, histone modification binding proteins, the DNA methylation machinery and other chromatin proteins have also been documented (Du et al., 2015; Zhang et al., 2015). However, the relationships of histone modification patterns in chromatin to the developmental decisions taken by DNA-binding transcription factors to activate or repress specific genes and to the functions of chromatin remodelers that make DNA sequences more

Table 1 Selected posttranslational modifications of residues in the tails of core histones known to be Involved in regulation of gene transcription

Core Histone	Modification	DNA Element	Role in Transcription	Modifier	Binder	Remover	Detected in Zebrafish? (References)
H2A	K4/5/7ac		Activation				
	K119uq	Promoter	Repression	RING1A			
H2B	S33P		Activation				
	K5/11/12ac		Activation				
	K15/16/20ac		Activation				
	K120uq	TSS Gene Body	Activation				
H3	K4me1	Enhancer TSS	Activation	MLL3/4 Set7/9		KDM1A	Yes [1,2,3,4,5,6]
	K4me2	Promoter Gene Body	Activation	MLL PRDM9	BPTF KDM5	KDM1A	Yes [7]
	K4me3	Promoter TSS Gene Body	Activation	Set1A/B MLL	JMJD2 BPTF ING2	KDM5	Yes [1,8,7,2,9,4, 5,6,10, 11,12, 13,14,15,16, 17,18,19]
	K9me1		Repression	G9a, Prdm3/16		KDM1A KDM3B PHF8	Yes [20]
	K9me2		Repression	G9a	HP1	KDM1A KDM3B PHF8	Yes [20,14]
	K9me3	Promoter	Repression	G9a Suv39h1	HP1	KDM4	Yes [9,20,13,15, 21,17]
	K27me1	Gene Body	Activation				
	K27me2	All Elements	Repression			KDM6	
	K27me3	Promoter TSS Gene Body	Repression	EZH2	Pc	KDM6	Yes [8,7,9,4,5, 6,11,13,15, 16,17,19,22]

	K36me3	Gene Body	Elongation	SETD2	PHF1, PHF16	KDM4	Yes [8,7,17,19, 23]
	R17me		Activation	PRMT4			Yes [24]
	K4ac	Promoter TSS	Activation				
	K9ac	Promoter TSS	Activation	MOZ	BRD7	HDAC1/2	Yes [14,15]
	K14ac	Promoter TSS	Activation	p300/CBP Tip60	CBP, BRD4 SMARCA4	HDAC1	Yes [7]
	K18ac	Promoter TSS	Activation	p300/CBP			
	K23ac	Promoter TSS	Activation	p300/CBP			
	K27ac	Promoter TSS Enhancer	Activation	CBP MOZ	p300	HDAC1/2/3	Yes [1,5,6,10,11]
	S10P		Activation	MSK1	14-3-3	PP1α	
	S28P		Activation	MSK1	14-3-3	PP1α	
H4	K20me1		Activation & Repression	Pr-SET7		PHF8 Suv420h1	Yes [21]
	K20me3	All Elements	Activation & Repression	SMYD3/5	JMJD2		
	K5ac		Activation	p300/CBP Tip60	BRD2/4	HDAC1/2	
	K8ac		Activation	p300/CBP Tip60	BRD3	HDAC1/2	
	K12ac		Activation	p300/CBP Tip60	BRD2	HDAC1/2	
	K16ac		Activation	p300/CBP Tip60	TAF1	HDAC1/2	

Core histones H2A, H2B, H3 and H4 are substrates for acetylation, methylation and phosphorylation on specific Lysine (K), Arginine (R) and Serine (S) residues in their N-terminal tails. The table indicates some of the known properties of these histone modifications, including their association with distinct gene regulatory and coding elements, their roles in transcription, the identities of enzymes that add or remove modifications, as well as proteins that bind to each modification. The histone modifications that have been detected in zebrafish are also identified (relevant references are indicated); [1] Lee et al. (2015); [2] Aday et al. (2011); [3] Gorkin et al. (2012); [4] Irimia et al. (2012); [5] Bogdanović et al. (2012); [6] Bogdanovic et al. (2013); [7] Wu et al. (2011); [8] Vastenhouw et al. (2010); [9] Andersen et al. (2012); [10] Tena et al. (2014); [11] Choe, Ladam, and Sagerström (2014); [12] Wardle et al. (2006); [13] Ostrup et al. (2014); [14] Harrison et al. (2011); [15] Lindeman, Winata, et al. (2010); [16] Stewart, Tsun, and Izpisua Belmonte (2009); [17] Lindeman et al. (2011); [18] Haberle et al. (2014); [19] Zhang et al. (2014); [20] Ferreira et al. (2014); [21] Li et al. (2011); [22] A.H. Wang et al. (2013); [23] Aanes et al. (2013); [24] Selvi et al. (2015).

or less accessible, remain much less well understood. Despite the large number of studies undertaken to date, there is relatively little evidence to support the idea that specific patterns of histone modifications represent an instructive epigenetic code that embodies precise information defining the biological circumstances under which a gene may or may not be transcribed. Rather, the emerging consensus is that epigenetic modifications and their associated proteins represent dynamic features within the physical landscape of chromatin, which facilitate or impede movement of DNA-binding transcription factors and RNA polymerases toward their target sites in DNA, through interactions with chromatin remodeling machinery (Bartholomew, 2014; Henikoff & Shilatifard, 2011). At any point in time, therefore, the epigenetic landscape includes a genome-wide pattern of restrictive and permissive histone modifications, which reflects the cell's current status and includes traces of both its developmental history and future potential. Zebrafish offer unique opportunities to investigate the structure, function, and dynamic regulation of this epigenetic landscape during development. After external fertilization, zebrafish embryos develop rapidly and synchronously in large clutches that are highly amenable to molecular analysis. Together with its genetic tractability, the embryo's optical clarity further enables gene function, gene expression, and cell behaviors to be investigated within the context of the whole organism, using a wide variety of molecular approaches and in vivo imaging techniques. With the recent development of cost-effective, high-throughput whole genome sequencing technology, experimental approaches that take full advantage of these particularly valuable attributes of the zebrafish, can now be deployed to elucidate the roles of epigenetic mechanisms in a range of different developmental settings.

1. RECENT PROGRESS IN UNDERSTANDING THE ROLES OF HISTONE MODIFICATIONS IN ZEBRAFISH DEVELOPMENT

Over the last 5 years, a growing number of studies taking genome-wide approaches have documented the changing patterns of histone modifications across the zebrafish genome during embryonic development. While initial studies employed genome tiling and defined promoter microarrays (Lindeman, Reiner, et al., 2010; Vastenhouw et al., 2010; Wardle et al., 2006), the sensitivity of these technologies is relatively limited, and they have since been superseded by whole genome shotgun sequence determination of DNA immunoprecipitates, using Illumina Genome Analyzer and more recently Illumina HiSeq technology.

1.1 CHROMATIN MODIFICATIONS AT THE PROMOTERS OF ZYGOTICALLY ACTIVE GENES

The process of zygotic genome activation (ZGA) has been particularly intensively investigated (Aanes et al., 2013; Andersen et al., 2012; Haberle et al., 2014;

Lindeman et al., 2011; Potok et al., 2013; Vastenhouw et al., 2010; Zhang et al., 2014). This event is the first major developmental transition where regulated gene transcription plays a deterministic role in zebrafish embryogenesis, with RNA polymerases initiating transcription coordinately in all embryonic blastomeres to generate their pluripotent phenotype. The first genome-wide studies of the epigenome at these stages used a commercially produced promoter microarray to identify a consistent pattern of colocalized "bivalent" enrichment for H3K4me3 and H3K27me3 at the promoters and transcription start sites (TSSs) of many "bivalent" genes, shortly before the onset of ZGA (Vastenhouw et al., 2010). By contrast, a similar number of other genes are marked "monovalently" with H3K4me3 only at this same pre-ZGA stage. These results provided in vivo confirmation of observations first reported in mammalian embryonic stem cells (ESC), that the H3K4me3 and H3K27me3 combination of histone modifications marks the promoters of genes involved in conferring pluripotency to embryonic blastomeres (Azuara et al., 2006; Bernstein et al., 2006; Harikumar & Meshorer, 2015; Vastenhouw et al., 2012). At the time of ZGA, only a subset of genes associated with either the monovalent H3K4me3 or bivalent H3K4me3/H3K27me3 modifications are actively transcribed, indicating that these chromatin modifications identify loci that are competent for induction and thus "poised" for transcriptional activation, as well as transcriptionally active genes. These findings have been confirmed in zebrafish ZF4 fibroblasts, in a study which also identified large numbers of H3K4me3/H3K27me3 bivalently marked genes, as well as H3K4me3 and H3K27me3 monovalent loci (Lindeman, Reiner, et al., 2010).

An important step forward in epigenome analysis has been to develop methods for integrating genome-wide histone modification data with transcriptomic data. A report describing such an approach revealed that high levels of the H3K4me1 monomethylation mark are found in combination with H3K4me3 in zebrafish embryos at a majority of transcriptionally active tissue-specific TSSs (Aday et al., 2011). Another aspect of early development in which the roles of histone modifications have been explored is the transformation of sperm chromatin into zygotic chromatin before it becomes transcriptionally active. A customized zebrafish promoter microarray was used to discover that sperm chromatin is enriched for H3K4me3 and H3K27me3, as well as the variant histone H2AFV and H3K36me3, in specific regions of the genome encoding developmentally regulated transcription factors and intercellular signaling proteins, which only become transcriptionally active during embryonic axis formation, regional specification, and organogenesis (Wu, Zhang, & Cairns, 2011). Similar embedding of the promoters of a variety of transcriptionally inactive developmental regulatory genes in multivalent chromatin comprising combinations of H3K4me3 with H3K27me3, H3K36me3 and/or H3K9me3, was also observed in 256-cell embryos long before onset of ZGA (Lindeman et al., 2011). A large class of promoters are localized within chromatin that is marked only by H3K4me3 in both sperm and pre-ZGA embryonic chromatin, some of which then become bivalent with either H3K27me3 or H3K9me3 after the onset of ZGA. Interestingly, those promoters which persist within chromatin that is monovalently marked with H3K4me3 have a greater propensity to become transcriptionally active post-ZGA than do bivalent promoters (Lindeman et al., 2011). While these results are consistent with inhibitory roles for H3K27me3 and H3K9me3, they raise the question of how monovalent H3K4me3-marked loci remain silent until ZGA. Moreover, these studies confirm that the ZGA is a developmental transition characterized by major remodeling of chromatin, but they also point to the existence of patterned histone modifications in sperm and the early zygote, which persist until ZGA. The theme of a persisting paternal epigenotype which provides a template for remodeling the maternal epigenotype, and transcriptome at ZGA, has been explored further in a study that used an integrative approach to align previously published histone modification data with whole genome deep sequences of transcriptomes and bisulfite-converted whole genomic DNA in sperm, pre-ZGA embryos, and post-ZGA embryos (Potok et al., 2013). The results confirmed that genes with multivalent histone modifications post-ZGA exhibit much lower levels of these modifications in sperm and that the ZGA is accompanied by major changes in the patterns of zygotic histone methylation. However, the patterns of DNA methylation in the sperm and post-ZGA embryonic epigenomes are extremely similar, because the oocyte methylome is progressively, radically remodeled to resemble that of the paternal methylome (which persists relatively unchanged) during the transition from fertilized egg to transcriptionally active sphere stage embryo (Potok et al., 2013). How this targeted remodeling is achieved is not at all well understood, but some regions of the maternal epigenome are less susceptible than other. For example, maternally transcribed genes are persistently preferentially hypomethylated and marked by H3K4me3 in embryos before and after ZGA (Andersen et al., 2012). Thus, H3K4me3-marking may help to suppress conversion of hypomethylated regions of the maternal methylome toward the patterns of methylation exhibited by these regions within the paternal methylome during embryonic development. In this way, patterns of histone modifications could protect genes from methylation that are involved in generating the maternal transcriptome in oocytes, by maintaining them in a transcriptionally competent state.

1.2 INTERPLAY BETWEEN HISTONE MODIFICATIONS AND CHROMATIN REMODELING MACHINERY

Nucleosomal mapping in chromatin, using micrococcal nuclease and high-throughput sequencing of digested DNA fragments, has recently been used to determine the genomic locations of nucleosomes with embryonic chromatin (Zhang et al., 2014). Prior to the onset of ZGA, the distribution of nucleosomes is relatively irregular with respect to DNA sequence. However, during ZGA, nucleosomes are organized into regular, precisely positioned arrays within embryonic chromatin, in a manner that is closely correlated with the emergence of H3K4me3-marked nucleosomes in gene promoter regions (Zhang et al., 2014). The mechanistic impetus for ZGA may thus be based on a global reorganization of chromatin that involves both the concerted recruitment of histone-modifying enzymes and chromatin remodeling factors to specific locations within the genome.

Further insights into the chromatin mechanisms that prime the maternal-zygotic switch underlying ZGA have emerged from studying chromatin domains harboring genes with distinct maternal and zygotic promoter architectures and TSS patterns (Haberle et al., 2014). Cap Analysis of Gene Expression (CAGE) was used for a high-resolution genome-wide, deep sequencing analysis of transcription start sites (TSS), leading to the identification of two distinct classes of TSS and promoter DNA sequence architectures in these zebrafish genes, ie, maternal and zygotic promoter architectures. Genes that are both maternally and zygotically expressed produce transcripts with TSSs that are characteristic of either the maternal or the zygotic classes, demonstrating the coexistence of distinct sequence elements within the same promoter that define each of the two types of TSS. Using chromatin immunoprecipitation (ChIP) to detect H3K4me3 and micrococcal nuclease chromatin digestion to map nucleosome positioning, it was demonstrated that H3K4me3-marked nucleosomes are precisely positioned ~50 bp downstream of zygotic TSSs at the onset of ZGA, before transcription is initiated (Haberle et al., 2014), which is consistent with other published findings (Zhang et al., 2014). By contrast, there is no association between maternal TSS selection and nucleosome positioning in the oocyte, indicating that the regular positioning of nucleosomes is an emergent, dynamic property of chromatin remodeling processes in the embryo that precedes the use of TSSs and targets chromatin domains that are H3K4me3 marked for ZGA.

1.3 ASSOCIATION OF COVALENTLY MODIFIED HISTONES WITH DEVELOPMENTALLY REGULATED ENHANCER SEQUENCES

Once lineage specification is underway, then the growing complexity and diversity of embryonic tissues makes the analysis of tissue-specific transcriptional regulation particularly dependent on having sufficient capacity to generate, analyze, and integrate large quantities of high-throughput, deep sequencing data of multiple types, such as transcription factor ChIP-seq, histone ChIP-seq, Bisulfite-seq, CAGE, and RNA-seq datasets. Cell type–specific enhancers are known to be key components of lineage-specific transcription mechanisms, but unlike promoters, their identities cannot be inferred on the basis of their position relative to the TSS and they may be located many tens or hundreds of kb upstream or downstream of their cognate transcription unit. However, many enhancers comprise DNA sequences that are evolutionarily highly conserved noncoding elements (CNEs; Goode et al., 2011) and moreover, evidence from multiple species indicates that active enhancers are frequently localized within domains of chromatin marked with H3K4me1 and H3K27ac (Creyghton et al., 2010; Heintzman et al., 2009). These properties have facilitated several studies that identified candidate enhancers in zebrafish (Aday et al., 2011; Bogdanović et al., 2012; Irimia et al., 2012; Lee et al., 2015; Tena et al., 2014). Enhancers that are associated with transcriptionally active genes involved in lineage specification and cell differentiation are enriched for DNA sequences associated with H3K4me1- and H3K27ac-marked histones during the first 48 hours postfertilization (hpf) (Bogdanović et al., 2012). Some evolutionarily highly conserved DNA sequences known as genomic regulatory blocks (GRBs), found within the introns of bystander genes lying close to tissue-specific developmental regulatory genes, also exhibit functional properties of enhancers and are specifically associated with H3K4me1-marked histones when the developmental regulatory gene is transcribed, even if the bystander gene in which they are located is transcriptionally inactive (Aday et al., 2011; Irimia et al., 2012). Moreover, chromatin conformation capture experiments show that an H3K4me1-associated, evolutionarily conserved GRB can physically interact with the promoter of an adjacent developmental regulatory gene, further indicating a likely role for GRBs as enhancers (Irimia et al., 2012). The existence of evolutionarily conserved regulatory DNA sequence elements in the genomes of related species might be predicted to underlie the existence of evolutionarily conserved patterns of histone modifications within the epigenomes of those species. In line with this idea, recent experiments have identified evolutionarily conserved patterns of H3K27ac and H3K4me3 associated with conserved developmental regulatory genes that are expressed in the phylotypic, pharyngula stages of zebrafish and Medaka. Moreover, whereas promoter sequences from these genes are associated with evolutionarily conserved patterns of both H3K27ac and H3K4me3 modifications, more distally located, putative enhancer elements are associated with H3K27ac but not H3K4me3 (Tena et al., 2014). Taken together, these results indicate that the patterns of both histone modifications and the regulatory DNA sequences with which they are physically associated exhibit evolutionary conservation that reflects the combined functions of both chromatin proteins and the associated DNA in transcription. An even larger number of zebrafish enhancer elements were recently identified on the basis of their differential CpG methylation patterns in different developmental stages (Lee et al., 2015). ChIP-seq analysis of these differentially methylated regions (DMRs) during development revealed enrichment of H3K4me1 and H3K27ac centered on each DMR, consistent with putative enhancer functions. Gene Ontology (GO) analysis further confirmed that these DNA sequences are enriched in genomic regions encoding developmental regulatory genes, and transcription factor binding motif analysis, using the HOMER and DAVID algorithms, identified regulatory roles for SOX family, and nuclear receptor proteins in the functions of the DMRs themselves (Lee et al., 2015).

2. CHROMATIN IMMUNOPRECIPITATION FROM ZEBRAFISH EMBRYOS

2.1 RATIONALE

ChIP is a remarkably powerful technique for investigating the structure and function of chromatin components. The distribution and regulation of specific histone modifications in chromatin can be analyzed using antibodies that recognize individual modifications themselves or the proteins that generate, bind to, or remove them. To analyze histone modifications, their regulators and effectors, using immunoprecipitation, chromatin extracts must be prepared in which the modified histones and regulatory proteins remain intact and soluble, and their physical associations with other components of chromatin are largely undisturbed. To stabilize these interactions during the extract preparation and immunoprecipitation steps, chromatin is chemically cross-linked with chemicals such as ethylene glycol bis(succinic acid N-hydroxysuccinimide ester) (EGS) and formaldehyde. Formaldehyde is an effective cross-linker for covalently coupling proteins to DNA sequences with which they are in direct physical contact, and EGS has a long spacer between its reactive moieties, which allows chromatin proteins that do not contact DNA directly to be cross-linked to the DNA sequences with which they are associated. After separation and removal of sticky yolk platelets, embryos are homogenized and a soluble chromatin extract is prepared that is randomly sheared by sonication. Sonicated chromatin extracts are then incubated with antibodies that are specific for the chromatin proteins and/or covalent modifications of interest and which have been validated for immunoprecipitation. A wide range of validated antibodies is now available from a range of commercial suppliers, such as Abcam, Covance, Diagenode, Millipore, or Sigma. Once chromatin immunoprecipitates have been purified, the chemical cross-links are reversed and the DNA is subjected to end-repair, adapter ligation, and PCR amplification, and the fragments are then sequenced on an Illumina platform.

The following protocol is a versatile method for preparing chromatin immunoprecipitates from zebrafish embryos that may be used to detect specific transcription factors, histones, and other chromatin proteins, and analyze their interactions with DNA. The protocol was originally developed to investigate the functions of histone deacetylase Hdac1 in zebrafish embryonic chromatin (Harrison et al., 2011).

2.2 CHROMATIN IMMUNOPRECIPITATION METHOD

2.2.1 Dechorionation of embryos and cross-linking of chromatin with EGS

1. Collect zebrafish embryos and dispense 50 embryos per 80 mm petri dish. Incubate in E3 embryo medium until embryos reach the desired stage of development.
2. Collect 300 embryos into a 5 mL flat-bottomed plastic universal container. Add 5 mL E3 medium containing 10 mg/mL Pronase. Place universal on a benchtop rotator and rotate gently at 16 rpm for 30 min at room temperature. In the meantime, prepare a fresh solution of the chemical cross-linking agent Ethylene glycol bis(succinic acid N-hydroxysuccinimide ester) (EGS; Sigma) as set out below. If EGS treatment is not required, then proceed directly to cross-linking with formaldehyde in Section 2.2.2.
3. Dissolve 10 mg EGS in 200 μL DMSO in a 1.5 mL Eppendorf tube. Vortex to dissolve.

4. Add 150 μL (EGS in DMSO) to 10 mL phosphate-buffered saline (PBS) to a final concentration of 1.5 mM.
5. Remove the E3-Pronase solution from the embryos and rinse twice with PBS, pipetting up and down carefully to remove the chorions.
6. Remove PBS and add 5 mL EGS-PBS solution. Return the tube to the rotator and rotate at 16 rpm for a further 20 min at room temperature.

2.2.2 Cross-linking of chromatin with formaldehyde

1. Remove EGS-PBS solution and replace with 5 mL PBS. Add 200 μL 37% formaldehyde solution.
2. Return the tube to the rotator and rotate at 16 rpm for a further 20 min at room temperature.
3. Remove the solution and replace with PBS and wash once.
4. Replace PBS wash with a fresh 5 mL of PBS and add 0.625 mL of 1.25M Glycine to quench the cross-linking process. Return the tube to the rotator and rotate at 16 rpm for a further 5 min at room temperature.
5. Transfer the embryos in an aliquot of PBS-glycine to a 1.5 mL Eppendorf tube.

2.2.3 Deyolking of embryos

1. Remove the solution above the settled clump of embryos and add 1 mL of Deyolking Buffer.
2. Pipette the buffer up and down to disrupt the embryonic yolk masses.
3. Tape each tube to the top of a vortexer and shake the embryos for 5 min at a relatively low setting (eg, number 3).
4. Spin the tube in a microfuge at 2000 rpm for 1 min.
5. Remove supernatant and wash pellet with 1 mL Wash Buffer for 2 min on the vortexer at setting number 3.
6. Spin again in a microfuge at 2000 rpm for 1 min. Remove the supernatant and spin again with fresh 1 mL of Wash Buffer. Discard supernatant.

2.2.4 Chromatin preparation from embryos

1. Add 1 μL of each of the Sigma-supplied stocks of pepstatin, leupeptin, and PMSF proteinase inhibitors to 1 mL of DNA Extraction Buffer. Add 600 μL of this solution to the pellet of embryos. Using a plastic pestle (disposable, but reautoclavable), fitted to the Eppendorf tube, homogenize the embryos. Cap the Eppendorf tube, place it into a 50 mL Falcon tube, fill with tissue paper, and place on roller in the cold room. Rotate for 2 h at 4°C.

2.2.5 Sonication

1. Place Eppendorf tube containing homogenate into a small beaker filled with ice.
2. Place the beaker in the Beckman Soniprep, and insert the probe into the Eppendorf tube, lifting the tube off ice so that the probe tip is placed deep within the homogenate but not touching the bottom of the Eppendorf tube.

3. Wearing ear protectors, sonicate as follows: 20 s at 16 μm, then 6× for 15 s each time at 14 μm, with a period of 45 s cooling off in between each of the six bursts, by placing the tube on ice without sonication.
4. After sonication, place the tube on ice and return to the cold room. Microfuge at 14,000 rpm for 30 min at 4°C.
5. Remove the supernatant cleanly without disturbing the pellet and transfer it to a clean Eppendorf tube on ice. This is the Chromatin Extract.
6. Remove ~10% of the extract (~60 μL) and save it at 4°C. This is the Input Sample.
7. Take 20 μL of the Chromatin Extract to test the effectiveness of the sonication procedure. Add 280 μL Digestion Buffer, 2 μL Ribonuclease A (10 mg/mL) and incubate overnight at 65°C.
8. Store the remaining Chromatin Extract at 4°C.

2.2.6 Sonication test

1. Next day, add 2 μL of 10 mg/mL Proteinase K to the 300 μL sample of Chromatin Extract in Digestion Buffer. Incubate 2 h at 42°C. This treatment will reverse the protein–DNA cross-links.
2. Add 98 μL Digestion Buffer, then 400 μL phenol-chloroform-isoamyl alcohol (PCIA) and vortex to form an emulsion. Spin at room temperature in microfuge at 12,000 rpm for 5 min, then remove aqueous layer (400 μL), and add 40 μL 3M NaAc pH5.2. Add 1 mL absolute ethanol, mix, and place on dry ice for 20 min or at −20°C for a minimum of 2 h.
3. Microfuge in cold room for 30 min at 12,000 rpm, wash pellet with 1 mL 75% ethanol, spin again for 5 min in microfuge, and then remove ethanol carefully and discard.
4. Air dry pellet and resuspend in 20 μL distilled water.
5. Add loading dyes/buffer and load sample onto 1% agarose gel (that does not contain ethidium bromide) to check the size distribution of the sonicated DNA fragments. Run 1 kb ladder alongside sample.
6. After electrophoresis at 80 V for an hour, transfer the gel to a staining tank containing 100 mL 1× TAE, and add 5 μL ethidium bromide stock solution (10 mg/mL). Stain at room temperature. If the population of DNA fragments is satisfactory, proceed with ChIP protocol.

2.2.7 Incubation of chromatin extract with antibodies

1. Split the remaining Chromatin Extract into two samples of ~260 μL. To one sample, add 6 μg of primary antibody specific for the chromatin protein, chromatin modification, or transcription factor of interest. To the other sample, add 6 μg control IgG. Mix samples carefully, place Eppendorf tubes in 50 mL Falcon tubes containing tissues to cradle Eppendorfs, and rotate gently in cold room for 2–3 h.

2. Transfer 80 μL aliquots of Protein G magnetic beads into Eppendorf tubes. Wash beads 3 times with 1 mL PBS containing 0.5% BSA, using magnetic stand to collect beads.
3. Add the washed beads to the Chromatin Extract samples containing added IgGs.
4. Return Eppendorfs to 50 mL Falcon tubes containing tissues to cradle Eppendorfs, and rotate gently in cold room overnight.

2.2.8 Chromatin-antibody complex washes

1. Retrieve the Eppendorf tubes and briefly spin them down to collect the samples in the bottom of each tube. Place the tubes in the magnetic stand to pellet the Protein G beads.
2. Remove supernatant, discard, and wash the beads, in the cold room, as follows:
3. 3× washes with 0.5–1 mL 1× DNA Extraction Buffer.
4. 3× washes with 0.5–1 mL 5× RIPA Buffer.
5. 1× wash with 0.5–1 mL LiCl Buffer.
6. 1× wash with 0.5–1 mL 1× DNA Extraction Buffer.
   For each wash: place tubes in magnetic stand for 1 min to let beads settle. Remove supernatant. Place tubes on ice then add the wash buffer. Pipette up and down until beads are resuspended. Place tubes on rotator in cold room for 5 min. Place tubes in magnetic stand and let beads settle for 1 min. Begin next wash.
7. After last wash, transfer beads in 1× DNA Extraction Buffer equally to two new Eppendorf tubes, place in magnetic stand, and remove 1× DNA Extraction Buffer once beads have settled.
8. Add 100 μL Elution Buffer and incubate at 65°C for 20 min in a hot block.
9. Microfuge at 6000 rpm at room temperature for 3 min. Transfer supernatant to a fresh tube.

2.2.9 Purification of immunoprecipitated DNA

1. Add 200 μL Digestion Buffer, and then add 2 μL 10 mg/mL Ribonuclease A and incubate at 65°C for 6 h to overnight, to reverse the protein–DNA cross-links.
2. Next day, add 2 μL 10 mg/mL Proteinase K and incubate samples at 42°C for 2 h.
3. Extract samples with PCIA and precipitate DNA with absolute ethanol. Wash pelleted DNA with 75% Ethanol, air dry, and resuspend in 5–20 μL distilled water.
4. To proceed to high-throughput sequencing, an Agilent Tapestation System or equivalent should be used to check that DNA samples are in the required size range of 200–300 bp, and samples are accurately quantified using a fluorimeter such as the Qubit. DNA samples are then end-repaired with NEBNext, purified A-tailed, ligated to Illumina adapters, and amplified by PCR, prior to DNA sequence analysis.

2.2.10 Buffers and solutions

Pronase Solution

Add 200 μL of Pronase stock (5–10 mg/mL) to 5 mL E3 medium (without methylene blue).

1× DNA Extraction Buffer (50 mL)

Stock Conc.	Volume added	Final Conc.
1M Tris pH 8.2 (100×)	500 μL	10 mM Tris pH 8.2
0.5M EDTA (50×)	1 mL	10 mM EDTA
5M NaCl (25×)	2 mL	200 mM NaCl
NP40	250 μL	0.5%

5× DNA Extraction Buffer (50 mL)

Stock Conc.	Volume added	Final Conc.
1M Tris HCl pH 8.2 (100×)	500 μL	10 mM Tris pH 8.2
0.5M Na EDTA (50×)	1 mL	10 mM EDTA
5M NaCl (25×)	2 mL	200 mM NaCl
NP40	250 μL	0.5%
10% SDS	2.5 mL	0.5% SDS

5× RIPA Buffer (50 mL)

Stock Conc.	Volume added	Final Conc.
1M Tris HCl pH 8.0	500 μL	10 mM Tris
0.5M Na EDTA	100 μL	1 mM EDTA
50 mM EGTA	500 μL	0.5 mM EGTA
5M NaCl	5 mL	500 mM NaCl
10% SDS	500 μL	0.1% SDS
Triton X-100	500 μL	1% Triton
10% Sodium deoxycholate	5 mL	0.1% cholate

Digestion Buffer (50 mL)

Stock Conc.	Volume added	Final Conc.
1M Tris HCl pH 8.0	2.5 mL	50 mM Tris
500 mM Na EDTA	100 μL	1 mM EDTA
5M NaCl	1 mL	100 mM NaCl
10% SDS	2.5 mL	0.5% SDS

LiCl Buffer (50 mL)

Stock Conc.	Volume added	Final Conc.
1M Tris HCl pH 8.0	500 μL	10 mM Tris
500 mM Na EDTA	100 μL	1 mM EDTA
50 mM Na EGTA	500 μL	0.5 mM EGTA
1M LiCl	12.5 mL	250 mM LiCl
Triton X-100	500 μL	1% Triton
10% Sodium deoxycholate	5 mL	1% cholate

Elution Buffer (15 mL)

Stock Conc.	Volume added	Final Conc.
1M Tris HCl pH 7.5	375 μL	25 mM Tris
500 mM Na EDTA	300 μL	10 mM EDTA
10% SDS	750 μL	0.5% SDS

Deyolking Buffer (100 mL)

Final Conc.	Mass per 100 mL H_2O
55 mM NaCl	0.32 g
1.8 mM KCl	0.01 g
1.25 mM $NaHCO_3$	0.01 g
2.7 mM $CaCl_2$	0.04 g

Wash Buffer (100 mL)

Final Conc.	Mass/Vol per 100 mL H_2O
110 mM NaCl	0.64 g
3.5 mM KCl	0.03 g
2.7 mM$CaCl_2$	0.04 g
10 mM Tris pH 8.5	1 mL 1M stock

3. HIGH-THROUGHPUT WHOLE-GENOME DNA SEQUENCING AND INTEGRATIVE ANALYSIS OF EPIGENOMIC DATA

Specific studies of particular genes and the local histone modification context within which their transcription is regulated can provide important mechanistic insights into the regulatory mechanisms for those genes in particular cells or tissues and illustrate the principles of their operation. However, when new knowledge about the broader role of the epigenome in a particular biological process is sought without bias toward a particular gene or genomic region, then genome-wide approaches that are able to identify and analyze epigenetic modifications across the entire genome are highly desirable. Before reliable approaches were available for high-throughput analysis of entire genome sequences, microarray-based methods for interrogating collections of experimentally annotated sequences that were associated with or which included specific genes, such as promoter arrays, were used to provide initial insights into the roles of epigenetic mechanisms in regulation of gene transcription. However, comprehensive DNA genome-wide sequence analysis of chromatin immunoprecipitates, using technology platforms such as the Illumina HiSeq is now the method of choice for a growing range of genomic, epigenomic, and transcriptomic applications. Short-read sequence files are generated from fragment libraries as fastq files, which are checked for quality using the FastQC program, then trimmed to remove low-quality base calls. Sequences are mapped to a reference genome sequence using one of several alignment algorithms that have been validated for zebrafish, such as Bowtie/Bowtie 2.0 (Langmead & Salzberg, 2012; Langmead et al., 2009), BWA (Li & Durbin, 2009), SOAP2 (Yu et al., 2012), or Novoalign (Yu et al., 2012). These and other analysis tools are available within web browser–based platforms such as the Galaxy Project (https://galaxyproject.org/) or the Cistrome Analysis Pipeline (http://cistrome.org/Cistrome/Cistrome_Project.html). Once mapped, peaks of histone modifications or transcription factor binding are called using software such as model-based analysis of ChIP-Seq (MACS) (Zhang et al., 2008), which measures the level of enrichment of short stretches of DNA sequences in the immunoprecipitated DNA sample, relative to their occurrence in the whole genome sequence (Fig. 1). MACS generates BED files, which give the locations of the histone modification/transcription factor binding peaks, and WIG files, which specify the properties of the signal profiles. BED and WIG files are then analyzed with a suite of tools within platforms, such as the Galaxy Project or the Cistrome Analysis Pipeline, to identify overlaps and complementarities between different patterns of chromatin modifications, map transcription factor binding profiles, and integrate new data with other annotated tracks that are aligned to the reference genome sequence (Blankenberg, Taylor, & Nekrutenko, 2015; Liu et al., 2011; Quinlan, 2014; Quinlan & Hall, 2010). A wide range of analytical and visualization tools are available within these platforms, which facilitate comparative analysis of ChIP-seq data with transcriptomic, methylomic, transcription factor binding and other types of data (eg, the Kyoto Encyclopedia of Genes and Genomes (KEGG) database

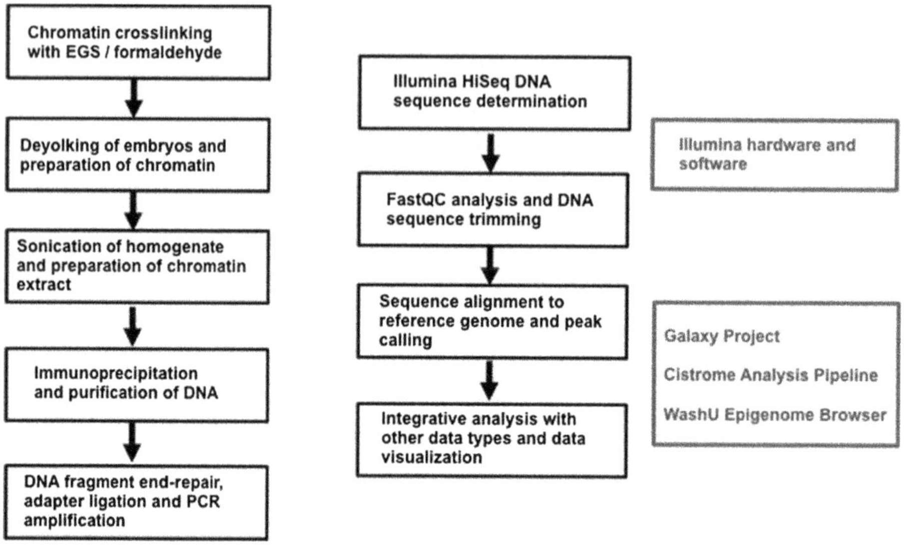

FIGURE 1

Schematic outline of the methods described for high-throughput epigenome analysis in zebrafish, using ChIP-seq to generate whole epigenome maps of histone modification patterns and/or transcription factor or other chromatin protein-binding profiles. The *blue box* (gray in print versions) indicates the bioinformatics pipelines and web browsers that are currently used for genome wide visualization and integrative analysis of epigenomic datasets.

contents), aligned to reference genomes in the UCSC genome browsers or the Washington University Epigenome Browser (Bailey et al., 2013; Gorkin et al., 2012; S. Wang et al., 2013; Zhou & Wang, 2012; Zhou et al., 2011). GO enrichment for GO terms can be performed with GREAT, DAVID, or PANTHER algorithms (Lee et al., 2015; Tena et al., 2014). An illustration of the type of output from the Washington University EpiGenome Browser is shown in Fig. 2, from the public track hub containing a wealth of published zebrafish epigenome data (http://epigenomegateway.wustl.edu/browser/). The RefSeq gene *pax2a* is aligned with ChIP-seq data for a selected range of whole embryo and whole larva histone modification tracks for several different developmental stages. A methylated DNA immunoprecipitation sequencing (MeDIP-seq) track from 24 hpf embryos and RNA-seq data from seven different developmental stages are also included. Of particular interest is the methylation-free CpG island encompassing the *pax2a* promoter and TSS, which is enriched in H3K27ac, H3K4me1, H3K4me3, and H3K27me3 modifications at multiple developmental stages. The *pax2a* gene body is strongly associated with the transcriptional elongation modification, H3K36me3, in 24 hpf embryos and 72 hpf larvae, when *pax2a* is strongly transcribed. By contrast, in shield and high stages, when *pax2a* is transcriptionally inactive, the *pax2a* promoter and TSS are enriched in the repressive H3K27me3 modification and deficient in H3K36me3 and H3K9me3.

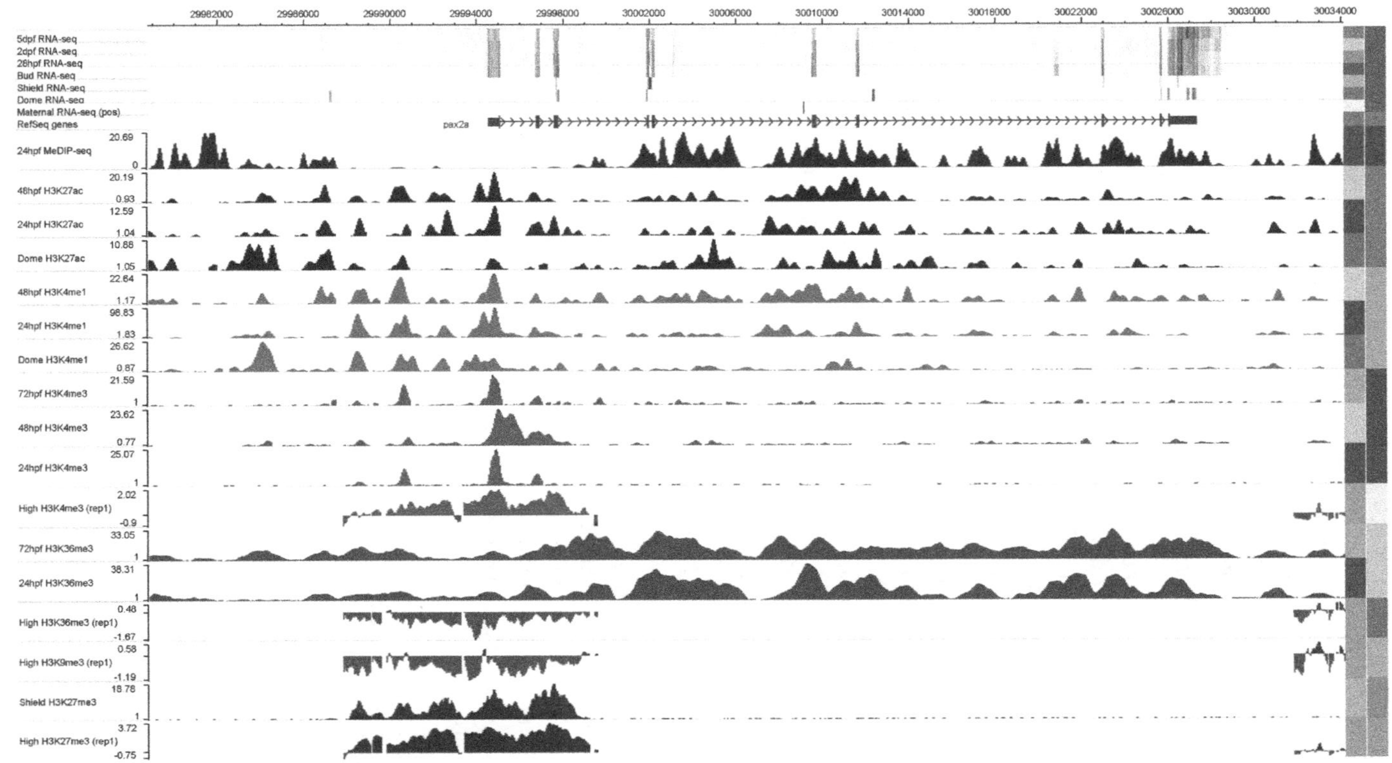

FIGURE 2

An example of the Washington University Epigenome Browser output for a specific region of chromatin encompassing the zebrafish *pax2a* gene, in whole embryos, analyzed at multiple stages of development. A selection of publicly accessible data tracks showing selected histone modification patterns is aligned with the *pax2a* gene, along with RNA seq data on *pax2a* transcripts at multiple stages of development (located above the RefSeq gene track), and MeDIP-seq data illustrating DNA methylation patterns around the *pax2a* locus (located immediately below the RefSeq gene track). (See color plate)

CONCLUSIONS AND PROSPECTS

Recent advances in the development of methods for comprehensive genome-wide analysis of histone modifications in the zebrafish embryo have provided valuable insights into important processes that regulate early development. Histone modifications provide distinct signatures for promoters that become transcriptionally active at the onset of ZGA, and H3K4me3 seems to play a critical role in facilitating the remodeling of chromatin so that the first nucleosome in the first exon of genes whose transcription is initiated at the ZGA is located ~50 bp downstream of its TSS (Haberle et al., 2014). Epigenomic analysis of developmentally regulated, lineage-specific enhancer elements further indicates that enhancer activation is accompanied by enrichment of the H3K4me1 and H3K27ac modifications in the enhancer centers, and that these modification patterns, like their cognate DNA sequences, include evolutionarily conserved features. The molecular methods now available for high-resolution global epigenome analysis in zebrafish are extremely powerful and include chromatin conformation capture to map *trans*-interactions between different chromatin domains (Ay & Noble, 2015; Irimia et al., 2012). In combination with the development of open-source computational platforms for integrative analysis and visualization of epigenomic data, such as the Galaxy Project, the Cistrome Analysis Pipeline, and the Washington University EpiGenome Browser, these resources are helping to create a readily accessible, data-rich scientific infrastructure that will be a fertile ground for hypothesis-driven studies of developmental regulatory mechanisms. While the zebrafish has proved to be an exceptionally tractable model organism for developmental biology, it also promises to be of considerable value to biologists seeking to advance understanding of many aspects of physiology and behavior, such as how extrinsic factors influence epigenetic regulation of phenotypic plasticity across the life course, including the molecular basis of adaptations that underlie healthy aging and changes that engender vulnerability to chronic diseases (Cunliffe, 2015). The methods described in this chapter are likely to be invaluable in advancing such studies in the foreseeable future.

ACKNOWLEDGMENTS

I am grateful to Michael Harrison and Aristophanes Georgiou for their contributions in developing and optimizing immunoprecipitation methods, to Sheffield aquarium staff for excellent fish husbandry, and to MRC, BBSRC, ESRC, and the Wellcome Trust for the funding to support my laboratory's research in this area.

REFERENCES

Aanes, H., et al. (2013). Differential transcript isoform usage pre-and post-zygotic genome activation in zebrafish. *BMS Genomics, 14*, 331.

Aday, A. W., et al. (2011). Identification of cis regulatory features in the embryonic zebrafish genome through large-scale profiling of H3K4me1 and H3K4me3 binding sites. *Developmental Biology, 357*, 450–462.

Andersen, I. S., et al. (2012). Developmental features of DNA methylation during activation of the embryonic zebrafish genome. *Genome Biology, 13*, R65.

Ay, F., & Noble, W. S. (2015). Analysis methods for studying the 3D architecture of the genome. *Genome Biology, 16*(1), 183. Available at: http://genomebiology.com/2015/16/1/183.

Azuara, V., et al. (2006). Chromatin signatures of pluripotent cell lines. *Nature Cell Biology, 8*(5), 532–538. Available at: http://www.ncbi.nlm.nih.gov/pubmed/16570078.

Badeaux, A. I., & Shi, Y. (2013). Emerging roles for chromatin as a signal integration and storage platform. *Nature Reviews Molecular Cell Biology, 14*(4), 211–224. Available at: http://www.pubmedcentral.nih.gov/articlerender.fcgi?artid=4082330&tool=pmcentrez&rendertype=abstract.

Bailey, T., et al. (2013). Practical guidelines for the comprehensive analysis of ChIP-seq data. *PLoS Computational Biology, 9*(11), 5–12.

Bartholomew, B. (February 2014). Regulating the chromatin landscape: structural and mechanistic perspectives. *Annual Review of Biochemistry, 83*, 671–696. Available at: http://www.ncbi.nlm.nih.gov/pubmed/24606138.

Bernstein, B. E., et al. (2006). A bivalent chromatin structure marks key developmental genes in embryonic stem cells. *Cell, 125*(2), 315–326.

Blankenberg, D., Taylor, J., & Nekrutenko, A. (2015). Online resources for genomic analysis using high-throughput sequencing. *Cold Spring Harbor Protocols*, 324–336. Available at: http://www.ncbi.nlm.nih.gov/pubmed/25655493.

Bogdanović, O., et al. (2012). Dynamics of enhancer chromatin signatures mark the transition from pluripotency to cell specification during embryogenesis. *Genome Research, 22*(10), 2043–2053.

Bogdanovic, O., et al. (2013). The developmental epigenomics toolbox: ChIP-seq and MethylCap-seq profiling of early zebrafish embryos. *Methods, 62*, 207–215.

Chen, T., & Dent, S. (2014). Chromatin modifiers: regulators of cellular differentiation. *Nature Reviews Genetics, 15*(2), 93–106.

Choe, S.-K., Ladam, F., & Sagerström, C. G. (2014). Tale factors poise promoters for activation by Hox proteins. *Developmental Cell, 28*, 203–211. Available at: http://dx.doi.org/10.1016/j.devcel.2013.12.011.

Creyghton, M. P., et al. (2010). Histone H3K27ac separates active from poised enhancers and predicts developmental state. *Proceedings of the National Academy of Sciences of the United States of America, 107*(50), 21931–21936. Available at: http://www.pnas.org/cgi/doi/10.1073/pnas.1016071107.

Cunliffe, V. T. (2015). Experience-sensitive epigenetic mechanisms, developmental plasticity, and the biological embedding of chronic disease risk. *Wiley Interdisciplinary Reviews: Systems Biology and Medicine, 7*(2), 53–71. Available at: http://dx.doi.org/10.1002/wsbm.1291.

Du, J., et al. (2015). DNA methylation pathways and their crosstalk with histone methylation. *Nature Reviews Molecular Cell Biology, 16*(9), 519–532. Available at: http://dx.doi.org/10.1038/nrm4043.

Ferreira, T., et al. (2014). Silencing of odorant receptor genes by G protein βγ signaling ensures the expression of one odorant receptor per olfactory sensory neuron. *Neuron, 81*, 847–859. Available at: http://dx.doi.org/10.1016/j.neuron.2014.01.001.

Goode, D. K., et al. (2011). Minor change, major difference: divergent functions of highly conserved cis-regulatory elements subsequent to whole genome duplication events. *Development (Cambridge, England), 884*, 879–884.

Gorkin, D. U., et al. (2012). Integration of ChIP-seq and machine learning reveals enhancers and a predictive regulatory sequence vocabulary in melanocytes. *Genome Research, 22*, 2290–2301.

Haberle, V., et al. (2014). Two independent transcription initiation codes overlap on vertebrate core promoters. *Nature, 507*, 381–385.

Harikumar, A., & Meshorer, E. (2015). Chromatin remodeling and bivalent histone modifications in embryonic stem cells. *EMBO Reports, 16*(12), 1609–1619.

Harrison, M. R. M., et al. (2011). The epigenetic regulator histone deacetylase 1 promotes transcription of a core neurogenic programme in zebrafish embryos. *BMC Genomics, 12*(1), 24. Available at: http://www.biomedcentral.com/1471-2164/12/24.

Heintzman, N. D., et al. (2009). Histone modification at human enhancers reflect global cell-type specific gene expression. *Nature, 459*(7243), 108–112.

Henikoff, S., & Shilatifard, A. (2011). Histone modification: cause or cog? *Trends in Genetics, 27*(10), 389–396. Available at: http://linkinghub.elsevier.com/retrieve/pii/S0168952511000977.

Irimia, M., et al. (2012). Extensive conservation of ancient microsynteny across metazoans due to cis-regulatory constraints. *Genome Research, 22*, 2356–2367.

Kouzarides, T. (2007). Chromatin modifications and their function. *Cell, 128*(4), 693–705. Available at: http://www.sciencedirect.com/science/article/pii/S0092867407001845.

Langmead, B., & Salzberg, S. L. (2012). Fast gapped-read alignment with Bowtie 2. *Nature Methods, 9*(4), 357–359. Available at: http://dx.doi.org/10.1038/nmeth.1923.

Langmead, B., et al. (2009). Ultrafast and memory-efficient alignment of short DNA sequences to the human genome. *Genome Biology, 10*(3), R25. Available at: http://www.ncbi.nlm.nih.gov/entrez/query.fcgi?cmd=Retrieve&db=PubMed&dopt=Citation&list_uids=19261174.

Lee, H. J., et al. (2015). Developmental enhancers revealed by extensive DNA methylome maps of zebrafish early embryos. *Nature Communications, 6*, 6315. Available at: http://www.ncbi.nlm.nih.gov/pubmed/25697895.

Li, H., & Durbin, R. (2009). Fast and accurate short read alignment with Burrows – Wheeler transform. *Bioinformatics (Oxford, England), 25*(14), 1754–1760.

Li, Z., et al. (2011). Histone H4 Lys 20 monomethylation by histone methylase SET8 mediates Wnt target gene activation. *Proceedings of the National Academy of Sciences of the United States of America, 108*(8), 3116–3123.

Lindeman, L. C., Reiner, A. H., et al. (2010). Tiling histone H3 lysine 4 and 27 methylation in zebrafish using high-density microarrays. *PLoS One, 5*(12), e15651.

Lindeman, L. C., Winata, C. L., et al. (2010). Chromatin states of developmentally-regulated genes revealed by DNA and histone methylation patterns in zebrafish embryos. *International Journal of Developmental Biology, 54*, 803–813.

Lindeman, L. C., et al. (2011). Prepatterning of developmental gene expression by modified histones before zygotic genome activation. *Developmental Cell, 21*, 993–1004.

Liu, T., et al. (2011). Cistrome: an integrative platform for transcriptional regulation studies. *Genome Biology, 12*(8), R83. Available at: http://genomebiology.com/2011/12/8/R83.

Ostrup, O., et al. (2014). The specific alteration of histone methylation profiles by DZNep during early zebrafish development. *Biochimica et Biophysica Acta, 1839*, 1307–1315.

Patel, D., & Wang, Z. (2013). Readout of epigenetic modifications. *Annual Review of Biochemistry, 82*, 81–118.

Potok, M. E., et al. (2013). Reprogramming the maternal zebrafish genome after fertilization to match the paternal methylation pattern. *Cell, 153*, 759–772. Available at: http://dx.doi.org/10.1016/j.cell.2013.04.030.

Quinlan, A. R. (2014). BEDTools: the Swiss-army tool for genome feature analysis. *Curr. Protoc. Bioinformatics, 47*, 11.12.1–11.12.34. Available at: http://www.ncbi.nlm.nih.gov/pubmed/25199790.

Quinlan, A. R., & Hall, I. M. (2010). BEDTools: a flexible suite of utilities for comparing genomic features. *Bioinformatics (Oxford, England), 26*(6), 841–842. Available at: http://www.pubmedcentral.nih.gov/articlerender.fcgi?artid=2832824&tool=pmcentrez&rendertype=abstract/n http://bioinformatics.oxfordjournals.org/content/26/6/841.short.

Selvi, B. R., et al. (2015). CARM1 regulates astroglial lineage through transcriptional regulation of Nanog and posttranscriptional regulation by miR92a. *Molecular Biology of the Cell, 26*(2), 316–326. Available at: http://www.pubmedcentral.nih.gov/articlerender.fcgi?artid=4294678&tool=pmcentrez&rendertype=abstract.

Stewart, S., Tsun, Z.-Y., & Izpisua Belmonte, J. C. (2009). A histone demethylase is necessary for regeneration in zebrafish. *Proceedings of the National Academy of Sciences of the United States of America, 106*(47), 19889–19894.

Tena, J. J., et al. (2014). Comparative epigenomics in distantly related teleost species identifies conserved cis-regulatory nodes active during the vertebrate phylotypic period. *Genome Research, 24*, 1075–1085.

Vastenhouw, N. L., et al. (2010). Chromatin signature of embryonic pluripotency is established during genome activation. *Nature, 464*.

Vastenhouw, N. L., et al. (2012). Bivalent histone modifications in early embryogenesis this review comes from a themed issue on nucleus and gene expression edited. *Current Opinion in Cell Biology, 24*, 374–386.

Wang, A. H., et al. (2013). The histone chaperone Spt6 coordinates histone H3K27 demethylation and myogenesis. *EMBO Journal, 32*, 1075–1086.

Wang, S., et al. (2013). Target analysis by integration of transcriptome and ChIP-seq data with BETA. *Nature Protocols, 8*(12), 2502–2515. Available at: http://www.nature.com/doifinder/10.1038/nprot.2013.150.

Wardle, F. C., et al. (2006). Zebrafish promoter microarrays identify actively transcribed embryonic genes. *Genome Biology, 7*(8), R71.

Wu, S.-F., Zhang, H., & Cairns, B. R. (2011). Genes for embryo development are packaged in blocks of multivalent chromatin in zebrafish sperm. *Genome Research, 21*, 578–589.

Yu, X., et al. (2012). How do alignment programs perform on sequencing data with varying qualities and from repetitive regions? *BioData Mining, 5*(1), 6.

Zhang, Y., et al. (2008). Model-based analysis of ChIP-seq (MACS). *Genome Biology, 9*(9), R137. Available at: http://genomebiology.com/2008/9/9/R137.

Zhang, Y., et al. (2014). Canonical nucleosome organization at promoters forms during genome activation. *Genome Research, 24*, 260–266.

Zhang, T., et al. (2015). The interplay of histone modifications – writers that read. *EMBO Reports, 16*(11), 1467–1481.

Zhou, X., & Wang, T. (2012). Using the wash U epigenome browser to examine genome-wide sequencing data. *Current Protocols in Bioinformatics* (Chapter 10), p. Unit10.10. Available at: http://www.pubmedcentral.nih.gov/articlerender.fcgi?artid=3643794&tool=pmcentrez&rendertype=abstract.

Zhou, X., et al. (2011). The human epigenome browser at Washington University. *Nature Methods, 8*(12), 989–990. Available at: http://dx.doi.org/10.1038/nmeth.1772.

CHAPTER

Chromatin immunoprecipitation and an open chromatin assay in zebrafish erythrocytes

20

S. Yang*[§,¶], C.J. Ott[§], M.P. Rossmann[||], M. Superdock*[§,¶], L.I. Zon*[§,¶,||], Y. Zhou*[§,¶,||,1]

**Boston Children's Hospital, Boston, MA, United States*

[§]Dana Farber Cancer Institute, Harvard Stem Cell Institute, Boston, MA, United States

[¶]Harvard Medical School and Howard Hughes Medical Institute, Boston, MA, United States

[||]Harvard University, Harvard, Cambridge, MA, United States

[1]Corresponding author: E-mail: yzhou@enders.tch.harvard.edu

CHAPTER OUTLINE

Methods in Cell Biology, Volume 135, ISSN 0091-679X, http://dx.doi.org/10.1016/bs.mcb.2016.04.019

Abstract

Zebrafish is an excellent genetic and developmental model for the study of vertebrate development and disease. Its ability to produce an abundance of transparent, externally developed embryos has facilitated large-scale genetic and chemical screens for the identification of critical genes and chemical factors that modulate developmental pathways. These studies can have profound implications for the diagnosis and treatment of a variety of human diseases. Recent advancements in molecular and genomic studies have provided valuable tools and resources for comprehensive and high-resolution analysis of epigenomes during cell specification and lineage differentiation throughout development. In this chapter, we describe two simple methods to evaluate protein–DNA interaction and chromatin architecture in erythrocytes from adult zebrafish. These are chromatin immunoprecipitation coupled with next-generation sequencing (ChIP-seq) and an assay for transposase-accessible chromatin with high-throughput sequencing (ATAC-seq). These techniques, together with gene expression profiling, are useful for analyzing epigenomic regulation of cell specification, differentiation, and function during zebrafish development in both normal and disease models

INTRODUCTION

As a well-established model for studies concerning vertebrate developmental biology and genetics, including areas of developmental hematopoiesis (Paik & Zon, 2010) and organ formation and function (Cox & Goessling, 2015; Jaźwińska & Sallin, 2016; Lagendijk, Yap, & Hogan, 2014; Quiros-Gonzalez & Yadav, 2014; Schlegel & Gut, 2015; Takeuchi, 2014), zebrafish has been reliably utilized to model human biology and disease and develop therapies and drugs for human disease treatment (Asnani & Peterson, 2014; Carroll & North, 2014; Goessling & Sadler, 2015; Gurevich, Siegel, & Currie, 2015; Haesemeyer & Schier, 2015; van Houcke, De Groef, Dekeyster, & Moons, 2015; Jörgens et al., 2015; McCammon & Sive, 2015; Mohseny & Hogendoorn, 2014; Mort, Jackson, & Patton, 2015; Phillips & Westerfield, 2014; Plantié, Migocka-Patrzałek, Daczewska, & Jagla, 2015; Powles-Glover, 2014; Saralahti & Rämet, 2015; Wager, Mahmood, & Russell, 2014; White, 2015; Wilkinson, Jopling, & van Eeden, 2014; Wilkinson & van Eeden, 2014). Zebrafish offers a number of unique advantages for studying development in particular, including the transparency of zebrafish embryos, which facilitates the use of in vivo microscopy coupled with fluorescent labeling for the direct visualization of organ development and function. A major breakthrough that has extended the advantages of transparency to adult fish is the creation of the "casper" fish (White et al., 2008), which lacks melanocytes and iridophores. These traits cause the "casper" fish to remain transparent through adulthood. An additional benefit of zebrafish is its fecundity, which facilitates large-scale phenotype-based forward genetic and chemical genetic screens. These classic advantages of zebrafish make it ideal for developmental and genetic studies.

Many advantageous tools have been created to facilitate the use of this powerful system. For example, the introduction of a high-quality reference genome has supported many comprehensive genomic and epigenomic studies, including genome-wide transcriptome analyses and detailed studies of the effect of transcription factors and chromatin architecture on transcription regulation. These tools have accelerated in vivo molecular mechanism studies regarding stem cell specification, maintenance, and subsequent lineage differentiation throughout vertebrate development. However, many molecular and genomic tools still remain largely underutilized in zebrafish studies.

Despite the relative abundance of gene expression data, limited chromatin immunoprecipitation (ChIP) experiments have been used to examine the changes in histone marks that occur during zebrafish embryogenesis (Lindeman, Vogt-Kielland, Alestrom, & Collas, 2009; Vastenhouw et al., 2010). Using ChIP to examine histone modifications provides a better molecular understanding of the alterations to chromatin architecture that accompany gene expression changes. ChIP can also be employed to define the specific DNA binding sites of transcription factors and infer which genes they directly regulate. The assay for transposase-accessible chromatin with high-throughput sequencing (ATAC-seq) has been similarly underutilized in zebrafish studies. ATAC-seq is an assay that is used to identify regions of open chromatin that are more accessible to DNA-interacting factors at a genome-wide level. It functions similarly to assays of DNase I hypersensitivity coupled with next-generation sequencing (DNase-seq) but requires many fewer cells. This method supports the detection of known and novel DNA regulatory elements, such as promoters and enhancers. It also allows for the confirmation or inference of known and novel cognate transcription factors and epigenetic factors through the analysis of the underlying DNA sequence motifs in these regulatory regions. Here we describe both protocols for ChIP-seq and ATAC-seq analysis of adult zebrafish erythrocytes.

1. RATIONALE

ChIP has been used extensively for studying the chromosome landscape, including histone modifications such as methylation and acetylation as well as the various transcription and chromatin factors that bind to DNA. This method describes the isolation of the zebrafish cell type of interest, the ChIP procedure, and the subsequent analysis of sequence datasets. We describe this method here briefly. DNA and proteins are cross-linked to maintain the chromatin architecture and fix the interaction between factors bound to chromatin and DNA. Antibodies against specific histone modifications or transcription and chromatin factors are conjugated to solid beads and used to bind and subsequently precipitate the modified histones or transcription and chromatin factors of interest, respectively. After immunoprecipitation, the beads are thoroughly washed and the cross-links are reversed. Isolated DNA fragments from a specific chromosome region of interest can then be detected by quantitative PCR. Binding to all immunoprecipitated sites across the genome or specific sites of interest can be identified in a genome-wide manner by next-generation sequencing.

For ATAC-seq, cells are permeabilized to allow the modified bacterial Tn5 transposase to effectively create double-stranded breaks and add a specific short sequencing oligonucleotide to both 5′-ends of the DNA at each break. The "tagged" DNA fragments are then purified from their interacting proteins. DNA fragments from each sample are amplified and indexed for high-throughput sequencing. These techniques allow the sequences from precipitated or "tagged" fragments to be mapped to a reference genome, at which point comprehensive bioinformatics analysis can be applied to evaluate the function of these target regions by the identification of specific enriched genomic locations, binding motifs, and associated functional genes.

2. ADULT ZEBRAFISH EXSANGUINATION

For each ChIP, a minimum of 10^7 cells should be sufficient to avoid an amplification step post ChIP. Using the following exsanguination protocol, approximately 10^6 red blood cells can be acquired from one adult fish. Therefore, at least 10 adult zebrafish should be processed per ChIP. Fish are first euthanized in an overdose of Tricaine-S (MS-222, TMS, tricaine methanesulfonate, 200−300 mg/L), and they are bled quickly using a heparinized 200-μL tip attached to a pipette. Blood cells are collected by puncturing each fish through the gills deeply enough to penetrate the heart. To aid bleeding and prevent clotting, 200 μL of red cell isolation buffer (0.9% PBS, 5% fetal bovine serum, and 0.3-1 U heparin/mL) should be applied near the puncture site before the initial puncture. Using the same heparinized pipette tip, the blood is extracted by repeated pipetting and isolated blood is transferred to an Eppendorf tube containing the red cell isolation buffer. Samples should be kept at room temperature, since zebrafish erythrocytes are fragile and will lyse at lower temperatures.

Another easier method for exsanguination is by decapitation. Euthanized fish are placed in a Petri dish containing several milliliters of heparin solution (0.3-1 U heparin/mL). Zebrafish are then decapitated with a sharp razor blade. This procedure needs to be approved by the Institutional Animal Care and Use Committee (IACUC) (Fan & Yazulla, 1997) at individual institutions. The blood will flow freely out of the animal into the heparin solution. To collect the cells, use a 200-μL heparinized pipette tip and transfer cells to a 50-mL canonical tube containing the red cell isolation buffer.

3. CHROMATIN IMMUNOPRECIPITATION AND ChIP-seq LIBRARY CONSTRUCTION (LEE, JOHNSTONE, & YOUNG, 2006)

3.1 CROSS-LINKING

After isolation, the DNA and proteins should be immediately cross-linked using the following protocol:

1. Add formaldehyde solution (50 mM HEPES-KOH pH 7.5, 100 mM NaCl, 1 mM EDTA, 0.5 mM EGTA, and 11% formaldehyde) at 1/10 of the volume of the red cell isolation buffer. For example, 1 mL of formaldehyde solution should be added to 10^7 cells in 10 mL of the red cell isolation buffer.
2. Place samples on a shaker at room temperature for 5 min.
3. Immediately after the 5-min incubation, add 1/20 volume of 2.5 M glycine to quench the formaldehyde.
4. Incubate the cells for 5 min at room temperature.
5. Pellet the cells by centrifugation at 1100 × g for 5 min at 4°C.
6. Wash the cells twice with excess amounts of PBS (for example, 50 mL). This should be done as fast as possible to prevent excessive lysis of the erythrocytes. The researcher will notice the supernatant turning red, which is an indication of free hemoglobin from the lysed cells when significant cell lysis occurs.
7. After the washes, use the cell pellets immediately or flash freeze them in liquid nitrogen. Flash frozen cell pellets can be stored for months at −80°C.

Cross-linking is crucial for an effective ChIP as inefficient cross-linking can lead to loss of interactions, whereas too much cross-linking can result in identifying false positive protein–protein and protein–DNA interactions. Cross-linking times may need to be experimentally determined for a specific antibody and/or a particular cell type.

In previously published protocols detailing ChIP analysis of mouse embryonic stem cells (ESCs), a limited amount of cells of interest were mixed with a large number of cells from other species, such as *Drosophila* cells, to minimize material loss in the process due to limited starting material (carrier ChIP or CChIP) (O'Neill, VerMilyea, & Turner, 2006). A new carrier ChIP-seq protocol was developed by Zheng et al. (2015). It employed not only bacterial cell carriers at the cross-linking and sonication steps but also DNA carriers at the DNA isolation and amplification steps. A single cell ChIP-seq analysis was also used to define subpopulations in ESCs by chromatin state (Rotem et al., 2015). It will be very interesting to determine whether or not these protocols can be adopted for ChIP-seq analysis of rare chromatin and transcription factors. However, for the protocol presented here, these modifications are unnecessary, as an adequate number of adult zebrafish erythrocytes can be isolated for performing conventional ChIP. Nevertheless, in the case that only a limited number of cells of interest can be isolated, these modified protocols could serve as useful alternatives.

3.2 SONICATION

The sonication conditions are crucial for the success of the ChIP. Sonication permits the fragmentation and solubilization of chromatin. The size of the fragments determines the precision of the identification of protein factor binding regions. Smaller fragments will allow for more precise localization, but too much sonication can lead to lower signal-to-noise ratios as a result of the loss of antibody epitopes due to protein or DNA degradation. The ideal fragment size should be around 200 bp.

Prior to sonication, fixed cells need to be lysed. This can be done using the following protocol:

1. For 10^7 erythrocytes, resuspend the cells in 10 mL of lysis buffer 1 (50 mM HEPES-KOH pH 7.5, 140 mM NaCl, 1 mM EDTA, 10% glycerol, 0.5% NP-40, 0.25% Triton X-100, 1× protease inhibitors).
2. Incubate the cells for 10 min at 4°C.
3. Pellet the cells by centrifugation at 1350 × g for 5 min at 4°C.
4. Resuspend the pellets in 10 mL of lysis buffer 2 (10 mM Tris–HCl pH 8.0, 200 mM NaCl, 1 mM EDTA, 0.5 mM EGTA, 1× protease inhibitors)
5. Incubate the cells for 10 min at room temperature.
6. Pellet the cells by centrifugation at 1350 × g for 5 min at 4°C.
7. Resuspend the pellets in 1 mL of lysis buffer 3/sonication buffer (10 mM Tris–HCl pH 8.0, 100 mM NaCl, 1 mM EDTA, 0.5 mM EGTA, 0.1% Na-Deoxycholate, 0.5% *N*-lauroylsarcosine, 1× protease inhibitors).

Once the cells have been lysed, the samples can be sonicated in a BioRuptor sonicator for 24 cycles of 30-s sonication with a 59-s resting interval after each cycle. The sonicated material should be checked for the appropriate fragment sizes before proceeding. Fragment sizes can be estimated using the following protocol:

1. Combine 10 μL of sample with 20 μL TE buffer (10 mM Tris–HCl pH 8.0, 1 mM EDTA pH 8.0) containing 0.1% SDS and boiled for 15 min.
2. Mix the samples with a proper amount of loading dye and run on a 1.5% agarose gel to evaluate the size of the fragments. If the mean fragment size is larger than 200 bp, perform additional cycles of sonication in increments of two to six cycles.
3. Check fragment sizes on an agarose gel as described earlier until the desired size has been achieved.

This boiling method of denaturing chromatin-bound proteins for the visualization of sonicated DNA on an agarose gel is a quick way to approximate the fragment size after sonication. To assess the sonication efficiency more accurately, the proper reversal of cross-linking is necessary (see later discussion).

After sonication, Triton X-100 is added to sonicated lysates at 1% final concentration, after which they are centrifuged for 10 min at 18,000 × g to pellet cell debris. Following this centrifugation step, 50 μL of the supernatant should be set aside as a total input control for a ChIP starting with approximately 10^7 cells and kept at −20°C until step F.

As an alternative to the protocol described previously, a single buffer can be used for cell lysis and sonication so as to minimize loss of cells. Collected cell pellets should be resuspended in an SDS ChIP buffer (20 mM Tris–HCl pH 8.1, 150 mM NaCl, 2 mM EDTA, 0.1% SDS, 1% Triton X-100, 1× protease inhibitors) (Kim, Cantor, Orkin, & Wang, 2009). Sonication conditions will be different using this buffer, so fragmented chromatin should be checked on an agarose gel as described earlier to determine the optimal number of sonication cycles.

3.3 PRETREATMENT OF BEADS AND ANTIBODIES

Prior to immunoprecipitation, the selected antibody should be attached to magnetic protein A or protein G Dynabeads. Fifty microliters of beads is used for each immunoprecipitation. To prepare the beads, they are washed twice with 1 mL blocking buffer (1× PBS, 0.5% BSA) and then resuspended in 250 μL blocking buffer plus 5 μg of antibody. The amount of antibody can be adjusted. All washes are performed using the Dynal MPC magnet for bead/antibody collection. Beads are collected by a 30-s incubation on the magnet, the supernatant is removed, and then the next buffer is added immediately. The beads are rotated with the antibody for 6 h to overnight at 4°C. After the incubation, the beads are washed twice with 1 mL blocking buffer to remove any excess antibody. It is crucial for the beads not to get dry between washes. The beads are now ready to be used for immunoprecipitation.

3.4 CHROMATIN IMMUNOPRECIPITATION

The sonicated material from step B (after the addition of Triton X-100) can be directly added to the antibody-conjugated beads from step C. If the alternative buffer for sonication, which does not require nuclear extraction, is used, then the sonicated material can be used for immunoprecipitation directly in the same buffer. The beads and lysates are incubated together with agitation for a time ranging from 4 h to overnight at 4°C.

It is important to perform a control immunoprecipitation in parallel. Two standard controls commonly used are an IgG isotype control that is the same isotype as the antibody used for the ChIP or a control with the beads alone (Sandmann, Jakobsen, & Furlong, 2006). These controls will help eliminate false positive results caused by nonspecific binding to antibodies or beads. In addition to these two types of nonspecific binding, chromatin may also adhere nonspecifically to the surface of plastic tubes. To minimize this problem, siliconized or manufacturer-certified low-retention tubes can be used, especially when the amount of starting material is less than optimal.

3.5 CHROMATIN IMMUNOPRECIPITATION WASHES

Following the immunoprecipitation, several washes are employed to remove the nonbound chromatin from the bead/antibody mixture. Beads are washed once with wash buffer 1 (20 mM Tris–HCl pH 8.0, 150 mM NaCl, 2 mM EDTA, 0.1% SDS, 1% Triton X-100), once with wash buffer 2 (20 mM Tris–HCl pH 8.0, 500 mM NaCl, 2 mM EDTA, 0.1% SDS, 1% Triton X-100), once with wash buffer 3 (10 mM Tris–HCl pH 8.0, 250 mM LiCl, 2 mM EDTA, 1% NP40), and once with TE buffer. All washes are performed at 4°C and using the Dynal magnet as described in part C. For each wash, the supernatant is removed using a pipette and 1 mL of the next solution is added. Tubes are shaken vigorously until all the beads are resuspended and then placed on a rocker for 3-min incubations at 4°C. If siliconized or low-retention tubes are not used, the beads are transferred to new tubes after each wash to minimize nonspecific binding of chromatin fragments to plastic tubes. After the last wash with TE buffer, the supernatant is removed, and the tubes are centrifuged at 960 × g for 3 min at 4°C. Residual TE is then removed.

3.6 ELUTION AND REVERSAL OF CROSS-LINKING

The beads from step E are resuspended in 200 μL elution buffer (50 mM Tris–HCl pH 8.0, 10 mM EDTA, and 1% SDS) by heating for 30 min at 65°C in a shaking heat block. The supernatant is then transferred to a new tube for cross-link reversal that is done for 6 h to overnight at 65°C. Longer incubations may result in nosier data. It is important to also reverse the cross-links of the total input controls saved in step B. For this, 150 μL elution buffer is added to each 50 μL of the total input controls. They are then incubated at 65°C along with the ChIP samples.

3.7 DNA RECOVERY AND QUALITY CONTROL

After the reversal of cross-linking, the samples are treated with RNase for 2 h at 37°C, followed by proteinase K for 2 h to overnight at 55°C. Then the DNA is extracted using phenol/chloroform/isoamyl alcohol followed by an ethanol precipitation. The ChIP DNA is resuspended in 30 μL TE buffer and the total input controls in 150 μL TE buffer. The concentration of the ChIP DNA samples should be around 1 ng/μL.

To validate the ChIP, quantitative PCR is used. The primers are designed based on known information about the binding sites of the transcription factor. Standard qPCR conditions are utilized. If available, positive and negative primer sets should be used. If a 384-well qPCR platform is available, less than 0.5 μL of the ChIP sample can be used in a 5-μL total reaction volume. Primer sets for a housekeeping gene are included to normalize different ChIP samples since there is always residual nonspecific binding. The enrichment is quantified in comparison to the respective total input control samples and the negative controls ChIP (IgG isotype or beads alone) using the ΔΔCt method. Alternatively, equal amounts of the total input control and ChIP material can be used, and the enrichment can be measured in comparison to the respective total input control.

3.8 ChIP-seq LIBRARY PREPARATION

In general, the outlined ChIP-seq library preparation protocol is based on what has been established by Guenther et al. (2008). The sequencing libraries can be prepared with the NEBNext ChIP-Seq Library Prep Master Mix Set for Illumina (NEB, E6240) according to the manufacturer's instructions. 200–1000 ng of input DNA and as little as 5 ng of ChIP DNA have been used successfully to make high-quality libraries. DNA overhangs are turned into phosphorylated blunt ends and the samples are purified using Agencourt AMPure XP beads (Beckman Coulter, A63880; 1.8× of the sample volume). Library preparation continues by the addition of a single dA to the 3′ end to allow for directional adaptor ligation. Samples are purified again as before using the AMPure XP beads at 1.8× and Illumina adapter oligonucleotides (1/10 dilution) are added to the samples. The adapter-ligated DNA is cleaned up by bead purification of the samples using the AMPure XP beads at 1.8× followed by a size selection step using AMPure XP beads (0.9× followed by 0.2× of the sample volume), which selects for library fragment sizes of around 150 bp. The samples are amplified by PCR (limited amplification of up to 18 cycles) that adds additional linker sequence to the fragments to prepare them for annealing to binding sites in a proper Illumina sequencing flow cell. Lastly, the amplified libraries are again purified with AMPure XP beads (1× of the sample volume). The purified libraries can then be stored at −20°C. Libraries are quantified using the QuBit dsDNA HS assay kit (Thermo Fisher Scientific), and their quality is assessed by running 1 μL of appropriately diluted sample on a 2100 Bioanalyzer (Agilent Technologies), a TapeStation, or an AATI Fragment Analyzer. Using the protocol with size selection described previously, the peak fragment size usually runs at 250–300 bp. In our experience, size selection using beads is sufficient to generate libraries with this fragment range. If that is not the case, one can add another size selection step after library amplification (instead of the final bead purification), in which the libraries are separated on a 2% agarose gel and products between 150 and 350 bp are selected and extracted from the gel (the products include fragments of 50–250 bp plus approximately 100 bp of primer sequence). It is worth noting that gel purification leads to a greater loss of material than bead purification.

As an alternative to using a QuBit for quantifying the ChIP-seq libraries, library concentration can be determined using the KAPA Library Quantification Kit (Kapa Biosystems, kit varies depending on PCR machine used). However, very high dilutions (1:2,500,000 and higher) are required to accurately measure the library concentration by qPCR, which can lead to error. In our experience, the library concentration measured by a QuBit assay in combination with the fragment size, as determined by the Bioanalyzer or AATI runs, has yielded sufficiently accurate information for the calculation of molarity for pooling 12 libraries into one Illumina Rapid SE flow cell run on a HiSeq 2500 system.

All protocols for Illumina/Solexa sequence preparation, sequencing, and quality control are also available on the Illumina website (http://www.illumina.com/pages.ilmn?ID=203). Different platforms can also be used such as the Ion Torrent, or the

Roche 454 platform (Harismendy et al., 2009; Kircher & Kelso, 2010). Distinct experiments may require diverse sequencing platforms and thus appropriate sample preparations.

4. ATAC-seq ANALYSIS OF OPEN CHROMATIN REGIONS

A modified bacterial Tn5 transposase (Adey et al., 2010) has been engineered to effectively induce random double-stranded breaks into genomic DNA both in vitro and in vivo, followed by the ligation of specific short nucleotide sequences to the 5′-ends of either side of the breaks. The assay for transposase-accessible chromatin with high-throughput sequencing (ATAC-seq) (Buenrostro, Giresi, Zaba, Chang, and Greenleaf (2013)) utilizes the ability of this modified Tn5 transposase for "tagmentation" of chromatin in permeabilized cells, which tags DNA ends within open chromatin regions with unique Illumina library adaptors. The assumption is that there are higher chances of breaking and tagging chromatin regions that are more open and accessible, which are probable sites of genomic regulatory elements. Coupled with gene expression profiling and ChIP-seq analysis of histone modifications, chromatin factors, and transcription factors, this assay can offer a confirmatory and complementary view of the epigenomic landscape of particular cell types. Moreover, ATAC-seq works robustly on samples with limited cell inputs (500-50,000 cells), making it a useful tool for epigenomic analysis of rare cell populations.

As the epigenomic landscape of a given cell type is unique, it is important to isolate a pure population of cells for ATAC-seq analysis to achieve an adequate signal-to-noise ratio. Given the relatively low number of cells needed for ATAC-seq, one can afford to sacrifice cell quantity for purity during cell isolation procedures from zebrafish embryos or organs. Here we outline an example protocol for ATAC-seq analysis of zebrafish peripheral red blood cells. This is an adaptation of the original protocol published in Nature Methods by Buenrostro et al. (2013) with slight modifications that we have found to work best for analyzing freshly harvested zebrafish red blood cells. This assay can also work with viable frozen cells. Cells after flash freezing treatment also have worked for some cell lines and primary human T cells purified from frozen peripheral blood.

Zebrafish adult peripheral red cells are harvested by heart puncture or decapitation as described in Section 1. As mentioned earlier, approval from institutional IACUC is required before carrying out any of the described handling of zebrafish adults. Since the ATAC-seq analysis requires only small numbers of cells, from 500 to 50,000, a small fraction of about 1 million peripheral red cells collected from one adult zebrafish is sufficient for analysis by ATAC-seq. Other cells can be processed similarly by following the protocol outlined later. However, it is important to optimize the transposition step by optimizing the amount of enzyme mix, number of cells, and incubation time (30–90 min) used in each transposition reaction.

4.1 CELL TREATMENT AND NUCLEUS ISOLATION

Collected cells are spun down in the cell collection medium at 500 × g for 5 min at 4°C. Excess cold PBS or cell collection medium can be used to achieve a balance between different sample volumes. Extreme care should be taken not to disturb (almost) invisible cell pellets. Less than 10,000 cells can be used to produce meaningful results, but they require a careful titration of the amount of transposase, incubation time, and reaction volume that can vary significantly between different types and numbers of cells. Additionally, no purification after a transposition reaction is recommended for cell numbers between 500 and 5000. Samples with low cell numbers require immediate library PCR amplification and indexing after the transposition reaction (see later discussion).

Cell pellets are gently washed each with 50 μL of cold 1× PBS and spun down at 500 × g for 5 min at 4°C. After removal of PBS, 50 μL of cold cell lysis buffer (10 mM Tris–HCl pH 7.4, 10 mM NaCl, 3 mM $MgCl_2$, 0.1% IGEPAL CA-360) is gently added, and cells are resuspended by gentle pipetting followed by an immediate spin at 500 × g for 5 min at 4°C. The transposition reaction mix can be set up during the spin. For each sample, the following mix is prepared: 25 μL of 2× TD Buffer (in the Nextera DNA Library Preparation Kit, Illumina, FC-121-1031), 2.5 μL of transposase enzyme (in the same kit), and 22.5 μL of nuclease-free water for a total of 50 μL.

4.2 TRANSPOSITION OF OPEN CHROMATIN REGIONS

The supernatant of the cell lysis reaction buffer is carefully pipetted off from all samples, and the cell pellets are left on ice. Immediately the transposition reaction mix is carefully added, and the cell pellet is resuspended by pipetting up and down very gently. Since the incubation time, amount of transposition enzyme and transposition reaction volume can vary significantly between different cell types, it is a good practice to run an optimization experiment by varying the time of incubation and the amount of transposase. Normally, an incubation time of 30–90 min at 37°C can work well for a 50,000-cell sample.

4.3 DNA ISOLATION AND LIBRARY PREPARATION

Total DNA from each transposition reaction is isolated using the Qiagen MinElute PCR purification kit and eluted into 10 μL of the elution buffer provided in the kit. The isolated DNA samples can be stored at −20°C before library amplification and indexing or used immediately for the next step. If 50,000 cells are used, it is also possible to run a fraction (10%) of the DNA samples (in a nano-gram range) on a TapeStation, Bioanalyzer DNA chip, or AATI fragment analyzer before library amplification. This can help with a quick optimization of the transposition reactions for new cell types. This is not recommended for samples with a fewer number of cells since the amount of purified DNA will be below the detection limits of the aforementioned DNA analysis systems.

4.4 LIBRARY AMPLIFICATION AND INDEXING

Any high-fidelity PCR master mixes can be used for the outlined initial library amplification. Here we use the high-fidelity PCR master mix (New England Biolabs, M0541) as an example. This PCR reaction can be carried out according to the following protocol:

1. Make a PCR reaction mix consisting of 7 μL nuclease-free water (adjustable), 2.5 μL Nextera primer Ad1 (at 25 μM), 2.5 μL Nextera primer Ad2.[x] (at 25 μM), 3 μL 10× SYBR Green I (a 1:1000 dilution of Invitrogen 10,000× SYBR Green I, S-7563, in 10 mM Tris−HCl), and 25 μL NEBNext High-Fidelity 2× PCR master mix (NEB, M0541). Be sure to prepare some extra mix to account for loss in aliquots in each sample.
2. Add 40 μL of the above mix to each of the 10 μL transposed DNA samples and mix thoroughly on ice.
3. Perform an initial PCR amplification of the libraries:
   a. 72°C, 5 min
   b. 98°C, 30 s
   c. 98°C, 10 s
   d. 63°C, 30 s
   e. 72°C, 1 min
   f. Repeat steps 3−5, 4 times
   g. Hold at 4°C

It is recommended to synthesize a set of custom Greenleaf ATAC primers (http://seq.liai.org/204-2/and Table 1) that includes one common primer ATAC_Ad1_NoMX and 24 index primers ATAC_Ad2.1-Ad2.24 (from Buenrostro et al., 2013). We have not tried to use the primers from the Nextera Index Kits (24 or 96 indexes) for amplification and indexing of ATAC-seq libraries.

A side qPCR reaction is performed to determine how many additional PCR cycles described previously are needed to obtain enough library material for sequencing without overamplification, which can lead to GC-rich sequence and size bias during the library amplification and indexing. While the initial library amplification is being performed, a qPCR side reaction is prepared in a 15-μL total volume. This can be done using the following procedure:

1. Make a qPCR reaction mix consisting of 3.9 μL nuclease-free water (adjustable), 0.25 μL Nextera primer Ad1 (at 25 uM), 0.25 μL Nextera primer Ad2.[x] (at 25 uM), 0.6 μL 10× SYBR Green I, and 5 μL 2× High-Fidelity NEB PCR Master Mix. Be sure to prepare some extra mix to account for loss in aliquots in each sample.
2. Combine 10 μL of the qPCR reaction mix with 5 μL of each of the "5-cycle" initial library amplification and indexing reactions. Mix thoroughly by gentle pipetting.
3. Store the remaining 45 μL of the initial library amplification and indexing reactions at 4°C.

Table 1 ATAC-seq Index Primers

Greenleaf ATAC Primers		
Primer name	**Index**	**Primer Sequence to Order**
ATAC_Ad1_noMX:	N/A	AATGATACGGCGACCACCGAGATCTACACTCGTCGGCAGCGTCAGATGTG
ATAC_Ad2.1_TAAGGCGA	TAAGGCGA	CAAGCAGAAGACGGCATACGAGATTCGCCTTAGTCTCGTGGGCTCGGAGATGT
ATAC_Ad2.2_CGTACTAG	CGTACTAG	CAAGCAGAAGACGGCATACGAGATCTAGTACGGTCTCGTGGGCTCGGAGATGT
ATAC_Ad2.3_AGGCAGAA	AGGCAGAA	CAAGCAGAAGACGGCATACGAGATTTCTGCCTGTCTCGTGGGCTCGGAGATGT
ATAC_Ad2.4_TCCTGAGC	TCCTGAGC	CAAGCAGAAGACGGCATACGAGATGCTCAGGAGTCTCGTGGGCTCGGAGATGT
ATAC_Ad2.5_GGACTCCT	GGACTCCT	CAAGCAGAAGACGGCATACGAGATAGGAGTCCGTCTCGTGGGCTCGGAGATGT
ATAC_Ad2.6_TAGGCATG	TAGGCATG	CAAGCAGAAGACGGCATACGAGATCATGCCTAGTCTCGTGGGCTCGGAGATGT
ATAC_Ad2.7_CTCTCTAC	CTCTCTAC	CAAGCAGAAGACGGCATACGAGATGTAGAGAGGTCTCGTGGGCTCGGAGATGT
ATAC_Ad2.8_CAGAGAGG	CAGAGAGG	CAAGCAGAAGACGGCATACGAGATCCTCTCTGGTCTCGTGGGCTCGGAGATGT
ATAC_Ad2.9_GCTACGCT	GCTACGCT	CAAGCAGAAGACGGCATACGAGATAGCGTAGCGTCTCGTGGGCTCGGAGATGT
ATAC_Ad2.10_CGAGGCTG	CGAGGCTG	CAAGCAGAAGACGGCATACGAGATCAGCCTCGGTCTCGTGGGCTCGGAGATGT
ATAC_Ad2.11_AAGAGGCA	AAGAGGCA	CAAGCAGAAGACGGCATACGAGATTGCCTCTTGTCTCGTGGGCTCGGAGATGT
ATAC_Ad2.12_GTAGAGGA	GTAGAGGA	CAAGCAGAAGACGGCATACGAGATTCCTCTACGTCTCGTGGGCTCGGAGATGT
ATAC_Ad2.13_GTCGTGAT	GTCGTGAT	CAAGCAGAAGACGGCATACGAGATATCACGACGTCTCGTGGGCTCGGAGATGT
ATAC_Ad2.14_ACCACTGT	ACCACTGT	CAAGCAGAAGACGGCATACGAGATACAGTGGTGTCTCGTGGGCTCGGAGATGT
ATAC_Ad2.15_TGGATCTG	TGGATCTG	CAAGCAGAAGACGGCATACGAGATCAGATCCAGTCTCGTGGGCTCGGAGATGT
ATAC_Ad2.16_CCGTTTGT	CCGTTTGT	CAAGCAGAAGACGGCATACGAGATACAAACGGGTCTCGTGGGCTCGGAGATGT
ATAC_Ad2.17_TGCTGGGT	TGCTGGGT	CAAGCAGAAGACGGCATACGAGATACCCAGCAGTCTCGTGGGCTCGGAGATGT
ATAC_Ad2.18_GAGGGGTT	GAGGGGTT	CAAGCAGAAGACGGCATACGAGATAACCCCTCGTCTCGTGGGCTCGGAGATGT
ATAC_Ad2.19_AGGTTGGG	AGGTTGGG	CAAGCAGAAGACGGCATACGAGATCCCAACCTGTCTCGTGGGCTCGGAGATGT
ATAC_Ad2.20_GTGTGGTG	GTGTGGTG	CAAGCAGAAGACGGCATACGAGATCACCACACGTCTCGTGGGCTCGGAGATGT
ATAC_Ad2.21_TGGGTTTC	TGGGTTTC	CAAGCAGAAGACGGCATACGAGATGAAACCCAGTCTCGTGGGCTCGGAGATGT
ATAC_Ad2.22_TGGTCACA	TGGTCACA	CAAGCAGAAGACGGCATACGAGATTGTGACCAGTCTCGTGGGCTCGGAGATGT
ATAC_Ad2.23_TTGACCCT	TTGACCCT	CAAGCAGAAGACGGCATACGAGATAGGGTCAAGTCTCGTGGGCTCGGAGATGT
ATAC_Ad2.24_CCACTCCT	CCACTCCT	CAAGCAGAAGACGGCATACGAGATAGGAGTGGGTCTCGTGGGCTCGGAGATGT

4. Perform the side qPCR amplification with the following cycles:
   a. 98°C, 30 s
   b. 98°C, 10 s
   c. 63°C, 30 s
   d. 72°C, 1 min
   e. repeat steps 2–4, 19 times
   f. hold at 4°C

As shown in Fig. 1, fluorescent signals from the SYBR Green that are bound to the amplified products are plotted against cycle numbers. After determining the maxima of SYBR fluorescence for the plateau of each sample, plot and calculate the cycle number on the x-axis that corresponds to a quarter of the maxima. This quarter-maxima cycle number is used to perform additional library amplification and indexing for a particular sample. If the quarter-maxima cycle number falls between two cycle numbers, it is recommended to take the smaller number for additional PCR cycles.

Additional PCR cycles should be run with identical PCR settings as described previously across all samples, except for the number of cycles, which should vary based on the specific quarter-maxima value for each sample. It is more convenient to set the PCR machine to perform the highest number of cycles for a group of

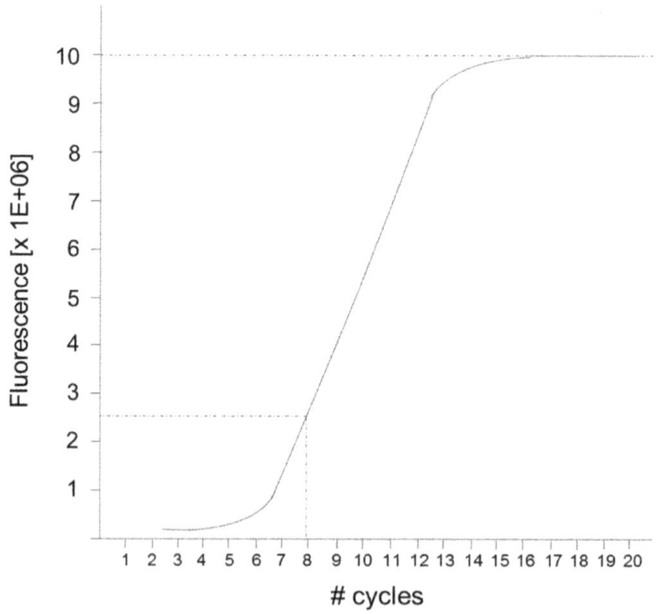

FIGURE 1

Determination of additional number of PCR cycles needed for library amplification and indexing.

samples and to remove individual samples quickly once they have completed the appropriate number of cycles. The final amplified libraries are purified again using the Qiagen MinElute PCR purification kit and eluted into 20 μL of the elution buffer. The purified libraries can then be stored at −20°C.

4.5 LIBRARY QUALITY CONTROL AND QUANTIFICATION

Accurate measurement of library concentration and quality is critical for successful sequencing and subsequent data analysis. Library concentration can be measured using the QuBit dsDNA HS Assay Kit. Alternatively, a qPCR library quantification kit (eg, from Kapa Biosystems, but see comments in Section 3, step H) can be used. A proper sensitivity chip and/or run setup of TapeStation, Bioanalyzer, or AATI Fragment Analysis is used to check for library quality and average fragment size information (Fig. 2). Successfully tagmented samples will display a characteristic nucleosomal "laddering" pattern. The molar concentration of a library is calculated for sequencing flow cell cluster generation and subsequent sequencing. The dark smear peak below 100 bp is not included in the calculation of molar concentration of libraries and their average fragment sizes. This smear is believed to be DNA fragments from mitochondrial DNA without proper tagging for clustering and sequencing on an Illumina next-generation sequencing platform.

4.6 SEQUENCING CONSIDERATIONS

Because a low number of cells are sufficient for ATAC-seq analysis, the sequencing of libraries with relatively higher amplification normally affords a lower mapping rate (50−70%) of the resulting sequence reads. We typically aim to obtain 40−50 million reads from each library for a good representation of genomic complexity. We typically pool 12 libraries into one Rapid PE flow cell for a 75- or 100-bp paired-end sequencing run on a HiSeq 2500 platform. Longer reads and paired-end sequence reads typically result in more accurate mapping and higher mapping rates.

5. ChIP-seq and ATAC-seq DATA ANALYSIS

As a method to detect DNA-protein binding in vivo, ChIP-seq has become the most widely used procedure for investigation of genome-wide transcriptional regulation and epigenetic modifications (Bailey et al., 2013). The typical ChIP-seq raw reads from a sequencing machine are short DNA fragments from the DNA-protein binding regions. ATAC-seq produces a library of DNA read tags that map to genomic regions that are relatively more accessible to transposase activity due to a looser organization of nucleosomes. Since both approaches aim to find enriched genomic regions from short DNA fragments, the computational analysis procedures of ChIP-seq and ATAC-seq are similar. Most of the tools discussed here are command-line tools. However, user-friendly integrated online ChIP-seq and ATAC-seq pipelines have

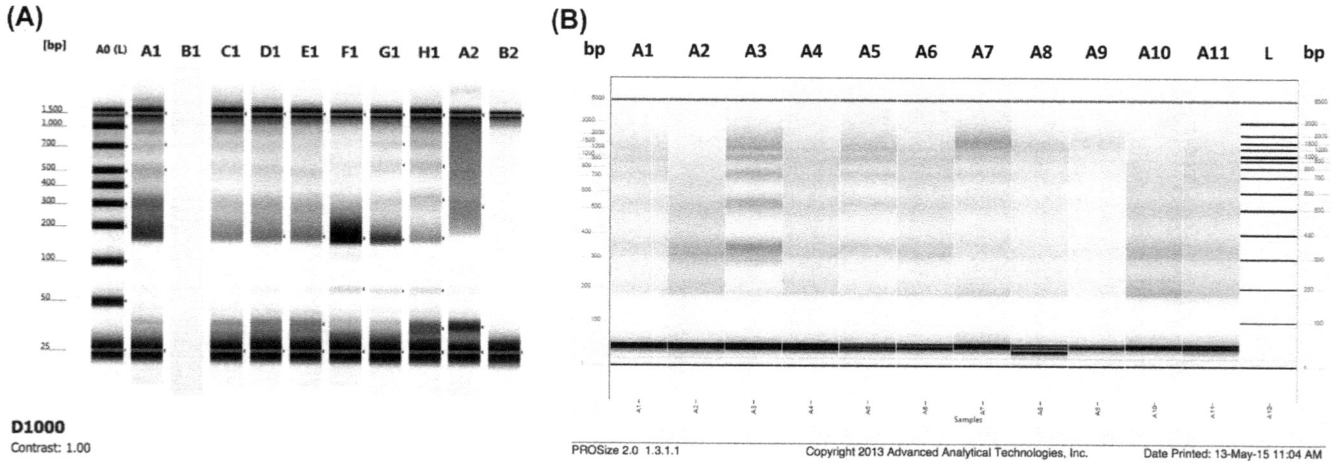

FIGURE 2

Example results from Agilent TapeStation and AATI Fragment Analysis of ATAC-seq libraries. (A) Agilent TapeStation results showing the size and laddering pattern of a set of ATAC-seq libraries. (B) AATI Fragment Analyzer results showing the size and laddering pattern of a different set of ATAC-seq libraries.

been developed for life science researchers who do not have extensive computational experience. Galaxy (https://usegalaxy.org) and GenePattern (http://www.broadinstitute.org/cancer/software/genepattern/) are alternative tools that are commonly used.

Sufficient sequencing depth in ChIP-seq and ATAC-seq is required for effective data analysis. The necessary sequencing depth depends on the genome size and the size of the genomic regions targeted for investigation. For typical zebrafish transcription factor binding sites and ATAC-seq regions, a total of 20 million mappable reads per sample are sufficient. For most histone marks and for proteins with many binding sites, more sequence reads (40–50 million) can help identify binding sites with greater confidence. An abundance of reads can also help to reveal weaker binding sites that may be harder to detect. Similarly, for ATAC-seq, open chromatin regions can be located more accurately and with greater certainty when more reads are available from a library.

Preprocessing and quality examination of the raw sequencing data are necessary for assessing the quality of the raw reads and identifying sources of error or bias. FastQC is used to provide an efficient and comprehensive evaluation of data quality. This tool can be incorporated into the whole data processing pipeline to ensure the reliability of the downstream analysis. Key parameters to examine include the Phred score (http://www.illumina.com/documents/products/technotes/technote_Q-Scores.pdf), which measures the error probability at each base (eg, a Q30 value corresponds to a 1-in-1000 base error rate), the overrepresented sequences that are possibly sequencing adapters or other contaminants, and the overall GC content. Cutadapt (http://cutadapt.readthedocs.org/en/stable/index.html) can be used to trim the raw reads based on quality scores, adapter contamination, or other criteria, such as read length to remove short reads.

After quality filtration, the sequence reads are mapped to the zebrafish reference genome using available mapping programs, such as Bowtie (Langmead, Trapnell, Pop, & Salzberg, 2009) and BWA (Li & Durbin, 2009). Sequence variations are more common in the zebrafish genome than they are in the human genome, so a higher mismatch rate (2 in 100 bp) is allowed for individual sequence alignment. These mapping programs output all mapped reads in a SAM or BAM format. After these files are sorted by genomic locations and indexes—procedures which can both be done using either SAMtools by Li et al. (2009) and Li (2011). (http://samtools.sourceforge.net) or Picard Tools by the Broad Institute (http://broadinstitute.github.io/picard/)—all mapped reads can be visualized in IGV by the Broad Institute (https://www.broadinstitute.org/igv/) and/or UCSC genome browser by Kent et al. (2002) (https://genome.ucsc.edu).

The next step is peak calling, the prediction of genomic regions characterized by an enrichment of mapped reads. Several peak-calling algorithms and tools are available, such as MACS (Zhang et al., 2008), SPP (Kharchenko, Tolstorukov, & Park, 2008), and ZINBA (Rashid, Giresi, Ibrahim, Sun, & Lieb, 2011). Peak-calling steps are generally similar between ChIP-seq and ATAC-seq, but there are a few differences: 1. ChIP-seq of input DNA or mock-IP are used as controls in ChIP-seq analysis, but not in ATAC-seq; 2. In general, ChIP-seq uses single-end reads and ATAC-seq uses paired-end reads; 3. When calling ATAC-seq peaks using MACS2, three parameters (−nomodel −shift −100 (or 75) −extsize 200 (or 150)) are added to the default MACS2 arguments, which are initially designed for ChIP-seq analysis. These different parameters are used so that called ATAC-seq peaks are centered on the cutting sites created by the modified transposase, Tn5, rather than on the locations of the reads themselves. Common enrichment metrics such as p-value or q-value (FDR) are used as statistical threshold for calling peaks in both ChIP-seq and ATAC-seq analyses. It is worth noting that the number of detected peaks can vary significantly between samples from ChIP-seq and ATAC-seq analyses. Peak numbers can also vary significantly between ChIPs when different antibodies are used to analyze the same cells, even when the different antibodies are against the same protein in the same cells.

It is of great interest to identify similar and different genomic regions defined by ChIP-seq and ATAC-seq in different but related cell types. It is also illuminating to compare the peaks either within ChIP-seq and ATAC-seq results or between ChIP-seq and ATAC-seq results for the same cells under different treatment conditions or for entirely different cells. Comparison of protein-bound or open chromatin regions across different conditions or cell types can be simply performed by peak overlap using BEDTools (Quinlan & Hall, 2010) and can provide useful information about similar and different mechanisms that are used to regulate cell function and define cell identity. However, reliable quantitative and differential binding analyses rely on an adequate number of reads at the relevant peaks, which can be determined using MAnorm (Shao, Zhang, Yuan, Orkin, & Waxman, 2012) or DBChIP (Liang & Keles, 2012). It may be necessary to slightly modify parameters for each individual analysis to evaluate more accurately the peaks of interest in different experiments. Experimental and functional verifications of called peaks are also invaluable whenever possible.

The called peaks for both ChIP-seq and ATAC-seq experiments are presented in a BED file, which highlights the genomic positions of the peaks, and a bigWig file, which displays the genomic profile of the mapped reads, which can both be visualized with the IGV or UCSC genome browser tools. Using these visualization tools, the genomic annotation of peaks, such as whether they occupy a promoter region, gene body, etc., is made readily visible. A genome-wide peak annotation can be achieved by GREAT (McLean et al., 2010) or other methods that associate genome regions with annotated genes and their structures. Then functional analysis of these annotated peaks, made possible through tools such as DAVID (Huang da, Sherman, & Lempicki, 2009), GREAT, and GSEA (Subramanian et al., 2005), can provide information about gene ontology enrichment or gene set enrichment of the genes associated with called peaks.

DNA-protein binding motifs enriched in ChIP-seq peaks can be found using motif discovery algorithms that are designed for analyzing ChIP-seq data such as MEME-ChIP (Machanick & Bailey, 2011). Motif analysis can be used to validate the ChIP-seq experiment if the motif is known or for the discovery of novel motifs

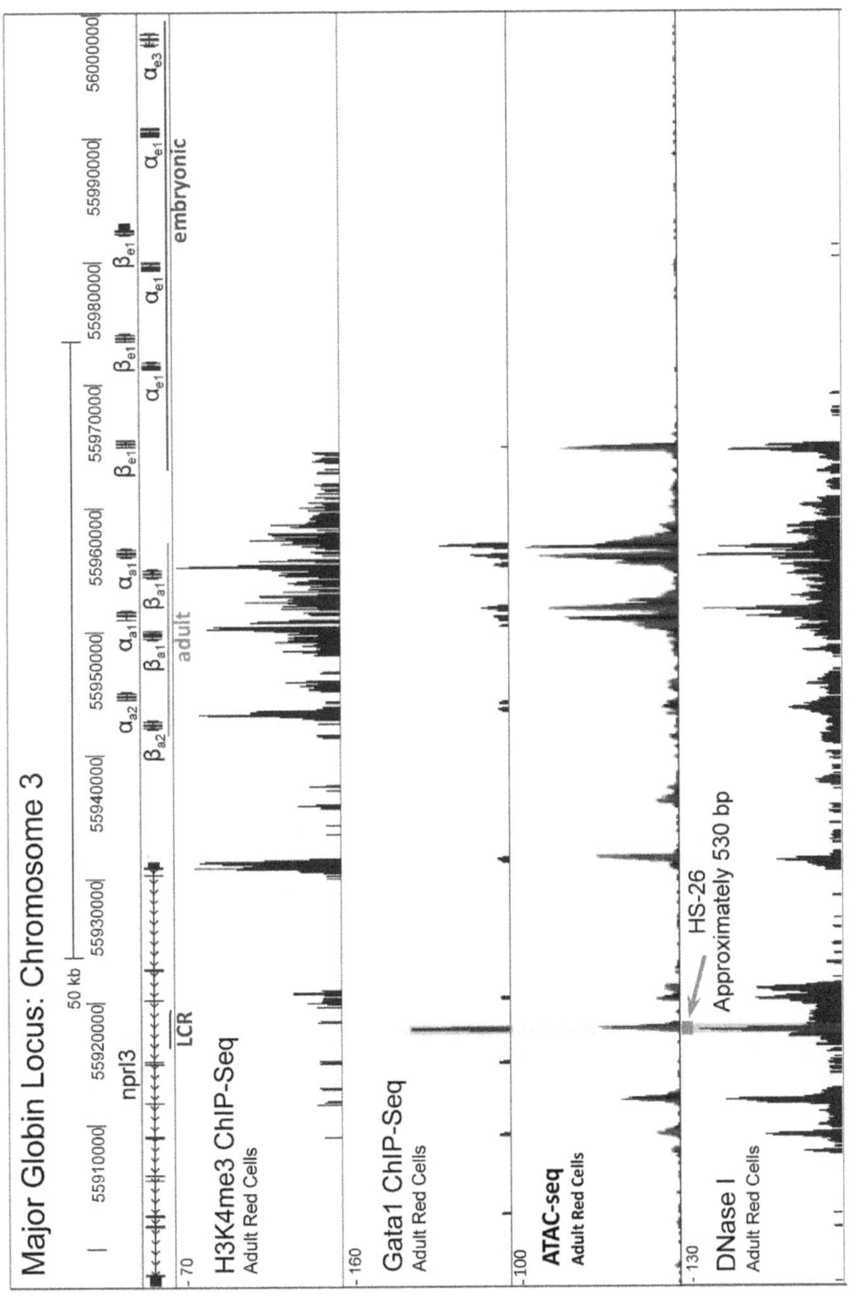

FIGURE 3

A screen shot of H3K4me3 ChIP-seq, gata1 ChIP-seq, ATAC-seq, and DNase-seq tracks at the zebrafish major *globin* locus. The *red line* (light gray in print versions) marks the adult globin cluster, the *blue line* (black in print versions) indicates the embryonic globin cluster, the *short purple line* (gray in print versions) dictates the LCR, and the *red bar* (light gray in print versions) on the DNase I track shows the location of the LCR core site. These tracks demonstrate the coexistence of binding by the active chromatin mark (H3K4me3) and erythroid master transcription factor (*gata1*) in the adult globin gene cluster and the LCR. They also highlight an open chromatin conformation in the corresponding genomic regions, as evidenced by ATAC-seq and DNase-seq peaks.

and binding partners that are associated with ChIPed proteins or histone modifications. Likewise, motifs underneath and around ATAC-seq peaks provide useful information about protein factors that interact with these DNA segments. All of these analyses will facilitate the studies of molecular mechanisms that modulate transcriptional control of cell fate and function.

6. RESULTS

We applied ChIP-seq and ATAC-seq to analyze the molecular mechanisms that modulate transcriptional regulation and to reveal the corresponding epigenetic landscape in the major globin locus. Globin switch is a biological process that is conserved from zebrafish to mammals. The current understanding of the mechanism that governs the globin switch has important implications in the diagnosis and treatment of anemia, especially thalassemia, which results from an imbalance in the production of alpha- and beta-globin chains. Zebrafish has been shown to be an excellent model for the understanding of vertebrate globin switch and transcriptional control of globin gene expression (Brownie et al., 2003; Ganis et al., 2012; Song et al., 2004). There is little or no embryonic *globin* mRNA expression in adult erythroid cells, even though both embryonic and adult *globin* genes are located near each other at the major globin locus and are under the control of the same locus control region (LCR) that is located in the intron 4 of its neighboring gene *nprl3*. ChIP-seq analyses of the H3K4me3 mark, an indicator of active transcription, and *gata1*, the erythroid master transcription regulator, reveal specific binding at the promoter sites of adult *globin* genes, while there is no binding at the embryonic *globin* promoter sites (Fig. 3). These evaluations at a molecular and genomic level exhibit cell-specific histone modifications and molecular interactions of the cell-specific transcription factor. They are consistent with cell-specific transcription in adult erythroid cells revealed by qPCR and gene expression profiling.

It is also interesting that gata1 interacts only with promoters of adult globins and the LCR. We examined the chromatin landscape in adult erythroid cells by ATAC-seq and DNase-seq to see if the adult globin cluster and LCR have a different epigenetic organization in adult erythrocytes. As expected, the adult globin cluster is open while the embryonic globin cluster stays closed, as evidenced by the peaks from ATAC-seq and DNase-seq analyses. Together, these comprehensive analyses demonstrate that there is an adult erythroid epigenetic organization that supports adult erythroid specific gene transcription and therefore cell identity and function.

SUMMARY AND CONCLUSION

This chapter describes effective methods for the identification of transcription factor binding sites and open chromatin regions of the zebrafish epigenome in a genome-wide manner with the use of ChIP-seq and ATAC-seq technologies. These methods can be easily extended to other abundant and relatively pure cell types besides adult erythrocytes. However, as with many other molecular and biochemical assays, the cell populations analyzed need to be as homogenous as possible to achieve consistent and easily interpretable results. It should also be noted that the choice of antibody for ChIP-seq studies is crucial and that not all antibodies can be used successfully in ChIP, even though they may work well in other methods such as Western blotting or immunohistochemistry. The conditions of the experiment may need to be optimized for the number of cells, the abundance of the protein factors of interest, and the available antibodies. The number of cells required for open chromatin detection by ATAC-seq is much smaller. Furthermore, we recommend trying to optimize the methods presented here to achieve the best outcomes, especially when they are applied to different cell types. In conclusion, these epigenomic analyses can reveal the landscape of transcription and chromatin factors as well as the chromatin states of a variety of zebrafish cell types. The in-depth and comprehensive bioinformatics analyses described here, especially in conjunction with gene expression profiling information, are essential for our understanding of the molecular mechanisms that regulate zebrafish cell function and fate.

ACKNOWLEDGMENTS

This work was supported by NIH #5PO1HL32262-34 (to S.H.O), 3P30DK049216-21S1 (to S. H. O), #5R01HL048801-23 (to L.I.Z), and NIH #5K99CA190861-01 (to C.J.O).

Conflicts of interest: L.I.Z. is a founder and stock holder of Fate Therapeutics Inc. and Scholar Rock, and a scientific advisor for Stemgent.

REFERENCES

Adey, A., Morrison, H. G., Asan, Xun, X., Kitzman, J. O., Turner, E. H., ... Shendure, J. (2010). Rapid, low-input, low-bias construction of shotgun fragment libraries by high-density in vitro transposition. *Genome Biology, 11*(12), R119. http://dx.doi.org/10.1186/gb-2010-11-12-r119. Epub 2010 December 8. PubMed PMID:21143862; PubMed Central PMCID:PMC3046479.

Asnani, A., & Peterson, R. T. (July 2014). The zebrafish as a tool to identify novel therapies for human cardiovascular disease. *Disease Models & Mechanisms, 7*(7), 763–767. http://dx.doi.org/10.1242/dmm.016170. Review. PubMed PMID:24973746; PubMed Central PMCID:PMC4073266.

Bailey, T., Krajewski, P., Ladunga, I., Lefebvre, C., Li, Q., Liu, T., ... Zhang, J. (2013). Practical guidelines for the comprehensive analysis of chip-seq data. *PLoS Computational Biology, 9*(11), e1003326. http://dx.doi.org/10.1371/journal.pcbi.1003326.

Brownlie, A., Hersey, C., Oates, A. C., Paw, B. H., Falick, A. M., Witkowska, H. E., … Zon, L. (March 1, 2003). Characterization of embryonic globin genes of the zebrafish. *Developmental Biology, 255*(1), 48–61. PubMed PMID:12618133.

Buenrostro, J. D., Giresi, P. G., Zaba, L. C., Chang, H. Y., & Greenleaf, W. J. (October 6, 2013). Transposition of native chromatin for fast and sensitive epigenomic profiling of open chromatin, DNA-binding proteins and nucleosome position. *Nature Methods, 10*(12), 1213–1218. http://dx.doi.org/10.1038/nmeth.2688.

Carroll, K. J., & North, T. E. (August 2014). Oceans of opportunity: exploring vertebrate hematopoiesis in zebrafish. *Experimental Hematology, 42*(8), 684–696. http://dx.doi.org/10.1016/j.exphem.2014.05.002. Epub 2014 May 9. Review. PubMed PMID:24816275; PubMed Central PMCID:PMC4461861.

Cox, A. G., & Goessling, W. (June 2015). The lure of zebrafish in liver research: regulation of hepatic growth in development and regeneration. *Current Opinion in Genetics & Development, 32*, 153–161. http://dx.doi.org/10.1016/j.gde.2015.03.002. Epub 2015 April 6. Review. PubMed PMID:25863341.

Fan, S. F., & Yazulla, S. (1997). Electrogenic hyperpolarization-elicited chloride transporter current in blue cones of zebrafish retinal slices. *Journal of Neurophysiology, 77*, 1447–1459.

Ganis, J. J., Hsia, N., Trompouki, E., de Jong, J. L., DiBiase, A., Lambert, J. S., … Zon, L. I. (June 15, 2012). Zebrafish globin switching occurs in two developmental stages and is controlled by the LCR. *Developmental Biology, 366*(2), 185–194. http://dx.doi.org/10.1016/j.ydbio.2012.03.021. Epub 2012 April 19. PubMed PMID:22537494; PubMed Central PMCID:PMC3378398.

Goessling, W., & Sadler, K. C. (November 2015). Zebrafish: an important tool for liver disease research. *Gastroenterology, 149*(6), 1361–1377. http://dx.doi.org/10.1053/j.gastro.2015.08.034. Epub 2015 August 28. Review. PubMed PMID:26319012.

Guenther, M. G., Lawton, L. N., Rozovskaia, T., Frampton, G. M., Levine, S. S., Volkert, T. L., … Young, R. A. (2008). Aberrant chromatin at genes encoding stem cell regulators in human mixed-lineage leukemia. *Genes Development, 22*, 3403–3408.

Gurevich, D., Siegel, A., & Currie, P. D. (2015). Skeletal myogenesis in the zebrafish and its implications for muscle disease modelling. *Results and Problems in Cell Differentiation, 56*, 49–76. http://dx.doi.org/10.1007/978-3-662-44608-9_3. Review. PubMed PMID:25344666.

Haesemeyer, M., & Schier, A. F. (February 2015). The study of psychiatric disease genes and drugs in zebrafish. *Current Opinion in Neurobiology, 30*, 122–130. http://dx.doi.org/10.1016/j.conb.2014.12.002. Epub 2014 December 16. Review. PubMed PMID:25523356; PubMed Central PMCID:PMC4294547.

Harismendy, O., Ng, P. C., Strausberg, R. L., Wang, X., Stockwell, T. B., Beeson, K. Y., … Frazer, K. A. (2009). Evaluation of next generation sequencing platforms for population targeted sequencing studies. *Genome Biology, 10*, R32.

van Houcke, J., De Groef, L., Dekeyster, E., & Moons, L. (November 2015). The zebrafish as a gerontology model in nervous system aging, disease, and repair. *Ageing Research Reviews, 24*(Pt B), 358–368. http://dx.doi.org/10.1016/j.arr.2015.10.004. Epub 2015 October 29. Review. PubMed PMID:26538520.

Huang da, W., Sherman, B. T., & Lempicki, R. A. (2009). Systematic and integrative analysis of large gene lists using DAVID bioinformatics resources. *Nature Protocols Other Titles: Protocols, 4*, 44–57. http://dx.doi.org/10.1038/nprot.2008.211.

Jaźwińska, A., & Sallin, P. (January 2016). Regeneration versus scarring in vertebrate appendages and heart. *The Journal of Pathology, 238*(2), 233–246. http://dx.doi.org/10.1002/path. 4644. Epub 2015 October 27. Review. PubMed PMID:26414617.

Jörgens, K., Stoll, S. J., Pohl, J., Fleming, T. H., Sticht, C., Nawroth, P. P., Hammes, H. P., & Kroll, J. (Jan 2015). High tissue glucose alters intersomitic blood vessels in zebrafish via methylglyoxal targeting the VEGF receptor signaling cascade. *Diabetes, 64*(1), 213–225. http://dx.doi.org/10.2337/db14-0352. Epub 2014 Aug 4. PubMed PMID: 25092676.

Kent, W. J., Sugnet, C. W., Furey, T. S., Roskin, K. M., Pringle, T. H., Zahler, A. M., & Haussler, D. (June 2002). The human genome browser at UCSC. *Genome Research, 12*(6), 996–1006.

Kharchenko, P. V., Tolstorukov, M. Y., & Park, P. J. (2008). Design and analysis of ChIP-seq experiments for DNA-binding proteins. *Nature Biotechnology, 26*, 1351–1359. http://dx.doi.org/10.1038/nbt.1508.

Kim, J., Cantor, A. B., Orkin, S. H., & Wang, J. (2009). Use of in vivo biotinylation to study protein-protein and protein-DNA interactions in mouse embryonic stem cells. *Nature Protocols Other Titles: Protocols, 4*, 506–517.

Kircher, M., & Kelso, J. (2010). High-throughput DNA sequencing–concepts and limitations. *BioEssays, 32*, 524–536.

Lagendijk, A. K., Yap, A. S., & Hogan, B. M. (2014). Endothelial cell-cell adhesion during zebrafish vascular development. *Cell Adhesion & Migration, 8*(2), 136–145. Review. PubMed PMID:24621476; PubMed Central PMCID:PMC4049859.

Langmead, B., Trapnell, C., Pop, M., & Salzberg, S. L. (2009). Ultrafast and memory-efficient alignment of short DNA sequences to the human genome. *Genome Biology, 10*, R25. http://dx.doi.org/10.1186/gb-2009-10-3-r25.

Lee, T. I., Johnstone, S. E., & Young, R. A. (2006). Chromatin immunoprecipitation and microarray-based analysis of protein location. *Nature Protocols Other Titles: Protocols, 1*, 729–748.

Li, H., & Durbin, R. (2009). Fast and accurate short read alignment with Burrows-Wheeler transform. *Bioinformatics, 25*, 1754–1760. http://dx.doi.org/10.1093/bioinformatics/btp324.

Li, H., Handsaker, B., Wysoker, A., Fennell, T., Ruan, J., Homer, N., … 1000 Genome Project Data Processing Subgroup. (2009). The Sequence alignment/map (SAM) format and SAMtools. *Bioinformatics, 25*, 2078–2079. PMID:19505943.

Li, H. A. (November 1, 2011). Statistical framework for SNP calling, mutation discovery, association mapping and population genetical parameter estimation from sequencing data. *Bioinformatics, 27*(21), 2987–2993. Epub 2011 September 8. (PMID:21903627).

Liang, K., & Keles, S. (2012). Detecting differential binding of transcription factors with ChIP-seq. *Bioinformatics, 28*, 121–122. http://dx.doi.org/10.1093/bioinformatics/btr605.

Lindeman, L. C., Vogt-Kielland, L. T., Alestrom, P., & Collas, P. (2009). Fish'n ChIPs: chromatin immunoprecipitation in the zebrafish embryo. *Methods in Molecular Biology, 567*, 75–86.

Machanick, P., & Bailey, T. L. (2011). MEME-ChIP: motif analysis of large DNA datasets. *Bioinformatics, 27*, 1696–1697. http://dx.doi.org/10.1093/bioinformatics/btr189.

McCammon, J. M., & Sive, H. (July 1, 2015). Challenges in understanding psychiatric disorders and developing therapeutics: a role for zebrafish. *Disease Models & Mechanisms, 8*(7), 647–656. http://dx.doi.org/10.1242/dmm.019620. Review. PubMed PMID:26092527; PubMed Central PMCID:PMC4486859.

McLean, C. Y., Bristor, D., Hiller, M., Clarke, S. L., Schaar, B. T., … Bejerano, G. (2010). GREAT improves functional interpretation of cis-regulatory regions. *Nature Biotechnology, 28*, 495–501. http://dx.doi.org/10.1038/nbt.1630.

Mohseny, A. B., & Hogendoorn, P. C. (2014). Zebrafish as a model for human osteosarcoma. *Advances in Experimental Medicine and Biology, 804*, 221–236. http://dx.doi.org/10.1007/978-3-319-04843-7_12. Review. PubMed PMID:24924177.

Mort, R. L., Jackson, I. J., & Patton, E. E. (February 15, 2015). The melanocyte lineage in development and disease. *Development, 142*(4), 620–632. http://dx.doi.org/10.1242/dev.106567. Review. Erratum in: Development. 2015 April 1;142(7):1387. PubMed PMID:25670789; PubMed Central PMCID:PMC4325379.

O'Neill, L. P., VerMilyea, M. D., & Turner, B. M. (2006). Epigenetic characterization of the early embryo with a chromatin immunoprecipitation protocol applicable to small cell populations. *Nature Genetics, 38*, 835–841.

Paik, E. J., & Zon, L. I. (2010). Hematopoietic development in the zebrafish. *International Journal of Developmental Biology, 54*, 1127–1137.

Phillips, J. B., & Westerfield, M. (July 2014). Zebrafish models in translational research: tipping the scales toward advancements in human health. *Disease Models & Mechanisms, 7*(7), 739–743. http://dx.doi.org/10.1242/dmm.015545. Review. PubMed PMID:24973743; PubMed Central PMCID:PMC4073263.

Plantié, E., Migocka-Patrzałek, M., Daczewska, M., & Jagla, K. (April 9, 2015). Model organisms in the fight against muscular dystrophy: lessons from drosophila and zebrafish. *Molecules: a Journal of Synthetic Chemistry and Natural Product Chemistry, 20*(4), 6237–6253. http://dx.doi.org/10.3390/molecules20046237. Review. PubMed PMID:25859781.

Powles-Glover, N. (September 2014). Cilia and ciliopathies: classic examples linking phenotype and genotype-an overview. *Reproductive Toxicology, 48*, 98–105. http://dx.doi.org/10.1016/j.reprotox.2014.05.005. Epub 2014 May 23. Review. PubMed PMID:24859270.

Quinlan, A. R., & Hall, I. M. (2010). BEDTools: a flexible suite of utilities for comparing genomic features. *Bioinformatics, 26*, 841–842. http://dx.doi.org/10.1093/bioinformatics/btq033.

Quiros-Gonzalez, I., & Yadav, V. K. (November 1, 2014). Central genes, pathways and modules that regulate bone mass. *Archives of Biochemistry and Biophysics, 561*, 130–136. http://dx.doi.org/10.1016/j.abb.2014.06.005. Epub 2014 June 17. Review. PubMed PMID:24950023.

Rashid, N. U., Giresi, P. G., Ibrahim, J. G., Sun, W., & Lieb, J. D. (July 25, 2011). ZINBA integrates local covariates with DNA-seq data to identify broad and narrow regions of enrichment, even within amplified genomic regions. *Genome Biology, 12*(7), R67. http://dx.doi.org/10.1186/gb-2011-12-7-r67. PubMed PMID:21787385; PubMed Central PMCID:PMC3218829.

Rotem, A., Ram, O., Shoresh, N., Sperling, R. A., Goren, A., Weitz, D. A., & Bernstein, B. E. (November 2015). Single-cell ChIP-seq reveals cell subpopulations defined by chromatin state. *Nature Biotechnology, 33*(11), 1165–1172. http://dx.doi.org/10.1038/nbt. 3383. Epub 2015 October 12. PubMed PMID:26458175; PubMed Central PMCID:PMC4636926.

Sandmann, T., Jakobsen, J. S., & Furlong, E. E. (2006). ChIP-on-chip protocol for genome-wide analysis of transcription factor binding in *Drosophila melanogaster* embryos. *Nature Protocols Other Titles: Protocols, 1*, 2839–2855.

Saralahti, A., & Rämet, M. (September 2015). Zebrafish and streptococcal infections. *Scandinavian Journal of Immunology, 82*(3), 174–183. http://dx.doi.org/10.1111/sji.12320. Review. PubMed PMID:26095827.

Schlegel, A., & Gut, P. (June 2015). Metabolic insights from zebrafish genetics, physiology, and chemical biology. *Cellular and Molecular Life Sciences: CMLS, 72*(12), 2249–2260. http://dx.doi.org/10.1007/s00018-014-1816-8. Epub 2015 January 4. Review. PubMed PMID:25556679; PubMed Central PMCID:PMC4439526.

Shao, Z., Zhang, Y., Yuan, G. C., Orkin, S. H., & Waxman, D. J. (2012). MAnorm: a robust model for quantitative comparison of ChIP-Seq data sets. *Genome Biology, 13*, R16. http://dx.doi.org/10.1186/gb-2012-13-3-r16.

Song, H. D., Sun, X. J., Deng, M., Zhang, G. W., Zhou, Y., Wu, X. Y., … Chen, Z. (November 16, 2004). Hematopoietic gene expression profile in zebrafish kidney marrow. *Proceedings of the National Academy of Sciences of the United States of America, 101*(46), 16240–16245. Epub 2004 November 1. PubMed PMID:15520368; PubMed Central PMCID:PMC528959.

Subramanian, A., Tamayo, P., Mootha, V. K., Mukherjee, S., Ebert, B. L., … Mesirov, J. P. (2005). Gene set enrichment analysis: a knowledge-based approach for interpreting genome-wide expression profiles. *Proceedings of the National Academy of Sciences of the United States of America, 102*, 15545–15550. http://dx.doi.org/10.1073/pnas.0506580102.

Takeuchi, T. (June 2014). Regulation of cardiomyocyte proliferation during development and regeneration. *Development, Growth & Differentiation, 56*(5), 402–409. http://dx.doi.org/10.1111/dgd.12134. Epub 2014 April 16. Review. PubMed PMID:24738847.

Vastenhouw, N. L., Zhang, Y., Woods, I. G., Imam, F., Regev, A., Liu, X. S., … Schier, A. F. (2010). Chromatin signature of embryonic pluripotency is established during genome activation. *Nature, 464*, 922–926.

Wager, K., Mahmood, F., & Russell, C. (July 2014). Modelling inborn errors of metabolism in zebrafish. *Journal of Inherited Metabolic Disease, 37*(4), 483–495. http://dx.doi.org/10.1007/s10545-014-9696-5. Epub 2014 May 6. Review. PubMed PMID:24797558.

White, R. M. (February 2015). Cross-species oncogenomics using zebrafish models of cancer. *Current Opinion in Genetics & Development, 30*, 73–79. http://dx.doi.org/10.1016/j.gde.2015.04.006. Epub 2015 June 9. Review. PubMed PMID:26070506; PubMed Central PMCID:PMC4603543.

White, R. M., Sessa, A., Burke, C., Bowman, T., LeBlanc, J., Ceol, C., … Zon, L. I. (2008). Transparent adult zebrafish as a tool for in vivo transplantation analysis. *Cell Stem Cell, 2*, 183–189.

Wilkinson, R. N., & van Eeden, F. J. (2014). The zebrafish as a model of vascular development and disease. *Progress in Molecular Biology and Translational Science, 124*, 93–122. http://dx.doi.org/10.1016/B978-0-12-386930-2.00005-7. Review. PubMed PMID: 24751428.

Wilkinson, R. N., Jopling, C., & van Eeden, F. J. (2014). Zebrafish as a model of cardiac disease. *Progress in Molecular Biology and Translational Science, 124*, 65–91. http://dx.doi.org/10.1016/B978-0-12-386930-2.00004-5. Review. PubMed PMID:24751427.

Zhang, Y., Liu, T., Meyer, C. A., Eeckhoute, J., Johnson, D. S., Bernstein, B. E., … Liu, X. S. (2008). Model-based analysis of chip-seq (MACS). *Genome Biology, 9*, R137.

Zheng, X., Yue, S., Chen, H., Weber, B., Jia, J., & Zheng, Y. (November 17, 2015). Low-cell-number epigenome profiling aids the study of lens aging and hematopoiesis. *Cell Reports, 13*(7), 1505–1518. http://dx.doi.org/10.1016/j.celrep.2015.10.004. Epub 2015 November 5. PubMed PMID:26549448.

CHAPTER

21

Assay for transposase-accessible chromatin and circularized chromosome conformation capture, two methods to explore the regulatory landscapes of genes in zebrafish

A. Fernández-Miñán*[,a], J. Bessa[§,¶,a], J.J. Tena*, J.L. Gómez-Skarmeta*[,1]

**Centro Andaluz de Biología del Desarrollo (CABD), Consejo Superior de Investigaciones Científicas/Universidad Pablo de Olavide/Junta de Andalucía, Sevilla, Spain*

[§]IBMC-Instituto de Biologia Molecular e Celular, Porto, Portugal

[¶]Universidade do Porto, Porto, Portugal

[1]Corresponding author: E-mail: jlgomska@upo.es

CHAPTER OUTLINE

[a]These authors contributed equally to this work.

Methods in Cell Biology, Volume 135, ISSN 0091-679X, http://dx.doi.org/10.1016/bs.mcb.2016.02.008

Abstract

Accurate transcriptional control of genes is fundamental for the correct functioning of organs and developmental processes. This control depends on the interplay between the promoter of genes and other noncoding sequences, whose interaction is mediated by 3D chromatin arrangements. Thus, the detailed description of transcriptional regulatory landscapes is essential to understand the mechanisms of transcriptional regulation. However, to achieve that, two important challenges have to be faced: (1) the identification of the noncoding sequences that contribute to gene transcription and (2) the association of these sequences to the respective genes they control. In this chapter, we describe two protocols that allow overcoming these important challenges: the assay for transposase-accessible chromatin using sequencing (ATAC-seq) and circularized chromosome conformation capture (4C-seq). ATAC-seq is a very efficient technique that, using a very low number of cells as starting material, allows the identification of active chromatin regions genome wide, whereas 4C-seq detects the subset of sequences that interact specifically with the promoter of a given gene. When combined, both techniques provide a comprehensive snapshot of the regulatory landscapes of developmental genes. The protocols we present here have been optimized for teleost fish samples, zebrafish and medaka, allowing the in-depth study of transcriptional regulation in these two emerging animal models. Given the amenability and easy genetic manipulation of these two experimental systems, we anticipate that they will be important in revealing general principles of the vertebrate regulatory genome.

INTRODUCTION

The precise control of gene expression is crucial for the development and homeostasis of multicellular organisms. Indeed, alteration of this control is the cause of most human genetic-related diseases (Maurano et al., 2012). Gene expression is controlled by proximal and distal noncoding regulatory sequences. Proximally, at the 5′ of the coding region, the promoters act as regulatory sequences essential for the assembly of the basal transcription machinery that recruits RNA polymerase II (PolII), triggering transcription. For many genes, promoters contain most of the information that controls their expression. However, for those genes with complex and dynamic expression patterns, in addition to promoters, distal *cis*-regulatory regions play essential functions to control transcription precisely in time and space (Danino, Even, Ideses, & Juven-Gershon, 2015). The most widely studied regulatory sequences are transcriptional enhancers. Enhancers contain binding sites for transcription factors that mediate the activation of PolII at the core of the gene's promoter to trigger transcription (Ong & Corces, 2011; Tjian & Maniatis, 1994).

One of the major functions of chromatin is to pack DNA tightly in the nucleus. This highly organized macromolecule is mostly composed by a successive repetition of basic structural units composed of DNA wrapped around eight histones, the nucleosomes. At promoters and regulatory regions, nucleosomes are dismantled to facilitate the accessibility of proteins to DNA to control gene transcription properly. Different techniques have been developed to identify these accessible open chromatin regions. DNAse assays (John et al., 2013; Song & Crawford, 2010) or formaldehyde-assisted isolation of regulatory elements with sequencing (FAIRE-seq; (Simon, Giresi, Davis, & Lieb, 2012)) have been widely used to identify open chromatin. Although the principles of both techniques are different, the first uses DNAse I to detect DNA accessibility and the second uses isolation of cross-linked DNA, both techniques involve several steps prone to DNA loss, limiting their efficiency and requiring large number of cells. A more efficient alternative, an assay for transposase-accessible chromatin using sequencing (ATAC-seq) has been recently developed. This technique consists of the in vitro transposition of sequencing adapters into native accessible chromatin, requiring a minimum of 500 cells to generate a robust and reproducible genome-wide profile of open chromatin (Buenrostro, Giresi, Zaba, Chang, & Greenleaf, 2013). This strategy has been refined and has been combined with the use of a microfluidics platform, so it can be applied to single cells (Buenrostro et al., 2015). ATAC-seq was originally used to identify open chromatin in a human cell line (GM12878; (Buenrostro et al., 2013)), but has been shown to be also very efficient in multicellular tissues of *Drosophila* or zebrafish (Davie et al., 2015; Gehrke et al., 2015). In this chapter, we will describe a detailed ATAC-seq protocol for zebrafish and medaka embryos.

Chromatin immunoprecipitation followed by sequencing (ChIP-seq) is also widely employed to identify different histone modifications associated with different chromatin states. Monomethylation of lysine 4 in histone 3 (H3K4me1) and acetylation of lysine 27 in histone 3 (H3K27ac) are modifications associated with enhancer sequences. Active enhancers are characterized by the presence of H3K27ac and H3K4me1 while poised enhancers are only decorated with H3K4me1 (Creyghton et al., 2010; Hawkins et al., 2011; Rada-Iglesias et al., 2011). It has been shown that ATAC-seq correlates with H3K4me1, therefore ATAC signal can be found in both active and poised enhancers (Lara-Astiaso et al., 2014). Promoters can also be identified by specific chromatin modifications, and trimethylation of lysine 4 in histone 3 (H3K4me3) is the most common mark used to identify them (Santos-Rosa et al., 2002). We have profiled these three marks at different stages of zebrafish development, allowing us to determine the developmental dynamics of promoters and enhancers during early embryogenesis (Bogdanovic et al., 2012). These data are available through our Web page (https://www.upo.es/CABD/GomezSkarmetaLab/epigenomics.html) and a detailed ChIP-seq protocol for zebrafish embryos can be found elsewhere (Bogdanovic et al., 2013; Trompouki et al., 2011).

These techniques facilitate the genome-wide identification of the noncoding *cis*-regulatory regions, but they do not provide information about the genes they regulate. Since *cis*-regulatory regions can be located hundred of kilobases from the genes they regulate, even at the introns of neighboring genes, the identification of their target promoters is not a straightforward task (Irimia et al., 2012; de Laat & Duboule, 2013). *Cis*-regulatory regions are placed in the vicinity of their target promoters through chromatin loops. In this 3D configuration of the chromatin, the proteins assembled at enhancers can interact with the proteins present at promoters, ultimately controlling the activity of PolII (Ong & Corces, 2011). This rationale is at the base of the chromosome conformation capture (3C) technique, which allows monitoring the proximity of two candidate regions of the chromatin in its 3D structure (Dekker, Rippe, Dekker, & Kleckner, 2002). 3C is performed by cross-linking the chromatin to fix its 3D organization, followed by a digestion with a restriction enzyme, a ligation, reversion of the cross-link, and finally a PCR. In these steps, two genomic regions that are in close proximity in the 3D structure of the chromatin will ligate more efficiently than other sequences located farther apart. Therefore, a PCR reaction using this ligation product will show more efficient amplifications for the chimeric products obtained from genomic regions that are in close proximity than for those resulting from less frequent ligations of distant regions. One important limitation of this technique is that it can only be used in a candidate-based approach, because to design appropriate primers that cover the ligations events, the sequences whose interactions are going to be interrogated have to be selected in advance. An improved 3C based protocol, circularized chromosome conformation capture followed by sequencing (4C-seq; (Zhao et al., 2006)), can overcome this limitation and profile all possible interactions from a region of interest (viewpoint). In 4C-seq, the PCR step is replaced by an inverse PCR anchored to a selected genomic locus, using primers with adapters for next generation sequencing (NGS). Overrepresented NGS reads indicate genomic regions that are in the vicinity of the sequence of interest, as enhancers that may interact with a given promoter. Therefore, this approach can help to identify the genomic territory in contact with a particular promoter, in other words, the regulatory landscape read by a specific gene.

Here we provide detailed protocols for ATAC-seq and 4C-seq, optimized for zebrafish and medaka samples, that when combined allow exploring the regulatory landscapes of genes, as we have shown for several developmental genes in zebrafish (Fig. 1) (Gomez-Marin et al., 2015; Smemo et al., 2014). Together, these two techniques will be of great utility in functional genomic analysis of teleost embryogenesis.

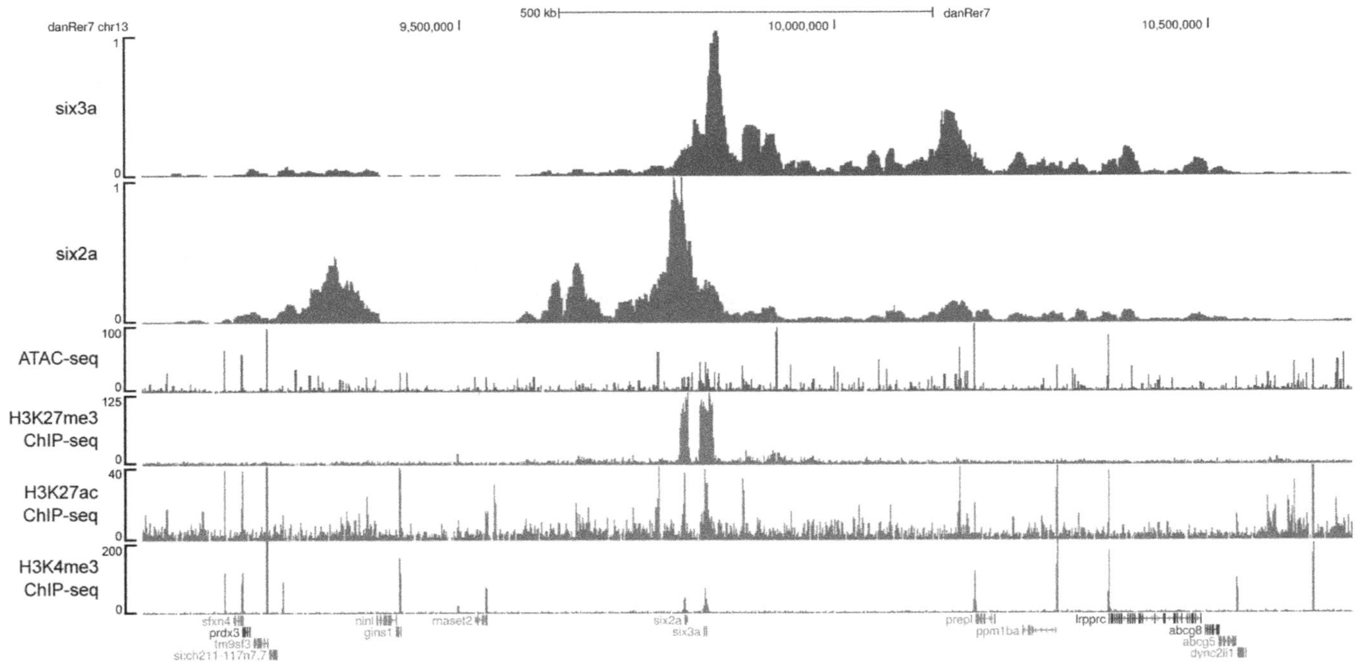

FIGURE 1 Genomic landscape around six2a/six3a cluster in zebrafish.

UCSC Genome Browser tracks showing different data sets in the region of *six2a* and *six3a* genes in zebrafish. These two genes have largely different expression patterns, which support the complementary contact maps revealed by circularized chromosome conformation capture using sequencing (4C-seq) from these two genes promoters. The 4C-seq signal from *six3a* promoter (in green) shows predominant contacts with the right half of the cluster, while the *six2a* promoter (in blue) is more prone to contact with the left side. Open chromatin regions detected with transposase-accessible chromatin using sequencing (*orange track*) overlap with other epigenomic marks such as H3K4me4, H3K27ac, and H3K27me3 (light gray tracks). (See color plate)

1. 4C-SEQ PROTOCOL

Here, we provide a step-by-step protocol for template preparation that can be used for quantitative PCR (3CqPCR) or deep sequencing (4C-seq) based on previous published protocols (Splinter, de Wit, van de Werken, Klous, & de Laat, 2012).

1.1 CELL ISOLATION

All the steps in this protocol are optimized for 1×10^7 cells. For 3C, we recommend using large numbers of embryos to end up with enough DNA to perform nonbiological replicates of qPCR, if necessary. For reference, we usually use 2000 embryos at 80% epiboly and 500 embryos at 24 hpf and 48 hpf. For 4C-seq, fewer cells can be used. We have successfully prepared 4C samples from 5×10^5 cells.

1.1.1 Day 1

1.1.1.1 Zebrafish embryos

1a. Zebrafish embryos are kept at 28°C until the desired stage. Synchronized embryos are collected in a glass beaker in embryo medium (E3 medium: 5 mM NaCl, 0.17 mM KCl, 0.4 mM $CaCl_2$, 0.16 mM $MgSO_4$). Use pronase (Roche #10165921001) to remove the chorion enzymatically. Add 500 μL of pronase (30 mg/mL) to 50 mL of E3. Shake the embryos gently until a few embryos without chorions are observed under the stereoscope. This usually takes a few minutes (15–20 min). Chorion softening can be speeded up by placing the beaker at 28°C (more details in (Westerfield, 1995)).
Note: Before gastrulation embryos without chorions should be kept always in glass or agarose-coated plates since they are fragile and stick to the plastic.

2a. Wash the embryos immediately three to five times with E3 to remove the pronase completely. The washes will help to release most of the embryos from their chorions. Alternatively, after the washes, use a Pasteur pipette and carefully pipette them several times in E3 medium. Once all the embryos are dechorionated, transfer them to a 1.5-mL safe-lock tube. Remove as much E3 as possible. Proceed with step 3.

1.1.1.2 Medaka embryos

1b. Medaka embryos are kept at 28°C until the desired stage. Synchronized embryos are dechorionated manually. Because manual dechorionation is time-consuming, it is advisable to maintain the embryos on a Petri dish placed on ice and dechorionate embryos in groups of 10, to prevent the embryos from getting out of development synchrony. Depending on the number of embryos required, cell isolation must be performed in batches (see step 10). Dechorionate the embryos with sharp forceps in a Petri dish containing ice-cold 1X PBS (phosphate buffer). Chorion removal results in yolk content release.
Note: Manual dechorionation is not feasible at early stages. In that case, follow the hatching enzyme protocol (see Rembold & Wittbrodt, 2004).

2b. Embryos are transferred with forceps to an Eppendorf tube placed on ice and containing ice-cold 1X PBS as soon as they are dechorionated. Once all the embryos are dechorionated, remove as much PBS as possible and proceed with step 3.

3. Add 1 mL of deyolking buffer (Ginzburg Fish Ringer without calcium: 55 mM NaCl, 1.8 mM KCl, 1.25 mM $NaHCO_3$).

4. To disrupt the yolks, pipette up and down with a P200 micropipette and then place the tube in a shaker at 1100 rpm for 5 min.
Note: The deyolking step is critical, since yolk contents impair the activity of restriction enzymes and therefore DNA remains undigested.

5. Place the tube in a microcentrifuge and spin at 300 g for 30 s to pellet the cells.

6. Carefully remove the supernatant with a P200 micropipette. Resuspend the pellet in 500 μL 1X PBS.

7. Transfer the cells to a 15 mL Falcon tube and add 4.5 mL 1X PBS. Proceed with chromatin cross-linking.

1.2 CHROMATIN CROSS-LINKING

8. Cross-link the chromatin with paraformaldehyde (PFA) at 2% final concentration. To that end, add 5 mL PFA 4% (in phosphate buffer 20 mM, pH 7.4) to the 15 mL Falcon containing the cells and incubate for 10 min at room temperature while tumbling.

9. To quench the cross-linking reaction, add 1.425 mL 1 M glycine (can be kept filtered at 4°C) to the tube, mix and place the tubes on ice immediately.

10. Directly pellet cells by centrifuging 8 min at 1300 rpm at 4°C. Remove supernatant. Proceed with cell lysis (step 11).
Note: After centrifugation, the pelleted cells can be shock frozen in liquid nitrogen and stored at −80°C. This allows processing of samples in batches if necessary by storing them at −80°C until enough material is achieved. These batches can then be processed together.

1.3 CELL LYSIS

11. Resuspend the cells gently in 5 mL cold lysis buffer (50 mM Tris−HCl pH 7.5, 150 mM NaCl, 5 mM EDTA, 0.5% NP-40 (Sigma #I8896), 1% TX-100 (Sigma #T8532) and 1X CØmplete protease inhibitors (Roche #11245200)). Transfer them to a tissue grinder (15 mL Tenbroeck Tissue Grinder, Wheaton 35,7426) and homogenize on ice every 5−10 min to ensure that cells are completely lysed.
Methyl green pyronin (Sigma #HT70116) is used to assess lysis efficiency since it stains the cytoplasm pink and the nuclei blue/green. Mix 3 μL of cells with 3 μL of methyl green pyronin staining on a microscope slide, overlay with a coverslip and check under a microscope. When cells are lysed, either green nuclei without cytoplasm or green nuclei with a collapsed cytoplasm are observed (Fig. 2A).

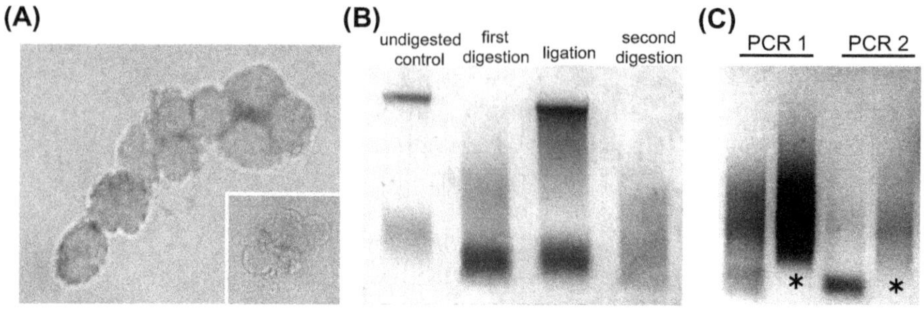

FIGURE 2 Circularized chromosome conformation capture using sequencing (4C-seq) sample preparation.

Different steps during 4C-seq sample preparation. (A) Cell lysis efficiency assessed by methyl green pyronin staining that stains the cytoplasm pink (gray in print versions) and the nuclei blue/green (dark gray in print versions). Inset: example of lysed cells with the cytoplasms completely disrupted. (B) Samples of the chromatin at different stages of the protocol visualized in an agarose gel. A clear shift is observed in the average sizes after both steps of digestion. (C) **Two 4C-seq libraries** before and after the purification step that increases chromatin concentration and removes primer-dimers (*asterisks*).

Note: It is critical to ensure proper cell lysis so the subsequent digestion step works efficiently. However, we have also noticed that leaving the cell to lysis for too long, ie, overnight, results in DNA degradation. For reference, 3 h are enough for lysis of zebrafish and medaka cells.

12. Transfer the sample back to the 15 mL Falcon and centrifuge 5 min at 1800 rpm at 4°C. Remove supernatant and proceed with digestion.
    Note: At this point, nuclei can be shock frozen in liquid nitrogen and stored at −80°C.

1.4 DIGESTION AND LIGATION OF THE CROSS-LINKED DNA

13. Resuspend the nuclei in 450 μL of H_2OmQ and transfer them to a 1.5-mL safe-lock tube.
14. Add 60 μL of DpnII 10X buffer (NEB #R0543M).
    Note: In our hands, the following enzymes work properly in the first (cross-linked chromatin) and second digestions: DpnII, Csp6I, HindIII, EcoRI, and NlaIII; nevertheless, other enzymes as BglII have been reported to work as well (Hagege et al., 2007).
15. Place the tube in a thermomixer at 37°C and add 15 μL of 10% sodium dodecyl sulfate (SDS). Incubate 1 h at 37°C while shaking at 900 rpm.
16. Add 75 μL 20% Triton X-100 and incubate 1 h at 37°C while shaking at 900 rpm.
17. Take a 5 μL aliquot of the sample as "undigested control" and store it at 4°C until step 21.
18. Add 400 units of DpnII and incubate overnight at 37°C while shaking at 900 rpm.

1.4.1 Day 2

19. Add 200 units of DpnII and incubate 4 h at 37°C while shaking at 900 rpm.
20. Take a 5 μL aliquot of the sample as "digested control."
21. Leave digestion running and determine digestion efficiency. To that end:
  21.1 Add 90 μL 10 mM Tris–HCl pH 7.5 to the 5 μL control samples from steps 17 to 20.
  21.2 Add 5 μL Proteinase K (10 mg/mL) and incubate for 1 h at 65°C in a thermoblock or a bath.
  21.3 Add 100 μL phenol-chloroform (Amresco, 11732668001) to the samples and mix vigorously.
  21.4 Spin at 13,000 rpm for 5 min at room temperature.
  21.5 Transfer the upper phase to a clean tube and load 20 μL on a 0.6% agarose gel (Fig. 2B).
22. If DNA was properly digested proceed with the first ligation. Otherwise, repeat from step 18.
23. Heat-inactivate the restriction enzyme by incubating 20 min at 65°C and proceed with step 24.
Alternatively, when the selected restriction enzyme is not sensitive to heat inactivation, proceed as follows:
  23.1 Add 80 μL 10% SDS and incubate 30 min at 65°C.
  23.2 Transfer the sample to a 50 mL Falcon tube and add 5.4 mL H_2OmQ.
  23.3 Add 700 μL 10X Promega T4 Ligase buffer (300 mM Tris–HCl pH 7.8, 100 mM $MgCl_2$, 100 mM DTT and 10 mM ATP). Add 375 μL 20% Triton X-100 and incubate 1 h at 37°C.
  23.4 Continue with step 25.
24. Transfer the sample to a 50 mL Falcon tube and add 5.7 mL H_2OmQ and 700 μL 10X Promega T4 Ligase buffer.
25. Add 60 units T4 DNA Ligase (Promega, 10481220001), mix by swirling and incubate overnight at 16°C. A water bath located in the cold room can be used since shaking is not necessary.

1.4.2 Day 3

26. Take a 100 μL aliquot of the sample as "ligation control."
27. Leave ligation running and determine ligation efficiency. To that end:
  27.1 Add 5 μL Proteinase K (10 mg/mL) and incubate for 1 h at 65°C in a thermoblock or a bath.
  27.2 Add 100 μL phenol-chloroform to the sample and mix vigorously.
  27.3 Spin at 13,000 rpm for 5 min at room temperature.
  27.4 Transfer the upper phase to a clean tube and load 20 μL on a 0.6% agarose gel together with the undigested and digested controls from step 21 (Fig. 2B).
28. If ligation occurred properly, proceed with step 29. If not, add fresh ATP (final concentration of 1 mM) and repeat from step 25.

1.5 REVERSE CROSS-LINKING AND PRECIPITATION

29. Add 30 μL Proteinase K (10 mg/mL) and reverse cross-link overnight at 65°C in a water bath.

1.5.1 *Day 4*

30. Add 30 μL RNase A (10 mg/mL) and incubate 45 min at 37°C in a water bath.
31. Add 7 mL phenol-chloroform and mix vigorously. Centrifuge for 15 min at 3270 g at room temperature.
32. Transfer the aqueous phase to a new 50 mL Falcon tube and add: 7.5 mL H_2OmQ, 1 mL 3M NaAC pH 5.2, 7 μL glycogen (1 mg/mL) (Roche #10901393001) and 35 mL 100% EtOH.
33. Mix vigorously and incubate at −80°C until the sample is frozen solid. This can take several hours.
34. Centrifuge for 60 min at 3270 g at 4°C.
35. Remove supernatant and add 10 mL cold 70% ethanol. Centrifuge at 4°C for 30 min at 3270 g.
36. Remove the supernatant and briefly dry the pellet at room temperature.
37. Dissolve the pellet in 150 μL 10 mM Tris−HCl pH 7.5 at 37°C for 30 min. The 3C template is now ready to be used for quantitative PCRs (pPCR) or it can be stored at −20°C. Analysis of 3C assays by qPCR has been already discussed in Hagege et al. (2007).

1.6 SECOND ROUND OF DIGESTION AND LIGATION

38. To the 150 μL 3C sample add: 50 μL Csp6I 10X buffer, H_2OmQ up to 500 μL and 50 units Csp6I (Fermentas #ER0211). Incubate overnight at 37°C.

1.6.1 *Day 5*

39. Take a 5 μL aliquot of the sample as "digestion control."
40. Leave digestion running and determine digestion efficiency. To that end:
    40.1 Add 95 μL 10 mM Tris−HCl pH 7.5 to the 5 μL sample from step 39.
    40.2 Load 20 μL on a 0.6% agarose gel next to the "ligation control" from step 27 (Fig. 2B).
41. If digestion occurred properly, proceed with step 42. If not, add fresh restriction enzyme (100 units) and repeat from step 38. Alternatively, the sample can be repurified with phenol-chloroform to facilitate efficient digestion.
42. Heat-inactivate the restriction enzyme by incubating 20 min at 65°C and proceed with step 43.
    Alternatively, when the restriction enzyme is not sensitive to heat inactivation, it can be inactivated by sample purification, see step 42.1:
    42.1 Add 500 μL phenol-chloroform and mix vigorously.
    42.2 Centrifuge 10 min at 16400 g at room temperature.

42.3 Transfer the aqueous phase to a fresh tube and add 50 μL 2M NaAc pH 5.2 and 950 μL 100% EtOH.
42.4 Incubate at −80°C until completely frozen.
42.5 Centrifuge 20 min at 16400 g at 4°C.
42.6 Remove supernatant and add 150 μL cold 70% ethanol.
42.7 Centrifuge 10 min 16400 g at 4°C.
42.8 Resuspend the pellet in 500 μL 10 mM Tris–HCl pH 7.5 and proceed with step 43.

43. Transfer the sample to a 50 mL Falcon and add 12.1 mL H_2OmQ, 1.4 mL 10X Promega T4 ligase buffer and 100 units T4 DNA ligase.
44. Mix by swirling and incubate overnight at 16°C. A water bath located in the cold room can be used since shaking is not necessary.

1.7 SAMPLE PURIFICATION

1.7.1 Day 6

45. Transfer the sample to an Amicon Ultra-15 Centrifugal Filter Device (Millipore) and centrifuge at room temperature for 30 min at 3270 g.
46. Remove the filter device from the Amicon Filter Falcon and discard filtrate in centrifuge tube. Place the filter device back in the centrifuge tube and wash by adding 10 mL Tris–HCl pH 7.5. Centrifuge for 30 min at 3270 g at room temperature.
47. Repeat previous step.
48. Recover the concentrated solute, approximately 200 μL with a P200 micropipette and transfer it to a safe-lock tube.
    Note: Longer centrifugation times (up to 1 h) can be necessary to end up with approximately 200 μL of concentrated solute.
49. Use 1 μL to determine sample quantity using Qubit dsDNA BR Assay Kit (Molecular Probes #Q32850). Follow instructions provided with the kit.
50. The 4C template is now ready and can be stored at −20°C or used directly in step 51.

1.8 PREPARATION OF 4C LIBRARIES FOR ILLUMINA SEQUENCING

Details on experimental design, ie, primers design, choice of region of interest (viewpoint), and restriction enzyme selection have been already discussed in detail (see Splinter et al., 2012).

51. Once the viewpoint is selected, we use Primer3 (http://bioinfo.ut.ee/primer3/) (Koressaar & Remm, 2007; Untergasser et al., 2012) to design the short primers with the following conditions: primer size between 18 and 27 bp, primer Tm between 59°C and 61°C, primer GC% content between 40% and 60%, Max-Poly X is set to 3 and GC clamp to 2. Primers are ordered desalted and then resuspended in Tris–HCl pH 8 10 mM to a final concentration of 1 μg/μL.

52. 4C template dilutions of 12.5, 25, 50, and 100 ng/μL are used to determine the linear range of amplification. To that end mix the following in a PCR tube:
    a. 1 μL 4C DNA template
    b. 2.5 μL 10X PCR buffer 1
    c. 0.5 μL dNTP (10 mM)
    d. 2 μL forward + reverse short primers (1/7 dilution mix)
    e. 0.35 μL Expand Long Template Polymerase (Roche, #11759060001)
    f. 18.65 μL H_2OmQ
    Final volume is 25 μL.

4C PCR program: 2 min 94°C; 30 cycles of 10 s 94°C, 1 min 55°C, and 3 min 68°C; 5 min 68°C; hold at 4°C.

53. To evaluate linear amplification run 15 μL of the PCR products on a 1.5% agarose gel. Keep the remaining 10 μL to compare with the long primers PCR product.
54. Order the long primers with the Illumina adapters. An example of long primers, where adapters are highlighted in uppercase:
    Reading long primer:
    AATGATACGGCGACCACCGAACACTCTTTCCCTACACGACGCTCTTCCGATCTgccgcttccttctgtttgatc
    Nonreading long primer:
    CAAGCAGAAGACGGCATACGAgaagttgcggctctttcc
    Primers are ordered HPLC-purified and resuspended in Tris–HCl pH 8 10 mM to a final concentration of 1 μg/μL.
55. Perform a 4C PCR with the long primers. Mix in a PCR tube:
    a. 1 μL 4C DNA template (concentration depends on PCR from step 52)
    b. 2.5 μL 10X PCR buffer 1
    c. 0.5 μL dNTP (10 mM)
    d. 4.5 μL Reading long primer (1/7 dilution)
    e. 3 μL nonreading long primer (1/7 dilution)
    f. 0.35 μL Expand Long Template Polymerase
    g. 13.15 μL H_2OmQ
    Final volume is 25 μL.

Run 15 μL of the PCR product on a 1.5% agarose gel together with the 10 μL from step 53. Due to the extra nucleotides of the adapters, 75 nt for the reading primer and the 40 nt for the nonreading primer, there is a shift in the PCR product length.

Note: When there is low quantity of the PCR product and high amount of unused primer, you can reduce primer volumes to half. It may also be necessary to increase the number of cycles up to 32 and perform PCR with different annealing temperatures.

56. Once the optimal conditions are established for the long primers, proceed with the 4C library generation. Mix in a tube the following:
    a. 8 μL 4C DNA template
    b. 20 μL 10X PCR buffer 1
    c. 4 μL dNTP (10 mM)

d. 6 μL reading long primer (1 μg/μL stock)
e. 4 μL nonreading long primer (1 μg/μL stock)
f. 2.8 μL Expand Long Template Polymerase
g. 155.2 μL H_2OmQ
Final volume is 200 μL.

Mix and divide into eight PCR tubes of 25 μL each and run a PCR with the conditions established in step 55.

57. Pool all the PCRs in an Eppendorf tube, mix and take a 20 μL sample. Run 10 μL on a 1.5% agarose gel. Keep 10 μL to compare with the purified library from step 60.
58. Purify the sample using the High Pure PCR Product Purification Kit (Roche, 28104) to remove the nonused adapter primers from the PCR product. Follow instructions provided with the kit. Add 500 μL of Binding Buffer to the 200 μL PCR and use one column. Elute in 40 μL elution buffer.
59. We use AMPure XP PCR purification kit (Agencourt AMPure XP) to further purify the 4C library. Mix the sample with the AMPure XP reagent at a 1:1 ratio and follow instructions provided with the kit. Elute in 40 μL 10 mM Tris pH 8.5.
60. Use 1 μL of the purified library to determine sample quantity using Qubit dsDNA BR Assay Kit (Molecular Probes #Q32850).
61. Run 2–4 μL of the purified PCR together with the 10 μL of nonpurified PCR from step 57 in a 1.5% agarose gel. Primer bands should be absent in the purified sample (Fig. 2C).
62. 4C library is now ready. Minimum concentration required for NGS is 10 ng/μL. The genomic area labeled by a 4C-seq signal is usually limited to several megabases. As a consequence, it requires many fewer reads than other genome-wide techniques and allows pooling of several libraries within one lane.

2. ATAC-SEQ PROTOCOL

Here we have adapted the reported protocol (Buenrostro et al., 2013) to zebrafish and medaka embryos. The number of embryos used for ATAC-seq at different stages is optimized to obtain enough cells (120000–160000 cells) to use 60000–80000 cells for the fragmentation and tagging of the genome with sequencing adapters (tagmentation) (step 9) and the rest to confirm cell number (step 8). See the following table:

Stage	Number of Embryos	Total Number of Cells	μL Used for Tagmentation
256 cells	1500–2000	1700 cells/μL	50 μL
Dome	100	2700 cells/μL	30 μL
Shield	50	4200 cells/μL	15 μL
24 hpf	5	4300 cells/μL	20 μL
48 hpf	5	7600 cells/μL	10 μL

2.1 CELL LYSIS

2.1.1 Zebrafish

1a. Embryos are kept at 28°C until the desired stage. 24 hpf and 48 hpf embryos are dechorionated manually. For earlier stages, synchronized embryos are collected in a glass beaker in embryo medium (E3 medium: 5 mM NaCl, 0.17 mM KCl, 0.4 mM $CaCl_2$, 0.16 mM $MgSO_4$) and pronase (Roche #10165921001) is used to enzymatically remove the chorion. To that end, add 500 μL of pronase (30 mg/mL) to 50 mL of E3. Shake the embryos gently until a few embryos without chorion are observed under the stereoscope. This usually takes a few minutes (15–20 min). Chorion softening can be speed up by placing the beaker at 28°C (more details in (Westerfield, 1995)).
Note: Before gastrulation embryos without chorion should be kept always in glass or agarose-coated plates since they stick to plastic and die.

2a. Wash the embryos immediately three to five times with E3 to completely remove the pronase. The washes will release most of the embryos from their chorion. Alternatively, after the washes use a Pasteur pipette and carefully pipette them several times in the E3 medium. Proceed with step 3.

2.1.2 Medaka

1b. Medaka embryos are kept at 28°C until the desired stage. Synchronized embryos are dechorionated manually. Dechorionate the embryos with sharp forceps in a petri dish containing ice-cold 1X PBS. Chorion removal results in yolk content release.
Note: Manual dechorionation is not feasible at early stages. In that case, follow the hatching enzyme protocol previously reported (Rembold & Wittbrodt, 2004).

2b. Embryos are transferred with forceps to an Eppendorf tube placed on ice and containing ice-cold 1X PBS as soon as they are dechorionated. Proceed with step 3.

3. Once all the zebrafish or medaka embryos are dechorionated, transfer them to a 1.5 mL safe-lock tube. Remove as much medium as possible and add 1 mL of deyolking buffer (Ginzburg Fish Ringer without calcium: 55 mM NaCl, 1.8 mM KCl, 1.25 mM $NaHCO_3$).

4. To disrupt the yolks, pipette up and down with a P200 micropipette and then place the tube in a shaker at 1100 rpm for 5 min.

5. Place the tube in a microcentrifuge at 4°C and spin at 500g for 5 min to pellet the cells. Carefully remove the supernatant with a P200 micropipette.

6. Wash with ice-cold 1X PBS and repeat step 5.

7. Resuspend the pellet in 50 μL lysis buffer (10 mM Tris–HCl, pH 7.4, 10 mM NaCl, 3 mM $MgCl_2$, 0.1% NP-40) and pipette up and down to lyse the cells. Immediately proceed with step 8.

8. Take the corresponding volume of cell lysate (see Table above) and centrifuge at 500 g for 10 min at 4°C. Meanwhile, calculate the total number of cells with the help of a Neubauer chamber. To that end, add 1 μL of Hoechst (0.01 mg/mL) to 10 μL of the remaining sample.

2.2 TAGMENTATION

9. Carefully remove supernatant with a P200 micropipette and add 50 μL of the transposition reaction mix:
   a. 25 μL 2X TD buffer (Illumina Cat#FC-121-1030)
   b. 2.5 μL Tn5 transposase (Illumina Cat#FC-121-1030)
   c. 22.5 μL nuclease-free H_2O

Resuspend the nuclei gently by pipetting up and down.

10. Incubate the transposition reaction at 37°C for 30 min.
11. Immediately after transposition, purify using a Qiagen MinElute Kit (Qiagen, 28004). Follow instructions provided with the kit. Elute in 21 μL Elution Buffer (10 mM Tris buffer, pH 8).
12. At this point, purified DNA can be stored at −20°C or proceed with step 13.

2.3 AMPLIFICATION AND SEQUENCING OF TAGMENTATION LIBRARIES

13. The complete list of primers is available in Supp. Table 1 from Buenrostro et al. (2013).
    Determine the optimum number of enrichment cycles. Use 1 μL DNA from step 12 in the following quantitative PCR:
   a. 1 μL DNA
   b. 0.15 μL Ad1 (10 μM)
   c. 0.15 μL Ad2.1 (10 μM)
   d. 1 μL 10X Sybr Green
   e. 5 μL Kapa Hifi HotStart ReadyMix (Kapa Biosystems)
   f. 2.70 μL H_2OmQ
      Final volume 10 μL.

A typical tagmentation program consist of 98°C for 45 s (hot-start activation step); 72°C for 5 min; 98°C for 30 s; 24 cycles of 98°C for 10 s, 63°C for 30 s, and 72°C for 30 s; 72°C for 1 min.

14. To generate the ATAC library, mix the following in a PCR tube:
   a. 20 μL transposed DNA
   b. 2.5 μL Ad1 (25 μM)
   c. 2.5 μL Ad2.1 (25 μM)
   d. 25 μL Kapa Hifi HotStart ReadyMix (Kapa Biosystems)
      Final volume 50 μL.

A typical ATAC library PCR program consist of 98°C 45 s (hot-start activation step); 72°C for 5 min; 98°C for 30 s; N cycles (N = Ct value determined in qPCR from step 13) of 98°C for 10 s, 63°C for 30 s, 72°C for 30 s; 72°C for 1 min; hold at 4°C.

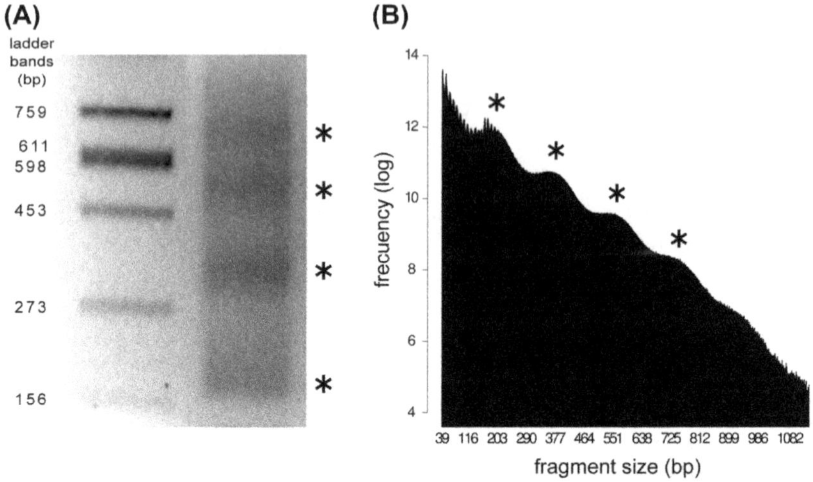

FIGURE 3 Nucleosome-associated bands in transposase-accessible chromatin using sequencing samples.

(A) After tagmentation of the chromatin, clear bands (*asterisks*) can be observed in a sample loaded in an agarose gel. Band sizes correspond to the first few linkers between nucleosomes close to the tagmentation site. (B) Mapping the pair-end reads to the zebrafish genome results in a similar fragment size distribution pattern. In this case, the nearest nucleosomes positions match with valleys in the graph, while linkers positions correspond to peaks (the first four are marked with *asterisks*).

15. Purify the amplified library using Qiagen PCR Cleanup Kit (Qiagen 28104). Follow instructions provided with the kit. Elute in 20 μL Elution Buffer (10 mM Tris buffer, pH 8).
16. Use 1 μL to determine concentration using the Qubit dsDNA BR Assay Kit (Molecular Probes #Q32850). Follow instructions provided with the kit.
17. Run 2–5 μL of the amplified library on a 2% agarose gel. Clear bands corresponding to the first few linkers between nucleosomes close to the tagmentation site, which are more accessible than nucleosome cores, should be observed (see Fig. 3).
18. A minimum concentration of 30 ng/μL is preferred for NGS. Usually at least 100 million paired-end reads per library are needed to obtain a genome-wide profile with enough footprinting signal.

ACKNOWLEDGMENTS

We thank Elisa de la Calle-Mustienes and Ignacio Maeso for helpful discussions. Work was funded by grants from Ministerio de Economía y Competitividad (BFU2013-41322-P and BFU2014-58449-JIN) and the Andalusian Government (BIO-396). JB was supported by an FCT Investigator grant (IF/00654/2013).

REFERENCES

Bogdanovic, O., Fernandez-Minan, A., Tena, J. J., de la Calle-Mustienes, E., & Gomez-Skarmeta, J. L. (2013). The developmental epigenomics toolbox: ChIP-seq and MethylCap-seq profiling of early zebrafish embryos. *Methods, 62*, 207–215.

Bogdanovic, O., Fernandez-Minan, A., Tena, J. J., de la Calle-Mustienes, E., Hidalgo, C., van Kruysbergen, I. … Gómez-Skarmeta, J. L. (2012). Dynamics of enhancer chromatin signatures mark the transition from pluripotency to cell specification during embryogenesis. *Genome Research, 22*, 2043–2053.

Buenrostro, J. D., Giresi, P. G., Zaba, L. C., Chang, H. Y., & Greenleaf, W. J. (2013). Transposition of native chromatin for fast and sensitive epigenomic profiling of open chromatin, DNA-binding proteins and nucleosome position. *Nature Methods, 10*, 1213–1218.

Buenrostro, J. D., Wu, B., Litzenburger, U. M., Ruff, D., Gonzales, M. L., Snyder, M. P. … Greenleaf, W. J. (2015). Single-cell chromatin accessibility reveals principles of regulatory variation. *Nature, 523*, 486–490.

Creyghton, M. P., Cheng, A. W., Welstead, G. G., Kooistra, T., Carey, B. W., Steine, E. J. … Jaenisch, R. (2010). Histone H3K27ac separates active from poised enhancers and predicts developmental state. *Proceedings of the National Academy of Sciences of the United States of America, 107*(50), 21931–21936.

Danino, Y. M., Even, D., Ideses, D., & Juven-Gershon, T. (2015). The core promoter: at the heart of gene expression. *Biochimica et Biophysica Acta, 1849*, 1116–1131.

Davie, K., Jacobs, J., Atkins, M., Potier, D., Christiaens, V., Halder, G., & Aerts, S. (2015). Discovery of transcription factors and regulatory regions driving in vivo tumor development by ATAC-seq and FAIRE-seq open chromatin profiling. *PLoS Genetics, 11*, e1004994.

Dekker, J., Rippe, K., Dekker, M., & Kleckner, N. (2002). Capturing chromosome conformation. *Science, 295*, 1306–1311.

Gehrke, A. R., Schneider, I., de la Calle-Mustienes, E., Tena, J. J., Gomez-Marin, C., Chandran, M. … Shubin, N. H. (2015). Deep conservation of wrist and digit enhancers in fish. *Proceedings of the National Academy of Sciences of the United States of America, 112*, 803–808.

Gomez-Marin, C., Tena, J. J., Acemel, R. D., Lopez-Mayorga, M., Naranjo, S., de la Calle-Mustienes, E. … Gómez-Skarmeta, J. L. (2015). Evolutionary comparison reveals that diverging CTCF sites are signatures of ancestral topological associating domains borders. *Proceedings of the National Academy of Sciences of the United States of America, 112*, 7542–7547.

Hagege, H., Klous, P., Braem, C., Splinter, E., Dekker, J., Cathala, G. … Forné, T. (2007). Quantitative analysis of chromosome conformation capture assays (3C-qPCR). *Nature Protocols, 2*, 1722–1733.

Hawkins, R. D., Hon, G. C., Yang, C., Antosiewicz-Bourget, J. E., Lee, L. K., Ngo, Q. M. … Ren, B. (2011). Dynamic chromatin states in human ES cells reveal potential regulatory sequences and genes involved in pluripotency. *Cell Research, 21*, 1393–1409.

Irimia, M., Tena, J. J., Alexis, M. S., Fernandez-Minan, A., Maeso, I., Bogdanovic, O. … Fraser, H. B. (2012). Extensive conservation of ancient microsynteny across metazoans due to cis-regulatory constraints. *Genome Research, 22*, 2356–2367.

John, S., Sabo, P. J., Canfield, T. K., Lee, K., Vong, S., Weaver, M. … Stamatoyannopoulos, J. A. (July 2013). Genome-scale mapping of DNase I hypersensitivity. *Current Protocols in Molecular Biology*. http://dx.doi.org/10.1002/0471142727.mb2127s103 (Chapter 27) Unit 21.27.

Koressaar, T., & Remm, M. (2007). Enhancements and modifications of primer design program Primer3. *Bioinformatics, 23*, 1289–1291.

de Laat, W., & Duboule, D. (2013). Topology of mammalian developmental enhancers and their regulatory landscapes. *Nature, 502*, 499–506.

Lara-Astiaso, D., Weiner, A., Lorenzo-Vivas, E., Zaretsky, I., Jaitin, D. A., David, E.… Amit, I. (2014). Immunogenetics. Chromatin state dynamics during blood formation. *Science, 345*, 943–949.

Maurano, M. T., Humbert, R., Rynes, E., Thurman, R. E., Haugen, E., Wang, H.… Stamatoyannopoulos, J. A. (2012). Systematic localization of common disease-associated variation in regulatory DNA. *Science, 337*, 1190–1195.

Ong, C. T., & Corces, V. G. (2011). Enhancer function: new insights into the regulation of tissue-specific gene expression. *Nature Reviews Genetics, 12*, 283–293.

Rada-Iglesias, A., Bajpai, R., Swigut, T., Brugmann, S. A., Flynn, R. A., & Wysocka, J. (2011). A unique chromatin signature uncovers early developmental enhancers in humans. *Nature, 470*, 279–283.

Rembold, M., & Wittbrodt, J. (July 2004). In vivo time-lapse imaging in medaka—n-heptanol blocks contractile rhythmical movements. *Mechanisms of Development, 121*, 965–970.

Santos-Rosa, H., Schneider, R., Bannister, A. J., Sherriff, J., Bernstein, B. E., Emre, N. C.… Kouzarides, T. (2002). Active genes are tri-methylated at K4 of histone H3. *Nature, 419*, 407–411.

Simon, J. M., Giresi, P. G., Davis, I. J., & Lieb, J. D. (2012). Using formaldehyde-assisted isolation of regulatory elements (FAIRE) to isolate active regulatory DNA. *Nature Protocols, 7*, 256–267.

Smemo, S., Tena, J. J., Kim, K. H., Gamazon, E. R., Sakabe, N. J., Gomez-Marin, C.… Nóbrega, M. A. (2014). Obesity-associated variants within FTO form long-range functional connections with IRX3. *Nature, 507*, 371–375.

Song, L., & Crawford, G. E. (2010). DNase-seq: a high-resolution technique for mapping active gene regulatory elements across the genome from mammalian cells. *Cold Spring Harbor Protocols*. pdb prot5384.

Splinter, E., de Wit, E., van de Werken, H. J., Klous, P., & de Laat, W. (2012). Determining long-range chromatin interactions for selected genomic sites using 4C-seq technology: from fixation to computation. *Methods, 58*(3), 221–230.

Tjian, R., & Maniatis, T. (1994). Transcriptional activation: a complex puzzle with few easy pieces. *Cell, 77*, 5–8.

Trompouki, E., Bowman, T. V., Dibiase, A., Zhou, Y., & Zon, L. I. (2011). Chromatin immunoprecipitation in adult zebrafish red cells. *Methods in Cell Biology, 104*, 341–352.

Untergasser, A., Cutcutache, I., Koressaar, T., Ye, J., Faircloth, B. C., Remm, M., & Rozen, S. G. (2012). Primer3—new capabilities and interfaces. *Nucleic Acids Research, 40*, e115.

Westerfield, M. (1995). *The zebrafish book*. Eugene: University of Oregon Press.

Zhao, Z., Tavoosidana, G., Sjolinder, M., Gondor, A., Mariano, P., Wang, S.… Ohlsson, R. (2006). Circular chromosome conformation capture (4C) uncovers extensive networks of epigenetically regulated intra- and interchromosomal interactions. *Nature Genetics, 38*, 1341–1347.

CHAPTER

Epigenetic regulation of hematopoietic stem cell development

22

C. Li[*,§], **T. Evans**[¶,1], **M.G. Goll**[*,1]

**Memorial Sloan Kettering Cancer Center, New York, NY, United States*

[§]Weill Cornell Graduate School of Medical Sciences, New York, NY, United States

[¶]Weill Cornell Medical College, New York, NY, United States

[1]Corresponding authors: E-mail: tre2003@med.cornell.edu; gollm@mskcc.org

CHAPTER OUTLINE

Abstract

Hematopoietic stem cells (HSCs) are multipotent self-renewing precursors with the capacity to differentiate into all adult blood cell lineages. HSC development is a highly orchestrated process regulated by multiple transcription factors and signaling pathways.

Methods in Cell Biology, Volume 135, ISSN 0091-679X, http://dx.doi.org/10.1016/bs.mcb.2016.01.010

Emerging evidence suggests that epigenetic regulation is an additional essential component of HSC development. Powerful genetic and imaging approaches, combined with conservation of mammalian programs, have made zebrafish a prominent model for the study of HSC production. This chapter summarizes approaches that have been used to identify epigenetic regulators of HSC development in zebrafish and highlights additional strategies that are likely to facilitate progress in this promising field.

INTRODUCTION

Hematopoietic stem cells (HSCs) are self-renewing precursors from which all adult blood cell types are derived. The production of HSCs during embryonic development is essential for the establishment of the adult blood system (Orkin & Zon, 2008). Transplantation of HSCs derived from bone marrow or cord blood is a true stem cell therapy, used to restore the hematopoietic system in patients with diseased or defective bone marrow. This important therapeutic application has raised significant interest in understanding the mechanisms that regulate the de novo production and maintenance of HSCs in vivo.

The acquisition of cellular identity relies on a combination of genetic and epigenetic information. Genetic information encodes the transcription factors and effector genes that drive progenitor fate and differentiation. In contrast, epigenetic information shapes the accessibility of transcription factors to DNA, providing a heritable chromatin landscape that can direct or reinforce cell type–specific transcription programs. Studies in mice, cell culture, and zebrafish have identified a number of transcription factors and signaling pathways that are important for HSC development (Jagannathan-Bogdan & Zon, 2013). However, much less is known about how epigenetic regulation contributes to the production and maintenance of HSCs in the developing embryo. Zebrafish are well suited to addressing this topic. In this chapter, we provide an overview of approaches that have been used to identify epigenetic regulators of HSC development in zebrafish and we discuss additional strategies that may accelerate progress in this field.

1. MECHANISMS OF EPIGENETIC REGULATION

Epigenetic mechanisms control the accessibility of transcription factors to DNA through the manipulation of chromatin structure. A principle component of chromatin is the nucleosome, which consists of a segment of DNA wound around a core histone octamer. Covalent modifications of DNA and histone tails provide two mechanisms to regulate chromatin compaction; these modifications contribute to a heritable landscape that can influence recruitment of transcriptional machinery through short- and long-range interactions at *cis*-regulatory regions.

1.1 5-METHYLCYTOSINE

In vertebrate genomes, DNA can be modified through methylation of cytosine residues (5-methylcytosine, 5mC). DNA methylation is associated with transcriptional repression and is required for normal development (Goll & Bestor, 2005). The methylation reaction is executed by two groups of DNA methyltransferases: the maintenance methyltransferase, Dnmt1, and the de novo methyltransferases of the Dnmt3 family. De novo methyltransferases methylate unmodified DNA to establish methylation patterns. During maintenance of DNA methylation, the cofactor Uhrf1 recruits Dnmt1 to hemimethylated CpG dinucleotides at the replication fork, allowing faithful propagation of methylation patterns to the newly synthesized DNA strand (Bostick et al., 2007; Sharif et al., 2007). The zebrafish genome encodes single well-conserved orthologs of Dnmt1 and Uhrf1. During the first 2 days of development, *uhrf1*- and *dnmt1*-mutant zebrafish larvae appear overtly normal and genome-wide methylation patterns are largely unaffected, perhaps due to significant maternal deposition of mRNA encoding Dnmt1 and Uhrf1 (Chu et al., 2012; Martin, Laforest, Akimenko, & Ekker, 1999). However, homozygous mutations in either gene result in reduced genome-wide levels of DNA methylation and larval lethality by 7 days post fertilization (dpf) (Anderson et al., 2009; Sadler, Krahn, Gaur, & Ukomadu, 2007; Tittle et al., 2011). Mammalian genomes encode two de novo DNA methyltransferases (Dnmt3A and Dnmt3b), whereas the zebrafish genome encodes six Dnmt3 orthologs (Goll & Halpern, 2011; Shimoda, Yamakoshi, Miyake, & Takeda, 2005; Smith, Dueck, Mhanni, & McGowan, 2005). Mutations in individual zebrafish *dnmt3* genes have not been uncovered in forward genetic screens. It is possible that functional redundancies among these genes have prevented their detection using unbiased approaches.

1.2 5-HYDROXYMETHYLCYTOSINE

While the mechanisms that establish and maintain DNA methylation are well studied, the pathways that regulate removal of 5mC have been more elusive (Wu & Zhang, 2011). In 2009, the Tet proteins (Tet1, Tet2 and Tet3) were identified as a family of cytosine dioxygenases that are capable of converting 5mC to 5-hydroxymethylcytosine (5hmC) and its derivatives 5-formylcytosine and 5-carboxylcytosine (He et al., 2011; Tahiliani et al., 2009). Conversion of 5mC to 5hmC and its derivatives appears to drive DNA demethylation through the active excision or passive dilution of oxidized bases (Wu & Zhang, 2011). Tet-mediated DNA demethylation provides one mechanism for reactivation of genes that have been transcriptionally silenced by 5mC. It is also possible that in some contexts 5hmC may act as a unique DNA modification that imparts distinct epigenetic information on the underlying genome. The zebrafish genome encodes single orthologs of Tet1, Tet2, and Tet3, and combinatorial mutation of these three genes reduces 5hmC in total genomic DNA to levels below the threshold for antibody detection (Almeida et al., 2012; Li et al., 2015).

1.3 HISTONE MODIFICATION

The tails of individual histones within nucleosome are also subject to a variety of covalent modifications that regulate chromatin compaction (Bannister & Kouzarides, 2011). Histone tails are well conserved between zebrafish and other species and the zebrafish genome encodes orthologs of many histone-modifying enzymes identified in mammals. Among the most well-studied histone tail modifications are histone acetylation and methylation. Histone acetylation is regulated by the antagonistic action of histone acetyltransferases and deacetylaces. Acetylation is generally associated with open chromatin structure and active gene expression, whereas histone deacetylation is correlated with gene repression (Eberharter & Becker, 2002). Histone methylation represents a more complex level of regulation as residues can be mono-, di-, or trimethylated and methylation can facilitate gene activation or repression depending on which histone tail residues are modified and to what degree. Polycomb and trithorax complexes represent two groups of proteins that are important for the regulation of histone tail methylation. Trithorax group proteins are important mediators of H3K4 di- and trimethylation, marks typically associated with transcriptional activation (Schuettengruber, Chourrout, Vervoort, Leblanc, & Cavalli, 2007). In contrast, polycomb group proteins of the PRC2 complex are responsible for trimethylation of H3K27, a mark of transcriptional inactivation. H3K27me3 further mediates the recruitment of a second polycomb complex, PRC1, which monoubiquitinates histone H2A on lysine 119 and stabilizes the repressive chromatin state (de Napoles et al., 2004; Wang et al., 2004).

2. HEMATOPOIETIC STEM CELL DEVELOPMENT

As for all vertebrate species, the earliest hematopoietic cells in zebrafish represent "primitive" lineages that support embryonic development but are derived from relatively short-lived progenitors that are eventually replaced by HSC-derived cells (de Jong & Zon, 2005). At the molecular level, early HSC development is regulated by a core set of transcription factors including *scl*, *gata2*, and *runx1* (Jagannathan-Bogdan & Zon, 2013). Upstream of these factors, inputs from major signaling pathways including Bmp, Wnt, and Notch contribute to HSC development (Clements & Traver, 2013). In mammals, a major site where HSCs are born is the aorta-gonad-mesonephros region. Within this region, nascent HSCs emerge from a specialized subset of endothelial cells within the dorsal aorta through a process termed the endothelial-to-hematopoietic transition (EHT). During this process, cells of the hemogenic endothelium transition to a hematopoietic fate in the absence of cell division (Bertrand et al., 2010; Boisset et al., 2010; Kissa & Herbomel, 2010). Later, these cells migrate to the fetal liver and eventually to the bone marrow. Evidence suggests that the de novo production of HSCs only occurs during a defined window in the embryo, and self-renewing divisions of this initial pool provide for the expansion and maintenance of HSCs through adulthood (Chen, Yokomizo, Zeigler,

Dzierzak, & Speck, 2009; Göthert et al., 2005). Although more challenging than in vitro approaches, the exquisite spatial and temporal regulation of this process makes the in vivo study of HSC development important.

The rapid development of the hematopoietic system combined with external fertilization and larval clarity make zebrafish an appealing model for the in vivo study of HSC development. In zebrafish, the earliest HSCs emerge from the ventral wall of the dorsal aorta at around 30 h post fertilization (hpf) and HSC production continues until roughly 48 hpf (Bertrand et al., 2010; Kissa & Herbomel, 2010). Exploiting the ability to visualize the process in zebrafish embryos, recent studies identified additional signals required for HSC birth, including Cxcl12 from somite-derived endothelial cells (Nguyen et al., 2014), TNFα from primitive neutrophils (Espín-Palazón et al., 2014), and INF-γ from the hemogenic endothelium itself (Sawamiphak, Kontarakis, & Stainier, 2014). Following emergence, these newly born HSCs rapidly enter the subaortic space and subsequently migrate either directly to the thymus or to the caudal hematopoietic tissue. By 4 dpf, HSCs seed the kidney marrow, which is the site of adult hematopoiesis in zebrafish (Jin, Xu, & Wen, 2007; Murayama et al., 2006).

3. APPROACHES

The dramatic shift in transcriptional profiles that occurs during EHT is likely to require significant changes in the chromatin landscape. However, our current understanding of how epigenetic regulation contributes to the production and early maintenance of HSCs is limited. The accessibility of early stages of HSC development and large brood sizes makes zebrafish well suited to genetic approaches aimed at identifying epigenetic regulators of HSC development. The application of forward and reverse genetic approaches in zebrafish is beginning to unravel the relevant epigenetic pathways. It is important to note that because HSCs are produced from specialized cells within the dorsal aorta, vascular development and artery specification are prerequisites for HSC emergence. In this chapter, we focus our discussion on identified genes that impact early HSC development but do not overtly impair development of the vascular system.

3.1 UNBIASED FORWARD GENETIC APPROACHES

Small size and large brood sizes make zebrafish amenable to unbiased genetic screening approaches. Based on these strengths, several genetic screens have been performed to identify zebrafish mutants with impaired HSC development (Burns et al., 2009; Du et al., 2011; Liu et al., 2015). These screens were not designed to specifically identify epigenetic regulators of HSC development; nonetheless, two known chromatin modifiers were among the identified genes.

Burns et al. (2009) screened a collection of 194 zebrafish lines harboring retroviral insertions in known genes for reduced expression of the HSC marker *c-myb*.

The histone deacetylase enzyme, *hdac1* was identified as one of nine genes in this collection that was required for HSC production. Epistasis analysis demonstrated that *hdac1* is required downstream of Notch signaling and arterial formation. Expression of *runx1*, an essential regulator of EHT, was reduced in *hdac1* mutants and HSCs could be rescued by injection of mRNA encoding this transcription factor. Although direct targets were not identified, these results suggest that *hdac1* acts upstream or in parallel to *runx1* to promote HSC specific transcriptional programs. The identification of *hdac1* in this retroviral screen provided a first indication that epigenetic regulation of chromatin states contributes to the control of HSC emergence.

More recently, an N-ethyl-N-nitrosourea-based screen for genes that disrupt *c-myb* expression identified a requirement for the DNA methyltranserase, *dnmt1*, in the early maintenance of HSCs (Liu et al., 2015). *Dnmt1* mutant larvae had overtly normal vascular development and arterial specification, but exhibited a clear reduction in *c-myb* expression along the dorsal aorta, attributed to a diminished capacity for HSC proliferation in the mutants. Consistent with this hypothesis, *cebpa*, a transcription factor known to negatively regulate HSC proliferation, was upregulated in the mutants, and combined mutation of *cebpa* and *dnmt1* rescued expression of HSC markers. Studies in mouse have also identified a requirement for *Dnmt1* in the self-renewal adult HSCs derived from the bone marrow, suggesting a conserved role for 5mC in HSC maintenance at multiple developmental stages (Bröske et al., 2009; Trowbridge, Snow, Kim, & Orkin, 2009). It remains possible that Dnmt1 also has roles in the de novo generation of HSCs that are masked by maternally deposited mRNA in zebrafish *dnmt1* mutants. Additional analysis of maternal/zygotic *dnmt1* mutant embryos will be required to definitively address requirements for *dnmt1* in early steps of HSC production.

3.2 REVERSE GENETIC APPROACHES

Forward genetic screens provide a powerful approach to identify novel gene functions and pathways. However, they can be labor intensive and screens to date have not been specifically designed to identify epigenetic regulators of HSC development. Reverse genetic approaches for targeted elimination of specific gene products offer an alternative strategy to directly test known epigenetic regulators for their role in HSC development. Reverse genetic approaches also facilitate testing of gene families for functionally overlapping requirements in HSC development. In zebrafish, antisense morpholinos that block translation or splicing of target mRNAs have historically been used as the primary tool for reverse genetics (Bill, Petzold, Clark, Schimmenti, & Ekker, 2009). An advantage of this approach is the ability to rapidly assess gene function, including for more than one gene, in morpholino-injected embryos. However, there has been growing concern over the potential for nonspecific phenotypes in morphant embryos (Kok et al., 2014; Robu et al., 2007). Genome-editing approaches have now emerged as an attractive alternative, allowing for the selective mutation of any gene of interest within the zebrafish genome (Bedell et al., 2012; Hwang et al., 2013).

To date, several groups have provided data implicating polycomb and trithorax group proteins in HSC development using antisense morpholinos. Depletion of the PRC1 genes *Bmi1/1b* or *Ring1b*, led to decreased expression of the HSC markers *runx1* and *c-myb* in the dorsal aorta (Yu et al., 2012). Similarly, morpholino depletion of the trithorax gene *mll* resulted in reduced expression of HSC markers (Wan, Hu, Liu, Feng, & Xiao, 2011). However, rescue of HSCs by coinjection of mRNA encoding wild-type proteins was not demonstrated for any of these morphants and upstream defects in vascular development were not investigated, making it difficult to definitively assess the specificity of these phenotypes. More recently, morpholino-knockdown of the 5mC dioxygenase *tet2* gene in zebrafish was reported to cause reduced expression of the HSC marker *c-myb* (Ge et al., 2014). However, similar reductions in *c-myb* expression were not observed in *tet2* mutant zebrafish, which are viable to adulthood and show no obvious defects in embryonic HSC production (Gjini et al., 2015; Li et al., 2015).

A major advance for the zebrafish field has been the development of tools for the targeted introduction of mutations. Transcription activator-like effector nucleases (TALENs) (Cermak et al., 2011) and clustered, regularly interspaced short palindromic repeats combined with CRISPR-associated proteins (CRISPR/Cas9) (Cong et al., 2013; Mali et al., 2013) can now be used to direct double-strand breaks to specific genomic loci. The subsequent repair of these breaks by the nonhomologous end joining machinery is error-prone and often introduces small insertions or deletions into the targeted gene. Both TALENs and CRISPR/Cas9 systems have been successfully applied in zebrafish with high efficiency (Bedell et al., 2012; Hwang et al., 2013).

Because genome-editing techniques are relatively new to the zebrafish system, they have yet to be extensively exploited for the mutation of known epigenetic regulators. However, the potential of these approaches is illustrated by a recent study identifying overlapping requirements for the 5mC dioxygenases Tet2 and Tet3 in HSC emergence. Using TALENs, Li et al. generated catalytic loss of function alleles for each of the three zebrafish *tet* orthologs and systematically examined all mutant combinations (Li et al., 2015). Mutation of either *tet2* or *tet3* alone caused only modest reductions in 5hmC and was not associated with defective HSC development. However, combinatorial mutation of *tet2* and *tet3* led to a greater than 30-fold reduction in genome-wide 5hmC levels and a significant reduction in the number of successful EHT events observed by time-lapse confocal microscopy (Fig. 1A and B). Mutation of *tet2/3* did not affect vascular development or arterial specification; however, Notch signaling in the hemogenic endothelium and the downstream expression of the *scl/gata2b/runx1* hematopoietic transcription factor network was disrupted in the double mutants (Fig. 1C and D). The results of this study uncover a requirement for Tet enzymes in the function of the hemogenic endothelium and identify dynamic regulation of 5mC as a previously unappreciated aspect of HSC production. The combinatorial requirement for Tets in HSC emergence identified in this study is unlikely to have been detected using forward genetic approaches.

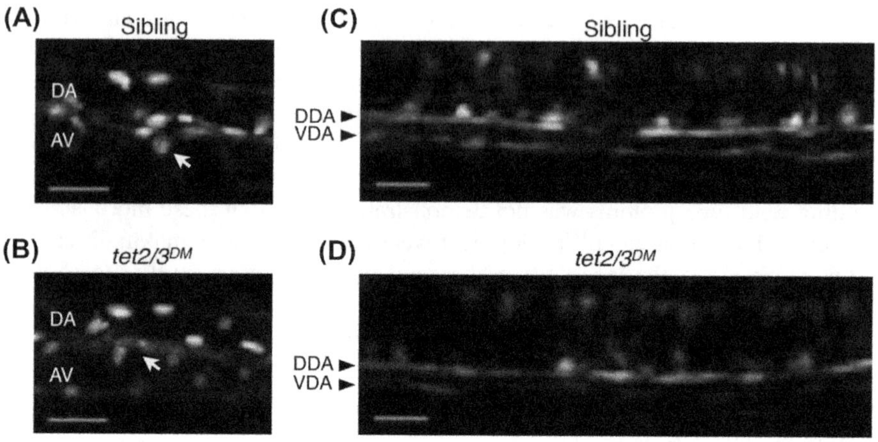

FIGURE 1 Tet2/3 are required for hematopoietic stem cell emergence and Notch signaling in the hemogenic endothelium.

(A, B) Time-lapse confocal images of *Tg(kdrl:Ras-mCherry)*, *Tg(kdrl:H2B-EGFP) tet2/3DM* (double mutant), and control siblings from 30 to 46 hpf. The *yellow arrow* indicates a cell that has just completed endothelial-to-hematopoietic transition in a sibling larva (A) and a cell undergoing nuclear fragmentation in a *tet2/3DM* larva (B).
DA indicates the dorsal aorta. *AV* indicates the axial vein. Scale bars indicate 30 μM.
(C, D) Images of *Tg(tp1:GFP) tet2/3DM* and control siblings at 26 hpf. *DDA* indicates the dorsal wall of the dorsal aorta. *VDA* indicates the ventral wall of the dorsal aorta. Scale bars indicate 60 μM. (See color plate)

3.3 CANDIDATE SCREENS

Reverse genetic approaches provide a straightforward mechanism to evaluate the role of known epigenetic regulators in HSC development. However, this strategy is typically applied to one or a few genes with anticipated phenotypes. Candidate screens essentially leverage advantages of both forward and reverse genetic approaches. A recent candidate screen for chromatin factors that regulate zebrafish hematopoiesis substantially expanded the number of epigenetic regulators implicated in HSC development. Screening of a panel of 425 morpholinos that depleted proteins containing amino acid motifs associated with chromatin or nucleic acid binding revealed 31 candidates that caused strong alterations in *runx1/c-myb* expression levels (Huang et al., 2013). Among these candidates, 20 morpholinos impacted HSC development without negative effects on vascular or arterial development (Tables 1 and 2). Consistent with previous studies, this candidate screen implicated *hdac1* and components of the PRC1 and SET1/trithorax complexes in HSC production. In addition, several genes that had not been previously implicated in HSC development were uncovered, including *brd8a*, *jmjd1*, and *napll4a*. An added advantage of the large scale–candidate approach is that it allows for identification

Table 1 Epigenetic Regulators With Known Requirements for Hematopoietic Stem Cell Development

Pathway	Gene	Gene Function	Mutant or Morphant	Rescue	References
DNA methylation	*dnmt1*	DNA methyltransferase	Mutant	Yes	Liu et al. (2015)
	tet2	Methylcytosine dioxygenase	Morphant	Yes	Ge et al. (2014)
	tet2; tet3	Methylcytosine dioxygenase	Mutant	Yes	Li et al. (2015)
Polycomb regulation	*bmi1*	Polycomb repressive complex 1	Morphant	No	Yu et al. (2012)
	ring1b	Polycomb repressive complex 1	Morphant	No	Yu et al. (2012)
	cbx6b	Polycomb repressive complex 1	Morphant	No	Huang et al. (2013)
	cbx8b	Polycomb repressive complex 1	Morphant	No	Huang et al. (2013)
Histone methylation	*mll*	SET1 histone methyltransferase complex	Morpahnt	No	Wan et al. (2011)
	ash2l	SET1 histone methyltransferase complex	Morphant	No	Huang et al. (2013)
	cxxc1l	SET1 histone methyltransferase complex	Morphant	No	Huang et al. (2013)
	setd1ba	SET1 histone methyltransferase complex	Morphant	No	Huang et al. (2013)
	prdm12	Histone methyltransferase	Morphant	No	Huang et al. (2013)
	prdm16	Histone methyltransferase	Morphant	No	Huang et al. (2013)
Histone deacetylation	*hdac1*	Histone deacetylase	Mutant	No	Burns et al. (2009)
	hdac1	Histone deacetylase	Morphant	No	Huang et al. (2013)
	hdac6	Histone deacetylase	Morphant	No	Huang et al. (2013)
	hdac9a	Histone deacetylase	Morphant	No	Huang et al. (2013)
	sirt7	Histone deacetylase	Morphant	No	Huang et al. (2013)
Chromatin remodeling	*smarcd1*	SWI/SNF chromatin–remodeling complex	Morphant	No	Huang et al. (2013)
	smarcd2	SWI/SNF chromatin–remodeling complex	Morphant	No	Huang et al. (2013)
	nap1l4a	Nucleosome assembly factor	Morphant	No	Huang et al. (2013)
	brd8a	Bromodomain-containing protein	Morphant	No	Huang et al. (2013)
Histone acetylation	*p300*	p300/CBP transcriptional coactivator complex	Morphant	No	Huang et al. (2013)

Table 2 Zebrafish Morphants That Cause an Expansion of Hematopoietic Stem Cell Populations

Pathway	Gene	Gene Function	Mutant or Morphant	Rescue	References
Chromatin remodeling	*cecr2*	Bromodomain-containing protein	Morphant	No	Huang et al. (2013)
	chd7	Chromodomain helicase DNA–binding protein	Morphant	No	Huang et al. (2013)
Histone acetylation	*crebbpa*	CREB-binding protein	Morphant	No	Huang et al. (2013)
Unknown	*mbd6*	Methyl-CpG-binding protein	Morphant	No	Huang et al. (2013)

of multisubunit complexes that include more than one gene implicated by the screen. This analysis revealed likely requirements for the SWI/SNF, ISWI, SET, and HAT complexes in regulating HSC development. The advent and adaptation of Crispr/Cas9 in zebrafish now allows for larger candidate screens that can overcome limitations of high cost and nonspecific effects encountered with morpholinos (Shah, Davey, Whitebirch, Miller, & Moens, 2015). This approach can be used to further evaluate epigenetic components in HSC development.

4. DISEASE MODELS

There is strong evidence that disruption of the chromatin environment contributes to malignancies of the hematopoietic system, including both myeloid (Shih, Abdel-Wahab, Patel, & Levine, 2015) and lymphoid (Jiang & Melnick, 2015) malignancies. Parallels between the hematopoietic system in humans and zebrafish have inspired the use of zebrafish to model human leukemia (White, Rose, & Zon, 2013). In particular, efforts to model myeloid malignancies, a class of clonal diseases arising in hematopoietic stem or progenitor cells, have revealed new insights into how epigenetic mechanisms promote normal and diseased states.

In humans, the *NUP98-HOXA9* (*NHA9*) fusion oncogene is detected in high-risk acute myeloid leukemia. This oncogene results from a chromosomal translocation that fuses nucleoporin 98 kDa (*NUP98*) to homeobox A9 (*HOXA9*) (Nakamura et al., 1996). In zebrafish, expression of the NAH9 fusion protein from a transgenic *pu.1* promoter causes fish to develop myeloproliferative neoplasms in adulthood (Forrester et al., 2011). This same fusion also causes overproduction of HSCs in the zebrafish embryo (Deveau et al., 2015). Microarray analysis revealed that the DNA methyltransferase, *dnmt1*, is one of the most upregulated genes in embryos expressing NHA9 (Deveau et al., 2015). Moreover, chemical inhibition of Dnmt1 by the DNA methyltransferase inhibitor decitabine rescued overproduction of HSCs in embryos carrying the NHA9 transgene, suggesting that Dnmt1 overexpression plays a causative role in promoting HSC expansion. This observation is consistent with the role for *dnmt1* in HSC proliferation that was reported by Liu et al. (2015). Intriguingly, combining sub-monotherapeutic doses of the histone deacetylase inhibitor valproic acid and decitabine also blocked the effects of NHA9 on zebrafish blood development. These results link expression of the NAH9 fusion to changes in the chromatin landscape and highlight the potential of combinatorial epigenetic therapies in the treatment of NHA9-induced myeloid disease.

More recently, the first zebrafish cancer model driven by mutation of a known chromatin regulator was described (Gjini et al., 2015). Mutations in the 5mC dioxygenase *TET2* are detected with high frequencies in human myeloid malignancies and mutations in *Tet2* cause similar malignancies in mouse (Delhommeau et al., 2009; Ko et al., 2011; Li et al., 2011; Moran-Crusio et al., 2011). As previously discussed, homozygous mutation of *tet2* does not impact HSC development in the zebrafish embryo. However, as they age, *tet2* mutant zebrafish adults develop progressive clonal myelodysplasia, culminating in myelodysplastic syndromes by 24 months. Decreases in the modified base 5hmC were observed in hematopoietic cells of the kidney marrow of *tet2* mutants but not in other cell types, suggesting that other Tet family members may compensate for *tet2* outside of the hematopoietic system. The lack of Tet redundancy in HSCs provides one likely explanation for the high oncogenic potential of *TET2* mutation in myeloid malignancies. Because *tet2* mutants are viable and fertile, they provide a powerful model to dissect the pathways that are regulated by Tet2 in the adult hematopoietic system. In addition, *tet2* mutant embryos offer an opportunity for small-molecule screens to identify compounds that selectively kill *tet2* mutant HSCs. Such compounds may have therapeutic potential for the treatment of myeloid malignancies.

5. CURRENT CHALLENGES AND FUTURE DIRECTIONS

To date, genetic approaches in zebrafish have implicated a number of important epigenetic pathways in the regulation of HSC development. However, the generation of additional tools and approaches will be required to fully realize the potential of the zebrafish system.

5.1 TOOLS FOR UNBIASED SCREENING OF EPIGENETIC STATES IN VIVO

To date, screens for regulators of HSC development in zebrafish have relied on expression of HSC associated markers, leading to identification of a broad spectrum of genes with diverse functions. More narrowly focused screens for epigenetic regulators of HSC development have not been performed, likely due to the lack of tools to rapidly monitor epigenetic changes at the enhancers or promoters of relevant hematopoietic genes. Tools to monitor DNA methylation changes in zebrafish using fluorescent reporters under the control of the yeast Gal4/UAS system have previously been described (Akitake, Macurak, Halpern, & Goll, 2011; Goll, Anderson, Stainier, Spradling, & Halpern, 2009). However, the use of exogenous UAS sequences to monitor DNA methylation makes this approach better suited for the identification of global regulators of chromatin, rather than to those involved in a particular developmental process.

A recently described variation on the Gal4/UAS approach has the potential to be adapted for screening in zebrafish. The minimal *Snrpn* gene promoter is a methylation-sensitive promoter that is influenced by the DNA methylation status of adjacent sequences. By placing this promoter upstream of a fluorescent reporter, Stelzer and colleagues were able to visually track methylation changes at nearby enhancer sequences in mouse embryonic stem cells based on fluorophore expression (Stelzer et al., 2015). Placing the Snrpn reporter adjacent to enhancers of hematopoietic transcription factors could provide one approach to monitor the methylation status of HSC-specific enhancers in the developing zebrafish embryo. If successful, this approach would allow for unbiased screens specifically designed to identify genes involved in the epigenetic control of the HSC transcriptional program.

5.2 TOOLS FOR TISSUE-SPECIFIC MUTATION

Another major challenge is that current approaches have relied on whole embryo depletion of gene products. However, many epigenetic regulators are likely to be important in multiple tissues and mutant embryos may not survive long enough for analysis of HSC development. In other cases, upstream defects in mesoderm patterning, vascular development, or arterial identify may mask later specific requirements for epigenetic regulators in HSC development. These challenges will require the use of tissue-specific approaches for mutation of candidate genes. The Cre/loxP system has been widely used in mouse for the conditional mutation of specific genes (Nagy, 2000). This system is also functional in zebrafish (Hans, Kaslin, Freudenreich, & Brand, 2009; Langenau et al., 2005). A recently described CRISPR/Cas-based vector system that enables tissue-specific gene inactivation in zebrafish provides one exciting alternative that could be applied to test a large number of candidate epigenetic regulators for their requirements in HSCs or their endothelial precursors (Ablain, Durand, Yang, Zhou, & Zon, 2015).

CONCLUSIONS

It is perhaps not surprising that epigenetic regulation plays a major role in HSC generation, as the transition from endothelial to hematopoietic identity represents a dramatic reprogramming of fate. The accessibility of zebrafish embryos from the time of fertilization, coupled with larval clarity and large brood sizes has facilitated the discovery of signals and genes that regulate the generation of HSCs. Identifying mechanisms to manipulate HSC fate and proliferation has important implications for the development of therapeutics to treat hematopoietic disorders. The translational potential of the zebrafish system is exemplified by the development of a PGE2 derivative as a drug to expand transplantable HSCs from human cord blood. A chemical screen identified the capacity of PGE2 for impacting production of zebrafish HSCs (North et al., 2007), and within a few years the concept was being tested in human clinical trials (Hagedorn, Durand, Fast, & Zon, 2014). In addition to chemical screens, and the genetic strategies described above, new gene-editing techniques offer opportunities to build zebrafish models for hematopoietic disease. For example, mutations in genes known to be associated with leukemia, including patient-specific alleles, can be created and combined with *tet2* mutations to investigate the impact on myeloid dysplasia. While the DNA and chromatin modification machineries that are key to HSC biology are now beginning to be defined, the next challenges will be to understand their target genes and to distinguish epigenetic marks that are drivers of HSC phenotype from those that are consequences of altered transcription. Again, emerging technologies provide exciting avenues of inquiry, limited only by investigators' imagination. For example, modified Crispr/Cas9 technologies can now be developed to target not only mutations, but epigenetic modifications (Hilton et al., 2015; Maeder et al., 2013).

REFERENCES

Ablain, J., Durand, E. M., Yang, S., Zhou, Y., & Zon, L. I. (2015). A CRISPR/Cas9 vector system for tissue-specific gene disruption in zebrafish. *Developmental Cell, 32*(6), 756–764.

Akitake, C. M., Macurak, M., Halpern, M. E., & Goll, M. G. (2011). Transgenerational analysis of transcriptional silencing in zebrafish. *Developmental Biology, 352*(2), 191–201.

Almeida, R. D., Loose, M., Sottile, V., Matsa, E., Denning, C., Young, L. … Ruzov, A. (2012). 5-hydroxymethyl-cytosine enrichment of non-committed cells is not a universal feature of vertebrate development. *Epigenetics: Official Journal of the DNA Methylation Society, 7*(4), 383–389.

Anderson, R. M., Bosch, J. A., Goll, M. G., Hesselson, D., Dong, P. D. S., Shin, D. … Stainier, D. Y. R. (2009). Loss of Dnmt1 catalytic activity reveals multiple roles for DNA methylation during pancreas development and regeneration. *Developmental Biology, 334*(1), 213–223.

Bannister, A. J., & Kouzarides, T. (2011). Regulation of chromatin by histone modifications. *Cell Research, 21*(3), 381–395.

Bedell, V. M., Wang, Y., Campbell, J. M., Poshusta, T. L., Starker, C. G., Krug, R. G. … Ekker, S. C. (2012). In vivo genome editing using a high-efficiency TALEN system. *Nature, 491*(7422), 114–118.

Bertrand, J. Y., Chi, N. C., Santoso, B., Teng, S., Stainier, D. Y. R., & Traver, D. (2010). Haematopoietic stem cells derive directly from aortic endothelium during development. *Nature, 464*(7285), 108–111.

Bill, B. R., Petzold, A. M., Clark, K. J., Schimmenti, L. A., & Ekker, S. C. (2009). A primer for morpholino use in zebrafish. *Zebrafish, 6*(1), 69–77.

Boisset, J.-C., van Cappellen, W., Andrieu-Soler, C., Galjart, N., Dzierzak, E., & Robin, C. (2010). In vivo imaging of haematopoietic cells emerging from the mouse aortic endothelium. *Nature, 464*(7285), 116–120.

Bostick, M., Kim, J. K., Estève, P.-O., Clark, A., Pradhan, S., & Jacobsen, S. E. (2007). UHRF1 plays a role in maintaining DNA methylation in mammalian cells. *Science (New York, NY), 317*(5845), 1760–1764.

Bröske, A.-M., Vockentanz, L., Kharazi, S., Huska, M. R., Mancini, E., Scheller, M. … Rosenbauer, F. (2009). DNA methylation protects hematopoietic stem cell multipotency from myeloerythroid restriction. *Nature Genetics, 41*(11), 1207–1215.

Burns, C. E., Galloway, J. L., Smith, A. C. H., Keefe, M. D., Cashman, T. J., Paik, E. J. … Zon, L. I. (2009). A genetic screen in zebrafish defines a hierarchical network of pathways required for hematopoietic stem cell emergence. *Blood, 113*(23), 5776–5782.

Cermak, T., Doyle, E. L., Christian, M., Wang, L., Zhang, Y., Schmidt, C. … Voytas, D. F. (2011). Efficient design and assembly of custom TALEN and other TAL effector-based constructs for DNA targeting. *Nucleic Acids Research, 39*(12), e82.

Chen, M. J., Yokomizo, T., Zeigler, B. M., Dzierzak, E., & Speck, N. A. (2009). Runx1 is required for the endothelial to haematopoietic cell transition but not thereafter. *Nature, 457*(7231), 887–891.

Chu, J., Loughlin, E. A., Gaur, N. A., SenBanerjee, S., Jacob, V., Monson, C. … Sadler, K. C. (2012). UHRF1 phosphorylation by cyclin A2/cyclin-dependent kinase 2 is required for zebrafish embryogenesis. *Molecular Biology of the Cell, 23*, 59–70.

Clements, W. K., & Traver, D. (2013). Signalling pathways that control vertebrate haematopoietic stem cell specification. *Nature Reviews Immunology, 13*(5), 336–348.

Cong, L., Ran, F. A., Cox, D., Lin, S., Barretto, R., Habib, N. … Zhang, F. (2013). Multiplex genome engineering using CRISPR/Cas systems. *Science (New York, NY), 339*(6121), 819–823.

Delhommeau, F., Dupont, S., Della Valle, V., James, C., Trannoy, S., Massé, A. … Bernard, O. A. (2009). Mutation in TET2 in myeloid cancers. *The New England Journal of Medicine, 360*(22), 2289–2301.

Deveau, A. P., Forrester, A. M., Coombs, A. J., Wagner, G. S., Grabher, C., Chute, I. C. … Berman, J. N. (2015). Epigenetic therapy restores normal hematopoiesis in a zebrafish model of NUP98-HOXA9-induced myeloid disease. *Leukemia, 29*(10). April.

Du, L., Xu, J., Li, X., Ma, N., Liu, Y., Peng, J. … Wen, Z. (2011). Rumba and Haus3 are essential factors for the maintenance of hematopoietic stem/progenitor cells during zebrafish hematopoiesis. *Development (Cambridge, England), 138*(4), 619–629.

Eberharter, A., & Becker, P. B. (2002). Histone acetylation: a switch between repressive and permissive chromatin. Second in review on chromatin dynamics. *EMBO Reports, 3*(3), 224–229.

Espín-Palazón, R., Stachura, D. L., Campbell, C. A., García-Moreno, D., Del Cid, N., Kim, A. D. … Traver, D. (2014). Proinflammatory signaling regulates hematopoietic stem cell emergence. *Cell, 159*(5), 1070–1085.

Forrester, A. M., Grabher, C., Mcbride, E. R., Boyd, E. R., Vigerstad, M. H., Edgar, A. … Berman, J. N. (2011). NUP98-HOXA9-transgenic zebrafish develop a myeloproliferative neoplasm and provide new insight into mechanisms of myeloid leukaemogenesis. *British Journal of Haematology, 155*, 167–181. August.

Ge, L., Zhang, R.-P., Wan, F., Guo, D.-Y., Wang, P., Xiang, L.-X., & Shao, J.-Z. (2014). TET2 plays an essential role in erythropoiesis by regulating lineage-specific genes via DNA oxidative demethylation in a zebrafish model. *Molecular and Cellular Biology, 34*(6), 989–1002.

Gjini, E., Mansour, M. R., Sander, J. D., Moritz, N., Nguyen, A. T., Kesarsing, M. … Look, A. T. (2015). A zebrafish model of myelodysplastic syndrome produced through tet2 genomic editing. *Molecular and Cellular Biology, 35*(5), 789–804.

Goll, M. G., Anderson, R., Stainier, D. Y. R., Spradling, A. C., & Halpern, M. E. (2009). Transcriptional silencing and reactivation in transgenic zebrafish. *Genetics, 182*(3), 747–755.

Goll, M. G., & Bestor, T. H. (2005). Eukaryotic cytosine methyltransferases. *Annual Review of Biochemistry, 74*, 481–514.

Goll, M. G., & Halpern, M. E. (2011). DNA methylation in zebrafish. *Progress in Molecular Biology and Translational Science, 101*, 193–218.

Göthert, J. R., Gustin, S. E., Hall, M. A., Green, A. R., Göttgens, B., Izon, D. J., & Begley, C. G. (2005). In vivo fate-tracing studies using the Scl stem cell enhancer: embryonic hematopoietic stem cells significantly contribute to adult hematopoiesis. *Blood, 105*(7), 2724–2732.

Hagedorn, E. J., Durand, E. M., Fast, E. M., & Zon, L. I. (2014). Getting more for your marrow: boosting hematopoietic stem cell numbers with PGE2. *Experimental Cell Research, 329*(2), 220–226.

Hans, S., Kaslin, J., Freudenreich, D., & Brand, M. (2009). Temporally-controlled site-specific recombination in zebrafish. *PLoS One, 4*(2), e4640.

He, Y.-F., Li, B.-Z., Li, Z., Liu, P., Wang, Y., Tang, Q. ... Xu, G.-L. (2011). Tet-mediated formation of 5-carboxylcytosine and its excision by TDG in mammalian DNA. *Science (New York, NY), 333*(6047), 1303–1307.

Hilton, I. B., D'Ippolito, A. M., Vockley, C. M., Thakore, P. I., Crawford, G. E., Reddy, T. E., & Gersbach, C. A. (2015). Epigenome editing by a CRISPR-Cas9 based acetyltransferase activates genes from promoters and enhancers. *Nature Biotechnology, 33*(5), 510–519.

Huang, H.-T., Kathrein, K. L., Barton, A., Gitlin, Z., Huang, Y.-H., Ward, T. P. ... Zon, L. I. (2013). A network of epigenetic regulators guides developmental haematopoiesis in vivo. *Nature Cell Biology, 15*(12), 1516–1525.

Hwang, W. Y., Fu, Y., Reyon, D., Maeder, M. L., Tsai, S. Q., Sander, J. D. ... Joung, J. K. (2013). Efficient genome editing in zebrafish using a CRISPR-Cas system. *Nature Biotechnology, 31*(3), 227–229.

Jagannathan-Bogdan, M., & Zon, L. I. (2013). Hematopoiesis. *Development (Cambridge, England), 140*(12), 2463–2467.

Jiang, Y., & Melnick, A. (2015). The epigenetic basis of diffuse large B-cell lymphoma. *Seminars in Hematology, 52*(2), 86–96.

Jin, H., Xu, J., & Wen, Z. (2007). Migratory path of definitive hematopoietic stem/progenitor cells during zebrafish development. *Blood, 109*(12), 5208–5214.

de Jong, J. L. O., & Zon, L. I. (2005). Use of the zebrafish system to study primitive and definitive hematopoiesis. *Annual Review of Genetics, 39*, 481–501.

Kissa, K., & Herbomel, P. (2010). Blood stem cells emerge from aortic endothelium by a novel type of cell transition. *Nature, 464*(7285), 112–115.

Ko, M., Bandukwala, H. S., An, J., Lamperti, E. D., Thompson, E. C., Hastie, R. ... Rao, A. (2011). Ten-Eleven-Translocation 2 (TET2) negatively regulates homeostasis and differentiation of hematopoietic stem cells in mice. *Proceedings of the National Academy of Sciences of the United States of America, 108*(35), 14566–14571.

Kok, F. O., Shin, M., Ni, C.-W., Gupta, A., Grosse, A. S., van Impel, A. ... Lawson, N. D. (2014). Reverse genetic screening reveals poor correlation between morpholino-induced and mutant phenotypes in zebrafish. *Developmental Cell, 32*(1), 97–108.

Langenau, D. M., Feng, H., Berghmans, S., Kanki, J. P., Kutok, J. L., & Look, A. T. (2005). Cre/lox-regulated transgenic zebrafish model with conditional myc-induced T cell acute lymphoblastic leukemia. *Proceedings of the National Academy of Sciences of the United States of America, 102*(17), 6068–6073.

Li, C., Lan, Y., Schwartz-Orbach, L., Korol, E., Tahiliani, M., Evans, T., & Goll, M. G. (2015). Overlapping requirements for Tet2 and Tet3 in normal development and hematopoietic stem cell emergence. *Cell Reports, 12*(7), 1133–1143.

Li, Z., Cai, X., Cai, C. L., Wang, J., Zhang, W., Petersen, B. E. ... Xu, M. (2011). Deletion of Tet2 in mice leads to dysregulated hematopoietic stem cells and subsequent development of myeloid malignancies. *Blood, 118*(17), 4509–4518.

Liu, X., Jia, X., Yuan, H., Ma, K., Chen, Y., Jin, Y. ... Zhu, J. (2015). DNA methyltransferase 1 functions through C/ebpa to maintain hematopoietic stem and progenitor cells in zebrafish. *Journal of Hematology & Oncology, 8*(1), 15.

Maeder, M. L., Linder, S. J., Cascio, V. M., Fu, Y., Ho, Q. H., & Joung, J. K. (2013). CRISPR RNA-guided activation of endogenous human genes. *Nature Methods, 10*(10), 977–979.

Mali, P., Yang, L., Esvelt, K. M., Aach, J., Guell, M., DiCarlo, J. E. ... Church, G. M. (2013). RNA-guided human genome engineering via Cas9. *Science (New York, NY), 339*(6121), 823–826.

Martin, C. C., Laforest, L., Akimenko, M. A., & Ekker, M. (1999). A role for DNA methylation in gastrulation and somite patterning. *Developmental Biology, 206*(2), 189–205.

Moran-Crusio, K., Reavie, L., Shih, A., Abdel-Wahab, O., Ndiaye-Lobry, D., Lobry, C.… Levine, R. L. (2011). Tet2 loss leads to increased hematopoietic stem cell self-renewal and myeloid transformation. *Cancer Cell, 20*(1), 11–24.

Murayama, E., Kissa, K., Zapata, A., Mordelet, E., Briolat, V., Lin, H. F.… Herbomel, P. (2006). Tracing hematopoietic precursor migration to successive hematopoietic organs during zebrafish development. *Immunity, 25*(6), 963–975.

Nagy, A. (2000). Cre recombinase: the universal reagent for genome tailoring. *Genesis, 26*(2), 99–109.

Nakamura, T., Largaespada, D. A., Lee, M. P., Johnson, L. A., Ohyashiki, K., Toyama, K.… Shaughnessy, J. D. (1996). Fusion of the nucleoporin gene NUP98 to HOXA9 by the chromosome translocation t(7;11)(p15;p15) in human myeloid leukaemia. *Nature Genetics, 12*(2), 154–158.

de Napoles, M., Mermoud, J. E., Wakao, R., Tang, Y. A., Endoh, M., Appanah, R.… Brockdorff, N. (2004). Polycomb group proteins ring1A/B link ubiquitylation of histone H2A to heritable gene silencing and X inactivation. *Developmental Cell, 7*(5), 663–676.

Nguyen, P. D., Hollway, G. E., Sonntag, C., Miles, L. B., Hall, T. E., Berger, S.… Currie, P. D. (2014). Haematopoietic stem cell induction by somite-derived endothelial cells controlled by meox1. *Nature, 512*(7514), 314–318.

North, T. E., Goessling, W., Walkley, C. R., Lengerke, C., Kopani, K. R., Lord, A. M.… Zon, L. I. (2007). Prostaglandin E2 regulates vertebrate haematopoietic stem cell homeostasis. *Nature, 447*(7147), 1007–1011.

Orkin, S. H., & Zon, L. I. (2008). Hematopoiesis: an evolving paradigm for stem cell biology. *Cell, 132*(4), 631–644.

Robu, M. E., Larson, J. D., Nasevicius, A., Beiraghi, S., Brenner, C., Farber, S. A., & Ekker, S. C. (2007). P53 activation by knockdown technologies. *PLoS Genetics, 3*(5), 787–801.

Sadler, K. C., Krahn, K. N., Gaur, N. A., & Ukomadu, C. (2007). Liver growth in the embryo and during liver regeneration in zebrafish requires the cell cycle regulator, uhrf1. *Proceedings of the National Academy of Sciences of the United States of America, 104*(5), 1570–1575.

Sawamiphak, S., Kontarakis, Z., & Stainier, D. Y. R. (2014). Interferon gamma signaling-positively regulates hematopoietic stem cell emergence. *Developmental Cell, 31*(5), 640–653.

Schuettengruber, B., Chourrout, D., Vervoort, M., Leblanc, B., & Cavalli, G. (2007). Genome regulation by polycomb and trithorax proteins. *Cell, 128*(4), 735–745.

Shah, A. N., Davey, C. F., Whitebirch, A. C., Miller, A. C., & Moens, C. B. (2015). Rapid reverse genetic screening using CRISPR in zebrafish. *Nature Methods, 12*(6), 535–540.

Sharif, J., Muto, M., Takebayashi, S., Suetake, I., Iwamatsu, A., Endo, T. A.… Koseki, H. (2007). The SRA protein Np95 mediates epigenetic inheritance by recruiting Dnmt1 to methylated DNA. *Nature, 450*(7171), 908–912.

Shih, A. H., Abdel-Wahab, O., Patel, J. P., & Levine, R. L. (2015). The role of mutations in epigenetic regulators in myeloid malignancies. *Immunological Review, 263*, 22–35.

Shimoda, N., Yamakoshi, K., Miyake, A., & Takeda, H. (2005). Identification of a gene required for de novo DNA methylation of the zebrafish no tail gene. *Developmental Dynamics: An Official Publication of the American Association of Anatomists, 233*(4), 1509–1516.

Smith, T. H. L., Dueck, C. C., Mhanni, A. A., & McGowan, R. A. (2005). Novel splice variants associated with one of the zebrafish dnmt3 genes. *BMC Developmental Biology, 5*, 23.

Stelzer, Y., Shivalila, C. S., Soldner, F., Markoulaki, S., Stelzer, Y., Shivalila, C. S.... Jaenisch, R. (2015). Tracing dynamic changes of DNA methylation at resource tracing dynamic changes of DNA methylation at single-cell resolution. *Cell, 163*(1), 218–229.

Tahiliani, M., Koh, K. P., Shen, Y., Pastor, W. A., Bandukwala, H., Brudno, Y.... Rao, A. (2009). Conversion of 5-methylcytosine to 5-hydroxymethylcytosine in mammalian DNA by MLL partner TET1. *Science (New York, NY), 324*(5929), 930–935.

Tittle, R. K., Sze, R., Ng, A., Nuckels, R. J., Swartz, M. E., Anderson, R. M.... Gross, J. M. (2011). Uhrf1 and Dnmt1 are required for development and maintenance of the zebrafish lens. *Developmental Biology, 350*(1), 50–63.

Trowbridge, J. J., Snow, J. W., Kim, J., & Orkin, S. H. (2009). DNA methyltransferase 1 is essential for and uniquely regulates hematopoietic stem and progenitor cells. *Cell Stem Cell, 5*(4), 442–449.

Wan, X., Hu, B., Liu, J., Feng, X., & Xiao, W. (2011). Zebrafish mll gene is essential for hematopoiesis. *The Journal of Biological Chemistry, 286*(38), 33345–33357.

Wang, H., Wang, L., Erdjument-Bromage, H., Vidal, M., Tempst, P., Jones, R. S., & Zhang, Y. (2004). Role of histone H2A ubiquitination in Polycomb silencing. *Nature, 431*(7010), 873–878.

White, R., Rose, K., & Zon, L. (2013). Zebrafish cancer: the state of the art and the path forward. *Nature Reviews Cancer, 13*(9), 624–636.

Wu, H., & Zhang, Y. (2011). Mechanisms and functions of Tet protein-mediated 5-methylcytosine oxidation. *Genes & Development, 25*(23), 2436–2452.

Yu, M., Mazor, T., Huang, H., Huang, H. T., Kathrein, K. L., Woo, A. J.... Cantor, A. B. (2012). Direct recruitment of polycomb repressive complex 1 to chromatin by core binding transcription factors. *Molecular Cell, 45*(3), 330–343.

PART 5

Zebrafish Infrastructure

CHAPTER

A scientist's guide for submitting data to ZFIN

23

D.G. Howe[1], Y.M. Bradford, A. Eagle, D. Fashena, K. Frazer, P. Kalita, P. Mani, R. Martin, S.T. Moxon, H. Paddock, C. Pich, S. Ramachandran, L. Ruzicka, K. Schaper, X. Shao, A. Singer, S. Toro, C. Van Slyke, M. Westerfield

University of Oregon, Eugene, OR, United States

[1]*Corresponding author: E-mail: doughowe@uoregon.edu*

CHAPTER OUTLINE

Methods in Cell Biology, Volume 135, ISSN 0091-679X, http://dx.doi.org/10.1016/bs.mcb.2016.04.010

Abstract

The Zebrafish Model Organism Database (ZFIN; zfin.org) serves as the central repository for genetic and genomic data produced using zebrafish (*Danio rerio*). Data in ZFIN are either manually curated from peer-reviewed publications or submitted directly to ZFIN from various data repositories. Data types currently supported include mutants, transgenic lines, DNA constructs, gene expression, phenotypes, antibodies, morpholinos, TALENs, CRISPRs, disease models, movies, and images. The rapidly changing methods of genomic science have increased the production of data that cannot readily be represented in standard journal publications. These large data sets require web-based presentation. As the central repository for zebrafish research data, it has become increasingly important for ZFIN to provide the zebrafish research community with support for their data sets and guidance on what is required to submit these data to ZFIN. Regardless of their volume, all data that are submitted for inclusion in ZFIN must include a minimum set of information that describes the data. The aim of this chapter is to identify data types that fit into the current ZFIN database and explain how to provide those data in the optimal format for integration. We identify the required and optional data elements, define jargon, and present tools and templates that can help with the acquisition and organization of data as they are being prepared for submission to ZFIN. This information will also appear in the ZFIN wiki, where it will be updated as our services evolve over time.

INTRODUCTION

ZFIN is the central repository of genetic and genomic information for the zebrafish research community. Granting agencies increasingly require that large data sets be submitted to an appropriate database repository such as ZFIN. In light of this, we aim to integrate as much of the zebrafish mutant, transgenic, expression, and phenotype data from the research community as possible and to provide links back to source databases when possible. To accomplish that, there must be clear guidance and documentation of the process and requirements for adding data to the ZFIN database. The process of preparing, submitting, and loading data into the ZFIN database is called a "data submission." Planning ahead for data submission to ZFIN will result in more efficient and timely addition of data to the database. This publication, the associated ZFIN wiki pages, and data submission templates referenced herein provide an up-to-date reference resource to support submission of large and small data sets to ZFIN.

1. WHY LOAD DATA INTO ZFIN?

In this era of big data it is increasingly important to integrate data from multiple sources to provide a unified view of data at a single location. This integration maximizes the value of each piece of data by allowing queries to return accurate and more complete results, accelerating research, and reducing redundant effort and research cost. ZFIN supports a diverse collection of data types including mutants, transgenic lines, expression, phenotypes, constructs, morpholinos, transcription activator-like effector nucleases (TALENs), clustered regularly interspaced short palindromic repeats (CRISPRs), antibodies, and disease models. Data curated from publications and from prior data loads are integrated to provide as complete a picture as possible of the role and function of each gene based on all the information available in the ZFIN database. Often, data stored in lab-specific databases lack the long-term stability, accessibility, and data integration that they will have in ZFIN. The goal at ZFIN is to capture the essential core of the data and any additional details the ZFIN database is able to support. In many cases it is possible to link from ZFIN back to the laboratory web pages, providing easy access to any further details that are not currently included at ZFIN. In the long term, the most important services ZFIN can provide related to data loads are to integrate data from disparate resources, provide a central location presenting a complete picture of what is known about a topic of interest, and provide critical long-term data stability and accessibility.

2. THE STRUCTURE OF THE ZFIN DATABASE

Data at ZFIN are stored in a complex relational database consisting of over 300 database tables (http://zfin.org/schemaSpy). This database structure allows disparate pieces of data to relate to each other and to be presented in an integrated format. Aligning incoming data with existing ZFIN data and associated database constraints is one of the major challenges for data loads, particularly if alignment is considered only after data are collected. In contrast, if the structure of a potential data submission is understood early in the data gathering process, data collection can be optimized to facilitate a smooth data submission process. Below we describe the data submission process and each major data type we currently support, as well as their components, and we identify which components are required and which are optional.

3. THE DATA SUBMISSION PROCESS

Data submission requests are typically initiated by an inquiry from a researcher. Once submission of data has been agreed upon, there are several steps to a typical data submission process (Fig. 1). A curator assigned to the data load will provide guidance on data gathering, establish necessary records in ZFIN, and assist in getting the data into a format that can be loaded into the ZFIN database.

Once submitted, data are subject to a number of quality control and data validation steps. Inconsistencies in the data are resolved through discussion with the data submitter. Once the data are free of errors, they are loaded into a test database for final review by the submitter. Once approved by the submitter, the data are loaded into the ZFIN database.

ZFIN does not hold private data. Once submitted, the data load will proceed as part of the normal software release cycle. When there are data that cannot be released to the public until they are published, an initial submission of high-quality data from the same experiments that is not part of the publication can be considered. This may include data such as mutations with no obvious or early lethal phenotypes, gene expression where there is either no expression at a particular developmental stage or expression is ubiquitous, or enhancer traps that trap already characterized enhancers. This allows the researcher to become familiar with the data submission process and to validate the data submission file format while protecting the integrity of the unpublished data. Once published, those data can be submitted with confidence, knowing that similar data have already been integrated into ZFIN.

4. DATA SUBMISSIONS

The data required for a submission often involve multiple files that together provide the information needed to represent the data fully at ZFIN. In this section each data type that can be loaded into the ZFIN database is described along with its required and optional components.

4.1 MUTANT AND TRANSGENIC LINE SUBMISSION

Mutant features are genomic alterations often generated by an applied mutagen, whereas transgenic features are genomic alterations generated by insertion of one

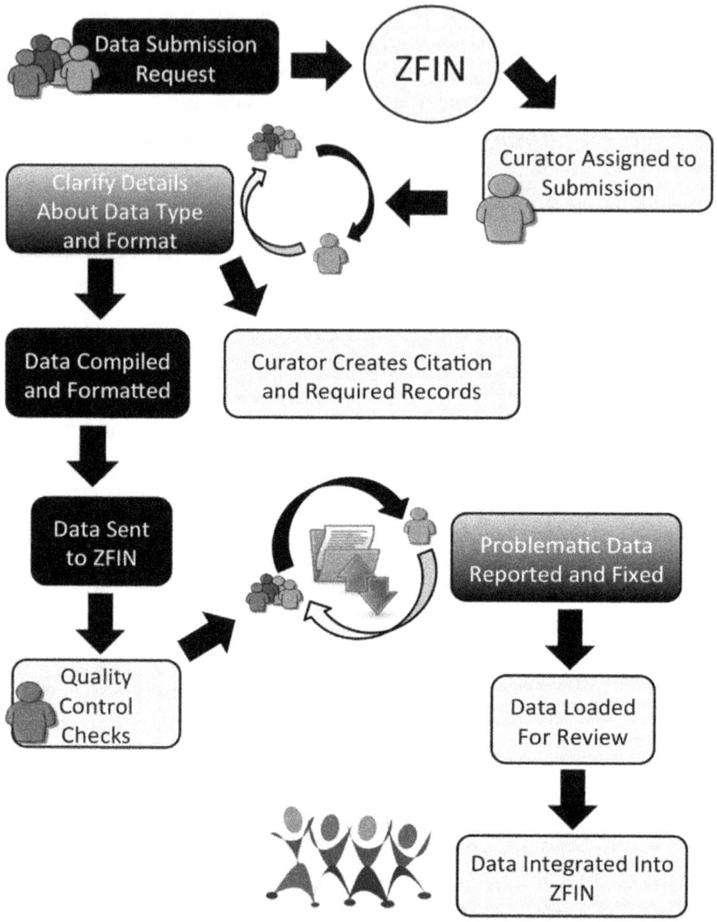

FIGURE 1 Summary of the data submission process

Black boxes indicate work done by the data submitter, light gray boxes indicate work done by ZFIN, gradient-filled boxes indicate iterative steps where work is shared between ZFIN and the data submitter.

or more copies of a transgenic construct. These may or may not result in alleles of genes. Mutant and transgenic lines are strains of fish that contain one or more heritable transgenic or mutant features. Each transgenic feature may contain one or more transgenic constructs. The exact insertion site or sites may or may not be known. Lines that contain multiple distinct known transgenic insertion loci will have a distinct genomic feature designated for each insertion site. The transgenic line is then composed of a combination of those distinct insertions. Data submissions containing genomic features not yet in ZFIN include information to create new feature records. Below we describe the various elements of data that can be included in a mutant or transgenic line data submission.

4.1.1 Genotypes

The genotype represents the primary genomic sequence alterations present in a fish. The genotype conveys zygosity information about specific loci having known sequence variants or transgenic insertions as well as the genetic background. To define a specific genotype in ZFIN completely and uniquely, each genomic feature and the zygosity of the locus where it resides are required. To define the genotype further, information about the parental zygosity of each locus and the genetic background can also optionally be provided (Table 1). The required and optional data elements for transgenic line submission (Table 2) and for mutant submission (Table 3) are similar but distinct.

4.1.2 Genomic feature

Each genomic feature has a unique name representing a specific genomic alteration or transgenic insertion event. Genomic feature names are composed of the line designation and a unique number identifier. Transgenic features have the suffix Tg, Et, Pt, or Gt depending on whether they are standard transgenic, enhancer trap, promoter trap, or gene trap features, respectively. Line designations are institution specific and can be obtained from the ZFIN nomenclature coordinator (nomenclature@zfin.org), and the researcher supplies the number identifier.

4.1.3 Feature zygosity

Feature zygosity is the zygosity of the locus with respect to the specific genomic feature. Valid choices include homozygous, heterozygous, and unknown. Each genomic feature in a genotype will have zygosity information associated with it. Transgenes that are present on only a single copy of the locus are considered heterozygous in ZFIN.

4.1.4 Feature maternal and paternal zygosity

The male and female parental zygosity of the locus at which there is a known genomic feature can be included to provide information about how the genomic feature was inherited. Parental zygosity of each mutant or Tg locus is reported as either homozygous, heterozygous, wild type, or unknown. Transgenes that are only present on a single copy of the male or female parental locus are considered heterozygous for the parent in ZFIN.

Table 1 Data for Submitting Genotypes to ZFIN

Required	Optional
Genomic feature	Feature maternal zygosity
Feature zygosity	Feature paternal zygosity
Genetic background	

Table 2 Data for Submitting Transgenic Lines to ZFIN

Required	Optional
Genomic feature	Link to alternate resource
Affected gene symbol	Insertion accession
Affected gene accession	Note
Transgene type	
Mutagen	
Subject	
Construct	
Laboratory of origin	
Citation	

Table 3 Data for Submitting Mutants to ZFIN

Required	Optional
Genomic feature	Link to alternate resource
Affected gene symbol	Mutant sequence accession
Affected gene accession	Note
Mutation type	
Mutagen	
Subject	
Laboratory of origin	
Citation	

4.1.5 Genetic background

The genetic background specifies on which standard or wild-type line (Table 4) the genotype is being carried. If there are no mutations involved, the genotype may be one of these standard lines. More detailed information about these standard lines can be found on the ZFIN Wild-Type Lines web page at http://zfin.org/action/feature/wildtype-list.

4.1.6 Affected gene symbol

If the genomic alteration is a point mutation, small insertion, small deletion, indel, transgene insertion, or results in a change in the coding region of one gene, that gene should be listed as an affected gene to indicate that the genomic alteration is an allele of that gene. For other mutation types (translocations, inversions, deficiencies), which may affect multiple genes, a separate file is used to specify the affected genes and how they relate to the genomic feature (Table 5). Genomic features with multiple affected genes are provided with one row per affected gene in this data file.

Table 4 Standard Lines in ZFIN

Abbreviation	Full Name	Description
AB	AB	Any of the inbred lines derived from the original Streisinger A and B in crosses. Includes AB* and ABC.
AB/TL	AB/Tupfel long fin	Mixed AB/Tupfel long-fin line either a maintained inbred line or a novel cross.
AB/TU	AB/Tuebingen	Mixed AB/Tuebingen line either a maintained inbred line or a novel cross.
C32	C32	Either C32 derivatives from the Steve Johnson laboratory or the Kimmel lab.
KOLN	Cologne	A wild-type line originally from the Campos-Ortega laboratory that has short fins.
DAR	Darjeeling	Wild-type line collected in Darjeeling, India, by Heiko Bleher in 1987. Line maintained by inbreeding.
EKW	Ekkwill	Wild-type line from Ekkwill Breeders in Florida.
HK	Hong Kong	Stock obtained from Hong Kong fish dealer.
IND	India	Stock obtained from expedition to Darjeeling (wild isolate).
NA	Nadia	Wild-type line from the Nadia district. Original stock collected from stagnant ponds and flood plain. Inbred.
NHGR-1	NHGR-1	Fully sequenced inbred line derived from a Tuebingen/AB cross (LaFave, Varshney, Vemulapalli, Mullikin, & Burgess, 2014)
RW	RIKEN WT	Wild-type line distributed by RIKEN.
SAT	Sanger AB Tuebingen	The AB/Tuebingen line derived from double haploid fish used by Sanger for genomic sequencing.
SJA	SJA	AB-derived line that is bred to reduce polymorphism.
SJD	SJD	Sibling line to Darjeeling. Inbred to reduce polymorphisms.
TU	Tuebingen	Short fins, original stock from a Tuebingen pet shop.
TL	Tupfel long fin	Homozygous for leo^{t1} and lof^{dt2}.
TLN	Tupfel long-fin nacre	The TL-derived TLN wild-type strain carries a mix of molecularly uncharacterized $mitfa(nacre)^{s170}$ and $mitfa(nacre)^{s184}$ in the background. TL is homozygous for $cx41.8(leo)^{t1}$ and lof^{dt2}.
WIK	WIK	The WIK line is very polymorphic relative to the TU line.
WT	Wild type	Used to denote any wild type not listed above.

Table 5 Data Required for Specifying Multiple Affected Genes

Required Data	Description
Genomic feature	The unique identifier for the mutant.
Affected gene	The symbol for the affected gene.
Affected gene accession	A ZDB-GENE ID or sequence accession number for the affected gene.
Relationship	The relationship between the genomic feature and the affected gene. One of gene missing, gene present, gene moved, is allele of gene.

4.1.7 *Affected gene accession*

When submitting a mutant or transgenic line with an affected gene it is essential to supply a unique identifier for the gene to remove any ambiguity in gene identification. ZFIN gene record IDs (ZDB-GENE IDs) are the best identifiers for unambiguous identification of genes in ZFIN, so they are preferred whenever possible. If ZDB-GENE IDs are difficult to obtain for your data, Ensembl IDs (OTTDART or ENSDARG) are the next best choice because they fit most easily into existing gene identification pipelines at ZFIN. Minimally, a sequence accession number, such as the GenBank ID for the longest transcript of the gene, must be provided to ensure accurate identification of genes for the incoming data. Without such gene-specific identifiers, it may not be possible to load all data in a submitted data set.

4.1.8 *Transgene type*

In cases where trapping has occurred, transgene type abbreviations are added to the end of the line designation to denote the type of transgene (Et for enhancer trap, Gt for gene trap, and Pt for promoter trap). It is acceptable to use Tg to name trap lines, but the more specific name using Et, Pt, or Gt is helpful in denoting their function.

4.1.9 *Mutation type*

ZFIN supports many types of genomic sequence alterations derived from the sequence ontology (SO) (Eilbeck et al., 2005) (Table 6). One of these types must be assigned to each mutant submitted to ZFIN.

4.1.10 *Mutagen*

The mutagen used to generate the genomic feature or transgenic line must be provided. Valid mutagens come from a constrained set of terms (Table 7). If a TALEN or CRISPR was the mutagen, then the name of the specific TALEN or CRISPR should be provided as the mutagen. For transgenic lines using only a construct, the mutagen is "DNA." If a TALEN or CRISPR was also used during creation of a transgenic insertion, the TALEN/CRISPR name should be provided as the mutagen. The construct name is provided in a separate column (see the later discussion). Details for TALENs, CRISPRs, and constructs not already in ZFIN must be

Table 6 Mutation Types Supported in ZFIN

Mutation Type	Definition	Notes	Data to Provide	SO ID
Point mutation	A single nucleotide change which has occurred at the same position of a corresponding nucleotide in a reference sequence.	Can be an allele of a single gene.	Affected gene and its ZDB-GENE ID.	SO:1000008
Small deletion	The point at which one or more contiguous nucleotides were excised.	The excision is within a single gene.	Affected gene and its ZDB-GENE ID.	SO:0000159
Insertion	The sequence of one or more nucleotides added between two adjacent nucleotides in the sequence.	Usually an allele of a single gene.	Affected gene and its ZDB-GENE ID.	SO:0000667
Indel	A sequence alteration which includes an insertion and a deletion, affecting two or more bases.	Usually an allele of a single gene.	Affected gene and its ZDB-GENE ID.	SO:1000032
Translocation	A region of nucleotide sequence that has translocated to a new position. The observed adjacency of two previously separated regions.	Has at least one breakpoint, frequently within a gene. Has genes that are in a new genomic context.	Genes at breakpoint and their ZDB-GENE IDs. Genes that have been relocated and their associated ZDB-GENE IDs.	SO:0000199
Inversion	A continuous nucleotide sequence is inverted in the same position.	Has two breakpoints which may occur in one or more genes and may have additional genes within the inverted sequence.	Genes at breakpoint and their ZDB-GENE IDs. Genes that have been relocated and their associated ZDB-GENE IDs.	SO:1000036
Deficiency	An incomplete chromosome.	The chromosome is missing more than a single gene. Has two breakpoints. Other genes existing between the breakpoints may also have been lost.	Genes at breakpoint and their ZDB-GENE IDs. Genes that have been lost and their associated ZDB-GENE IDs.	SO:1000029
Unknown	A mutation where the lesion type is unknown.	May be an allele of a gene or in an unknown location.	Affected gene and its ZDB-GENE ID if known.	NA

Table 7 Mutagen Types Supported in ZFIN

Mutagen Type	Description
TALEN	Transcription activator-like effector nucleases (TALENs) are nucleases specifically designed to cleave a DNA sequence of interest. Provide the name of the TALEN.
CRISPR	Clustered regularly interspaced short palindromic repeats (CRISPRs) are specifically designed to recruit the Cas9 enzyme to cleave DNA at a targeted DNA locus. Mutations often result from the subsequent DNA repair event. Provide the name of the CRISPR.
ENU	N-ethyl-N-nitrosourea, ENU, is a chemical alkylating agent and mutagen when applied to animals.
TMP	4,5′,8-Trimethylpsoralen is a DNA cross-linking agent which often produces deletion mutations.
Gamma rays	Ionizing electromagnetic radiation used to induce mutations. Often produces large deletions and chromosomal aberrations.
Spontaneous	De novo mutations not generated by the application of an external mutagen.
Zinc finger nuclease	Zinc finger nucleases are artificial restriction enzymes designed to target and cleave specific DNA sequences.
DNA	DNA sequence, usually a transgenic construct, injected into embryos to create heritable transgenic insertions.

specified in a separate file dedicated to these new record types. See Sections 4.2 and 4.7.3 for details on supplying these data.

4.1.11 Subject

This is the type of original recipient animal of the mutagen treatment. Subject is used in combination with "mutagen" data to capture how the genomic alteration was generated. Valid choices for the subject come from the following constrained set: adult males, adult females, embryos, sperm.

4.1.12 Construct

The name of the transgenic construct used to make a transgenic line should be specified. If the construct is already in ZFIN, the ZFIN construct name should be provided. When multiple constructs are inserted simultaneously there should be multiple rows of data for the line designation specifying only one construct per row. Details for submission of new construct records to ZFIN is described in Section 4.2 of this chapter.

4.1.13 Laboratory of origin

All genomic features require a laboratory of origin, the laboratory from which the feature originated. This laboratory must have a laboratory record in ZFIN. New laboratory records can be created upon request at zfinadmn@zfin.org.

4.1.14 Sequence accession

When genomic features are located by sequencing, a GenBank accession number may be provided to describe the altered genomic sequence or transgene insertion site. The GenBank number is included on the genomic feature page in ZFIN and may also be used to locate features on a genome browser.

4.1.15 Link to alternate resource

Some laboratories provide access to their data via a dedicated laboratory Website. A URL pointing to such an alternate web-based resource about the mutant or transgenic line can be provided. This URL is used to link from ZFIN back to the original resource web page where additional information about the feature may be available.

4.1.16 Citations

All data added to ZFIN must be associated with a publication record specifically created at ZFIN to represent the data load. The publication record created at ZFIN must minimally have a list of authors, a title, and an abstract. Data that have been published in a peer-reviewed journal article may also be attributed to that publication by providing the PubMed ID (Table 8).

4.1.17 Note

A free text note with further details about the mutant or transgenic line can be included. For a mutant line these details may include transition/transversion, location info, null/hypomorph/gain of function, amino acid changes, etc. These details clarify the nature and location of the mutation, particularly for details that cannot be stored in a dedicated field in the ZFIN database. For a transgenic line the note may include information such as the insertion site construct copy number, the consequence of the insertion (null, hypomorph, etc.), or specific details about the function or composition of the line.

4.1.18 Sperm samples

When features are only available as sperm samples, ZFIN only accepts data for those features that have been accepted by a resource center such as the Zebrafish International Resource Center. Data loads involving sperm samples will be arranged after the resource center contacts ZFIN regarding the sperm samples they expect to receive.

Table 8 Data for Citation of a Data Submission

Required	Optional
Author list Publication title Abstract describing data and methods	PubMed ID

4.2 TRANSGENIC CONSTRUCTS

Transgenic constructs are engineered nucleotide sequences injected into fish resulting in heritable transgenic insertions into the genome. ZFIN creates construct records only for constructs used to make stable and heritable transgenic insertions. Information about the functional portions of constructs is collected, but information about plasmid propagation is not. The required and optional data elements for creating new transgenic construct records in ZFIN are described later (Table 9). Transgenic construct design varies widely and, as a result, multiple construct submission formats are required. The information requested here will support only basic constructs with a single functional cassette. When construct data do not fit into this simple set of fields, we are happy to work with data submitters to accept construct data as completely as possible. This can typically be accomplished by submitting an extended version of the simple data we typically request.

4.2.1 Construct name

The name ZFIN uses to label constructs is composed of the functional parts of the construct, mainly the promoter elements and expressed genes. A single colon is used to separate the promoter element from the expressed gene. Basal promoters are not recorded, but any gene-specific promoter elements should be listed in the promoter section of the name (Fig. 2). Additionally, the type of transgenic construct is encoded in the name by prepending the name with Tg if the construct is a general transgenic construct, Gt if the construct is a gene trap, Et if the construct is an enhancer trap, and Pt if the construct is a promoter trap. Additional names for constructs can be provided as aliases if desired. Assistance with naming

Table 9 Data for Submitting Transgenic Constructs to ZFIN

Required	Optional
Construct name	Link to alternate resource
Promoter gene symbol	Construct accession
Promoter accession	Construct map image name
Coding sequence gene symbol	Note
Coding sequence accession	
Engineered region name	
Citation	

FIGURE 2

The basic elements of a transgenic construct.

transgenic constructs is provided by the ZFIN nomenclature coordinator (nomenclature@zfin.org).

4.2.2 Promoter gene symbol

Many transgenic constructs drive gene expression using the promoter of a gene from zebrafish or another species. The symbol for the promoter gene should be provided, including an abbreviation for the species from which the sequence was derived if it is not a zebrafish gene (Hsa for human, Mmu for mouse, etc.). In some cases, such as promoter trap constructs, there may not be a regulatory region to include in the construct name.

4.2.3 Promoter gene accession

When submitting data specifying a gene it is essential to supply a stable and unique identifier for the gene to remove any ambiguity in gene identification. ZDB-GENE IDs are the best identifiers for this purpose. Please see Section 4.1.7 of this chapter for more details on providing unique gene identifiers.

4.2.4 Coding sequence gene symbol

Most transgenic constructs drive expression of a gene or a reporter in the derived transgenic fish. The symbol for the product that will be expressed from the integrated construct should be provided here, including an abbreviation for the species from which it was derived if it is not a zebrafish gene.

4.2.5 Coding sequence gene accession

When submitting data specifying a gene it is essential to supply a unique and immutable identifier for the gene to remove any ambiguity in gene identification. ZDB-GENE IDs are the best identifiers for this purpose. Please see Section 4.1.7 of this chapter for more details on providing unique gene identifiers.

4.2.6 Engineered region name

Constructs often include additional elements in them such as binding or cleavage sites, internal ribosomal entry sites, etc. ZFIN refers to these as Engineered Regions (ERs). ERs are typically not included in the name of the construct. Instead, they are linked to the construct record using a "construct contains" relationship, making the ER details available on the construct page. The complete list of ERs currently supported at ZFIN is available via a marker search at ZFIN. New ER records can be requested through the nomenclature coordinator (nomenclature@zfin.org).

4.2.7 Construct sequence accession

The GenBank number for a construct sequence can optionally be provided. This GenBank number will be visible on the construct record in ZFIN. If a BAC or PAC clone was used to generate the construct, the accession number of the BAC or PAC should be provided.

4.2.8 Construct map image name

The name of an image file (jpg or png) can be provided showing a graphical representation of the construct map. The image file itself can be provided via email, ftp, or a file-sharing service. The image of the construct map will be included on the ZFIN construct page.

4.2.9 Link to alternate construct resource

Laboratories may have a Website where more details about the construct are provided. A URL for such a web-based resource can be optionally provided. This is used to support a link to the alternate construct page from the ZFIN construct page.

4.2.10 Citation

All data loaded into ZFIN require a data load citation to which the data are attributed. This will be a ZDB-PUB ID for the data load publication and an optional PubMed ID for any related journal article. Please see Section 4.1.16 of this chapter for further details.

4.2.11 Construct note

Unfortunately, the ZFIN database cannot support every detail for every construct. Sometimes there are additional pieces of information that further illuminate the nature of a construct. A note can be included here to provide salient details.

4.3 MORPHOLINOS, TALENS, AND CRISPRS

Morpholinos, TALENs, and CRISPRs are sequence-specific regents used to disrupt gene translation or splicing, or to introduce sequence alterations at sequence-targeted locations. Once these tools are designed and tested, data describing them can be submitted to ZFIN. The required and optional data elements for submitting these records to ZFIN are described in Table 10.

4.3.1 MO/TALEN/CRISPR name

Morpholinos, TALENs, and CRISPRs are given unique names at ZFIN using a type-specific prefix with a unique numerical index and the gene symbol of the targeted gene appended. For example, TALEN1-pax2a would be the first TALEN added to

Table 10 Data for Submitting Morpholinos, TALENs, and CRISPRs to ZFIN

Required	Optional
MO/TALEN/CRISPR name	Link to alternate resource
Target sequence 1	Note
Target sequence 2 (TALEN only)	Citation PMID
Target gene symbol	
Target gene accession	
Data load citation	

ZFIN that targets the pax2a gene. These names are assigned in sequence as each new record is created in ZFIN. Any name can be included on the record as a synonym.

4.3.2 Target sequence 1

The target sequence reported is the 5′-3′ genomic sequence being targeted by the TALEN, CRISPR, or morpholino.

4.3.3 Target sequence 2

The target sequence reported is the 5′-3′ genomic sequence for the second genomic target of a TALEN. These data are not provided for CRISPRs or morpholinos, which only have a single target sequence.

4.3.4 Target gene symbol

The target gene symbol is for the gene being affected by the morpholino, TALEN, or CRISPR. If there are multiple target genes for a morpholino, TALEN, or CRISPR, there should be one row per target gene in the data submission file.

4.3.5 Target gene accession

When submitting data specifying a gene, it is essential to supply a unique and immutable identifier for the gene to remove any ambiguity in gene identification. ZDB-GENE IDs are the best identifiers for this purpose. Please see Section 4.1.7 of this chapter for more details on providing unique gene identifiers.

4.3.6 Link to alternate resource

Some laboratories maintain databases specifically to house their own TALEN, CRISPR, or morpholino data. ZFIN can link back to those resources if a URL with long-term stability is provided.

4.3.7 Citations

All data loaded into ZFIN require a data load citation to which the data are attributed. This will be a ZDB-PUB ID for the data load publication and an optional PubMed ID for any related journal article. Please see Section 4.1.16 of this chapter for further details.

4.4 EXPRESSION DATA

Zebrafish are used extensively to assay gene expression and antibody labeling patterns. The information for an expression data submission is described later (Table 11).

4.4.1 Expressed gene symbol

Expression data are often recorded for a specific gene. This could be a zebrafish gene or a reporter transgene. The symbol of the gene whose expression is measured should be included. In some cases, the labeled gene may not be known or the

Table 11 Data for Submitting Expression Data to ZFIN

Required	Optional
Expressed gene symbol	Image/movie file name
Expressed gene accession	Antibody name (for immunoassays)
Genotype	Probe GenBank accession number (for hybridization assays)
Morpholinos, TALENs, CRISPRs	Antibody name (for immunoassays)
Anatomical structure	
Developmental stage	
Experimental conditions	
Citation	
Assay type	

antibody may be made using immunogens other than gene products, such as lipids, or chemicals. In those cases, the expressed gene data are not included.

4.4.2 Expressed gene accession

When submitting data specifying a gene, it is essential to supply a unique and immutable identifier for the gene to remove any ambiguity in gene identification. ZDB-GENE IDs are the best identifiers for this purpose. Please see Section 4.1.7 of this chapter for more details on providing unique gene identifiers.

4.4.3 Genotype

The genotype of the fish in which the expression observation was made, including the genetic background, must be specified. See Section 4.1 in this chapter for details on reporting genotypes to ZFIN.

4.4.4 Morpholinos, TALENs, and CRISPRs

Morpholinos, TALENs, and CRISPRs can be injected into zygotes to induce anatomically localized modification of gene expression or gene structure. When used in this way during an expression analysis, the name of the morpholino, TALEN, or CRISPR should be reported. Morpholinos, TALENS, and CRISPRs that already have records in ZFIN can be listed by their ZDB ID or ZFIN name. When a morpholino, TALEN, or CRISPR is not already in ZFIN, a separate file must be submitted for creation of these new records. Please see Section 4.3 of this chapter for details on submitting these new records to ZFIN.

4.4.5 Anatomical structure

The anatomical structure where the expression observation was made must be specified. To use a common vocabulary, the anatomical term should be selected from the Zebrafish Anatomy Ontology (ZFA), which describes zebrafish anatomy from larvae to adult (Van Slyke, Bradford, Westerfield, & Haendel, 2014). Any anatomical terms that are not drawn from the ZFA will need to be mapped to equivalent

ZFA terms. This can be time-consuming work before your submission can be accepted, so it is advised that the ZFA terms be used from the outset of data gathering. The most recent update of the ZFA is available at GitHub: https://github.com/cerivs/zebrafish-anatomical-ontology. More complex compound anatomical terms, including organ parts or subcellular localization (such as "mitochondrion part of retinal ganglion cell"), can also be supported. Contact ZFIN if you would like to use such a term so a data format can be established for your particular data load.

4.4.6 Developmental stage

The developmental stage is the developmental time point at which the experimental observation was made. The developmental stage is reported in hours postfertilization and is selected from the Zebrafish Developmental Stages ontology (ZFS), which is based on developmental stages as described in Kimmel, Ballard, Kimmel, Ullmann, & Schilling, 1995. The current ZFS can be found from http://zfin.org/zf_info/zfbook/stages/index.html. Developmental stage terms that do not come from the ZFS will need to be mapped to equivalent ZFS terms. This can be time-consuming work before a data submission can be accepted, so it is advised that the ZFS terms be used from the outset of data gathering.

4.4.7 Experimental conditions

Experimental conditions provide details of the pertinent experimental treatments present at the time data were collected. Experimental conditions are represented in ZFIN by a constrained set of choices (Table 12). New conditions can be requested if necessary.

4.4.8 Experimental conditions note

A free text note can be used to capture further details about an experimental condition. If "chemical" is selected as the experimental condition the chemical name, supplier number, and a chemical identifier from ChEBI (Hastings et al., 2015) should be supplied in the note.

4.4.9 Citations

All data loaded into ZFIN require a data load citation to which the data are attributed. This will be a ZFIN publication ID (ZDB-PUB ID) for the data load publication and an optional PubMed ID for any related journal article. Please see Section 4.1.16 of this chapter for further details.

4.4.10 Assay type

The assay used to detect expression should be listed. Valid options are immunohistochemistry, western blot, mRNA in situ hybridization, DNA in situ hybridization, intrinsic fluorescence, northern blot, and reverse transcription PCR.

4.4.11 Antibody name

When expression data are gathered by immunohistochemistry, the specific antibody can be included. If the antibody is already found in ZFIN, then the ZFIN ID for that

Table 12 Experimental Conditions in ZFIN

Condition	Description
Standard	Experimental condition that is the standard environment for zebrafish husbandry, as described in The Zebrafish Book. In general the standard environment utilizes contaminant-free tank water, heated to 28.5°C, with the fish fed a normal contaminant-free diet, with standard osmolarity, pH, and normal light cycle of 14 h light/10 h dark.
Generic control	Experimental condition that is used as a reference point to compare with results of treated zebrafish. Generic experimental controls often use sham injections, injections of vehicle, injections of control MOs, etc. This environment is used for nonstandard conditions used in control treatments.
Chemical	Experimental condition in which the fish is treated in tank water, or by injection or consumption, with a chemical substance. The ChEBI ID for the chemical should be included in the data submission.
pH, acidic	Experimental condition in which the pH of the water is lower than the pH of the controlled conditions.
pH, basic	Experimental condition in which the pH of the water is higher than the pH of the controlled conditions.
Electric field	Experimental condition in which an electric field is applied to the fish, fish cells, or organs as compared to control conditions.
Gravity	Experimental condition in which the fish is exposed to forces that simulate low or high gravity as compared to earth's gravity.
Hyperoxia	Experimental condition in which the oxygen (O_2) concentration is higher than the one in controlled conditions.
Hypoxia	Experimental condition in which the oxygen (O_2) concentration is lower than the one in controlled conditions.
Light	Experimental condition in which the intensity, wavelength, and/or duration of illumination is (are) different from the one in controlled conditions.
Magnetic field	Experimental condition in which the fish is exposed to a magnetic field as compared to control conditions. A magnetic field is a region in which the force of magnetism is applied.
Mechanical stress	Experimental condition in which an external force is applied to the fish or part of the fish.
Radiation	Experimental condition in which the fish is exposed to ionizing and/or nonionizing radiation. The radiation could be ionizing such as gamma rays, alpha particles, UV, X-ray, and nonionizing such as infrared, microwaves, etc.
Bacterial infection	Experimental condition in which fish have been infected with bacteria. This infection can be done by addition of bacteria in the water or by injection of bacteria, (for example, in the brain ventricle, in the caudal vein, in the yolk sac), or ingestion, or other means.

Cancer	Experimental condition in which cancer cells are introduced to the fish via injection of tumor cells.
Fungal infection	Experimental condition in which fish have been infected with a fungus.
Germ free	Experimental condition in which fish were raised in the absence of bacteria.
High-calorie diet	Experimental condition in which fish are fed a high-calorie diet as compared to the normal diet.
Low-calorie diet	Experimental condition in which fish are fed a low-calorie diet as compared to the normal diet.
Organ culture	Experimental condition in which an organ is dissected/isolated/collected from the fish and placed in culture. The analysis of the experiment is done on this organ in culture.
Primary cell culture	Experimental condition in which an embryo or adult fish is dissociated to a single cell suspension. The analysis is made on this cell culture.
Regeneration/healing	Experimental condition in which fish's organ (eg, heart) or anatomical structure (eg, fin) was wounded or amputated.
Starvation	Experimental condition in which fish were deprived of food.
Salinity, hypertonic	Experimental condition in which the salt concentration is higher than the one in controlled conditions.
Salinity, hypotonic	Experimental condition in which the salt concentration is lower than the one in controlled conditions.
Temperature, cold shock	Experimental condition in which fish are subjected for a short period of time to temperature lower than the controlled temperature. The standard controlled temperature (according to The Zebrafish Book) is 28.5°C.
Temperature, heat shock	Experimental condition in which fish are subjected for a short period of time to temperature higher than the controlled temperature. The standard controlled temperature (according to The Zebrafish Book) is 28.5°C.
Temperature, stable	Experimental condition in which fish are raised in temperature different (lower or higher) than the controlled temperature. The standard controlled temperature (according to The Zebrafish Book) is 28.5°C.

antibody should be provided. However, if the antibody is not in ZFIN, then an antibody record can be created. A number of pieces of information are required to create the antibody record at ZFIN (Table 13). Information not listed as optional is required.

4.4.12 Probe GenBank accession number

Expression data are often gathered by in situ hybridization using a labeled sequence-specific nucleotide probe. If there is a GenBank number for the probe sequence, that information can be included here.

4.4.13 Images and movies

Images and movies are often the best medium to convey gene expression information. Submission of images or movies in conjunction with the expression data text files is encouraged. Image files should be in JPEG or PNG format. Movies should be in MP4 format with a 10 MB per file size limit. Media file names need to be unique within the load. Encoding information about the content of the image/movie in the file name can be helpful. For example, expressed gene, stage, and image number could all be in the file name. If expression was in a nonstandard background or if multiple alleles were used, this information could be included in the file name. For mutant phenotypes the allele and stage of observation could be included in the file name. If an antibody is used then the antibody could be encoded in the file name.

Table 13 Data Used to Create New Antibody Records

Data	Description
Host organism	Organism from which the antibody was made.
Immunogen organism	Species from which the immunogen was obtained. If it is a peptide based on a sequence from a particular organism, list that organism.
Antibody type	List whether antibody is polyclonal or monoclonal.
Antibody isotype	List isotype if known (optional).
Source	If the antibody was purchased from a commercial supplier, list the supplier.
Catalog number	If the antibody is from a commercial supplier please provide the catalog number (optional).
Name	Include clone names if known (optional).
Note	Include sequence of peptide or accession number of sequenced used to produce antibody if the antibody was custom made. Also include any usage notes here (optional).
Target gene	Provide ZDB-GENE ID or sequence accession number for the target gene if known (optional).
Citation for original source	If this is a previously published antibody, please provide a reference PubMed ID. Otherwise the antibody will be attributed to the data load publication.

4.5 PHENOTYPE DATA

Phenotypes at ZFIN are recorded in the bipartite E:Q syntax (Howe et al., 2013; Washington et al., 2009). The "E" is an entity [such as "eye" (ZFA:0000107)] in which the phenotype manifests. The "Q" is a phenotypic quality describing the nature of the phenotype ["decreased size" (PATO:0000587), for example]. Together, this EQ combination can be used to describe a phenotype of "small eyes." Phenotype data must be in this format before it can be loaded into ZFIN. The required and optional data elements for a phenotype data submission are listed in Table 14.

4.5.1 Genotype

The genotype of the fish in which the phenotype was observed, including the feature name, zygosity, and genetic background, must be specified. See Section 4.1.1 in this chapter for details on how to provide genotype information.

4.5.2 Morpholinos, TALENS, and CRISPRs

Morpholinos, TALENs, and CRISPRs can be injected into zygotes to induce anatomically localized modification of gene expression and/or gene structure. When used in this way during a phenotypic analysis, the morpholino, TALEN, or CRISPR name should be reported. If the reagent already has a record in ZFIN, it can be listed by its ZFIN name or ZDB ID. When the reagent is not already in ZFIN, a separate file needs to be submitted to support creation of these new records. Please see the Section 4.3 of this chapter for details on submitting these new records to ZFIN.

4.5.3 Developmental stage

The developmental stage is a term selected from the ZFS. This should report the developmental stage at which the phenotype observation was made. Developmental stage terms that do not come from the ZFS will need to be mapped to ZFS terms before the data submission can be loaded. This data mapping can be time-consuming, so it is recommended that the ZFS terms be used from the outset of data gathering.

Table 14 Data for Submitting EQ Phenotypes to ZFIN

Required	Optional
Genotype Morpholinos, TALENs, CRISPRs Developmental stage Experimental conditions Phenotype entity Phenotype quality Tag Citation	Image or movie file name

4.5.4 Experimental condition

All phenotype annotations need to have the experimental conditions specified. Valid values must come from the constrained set of experimental conditions supported at ZFIN (Table 12).

4.5.5 Phenotype entity

The phenotype entity can either be an anatomical term selected from the ZFA (eg, "heart"), or a process term selected from the Gene Ontology (GO) (eg, "cell migration") (Ashburner et al., 2000). Any phenotype entity terms that are not drawn from these two sources will need to be mapped to equivalent terms from these ontologies. This can be time-consuming work, so it is advised that the GO and ZFA terms be used from the outset of data gathering. Example terms include "eye" from the ZFA or "cell migration" from the GO. More complex phenotype entities can be supported by combining terms such as "cell migration occurs in heart." Contact ZFIN if you would like to use more complex entities so we can review the details together for your particular data load.

4.5.6 Phenotype quality

The phenotype quality is a term selected from the Phenotypic Quality Ontology (PATO) (Mungall et al., 2010). The current PATO can be found at BioPortal (http://purl.bioontology.org/ontology/PATO). The quality term describes the nature of the phenotype as it pertains to the phenotype entity. Any phenotype quality terms that are not drawn from PATO will need to be mapped to equivalent terms from PATO. This can be time-consuming work, so it is advised that the PATO terms be used from the outset of data gathering. Note that some PATO terms can be used only with anatomical structures [Example: "decreased size" (PATO:0000587)], and others are exclusively used with biological process terms [eg, "increased rate" (PATO:0000912)].

4.5.7 Tag

Phenotype annotations can capture observations of "abnormal" or "normal" processes or structures relative to control conditions. The "abnormal" tag is used when a phenotype involves an altered morphology or process compared to a control. For example, a mutant with small eyes relative to a control would be captured as Entity = "eye," PATO = "decreased size," Tag = "abnormal." If it is notable that no abnormal morphological phenotype is observed, that can be annotated using the "normal" tag to capture the phenotype as Entity = "whole organism," PATO = "morphology," Tag = "normal."

4.5.8 Media file name

Phenotype data can be submitted with accompanying images or movies to illustrate the phenotype. The image or movie file name is specified here. Please see Section 4.4.13 of this chapter for details about the media that can be provided.

4.5.9 Citations

All data loaded into ZFIN require a data load citation to which the data are attributed. This will be a ZDB-PUB ID for the data load publication and an optional PubMed ID for any related journal article. Please see Section 4.1.16 of this chapter for further details.

4.6 GENOME BROWSER TRACKS

It is now common for data to be placed into a genomic context by viewing them as a track in a genome browser. This can include gene-centric information, such as expression data, as well as epigenetic data such as DNA methylation status, specific binding sites, etc. Track files can be submitted to ZFIN and added to the ZFIN track hub (Raney et al., 2014). Tracks in the ZFIN track hub can be viewed in the UCSC genome browser (Kent et al., 2002) by adding the ZFIN track hub URL (http://trackhub.zfin.org/zfintracks/hub.txt) to the list of available hubs in the "My Data" tab at UCSC. The data necessary to submit a track for inclusion in the ZFIN track hub are described later.

4.6.1 Track files

Track hub files and configuration are well documented on the UCSC Website: https://genome.ucsc.edu/goldenPath/help/hgTrackHubHelp.html. Please consult that resource for detailed information about track hubs and file formats. Track files submitted to ZFIN must be in one of the compressed binary formats supported by the genome browser: bigBed, bigGenePred, bigWig, BAM, HAL, or VCF.

4.6.2 Track configuration and description

All tracks are generated using the coordinate system from a specific zebrafish reference genome such as Zv9 or GRCz10 (LaFave et al., 2014). Tracks from the ZFIN track hub will be visible in the UCSC genome browser only while viewing the reference genome for which the track was generated. Tracks generated using the Zv9 coordinate system will not be visible while viewing the GRCz10 build at UCSC. The track submitter provides track description and configuration details. The UCSC track hub documentation provides a comprehensive review of the options available for track configuration. The ZFIN track hub has the following minimum set of configuration information that must be provided by the track submitter (Table 15).

4.6.3 Citation

All data loaded into ZFIN require a data load citation to which the data are attributed. This will be a ZDB-PUB ID for the data load publication and an optional PubMed ID for any related journal article. Please see Section 4.1.16 of this chapter for further details.

4.6.4 Track maintenance

When new versions of genome builds are released, such as the transition from Zv9 to GRCz10, ZFIN will not port submitted tracks to the newer genome coordinate

Table 15 Required Track Hub Configuration Information

Label	Description
Track	The track file name.
shortLabel	A brief (17 character) label to describe the track in the genome browser. Visible to the left of the track in the genome browser. Example: 4-day methylome
longLabel	A longer label (76 character) to describe the track in the genome browser. Visible above the track in the genome browser. Provide enough detail to uniquely identify the track. Example: Chatterjee et al., 2013 adult brain methylome
Type	States the track file format (bigWig, bigBed, etc.).

system. New track files must be provided by the original track submitter. Previously submitted tracks will remain available in the ZFIN track hub when viewing the data source for which the track was made.

4.7 DISEASE MODELS

ZFIN recently added the ability to annotate disease models created using zebrafish. A disease model can be created using a mutant, a morpholino, TALEN, CRISPR, and/or manipulation of the experimental conditions such as application of a chemical or modification of diet. Mutants and experimental conditions can be used in any combination to generate a disease model. All of the following pieces of data are required to submit a new disease model to ZFIN.

4.7.1 Genotype

The genotype of the fish used in the disease model must be specified. See Section 4.1.1 of this chapter.

4.7.2 Experimental condition

A disease model may exist as a wild-type or mutant fish with a chemical treatment. Therefore, the experimental conditions for all disease model observations must be provided. ZFIN currently uses a constrained list of experimental conditions (Table 12).

4.7.3 Morpholino, TALEN, CRISPR

Some disease models are generated using morpholinos, TALENs, or CRISPRs. The name of the morpholino, TALEN, or CRISPR must be specified when one is used as an experimental treatment rather than as a germ line mutagen.

4.7.4 Disease term ID

The Disease Ontology (DO) (Schriml & Mitraka, 2015) is used to curate disease data in ZFIN. Any disease model data must therefore include the DO identifier for the DO term representing the intended disease. The DO can be searched at http://disease-ontology.org/. If your disease terms are not found in DO, let us know and we can get them added to the ontology.

4.7.5 Citation

All data loaded into ZFIN require a data load citation to which the data are attributed. This will be a ZDB-PUB ID for the data load publication and an optional PubMed ID for any related journal article. Please see Section 4.1.16 of this chapter for further details.

5. WHAT HAPPENS TO YOUR DATA AFTER SUBMISSION TO ZFIN?

Data that are submitted to ZFIN have many individual pieces of information that identify specific things such as genes, anatomical structures, antibodies, probes, constructs, genotypes, etc. Before submitted data can be loaded into ZFIN, it must be established whether records in the incoming data already exist in the ZFIN database. Quality control standards are also applied that may go beyond those used during data collection. Later we describe some of the primary data validation and quality control processes we use to treat data before entering them into the ZFIN database.

5.1 GENE IDENTIFICATION

When genes are specifically identified in a data submission, it is essential that the genes be correctly identified in the ZFIN database. Because gene symbols can change, we request a gene symbol and an additional stable accession number to identify the gene. This is very straightforward if ZDB-GENE IDs are provided. If another sequence accession number is provided, genes are identified first by matches to the provided accession number. Accession numbers for genes that cannot be identified by that method will be run through a set of scripts designed to help curators identify the correct gene in ZFIN using a BLAST analysis process known as the "redundancy pipeline." When the redundancy pipeline fails to locate an existing gene in ZFIN based on the provided sequence accession, a new gene record may be created depending on the nature of the data being submitted. A second BLAST analysis method is then used to help curators determine the best name possible for this new gene based on sequence similarity to mouse and human genes. This is known as the "nomenclature pipeline." These BLAST-based analyses can generate significant work and hence can significantly delay completion of a data load.

5.2 MORPHOLINO, TALEN, CRISPR IDENTIFICATION AND TARGET VALIDATION

Incoming sequences for morpholinos, TALENs, and CRISPRs will be used to identify any that may already have a record in ZFIN. When target genes are not provided in the data submission, morpholino, TALEN, and CRISPR sequences will be used to attempt unambiguous identification of target genes in the ZFIN database. If an exact match is not found for a target gene or a morpholino, TALEN, or CRISPR, a new record for it may be created to support the incoming data depending on the nature of the data being submitted.

5.3 ANATOMY TERM AND STAGE VALIDATION

Anatomy terms in the ZFA have developmental stages from the ZFS assigned to them to indicate when they develop (start stage) and when they disappear (end stage). For expression data, the ZFIN database requires that the developmental stage at which the expression data was collected overlaps with the developmental stages at which the labeled structure is present according to the ZFA and ZFS. This relationship is validated during expression loads and updates are requested for instances that violate this ZFS constraint on the ZFA.

5.4 EQ SYNTAX VALIDATION

There are quality control restrictions that govern which PATO qualities are valid for use with which phenotype entities. For example, it is not permitted to submit an EQ observation for "cell migration":"decreased size" because the PATO term "decreased size" is disallowed for use with GO biological process terms and is only allowed for use with physical entities such as "eye." Although many of these constraints follow common sense, strict validation ensures more consistent results for data integration, computed reasoning, and searches.

5.5 ANTIBODY IDENTIFICATION

When information about a new antibody is being submitted, the ZFIN antibody records are queried to determine whether the antibody already has a record in ZFIN and to confirm that the antibody details provided in the data submission agree with details for the antibody that already exists in ZFIN. When an antibody record in ZFIN cannot be unambiguously identified, a new antibody record may be created to support the data load depending on the nature of the data being submitted.

6. DATA SUBMISSION TEMPLATES

Several pieces of data included in data submissions come from ontologies or constrained lists of valid choices (Table 16). Submission of data that do not come from these term sets creates a challenge during data submissions.

To help promote use of correct ontology terms and the constrained term sets, a Google Spreadsheet has been produced that includes tabs for each of the data files that can be submitted (https://docs.google.com/spreadsheets/d/1p7e6LyxU1wSObD4q8Fon0f6Kf5w-O_xPF-QTmDZSwJc/edit?usp=sharing). To use the spreadsheet,

Table 16 Sources of Valid Values for Data Submissions

Data Type	Source of Valid Values
Anatomy	Zebrafish Anatomy Ontology
Developmental stage	Zebrafish Developmental Stages
Human disease	Human Disease Ontology
Biological processes	Gene Ontology
Experimental condition	Constrained list of experimental conditions (Table 12)
Mutagen	Constrained list of mutagens (Table 6)
Subject	Constrained list of subjects (Table 7)
Phenotype entity	Gene Ontology or Zebrafish Anatomy Ontology
Phenotype quality	Phenotypic Trait Ontology
Genetic background	The list of standard lines at ZFIN (Table 4)

Table 17 Data Submission Files for Each Data Type

Data Type Being Submitted	Data Sheets to Submit	Other Files
Mutants	Mutants/transgenics Genotypes Citations	
Transgenics	Constructs Mutants/transgenics Genotypes Citations	Construct image
Phenotype	Mutants/transgenics Genotypes Citations Constructs Phenotypes	Media files
Expression	Mutants/transgenics Genotypes Citations Constructs Expression	Media files
Morpholinos, TALENs, CRISPRs	MO/TAL/CRSP Citations	
Genome browser tracks	TrackInfo Citations	Track file
Disease models	Mutants/transgenics Genotypes Disease Models Citation	
Antibodies	Antibodies	

log in to a Google account and save a copy of the workbook for your data. Columns that accept values from a constrained set offer a pick list of valid choices to choose from and restrict entry to valid terms only. Columns that accept terms from specific ontologies use the OntoMaton (Maguire, González-Beltrán, Whetzel, Sansone, & Rocca-Serra, 2013) plug-in for Google Spreadsheets to restrict valid choices to terms from the current version of the correct ontology at BioPortal (Whetzel et al., 2011). The OntoMaton plug-in will not be available if the workbook is saved as an Excel file, and you may need to add the Ontomoton plug-in to your copy of the Google spreadsheet before you can access this feature. Once your data are gathered, you can share the files with ZFIN or export to Excel and send them to ZFIN. These basic data templates are offered to help researchers gather the correct data for a submission to ZFIN. Files that may be required for each type of data submission, regardless of how these are produced, are listed in Table 17.

Following these guidelines will help ensure a smooth experience for researchers who wish to submit data to ZFIN.

FUNDING INFORMATION

National Human Genome Research Institute (NIH) grant HG002659.

REFERENCES

Ashburner, M., Ball, C. A., Blake, J. A., Botstein, D., Butler, H., Cherry, J. M., … Sherlock, G. (2000). Gene ontology: tool for the unification of biology. *The Gene Ontology Consortium Nature Genetics, 25*, 25–29.

Chatterjee, A., Ozaki, Y., Stockwell, P. A., Horsfield, J. A., Morison, I. M., & Nakagawa, S. (2013). Mapping the zebrafish brain methylome using reduced representation bisulfitesequencing. *Epigenetics, 8*(9), 979–989.

Eilbeck, K., Lewis, S. E., Mungall, C. J., Yandell, M., Stein, L., Durbin, R., & Ashburner, M. (2005). The Sequence Ontology: a tool for the unification of genome annotations. *Genome Biology, 6*, R44.

Hastings, J., Owen, G., Dekker, A., Ennis, M., Kale, N., Muthukrishnan, V., … Steinbeck, C. (2015). ChEBI in 2016: Improved services and an expanding collection of metabolites. *Nucleic Acids Research, 44*(D1), D1214–D1219.

Howe, D. G., Bradford, Y. M., Conlin, T., Eagle, A. E., Fashena, D., Frazer, K., … Westerfield, M. (2013). ZFIN, the Zebrafish Model Organism Database: increased support for mutants and transgenics. *Nucleic Acids Research, 41*, D854–D860.

Kent, W. J., Sugnet, C. W., Furey, T. S., Roskin, K. M., Pringle, T. H., Zahler, A. M., & Haussler, D. (2002). The human genome browser at UCSC. *Genome Research, 12*, 996–1006.

Kimmel, C. B., Ballard, W. W., Kimmel, S. R., Ullmann, B., & Schilling, T. F. (1995). Stages of embryonic development of the zebrafish. *Developmental Dynamics, 203*, 253–310.

LaFave, M. C., Varshney, G. K., Vemulapalli, M., Mullikin, J. C., & Burgess, S. M. (2014). A defined zebrafish line for high-throughput genetics and genomics: NHGRI-1. *Genetics, 198*, 167–170.

Maguire, E., González-Beltrán, A., Whetzel, P. L., Sansone, S.-A., & Rocca-Serra, P. (2013). OntoMaton: a bioportal powered ontology widget for Google Spreadsheets. *Bioinformatics, 29*, 525–527.

Mungall, C. J., Gkoutos, G. V., Smith, C. L., Haendel, M. A., Lewis, S. E., & Ashburner, M. (2010). Integrating phenotype ontologies across multiple species. *Genome Biology, 11*, R2.

Raney, B. J., Dreszer, T. R., Barber, G. P., Clawson, H., Fujita, P. A., Wang, T., … Kent, W. J. (2014). Track data hubs enable visualization of user-defined genome-wide annotations on the UCSC Genome Browser. *Bioinformatics, 30*, 1003–1005.

Schriml, L. M., & Mitraka, E. (2015). The Disease Ontology: fostering interoperability between biological and clinical human disease-related data. *Mammalian Genome, 26*, 584–589.

Van Slyke, C. E., Bradford, Y. M., Westerfield, M., & Haendel, M. A. (2014). The zebrafish anatomy and stage ontologies: representing the anatomy and development of *Danio rerio*. *Journal of Biomedical Semantics, 5*, 12.

Washington, N. L., Haendel, M. A., Mungall, C. J., Ashburner, M., Westerfield, M., & Lewis, S. E. (2009). Linking human diseases to animal models using ontology-based phenotype annotation. *PLoS Biology, 7*, e1000247.

Whetzel, P. L., Noy, N. F., Shah, N. H., Alexander, P. R., Nyulas, C., Tudorache, T., & Musen, M. A. (2011). BioPortal: enhanced functionality via new Web services from the National Center for Biomedical Ontology to access and use ontologies in software applications. *Nucleic Acids Research, 39*, W541–W545.

CHAPTER

New frontiers for zebrafish management

24

C. Lawrence

Boston Children's Hospital, Boston, MA, United States

E-mail: christian.lawrence@childrens.harvard.edu

CHAPTER OUTLINE

Abstract

The zebrafish (*Danio rerio*) is a preeminent model organism with a wide and expanding utility for numerous scientific disciplines. The same features that once endeared this small freshwater minnow to developmental biologists combined with its relatively high genetic similarity to mammals and the advent of new, more efficient methods for genome editing are now helping to spur expanded growth in its usage in various fields, including toxicology, drug discovery, transplant biology, disease modeling, and even aquaculture. Continued maturation and adoption of the zebrafish model system in these and other fields of science will require that methods and approaches for husbandry and management of these fish in controlled settings be refined and improved to the extent that, ultimately, zebrafish research becomes more reproducible, defined, cost-effective, and accessible to the masses. Knowledge and technology transfer from laboratory animal science and commercial aquaculture will be a necessary part of this development.

Methods in Cell Biology, Volume 135, ISSN 0091-679X, http://dx.doi.org/10.1016/bs.mcb.2016.04.015

INTRODUCTION

The rise of the zebrafish over the past several decades to its place as a preeminent experimental model animal is well documented (Grunwald & Eisen, 2002; Lieschke & Currie, 2007; Santoriello & Zon, 2012). From its early days as an aquarium fish that was sometimes also studied in the laboratory (Laale, 1977), to its current status as an omnipresent complement and/or alternative to the laboratory mouse, the growth in the usage of this diminutive freshwater minnow in basic biomedical research has been driven—and to a large extent defined—by an ever-increasing array of advances in technology and molecular tools with which to study it. Among the most notable and recent of these developments include the completion of the zebrafish genome sequence (Howe et al., 2013), and the emergence of Clustered regularly-interspaced short palindromic repeats/CRISPR associated protein 9 (CRISPR/Cas9)-based genome editing techniques (Auer, Duroure, De Cian, Concordet, & Del Bene, 2014; Hwang, Fu, Reyon, Maeder, Kaini, et al., 2013; Hwang, Fu, Reyon, Maeder, Tsai, et al., 2013; Irion, Krauss, & Nusslein-Volhard, 2014), both of which are driving widespread adoption of the model across diverse scientific disciplines (Bailey, Oliveri, & Levin, 2013; Ribas & Piferrer, 2013; Steenbergen, Richardson, & Champagne, 2011; Tan & Zon, 2011).

Historically, the development of approaches for zebrafish management and husbandry have lagged far behind that of the research (Lawrence, 2007, 2011). Conventional management programs for zebrafish in research settings have long centered on the simple and straightforward goals of producing animals that are free from visible disease, breed well, and survive at rates high enough to complete experiments. The scientific literature, as it relates to published work done with the zebrafish, reflects this in the fact that an overwhelmingly large number of these articles still provide little or no reliable/reproducible information on the maintenance conditions of fish used in experiments. Many articles state that fish were maintained in accordance with "standard conditions" when in fact few to no standards actually exist.

Over the past several years, however, many working in the field have generally become aware of the need to correct these shortcomings, especially as the sophistication and scale of the science being done with the fish has moved beyond the traditional areas of developmental genetics and biology. The overall absence of defined conditions, standards, and control over the environment of laboratory zebrafish has also become more apparent as an increasing number of researchers accustomed to the much higher level of control and definition typical of mice, rats, and other traditional mammalian systems have looked for, but not found, when they begin working with zebrafish. These weaknesses have been pointed out with respect to nutrition (Penglase, Moren, & Hamre, 2012; Watts, Powell, & D'Abramo, 2012), genetic background (K. H. Brown et al., 2012), and pathogen control (Lawrence, Ennis, et al., 2012).

Thus, a paradigm shift is now underway in zebrafish research, one that revolves around the guiding premise that every aspect of the environment of the laboratory zebrafish has the potential to influence its physiology, and consequently, the results of studies conducted with it. Indeed, the concept of defining and limiting the so-called non–protocol-induced variation in experiments with laboratory mice and rats has long been appreciated (Crabbe, Wahlsten, & Dudek, 1999; Garner, Dufour, Gregg, Weisker, & Mench, 2004; Hufeldt, Nielsen, Vogensen, Midtvedt, & Hansen, 2010; Hutchinson, Avery, & Vandewoude, 2005; Schellinck, Cyr, & Brown, 2010; Wainwright, 2001). The same dynamics exists for those working with zebrafish, and some have pointed this fact out, especially with respect to the effect that underlying, subclinical infections may have on experiments (Kent, Harper, & Wolf, 2012). Therefore, the goals of zebrafish husbandry and management need to progress far beyond keeping fish visibly "healthy" and breeding toward the maintenance of similar levels of definition and control over the environment on par with that seen in laboratory mice and rats. The well-established field of laboratory animal science and medicine provides instructive approaches and technology that can help to guide and inform the fish research community in this endeavor.

Other driving forces are helping to shape a new landscape for zebrafish management and husbandry, most notably the increasing scale of infrastructure being constructed to support aquatic research at a growing number of institutions around the world. As the size of systems increases so, also does the demand for greater efficiency. To meet this need, the field is increasingly borrowing from the field of commercial aquaculture, not only for life support equipment but also for management systems and approaches. Indeed, the melding of commercial aquaculture with zebrafish biomedicine represents a space of dynamic exchange with a potential to drive improvements on both sides.

The purpose of this review is to highlight a number of recent advances in various fields as they relate to key aspects of zebrafish management that are helping to shape this new paradigm, and to point out areas where more work is needed.

1. PHYSICOCHEMICAL ENVIRONMENT

The physical and chemical environment, collectively referred to as water quality, is as pervasive a factor as there is in all of zebrafish management. A host of parameters encompass the external milieu of the laboratory zebrafish, including chemical—pH, salinity, alkalinity, hardness, carbon dioxide, dissolved oxygen, and phosphates, and physical—temperature, light intensity, and photoperiod. The collective provision and maintenance of these parameters within a range that supports the growth and development of zebrafish in the laboratory is perhaps the most elemental requirement of any fish research operation.

Yet, beyond this, it should be appreciated that each of the numerous factors that comprise water quality exert numerous and often interrelated impacts on the physiology of the target organism. Numerous examples of environmental factors influencing gene expression and physiology in zebrafish exist in the literature, including dissolved oxygen (Martinovic et al., 2009; Moore Hosey & Bagatto, 2006), temperature (Marvin & Lewis, 2013; McClelland, Craig, Dhekney, & Dipardo, 2006), mineral concentration (Craig, Wood, & Mcclelland, 2010), photoperiod (Ángeles Esteban,

Cuesta, Rodríguez, & Meseguer, 2006; Leonardi & Klempau, 2003; Marvin & Lewis, 2013), and light intensity (Bilotta, 2000; Kaneko & Cahill, 2005; Saszik & Bilotta, 1999; Villamizar, Vera, Foulkes, & Sánchez-Vázquez, 2013). Despite this fact, a great many studies in the literature do not include adequate descriptions of these basic conditions. This means that they are either not measured/known or are simply not being reported. Whenever this sort of underreporting occurs in any experimental situation where water quality has the potential to influence the physiology of the research subject, this can be a major weakness in the science.

The basic solution to this problem is for the authors to provide these data in the material and methods sections, and/or for the reviewers to ask for them during the editorial process. This necessitates that the water quality information is monitored, recorded, and regularly shared by facility staff with the research base, especially in instances where there is a separation between the laboratory doing the work and those maintaining the fish. In many cases, the modern housing systems used in many zebrafish laboratories are designed to continuously monitor and record a wide range of conditions within the systems (Lawrence & Mason, 2012), so considerable data are usually available if needed. The key is to connect this to the science.

One example of how this connection can be made comes from work at Boston Children's Hospital. In their published work on average growth curves and time to first reproduction in zebrafish, Lawrence, Adatto, Best, James, and Maloney (2012) included relatively expansive data on a full range of water quality conditions experienced by the fish during these trials. The power of this study, aside from the information it provided on its own merit, is that it also provides a ready reference for any work done by scientists with zebrafish at Boston Children's Hospital (BCH) going forward; ie, BCH scientists can cite this reference in the material and methods sections of their own articles such that these potentially important variables are known and accounted for in their work as well. This is an example of improved communication between aquaculture and the research program that results in the stronger science; if any parameter has the potential to influence results, it can be accounted for.

2. NUTRITION AND FEEDING

The nutritional state of the fish being used in research is as critical and elemental a factor for the consistency and quality of the science as water quality. As with the chemical and physical nature of the water, the nutritional state of the animal exerts far-reaching and pervasive effects on the fish as a research subject. The importance of limiting variability in the diet of experimental animals has long been understood in the context of murine work (Barnard, Lewis, Teter, & Thigpen, 2009), and there is ample evidence that the nutritional status of the zebrafish exerts profound, diverse, and far-reaching effects on their physiology (Craig, Galus, Wood, & McClelland, 2009; Jury et al., 2008; Lebold et al., 2013; Miller et al., 2014; Robison et al., 2008; Siccardi et al., 2010; Ulloa et al., 2013). Thus, there has been a recent shift in the focus on zebrafish nutrition and feeding from simply providing a diet that supports growth and reproductive function in the laboratory to one that meets those same goals with a much higher degree of efficiency while at the same time limiting variation associated with ingredients and nutrient profile.

Considerable gains in the first area, that of increased efficiency in zebrafish feeding programs, have been made over the past several years, drawing heavily on methodology developed for commercial aquaculture. The production of farmed fish for food production is a rapidly growing, multibillion dollar global industry (FAO, 2014) where a premium is placed on efficiency. This is particularly the case for feeding, since it represents such a major proportion of the typical operating budget of a fish culture operation. The same dynamics exists in zebrafish facilities; feed and particularly the labor involved with feed preparation and delivery represent a significant component of the annual expenses associated with running any zebrafish laboratory.

This fact has prompted reexamination of the traditional feeding protocols. In the past, a common approach involved feeding the fish, especially adults, a mixture of live and formulated feeds up to two to five times per day to support favorable growth rates and reproductive function (Castranova et al., 2011; Gonzales, 2012; Gonzales & Law, 2013; Lawrence, Adatto, et al., 2012; Varga, 2011). However, the drawbacks of this strategy are that it requires high labor inputs, especially in facilities with larger numbers of tanks, and the fact that the application admixtures of multiple diets, both live and formulated, add considerable variability and complexity to the nutritional status of the animals being fed.

There is mounting evidence that this general strategy is suboptimal. First, in one study on feed frequency, it was shown that presenting zebrafish a formulated feed at 5% of body weight once per day, from the subadult stage onward, supported statistically similar rates of growth and reproduction as when fish received the same amount of diet spread out over three or five feedings per day (Lawrence, Best, James, & Maloney, 2012).

The implications of this work are significant for several reasons. First, the demonstration that zebrafish—at least from the subadult stage onward—do not absolutely require more than one feeding per day represents a potentially significant labor savings, especially as the scale of a given operation is increased. Secondly, it also showed that the performance of fish fed a single formulated diet matches or even exceeds the performance of fish fed a more conventional mixture of live and formulated feeds, based on published results in the literature (Castranova et al., 2011; Gonzales, 2012; Markovich, Rizzuto, & Brown, 2007). The practice of using a single food source for zebrafish, which has since been further supported by other publications in the literature (Kaushik, Georga, & Koumoundouros, 2011; Siccardi et al., 2009) is notable because it simplifies nutritional inputs and avoids some of the problems associated with live diets, at least for adult fish. *Artemia nauplii*, the most common live feed used in zebrafish facilities, is expensive, variable in price, labor intensive to culture, and inherently unstable nutritionally because it is wild harvested. There is also a further efficiency gain resulting from a move toward using formulated feeds because they are typically much more dense than live prey items of a similar size.

The application of a single formulated feed is also a big step forward in terms of reducing unwanted variation in experiments associated with nutrition, ingredients, and in some cases, antinutritional factors. This is because a formulated diet is more easily (1) quantified in terms of the amount of feed going to the fish each day (ie, can be fed according to a percentage of body weight), and (2) quality controlled in terms of its ingredients and nutrient profiles. This approach promotes better definition and consistency of the nutritional state of the zebrafish used in experiments, which is a critical first step toward standardizing the diet for the model (Watts et al., 2012).

A necessary part of this movement toward greater efficiency and definition in diet is the determination of basic nutritional requirements for this species. Surprisingly, despite nearly 40 years of accumulated scientific work on the zebrafish, relatively little is known about its nutritional requirements (Lawrence, 2007; Siccardi et al., 2009). Again, information of this nature has been established for murine models and for many commercial aquaculture species, and is a critical and necessary component of standardization.

A collection of nutritional studies published over the past decade have provided the basic framework for determining zebrafish dietary requirements for protein (Hedrera et al., 2013; Smith et al., 2013), polyunsaturated fatty acids (Jaya-Ram, Kuah, Lim, Kolkovski, & Shu-Chien, 2008; Kaushik, Georga, & Koumoundouros, 2011; Powell et al., 2015), vitamins (Alsop, Matsumoto, Brown, & Van Der Kraak, 2008; Kirkwood et al., 2012; Lebold et al., 2013; Miller et al., 2012)', minerals (Craig et al., 2009; Siccardi et al., 2010), and carbohydrates (Robison et al., 2008). However, the field has yet to conduct the kind of systematic studies required to determine the daily nutritional requirements of the fish that will promote health and reproduction. Once this is done, these data can be utilized to help develop an appropriate open formulation standardized reference diet that can be used in the field.

3. BREEDING

Much of the zebrafish research being conducted today still depends much on the timely and consistent production of high-quality embryos for experiments. The growing understanding of the reproductive behaviors of the fish in nature and captivity (Darrow & Harris, 2004; Gerlach, 2006; Kurtzman, Craig, Grizzle, & Hove, 2010; Santos et al., 2007; Sessa et al., 2008; Spence, Fatema, et al., 2006; Spence, Jordan, & Smith, 2006; Spence & Smith, 2005; Uusi-Heikkil, Kuparinen, Wolter, Meinelt, & Arlinghaus, 2012) has fueled innovations in breeding techniques and equipment. One notable example is the development of the iSpawn (Adatto, Lawrence, Thompson, & Zon, 2011), a specialized spawning tank designed to capitalize on the natural preference of the fish to spawn in shallow water (Sessa et al., 2008). This has helped improve the throughput of certain kinds of experiments, such as chemical screens (Tan & Zon, 2011).

There is still much to learn about zebrafish reproductive behavior, and this information needs to be leveraged to make further improvements to breeding protocols and apparatuses. For example, there has been a long-standing debate in the field about the need for provision of environmental complexity in breeding setups, namely, by the addition of plastic plants. Arguments in favor of the practice were based only upon inferences made from observations that female fish in the wild prefer to oviposit in vegetated sites (Spence, Ashton, & Smith, 2007). New data now provide direct evidence of this relationship. Wafer and colleagues compared breeding performance (fecundity and fertility) between fish spawned in breeding setups with versus without plastic plants, and found that both parameters were significantly improved when plants were provided. Importantly, these authors also found that the type of plant material used had differential effects, depending upon the age of the fish being spawned. This work illustrates the complexity of behaviors that were long oversimplified by those working in the field. More data are needed to help further inform practices and design, such that the reproductive potential of the model is fully realized.

4. GENETIC MANAGEMENT

The genetic background of the research animal is yet another deeply pervasive variable that has the potential to influence experiments in myriad and often unforeseen ways. Intrinsic genetic forces, such as mutation and drift, tend to accumulate in animal populations over time, and can ultimately compromise the reproducibility of experimental data if they are not limited, defined, and controlled. The importance of this issue has long been identified and prioritized in the research community with respect to laboratory mice and rats, and has prompted the development of an international set of genetic standards for these species in research (Taft, Davisson, & Wiles, 2006). This has in turn spurred the prodigious growth of commercial suppliers that can readily supply research-grade, genetically defined animals to the research market.

Despite the rapid growth of the model system over the past several decades, no such options currently exist for the zebrafish research community. Indeed, the importance of genetic background in zebrafish work is still largely overlooked. This premise is supported by a cursory review of the literature; strain information is often only marginally provided in publications, and in the manner it is reported, there appears to be an implicit assumption that a given wild-type strain is the same, across time and space. In reality, because no true commercial suppliers of defined strains (inbred or outbred) exist, and wild-type strains that are the background of the thousands of new mutant and transgenic strains being created are propagated at individual institutions with little to no programmatic or defined interchange with others (Lawrence, 2011); what the field is essentially working with is hundreds, if not many thousands of small, isolated subpopulations that are only becoming more disparate from one another over time, even within the same "strain." A number of studies in the recent literature support this (Bowen, Henke, Siegfried, Warman, & Harris, 2012; Leshchiner et al., 2012; Obholzer et al., 2012).

Mounting evidence is beginning to help define the risk that the lack of genetic standardization poses for the integrity, scope, and future growth of zebrafish science. Brown and coauthors found significant levels of genomic diversity and substructuring between and within different laboratory strains of zebrafish in their analysis of copy number variants in zebrafish, and raised the possibility that such variability could compromise the utility of the zebrafish as a model for human disease if not accounted for and/or controlled (K. H. Brown et al., 2012). The impacts of strain differences on experimental outcomes are now well-characterized in toxicology (A. R. Brown et al., 2012; Coe et al., 2009), behavior (Vignet et al., 2013; Wright, Nakamichi, Krause, & Butlin, 2006), and other fields (Link, Gray, Smith, & John, 2004; Meyer, Froehlich, Galt, & Biga, 2013).

Recently, efforts have been made to address these shortcomings. Shinya and Sakai (2011) were able to establish a highly homogeneous inbred strain of zebrafish (IM) by 16 generations of a serial sib-pair mating scheme. Although it is unclear as to whether these authors were able to successfully complete the full 20-generation inbreeding scheme necessary to meet the definition of a classically inbred strain as defined in mouse and medaka, these efforts highlight that it might be possible to create additional and more robust versions of these strains in the future. This is a step above the homogeneous strains now available in the field created by gynogenesis (Mizgireuv & Revskoy, 2006; Mizgirev & Revskoy, 2010; Nechiporuk, Finney, Keating, & Johnson, 1999), in terms of genetic uniformity, but it is not yet clear that the classical problems of productivity, immune function, and sex ratio skewing has been solved.

In the meantime, LaFave, Varshney, Vemulapalli, Mullikin, and Burgess (2014) have come up with a potential solution to the problem of uncontrolled and unknown genetic diversity in strains by creating a new defined line of fish, NHGRI-1, that was derived from a cross founder of TAB-5 (TU × AB) parents and sequenced to a depth of 50×. The results of this analysis identified and documented regions of the genome in these fish that match the Zv9 reference sequence as well as over 10 million previously unreported single nucleotide variants. Consequently, the use of a strain with this level of genomic definition allows for an unprecedented level of control for any type of research that requires knowledge of possible sequence variation. Examples of this kind of work are numerous, but in particular are of benefit to genome targeting/modification technologies, such as CRISPRs.

The task ahead for the field is to find ways to maintain, expand, and distribute these resources for and across the field. To that end, it is likely that the time for commercial production of genetically defined and quality controlled research-grade animals is drawing near.

5. LARVICULTURE

The ability to grow fish from the embryonic stage to adulthood with high rates of survival and growth defines the success or failure of zebrafish research operations at the most elemental level. Larval rearing is perhaps the greatest operational challenge that managers face, especially those new to the field. Although a variety of "conventional" practices can be employed to rear fish to adulthood, the efficiency with which this can be done depends on a delicate balance of water quality and feeding. The age old challenge of zebrafish larviculture (and most cultured fish species), especially during the first feeding stages, is to balance the demands of the larval fish to have highly nutritious, digestible, prey items (live or inanimate) available nearly continuously in small amounts without adversely compromising water quality (Harper & Lawrence, 2010). Over the years, numerous methods have been employed that meet these demands with varying degrees of success (Carvalho & Arau, 2006; Hensley & Leung, 2010; Nusslein-Volhard & Dahm, 2002; Varga, 2011; Westerfield, 2007; Wilson, 2012). However, there is now a growing movement in the field to standardize zebrafish larviculture and make it more high throughput and efficient, drawing heavily upon methodologies and technology employed in commercial aquaculture.

The most prominent of these advances revolve around the usage of the marine rotifer (*Brachionus plicatilis*) as the primary feed item for first feeding fish. Rotifers are among the most important live feeds in all of aquaculture, due to a variety of favorable attributes, including small size, efficient gut loading, visual and chemical attractiveness, and ease of controlled mass culture (Lavens & Sorgeloos, 1996). For many years, conventional wisdom in the zebrafish community was that rotifers were too difficult to culture to be able to systemically employ them as a first feed in a typical zebrafish facility. However, this changed when Best and coauthors published what is now known as the zebrafish–rotifer polyculture method in 2010 (Best, Adatto, Cockington, James, & Lawrence, 2010). In this method, first feeding zebrafish and marine rotifers are grown together in static, brackish water enriched with microalgal paste. The environment created by this admixture is ideal for larval fish; the rotifers act as a continuous, on-demand source of nutrition for the fish as they grow without adversely compromising water quality. This results in very high rates of growth and survival, and rapidly transitions the fish to the point where they can be transitioned to environments with higher rates of recirculating flow and can be presented with larger prey items that help drive rapid growth to the juvenile stages (Best et al., 2010). The wide-scale adoption of this approach in the community has been bolstered by the development of rotifer production systems that can be readily set up in zebrafish facilities, regardless of the scale of the operation (Lawrence, Sanders, & Henry, 2012).

This methodology has been since been improved upon in different ways. In 2015, Aoyama and coauthors showed that the freshwater rotifer (*Brachionus calyciflorus*) can be used in place of saltwater rotifers in a similar culturing method, potentially offering an advantage over the use of saltwater rotifers because it does not require that the fish transition from brackish to freshwater conditions as they move beyond the first feeding stage (Aoyama et al., 2015).

Rotifer culturing methods have also been improved (Lawrence et al., 2016) to the extent that production levels of rotifers can be high enough to also sustain fish growth over the entire larval period of development, beyond the first few days of exogenous feeding. This allows for the effective elimination of *A. nauplii* from the diet, which reduces the costs and complexity of feeding protocols without compromising growth performance (Lawrence, James, & Mobley, 2015).

These developments in larviculture are now allowing an increasing number of laboratories to dramatically increase throughput in their nurseries in a cost-effective manner. This is important to help drive the production of new mutant and transgenic lines, especially in light of the emergence of enhanced genome modification techniques, such as CRISPRs and Transcription activator-like effector nucleases (TALENs).

6. HEALTH MANAGEMENT

Knowledge of the pathogen status of fish used in research has numerous and important implications for the science. Such detailed understanding of the health status of the fish populations in research facilities is useful on a number of levels. Most proximately, it helps managers to take steps to prevent disease outbreaks that could result in catastrophic losses of fish, which is an omnipresent risk in zebrafish facilities, due to the high rate of exchange of fish lines between laboratories worldwide, and the emergence of new pathogens, such as *Edwardsiella ictaluri*, which is highly pathogenic to zebrafish (Hawke et al., 2013). There is also an occupational health risk posed by certain pathogens known to infect zebrafish, such as *Mycobacterium marinum* (Ang, Rattana-Apiromyakij, & Goh, 2000; Ostland et al., 2008; Watral & Kent, 2007), and *Aeromonas hydrophila* (Cipriano, 2001; Pullium, Dillehay, & Webb, 1999) that have zoonotic potential. Finally, and perhaps most importantly, it is critical to define the health status of fish populations used in research because many diseases of zebrafish often present as subclinical infections but still have the potential to impact the physiology of the animals in ways that could compromise experiments (Kent et al., 2009, 2012; Lawrence, Ennis, et al., 2012).

Current understanding of the common diseases of zebrafish has dramatically increased over the past 5 years, concomitant with the expansion of the model system into new and disparate areas of research. A number of new agents have been identified and characterized, including *E. ictaluri* (Hawke et al., 2013), *Pleistophora hyphessobryconis* (Sanders et al., 2010), and several additional species of *Mycobacterium* (Whipps, Lieggi, & Wagner, 2012). At the same time, the most common and well-known diseases of zebrafish, such as *Pseudoloma neurophilia* (Murray et al., 2011; Sanders & Kent, 2011; Sanders, Watral, Clarkson, & Kent, 2013), *Pseudocapillaria tomentosa* (Collymore et al., 2014; Maley, Laird, Rinkwitz, & Becker, 2013), *M. marinum* (Broussard et al., 2009; Peterson et al., 2013; Tobin & Ramakrishnan, 2008) and *M. chelonae* (Peterson et al., 2013; Whipps, Matthews, & Kent, 2008) have been more fully characterized in their biology, pathogenesis, and treatability in research colonies of zebrafish.

Despite all these data, the community has been slow to adopt health management strategies that will help limit the spread of these agents within and across facilities and their impact on science. A relatively small number of facilities employ sentinel programs (Lawrence, Ennis, et al., 2012), and the overwhelming attitude toward disease is titled toward prevention against catastrophic losses, but not on its effect on research as a source of non–protocol-induced variation.

However, this dynamic is the one that poses the greatest risk to the future growth and utility of the zebrafish model system. One major difference between the zebrafish and murine research model systems is an appreciation of this fact. It is well known that subclinical infections have the potential to compromise results of experiments in murine models (Lipman & Homberger, 2003; Lipman & Perkins, 2002), and therefore, numerous and far-reaching efforts have been made to define, reduce, and eliminate rodent pathogens from research colonies. This movement is only in its infancy in zebrafish systems, but progress has been made.

The first step toward pushing the field in this direction is direct evidence that subclinical infections can influence the physiology of the fish. While publications in the literature have been generally been sparse, this is changing. In 2015, Spagnoli and coauthors demonstrated that subclinical infection of zebrafish by *P. neurophilia* resulted in quantifiable differences between the startle response of infected versus noninfected zebrafish (Spagnoli, Xue, & Kent, 2015). This work illustrates the potential negative impact that subclinical infection can have on research; if researchers working on zebrafish behavior are unaware that their fish may be infected with this agent, they are ignoring a potential source of variation that could influence their results and the reproducibility of these experiments in future work.

Therefore, it is clear that the community should be working toward the universal goal of defining, reducing, and ultimately eliminating pathogens from research populations of zebrafish. This is no easy proposition, but it is possible. It will first require whole-scale adoption of a set of international standards for health monitoring of zebrafish research facilities. This is now made more possible with the advent of polymerase chain reaction (PCR)–based screening techniques for many common pathogens (Whipps & Kent, 2006; Whipps et al., 2012). These assays can also be used to screen pooled embryos, as well as environmental samples, such as biofilms, sludge, water, and different feed-associated items. At the same time, a number of commercial diagnostic laboratories now offer readily available and rapid screening services. All these factors can be leveraged into the development of effective screening procedures that will help define the pathogen landscape of research populations.

Prevention of the spread of pathogens will also require improvements in quarantine procedures. The standard "bleached eggs only" policy utilized by many in the fish community has weaknesses, namely, that egg disinfection protocols are often ineffective at eliminating certain agents (Ferguson, Watral, Schwindt, & Kent, 2007; Mainous Smith, 2005; Sanders et al., 2013). New potentially more effective methods for embryo disinfection are being developed, as illustrated in the work done in 2015 by Chang and coauthors that showed that povidone-iodine was more effective at reducing certain *Mycobacterium* species than bleach (Chang, Colicino, DiPaola, Al-Hasnawi, & Whipps, 2015). The fact that quarantine fish and/or their embryos can also be screened by PCR also gives managers additional options to improve quarantine procedures so that the introduction of pathogens from outside sources is reduced.

Once pathogens have been identified in populations, it can be very difficult to eliminate them from existing zebrafish facilities, especially *Mycobacterium* species that reside in biofilms (Mohammad Hem, Shukla, Whitaker, Arnold, & Shahamat, 2007), and the protozoan parasites like *Pseudoloma* and *Pleistophora* with infective spores that are highly resistant to chlorine and other disinfectants (Ferguson et al., 2007). Still, treatment options do exist for some of the common agents. The nematode gut parasite, *P. tomentosa*, can be effectively treated and eliminated from laboratory zebrafish populations using antihelminthic drugs (Collymore et al., 2014; Maley et al., 2013). The field can look to commercial aquaculture for additional potential therapeutic options for zebrafish; a number of drugs and vaccines have been developed by the industry to treat common diseases in fish farms, and the potential to do the same for zebrafish populations should be explored where appropriate.

Ultimately, the field will require the establishment, maintenance, and distribution of pathogen-free stocks of animals to be used in experiments that require that subjects be free from specific or multiple diseases. For one pathogen at least, specific pathogen-free (SPF) stocks have been developed. Kent and coauthors developed a protocol for producing *P. neurophilia*-free stocks of fish (Kent et al., 2011). This program has been successful in supporting ongoing work in toxicology at the Oregon State University, and has been expanded to include a number of different strains of fish that have been made available to the community. This is so far the only example of SPF zebrafish in the field, and ultimately the broad-scale implementation of this approach will require involvement from commercial suppliers as well as improvement in surveillance, therapeutics, and housing infrastructure to maintain them over the long term.

7. HOUSING/AQUACULTURE

The advent of commercially produced aquaculture systems underpinned the growth of the zebrafish model system over the past several decades (Lawrence & Mason, 2012). In recent years, the industry has seen a shift in the nature and type of housing and filtration equipment less typical of hobbyist-level aquariums to that of commercial aquaculture, especially as the scale of systems have increased and the need for improved cost efficiency has risen. The advent of feeding practices borrowed from the aquaculture industry, such as the application of pelleted feeds at elevated feeding rates, has also pushed the industry in this direction.

One result of this movement is a shift toward a reduction in the use of consumable components for filtration, especially for the removal of suspended solids. Increasingly, the usage of bag filters and filter pads has been reduced, and replaced by backwashing sand/bead filters and/or rotating drum filters. These technologies are far more effective and efficient at removal of suspended solids from aquaculture systems. Both are also easily automated, reducing the amount of labor required for maintenance.

Automation in commercially produced systems has also been greatly expanded with respect to the monitoring of environmental parameters. Many commercial systems now include automated control and monitoring options that give managers far greater control over the environment within their systems, and also facilitates the collection of vast amounts of data concerning the chemical and physical parameters of the water. This is helping to contribute to far better definition of the water quality landscape in experiments.

Moving forward, the field will require further modification of system design necessitated by the employment of more intensive feeding and management programs. Solids removal has been improved at the level of systems, but has remained largely unchanged at the tank level. The traditional design of zebrafish tanks—facilitated to quickly remove solid wastes out along the bottom of tanks by the use of a siphoning baffle—has always been based on conventional feeding paradigms that consisted of a mixture of live diets and flake feeds. This approach, coupled with the square shape of zebrafish tanks (which is almost never seen anywhere else in aquaculture) is not necessarily compatible with the increase in the application of larger amounts of pelletized feeds now going into tanks, which results in an accumulation of solids behind the baffle and the consequent interruption of solids transport out of the tank and into the system. The industry will likely need to reexamine these designs as fish culturing techniques continue to become more intensive. Commercial aquaculture will once again prove to be a source of inspiration in this area.

The same concept also applies to pathogen control. The prevalent design of housing systems in the zebrafish market over the past two decades has increasingly been toward the construction of increasingly large recirculating systems with thousands of tanks and shared and integrated life support and filtration. This approach favors space and labor efficiency, but does not facilitate pathogen control. If a pathogen gets into a large, shared system, it automatically compromises the entire unit and once established, can be very difficult, if not impossible to eradicate. Now, especially with the increase in the prevalence of research on adult fish, especially with models of human disease that have variable immune status (ie, tumor-producing lines or transplantation models), there is a growing need for greater compartmentalization of systems. From this standpoint, smaller systems that can be isolated from one another to house discrete groups of animals used in different experiments would be preferable. This way, systems could be readily broken down and sanitized between experimental events. Employment of flow-through systems would also enhance flexibility in this regard. These developments would be critical in maintaining the pathogen-free status of fish for the duration of work, something that is typically not possible in conventional systems.

8. WELFARE

The dramatic rise in numbers of zebrafish used in science over the past decade has fueled increasing interest in their welfare in the laboratory (Reed & Jennings,

2011; Smith, 2014; Volpato, 2009). For many years, the behavior and welfare of zebrafish was largely overlooked and undermanaged, mostly because the conventional understanding of these aspects of the animal was based primarily on grossly oversimplified and anecdotal information derived from the hobbyist industry. Over the years, especially as the fish has become a more popular behavioral research subject, the realities have proven to be far more nuanced and complex; zebrafish are social animals with a rich and complex repertoire of behaviors that need to be understood and managed to promote their well-being in experimental settings.

Effective management of zebrafish welfare in the laboratory requires both a basic understanding of their normal behavior and its consequences, especially in captivity, and the development and employment of methods to measure their welfare (Lawrence, 2011). Current knowledge of zebrafish behavior comes from basic research in the laboratory, especially from behavioral researchers who have worked to develop models of anxiety, stress, and other syndromes (Filby, Paull, Hickmore, & Tyler, 2010; Maximino et al., 2010; Norton & Bally-Cuif, 2010; Steenbergen et al., 2011). At the same time, a small but growing number of field-based studies of the zebrafish in natural habitats have also contributed substantially to this knowledge (Arunachalam, Raja, Vijayakumar, Malaiammal, & Mayden, 2013; Engeszer, Patterson, Rao, & Parichy, 2007; Parichy, 2015; Suriyampola et al., 2015). Together, these data can collectively be used to construct a definition of what "normal behavior" is for this species.

Methods for assessing well-being in zebrafish range from simple, direct observations of behavior (Lawrence, 2011); to measures of various performance indices, such as growth, survival, and reproductive success (Castranova et al., 2011); to more complex biochemical and physiological analyses (Filby, Paull, Bartlett, Look, & Tyler, 2010). The most common of the third approach involves taking indirect measures of stress through quantification of cortisol, the primary stress hormone, in the subject animal or animals (Ellis et al., 2011). A number of recent studies have reported on the levels of circulating cortisol in zebrafish under a range of different experimental conditions (Canavello et al., 2011; Gronquist & Berges, 2013; Parker, Millington, Combe, & Brennan, 2012; Pavlidis et al., 2013; Ramsay et al., 2006). Although the results of cortisol-based studies are often variable, there is little question that this is the most reliable tool for measuring welfare, as long as the results are interpreted properly. Recent, noninvasive methods for collecting cortisol from water (Félix, Faustino, Cabral, & Oliveira, 2013), and from scales (Aerts et al., 2015), will likely lead to the development of more informative assays, especially the latter approach, which is currently the only way to measure stress over extended periods.

Importantly, this improved understanding of behavior and tools to measure welfare can be used to more effectively evaluate and inform practices and policy. There is a strong drive in the regulatory arena to provide environmental enrichment for laboratory zebrafish. Although the intentions behind this movement are good, the field needs to proceed with caution, since it has been demonstrated that enrichment has the potential to influence the science in unpredictable ways (Bayne, 2005).

9. EDUCATION AND TRAINING

The increasing complexity and scale of zebrafish research is driving not only the expansion of infrastructure to support it but also the variety and depth of knowledge and skills required to manage it. There is an acute and growing need in the field for specialists in "biomedical aquaculture" to staff the growing number of zebrafish facilities of today.

Although there are numerous people graduating from undergraduate and graduate programs in aquaculture, marine biology, and aquatic ecology, with requisite academic training to work in zebrafish facilities, very few of them posses sufficient understanding of both the aquaculture and the biomedical aspect of the zebrafish system. At the same time, there are thousands of people already staffing existing zebrafish facilities around the world who are still in need of continuing education and training to help keep them abreast of the latest technology and scientific developments.

There are now a growing number of resources that can meet the demand for training. For example, the joint initiative between the University of Alabama–Birmingham and Gadsden State Community college to develop on online certification program in biomedical aquaculture (http://www.gadsdenstate.edu/academics/science/an/zebrafish.php) is an example of a dedicated effort to provide virtual classroom training in many of the fundamentals of research aquaculture facility management for technicians, managers, investigators, and veterinarians working in this space. This program also offers a weeklong hands-on workshop that includes practical coursework and lectures on system design and maintenance, live feed production, necropsies, and fish breeding, taught by experts from around the United States and beyond. Other resources that provide online, on-demand training also exist, including the E-quarium course offered by a private company in Australia (http://www.theaquariumvet.com.au/pages/view/the_aquarium_vet_1), and the Experimental Fish course (http://www.theaquariumvet.com.au/pages/view/the_aquarium_vet_1) organized by the University of Prince Edward Island in Canada.

Weeklong training workshops in fish health and management are also offered at the MDI Biological Laboratory (https://mdibl.org/course/hcmlf-2016/) in Bar Harbor, Maine; by the International Zebrafish Husbandry Course in Milan, Italy; and by the International Zebrafish and Medaka Course organized by the Karlsruhe Institute of Technology in Germany (https://izmc.ezrc.kit.edu/english/index.php). These programs, along with other similar efforts are helping to train fish facility managers across the world, and are raising the standard of professionalism to match the growing demands of the field.

10. COMMERCIAL PROVISION OF MODEL INFRASTRUCTURAL SERVICES

The expansion of the zebrafish model system is also creating a demand for a wide range of goods and services to support it. Indeed, one of the primary differences between the murine and zebrafish model systems is that the former sector has an international, multibillion dollar industry of commercial entities built up around it that serves to provide infrastructural support to the research market. Collectively, these companies work to provide a wide range of goods and services that ultimately make mouse and rat research convenient, robust, and widely accessible to the scientific community. At this point, this type of commercially provided infrastructure is only marginally available for the zebrafish research market, but this is rapidly changing.

Over the past several years, a number of specialized companies have appeared on the scene to offer professional-level services to zebrafish researchers that up until this point have had to be produced organically, at the academic and institutional levels. As the field continues to expand, this internal provision of services will become less efficient and self-limiting. For example, the production of research-grade, genetically defined, pathogen-free animals will almost certainly not be possible without support from the industry. A wide range of products, ranging from housing systems, diagnostics, cryopreservation services, and many others, all have the potential to help the fish research model system grow and mature, but only if they are commercially produced. Indeed, many of the management challenges described in this review will be met only with contributions from this nascent services sector. One only needs to look to the mouse and rat to see how commercial entities have helped these model systems to grow and serve the needs of the scientific community, and how integral they are to their continued success. The next decade of growth in the zebrafish model system will undoubtedly be characterized by the same dynamics.

SUMMARY

These are exciting times for the zebrafish model system. The model is experiencing, and will continue to experience, rapid and wide-scale adoption across many fields of science in the next decade and beyond. The scale and rate of this expansion will depend on the continued development of improved management strategies that ultimately achieve greater levels of definition, control, and standardization of myriad environmental factors that influence the experimental landscape of the fish.

REFERENCES

Adatto, I., Lawrence, C., Thompson, M., & Zon, L. I. (2011). A new system for the rapid collection of large numbers of developmentally staged zebrafish embryos. *PLoS One, 6*(6), e21715. http://dx.doi.org/10.1371/journal.pone.0021715.

Aerts, J., Metz, J. R., Ampe, B., Decostere, A., Flik, G., & De Saeger, S. (2015). Scales tell a story on the stress history of fish. *PLoS One, 10*(4), e0123411. http://dx.doi.org/10.1371/journal.pone.0123411.

Alsop, D., Matsumoto, J., Brown, S., & Van Der Kraak, G. (2008). Retinoid requirements in the reproduction of zebrafish. *General and Comparative Endocrinology, 156*(1), 51–62. Retrieved from: http://www.ncbi.nlm.nih.gov/entrez/query.fcgi?cmd=Retrieve&db=PubMed&dopt=Citation&list_uids=18158153.

Ang, P., Rattana-Apiromyakij, N., & Goh, C. L. (2000). Retrospective study of *Mycobacterium marinum* skin infections. *International Journal of Dermatology, 39*(5), 343–347.

Ángeles Esteban, M., Cuesta, A., Rodríguez, A., & Meseguer, J. (2006). Effect of photoperiod on the fish innate immune system: a link between fish pineal gland and the immune system. *Journal of Pineal Research, 41*, 261–266. http://dx.doi.org/10.1111/j.1600-079X.2006.00362.x.

Aoyama, Y., Moriya, N., Tanaka, S., Taniguchi, T., Hosokawa, H., & Maegawa, S. (2015). A novel method for rearing zebrafish by using freshwater rotifers (*Brachionus calyciflorus*). *Zebrafish*. http://dx.doi.org/10.1089/zeb.2014.1032, 150504071741004.

Arunachalam, M., Raja, M., Vijayakumar, C., Malaiammal, P., & Mayden, R. L. (2013). Natural history of zebrafish (*Danio rerio*) in India. *Zebrafish, 10*, 1–14. http://dx.doi.org/10.1089/zeb.2012.0803.

Auer, T. O., Duroure, K., De Cian, A., Concordet, J.-P., & Del Bene, F. (2014). Highly efficient CRISPR/Cas9-mediated knock-in in zebrafish by homology-independent DNA repair. *Genome Research, 24*(1), 142–153. http://dx.doi.org/10.1101/gr.161638.113.

Bailey, J., Oliveri, A., & Levin, E. D. (2013). Zebrafish model systems for developmental neurobehavioral toxicology. *Birth Defects Research Part C — Embryo Today: Reviews*. http://dx.doi.org/10.1002/bdrc.21027.

Barnard, D. E., Lewis, S. M., Teter, B. B., & Thigpen, J. E. (2009). Open- and closed-formula laboratory animal diets and their importance to research. *Journal of the American Association for Laboratory Animal Science: JAALAS, 48*, 709–713.

Bayne, K. (2005). Potential for unintended consequences of environmental enrichment for laboratory animals and research results. *ILAR Journal, 46*(2), 129–139.

Best, J., Adatto, I., Cockington, J., James, A., & Lawrence, C. (2010). A novel method for rearing first-feeding larval zebrafish: polyculture with Type L saltwater rotifers (*Brachionus plicatilis*). *Zebrafish, 7*(3), 289–295. http://dx.doi.org/10.1089/zeb.2010.0667.

Bilotta, J. (2000). Effects of abnormal lighting on the development of zebrafish visual behavior. *Behavioural Brain Research, 116*(1), 81–87. http://dx.doi.org/10.1016/S0166-4328(00)00264-3.

Bowen, M. E., Henke, K., Siegfried, K. R., Warman, M. L., & Harris, M. P. (2012). Efficient mapping and cloning of mutations in zebrafish by low-coverage whole-genome sequencing. *Genetics, 190*, 1017–1024. http://dx.doi.org/10.1534/genetics.111.136069.

Broussard, G. W., Norris, M. B., Schwindt, A. R., Fournie, J. W., Winn, R. N., Kent, M. L., & Ennis, D. G. (2009). Chronic *Mycobacterium marinum* infection acts as a tumor promoter in Japanese Medaka (*Oryzias latipes*). *Comparative Biochemistry and Physiology. Toxicology & Pharmacology: CBP*, 152–160. http://dx.doi.org/10.1016/j.cbpc.2008.09.011.

Brown, A. R., Bickley, L. K., Ryan, T. A., Paull, G. C., Hamilton, P. B., Owen, S. F., … Tyler, C. R. (2012). Differences in sexual development in inbred and outbred zebrafish (*Danio rerio*) and implications for chemical testing. *Aquatic Toxicology (Amsterdam, Netherlands), 112–113*, 27–38. http://dx.doi.org/10.1016/j.aquatox.2012.01.017.

Brown, K. H., Dobrinski, K. P., Lee, A. S., Gokcumen, O., Mills, R. E., Shi, X., … Lee, C. (2012). Extensive genetic diversity and substructuring among zebrafish strains revealed through copy number variant analysis. *Proceedings of the National Academy of Sciences of the United States of America, 109*(2), 529–534. http://dx.doi.org/10.1073/pnas.1112163109.

Canavello, P. R., Cachat, J. M., Beeson, E. C., Laffoon, A. L., Grimes, C., Haymore, W. A. M., … Kalueff, A. V. (2011). Measuring endocrine (cortisol) responses of zebrafish to stress. *Neuromethods, 51*, 135–142. http://dx.doi.org/10.1007/978-1-60761-953-6_11.

Carvalho, P., & Arau, L. (2006). Rearing zebrafish (*Danio rerio*) larvae without live food: evaluation of a commercial, a practical and a purified starter diet on larval performance. *Aquaculture Research*, 1107–1111. http://dx.doi.org/10.1111/j.1365-2109.2006.01534.x.

Castranova, D., Lawton, A., Lawrence, C., Baumann, D. P., Best, J., Coscolla, J., … Weinstein, B. M. (2011). The effect of stocking densities on reproductive performance in laboratory zebrafish (*Danio rerio*). *Zebrafish, 8*(3), 141–146. http://dx.doi.org/10.1089/zeb.2011.0688.

Chang, C. T., Colicino, E. G., DiPaola, E. J., Al-Hasnawi, H. J., & Whipps, C. M. (2015). Evaluating the effectiveness of common disinfectants at preventing the propagation of *Mycobacterium* spp. isolated from zebrafish. *Comparative Biochemistry and Physiology. Toxicology & Pharmacology: CBP, 178*, 45–50. http://dx.doi.org/10.1016/j.cbpc.2015.09.008.

Cipriano, R. C. (2001). *Aeromonas hydrophila* and motile aeromonad septicemias of fish. In *Fish disease leaflet 68*. Washington: D. C. 20240 United States Department of the Interior Fish and Wildlife Service Division of Fishery Research.

Coe, T. S., Hamilton, P. B., Griffiths, A. M., Hodgson, D. J., Wahab, M. A., & Tyler, C. R. (2009). Genetic variation in strains of zebrafish (*Danio rerio*) and the implications for ecotoxicology studies. *Ecotoxicology, 18*(1), 144–150. http://dx.doi.org/10.1007/s10646-008-0267-0. Retrieved from: http://www.ncbi.nlm.nih.gov/entrez/query.fcgi?cmd=Retrieve&db=PubMed&dopt=Citation&list_uids=18795247.

Collymore, C., Watral, V., White, J. R., Colvin, M. E., Rasmussen, S., Tolwani, R. J., & Kent, M. L. (2014). Tolerance and efficacy of emamectin benzoate and ivermectin for the treatment of *Pseudocapillaria tomentosa* in laboratory zebrafish (*Danio rerio*). *Zebrafish, 11*(5), 490–497. http://dx.doi.org/10.1089/zeb.2014.1021.

Crabbe, J. C., Wahlsten, D., & Dudek, B. C. (1999). Genetics of mouse behavior: interactions with laboratory environment. *Science (New York, NY), 284*(5420), 1670–1672. http://dx.doi.org/10.1126/science.284.5420.1670.

Craig, P. M., Galus, M., Wood, C. M., & McClelland, G. B. (2009). Dietary iron alters waterborne copper-induced gene expression in soft water acclimated zebrafish (*Danio rerio*). *American Journal of Physiology-Regulatory Integrative and Comparative Physiology, 296*(2), R362–R373. http://dx.doi.org/10.1152/ajpregu.90581.2008.

Craig, P. M., Wood, C. M., & Mcclelland, G. B. (2010). Water chemistry alters gene expression and physiological end points of chronic waterborne copper exposure in zebrafish, *Danio rerio*. *Environmental Science & Technology, 44*(6), 2156–2162. http://dx.doi.org/10.1021/es902995q.

Darrow, K. O., & Harris, W. A. (2004). Characterization and development of courtship in zebrafish, *Danio rerio*. *Zebrafish, 1*(1), 40–45. Retrieved from: http://www.ncbi.nlm.nih.gov/entrez/query.fcgi?cmd=Retrieve&db=PubMed&dopt=Citation&list_uids=18248204.

Ellis, T., Yavuzcan, H., Lo, J., Tort, L., Øverli, Ø., & Martins, C. I. M. (2011). Cortisol and finfish welfare. *Fish Physiology and Biochemistry*. http://dx.doi.org/10.1007/s10695-011-9568-y.

Engeszer, R. E., Patterson, L. B., Rao, A. A., & Parichy, D. M. (2007). Zebrafish in the wild: a review of natural history and new notes from the field. *Zebrafish, 4*(1), 21–40. Retrieved from: http://www.ncbi.nlm.nih.gov/entrez/query.fcgi?cmd=Retrieve&db=PubMed&dopt=Citation&list_uids=18041940.

FAO. (2014). The state of world fisheries and aquaculture. *Food and Agriculture Oraganization of the United Nations, 2014*, 92-5-105177-1.

Félix, A. S., Faustino, A. I., Cabral, E. M., & Oliveira, R. F. (2013). Noninvasive measurement of steroid hormones in zebrafish holding-water. *Zebrafish, 10*, 110–115. http://dx.doi.org/10.1089/zeb.2012.0792.

Ferguson, J. A., Watral, V., Schwindt, A. R., & Kent, M. L. (2007). Spores of two fish microsporidia (*Pseudoloma neurophilia* and *Glugea anomala*) are highly resistant to chlorine. *Diseases of Aquatic Organisms, 76*(3), 205–214.

Filby, A. L., Paull, G. C., Bartlett, E. J., Van Look, K. J. W., & Tyler, C. R. (2010). Physiology & Behavior Physiological and health consequences of social status in zebra fish (*Danio rerio*). *Physiology & Behavior*. http://dx.doi.org/10.1016/j.physbeh.2010.09.004.

Filby, A. L., Paull, G. C., Hickmore, T. F., & Tyler, C. R. (2010). Unravelling the neurophysiological basis of aggression in a fish model. *BMC Genomics, 11*, 498.

Garner, J. P., Dufour, B., Gregg, L. E., Weisker, S. M., & Mench, J. A. (2004). Social and husbandry factors affecting the prevalence and severity of barbering ("whisker trimming") by laboratory mice. *Applied Animal Behaviour Science, 89*, 263–282. http://dx.doi.org/10.1016/j.applanim.2004.07.004.

Gerlach, G. (2006). Pheromonal regulation of reproductive success in female zebrafish: female suppression and male enhancement. *Animal Behaviour, 72*, 1119–1124. http://dx.doi.org/10.1016/j.anbehav.2006.03.009.

Gonzales, J. M. (2012). Preliminary evaluation on the effects of feeds on the growth and early reproductive performance of zebrafish (*Danio rerio*). *Journal of the American Association for Laboratory Animal Science : JAALAS, 51*, 412–417.

Gonzales, J. M., & Law, S. H. W. (2013). Feed and feeding regime affect growth rate and gonadosomatic index of adult zebrafish (*Danio rerio*). *Zebrafish, 10*, 532–540. http://dx.doi.org/10.1089/zeb.2013.0891.

Gronquist, D., & Berges, J. A. (2013). Effects of aquarium-related stressors on the zebrafish: a comparison of behavioral, physiological, and biochemical indicators. *Journal of Aquatic Animal Health, 25*(1), 53–65. http://dx.doi.org/10.1080/08997659.2012.747450.

Grunwald, D. J., & Eisen, J. S. (2002). Headwaters of the zebrafish emergence of a new model vertebrate. *Nature Reviews Genetics, 3*, 7–24.

Harper, C., & Lawrence, C. (2010). *The laboratory zebrafish (laboratory animal pocket reference)*. CRC Press.

Hawke, J. P., Kent, M., Rogge, M., Baumgartner, W., Wiles, J., Shelley, J., … Peterson, T. S. (2013). Edwardsiellosis caused by *Edwardsiella ictaluri* in laboratory populations of zebrafish *Danio rerio*. *Journal of Aquatic Animal Health, 25*(3), 171–183. http://dx.doi.org/10.1080/08997659.2013.782226.

Hedrera, M. I., Galdames, J. A., Jimenez-Reyes, M. F., Reyes, A. E., Avendaño-Herrera, R., Romero, J., & Feijóo, C. G. (2013). Soybean meal induces intestinal inflammation in zebrafish larvae. *PLoS One, 8*, e69983. http://dx.doi.org/10.1371/journal.pone.0069983.

Hensley, M. R., & Leung, Y. F. (2010). A convenient dry feed for raising zebrafish larvae. *Zebrafish, 7*, 219–231. http://dx.doi.org/10.1089/zeb.2010.0652.

Howe, K., Clark, M. D., Torroja, C. F., Torrance, J., Berthelot, C., Muffato, M., … Stemple, D. L. (2013). The zebrafish reference genome sequence and its relationship to the human genome. *Nature, 496*(7446), 498–503. http://dx.doi.org/10.1038/nature12111.

Hufeldt, M. R., Nielsen, D. S., Vogensen, F. K., Midtvedt, T., & Hansen, A. K. (2010). Variation in the gut microbiota of laboratory mice is related to both genetic and environmental factors. *Comparative Medicine, 60*(5), 336–342.

Hutchinson, E., Avery, A., & Vandewoude, S. (2005). Environmental enrichment for laboratory rodents. *ILAR Journal/National Research Council, Institute of Laboratory Animal Resources, 46*(2), 148–161. http://dx.doi.org/10.1093/ilar.46.2.148.

Hwang, W. Y., Fu, Y., Reyon, D., Maeder, M. L., Kaini, P., Sander, J. D., … Yeh, J. R. J. (2013). Heritable and precise zebrafish genome editing using a CRISPR-Cas system. *PLoS One, 8*(7), 1–9. http://dx.doi.org/10.1371/journal.pone.0068708.

Hwang, W. Y., Fu, Y., Reyon, D., Maeder, M. L., Tsai, S. Q., Sander, J. D., … Joung, J. K. (2013). Efficient genome editing in zebrafish using a CRISPR-Cas system. *Nature Biotechnology, 31*(3), 227–229. http://dx.doi.org/10.1038/nbt.2501.

Irion, U., Krauss, J., & Nusslein-Volhard, C. (2014). Precise and efficient genome editing in zebrafish using the CRISPR/Cas9 system. *Development, 141*(24), 4827–4830. http://dx.doi.org/10.1242/dev.115584.

Jaya-Ram, A., Kuah, M. K., Lim, P. S., Kolkovski, S., & Shu-Chien, A. C. (2008). Influence of dietary HUFA levels on reproductive performance, tissue fatty acid profile and desaturase and elongase mRNAs expression in female zebrafish *Danio rerio*. *Aquaculture, 277*(3–4), 275–281. http://dx.doi.org/10.1016/j.aquaculture.2008.02.027.

Jury, D. R., Kaveti, S., Duan, Z. H., Willard, B., Kinter, M., & Londraville, R. (2008). Effects of calorie restriction on the zebrafish liver proteome. *Comparative Biochemistry and Physiology – Part D: Genomics and Proteomics, 3*(4), 275–282. http://dx.doi.org/10.1016/j.cbd.2008.07.003.

Kaneko, M., & Cahill, G. M. (2005). Light-dependent development of circadian gene expression in transgenic zebrafish. *PLoS Biology, 3*(2), e34. http://dx.doi.org/10.1371/journal.pbio.0030034.

Kaushik, S., Georga, I., & Koumoundouros, G. (2011). Growth and body composition of zebrafish (*Danio rerio*) larvae fed a compound feed from first feeding onward: toward implications on nutrient requirements. *Zebrafish, 8*, 87–95. http://dx.doi.org/10.1089/zeb.2011.0696.

Kent, M. L., Buchner, C., Watral, V. G., Sanders, J. L., Ladu, J., Peterson, T. S., & Tanguay, R. L. (2011). Development and maintenance of a specific pathogen-free (SPF) zebrafish research facility for *Pseudoloma neurophilia*. *Diseases of Aquatic Organisms, 95*(1), 73–79. http://dx.doi.org/10.3354/dao02333.

Kent, M. L., Feist, S. W., Harper, C., Hoogstraten-Miller, S., Law, J. M., Sanchez-Morgado, J. M., … Whipps, C. M. (2009). Recommendations for control of pathogens and infectious diseases in fish research facilities. *Comparative Biochemistry and Physiology. Toxicology & Pharmacology: CBP, 149*(2), 240–248. http://dx.doi.org/10.1016/j.cbpc.2008.08.001. pii:S1532-0456(08)00148-8.

Kent, M. L., Harper, C., & Wolf, J. C. (2012). Documented and potential research impacts of subclinical diseases in zebrafish. *ILAR Journal/National Research Council, Institute of Laboratory Animal Resources, 53*(2), 126–134. http://dx.doi.org/10.1093/ilar.53.2.126.

Kirkwood, J. S., Lebold, K. M., Miranda, C. L., Wright, C. L., Miller, G. W., Tanguay, R. L., … Stevens, J. F. (2012). Vitamin C deficiency activates the purine nucleotide cycle in zebrafish. *Journal of Biological Chemistry*. http://dx.doi.org/10.1074/jbc.M111.316018.

Kurtzman, M. S., Craig, M. P., Grizzle, B. K., & Hove, J. R. (2010). Sexually segregated housing results in improved early larval survival in zebrafish. *Lab Animal, 39*, 183–189. http://dx.doi.org/10.1038/laban0610-183.

Laale, H. W. (1977). Biology and use of zebrafish, *Brachydanio rerio* in fisheries research - literature-review. *Journal of Fish Biology, 10*(2), 121.

LaFave, M. C., Varshney, G. K., Vemulapalli, M., Mullikin, J. C., & Burgess, S. M. (2014). A defined zebrafish line for high-throughput genetics and genomics: NHGRI-1. *Genetics, 198*, 167–170. http://dx.doi.org/10.1534/genetics.114.166769.

Lavens, P., & Sorgeloos, P. (1996). *Manual on the production and use of live food for aquaculture*. Vasa. Retrieved from: http://medcontent.metapress.com/index/A65RM03P4874243N.pdf http://www.cabdirect.org/abstracts/19971409717.html.

Lawrence, C. (2007). The husbandry of zebrafish (*Danio rerio*): a review. *Aquaculture, 269*(1–4), 1–20. http://dx.doi.org/10.1016/j.aquaculture.2007.04.077.

Lawrence, C. (2011). Advances in zebrafish husbandry and management. *Methods in Cell Biology, 104*, 429–451. http://dx.doi.org/10.1016/B978-0-12-374814-0.00023-9.

Lawrence, C. (2011). Environmental enrichment and the zebrafish. *The Enrichment Record*, 11–15.

Lawrence, C., Adatto, I., Best, J., James, A., & Maloney, K. (2012). Generation time of zebrafish (*Danio rerio*) and medakas (*Oryzias latipes*) housed in the same aquaculture facility. *Lab Animal, 41*(6), 158–165. http://dx.doi.org/10.1038/laban0612-158.

Lawrence, C., Best, J., James, A., & Maloney, K. (2012). The effects of feeding frequency on growth and reproduction in zebrafish (*Danio rerio*). *Aquaculture, 368–369*, 103–108. http://dx.doi.org/10.1016/j.aquaculture.2012.09.022.

Lawrence, C., Best, J., Cockington, J., Henry, E. C., Hurley, S., James, A., … Sanders, E. (2016). The complete and updated "Rotifer Polyculture Method" for rearing first feeding zebrafish. *Journal of Visualized Experiments: JoVE*, (107), e53629. http://dx.doi.org/10.3791/53629.

Lawrence, C., Ennis, D. G., Harper, C., Kent, M. L., Murray, K., & Sanders, G. E. (2012). The challenges of implementing pathogen control strategies for fishes used in biomedical research. *Comparative Biochemistry and Physiology – C Toxicology and Pharmacology, 155*, 160–166. http://dx.doi.org/10.1016/j.cbpc.2011.06.007.

Lawrence, C., James, A., & Mobley, S. (2015). Successful replacement of *Artemia salina* nauplii with marine rotifers (*Brachionus plicatilis*) in the diet of preadult zebrafish (*Danio rerio*). *Zebrafish, 12*(5), 366–371. http://dx.doi.org/10.1089/zeb.2015.1118.

Lawrence, C., & Mason, T. (2012). Zebrafish housing systems: a review of basic operating principles and considerations for design and functionality. *ILAR Journal*. http://dx.doi.org/10.1093/ilar.53.2.179.

Lawrence, C., Sanders, E., & Henry, E. (2012). Methods for culturing saltwater rotifers (*Brachionus plicatilis*) for rearing larval zebrafish. *Zebrafish, 9*, 140–146. http://dx.doi.org/10.1089/zeb.2012.0771.

Lebold, K. M., Lohr, C. V., Barton, C. L., Miller, G. W., Labut, E. M., Tanguay, R. L., & Traber, M. G. (2013). Chronic vitamin e deficiency promotes vitamin C deficiency in zebrafish leading to degenerative myopathy and impaired swimming behavior. *Comparative Biochemistry and Physiology – C Toxicology and Pharmacology, 157*, 382–389. http://dx.doi.org/10.1016/j.cbpc.2013.03.007.

Leonardi, M. O., & Klempau, A. E. (2003). Artificial photoperiod influence on the immune system of juvenile rainbow trout (*Oncorhynchus mykiss*) in the Southern Hemisphere. *Aquaculture, 221*, 581–591. http://dx.doi.org/10.1016/S0044-8486(03)00032-2.

Leshchiner, I., Alexa, K., Kelsey, P., Adzhubei, I., Austin-Tse, C. A., Cooney, J. D., … Sunyaev, S. R. (2012). Mutation mapping and identification by whole-genome sequencing. *Genome Research, 22*, 1541–1548. http://dx.doi.org/10.1101/gr.135541.111.

Lieschke, G. J., & Currie, P. D. (2007). Animal models of human disease: zebrafish swim into view. *Nature Reviews Genetics, 5*, 353–367.

Link, B. A., Gray, M. P., Smith, R. S., & John, S. W. (2004). Intraocular pressure in zebrafish: comparison of inbred strains and identification of a reduced melanin mutant with raised IOP. *Investigative Ophthalmology & Visual Science, 45*(12), 4415–4422. Retrieved from: http://www.ncbi.nlm.nih.gov/entrez/query.fcgi?cmd=Retrieve&db=PubMed&dopt=Citation&list_uids=15557450.

Lipman, N. S., & Homberger, F. R. (2003). Rodent quality assurance testing: use of sentinel animal systems. *Lab Animal, 32*(5), 36–43. http://dx.doi.org/10.1038/laban0503-36.

Lipman, N. S., & Perkins, S. N. (2002). Factors that may influence animal research. In L. C. Anderson, J. Fox, F. M. Loew, & F. W. Quimby (Eds.), *Laboratory animal medicine*. Orlando: Academic Press.

Mainous Smith, S. M. (2005). Efficacy of common disinfectants against *Mycobacterium marinum*. *Journal of Aquatic Health, 17*, 284–288.

Maley, D., Laird, A. S., Rinkwitz, S., & Becker, T. S. (2013). A simple and efficient protocol for the treatment of zebrafish colonies infected with parasitic nematodes. *Zebrafish, 10*(3), 447–450. http://dx.doi.org/10.1089/zeb.2013.0868.

Markovich, M. L., Rizzuto, N. V., & Brown, P. B. (2007). Diet affects spawning in zebrafish. *Zebrafish, 4*(1), 69–74. Retrieved from: http://www.ncbi.nlm.nih.gov/entrez/query.fcgi?cmd=Retrieve&db=PubMed&dopt=Citation&list_uids=18041944.

Martinovic, D., Villeneuve, D. L., Kahl, M. D., Blake, L. S., Brodin, J. D., & Ankley, G. T. (2009). Hypoxia alters gene expression in the gonads of zebrafish (*Danio rerio*). *Aquatic Toxicology, 95*(4), 258–272. Retrieved from: http://www.ncbi.nlm.nih.gov/entrez/query.fcgi?cmd=Retrieve&.

Marvin, G. A., & Lewis, M. (2013). Effect of temperature, photoperiod, and feeding on the rate of tail regeneration in a semiaquatic plethodontid salamander. *Journal of Thermal Biology, 38*(8), 548–552. http://dx.doi.org/10.1016/j.jtherbio.2013.09.008.

Maximino, C., de Brito, T. M., da Silva Batista, A. W., Herculano, A. M., Morato, S., & Gouveia, A. (2010). Measuring anxiety in zebrafish: a critical review. *Behavioural Brain Research*. http://dx.doi.org/10.1016/j.bbr.2010.05.031.

McClelland, G. B., Craig, P. M., Dhekney, K., & Dipardo, S. (2006). Temperature- and exercise-induced gene expression and metabolic enzyme changes in skeletal muscle of adult zebrafish (*Danio rerio*). *Journal of Physiology, 577*(Pt 2), 739–751. Retrieved from: http://www.ncbi.nlm.nih.gov/entrez/query.fcgi?cmd=Retrieve&db=PubMed&dopt=Citation&list_uids=16990399.

Meyer, B. M., Froehlich, J. M., Galt, N. J., & Biga, P. R. (2013). Inbred strains of zebrafish exhibit variation in growth performance and myostatin expression following fasting. *Comparative Biochemistry and Physiology: Part A, Molecular & Integrative Physiology, 164*, 1–9. http://dx.doi.org/10.1016/j.cbpa.2012.10.004.

Miller, G. W., Labut, E. M., Lebold, K. M., Floeter, A., Tanguay, R. L., & Traber, M. G. (2012). Zebrafish (*Danio rerio*) fed vitamin E-deficient diets produce embryos with increased morphologic abnormalities and mortality. *Journal of Nutritional Biochemistry, 23*, 478–486. http://dx.doi.org/10.1016/j.jnutbio.2011.02.002.

Miller, G. W., Truong, L., Barton, C. L., Labut, E. M., Lebold, K. M., Traber, M. G., & Tanguay, R. L. (2014). The influences of parental diet and vitamin E intake on the embryonic zebrafish transcriptome. *Comparative Biochemistry and Physiology – Part D: Genomics and Proteomics, 10*, 22–29. http://dx.doi.org/10.1016/j.cbd.2014.02.001.

Mizgireuv, I. V., & Revskoy, S. Y. (2006). Transplantable tumor lines generated in clonal zebrafish. *Cancer Research, 66*(6), 3120–3125. Retrieved from: http://www.ncbi.nlm.nih.gov/entrez/query.fcgi?cmd=Retrieve&db=PubMed&dopt=Citation&list_uids=16540662.

Mizgirev, I., & Revskoy, S. (2010). Generation of clonal zebrafish lines and transplantable hepatic tumors. *Nature Protocols, 5*, 383–394. http://dx.doi.org/10.1038/nprot.2010.8.

Mohammad Hem, A., Shukla, D., Whitaker, B., Arnold, J., & Shahamat, M. R. (2007). Attachment and biofilm formation of *Mycobacterium marinum* on a hydrophobic surface at the air interface. *World Journal of Microbiology & Biotechnology, 23*, 93–101.

Moore Hosey, M., & Bagatto, B. F. B. (2006). Cardiovascular system in larval zebrafish responds to developmental hypoxia in a family specific manner. *Frontiers in Zoology, 15*, 3–4.

Murray, K. N., Dreska, M., Nasiadka, A., Rinne, M., Matthews, J. L., Carmichael, C., … Westerfield, M. (2011). Transmission, diagnosis, and recommendations for control of *Pseudoloma neurophilia* infections in laboratory zebrafish (*Danio rerio*) facilities. *Comparative Medicine, 61*(4), 322–329.

Nechiporuk, A., Finney, J. E., Keating, M. T., & Johnson, S. L. (1999). Assessment of polymorphism in zebrafish mapping strains. *Genome Research, 9*, 1231–1238. http://dx.doi.org/10.1101/gr.9.12.1231.

Norton, W., & Bally-Cuif, L. (2010). Adult zebrafish as a model organism for behavioural genetics. *BMC Neuroscience, 11*, 90. http://dx.doi.org/10.1186/1471-2202-11-90.

Nusslein-Volhard, C., & Dahm, R. (2002). *Zebrafish, a practical approach*. Oxford: Oxford University Press.

Obholzer, N., Swinburne, I. A., Schwab, E., Nechiporuk, A. V., Nicolson, T., & Megason, S. G. (2012). Rapid positional cloning of zebrafish mutations by linkage and homozygosity mapping using whole-genome sequencing. *Development, 139*, 4280–4290. http://dx.doi.org/10.1242/dev.083931.

Ostland, V. E., Watral, V., Whipps, C. M., Austin, F. W., St-Hilaire, S., Westerman, M. E., & Kent, M. L. (2008). Biochemical, molecular, and virulence characteristics of select *Mycobacterium marinum* isolates in hybrid striped bass *Morone chrysops* × *M saxatilis* and zebrafish *Danio rerio*. *Diseases of Aquatic Organisms, 79*(2), 107–118. http://dx.doi.org/10.3354/dao01891.

Parichy, D. M. (2015). Advancing biology through a deeper understanding of zebrafish ecology and evolution. *eLife, 4*, e05635. http://dx.doi.org/10.7554/eLife.05635.

Parker, M. O., Millington, M. E., Combe, F. J., & Brennan, C. H. (2012). Housing conditions differentially affect physiological and behavioural stress responses of zebrafish, as well as the response to anxiolytics. *PLoS One, 7*. http://dx.doi.org/10.1371/journal.pone.0034992.

Pavlidis, M., Digka, N., Theodoridi, A., Campo, A., Barsakis, K., Skouradakis, G., … Tsalafouta, A. (2013). Husbandry of zebrafish, *Danio rerio*, and the cortisol stress response. *Zebrafish, 10*, 524–531. http://dx.doi.org/10.1089/zeb.2012.0819.

Penglase, S., Moren, M., & Hamre, K. (2012). Lab animals: standardize the diet for zebrafish model. *Nature*. http://dx.doi.org/10.1038/491333a.

Peterson, T. S., Ferguson, J. A., Watral, V. G., Mutoji, K. N., Ennis, D. G., & Kent, M. L. (2013). *Paramecium caudatum* enhances transmission and infectivity of *Mycobacterium marinum* and *M. chelonae* in zebrafish *Danio rerio*. *Diseases of Aquatic Organisms, 106*, 229–239. http://dx.doi.org/10.3354/dao02649.

Powell, M. L., Pegues, M. A., Szalai, A. J., Ghanta, V. K., Abramo, L. R. D., & Watts, S. A. (2015). Effects of the dietary ω 3: ω 6 fatty acid ratio on body fat and inflammation in zebrafish (*Danio rerio*). *Comparative Medicine, 65*(4), 289–294.

Pullium, J. K., Dillehay, D. L., & Webb, S. (1999). High mortality in zebrafish (*Danio rerio*). *Contemporary Topics in Laboratory Animal Science/American Association for Laboratory Animal Science, 38*(3), 80–83.

Ramsay, J. M., Feist, G. W., Varga, Z. M., Westerfield, M., Kent, M. L., & Schreck, C. B. (2006). Whole-body cortisol is an indicator of crowding stress in adult zebrafish, *Danio rerio*. *Aquaculture, 258*(1–4), 565–574. http://dx.doi.org/10.1016/j.aquaculture.2006.04.020.

Reed, B., & Jennings, M. (2011). *Guidance on the housing and care of zebrafish* Danio rerio (pp. 1–27). Research Animals Department, Science Group, RSPCA.

Ribas, L., & Piferrer, F. (2013). The zebrafish (*Danio rerio*) as a model organism, with emphasis on applications for finfish aquaculture research. *Reviews in Aquaculture*. http://dx.doi.org/10.1111/raq.12041.

Robison, B. D., Drew, R. E., Murdoch, G. K., Powell, M., Rodnick, K. J., Settles, M., … Hardy, R. W. (2008). Sexual dimorphism in hepatic gene expression and the response to dietary carbohydrate manipulation in the zebrafish (*Danio rerio*). *Comparative Biochemistry and Physiology: Part D, Genomics & Proteomics, 3*(2), 141–154. http://dx.doi.org/10.1016/j.cbd.2008.01.001.

Sanders, J. L., & Kent, M. L. (2011). Development of a sensitive assay for the detection of *Pseudoloma neurophilia* in laboratory populations of the zebrafish *Danio rerio*. *Diseases of Aquatic Organisms, 96*(2), 145–156. http://dx.doi.org/10.3354/dao02375.Development.

Sanders, J. L., Lawrence, C., Nichols, D. K., Brubaker, J. F., Peterson, T. S., Murray, K. N., & Kent, M. L. (2010). *Pleistophora hyphessobryconis* (Microsporidia) infecting zebrafish *Danio rerio* in research facilities. *Diseases of Aquatic Organisms, 91*(1), 47–56. http://dx.doi.org/10.3354/dao02245.

Sanders, J. L., Watral, V., Clarkson, K., & Kent, M. L. (2013). Verification of intraovum transmission of a microsporidium of vertebrates: *Pseudoloma neurophilia* infecting the zebrafish, *Danio rerio*. *PLoS One, 8*(9), e76064. http://dx.doi.org/10.1371/journal.pone.0076064.

Santoriello, C., & Zon, L. I. (2012). Hooked! modeling human disease in zebrafish. *Journal of Clinical Investigation*. http://dx.doi.org/10.1172/JCI60434.

Santos, E. M., Workman, V. L., Paull, G. C., Filby, A. L., Van Look, K. J. W., Kille, P., & Tyler, C. R. (2007). Molecular basis of sex and reproductive status in breeding zebrafish. *Physiological Genomics, 30*(2), 111–122. http://dx.doi.org/10.1152/physiolgenomics.00284.2006.

Saszik, S., & Bilotta, J. (1999). Effects of abnormal light-rearing conditions on retinal physiology in larvae zebrafish. *Investigative Ophthalmology & Visual Science, 40*(12), 3026–3031.

Schellinck, H. M., Cyr, D. P., & Brown, R. E. (2010). Chapter 7 – how many ways can mouse behavioral experiments go wrong? Confounding variables in mouse models of neurodegenerative diseases and how to control them. *Advances in the Study of Behavior, 41*. http://dx.doi.org/10.1016/S0065-3454(10)41007-4.

Sessa, A. K., White, R., Houvras, Y., Burke, C., Pugach, E., Baker, B., … Zon, L. I. (2008). The effect of a depth gradient on the mating behavior, oviposition site preference, and embryo production in the zebrafish, *Danio rerio*. *Zebrafish, 5*(4), 335–339. http://dx.doi.org/10.1089/zeb.2008.0535.

Shinya, M., & Sakai, N. (2011). Generation of highly homogeneous strains of zebrafish through full sib-pair mating. *G3 (Bethesda, MD), 1*, 377–386. http://dx.doi.org/10.1534/g3.111.000851.

Siccardi, A. J., Garris, H. W., Jones, W. T., Moseley, D. B., D'Abramo, L. R., & Watts, S. A. (2009). Growth and survival of zebrafish (*Danio rerio*) fed different commercial and laboratory diets. *Zebrafish, 6*, 275–280. http://dx.doi.org/10.1089/zeb.2008.0553. Retrieved from: http://www.ncbi.nlm.nih.gov/entrez/query.fcgi?cmd=Retrieve&db=PubMed&dopt=Citation&list_uids=19566408.

Siccardi, A. J., Padgett-Vasquez, S., Garris, H. W., Nagy, T. R., D'Abramo, L. R., & Watts, S. A. (2010). Dietary strontium increases bone mineral density in intact zebrafish (*Danio rerio*): a potential model system for bone research. *Zebrafish, 7*, 267–273. http://dx.doi.org/10.1089/zeb.2010.0654.

Smith, D. L., Barry, R. J., Powell, M. L., Nagy, T. R., D'Abramo, L. R., & Watts, S. A. (2013). Dietary protein source influence on body size and composition in growing zebrafish. *Zebrafish, 10*, 439–446. http://dx.doi.org/10.1089/zeb.2012.0864.

Smith, S. A. (2014). Welfare of laboratory fishes. *In Laboratory Animal Welfare*, 301–311.

Spagnoli, S., Xue, L., & Kent, M. L. (2015). The common neural parasite *Pseudoloma neurophilia* is associated with altered startle response habituation in adult zebrafish (*Danio rerio*): implications for the zebrafish as a model organism. *Behavioural Brain Research, 291*, 351–360. http://dx.doi.org/10.1016/j.bbr.2015.05.046.

Spence, R., Ashton, R., & Smith, C. (2007). Oviposition decisions are mediated by spawning site quality in wild and domesticated zebrafish, *Danio rerio*. *Behaviour, 144*, 953–966.

Spence, R., Fatema, M. K., Reichard, M., Huq, K. A., Wahab, M. A., Ahmed, Z. F., & Smith, C. (2006). The distribution and habitat preferences of the zebrafish in Bangladesh. *Journal of Fish Biology, 69*(5), 1435–1448. http://dx.doi.org/10.1111/j.1095-8649.2006.01206.x.

Spence, R., Jordan, W. C., & Smith, C. (2006). Genetic analysis of male reproductive success in relation to density in the zebrafish, *Danio rerio*. *Frontiers in Zoology, 3*, 5. Retrieved from: http://www.ncbi.nlm.nih.gov/entrez/query.fcgi?cmd=Retrieve&db=PubMed&dopt=Citation&list_uids=16597325.

Spence, R., & Smith, C. (2005). Male territoriality mediates density and sex ratio effects on oviposition in the zebrafish, *Danio rerio*. *Animal Behaviour, 69*, 1317–1323. http://dx.doi.org/10.1016/j.anbehav.2004.10.010.

Steenbergen, P. J., Richardson, M. K., & Champagne, D. L. (2011). The use of the zebrafish model in stress research. *Progress in Neuro-Psychopharmacology & Biological Psychiatry, 35*, 1432–1451. http://dx.doi.org/10.1016/j.pnpbp.2010.10.010.

Suriyampola, P. S., Shelton, D. S., Shukla, R., Roy, T., Bhat, A., & Martins, E. P. (2015). Zebrafish social behavior in the wild. *Zebrafish, 00*(00). http://dx.doi.org/10.1089/zeb.2015.1159. zeb.2015.1159.

Taft, R. A., Davisson, M., & Wiles, M. V. (2006). Know thy mouse. *Trends in Genetics: TIG, 22*, 649–653. http://dx.doi.org/10.1016/j.tig.2006.09.010.

Tan, J. L., & Zon, L. I. (2011). Chemical screening in zebrafish for novel biological and therapeutic discovery. *Methods in Cell Biology, 105*, 493–516. http://dx.doi.org/10.1016/B978-0-12-381320-6.00021-7.

Tobin, D. M., & Ramakrishnan, L. (2008). Comparative pathogenesis of *Mycobacterium marinum* and *Mycobacterium tuberculosis*. *Cellular Microbiology, 10*(5), 1027–1039. http://dx.doi.org/10.1111/j.1462-5822.2008.01133.x.

Ulloa, P. E., Peña, A. A., Lizama, C. D., Araneda, C., Iturra, P., Neira, R., & Medrano, J. F. (2013). Growth response and expression of muscle growth-related candidate genes in adult zebrafish fed plant and fishmeal protein-based diets. *Zebrafish, 10*, 99–109. http://dx.doi.org/10.1089/zeb.2012.0823.

Uusi-Heikkil, S., Kuparinen, A., Wolter, C., Meinelt, T., & Arlinghaus, R. (2012). Paternal body size affects reproductive success in laboratory-held zebrafish (*Danio rerio*). *Environmental Biology of Fishes, 93*(4), 461–474. http://dx.doi.org/10.1007/s10641-011-9937-5.

Varga, Z. M. (2011). Aquaculture and husbandry at the zebrafish international resource center. *Methods in Cell Biology, 104*, 453–478. http://dx.doi.org/10.1016/B978-0-12-374814-0.00024-0.

Vignet, C., Bégout, M.-L., Péan, S., Lyphout, L., Leguay, D., & Cousin, X. (2013). Systematic screening of behavioral responses in two zebrafish strains. *Zebrafish, 10*, 365–375. http://dx.doi.org/10.1089/zeb.2013.0871.

Villamizar, N., Vera, L. M., Foulkes, N. S., & Sánchez-Vázquez, F. J. (2013). Effect of lighting conditions on zebrafish growth and development. *Zebrafish, 11*(2). http://dx.doi.org/10.1089/zeb.2013.0926.

Volpato, G. L. (2009). Challenges in assessing fish welfare. *ILAR Journal, 50*(4), 329–337.

Wainwright, P. E. (2001). The role of nutritional factors in behavioural development in laboratory mice. *Behavioural Brain Research, 125*, 75–80. http://dx.doi.org/10.1016/S0166-4328(01)00286-8.

Watral, V., & Kent, M. L. (2007). Pathogenesis of *Mycobacterium* spp. in zebrafish (*Danio rerio*) from research facilities. *Comparative Biochemistry and Physiology. Toxicology & Pharmacology: CBP*, 55–60. http://dx.doi.org/10.1016/j.cbpc.2006.06.004.

Watts, S. A., Powell, M., & D'Abramo, L. R. (2012). Fundamental approaches to the study of zebrafish nutrition. *ILAR Journal, 53*, 144–160. http://dx.doi.org/10.1093/ilar.53.2.144.

Westerfield, M. (2007). *The zebrafish book. A guide for the laboratory use of zebrafish (*Danio rerio*)* (4th ed.). Eugene, OR: University of Oregon Press.

Whipps, C. M., & Kent, M. L. (2006). Polymerase chain reaction detection of *Pseudoloma neurophilia*, a common microsporidian of zebrafish (*Danio rerio*) reared in research laboratories. *Journal of the American Association for Laboratory Animal Science, 45*(1), 36–39.

Whipps, C. M., Lieggi, C., & Wagner, R. (2012). Mycobacteriosis in zebrafish colonies. *ILAR Journal/National Research Council, Institute of Laboratory Animal Resources, 53*(2), 95–105. http://dx.doi.org/10.1093/ilar.53.2.95.

Whipps, C. M., Matthews, J. L., & Kent, M. L. (2008). Distribution and genetic characterization of *Mycobacterium chelonae* in laboratory zebrafish *Danio rerio*. *Diseases of Aquatic Organisms, 82*(1), 45–54. http://dx.doi.org/10.3354/dao01967.

Wilson, C. (2012). Aspects of larval rearing. *ILAR Journal/National Research Council, Institute of Laboratory Animal Resources*. http://dx.doi.org/10.1093/ilar.53.2.169.

Wright, D., Nakamichi, R., Krause, J., & Butlin, R. K. (2006). QTL analysis of behavioral and morphological differentiation between wild and laboratory zebrafish (*Danio rerio*). *Behavior Genetics, 36*(2), 271–284. http://dx.doi.org/10.1007/s10519-005-9029-4.

CHAPTER

25 Aquaculture, husbandry, and shipping at the Zebrafish International Resource Center

Z.M. Varga

University of Oregon, Eugene, OR, United States

E-mail: zoltan@zebrafish.org

CHAPTER OUTLINE

Methods in Cell Biology, Volume 135, ISSN 0091-679X, http://dx.doi.org/10.1016/bs.mcb.2016.01.007

Abstract

This chapter provides an overview of the Zebrafish International Resource Center (ZIRC) aquaculture systems, husbandry, and live fish shipping methods. The ZIRC has a recirculating water system with mechanical and biological filtration in its main fish facility, and a flow-through water system for its quarantine room. I describe basic husbandry methods for breeding, rearing larvae, feeding and shipping of fish. Whereas the procedures presented here are merely an overview and only the most essential methods are included, they offer starting points to set up, develop, refine, or troubleshoot methods in other fish facilities.

INTRODUCTION

Zebrafish (*Danio rerio*) are popular biomedical research animals and many strains are considered suitable models for human medical conditions (Goessling, North, & Zon, 2007; Lieschke & Currie, 2007; Paik & Zon, 2010; Patton & Zon, 2005; Phillips & Westerfield, 2014). Key reasons for their popularity in research, in addition to many experimental advantages (Grunwald & Eisen, 2002), are the relative ease and cost-efficiency of maintaining a colony.

The increase of zebrafish use in research, and the increased diversity and specialization of zebrafish research also led to greater diversity in colony management styles and colony performance expectations. At the same time, there is a desire to develop reasonable husbandry standards to ensure that colony performance, compliance with animal care regulations, and research results can be established, compared, and maintained among different facilities (Lawrence et al., 2009). Basic husbandry standards for zebrafish maintenance, ranging from stand-alone static tanks (Trevarrow, 2004; Westerfield, 2007) to facilities with flowing and recirculating water systems have already been proposed (Brand & Nüsslein-Volhard, 2002; Harper & Lawrence, 2011; Lawrence, 2007; Trevarrow, 2004; Westerfield, 2007) and adapted by many laboratories. Initially, these standards were mainly based on general aquarium know-how, individual colony management experience, and limitations set by the particular aquaculture and aquaria equipment used at each facility. Recently however, more and more publications are available that inform about novel husbandry methods or expand on previous experience-based boundaries (see also chapter New frontiers for zebrafish management by, Lawrence, 2016).

Danio rerio is a tropical/subtropical cyprinid species with origins ranging from Pakistan, India, Bangladesh, Nepal, and Myanmar (Jhingran, 1991; Lawrence, 2007). Zebrafish have adapted to a variety of different habitats ranging from rivers, small streams, channels, stagnant or slow-moving pools, including bheels and rice paddies (Engeszer, Patterson, Rao, & Parichy, 2007; Jhingran, 1991; Fishbase.org). They can be found in relatively temperate Himalayan foothills along the Indian-Nepalese-Bangladesh border to subtropical zones of Southeast India near Bangalore (Engeszer et al., 2007; Whiteley et al., 2011; Wilson et al., 2014). The range of natural climates and habitats, and the relative hardiness and adaptability of zebrafish create opportunities for a variety of husbandry methods serving different research goals. There is also evidence that the laboratory conditions under which captive zebrafish populations have been maintained for decades may have contributed to domesticated substrains with genetic differences (Wilson et al., 2014). Examples include, Tübingen (TU) wild-type populations, approach the front screen of their aquaria when we approach. In contrast, AB wild types have not lost their flight response and swim away initially from the front screen. Strains derived from AB and TU hybrids such as TAB-14 show a mixed response. The difference can be explained with the aquaria types these populations were originally housed in and how they were selected for breeding (personal observation Z.V. David Lains). Similarly, there is evidence that these lines have also "evolved" different predispositions to some pathogens present in the research community (Murray et al., 2011). Lastly, unlike other wild-type strains, AB and TU both lack a normal sex-linked locus on chromosome 4, presumably because of the selection process used to establish embryonic lethal-free lines for mutagenesis screens. In these strains, environmental factors play a greater role in sex determination, whereas in other wild-type strains that have retained the sex-determining loci on chromosome 4, the environment plays a lesser or no role (Wilson et al., 2014).

Thus, husbandry standards, established for the welfare and experimental use of laboratory zebrafish should take into account their biology in natural habitats and the various adaptations to the facility environment of domesticated subpopulations.

The Zebrafish International Resource Center (ZIRC) is housed on the University of Oregon campus in Eugene. Its mission is to provide a central repository for wild-type, transgenic, and mutant lines of zebrafish and to distribute characterized, healthy, and breedable lines to the research community. Husbandry standards and procedures for maintaining healthy strains of zebrafish are continually evaluated, optimized, and developed (http://zebrafish.org/zirc/documents/protocols.php). ZIRC also provides a Veterinary Health Service for diagnosis of diseased fish. In addition, husbandry protocols and a manual for prevention, diagnosis, and treatment of diseases affecting zebrafish are available online: *Diseases of Zebrafish in Research Facilities* (http://zebrafish.org/zirc/health/diseaseManual.php).

Because ZIRC receives fish from facilities all over the world, biosecurity and the management or elimination of potential pathogens are important husbandry goals. ZIRC has developed stringent quarantine, husbandry, sentinel, and health surveillance policies for its colony, which are described in Chapter 27. This overview is designed to help promote basic best practices for most fish facilities. The husbandry and shipping methods can be adapted in other facilities; however, additional husbandry strategies may be needed to support specific research goals such as boosting reproductive rates or fecundity.

1. METHODS

1.1 ZIRC'S AQUATIC INFRASTRUCTURE

1.1.1 Overview

The ZIRC uses a flow-through water system in the quarantine room and recirculating water filtration systems for the main fish facility. To generate fish system water, ZIRC uses reverse osmosis (RO) filtered city water, and adds Instant Ocean (Aquatic Eco-Systems) and aragonite to restore freshwater conditions. For nitrification and removal of debris, return water is filtered in propeller wash bead filters and fluidized sand beds. The water is gas equilibrated in trickle columns after filtration, and also in each individual tank to ensure appropriate gas equilibrium and elimination of oxygen supersaturated water. To remove fine particulates from the water an optional bag filter loop and a nonoptional fine sediment filter system is coupled before the ultraviolet (UV) sterilizers, which eliminate pathogens from filtered water (Fig. 1).

1.1.2 Water source and volume

The Resource Center uses two sets of three RO pumps (CUNO 900) for its four recirculating filtration systems, which produce a maximum of 20,000 L RO water per day. Three out of four systems are in use to maintain a fish population of approximately 45,000 fish. The total water volume of each system with its associated tanks and pipes is approximately 56,800 L. To maintain stable water conditions and to remove nitrates, approximately 10% of the used water is replaced daily with fresh system water.

An additional RO pump (CUNO 1200) produces water for the flow-through system in the quarantine room. This system is spatially remote from the main facility; however, the salts and buffers to generate system water are the same. Because of space limitations only a single RO pump can be installed, and water production and flow in tanks is intermittent. Water quality is maintained with additional in-tank filter units during periods of static water.

1.1.3 Conductivity and buffering

The electrolytic conductivity of a solution (in μS/cm) is a useful value to monitor water salinity and total dissolved solids (TDS, in ppm or mg/L). RO removes between 95% and 99% of salts from water resulting in approximately 5–10 μS/cm conductivity. To restore freshwater conditions we prepare a concentrated stock solution of Instant Ocean in a large tub and use conductivity-controlled dosing pumps to add salts to prefiltration sumps (Fig. 1). The lower and upper thresholds for the pump are set to 450 and 550 μS/cm. A recent, comprehensive water analysis of our system

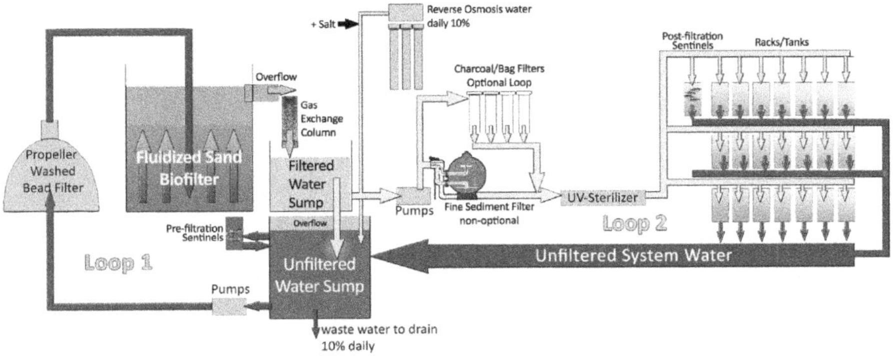

FIGURE 1 The ZIRC recirculating water system.

Loop1 produces and filters system water. Used fish water is collected in the prefiltration sump. Daily, 10% of the water is discarded to drain and is replaced by conditioned reverse osmosis water. Used water is pumped through a propeller washed bead filter that eliminates crude waste, and into the bottom of fluidized sand biofilters. Nitrifying bacteria populate the large surface area provided by sand and break ammonia down into less toxic nitrite and nitrate. From the sand filter, water overflows into a gas exchange column and into the filtered water sump. From here, water can reenter loop 1 by overflowing into the unfiltered water sump, or it is pumped into loop 2 as needed. Optional bag filters, fine sediment filters enhance the efficacy of the downstream ultraviolet (UV)-sterilizing units. Water is consumed in the main fish room aquaria, where used water enters a central drain and returns to prefiltration sumps (Loop 1). One set of sentinel fish receives water from the prefiltration sump. The other set is located after the UV sterilizer and receives filtered and sterilized clean water.

water (http://www.karlabs.com/) indicated that on average ZIRC's conditioned water has conductivity around 452 μS/cm.

Aragonite is a calcium carbonate substrate that functions as a buffer, provides additional trace elements to fish, and a carbon source for nitrifying bacteria. Aragonite also increases water hardness and pH, breaking down below pH 8.2 into calcium, to increase water hardness (120 mg/L) and carbonate, which buffers the water.

To maintain ZIRC's water pH in the range of pH 7.4–7.6, we manually regulate water to flow through an aragonite-reactor downstream from the fluidized sand filter. Because water flow can be adjusted as needed, this is an easy method to maintain stable pH. Alternatively, we can also add 250 mL of aragonite directly to the fluidized sand beds, which also makes use of the mineral's capacity to remove potential heavy metal pollutants from water (Kohler, Cubillas, Rodriguez-Blanco, Bauer, & Prieto, 2007).

1.1.4 Water quality and composition

The water parameters at ZIRC are maintained at a conductivity of 500 μS/cm, a pH in the range of 7.2–7.6, and total nitrates less than 5 ppm (mg/L; see also Table 1). These

Table 1 ZIRC Water Conditions

Water Parameter	ZIRC Target Value[a]	Actual Value 6/15/2015	Tolerated Range[b]
Conductivity (μS/cm)	500	452	Up to 2000
General hardness in dGH (mg/L $CaCO_3$)	6–7 (120–130)	3.2 57	1–19 (25–340)
pH	7.2–7.6	7.4	6.0–8.0
Temperature (°C)	28.5	28.0	16–32
Ammonia (NH_3/NH_4^+) mg/L	0	Not detected	Up to 0.05
Nitrite (NO_2^-) mg/L	0	<0.1	Up to 0.5
Nitrate (NO_3) mg/L	0–5	4.1	200

[a] *ZIRC water parameter values are based on Westerfield (2007).*
[b] *These values are based on the natural environments of zebrafish and recent publications (Engeszer et al., 2007; Lawrence, 2007; Trevarrow, 2004); Fishbase, http://www.fishbase.org/Summary/SpeciesSummary.php?id=4653.*

and the naturally occurring range of water quality values that zebrafish tolerate are indicated in Table 1 (also Westerfield, 2007). Water trace elements, salt and buffer composition based on a recent water sample analysis is listed in Table 2.

Table 2 ZIRC Water Composition Analysis 2015

Water Samples Analyzed for:	Units	"Facility Side A/System 3 + 4"	"Facility Side B/System 2"
		Result	
Boron, total	mg/L	0.08	0.06
Calcium, total	mg/L	10.9	12.6
Magnesium, total	mg/L	7.0	6.7
Potassium, total	mg/L	4.2	3.7
Sodium, total	mg/L	56.8	52.3
Sulfur, total, by ICP	mg/L	6.0	5.2
Alkalinity (as CaCO3)	mg/L	18	23
Bicarbonate (as CaCO3)	mg/L	18	22
Bromide	mg/L	0.48	0.45
Chloride	mg/L	109	98.9
Hardness	mg/L (as CaCO3)	56	59
Salinity	mg/L	232	221
Silica	mg/L	7.0	10.9
Sulfate	mg/L	16	15
TDS, estimated	mg/L	312	298

1.1.5 Filtration systems

The aquaculture system (Aquaneering Inc., Fig. 1) consists of mechanical propeller washed bead filters that remove coarse detritus and biological, fluidized sand-bed filters with an estimated 1300 km^2 of surface area per system and a water flow rate of 750 L/min. The sand filters harbor nitrifying bacteria such as *Nitrosomonas* and *Nitrobacter* spp. Uneaten food and fish waste increases ammonia (NH_3) levels, which is broken down by *Nitrosomonas* into nitrite (NO_2), and further oxidized by *Nitrobacter* to less toxic nitrate (NO_3). Ammonia, nitrite, and nitrate are toxic for fish; however, nitrate much less so than the first two. Daily 10% water replacement is sufficient to remove virtually all nitrates from the recirculating system. Both oxidizing processes lead to release of H^+ ions and hence the pH of the system water decreases. This in turn converts ammonia into ammonium ions (NH_4^+), which cannot be converted by the bacteria (Andrews, Exell, & Carrington, 2010). It is therefore important to keep the pH of the water system above pH 7 with adequate buffering.

Aragonite raises the pH and prevents formation of ammonium ions, hence supporting the nitrification process and providing a healthier environment for fish, whose gills are particularly susceptible to the ionized form of ammonia. It is also a porous material that provides a large surface area that supports bacterial metabolism of ammonia in our biofilters.

The ZIRC filtration system has recently been upgraded with fine sediment pool filters (Figs. 1 and 2 per water system) to remove particulates that could shield pathogens in the downstream UV sterilizers. These filters are rated to remove micrometer-sized particles, and need to be regularly back-flushed to remove sediment, further increasing the overall water exchange rate and improving the overall pollutant equilibrium, while optimizing exposure of microorganisms to UV in the sterilizers.

To further improve removal of sediment or other pollutants, an optional set of bag filters, with or without activated carbon, can be added via a bypass loop to the system, upstream of the UV sterilizers (Fig. 1). The mesh sizes range between 1 and 25 μm, which requires frequent changing to avoid clogging of the bags.

1.1.6 Ultraviolet water sterilization

UV sterilizers are among the most critical components for maintaining biosecurity in a recirculating water system by eliminating microorganisms from water. ZIRC's UV sterilizer units are Aquaneering 1000 W (5× 200 Watt amalgam bulbs per unit) that convert approximately 70% of electrical energy into germicidal UV-C output (peak output at 254 nm). We monitor the performance of bulbs closely by measuring total power output of each UV sterilizer and its five bulbs. We exchange bulbs approximately once every 1.5 years, or when the overall power output drops below 90%.

A key parameter describing the function of UV sterilizers is the UV-, kill-, or "zap-" dose produced by the units. It depends largely on

$$D_z = I \times T$$

FIGURE 2 ZIRC racks and tanks.

ZIRC uses two-faced rack systems with 2x49 3.6 L tanks. (A) Side view and (B) front view sketch of the rack system. *(Courtesy of Bob Drummond, Thoren Aquatics)* Letters and numbers indicate the Cartesian numbering system used in the ZIRC database to locate tanks. For example, A01G04 (B; bottom) refers to a tank in room side (A), on rack 01, in row (G), fourth position. (C) Front view of 3.6 L polycarbonate tank with feeding hole (top center), bar-code label, and label to indicate line and allele information. (D) Rear view of the same tank with coarse detritus baffle, siphon, and outflow valve. (E) In the juvenile grow-out section, the same tank is equipped with a fine mesh drain to prevent loss of fish. (F) Side view of standard crossing tank (Thoren Aquatics) with grated insert, artificial grass, fish, and lid. (G) The same setup with tilted inset, mimicking shallows of water bodies.

where D_z = zap-dose (in μWs/cm^2), I = UV intensity (in μW/cm^2), and T = dwell, or exposure time (in s). The effective maximal zap dose depends on the lowest possible water flow rate and additional facility specific data such as water UV transmissibility, diameter of UV sterilizers, bulb wattage, and number of bulbs in a unit, which all contribute to UV intensity. Thus, if the maximal UV intensity is still too low to eliminate a particular pathogen, the water flow can be reduced to increase dwell time and with it, the zap dose. According to manufacturer information, with an 8″ diameter vessel, each unit at ZIRC provides a dose of 180,000 μWs/cm^2 at the maximal flow rate of 755 L/min (200 gal/min).

1.1.7 System monitoring

Mechanical components and water quality of the main and quarantine room filtration systems are monitored via Internet using Sensaphone Web600 units and Icinga monitoring software. The software can send alerts via an SMS gateway. Customized alert thresholds trigger alert messages sent to staff via e-mail and/or SMS text message. Two ZIRC staff members are on call on a weekly rotating schedule to respond to system alarms after hours. They can review equipment and water status via SMS messaging or a web-based status page. The ZIRC's internal database displays monitored data and tracks and displays system information continuously. As needed, reports and system data can be downloaded as comma separated value (CSV) tables for further analysis or documentation. For each water system, continuously recorded data include the on/off status of pre- and postfiltration pumps, water pressure, sump water levels, water conductivity, pH, water and air temperature, salt dosing pump status, UV bulb intensity and UV failure alert, and floor "water bugs" to detect leaks. Other water parameters such as ammonia, nitrite, and nitrate tests are performed manually and recorded in weekly intervals. The Sensaphone/Icinga paging system can be extended to include additional monitoring in future. Other ZIRC equipment is also tied to the paging system such as fridges, freezers, ultra-low freezers, liquid nitrogen freezers, and liquid nitrogen fill levels.

1.2 TANKS, RACK SYSTEM, AND FISH MAINTENANCE

1.2.1 ZIRC tanks and rack systems

The Resource Center uses three types of aquaria. Seventy-five liter glass aquaria are used for large families of up to 300 fish, for example, wild-type and frequently requested transgenic lines. These are placed on self-made racks, which supply aeration and water through self-assembled manifolds and pipes.

For smaller groups of fish, Thoren Aquatics developed smaller polycarbonate tank types and their rack systems in collaboration with Resource Center staff (W. Trevarrow, A. Bailey at ZIRC). These consist of modular, low-maintenance tanks in the nursery (c. 1.3 L size) and 3.6 L tanks for adults and juveniles that receive individual aeration and adjustable water flow (Fig. 2; Table 3). The tanks have a V-shaped bottom and a baffle plus siphon on the backside to remove detritus. The racks hold tanks in a back-tilted angle to further facilitate tank cleaning. The ZIRC facility provides space for approximately 4100 3.6 L tanks.

1.2.2 Tank cleaning

Fish should be viewable from the front of the tank at all times to facilitate health assessments and ensure their well-being. An aquarium or tank requires cleaning when algae begin to form on its front or sides and/or debris begins to gather on the bottom of the tank.

Nursery tanks are used for approximately 21–28 days and are exchanged and cleaned when juveniles are moved to the grow-out section; 3.6 L juvenile and adult tanks need to be changed approximately once a month. To prevent mold buildup at the feeding hole, lids are changed more frequently, as needed. After removal from the rack, used tanks are scrubbed with a water-immersible automatic brush ("BarMaid," Electric Glass Washers, www.barmaidwashers.com) or with a water

Table 3 Tank Types, Fish Density, and Water Flow

Tank Type	Fish per Tank	Tank Volume (L)	Water flow (L × h^{-1})	Exchange rate/fish (L × fish^{-1} × h^{-1})	Tank volumes exchanged (h^{-1})
Nursery	50 Larvae	1.3	6.8	0.14	5.1
Grow-out/ adult	20 (±5)	3.6	10.2	0.51	2.8
Grow-out/ adult	188 (±20)	75	184.2	0.98	3

The ZIRC uses three tank types. In these, water flow is adjusted to three to five exchanges of the tank volume per hour. Nursery and 3.6 L tank flow rate is limited to a maximum of 20 L/h by the rack's water manifold design. Nursery tanks hold up to 39 larvae/L on dripping water and 0.1 L/h exchange rate. On average, 3.6 L tanks hold five to six fish/L, and 75 L tanks hold four to five fish/L.

jet and are then soaked in bleach (0.15% sodium hypochlorite, 1500 mg/L effective working concentration based on 8% Clorox) for at least 15 min, followed by a 0.003% (w/V) sodium thiosulfate (3.175 g/L; Western Chemical) neutralization dip or brief soak. Then, tanks receive a thorough, final rinse in an industrial tunnel washer at 82°C. Tank parts are also submerged in bleach, neutralized in sodium thiosulfate and rinsed in hot water. Bleach solutions are replaced every other week and sodium thiosulfate solutions weekly.

For work safety reasons, 75 L glass tanks must be cleaned in place. Glass aquaria are scrubbed in place using a sterilized scouring pad. A clean pair of shoulder length disposable gloves and new scouring pad is required for each tank to prevent the potential spread of pathogens. All sides and the bottom of aquaria are scrubbed, and the debris is flushed from the tank immediately by increasing the water flow to maximum for about 10 min. Water and air tubes are replaced in each tank after cleaning. Lids are replaced if any mold or food residue is visible around the feeding hole. Once a stock is discontinued and its aquarium is emptied, the air-stone is discarded. The tank is then drained and cleaned as described in the previous section.

1.2.3 Fish facility cleaning

The concrete floors are mopped once a week with a mild bleach solution (0.06% bleach in tap water, or 200 mL Clorox in 20 L). They are precleaned with bleach followed by a thorough cleansing with detergent. In addition, work surfaces and facility carts are cleaned daily (or after use) by spraying with 70% EtOH. Additional sterilization with Quatricide or bactericidal wipes is performed as needed.

All fish-related equipment is sterilized in a 0.15% bleach solution, followed by a sodium thiosulfate bath to neutralize residual chlorine. All items are then rinsed in the tunnel washer. Bleach and sodium thiosulfate solutions are replaced weekly. Equipment that cannot be bleached is sterilized in a laboratory-grade STERIS autoclave.

The efficiency of cleaning procedures is tested regularly. The tunnel washer and autoclave are tested weekly with temperature test strips to ensure that 82°C sterilization temperature is reached (Pharmacal Tri-Temp Tape; 3M Indicator Attest Biological for Steam Sterilizer). In addition, all cleaning systems are tested twice per year by sampling a representative list of aquaria, equipment, and surfaces with Replicate Organism Detection and Counting (RODAC) test plates.

1.2.4 Breeding fish and collecting embryos

Zebrafish breed most reliably between 3 months and 1 year of age. Breeding females and males are combined in mating tanks in the afternoon. Specialized mating tanks, or crossing cages, are needed to prevent cannibalization of eggs (Fig. 2F and G). Spawning is strongly controlled by the light cycle (Westerfield, 2007), and occurs after lights are turned on in the morning. If timing of breeding is crucial, females can be separated from males before spawning by placing a transparent polycarbonate screen diagonally into the grated insert. Eggs are collected in plastic tea sieves before noon, and rinsed into Petri dishes.

Depending on the line and on breeding goals, various combinations of males and females can be used for breeding. For example, individual pairs are used to identify carriers of mutations or transgenes. Group breeding is used to produce large numbers of embryos for a new generation or shipping, when the identity of adults is already known. We use 500 mL mating tanks (Fig. 2F and G) for single pairs or for "small group" crosses of 5–6 fish, typically two females and 3–4 males. For intermediate sized group crosses, up to 10–15 pairs, we use modified 3.6 L aquaria or transparent salad-bar trays with varying depth from a local restaurant supply store. For large group crosses, 25 or more pairs, modified polycarbonate rat cages are ideal (Tecniplast 2154F). All containers are modified by replacing the bottom of the insert tank with 3-mm diamond Trical grate (Aquatic Eco-systems, TN527) hot-glued to the sides. Stacking modified containers into unmodified ones provides breeding setups for all cross sizes similar to the one shown in Fig. 2F and G.

In addition to light, several environmental enrichment strategies can be used to stimulate breeding. Precut 1-inch strips of green mosquito netting (Hancock Fabrics) provide artificial "plants." Cutting or knotting of several strips can shape this material, which unlike plastic plants from pet stores, can be autoclaved (must be fire retardant-free!). Slanting the grated insert in the breeding setup mimics the edge of water bodies. This setup provides shallow areas for animals that are ready to breed, whereas others can withdraw in slightly deeper areas and hide in the netting material (Fig. 2F and G) (Sessa et al., 2008).

Breeding procedure:

1. Prepare mating tanks with grated inserts (Fig. 2F and G) in the afternoon; fill with fish water and two to three strips of mosquito netting. Leave approximately a 2–3 cm air space between the water surface and tank lid (See Fig 2F).
2. If accurate timing of spawning is necessary, place a divider in the mating tank to separate male(s) and female(s).

3. Net males and females into crossing cages. Treat fish gently!
4. Cover crossing cages with lids.
5. Label tanks with stock number, genotype or allele/lab designation, date, and caretaker on a Post-it flag or Time tape.
6. Fill in the breeding log, check in on fish before leaving for the day: assess behavior and water quality. Remove or separate fighting fish (indicated by scales in the tank).
7. Leave fish overnight.

Harvesting embryos:

1. If used, pull dividers immediately after lights are turned on.
2. If necessary, tilt grated cage insert to provide a sloped bottom (Fig 2G).
3. If necessary, place fish into freshwater, or into a container with less water (to provide a shallow breeding area throughout the tank (no slope).
4. Wait 2–3 h. Spawned eggs fall through the insert, protected from adults.
5. Transfer fish by placing the grated insert(s) into a new mating tank, or a 3.6 L holding tank, or back in their original aquaria on the water system.
6. To collect eggs, pour them into a fine-mesh plastic tea strainer.
7. Rinse eggs into a clean Petri dish with a squirt bottle (use fish water or embryo medium).
8. Label Petri dish with line name, stock or allele designation, date of birth, caretaker name, and other necessary information.
9. Return the fish to their original stock tank or tank location if you used identified adults.
10. If necessary, remove debris and feces from collected embryos.
11. If embryos need to be analyzed to characterize adult genotypes, maintain adults in isolated mating tanks until mutant/transgene carrier identification has been completed. Monitor isolated fish in static water daily; perform water changes as needed (look for cloudy discolored water or foul smell, and check water with an ammonia test kit every other day). Remove any dead fish immediately. Water in static tanks should be changed at least every 2–3 days and fish can be fed a few hours *before* water changes. Ideally, if fish need to be isolated for more than 5 days, transfer them back to a tank with normal water flow and feeding.

1.3 RAISING LARVAE

1.3.1 Tank system, water, and cleaning

The Resource Center aims to optimize time and survival rate of fish in the nursery, to enhance throughput and turnover of live lines in the facility.

Nursery tanks and racks. To provide an optimal rearing environment, we control water quality by adjusting water flow to each nursery tank. Nursery tanks have individual valves and drains that allow for switching from static conditions to intermittent, or constant water drip or flow. The mesh drain prevents larvae from entering the used water system, however, these can also clog with excess food. Therefore, we monitor and replace drains to prevent overflowing tanks.

Embryos are maintained at a density of 50 fish per Petri dish, at 28.5°C in 0.5× E2 medium and methylene blue (0.5 mg/L). At day 4 or 5, larvae are transferred to a nursery tank that contains a small amount of fish water (Fig. 3). If necessary, the transfer can be delayed up to day 7 (Fig. 3, *gray asterisk*).

E2 embryo medium contains 16 mM NaCl, 0.6 mM KCl, 1 mM $MgSO_4$, 0.15 mM KH_2PO_4, 0.05 mM Na_2HPO_4, 1 mM $CaCl_2$, and 0.7 mM $NaHCO_3$. We use 0.5× E2 in Petri dishes and during the static water phase in the nursery to provide a smoother transition, when system water is turned on in the nursery. We prepare E2 (Brand & Nüsslein-Volhard, 2002) from three stock solutions, E2A, E2B, and E2C, which are then used to make a large volume of the 0.5× E2. The detailed recipe has been described previously (Varga, 2011) and is available online: http://zebrafish.org/documents/protocols/pdf/Fish_Nursery/E2_solution.pdf.

1.3.2 Nursery schedule

Food. Larvae receive food that increases in size as the fish grow and their gape increases. The smallest food type, *Paramecium caudatum* (c. 150 μm length, 30 μm diameter) is introduced at day 5 post fertilization (5 days) and fed until 10 days (Fig. 3). Larvae are maintained in static water until 10 days; only system water containing paramecia is added during this time. Occasionally, when more fish need to be reared than the average paramecia production supports, we substitute paramecia with *Brachionus plicatilis* (L-type rotifer; Reed Mariculture). At 10 days, we start regular, 24-h water flow to the cages (0.1 L/h) and also start feeding Zeigler AP100 powdered food (see Table 4 for Food Mix Recipes) until 20 days.

Depending on nursery entry time and careful observation of growth rate and feeding behavior of larvae before transitions, water and food changes may be delayed by a day or two (indicated by *asterisks* in Fig. 3). Transfer of larvae from Petri dishes to static water nursery can occur as late as 7 days. Flake food, artemia, and water flow can be started as late as day 12 (Fig. 3, *black asterisk*). However,

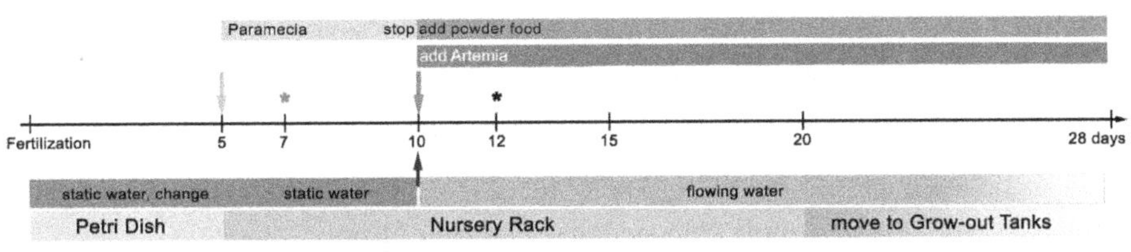

FIGURE 3 Timeline of ZIRC nursery. Water and feeding schedule.

Center, from left to right: time line. Top, *bars* above time line indicate food types. *Bars* below time line indicate embryos/larvae locations and water flow conditions. *Arrows* indicate times of food and water changes. *Asterisks* indicate optional, maximal delays for food and water flow transitions, based on larval growth and feeding observations.

Table 4 Dry Food Mix Recipes and Sources

Food Mix	Components	Source	Proportion (g)
Nursery (<21 days)	Zeigler Larval Diet, AP100, <100 μm	Aquatic Eco-Systems	Mix equal parts
	Zeigler Larval Diet, AP100, 100–150 μm	Aquatic Eco-Systems	Mix equal parts
	Rotifer diet	Reed Mariculture	
	Instructions: Use for fish up to 21 days post fertilization (day). Mix equal parts (by weight) of Zeigler Larval Diet powder. Date opened canisters and store in fish food freezer. Suspend 0.5 g of mix in 250 mL fish water. Feed each tank (50 fish) 5 mL with squirt bottle.		
Juvenile (21–90 days)	Zeigler Larval Diet, AP100, 150–250 μm	Aquatic Eco-Systems	500 g
	Zeigler Larval Diet, AP100, 250–450 μm	Aquatic Eco-Systems	500 g
	GP 200, Golden Pearls 200–300 μm	Artemia International LLC	250 g
	Instructions: Use for fish between 21 and 90 days. Weigh and mix all ingredients thoroughly. Date opened canisters and store in freezer.		
Adult (>90 days)	Zeigler Adult Zebrafish Diet	Aquatic Eco-Systems	1812 g (four 1lb bags)
	O.S.I. Spirulina Flake	Aquatic Eco-Systems	500 g
	GP 300, Golden Pearl 300–500 μm	Artemia International LLC	454 g (16 oz.)
	Instructions: Use for adult fish 90 days and older. Blend Spirulina flakes in blender to create more edible sizes. Avoid turning Spirulina into powder. Add ingredients in a large airtight bag and mix thoroughly. For extended storage, vacuum seal and freeze. Golden Pearl 300–500 μm		

change of food type and water flow should always coincide. At 20–28 days (Fig. 3), larvae are transferred to the juvenile or "grow-out" section of the facility, once they are large enough to feed on the juvenile food mix. Key for all transitions is to observe and ensure that larval/juvenile mouth gape is wide enough to permit feeding on larger food items.

Nursery Statistics. The ZIRC nursery has a capacity of approximately 4000 larvae per week and an approximately 90% survival rate (for 2015). Fertilization rates are recorded when embryos are harvested and cleaned, usually 2–3 h after egg laying. At 5–7 days, fish caretakers remove undeveloped, dead, dying, or mutant larvae, and sort 50 larvae per Petri dish. The number of larvae per stock that enter the nursery is recorded in the database. At 20–28 days larvae are counted again, before transfer to the juvenile grow-out section. Fifty juveniles are distributed to two 3.6 L tanks (small or less frequently requested strains) or up to 150–300 are pooled into a 75 L tank. For calculating survival rates, we divide the total number of larvae per stock leaving by the number entering the nursery.

Grow-out section. Adult 3.6or 75 L tanks are equipped with fine mesh drains to prevent loss of juveniles. Fish are fed juvenile diet three times a day and are maintained in the grow-out section until they reach adult size (around 3 months post fertilization). Fin clipping for genetic identification can occur as early as 2 months, otherwise the gender of the fish is identified at 3 months and they are bred to identify gene expression or phenotypes in their offspring. Once their genotype is identified, adults, or their offspring are made available to the public and can be requested through the ZIRC's site (zebrafish.org).

1.4 LIVE FOOD PRODUCTION

1.4.1 Live food types

Because of their small size, paramecia and rotifers (Best, Adatto, Cockington, James, & Lawrence, 2010) are ideal first feeding options for larvae and juveniles. At ZIRC, we use *Paramecia caudatum* and Great Salt Lake Artemia (Artemia International) for nursery, grow-out and adult fish (Westerfield, 2007). Whereas artemia offer only very basic nutritional value, they stimulate predatory behavior in fish and are therefore considered environmental enrichment. Detailed live food production methods have been described previously (Varga, 2011) and can be downloaded from the ZIRC site: http://zebrafish.org/documents/protocols.php#Fish%20Feeding.

1.4.2 Juvenile and adult feeding schedules

Juveniles receive three feedings per day, artemia in the morning, followed by two juvenile flake feedings at noon and in the afternoon (see also Nursery protocols and Table 4). On weekends and holidays, juveniles receive only two feedings in parallel with the adult feeding schedule.

Adults receive a morning feeding of brine shrimps at 9 am (Table 4). The 3 pm feeding consists of adult powdered food mix (Table 4). This feeding schedule maintains the fish in a healthy, breedable condition for more than a year and reduces the risk of obesity (Oka et al., 2010). Too much food can adversely affect water quality, requires increased tank cleaning effort, and might lead to occasional spinal deformities presumably stemming from vitamin or amino acid imbalances (Cahu, Infante, & Takeuchi, 2003; Lall & Lewis-McCrea, 2007).

We prefer feeding fish moderate portions more frequently, which are consumed in less than 5 min, rather than larger portions fewer times per day. This feeding regimen reduces food waste that contributes to ammonia buildup in the system water. In addition, the fish cannot overfeed, which can also lead to health problems. Several feedings also tend to reduce competition for food, which reduces stress and promotes better health (Ramsay et al., 2006). At the same time, we attempt to obtain a rapid growth rate so that sexual dimorphisms are recognizable at 2–3 months, and fish can be used for mating or distribution as early as possible.

1.5 LINE MANAGEMENT TOOLS

1.5.1 Record keeping

One of the most crucial tools to manage fish lines and stocks efficiently is a reliable database. ZIRC continues to develop its own web-based, PostgreSQL database to track live and cryopreserved fish line inventory and to support daily operations. The ZIRC database is extensively cross-linked to the Zebrafish Model Organism Database (ZFIN). Automated nightly data exchanges and frequent links from the Resource Center's online inventory pages to ZFIN database records ensure that inventory, scientific information, and nomenclature are up-to-date.

The public site is connected to the ZIRC database in real time and enables researchers to review the resource inventory, request and obtain fish, research materials, or request services. At the same time, the database helps fish caretakers to obtain a list of incoming requests and plan their daily and weekly activities. Each line imported to the Resource Center is assigned a unique ZIRC Line identification number (ZL). ZL numbers are background-specific, and a new number is assigned each time a mutation or a transgene is transferred onto a new genetic background. For every cross that establishes a new generation of a line, the database also generates a new, consecutive stock number to the offspring. In addition to line specific data, stock-specific husbandry data are also captured, such as date of birth, nursery survival rates, parental stock numbers, whether the stock was generated by an intercross or outcross, overall health status, tank locations, numbers of fish, fish removal numbers and reasons (ranging from distribution to health-related issues), labeling information, stock-specific notes, assigned animal caretaker, and planned work (eg, cryopreservation or breeding a next generation). The database also supports the Diagnostic Health Services ZIRC provides to the research community.

The database is organized into key areas of ZIRC activities such as general line specific information, line availability (for the display of lines to researchers on our public pages), phenotypes/genotypes (for line identification and propagation). The information on these "tabs" (at the line level) is extensively cross-linked with the ZFIN database and updated daily. In addition, stock-specific information is also stored (see the later discussion) including tank inventory, cryopreservation, and frozen sample inventory.

With the acquisition of multiallelic lines from large-scale screens (Kettleborough et al., 2013; Varshney et al., 2013) the database was reconfigured to track all alleles contained in sperm samples. Upon receiving requests for specific alleles, this facilitates the review of all samples that contain the requested alleles, and to retrieve and regenerate the most optimal samples for distribution. The database tracks and displays all associated alleles, in addition to the requested one(s), thus avoiding accidental elimination of unique alleles. In addition, if a particular allele is present in several samples, the staff can make use of the redundancy and delay or schedule amplification of alleles to better fit work plans. Thus, redundancies and unique alleles are equally represented and assist the decision-making of the ZIRC staff for various management strategies at the level of alleles, samples, or live fish lines.

1.5.2 Cryopreservation

Cryopreservation is the most cost-effective alternative for maintaining genetic resources because it reduces costs for animal and facility maintenance, personnel, and space. Cryopreservation is also the most efficient method for large-scale, long-term storage of important genetic materials because it prevents catastrophic loss of irreplaceable research lines and offers the possibility to maintain only the most important strains alive.

The freezing method we initially used was developed (Harvey, Kelley, & Ashwood-Smith, 1982) and adopted by Walker and Streisinger (Westerfield, 2007). The method was further modified and streamlined by Draper, McCallum, Stout, Slade, and Moens (2004) and Draper and Moens (2009). The Resource Center used this version of the protocol successfully with minor modifications as described (Carmichael, Westerfield, & Varga, 2009). To further improve and optimize frozen resource recovery, ZIRC staff researched and developed a new method recently. This method (personal communication Jen Matthews; http://zebrafish.org/documents/protocols.php) includes improved sperm acquisition and postthaw in vitro fertilization procedures. In addition, we improved cryopreservation by optimizing the extender, cell density, the cryoprotecting medium, the sample freeze rate, and by introducing several optional quality control steps into the process, such as sperm density and motility measurements. Moreover, the new method is flexible and can be performed by a single person or in pairs, sperm can be obtained by "squeezing" or testes dissection, and samples can be frozen from single males or can be pooled from many for more representative quality control assessments. In addition, sperm acquisition and freezing can occur at the same time, or can be uncoupled if high throughput is necessary by storing collected sperm on ice for hours, and bulk freezing all samples in a single step. With this new procedure, the average postthaw fertilization rate ranges between 50% and 60% per sample, which means virtually 100% recovery rate for most cryopreserved lines acquired or frozen by ZIRC.

After the Resource Center imports a new line into the quarantine room, we cryopreserve sperm as soon as possible from each male and store samples in our sperm bank as an immediate safety backup. ZIRC staff also collects finclips from each fish to confirm their genetic identity and to ensure that each fish carries the specific genetic modification. In addition, finclips also provide material to develop PCR genotyping protocols at ZIRC if necessary.

Later, once the line is established at the ZIRC, we rederive the line and amplify the number of samples so that they can be more effectively managed. For example, if there is a request for a line that is available only from the sperm bank, sperm are thawed and used to in vitro fertilize AB wild-type eggs. The resulting embryos are then sent to the requesting laboratory. We establish a threshold for the minimum number of samples necessary in our sperm bank and once this threshold is reached, we regenerate the line, raise the stock, replenish our frozen supplies, and supply the requesting laboratory with the line.

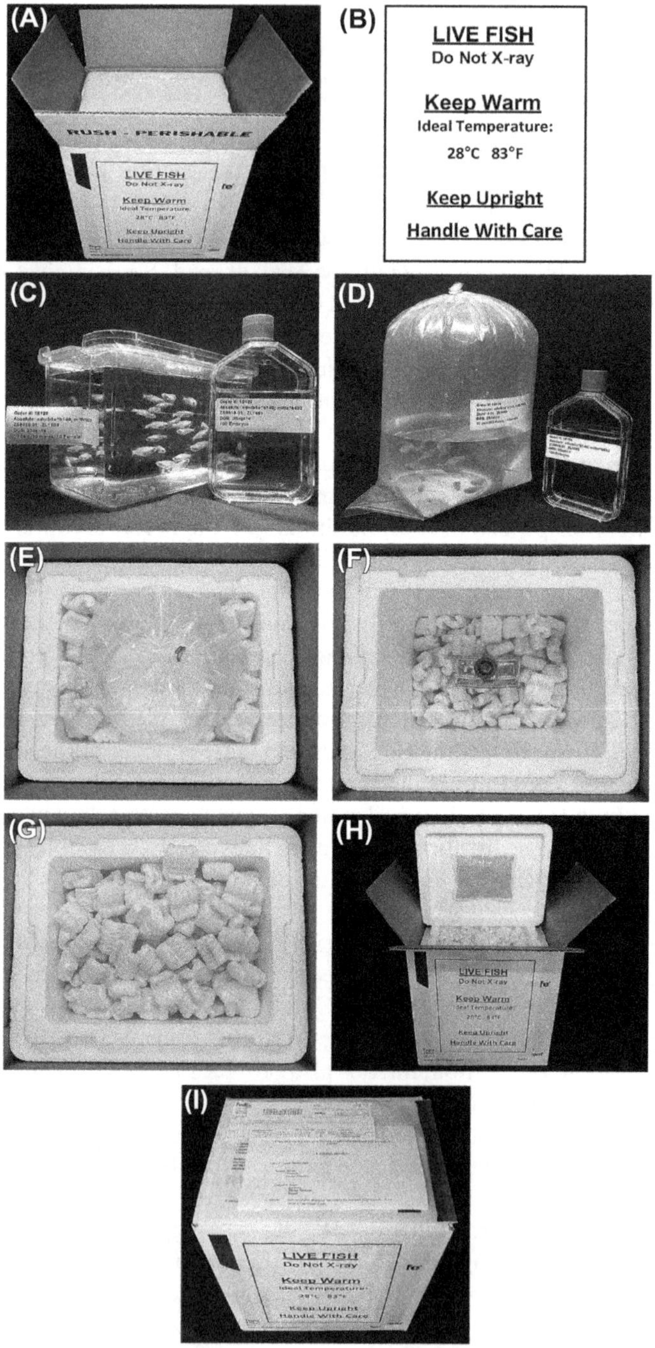

FIGURE 4 Packaging of embryos and adults for shipping.

(A) Thick-walled Styrofoam box inside cardboard box, labeled as indicated in panel (B). (C) Adults are prepared in labeled breeding cages, embryos in labeled 250 mL flasks. (D) Adults

Cryopreservation has also become a key tool to import large numbers of lines from individual laboratories. When a laboratory submits a large number of lines, we have found it more time- and cost-efficient to travel to the laboratory and cryopreserve sperm on site, rather than receiving live fish at ZIRC. Alternatively, ZIRC provides training for staff to cryopreserve lines in their own laboratory. In either case, the frozen samples are sent to ZIRC in liquid nitrogen shipping vessels and stored in the repository. We perform test-thaws to ensure that sperm from each line is viable.

For ZIRC, the turnover of lines between cryopreserved and live state has become *the* crucial line management tool. Limited tank space makes it necessary to cryopreserve lines that are not frequently requested and revive them if a request is placed by a researcher, allowing the Resource Center to operate as cost and space efficiently as possible (Carmichael et al., 2009).

1.6 SHIPPING FISH

Zebrafish shipments require special packaging to provide a controlled environment for fish while in transit and careful documentation to ensure expedient delivery. Because customs, animal welfare and biosafety regulations, vary from country to country, international shipments may require modifications of the labeling and documentation provided below. Packaging, however, must always conform to the International Air Transport Association (IATA) Live animal Regulations (container requirement 51; https://www.iata.org/whatwedo/cargo/live-animals/pages/index.aspx). International shipments should be sent Mondays or Fridays to allow time for customs and veterinary inspection clearances (1–3 days) and avoid weekend delivery. When shipping embryos, these should be bleached before packaging. For a detailed bleaching protocol see the Chapter 27 in this edition. Up-to-date modifications can also be found here: http://zebrafish.org/documents/protocols/pdf/Egg_Bleaching/Customer_Egg_Bleaching.pdf.

1.6.1 Packaging of embryos

1. Label tissue culture flasks with line specific information such as line name, allele designation, stock #, date of fertilization, and number of embryos.
2. Add 200 mL of 0.5× embryo medium with methylene blue to the flask(s).
3. Add 100 embryos per flask and close lid tightly, if in doubt, add parafilm around the cap (Fig. 4C and D).

and their label are transferred to two leak-proof polyethylene bags and closed with plenty of air or oxygen. The Styrofoam box is partially filled with Styrofoam packing peanuts before the bag or flask is added. (E,F). Completely filled Styrofoam box (G). Packing peanuts prevent movement of bag or flask and a heat pack (H) is added to the lid in the cold season or for cold-climate destinations. (I) Address labels and pouch with shipping documents (Invoice, Customs Statement, Copy of Import Permit, Health Declaration/Certification) are added in triplicate copy to the pouch.

4. Fill shipping box about 1/3 with Styrofoam packing peanuts, then place flask(s) with embryos in its center (Fig. 4F), and fill any remaining space with additional packing peanuts (Fig. 4G).

1.6.2 Packaging of adults

1. Write information on the plastic bag (line name and allele, number of fish, date of birth) or attach a label with the information (Fig. 4D).
2. Add the following to a square bottom plastic bag:
   a. Approximately 1/3 fish water
   b. ClorAM-X ammonia binder (1.5 mL ClorAM-X per liter of fish water)
   c. Fish. For the US or priority overnight shipments, place 20 fish/bag (1.5 L of water); for international or 2–3 days shipments, place 10 fish/bag (1.5 L water)
3. Close the bag trying to trap as much air as possible: Open the top of the bag wide, quickly grasp the top closing it off and then twist keeping pressure in the bag. The bag should contain approximately 2/3 air and 1/3 water when closed. If more air is needed, reopen the bag and try closing it again. Do not blow into the bag to add more air (CO_2!). If available, use pure oxygen, then the proportions can be inverted to 2/3 water and 1/3 oxygen.
4. Close bag by knotting or twisting the bag so the top doubles over and rubber-band this portion. Push the knot down to maintain pressure in the bag. The bag should remain tight and be leak free when turned upside down. Metal (Fig. 4D) or plastic clips can also be used instead of rubber bands.
5. Place the bag of fish inside another plastic bag! Line bags up (by their seams), so they are in the same orientation. Remove as much air as possible from the second bag, to provide no space for water if the inner bag spills. Close the outer bag as above. Ensure there are no leaks.
6. Place bagged fish in the shipping box and fill the remaining space with Styrofoam peanuts.

1.6.3 General packaging and labeling

7. During winter months, a heat pack should be placed in the box to avoid or minimize temperature drops when the package is exposed to outside temperatures. Punch a small hole in the center of the Styrofoam lid (not cardboard!) to allow oxygen to reach the heat pack. Activate heat pack and tape it to the inner side of the lid, covering the hole (Fig. 4H).
8. Place the Styrofoam lid on the box. Thick Styrofoam insulates the fish from outside temperature changes and also prevents water spills in case the plastic bags leak.
9. Close the cardboard box and seal with sturdy packing tape.
10. Place the following labels and documents on the outside of the box, including:
    a. Live fish label (Fig. 4B)
    b. Address label (on top of box)
    c. Pouch for customs and health certification documents for international shipments (usually provided by courier).

ZIRC currently uses World Courier for the following countries: Austria, Australia, China, Germany, Ireland, Israel, Italy, New Zealand, Spain, Saudi Arabia, Singapore, all South American countries, South Korea, Sweden, and the United Kingdom. Typical World Courier shipment cost is approximately $1500.

1.6.4 Shipping documentation

The paperwork for standard international distribution may vary due to country-specific requirements for customs and health declarations/certifications. In addition, many of these processes are still being developed between countries and information may be outdated or inaccurate. However, most international shipments must be accompanied by at least the following documents:

1. Customs Statement,
2. Customs Invoice (in triplicate),
3. Import Permit (of the receiving country), if required and
4. A Declaration of Health
   If the importing country requires a United States Department of Agriculture (USDA)-endorsed Health Certification, then the reciprocal import permit from that country is required and the sender must obtain the appropriate certificate form from the USDA website. Step (4) is not necessary in this case.

ZIRC provides templates of documents (1)–(4) upon request and on its site: http://zebrafish.org/documents/protocols.php. These documents should be copied onto your institute's official letterhead.

If a USDA endorsed Health Certificate is required, information can be obtained from the Animal & Plant Health Inspection Service (APHIS) at the USDA (http://www.aphis.usda.gov/regulations/vs/iregs/animals/). It is a good idea to get in touch with your USDA Area Office to ensure the appropriate steps are taken. Select the country of destination in the pull-down menu on the iRegs site. The guidelines detailed by USDA/APHIS and the receiving country must be met in order to obtain a USDA endorsed health certificate for export. Download the appropriate form for aquatic species and all supporting documents for the country of destination. If the country-specific form is not available, the USDA's APHIS Form 7001 must be used: http://www.aphis.usda.gov/library/forms/pdf/APHIS7001.pdf.

If a country requires certification that the exporting facility is free of a specific OIE reportable pathogen, you should contact the USDA Area Office. In this case a USDA Veterinary Medical Officer may need to schedule sampling and diagnostic testing of fish and possibly conduct a facility inspection.

Complete the health certificate for the shipment and add additional information as specified on the import permit. Note: this is a reciprocal process—the USDA will *only* issue a Health Certificate endorsement if a copy of the import permit from the recipient country is available, to ensure all country-specific information is provided. A USDA accredited veterinarian must sign and date the health certificate (or Form 7001), which then must be endorsed by the USDA/APHIS. To this end, mail the form to your local USDA Area Office with payment (currently $38).

Importantly, the health certificate is only valid for 30 days, and the importing country may limit shipment to a shorter time frame. The sender has to ensure that the shipment occurs before the earlier of the two dates expires. Hence it is important to schedule steps carefully and have a good working relationship with your local USDA office.

1.6.5 Zebrafish import to the United States

When receiving a shipment of zebrafish from abroad, a custom declaration (Form 3-177) must be filed with the US Fish and Wildlife Service (USFWS). Visit the http://www.fws.gov/le/declaration-form-3-177.html site, download the four pages, fill in the form and file it with the appropriate office (see instructions). Alternatively, you may with *eDecs*, an electronic declaration system offered by USFWS. In this case, create a new personal account on the eDecs site https://edecs.fws.gov/eDecsHome.cfm, create an Importer/Exporter Account and a new eDec file, and file it electronically. For academic institutions filing must occur, but a license is not needed and the fees are waived.

2. MATERIALS

1. ZIRC's Aquatic Infrastructure
   a. Instant Ocean IS 160 (Aquatic Eco-Systems, Inc.)
   b. Aragonite (Aragamax 00930, Sugar-sized, CaribSea Inc.)
   c. Filtration systems (Aquaneering Inc.)
   d. UV sterilizer (Aquaneering 1000)
2. Tanks, Rack System, and Fish Maintenance
   General:
   a. 3.6 and 1.3 L tanks and racks, breeding cages (Thoren Aquatic Systems, Inc.; http://thorenaquatics.com/)
   b. "BarMaid," water immersible brush (Electric Glass Washers, www.barmaidwashers.com)
   c. Ethanol (EtOH 190 Proof; Chemistry Stores UO)
   d. Sodium hypochlorite (NaOCl; Fisher Chemicals, SS290-1/L13934 and Clorox, Regular Bleach 5.38 L)
   e. Sodium thiosulfate (Western Chemical)
   f. RODAC plates (Pharmacal Research Labs Inc.)
   g. Polycarbonate mouse/rat cages (Tecniplast 2154F)
   h. Trical netting, 3 mm diamond shape (Aquatic Eco-systems, TN527)
   Spawning supplies:
   a. Small, intermediate, or large crossing cages with grated inserts or modified mesh bottoms (Tecniplast 2154F; Thoren Aquatic Systems, Inc.)
   b. Artificial grass (green mosquito netting strips: #2029239; Hancock Fabrics; http://hancockfabrics.com/)

c. Autoclavable fish net
d. 3.6 L tank (Thoren Aquatic Systems, Inc.)
e. Squirt bottles
f. Petri dishes
g. Plastic tea strainer (mesh size smaller than 0.4 mm)
h. Post-it flags or Time tape

3. Raising Larvae
    a. Methylene blue (Sigma, MB-1)
    b. E2 medium:
        i. Calcium chloride ($CaCl_2$) (Fisher Chemicals, C70-500)
        ii. Potassium chloride (KCl; Fisher Chemicals, P217-500)
        iii. Potassium phosphate (KH_2PO_4; Fisher Chemicals, P285-500)
        iv. Magnesium sulfate ($MgSO_4$; Fisher Chemicals, M65-500)
        v. Sodium chloride (NaCl; Mallinckrodt, 7581)
        vi. Sodium phosphate, dibasic anhydrous (Na_2HPO_4; Fisher Chemicals, S374-500)
        vii. Sodium bicarbonate ($NaHCO_3$; Fisher Chemicals, S233-500)
4. Fish Food
    a. Larval AP100, <100 Micron particle size (Zeigler Bros., 38-4710-02-11)
    b. Larval AP100, 100–150 Micron particle size (Zeigler Bros., 38-4709-03-11)
    c. Larval AP100, 150–250 Micron particle size (Zeigler Bros., 38-4710-04-11)
    d. San Francisco strain brine shrimp eggs (Brine Shrimp Direct)
    e. Premium brine shrimp eggs (Artemia International)
    f. Golden Pearls (Artemia International)
    g. Spirulina aquarium flake food (Ocean Star International (O.S.I.))
5. Shipping
    General Supplies:
    a. Styrofoam boxes, nested in cardboard box:
        VWR International, https://us.vwr.com/
        i. Small box dimensions 8″ L × 6″ W × 9″ H; Catalog number 14100-412
        ii. Medium box dimensions 12″ L × 10″ W × 7″ H; Catalog number 14100-402
        iii. Extra-large box dimensions 22″ L × 15″ W × 15″ H; Catalog number 15714-704

        Superior Shipping, http://superiorshippingsupplies.com/category and http://www.thermosafe.com/.
    b. Solutions:
        i. Ammonia binder: ClorAM-X, Reed Mariculture, http://reedmariculture.com/product_cloram_x.php
        ii. Sterile Fish System Water
    c. Styrofoam packing peanuts, do not use water dissolvable cellulose!
    d. Heat packs:
        i. Superior Shipping, http://superiorshippingsupplies.com/category
        ii. For the United Kingdom: http://www.heatpacksuk.co.uk/heat-packs.php

   iii. 40 h pack for domestic (overnight) shipments
   iv. 60–72 h packs for international (2–4 day) shipments
e. Live fish labels (Fig. 4B), address, courier labels, and document pouch (Fig. 4I)
f. Customs and veterinary health certification documents for international shipments

Embryo Packaging:

g. 250 mL tissue culture flasks
   i. Fisher Scientific, catalog # 10-126-5; http://www.fishersci.com
   ii. VWR International, catalog # 82050-854; https://us.vwr.com/
h. 0.5× embryo medium (E2)
i. Methylene blue (optional)

Adult Packaging:

j. Polyethylene plastic bags: at least 0.003 mm or 1.2×10^{-4} inches thick)
   i. Gulf Coast Plastics for bulk quantity, http://www.gulfcoastplastics.com/BottomSeal.htm
   ii. Superior Shipping, http://superiorshippingsupplies.com/category
   iii. Check any pet store for small quantities
   iv. Average size for holding 10–20 fish (10″ × 18″ (0.003 mil))
   v. Note: Square bottom bags can be made from the above using a plastic heat/bag sealer. You can also tape the corners of the bag to create square bottom.
k. Pure oxygen (not required, recommended)
l. Rubber bands, plastic ties, or metal clips

ACKNOWLEDGMENTS

I am grateful to all friends and colleagues at ZIRC for providing help and information. Special thanks to April Freeman, Katy Murray, Alex Schumann, and David Lains for reading and suggestions.

REFERENCES

Andrews, C., Exell, A., & Carrington, N. (2010). *The manual of fish health: Everything you need to know about aquarium fish, their environment and disease prevention.* (p. 208) Buffalo, New York: Firefly Books (U.S.) Inc.

Best, J., Adatto, I., Cockington, J., James, A., & Lawrence, C. (2010). A novel method for rearing first-feeding larval zebrafish: polyculture with Type L saltwater rotifers (*Brachionus plicatilis*). *Zebrafish, 7*, 289–295.

Brand, M., & Nüsslein-Volhard, C. (2002). Keeping and raising zebrafish. *Zebrafish* (Chapter 1 (pp. 7–37). New York: Oxford University Press Oxford.

Cahu, C., Infante, J. Z., & Takeuchi, T. (2003). Nutritional components affecting skeletal development in fish larvae. *Aquaculture, 227*, 245–258.

Carmichael, C., Westerfield, M., & Varga, Z. M. (2009). Cryopreservation and in vitro fertilization at the zebrafish international resource center. *Methods in Molecular Biology, 546*, 45–65.

Draper, B. W., McCallum, C. M., Stout, J. L., Slade, A. J., & Moens, C. B. (2004). A high-throughput method for identifying N-ethyl-N-nitrosourea (ENU)-induced point mutations in zebrafish. *Methods in Cell Biology, 77*, 91–112.

Draper, B. W., & Moens, C. B. (2009). A high-throughput method for zebrafish sperm cryopreservation and in vitro fertilization. *Journal of Visualized Experiments: JoVE*. http://www.jove.com/video/1395/a-high-throughput-method-for-zebrafish-sperm-cryopreservation-vitro.

Engeszer, R. E., Patterson, L. B., Rao, A. A., & Parichy, D. M. (2007). Zebrafish in the wild: a review of natural history and new notes from the field. *Zebrafish, 4*, 21–40.

Goessling, W., North, T. E., & Zon, L. I. (2007). New waves of discovery: modeling cancer in zebrafish. *Journal of Clinical Oncology, 25*, 2473–2479.

Grunwald, D. J., & Eisen, J. S. (2002). Headwaters of the zebrafish – emergence of a new model vertebrate. *Nature Reviews Genetics, 3*, 717–724.

Harper, C., & Lawrence, C. (2011). *The laboratory zebrafish*. Boca Raton: CRC Press.

Harvey, B., Kelley, R. N., & Ashwood-Smith, M. J. (1982). Cryopreservation of zebrafish (*Brachydanio rerio*) spermatozoa using methanol. *Canadian Journal of Zoology, 60*, 1867–1870.

Jhingran, A. (1991). *Inland fishes of India and adjacent countries*. books.google.com.

Kettleborough, R. N., Busch-Nentwich, E. M., Harvey, S. A., Dooley, C. M., de Bruijn, E., van Eeden, F. … Stemple, D. L. (2013). A systematic genome-wide analysis of zebrafish protein-coding gene function. *Nature, 496*, 494–497.

Kohler, S. J., Cubillas, P., Rodriguez-Blanco, J. D., Bauer, C., & Prieto, M. (2007). Removal of cadmium from wastewaters by aragonite shells and the influence of other divalent cations. *Environmental Science & Technology, 41*, 112–118.

Lall, S. P., & Lewis-McCrea, L. M. (2007). Role of nutrients in skeletal metabolism and pathology in fish – an overview. *Aquaculture, 267*, 3–19.

Lawrence, C. (2007). The husbandry of zebrafish (*Danio rerio*): a review. *Aquaculture, 269*, 1–20.

Lawrence, C. (2016). New Frontiers for Zebrafish Management. In H. W. Detrich, III, L. I. Zon, & M. Westerfield (Eds.), *The Zebrafish: Genetics, Genomics, and Transcriptomics. 135* (pp. 483–508).

Lawrence, C., Sanders, G. E., Varga, Z. M., Baumann, D. P., Freeman, A., Baur, B., & Francis, M. (2009). Regulatory compliance and the zebrafish. *Zebrafish, 6*, 453–456.

Lieschke, G. J., & Currie, P. D. (2007). Animal models of human disease: zebrafish swim into view. *Nature Reviews Genetics, 8*, 353–367.

Murray, K. N., Bauer, J., Tallen, A., Matthews, J. L., Westerfield, M., & Varga, Z. M. (2011). Characterization and management of asymptomatic Mycobacterium infections at the Zebrafish International Resource Center. *Journal of the American Association for Laboratory Animal Science, 50*, 675–679.

Oka, T., Nishimura, Y., Zang, L., Hirano, M., Shimada, Y., Wang, Z. … Tanaka, T. (2010). Diet-induced obesity in zebrafish shares common pathophysiological pathways with mammalian obesity. *BMC Physiology, 10*, 21.

Paik, E. J., & Zon, L. I. (2010). Hematopoietic development in the zebrafish. *International Journal of Developmental Biology, 54*, 1127–1137.

Patton, E. E., & Zon, L. I. (2005). Taking human cancer genes to the fish: a transgenic model of melanoma in zebrafish. *Zebrafish, 1*, 363–368.

Phillips, J. B., & Westerfield, M. (2014). Zebrafish models in translational research: tipping the scales toward advancements in human health. *Disease Models & Mechanisms, 7*, 739–743.

Ramsay, J. M., Feist, G. W., Varga, Z. M., Westerfield, M., Kent, M. L., & Schreck, C. B. (2006). Whole-body cortisol is an indicator of crowding stress in adult zebrafish, *Danio rerio*. *Aquaculture, 258*, 565–574.

Sessa, A. K., White, R., Houvras, Y., Burke, C., Pugach, E., Baker, B. … Zon, L. I. (2008). The effect of a depth gradient on the mating behavior, oviposition site preference, and embryo production in the zebrafish, *Danio rerio*. *Zebrafish, 5*, 335–339.

Trevarrow, B. (2004). Zebrafish facilities for small and large laboratories. *Methods in Cell Biology, 77*, 565–591.

Varga, Z. M. (2011). Aquaculture and husbandry at the zebrafish international resource center. *Methods in Cell Biology, 104*, 453–478.

Varshney, G. K., Lu, J., Gildea, D. E., Huang, H., Pei, W., Yang, Z. … Lin, S. (2013). A large-scale zebrafish gene knockout resource for the genome-wide study of gene function. *Genome Research, 23*, 727–735.

Westerfield, M. (2007). *The zebrafish book. A Guide for the laboratory use of zebrafish (*Danio rerio*)*. Eugene: University of Oregon Press.

Whiteley, A. R., Bhat, A., Martins, E. P., Mayden, R. L., Arunachalam, M., Uusi-Heikkila, S. … Bernatchez, L. (2011). Population genomics of wild and laboratory zebrafish (*Danio rerio*). *Molecular Ecology, 20*, 4259–4276.

Wilson, C. A., High, S. K., McCluskey, B. M., Amores, A., Yan, Y. L., Titus, T. A. … Postlethwait, J. H. (2014). Wild sex in zebrafish: loss of the natural sex determinant in domesticated strains. *Genetics, 198*, 1291–1308.

CHAPTER

Health monitoring and disease prevention at the Zebrafish International Resource Center

26

Z.M. Varga[1], K.N. Murray[1]

University of Oregon, Eugene, OR, United States

[1]*Corresponding authors: E-mail: zoltan@zebrafish.org; katy@zebrafish.org*

CHAPTER OUTLINE

Methods in Cell Biology, Volume 135, ISSN 0091-679X, http://dx.doi.org/10.1016/bs.mcb.2016.04.020

Abstract

In this chapter we review the components of the fish health program at the Zebrafish International Resource Center. We describe health-monitoring strategies to assess individual and colony health, practices to prevent the spread of pathogens within the fish colony, and a biosecurity program designed to prevent entry of new fish pathogens. While this program is designed for a facility on a recirculating water system with expectations of high volumes of import and export, many of the components can be directly applied or modified for application in facilities of different sizes and with other programmatic goals.

INTRODUCTION

The Zebrafish International Resource Center (ZIRC) is a closed facility that is physically separate from other aquatic facilities on the University of Oregon campus. ZIRC serves the biomedical research community by acquiring, maintaining, and redistributing wild-type, mutant, and transgenic zebrafish lines. In addition, ZIRC provides fish health consultations and a diagnostic pathology service. While providing these services we also maintain biosecurity and a healthy in-house fish population.

The ZIRC Main Fish Room is fed by recirculating mechanical and biological water filtration systems with UV sterilizers (see chapter 26: *Aquaculture, Husbandry, and Shipping at the Zebrafish International Resource Center* by Varga, 2016). Separate systems feed the two sides of the Main Fish Room. On average, 40,000 wild-type, mutant, and transgenic zebrafish are maintained in the facility at any time.

The ZIRC building also has a Quarantine Room, which functions on a separate flow-through water system. Water can be recirculated on one rack in the Quarantine Room, but the water originates from the flow-through system. New lines submitted to ZIRC are imported as live fish into the Quarantine Room or as cryopreserved sperm. After fish are acclimated in the Quarantine Room for at least 2 weeks, sperm is collected from males and cryopreserved for future regeneration of the line. When this is not possible, fish are spawned in the Quarantine Room, and their surface-sanitized embryos may be reared in the Main Fish Room or the Quarantine Room depending on their assigned biosecurity risk level.

New lines that enter the Main Fish Room come mainly from in vitro fertilization (IVF) using cryopreserved sperm and AB wild-type eggs from females in the Main Fish Room, and rarely from spawns in the Quarantine Room. Adult fish and embryos leave the main fish room to fulfill customer orders. Adult fish never leave the Quarantine Room and are euthanized after their genetic resource has been secured in the ZIRC collection.

As fish and gametes move into and around ZIRC, there is potential for fish pathogens to move with them. Here we describe procedures and policies that we have instituted to monitor individual, tank, and colony health status, to minimize the spread of previously identified pathogens within the facility, and to identify and prevent risks for new pathogen entry. This fish health program was designed to provide the highest level of animal welfare, to ensure redistribution of healthy fish to the research community, and to safeguard human health and safety.

1. METHODS

1.1 ASSESSING COLONY HEALTH

To determine the prevalence, spread, and success of control measures on pathogens in the Main Fish Room, we perform daily health monitoring and regular diagnostic testing of the groups described later. We have also identified background mutations and line-specific noninfectious pathologies with this approach. Finally, we believe this monitoring strategy will also facilitate rapid diagnosis of any new pathogens in the Main Fish Room.

1.1.1 Moribund fish

Fish are monitored daily for morbidity and mortality. Staff and student employees receive training in identification of signs of disease in zebrafish and the importance of removing these animals for overall colony health (Table 1). We stress that when feeding fish, cleaning a tank, or working with fish, one can also perform a health check. Employees isolate and label moribund fish in a 3.75-L (1-gallon) tank and inform the ZIRC veterinarian that they have identified a moribund fish. If they are in the middle of a task when they see a moribund fish, they mark the tank with a waterproof red sticky flag (Post-it's, 3M). They can remove the fish when they finish their task. Otherwise, an employee assigned to look for flagged fish will remove it during a facility health check.

The majority of fish displaying clinical signs of disease are euthanized, fixed in Dietrich's fixative, and submitted for histological processing, sectioning, and

Table 1 Physical and Behavioral Abnormalities That May Be Signs of Illness or Injury

Physical Abnormalities	Behavioral Abnormalities
Color change	Fish at surface or near water inlet
Weight loss	Fish on bottom of tank
Exophthalmia/pop-eye	Rapid operculum movements
Distended abdomen	Sluggish movements/lethargy
Skeletal deformity	Flashing/rubbing on tank surfaces
Mass/swelling	Circling, twirling, spinning
Hemorrhage/redness	Loss of equilibrium
Gas bubbles	Clamped fins
Protruding scales	
Fin erosion or lesion	
Skin ulceration or lesion	

staining. Immediate removal of moribund fish is key for early detection of pathogens. In the short term, removal of moribund fish prevents cannibalism and further spread of possible infectious agents. In the long term, removal and diagnostic testing of moribund fish allows us to implement appropriate control measures for infectious pathogens and make husbandry adjustments to improve nonpathogen associated morbidities at an early stage. However, to get an accurate picture of the prevalence of infectious and noninfectious pathologies in a colony, it is important to include clinically normal fish in a sampling strategy. Previous surveys at ZIRC have revealed a low prevalence of *Mycobacterium chelonae* infections in asymptomatic fish (Murray, Bauer, et al., 2011; Whipps, Matthews, & Kent, 2008). Therefore, we perform sampling and diagnostic testing of nonmoribund populations to assess the levels of preclinical and asymptomatic pathologies in the colony.

1.1.2 Sentinel fish

Along with testing moribund fish, a sentinel program can provide valuable information about pathogens present in a water system. It is also a useful tool in evaluating the efficacy of a filtration system and UV sterilizers in preventing the spread of pathogens within a facility. We use AB wild types as sentinels as this line is most representative of the colony. AB is our most popular wild-type line shipped to researchers and nearly all mutant, and transgenic lines at ZIRC are maintained on an AB background. Sentinel tanks are populated with 3- to 4-month-old AB fish from the same stock. A sample of fish from the source tank is tested for the microsporidian parasite *Pseudoloma neurophilia* before populating the sentinel tanks. ZIRC has sentinel tanks in pre- and postfiltration locations on both water systems feeding the Main Fish Room. Each location holds three tanks representing 3-months, 6-months, and 1-year exposure times. We maintain tanks representing three different exposure times to increase chances of catching pathogens requiring a minimum infective dose or time to develop disease and age-related pathologies. All exposure groups are sampled quarterly for histopathology.

However, reliance on sentinel fish testing alone can result in an underestimate of pathogen presence. Prefiltration sentinel fish become infected after receiving an infective dose from pooled system effluent. For a large water system, this can mean that several tanks or lines of fish become infected before sentinel fish receive an infective dose. Significant lag time may occur between when a pathogen is introduced in the colony, and when it is detected in sentinel fish, time that could be used to implement control measures to prevent an outbreak.

1.1.3 Wild-type lines

In addition to sentinel fish, we regularly perform diagnostic testing of clinically normal wild-type fish. Wild-type lines are maintained in 75-L (20-gallon) tanks of 150 − 300 fish. If a stock, which represents a specific breeding event, is split between multiple 20-gallon tanks, the tank populations remain separate. Fish from one tank do not move back and forth between tanks containing the same stock. However, if tank populations go below 100 fish, tanks of the same stock may be combined into one tank to maintain consistent animal densities and welfare conditions, and this is recorded in the database. In this way, we consider each tank a separate biosecure unit. At 8 months postfertilization, all tanks of wild-type fish are screened for *P. neurophilia* infection. For most lines, five fish are sampled, which gives 95% confidence of detecting a disease prevalence of 50% or higher (AFS-FHS, 2007; Simon & Schill, 1984). Because AB wild-type fish are used and spawned most frequently, we test 11 fish from all AB tanks, which provides 95% confidence of detecting a prevalence of at least 25%. Positive tanks are immediately removed and euthanized according to an IACUC-approved protocol.

The 8-month sample point was set after an initial screening of all wild-type tanks. During that screen and follow-up testing, we found false negative results for tanks tested before 8 months. The prevalence of *Pseudoloma* infection within a tank increases with age and accumulated exposure (Murray, Dreska, et al., 2011). Increasing the sample size for younger tanks could decrease the probability of a false negative. For example, by testing 28 fish per tank, one could detect a prevalence of 10% or more with 95% confidence. However, because we strive to isolate fish tanks procedurally, sacrificing fewer animals at a later time is acceptable considering the type of pathogen, nature of disease progression, animal welfare considerations to produce as few as possible animals for stocks maintenance, and the ZIRC mission as a genetic resource distribution center.

1.1.4 Additional sampling

Additional tanks and stocks are sampled for diagnostic testing depending on clinical signs, stock use, general health screening, and to determine prevalence of a trait or pathogen within a particular population. For example, in 2015 we sampled stocks that had exceeded our normal year-and-a-half lifespan due to extensive troubleshooting of the genotyping protocol and stocks that exhibited poor fecundity throughout their lifespan.

1.1.5 Diagnostic tests and results

The majority of fish sampled for diagnostic testing are euthanized, fixed in Dietrich's fixative, and submitted for histological processing, sectioning, and staining with hematoxylin and eosin (Humason, 1967). Histopathology is our diagnostic test of choice for broad screening because it allows for examination of all tissues for infectious pathogens, changes related to suboptimal water quality, neoplasia, and numerous other noninfectious pathologies. In fact, the initial diagnosis and descriptions of many of the major pathogens affecting zebrafish in research facilities, including *P. neurophilia* (Matthews et al., 2001), *M. chelonae* (Whipps et al., 2008), *Pleistophora hyphessobryconis* (Sanders et al., 2010), *Edwardsiella ictaluri* (Hawke et al., 2013), and *Myxidium streisingeri* (Whipps, Murray, & Kent, 2015) were made by screening histological sections of moribund zebrafish submitted to the ZIRC Diagnostic Pathology Service.

When screening wild-type tanks for *P. neurophilia*, fish are euthanized and submitted for histopathology with a Luna stain (Luna & Armed Forces Institute of

Pathology (US), 1968) or are dissected for an in-house PCR-based assay (Murray, Dreska, et al., 2011). While the Luna stain nicely highlights microsporidia spores in tissue sections, staining them brick red against a background of blue fish tissue (Peterson, Spitsbergen, Feist, & Kent, 2011), other pathologies can also be discerned in the sections, if present.

Other diagnostic tests, including bacterial culture, fin biopsy, skin scrapes, testing of tank debris, and PCR screening are performed as needed depending on clinical signs, histopathology results, and the number of moribund fish found in a particular tank or stock. Results of diagnostic tests are available in the ZIRC online Animal Health Report, which is updated quarterly at http://zebrafish.org/zirc/documents/health_report.php.

Diagnostic testing of fish from the ZIRC Main Fish Room has revealed *M. chelonae* and *P. neurophilia.* A survey of *Mycobacterium* spp. at ZIRC identified *M. chelonae* in biofilms and fish tissue (Whipps et al., 2008). *Mycobacterium chelonae* is considered an environmental, low-virulence species. At ZIRC, it is most often associated with subclinical infections (Murray, Bauer, et al., 2011). When mycobacteriosis is diagnosed in histological sections, the paraffin block of embedded tissue is sent to the Oregon State Veterinary Diagnostic Lab for a quantitative PCR assay to identify the species of *Mycobacterium* in the sections. *Mycobacterium chelonae* remains the only identified species in fish tissue from the ZIRC Main Fish Room.

The majority of morbidities observed at ZIRC and submitted for histopathology have noninfectious etiologies (Fig. 1A). *Pseudoloma neurophilia* and *M. chelonae* accounted for only 5% and 2%, respectively of the moribund fish examined from January through October of 2015. Clinical abnormalities can also be attributed to degenerative processes, physical anomalies, background genetic mutations, metabolic disturbances, neoplasia, noninfectious inflammation, and trauma. In addition, the majority of fish diagnosed with *P. neurophilia* and *M. chelonae*, appeared clinically normal (Fig. 1B), highlighting the importance of also sampling nonmoribund populations when assessing distribution and prevalence of pathogens.

1.2 PREVENTING PATHOGEN SPREAD WITHIN THE COLONY

We take a number of steps to prevent the spread of *P. neurophilia* and *M. chelonae* within and between tank populations. In addition, the procedures described later would allow us to isolate and prevent spread of any unintended introductions of new pathogens in the Main Fish Room.

1.2.1 UV sterilizers

UV sterilizers are an essential component for maintaining biosecurity and preventing pathogen spread in a recirculating water system. At ZIRC, the UV sterilizers are positioned at the end of the biological and mechanical filtration units in the water systems. Three sterilizers are in use, two on the water system that has historically utilized the higher water volume. Each sterilizer contains five 200-W bulbs (Aquaneering Inc.). At our maximum flow rate of 750 L/min (200 gal/min), each sterilizer

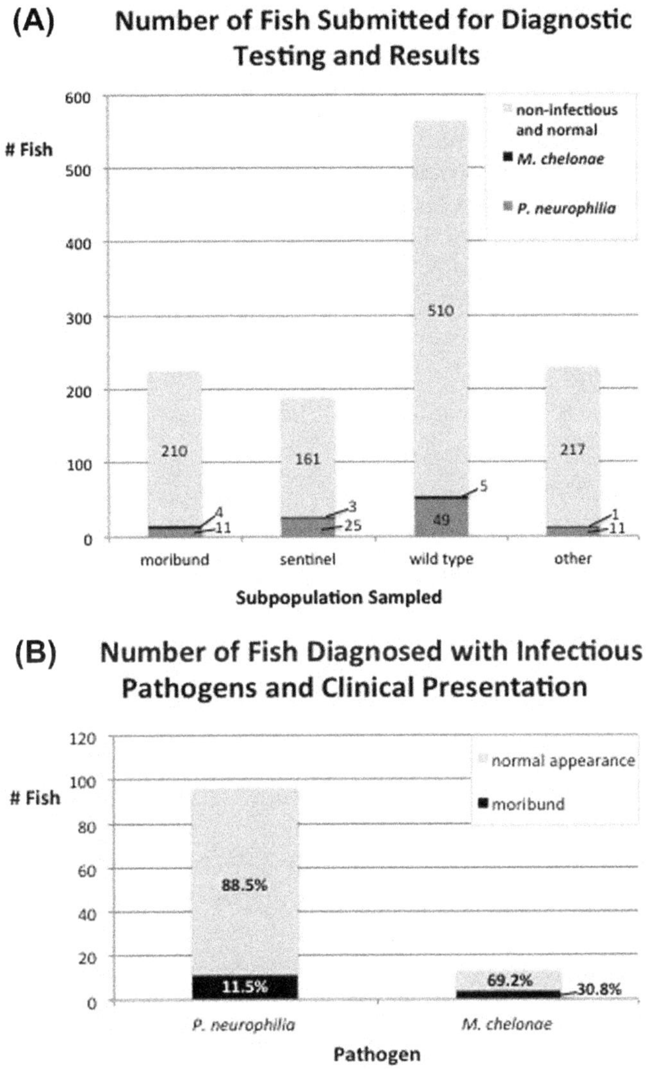

FIGURE 1 Fish sampled for diagnostic testing from January through October of 2015.

(A) Number of fish diagnosed with *Pseudoloma neurophilia*, *Mycobacterium chelonae*, and noninfectious/normal status for different sample groups. (B) The total number of fish diagnosed with *P. neurophilia* and *M. chelonae* with normal and moribund phenotypes.

delivers a UV dose of 165,000–132,000 μWsec/cm^2 over the first 20% decay in lamp life (assuming 98% water transmissibility). Slower flow rates result in greater dwell time in the sterilizer and hence a higher UV dose. UV output is monitored electronically. Bulbs are changed when output falls to 80% or after one-and-a-half years, whichever occurs first. The effect of the UV dose is dependent upon water transmissibility as fine particulates can shield pathogens from UV irradiation. To reduce particulates we have installed pressurized sand filters before the UV sterilizers. Bag filters, which can also be loaded with activated carbon, are on an optional bypass loop and are utilized as needed (see also chapter 26: *Aquaculture, Husbandry, and Shipping at the Zebrafish International Resource Center* by Varga, 2016). Because UV sterilization is directed at a short section of flowing water, it does not control pathogen proliferation and biofilm formation in the tanks and connected pipes. Thus, additional forms of disinfection strategies are required to control the pathogen load and maintain water quality within the aquaria.

1.2.2 Husbandry

We have identified several points during the fish lifespan where we have the opportunity to minimize potential pathogen transmission.

a. Embryo surface sanitization

All embryos produced at ZIRC, whether raised in-house or shipped to researchers, are surface sanitized in a 30 mg/L sodium hypochlorite solution. We bleach all wild-type lines individually in new solutions. Ten-minute exposure to 30 mg/L bleach solution has been shown to effectively neutralize *Mycobacterium marinum* on chorions when embryos are 24 h postfertilization or younger (Karen Guillemin, personal communication). It will not neutralize *M. chelonae* or *P. neurophilia* spores (Ferguson, Watral, Schwindt, & Kent, 2007). We include swirling and agitation of embryos in postbleach Rinse Solutions after bleaching to encourage dislodging *P. neurophilia* spores from the chorion surface, although the efficacy of agitation has not been documented. Our detailed embryo surface sanitization procedure is explained later. Exposure to bleach is toxic to fish and therefore embryos should not be bleached once they start to hatch (after 48 hpf).

a.1. Preparations:

1. Stock Solutions
   - 1.1. Methylene Blue Stock Solution: Dissolve 1 g methylene blue powder into 1 L RO water.
   - 1.2. Pronase Stock Solution: Dissolve 600 mg protease (from *Streptomyces griseus*) in 20 mL nanopure water. Stir very gently until dissolved. Keep on ice throughout preparation and aliquoting. Freeze 0.68 mL aliquots at −80°C for long-term storage or −20°C for short-term use. Do not thaw and refreeze.
2. Working Solutions
   - 2.1. First and Last Rinse Solution (Fig. 2; containers 1 and 5): Mix 10 mL methylene Blue Stock Solution with 20 L of 0.5× E2 embryo medium.
   - 2.2. Bleach Solution (Container 2): Mix 0.59 mL sodium hypochlorite (5.65–6% chemical grade bleach) with 1 L RO water (prepare fresh daily).

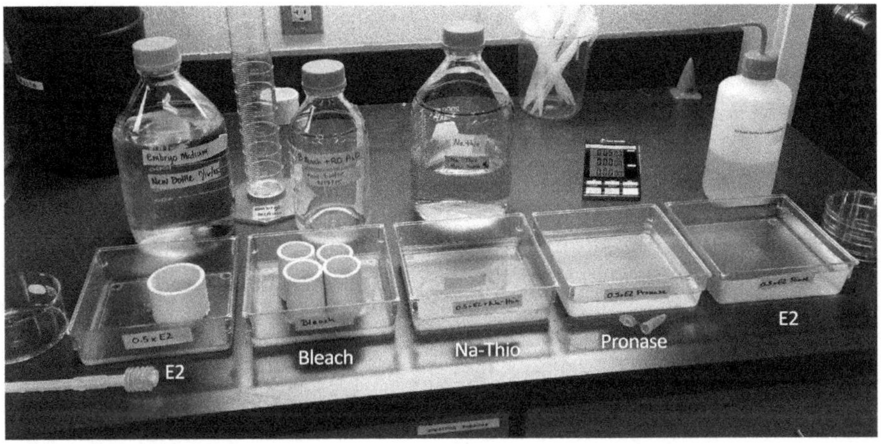

FIGURE 2 Embryo bleaching station.

2.3. Sodium Thiosulfate Solution (Container 3): Dissolve 1 g sodium thiosulfate in 2 L 0.5× E2 embryo medium.

2.4. Pronase Solution (Container 4): Add 0.68 mL Pronase Stock Solution to 500 mL 0.5× E2. Prepare fresh daily or every 2 h, as needed.

3. Prepare Stock and Working Solutions. Prepare mesh baskets (cut from PVC pipe with fine mesh hot-glued to the bottom) according to size and number of embryo batches. To bleach more than one batch, label each mesh basket to keep track of eggs, batches, and baskets. If necessary, note batches and corresponding basket labels on a piece of paper.
4. Important: Remove dead embryos and unfertilized eggs. Remove all debris and detritus by pipetting eggs into a clean Petri dish (es) with Rinse Solution.
5. Sanitize the work space with 70% alcohol (EtOH).
6. Set timer for 10 min.
7. Prepare five flat dish containers (Fig. 2):
   Container 1: Load/Rinse Solution
   Container 2: Bleach Solution
   Container 3: Sodium Thiosulfate Solution
   Container 4: Pronase Solution
   Container 5: Last Rinse Solution

   Containers 2, 3, 4, and 5 should contain progressively greater volumes to sufficiently neutralize or dilute the previous solution.

a.2 Bleaching Procedure:

1. Fill container 1 with Rinse Solution; use the smallest possible volume of Rinse Solution to cover the eggs.
2. Place mesh baskets into container 1.
3. Pipette up to 100 eggs into a small or up to 300 eggs into a large mesh basket.

4. Change pipettes between embryo batches.
5. Once all eggs are submersed in Rinse Solution, transfer the basket(s) to Container 2 with Bleach Solution. Start the timer and bleach for 10 min. Do not swirl or agitate embryos in the bleach solution.
6. Transfer baskets to Container 3, which contains the sodium thiosulfate. Gently swirl eggs in solution for 5 s.
7. Transfer eggs to Container 4. Swirl eggs in solution for 1 min.
   Note: Hatching occurs approximately 24 h following pronase treatment.
8. Transfer baskets to Container five and swirl in solution for 2 min.
9. After the final rinse, use a squirt bottle with Rinse Solution to wash the eggs out of the basket. Rinse the eggs into a clean finger bowl or Petri dish, and label.
10. Clean and rinse all equipment after use (following facility SOP).

b. Room and water system separation

Separate water systems supply sides A and B of the ZIRC Main Fish Room. Nursery, grow-out, and adult sections are present on each side. Surface-sanitized embryos are assigned to one side, where the stock remains until cryopreservation for long-term storage. AB and TAB wild-type stocks and mutant and transgenic lines that have been raised at ZIRC for several generations are maintained on side A. Other wild-type lines, stocks generated by IVF with thawed sperm from lines cryopreserved in the Quarantine Room or at an outside facility, and embryos from spawns in the Quarantine Room are all maintained on side B. On each side, all tanks of the same wild-type line are clustered together.

c. Maintenance, outcrossing, and retirement

Husbandry stress (chronic crowding and acute handling stressors) has been shown to affect susceptibility to and severity of infection with *P. neurophilia* and *Mycobacterium* spp. (Ramsay, Watral, Schreck, & Kent, 2009a, 2009b). Therefore, tank densities, feeding amounts and times, and time between spawning (see also chapter 26: *Aquaculture, Husbandry, and Shipping at the Zebrafish International Resource Center* by Varga, 2016) have been set to minimize stress on the fish.

Transmission of pathogens between tanks may occur particularly when populations are mixed for outcrossing. To minimize this potential, wild-type fish that have been outcrossed to mutant or transgenic carriers are not returned to their original tank. Once outcrossed, wild-type fish are euthanized, or if additional breeding is necessary, they are isolated in a separate 1-gallon tank and reused for breeding with that same mutant or transgenic stock.

Additional measures to minimize cross-contamination of stocks and tanks are used for example when fish are anesthetized for fin clipping or collection of gametes. For these procedures, separate anesthesia and recovery solutions and bowls are used for each tank of fish.

Because the likelihood of pathogen exposure and infection increases with age (Murray, Dreska, et al., 2011), removing aging stocks from the population decreases the risk of pathogen accumulation and transmission. ZIRC fish stocks are maintained for shipment to customers, collection of sperm for cryopreservation, outcrossing, and/or propagation of another generation. When these goals have been met, the stock is euthanized. Our goal is to stop maintaining stocks by 1 year-of-age. Depending on the number of fish carrying a particular allele, time needed for genotyping, and number of sperm samples needed to cryopreserve the line at the Resource Center, the lifespan of a stock may be extended, but generally not beyond a year-and-a-half.

1.2.3 Tank cleaning

The tank environment can harbor pathogen reservoirs beyond the fish population. *Mycobacterium* spp. exist in tank biofilms (Whipps et al., 2008) and *P. neurophilia* spores have been detected in spawning water (Sanders & Kent, 2011) and transmitted between tanks by sharing of tank debris (Murray, Dreska, et al., 2011). Eggs from *Pseudocapillaria tomentosa*, the intestinal nematode infecting zebrafish, can also be recovered from debris in tanks with infected fish (Murray & Peterson, 2015). Tank and equipment cleaning protocols are in place at ZIRC to prevent transmission of pathogens from these potential tank reservoirs. In addition, we treat each tank as an individual biosecure unit, regardless of genetic similarity to neighboring tanks.

A clean, autoclaved net is used on every tank. Twenty-gallon tanks are cleaned with the fish in the tank and a clean, autoclaved scrubber is used on every tank. Employees change gloves between cleaning different tanks. After scrubbing and dislodging debris and biofilm from the tank surfaces, the water inflow is increased to flush the debris out the tank drain, minimizing exposure time for fish and potential for ingestion (Murray, Bauer, et al., 2011). When 1-gallon tanks are dirty, fish are moved to a clean, new 1-gallon tank.

Debris in used/dirty tanks is manually removed, and tanks are soaked in a 0.15% bleach solution for a minimum of 20 min, well above the 500 mg/L bleach needed to kill *P. neurophilia* spores (Ferguson et al., 2007). After bleaching, tanks are soaked in 0.002% sodium thiosulfate to neutralize bleach residues, rinsed with hot water (82°C), and air-dried.

Tank and equipment sanitation procedures are routinely tested for effective disinfection with RODAC plates. The autoclave is tested using Attest Biological Indicator for Gravity Steam Sterilizer (3M).

1.2.4 Surface and equipment disinfection

At least once daily, work surfaces are sprayed with 70% ethanol, wiped to remove debris, then sprayed again and air-dried. After every use, embryo collection strainers are soaked in 70% ethanol for 10 min and air-dried. Tank scrubbers and nets are rinsed in tap water to remove obvious debris then autoclaved. Glass bowls used for embryo collection and anesthesia and mesh baskets that hold embryos during surface sanitation are cleaned in a laboratory grade dishwasher (with LabSolutions nonfoaming powder detergent and a 10 min rinse at 180°F). Floors are mopped twice weekly with a 600 ppm bleach solution.

1.2.5 Database tracking and record keeping

We record in our database whenever moribund fish are removed from a tank and the clinical signs observed. A stock or tank can also be assigned a particular health status in the database to reflect diagnosis of a pathogen, background mutation, an abnormality, or biosecurity risk. These records have allowed us to map tanks affected by these status changes and hypothesize about means of spread and transmission. In addition, the database helps to establish a baseline for normal levels of morbidity and mortality. Most of the morbidities that we have observed and tested are not related to infectious pathogens (Fig. 1A). Therefore, diagnostic test results should be regularly evaluated along with levels of morbidity and mortality to establish whether or not an increase in the latter should be regarded as a disease outbreak, an indication of suboptimal husbandry, or other etiology.

1.3 PREVENTING ENTRY OF NEW PATHOGENS

ZIRC receives new fish lines from facilities world wide and is expected to redistribute healthy fish. Identifying risks and preventing entry of new pathogens are key components of our biosecurity program. Policies and procedures are in place to minimize the risk of pathogen entry with imported live fish, cryopreserved sperm, and with personnel.

1.3.1 Request of health history from submitting laboratory

Before importing new zebrafish lines, we request health history information from the submitting laboratory. We ask for information about the water system, quarantine and introduction of new fish lines, frequency and type of health monitoring, sampling, and diagnostic testing that is performed, results of diagnostic tests, and the history of pathogens in the facility. This information is used to formulate a plan for handling the imported line(s) that will minimize biosecurity risks. For example, fish coming from a facility with a history of *P. tomentosa* could be monitored for the nematode infection by testing the tank debris for the capillarid eggs (Murray & Peterson, 2015). A fill-able form for facilities to use as a template for reporting their animal health results is available on our website, along with the ZIRC Animal Health Report.

1.3.2 Quarantine room

New lines imported as adult fish are housed in the Quarantine Room, which is isolated from the Main Fish Room by location, equipment, and procedures. When fish arrive at the Resource Center, shipping containers remain unopened until they are in the Quarantine Room. Once opened, we record water temperature, pH, and ammonia/nitrate/nitrite levels and transfer the fish to their new tanks. Genotype, date of birth, number and condition of fish, and water quality parameters are entered into the database. All shipping containers and packing materials are to disposed of immediately and are not reused because of potential contamination.

For the majority of imported lines, we request that only male zebrafish be submitted. Once acclimated to the Quarantine Room environment, the fish are anesthetized, sperm is obtained for cryopreservation, and fin clips may be obtained for later genetic identification. All imported fish are maintained in the Quarantine Room for their lifetime at ZIRC and euthanized after sufficient sperm samples have been acquired. When ordered by customers, these lines are regenerated by IVF using eggs from AB females reared in the Main Fish Room. Requiring that new lines go through cryopreserved sperm has cut down on the number of lines that we need to maintain in the Main Fish Room. In addition, it gives us greater potential to eliminate pathogens coming through the imported paternal stock compared to spawning and moving their surface-sanitized embryos into the Main Fish Room.

Quarantine Room tanks are monitored daily for morbidity and mortality. Moribund fish are euthanized, a piece of liver and spleen are removed and frozen at −80°C, and the rest of the body is fixed in Dietrich's fixative and submitted for histopathology. If mycobacteriosis is diagnosed in tissue sections, the frozen tissues are submitted to the Oregon State University Veterinary Diagnostic Lab (OSU VDL) for a PCR assay to identify the species of *Mycobacterium* in the tissues. All other imported fish in the Quarantine Room are euthanized, fixed, and submitted for histopathology after sufficient sperm samples are obtained. Paraffin blocks of fixed tissue may be sent to OSU VDL for follow-up PCR assays depending on histopathology findings. Results of diagnostic tests are used to determine the biosecurity risk posed by the next generation coming from cryopreserved sperm. If virulent pathogens, such as *M. marinum* (Watral & Kent, 2007), *Mycobacterium haemophilum* (Whipps, Dougan, & Kent, 2007), or *E. ictaluri* (Hawke et al., 2013), are identified, then the sperm samples from these fish are flagged in the database and the facility of origin may be contacted about reimporting the line. If reimportation is not possible, then the sperm samples may be used in IVF, but the progeny must be raised in the Quarantine Room. When they are mature, new sperm samples are obtained, replacing any remaining flagged samples, and all contributing males are again submitted for histopathology and follow-up PCR as needed. If no virulent pathogens are identified in males contributing to cryopreserved samples, then the sperm may be used for IVF and the progeny introduced to the B-side of the Main Fish Room.

Ideally, wild-type lines are imported when they can be the only fish in the Quarantine Room. This facilitates a complete sterilization of the Quarantine Room water system and equipment, if needed, before and/or after their importation. After the submitting laboratory health history has been evaluated, wild-type males and females are imported, acclimated, and then spawned in the Quarantine Room. Fish developing clinical disease are submitted for histopathology immediately, and portions of liver and spleen are frozen back for PCR. The rest of the fish are submitted for histopathology after sufficient numbers of viable embryos are obtained. The embryos are surface sanitized, isolated, and held in static water tanks, with regular water changes, pending histopathology results on the adults. If no virulent pathogens are detected, the progeny can be introduced on the B-side of the Main Fish Room. If virulent pathogens are diagnosed, the progeny are raised in the Quarantine Room, spawned, and tested again as described earlier.

1.3.3 Lines arriving as cryopreserved sperm

Some laboratories submitting a large number of lines to ZIRC have opted to send sperm cryopreserved at their facility. These samples are maintained in designated vapor-phase liquid freezers, separate from samples obtained at ZIRC. Along with the usual request for health history information, we also ask that these submitters send us 20 to 30 moribund, sump, or prefiltration sentinel fish, which we submit for histopathology. If no pathogens or low-virulence pathogens, such as *P. neurophilia* and *M. chelonae*, are identified in fish from the submitting facility, then the sperm samples are cleared for use in IVF, and the embryos can be raised on side B of the Main Fish Room. If high-virulence pathogens, such as *M. marinum, M. haemophilum*, and *E. icataluri*, are identified, then embryos generated from the sperm samples are raised in the Quarantine Room. The line is refrozen from this generation, and contributing males are submitted for histopathology. Sperm samples from facilities that do not send fixed fish for examination or are otherwise considered to have an unknown pathogen status are treated as potentially having high-virulence pathogens.

It is important to note that when lines are imported to ZIRC as frozen sperm, we receive multiple samples per line. To fulfill customer orders, samples may be used in IVF and embryos shipped. When samples are depleted, the line is raised at ZIRC, and new sperm samples are obtained to replenish frozen samples. Therefore lines generated by IVF using sperm samples from males not raised at ZIRC will be shipped to customers before we have a chance to evaluate the health status of their progeny at ZIRC. We recommend that zebrafish facilities quarantine all imported fish, adults and embryos, and that the health of the imported fish be evaluated before subsequent generations are moved into the main fish facility. Our quarantine recommendations can be found on the ZIRC Health Services website and in our online Animal Health Report.

1.3.4 People

ZIRC recognizes that people can act as fomites and inadvertently bring pathogens into a fish facility. Pathogens could be transferred from the Quarantine Room, the outside environment, home tanks, and other aquatic facilities.

The highest risk environment is likely the Quarantine Room. While it is separate from the Main Fish Room, it is in the same building and serviced by the same staff. We require that everyone entering the Quarantine Room don a set of shoes dedicated to that room. A variety of sizes of Crocs are available, which are routinely disinfected by soaking in bleach solution. Gloves are required to be worn for all tasks performed in the Quarantine Room. In addition, only a limited number of ZIRC staff members have duties in the Quarantine Room. These colleagues organize their work schedule such that they do not enter the Main Fish Room after performing tasks in the Quarantine Room.

Pathogens could also be introduced from outside of the ZIRC building. ZIRC is one of several aquatic facilities on the University of Oregon campus, each of which has a different disease status. When entering ZIRC, employees are required to change into dedicated facility shoes, which are disinfected monthly. Adhesive floor mats are also located throughout the facility to minimize tracking of debris between rooms. Employees also wash their hands upon entry and exit of the Main Fish Room.

We require that visitors not have been to another aquatic facility the same day before they visit ZIRC. Visitors also don shoe covers when entering the facility. A second set of shoe covers is added if a visitor enters the Quarantine Room.

Staff education is central to participation and compliance with the biosecurity program at ZIRC. We strive to create an environment where all employees feel invested in and responsible for fish health and biosecurity.

As a stock and distribution center for wild-type, mutant, and transgenic zebrafish, ZIRC handles a large volume of imports and exports. New lines are imported as adult fish or cryopreserved sperm and are exported to customers as embryos and adults. Staff members handle fish during cryopreservation and fin clipping procedures, spawning fish, cleaning tanks, and feeding. All staff members are trained to recognize moribund fish and appreciate the importance of husbandry and biosecurity protocols. Being able to identify moribund fish, understanding why we remove them and run diagnostic tests, and recognizing biosecurity risks on a daily basis help prevent transmission and introduction of pathogens. By securing the health of the fish at ZIRC, we simultaneously promote animal welfare, protect employee health and safety, and are able to supply healthy fish to the zebrafish research community.

2. MATERIALS

2.1 ASSESSING COLONY HEALTH

Dietrich's fixative	
30 mL Ethanol (95%)	Science Stores, UO
10 mL Formalin	Science Stores, UO
2 mL Glacial acetic acid	Science Stores, UO
58 mL Distilled water	
Store at room temperature.	

2.2 PREVENTING PATHOGEN SPREAD WITHIN THE COLONY

UV sterilizer	Aquaneering Inc.
Sodium hypochlorite (embryos)	Fisher Scientific, lab grade
Sodium hypochlorite (equipment)	AmSan, Pure Bright 3.75 L
Methylene blue	Sigma, MB-1

Sodium thiosulfate	Western Chemical
Pronase (Proteinase K)	Roche
Ethanol (EtOH 190 proof)	Science Stores, UO
RODAC plates	Pharmacal Research Labs Inc.
Attest Biological Indicator (3M)	Beesmed.com

2.3 PREVENTING ENTRY OF NEW PATHOGENS

Crocs	Amazon.com or Crocs.com
Adhesive multilayer floor mats	VWR
BacDown antibacterial handsoap	Fisher Scientific

REFERENCES

AFS-FHS. (2007). *FHS blue book: Suggested procedures for the detection and identification of certain finfish and shellfish pathogens*. Bethesda, MD: AFS-FHS.

Ferguson, J. A., Watral, V., Schwindt, A. R., & Kent, M. L. (2007). Spores of two fish microsporidia (*Pseudoloma neurophilia* and *Glugea anomala*) are highly resistant to chlorine. *Diseases of Aquatic Organisms, 76*(3), 205–214. Retrieved from: http://www.ncbi.nlm.nih.gov/entrez/query.fcgi?cmd=Retrieve&db=PubMed&dopt=Citation&list_uids=17803106.

Hawke, J. P., Kent, M., Rogge, M., Baumgartner, W., Wiles, J., Shelley, J., … Peterson, T. S. (2013). Edwardsiellosis caused by *Edwardsiella ictaluri* in laboratory populations of Zebrafish *Danio rerio*. *Journal of Aquatic Animal Health, 25*(3), 171–183. http://dx.doi.org/10.1080/08997659.2013.782226.

Humason, G. L. (1967). *Animal tissue techniques* (2nd ed.). San Francisco: W. H. Freeman.

Luna, L. G., & Armed Forces Institute of Pathology (US). (1968). *Manual of histologic staining methods of the Armed Forces Institute of Pathology* (3rd ed.). New York: Blakiston Division.

Matthews, J. L., Brown, A. M., Larison, K., Bishop-Stewart, J. K., Rogers, P., & Kent, M. L. (2001). *Pseudoloma neurophilia* n. g., n. sp., a new microsporidium from the central nervous system of the zebrafish (*Danio rerio*). *The Journal of Eukaryotic Microbiology, 48*(2), 227–233. Retrieved from: http://www.ncbi.nlm.nih.gov/entrez/query.fcgi?cmd=Retrieve&db=PubMed&dopt=Citation&list_uids=12095112.

Murray, K. N., Bauer, J., Tallen, A., Matthews, J. L., Westerfield, M., & Varga, Z. M. (2011). Characterization and management of asymptomatic *Mycobacterium* infections at the Zebrafish International Resource Center. *Journal of the American Association for Laboratory Animal Science: JAALAS, 50*(5), 675–679. Retrieved from: http://www.ncbi.nlm.nih.gov/entrez/query.fcgi?cmd=Retrieve&db=PubMed&dopt=Citation&list_uids=22330714.

Murray, K. N., Dreska, M., Nasiadka, A., Rinne, M., Matthews, J. L., Carmichael, C., ... Westerfield, M. (2011). Transmission, diagnosis, and recommendations for control of *Pseudoloma neurophilia* infections in laboratory zebrafish (*Danio rerio*) facilities. *Comparative Medicine, 61*(4), 322–329. Retrieved from: http://www.ncbi.nlm.nih.gov/pubmed/22330247.

Murray, K. N., & Peterson, T. S. (2015). Pathology in practice. *P. tomentosa* infection in zebrafish. *Journal of the American Veterinary Medical Association, 246*(2), 201–203. http://dx.doi.org/10.2460/javma.246.2.201.

Peterson, T. S., Spitsbergen, J. M., Feist, S. W., & Kent, M. L. (2011). The Luna stain, an improved selective stain for detection of microsporidian spores in histologic sections. *Diseases of Aquatic Organisms, 95*, 175–180.

Ramsay, J. M., Watral, V., Schreck, C. B., & Kent, M. L. (2009a). Husbandry stress exacerbates mycobacterial infections in adult zebrafish, *Danio rerio* (Hamilton). *Journal of Fish Diseases, 32*(11), 931–941. http://dx.doi.org/10.1111/j.1365–2761.2009.01074.x. pii: JFD1074.

Ramsay, J. M., Watral, V., Schreck, C. B., & Kent, M. L. (2009b). *Pseudoloma neurophilia* infections in zebrafish *Danio rerio*: effects of stress on survival, growth, and reproduction. *Diseases of Aquatic Organisms, 88*(1), 69–84. Retrieved from: http://www.ncbi.nlm.nih.gov/entrez/query.fcgi?cmd=Retrieve&db=PubMed&dopt=Citation&list_uids=20183967.

Sanders, J. L., & Kent, M. L. (2011). Development of a sensitive assay for the detection of *Pseudoloma neurophilia* in laboratory populations of the zebrafish *Danio rerio*. *Diseases of Aquatic Organisms, 96*(2), 145–156. http://dx.doi.org/10.3354/dao02375.

Sanders, J. L., Lawrence, C., Nichols, D. K., Brubaker, J. F., Peterson, T. S., Murray, K. N., & Kent, M. L. (2010). Pleistophora hyphessobryconis (Microsporidia) infecting zebrafish *Danio rerio* in research facilities. *Diseases of Aquatic Organisms, 91*(1), 47–56. Retrieved from: http://www.ncbi.nlm.nih.gov/entrez/query.fcgi?cmd=Retrieve&db=PubMed&dopt=Citation&list_uids=20853741.

Simon, R. C., & Schill, W. B. (1984). Tables of sample size requirements for detection of fish infected by pathogens: three confidence levels for different infection prevalence and various population sizes. *Journal of Fish Diseases, 7*(6), 6.

Varga, Z. M. (2016). Aquaculture, husbandry, and shipping at the Zebrafish International Resource Center. In H. W. Detrich, III, L. I. Zon, & M. Westerfield (Eds.), *135. The Zebrafish: Genetics, Genomics, and Transcriptomics* (pp. 509-534).

Watral, V., & Kent, M. L. (2007). Pathogenesis of *Mycobacterium* spp. in zebrafish (*Danio rerio*) from research facilities. *Comparative Biochemistry and Physiology. Toxicology & Pharmacology: CBP, 145*(1), 55–60. http://dx.doi.org/10.1016/j.cbpc.2006.06.004. pii:S1532-0456(06)00135-9.

Whipps, C. M., Dougan, S. T., & Kent, M. L. (2007). *Mycobacterium haemophilum* infections of zebrafish (*Danio rerio*) in research facilities. *FEMS Microbiology Letters, 270*(1), 21–26. http://dx.doi.org/10.1111/j.1574-6968.2007.00671.x.

Whipps, C. M., Matthews, J. L., & Kent, M. L. (2008). Distribution and genetic characterization of *Mycobacterium chelonae* in laboratory zebrafish *Danio rerio*. *Diseases of Aquatic Organisms, 82*(1), 45–54. Retrieved from: http://www.ncbi.nlm.nih.gov/entrez/query.fcgi?cmd=Retrieve&db=PubMed&dopt=Citation&list_uids=19062752.

Whipps, C. M., Murray, K. N., & Kent, M. L. (2015). Occurrence of a myxozoan parasite *Myxidium streisingeri* n. sp. in laboratory zebrafish *Danio rerio*. *The Journal of Parasitology, 101*(1), 86–90. http://dx.doi.org/10.1645/14-613.1.

Volumes in Series

Founding Series Editor
DAVID M. PRESCOTT

Volume 1 (1964)
Methods in Cell Physiology
Edited by David M. Prescott

Volume 2 (1966)
Methods in Cell Physiology
Edited by David M. Prescott

Volume 3 (1968)
Methods in Cell Physiology
Edited by David M. Prescott

Volume 4 (1970)
Methods in Cell Physiology
Edited by David M. Prescott

Volume 5 (1972)
Methods in Cell Physiology
Edited by David M. Prescott

Volume 6 (1973)
Methods in Cell Physiology
Edited by David M. Prescott

Volume 7 (1973)
Methods in Cell Biology
Edited by David M. Prescott

Volume 8 (1974)
Methods in Cell Biology
Edited by David M. Prescott

Volume 9 (1975)
Methods in Cell Biology
Edited by David M. Prescott

Volume 10 (1975)
Methods in Cell Biology
Edited by David M. Prescott

Volume 11 (1975)
Yeast Cells
Edited by David M. Prescott

Volume 12 (1975)
Yeast Cells
Edited by David M. Prescott

Volume 13 (1976)
Methods in Cell Biology
Edited by David M. Prescott

Volume 14 (1976)
Methods in Cell Biology
Edited by David M. Prescott

Volume 15 (1977)
Methods in Cell Biology
Edited by David M. Prescott

Volume 16 (1977)
Chromatin and Chromosomal Protein Research I
Edited by Gary Stein, Janet Stein, and Lewis J. Kleinsmith

Volume 17 (1978)
Chromatin and Chromosomal Protein Research II
Edited by Gary Stein, Janet Stein, and Lewis J. Kleinsmith

Volume 18 (1978)
Chromatin and Chromosomal Protein Research III
Edited by Gary Stein, Janet Stein, and Lewis J. Kleinsmith

Volume 19 (1978)
Chromatin and Chromosomal Protein Research IV
Edited by Gary Stein, Janet Stein, and Lewis J. Kleinsmith

Volume 20 (1978)
Methods in Cell Biology
Edited by David M. Prescott

Advisory Board Chairman
KEITH R. PORTER

Volume 21A (1980)
Normal Human Tissue and Cell Culture, Part A: Respiratory, Cardiovascular, and Integumentary Systems
Edited by Curtis C. Harris, Benjamin F. Trump, and Gary D. Stoner

Volume 21B (1980)
Normal Human Tissue and Cell Culture, Part B: Endocrine, Urogenital, and Gastrointestinal Systems
Edited by Curtis C. Harris, Benjamin F. Trump, and Gray D. Stoner

Volume 22 (1981)
Three-Dimensional Ultrastructure in Biology
Edited by James N. Turner

Volume 23 (1981)
Basic Mechanisms of Cellular Secretion
Edited by Arthur R. Hand and Constance Oliver

Volume 24 (1982)
The Cytoskeleton, Part A: Cytoskeletal Proteins, Isolation and Characterization
Edited by Leslie Wilson

Volume 25 (1982)
The Cytoskeleton, Part B: Biological Systems and In Vitro Models
Edited by Leslie Wilson

Volume 26 (1982)
Prenatal Diagnosis: Cell Biological Approaches
Edited by Samuel A. Latt and Gretchen J. Darlington

Series Editor
LESLIE WILSON

Volume 27 (1986)
Echinoderm Gametes and Embryos
Edited by Thomas E. Schroeder

Volume 28 (1987)
***Dictyostelium discoideum*: Molecular Approaches to Cell Biology**
Edited by James A. Spudich

Volume 29 (1989)
Fluorescence Microscopy of Living Cells in Culture, Part A: Fluorescent Analogs, Labeling Cells, and Basic Microscopy
Edited by Yu-Li Wang and D. Lansing Taylor

Volume 30 (1989)
Fluorescence Microscopy of Living Cells in Culture, Part B: Quantitative Fluorescence Microscopy—Imaging and Spectroscopy
Edited by D. Lansing Taylor and Yu-Li Wang

Volume 31 (1989)
Vesicular Transport, Part A
Edited by Alan M. Tartakoff

Volume 32 (1989)
Vesicular Transport, Part B
Edited by Alan M. Tartakoff

Volume 33 (1990)
Flow Cytometry
Edited by Zbigniew Darzynkiewicz and Harry A. Crissman

Volume 34 (1991)
Vectorial Transport of Proteins into and across Membranes
Edited by Alan M. Tartakoff

Selected from Volumes 31, 32, and 34 (1991)
Laboratory Methods for Vesicular and Vectorial Transport
Edited by Alan M. Tartakoff

Volume 35 (1991)
Functional Organization of the Nucleus: A Laboratory Guide
Edited by Barbara A. Hamkalo and Sarah C. R. Elgin

Volume 36 (1991)
***Xenopus laevis*: Practical Uses in Cell and Molecular Biology**
Edited by Brian K. Kay and H. Benjamin Peng

Series Editors
LESLIE WILSON AND PAUL MATSUDAIRA

Volume 37 (1993)
Antibodies in Cell Biology
Edited by David J. Asai

Volume 38 (1993)
Cell Biological Applications of Confocal Microscopy
Edited by Brian Matsumoto

Volume 39 (1993)
Motility Assays for Motor Proteins
Edited by Jonathan M. Scholey

Volume 40 (1994)
A Practical Guide to the Study of Calcium in Living Cells
Edited by Richard Nuccitelli

Volume 41 (1994)
Flow Cytometry, Second Edition, Part A
Edited by Zbigniew Darzynkiewicz, J. Paul Robinson, and Harry A. Crissman

Volume 42 (1994)
Flow Cytometry, Second Edition, Part B
Edited by Zbigniew Darzynkiewicz, J. Paul Robinson, and Harry A. Crissman

Volume 43 (1994)
Protein Expression in Animal Cells
Edited by Michael G. Roth

Volume 44 (1994)
***Drosophila melanogaster*: Practical Uses in Cell and Molecular Biology**
Edited by Lawrence S. B. Goldstein and Eric A. Fyrberg

Volume 45 (1994)
Microbes as Tools for Cell Biology
Edited by David G. Russell

Volume 46 (1995)
Cell Death
Edited by Lawrence M. Schwartz and Barbara A. Osborne

Volume 47 (1995)
Cilia and Flagella
Edited by William Dentler and George Witman

Volume 48 (1995)
Caenorhabditis elegans: Modern Biological Analysis of an Organism
Edited by Henry F. Epstein and Diane C. Shakes

Volume 49 (1995)
Methods in Plant Cell Biology, Part A
Edited by David W. Galbraith, Hans J. Bohnert, and Don P. Bourque

Volume 50 (1995)
Methods in Plant Cell Biology, Part B
Edited by David W. Galbraith, Don P. Bourque, and Hans J. Bohnert

Volume 51 (1996)
Methods in Avian Embryology
Edited by Marianne Bronner-Fraser

Volume 52 (1997)
Methods in Muscle Biology
Edited by Charles P. Emerson, Jr. and H. Lee Sweeney

Volume 53 (1997)
Nuclear Structure and Function
Edited by Miguel Berrios

Volume 54 (1997)
Cumulative Index

Volume 55 (1997)
Laser Tweezers in Cell Biology
Edited by Michael P. Sheetz

Volume 56 (1998)
Video Microscopy
Edited by Greenfield Sluder and David E. Wolf

Volume 57 (1998)
Animal Cell Culture Methods
Edited by Jennie P. Mather and David Barnes

Volume 58 (1998)
Green Fluorescent Protein
Edited by Kevin F. Sullivan and Steve A. Kay

Volume 59 (1998)
The Zebrafish: Biology
Edited by H. William Detrich III, Monte Westerfield, and Leonard I. Zon

Volume 60 (1998)
The Zebrafish: Genetics and Genomics
Edited by H. William Detrich III, Monte Westerfield, and Leonard I. Zon

Volume 61 (1998)
Mitosis and Meiosis
Edited by Conly L. Rieder

Volume 84 (2007)
Biophysical Tools for Biologists, Volume One: In Vitro Techniques
Edited by John J. Correia and H. William Detrich, III

Volume 85 (2008)
Fluorescent Proteins
Edited by Kevin F. Sullivan

Volume 86 (2008)
Stem Cell Culture
Edited by Dr. Jennie P. Mather

Volume 87 (2008)
Avian Embryology, 2nd Edition
Edited by Dr. Marianne Bronner-Fraser

Volume 88 (2008)
Introduction to Electron Microscopy for Biologists
Edited by Prof. Terence D. Allen

Volume 89 (2008)
Biophysical Tools for Biologists, Volume Two: In Vivo Techniques
Edited by Dr. John J. Correia and Dr. H. William Detrich, III

Volume 90 (2008)
Methods in Nano Cell Biology
Edited by Bhanu P. Jena

Volume 91 (2009)
Cilia: Structure and Motility
Edited by Stephen M. King and Gregory J. Pazour

Volume 92 (2009)
Cilia: Motors and Regulation
Edited by Stephen M. King and Gregory J. Pazour

Volume 93 (2009)
Cilia: Model Organisms and Intraflagellar Transport
Edited by Stephen M. King and Gregory J. Pazour

Volume 94 (2009)
Primary Cilia
Edited by Roger D. Sloboda

Volume 95 (2010)
Microtubules, in vitro
Edited by Leslie Wilson and John J. Correia

Volume 116 (2013)
Lipid Droplets
Edited by H. Robert Yang and Peng Li

Volume 117 (2013)
Receptor-Receptor Interactions
Edited by P. Michael Conn

Volume 118 (2013)
Methods for Analysis of Golgi Complex Function
Edited by Franck Perez and David J. Stephens

Volume 119 (2014)
Micropatterning in Cell Biology Part A
Edited by Matthieu Piel and Manuel Théry

Volume 120 (2014)
Micropatterning in Cell Biology Part B
Edited by Matthieu Piel and Manuel Théry

Volume 121 (2014)
Micropatterning in Cell Biology Part C
Edited by Matthieu Piel and Manuel Théry

Volume 122 (2014)
Nuclear Pore Complexes and Nucleocytoplasmic Transport - Methods
Edited by Valérie Doye

Volume 123 (2014)
Quantitative Imaging in Cell Biology
Edited by Jennifer C. Waters and Torsten Wittmann

Volume 124 (2014)
Correlative Light and Electron Microscopy II
Edited by Thomas Müller-Reichert and Paul Verkade

Volume 125 (2015)
Biophysical Methods in Cell Biology
Edited by Ewa K. Paluch

Volume 126 (2015)
Lysosomes and Lysosomal Diseases
Edited by Frances Platt and Nick Platt

Volume 127 (2015)
Methods in Cilia & Flagella
Edited by Renata Basto and Wallace F. Marshall

Volume 128 (2015)
Building a Cell from its Component Parts
Edited by Jennifer Ross and Wallace F. Marshall

Volume 129 (2015)
Centrosome & Centriole
Edited by Renata Basto and Karen Oegema

Volume 130 (2015)
Sorting and Recycling Endosomes
Edited by Wei Guo

Volume 131 (2016)
The Neuronal Cytoskeleton, Motor Proteins, and Organelle Trafficking in the Axon
Edited by K. Kevin Pfister

Volume 132 (2016)
G Protein-Coupled Receptors: Signaling, Trafficking and Regulation
Edited by Arun K. Shukla

Volume 133 (2016)
The Zebrafish: Cellular and Developmental Biology, Part A Cellular Biology
Edited by H. William Detrich, III, Monte Westerfield, and Leonard I. Zon

Volume 134 (2016)
The Zebrafish: Cellular and Developmental Biology, Part B Developmental Biology
Edited by H. William Detrich, III, Monte Westerfield, and Leonard I. Zon

Index

'*Note:* Page numbers followed by "f" indicate figures and "t" indicate tables.'

D

E

H

I

Q

R

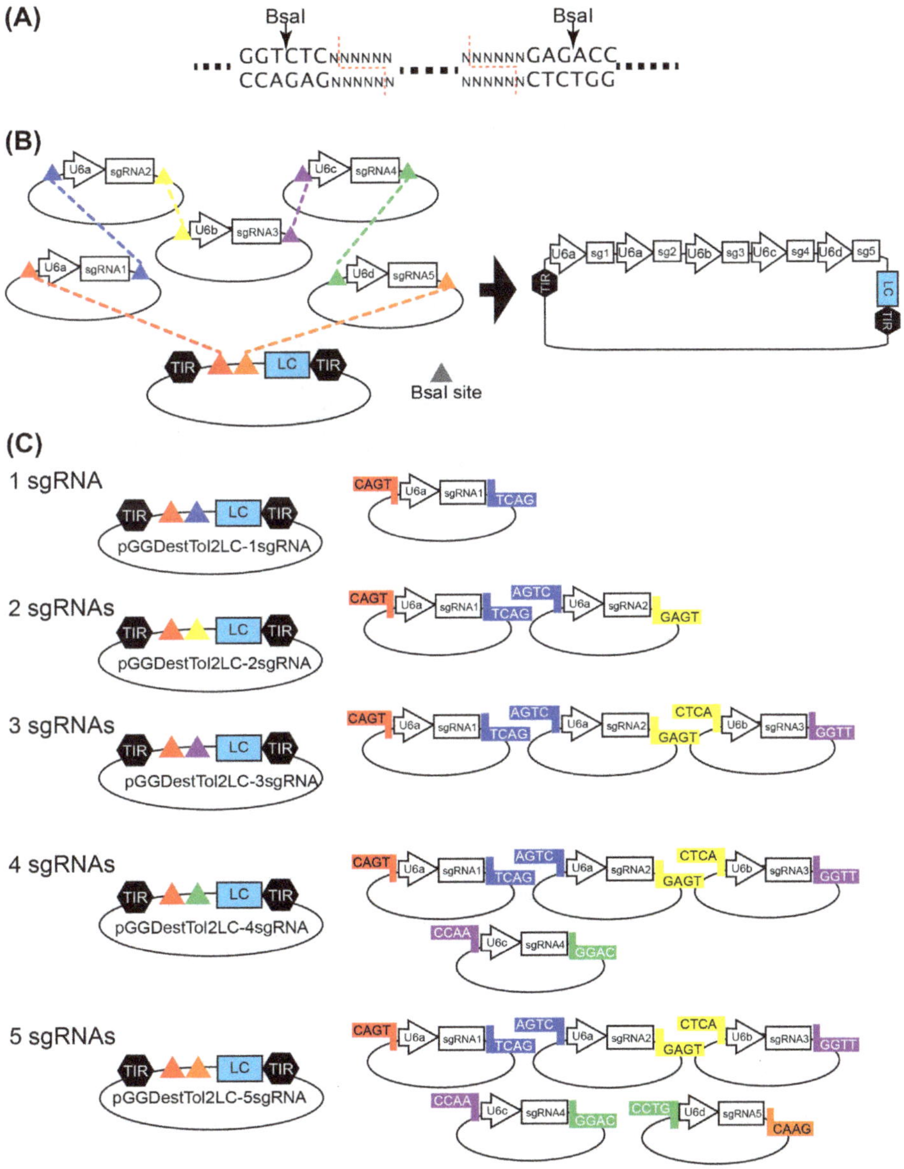

L. YIN *ET AL.*, FIGURE 3

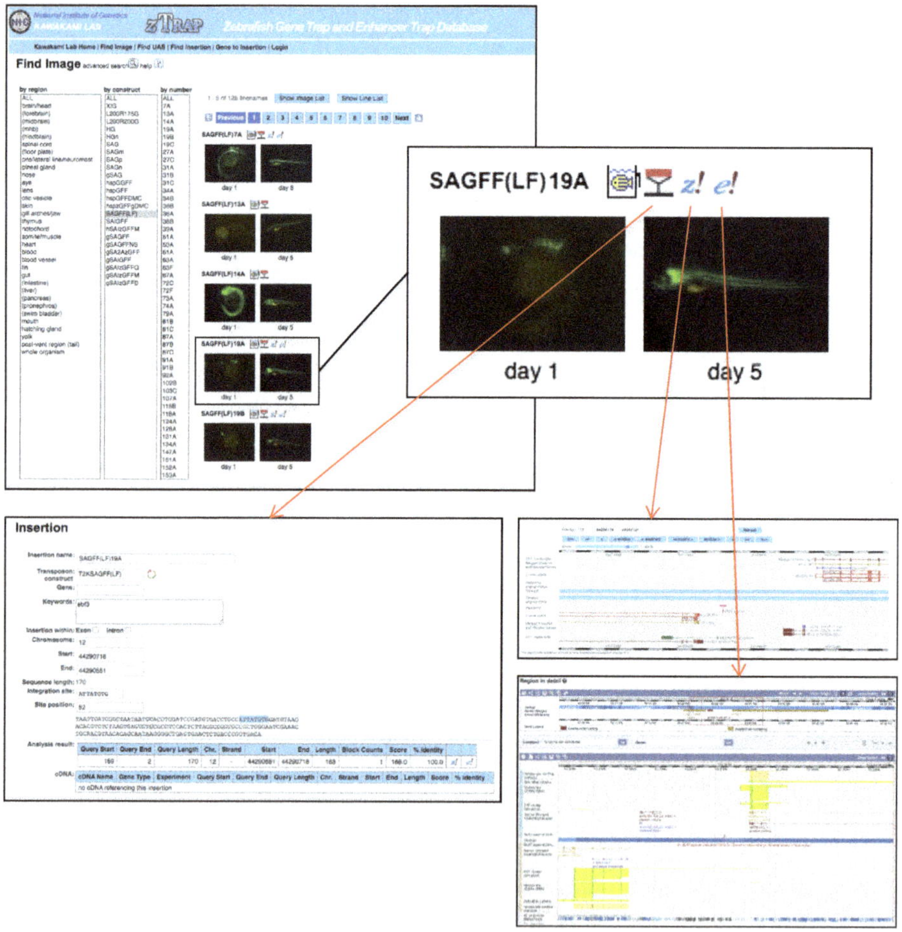

K. KAWAKAMI *ET AL.*, FIGURE 5

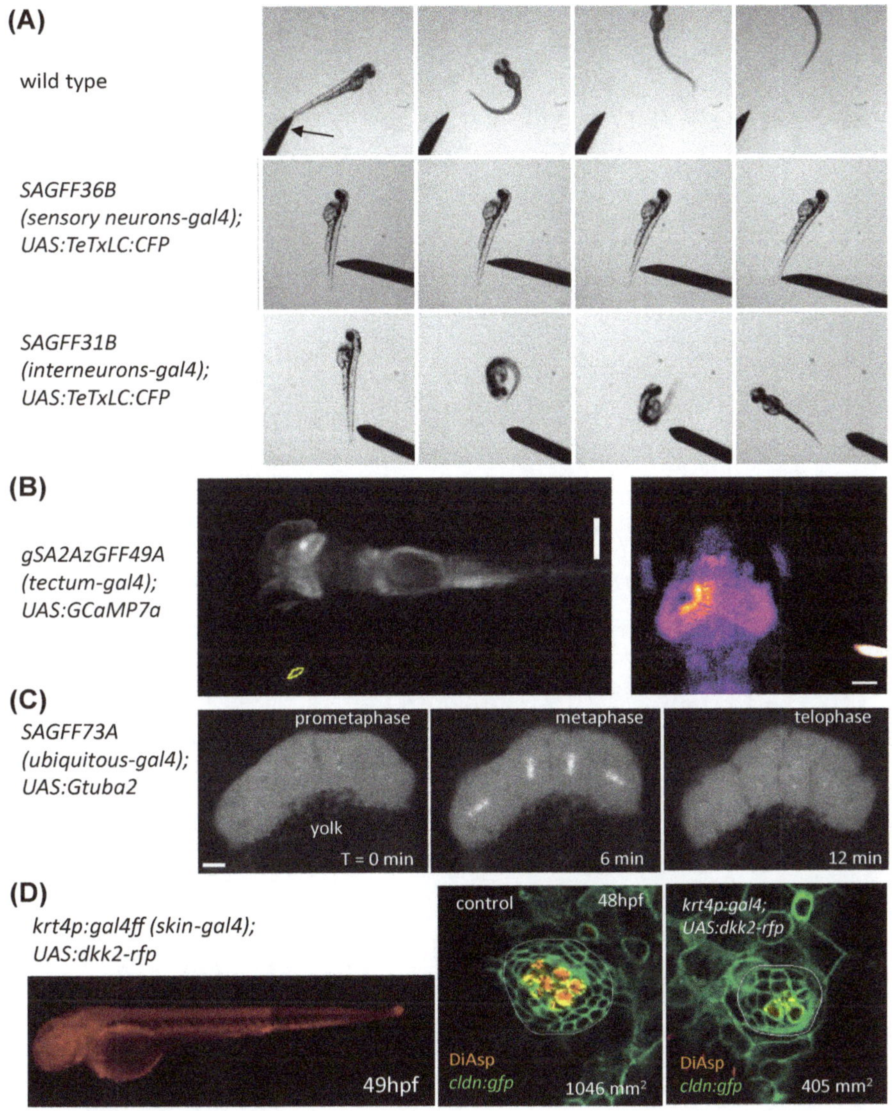

K. KAWAKAMI *ET AL.*, FIGURE 6

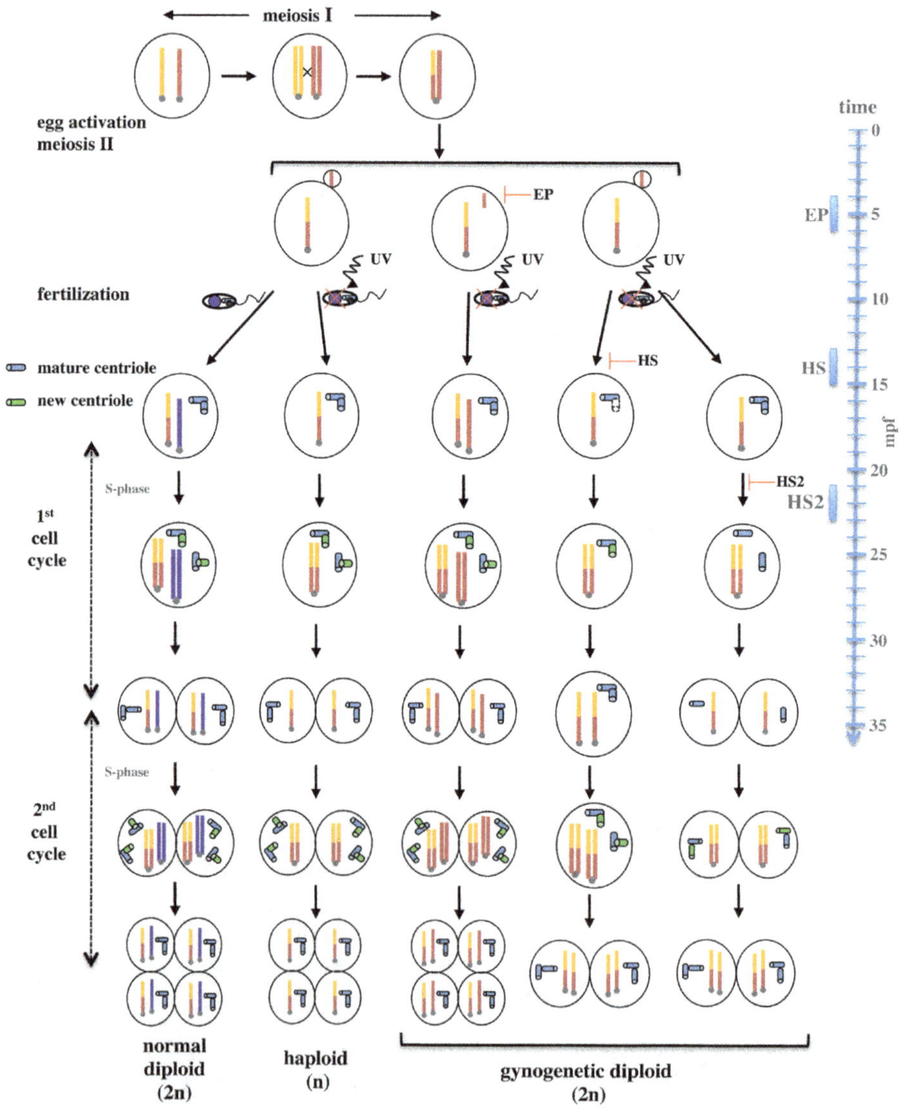

F. PELEGRI AND M.C. MULLINS, FIGURE 4

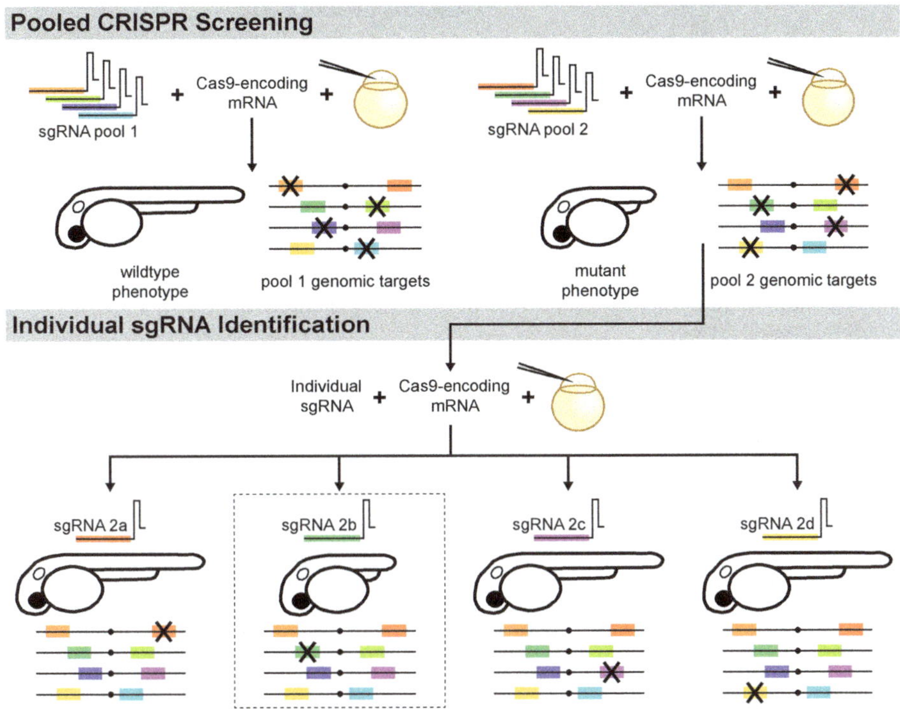

A.N. SHAH *ET AL.*, FIGURE 1

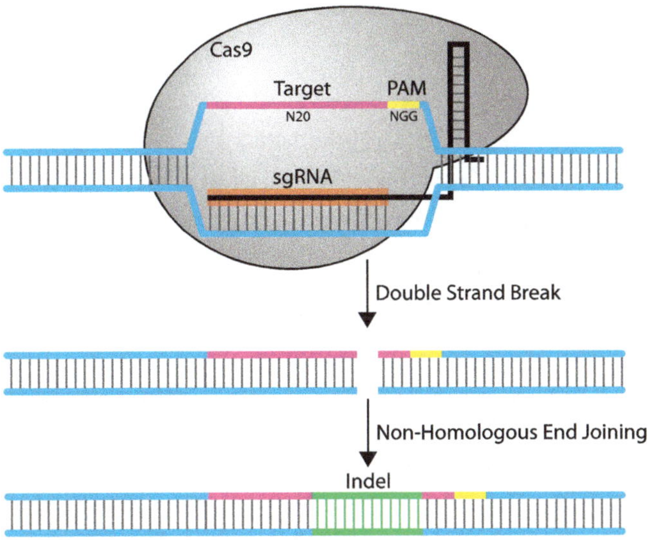

A.N. SHAH *ET AL.*, FIGURE 2

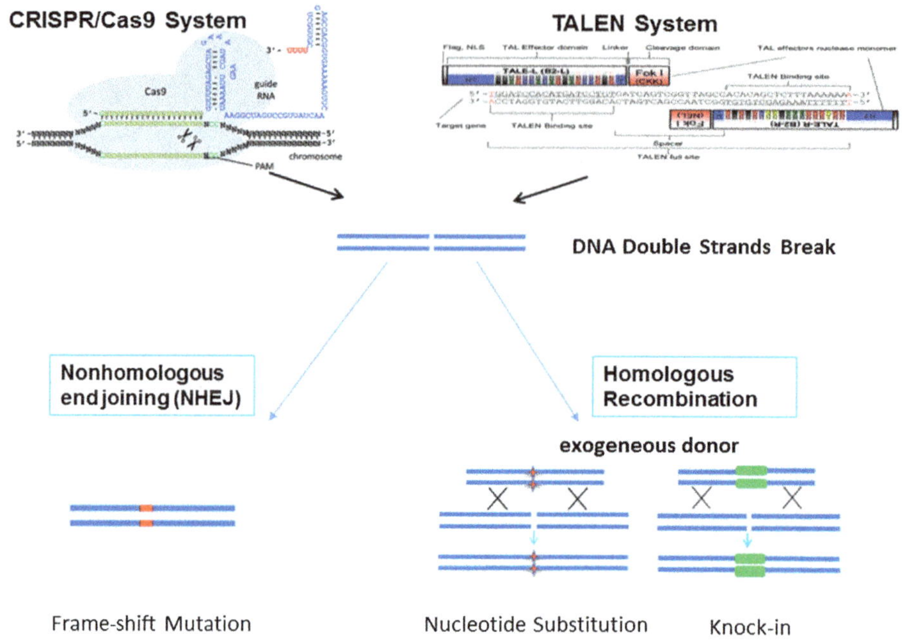

Y. ZHANG *ET AL.*, FIGURE 1

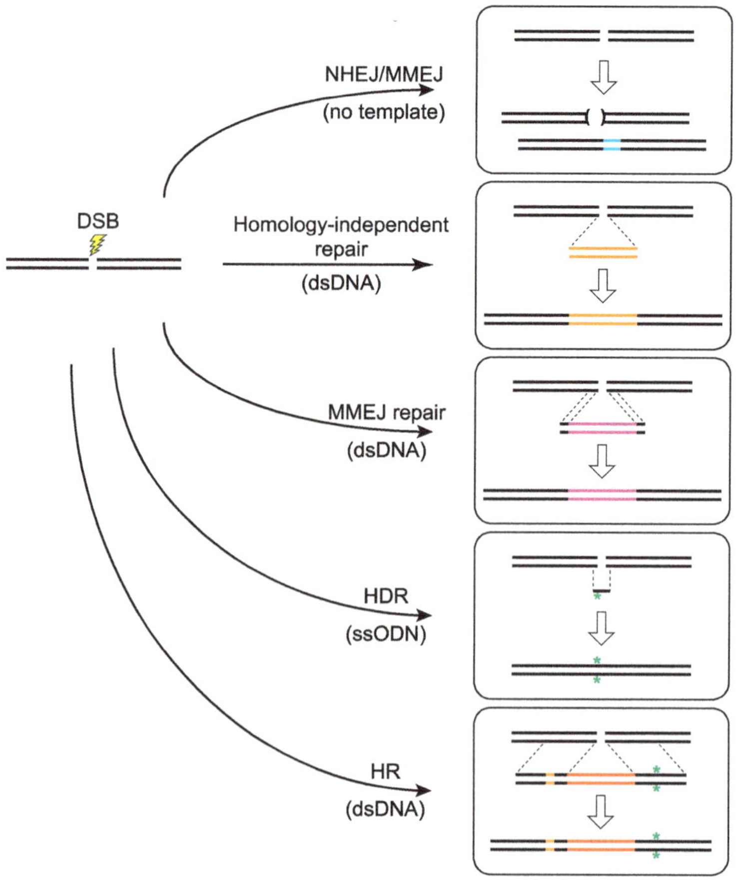

K. HOSHIJIMA *ET AL.*, FIGURE 1

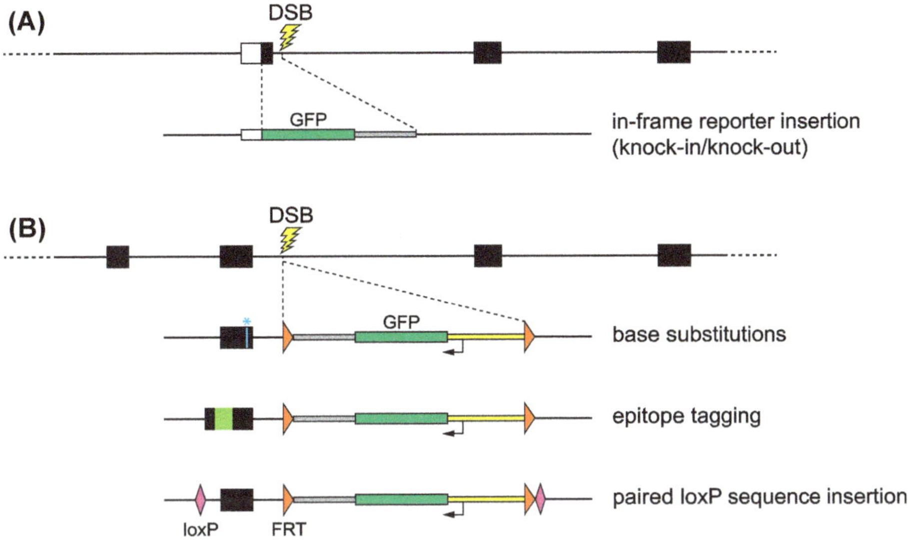

K. HOSHIJIMA *ET AL.*, FIGURE 2

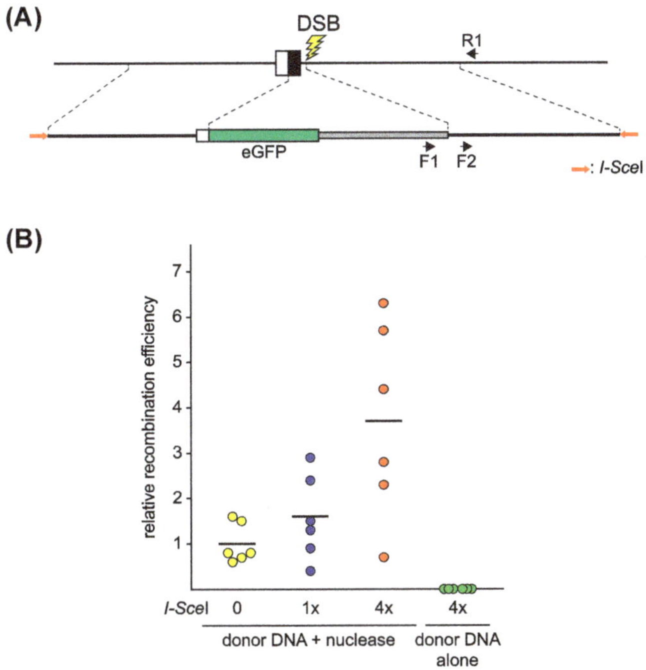

K. HOSHIJIMA *ET AL.*, FIGURE 5

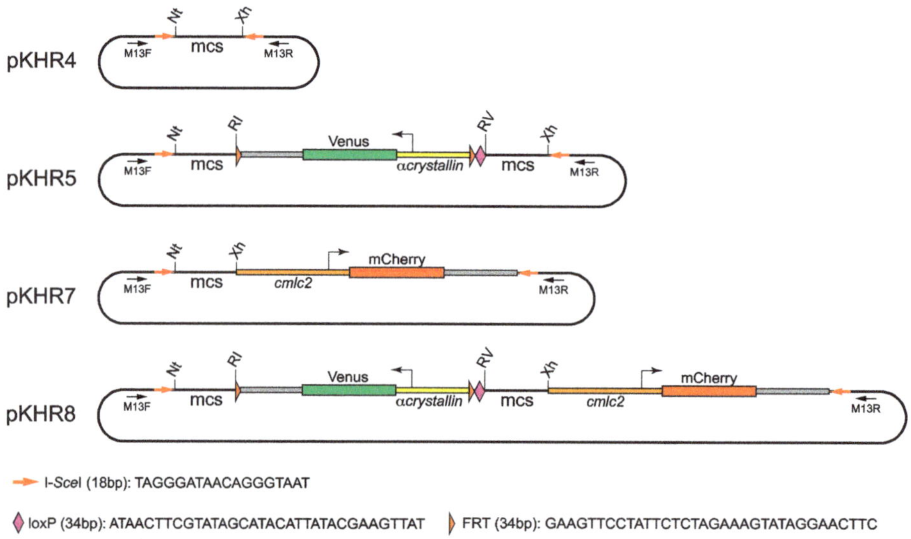

K. HOSHIJIMA *ET AL.*, FIGURE 6

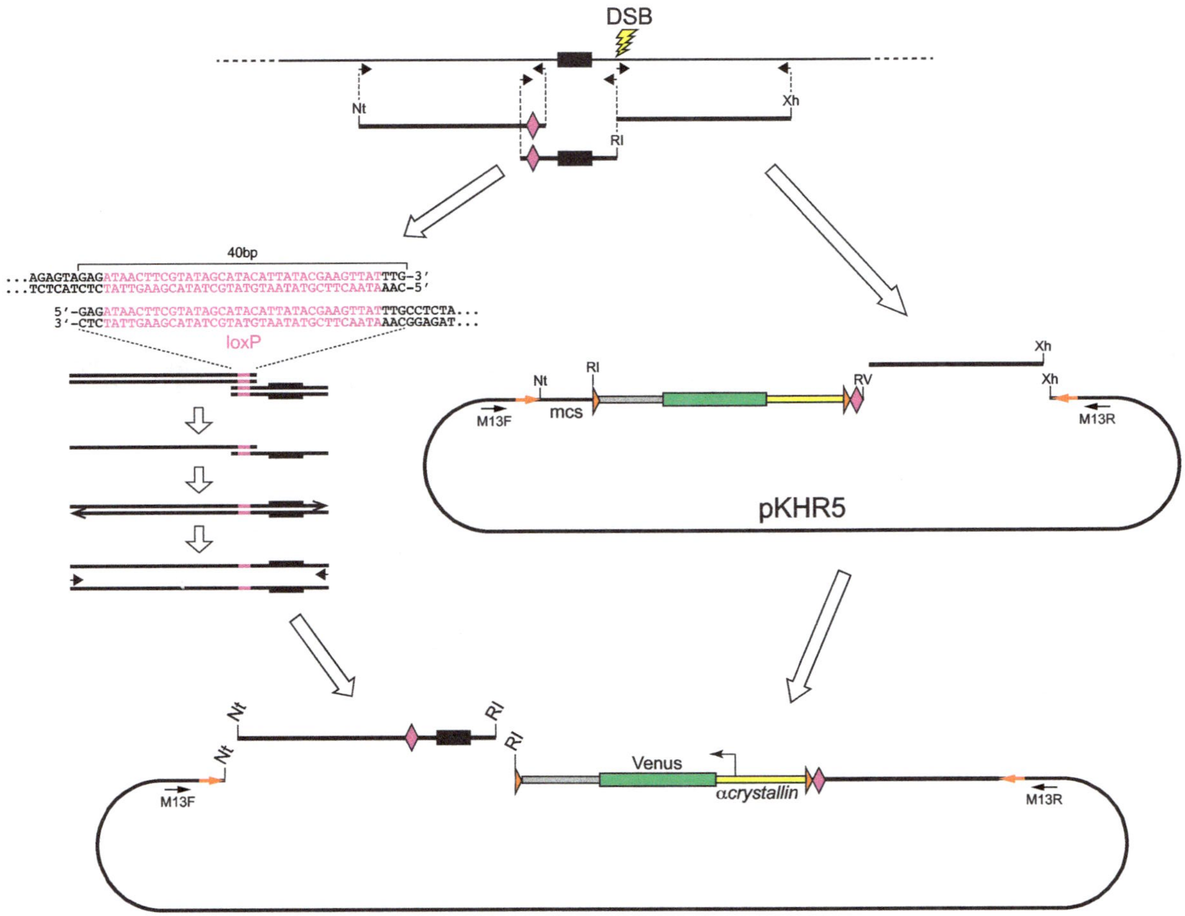

K. HOSHIJIMA *ET AL.*, FIGURE 8

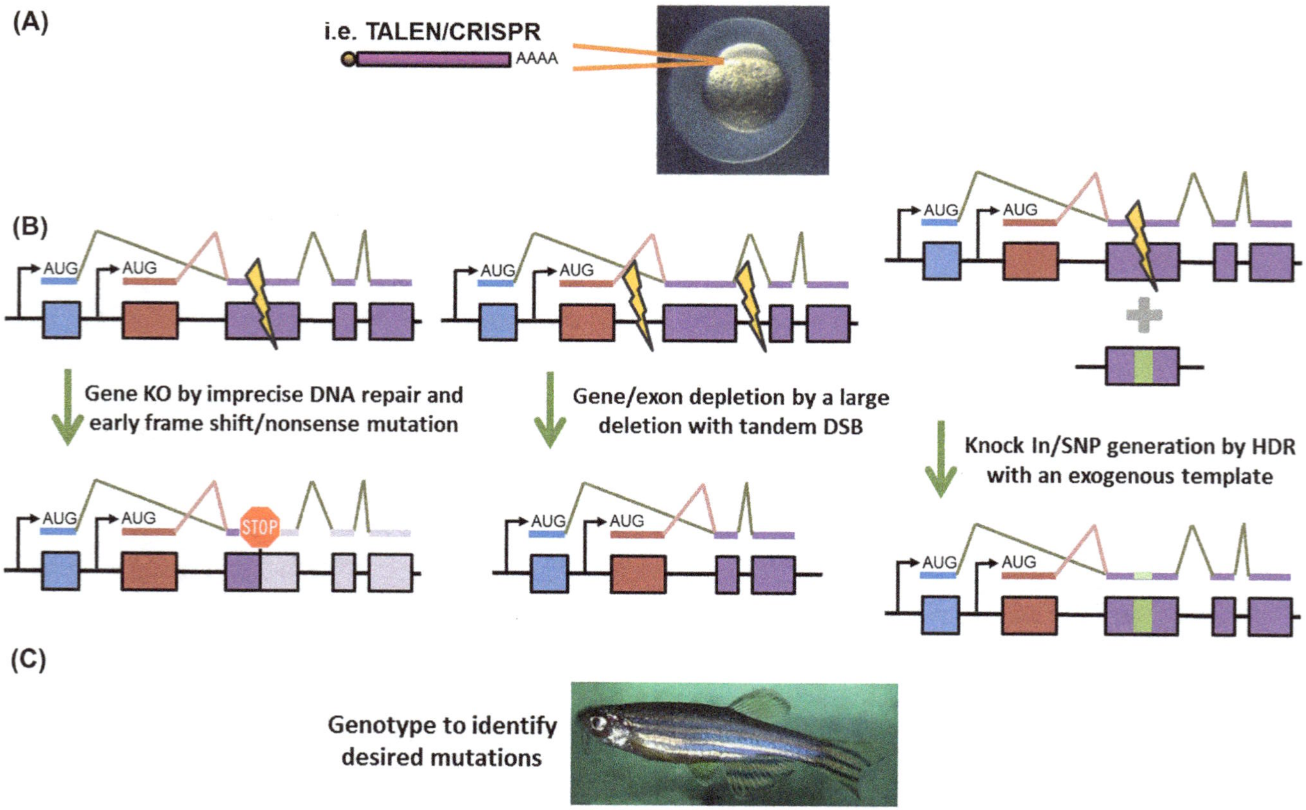

H. ATA *ET AL.*, FIGURE 1

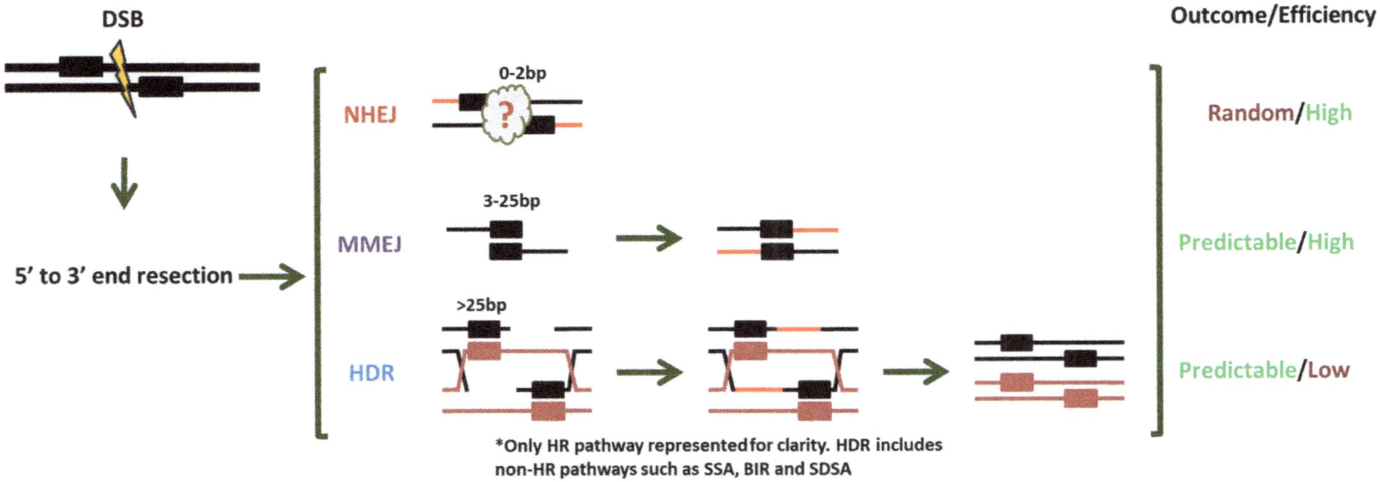

H. ATA *ET AL.*, FIGURE 2

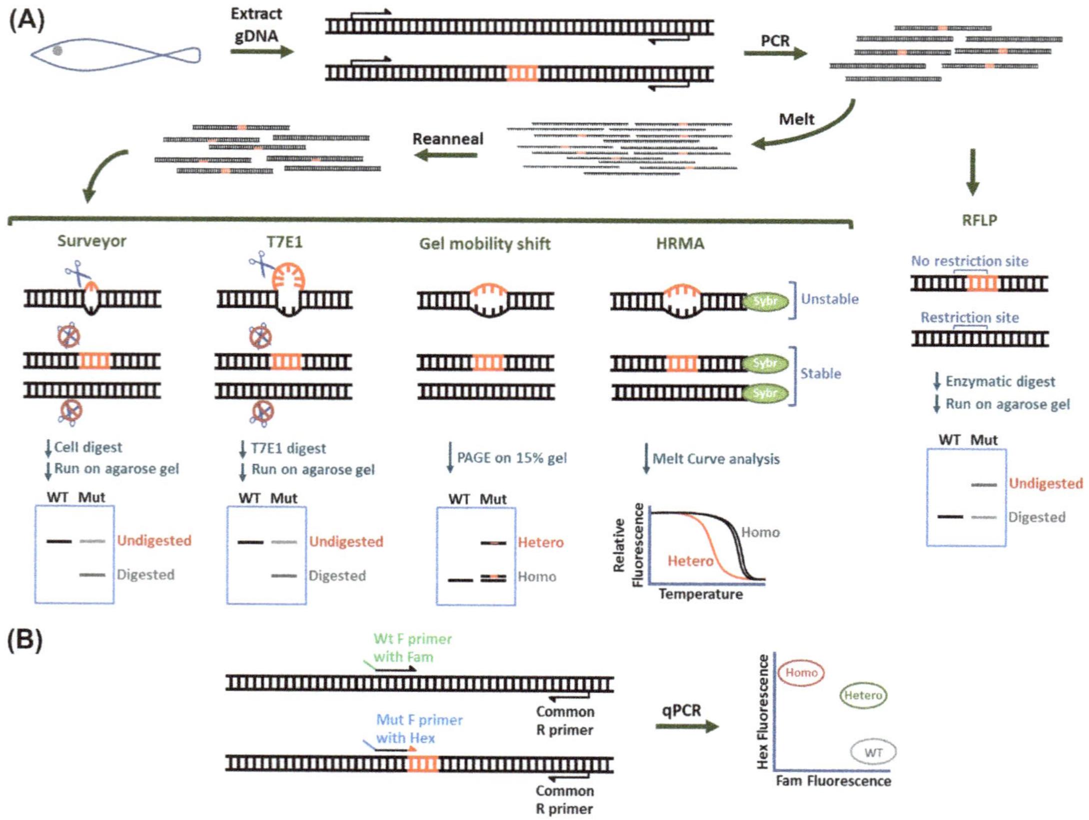

H. ATA *ET AL.*, FIGURE 3

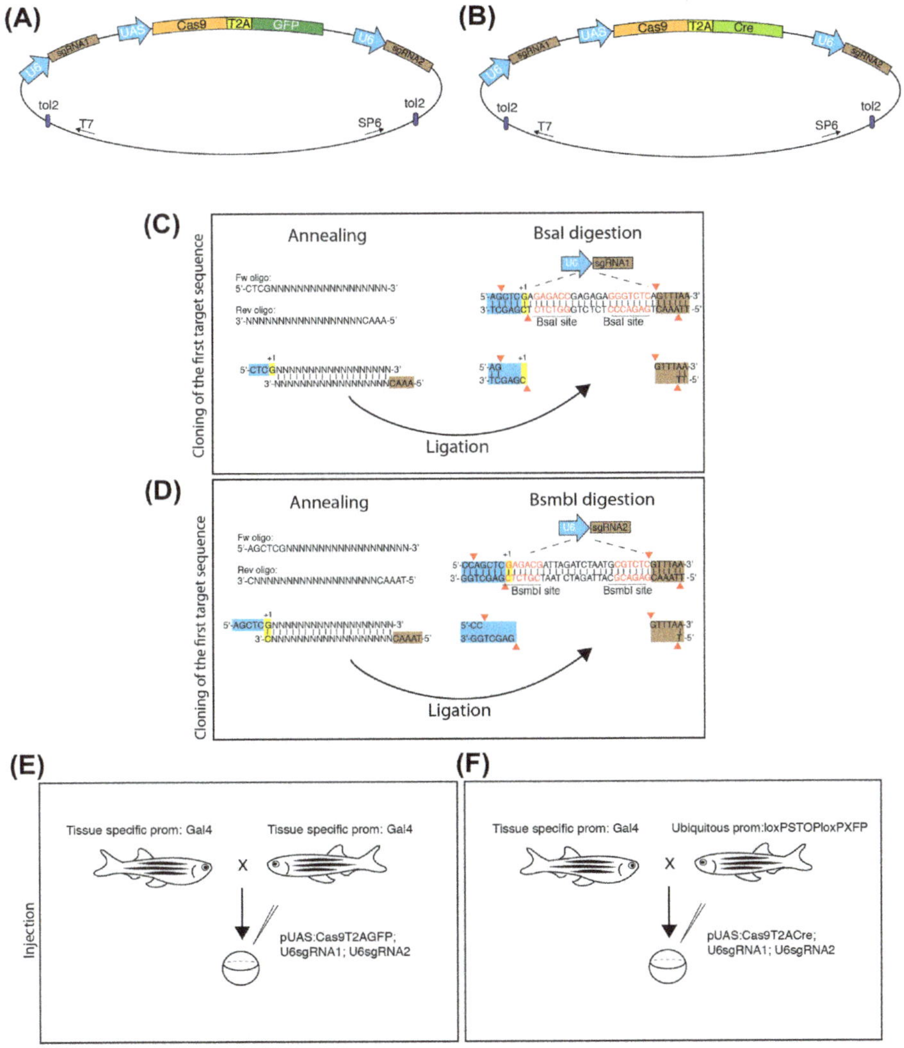

F. DE SANTIS *ET AL.*, FIGURE 1

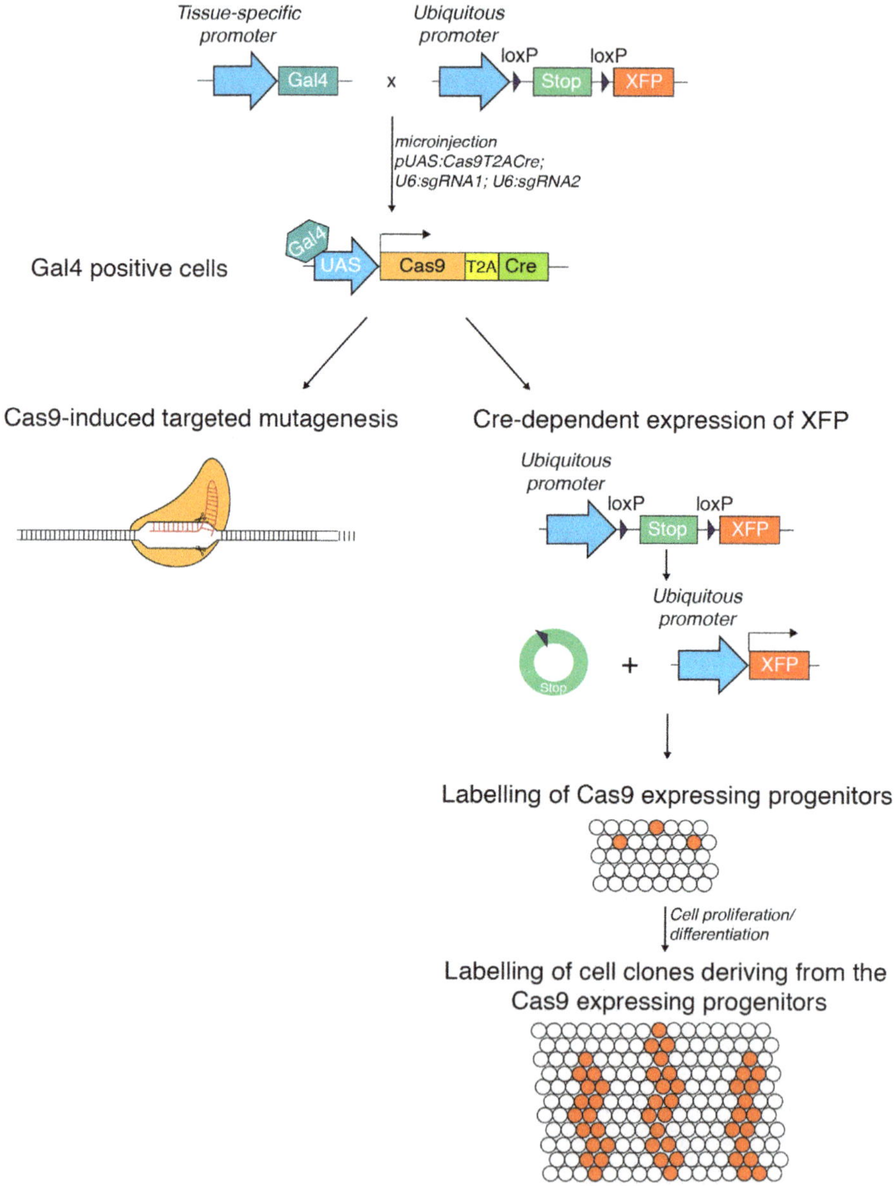

F. DE SANTIS *ET AL.*, FIGURE 3

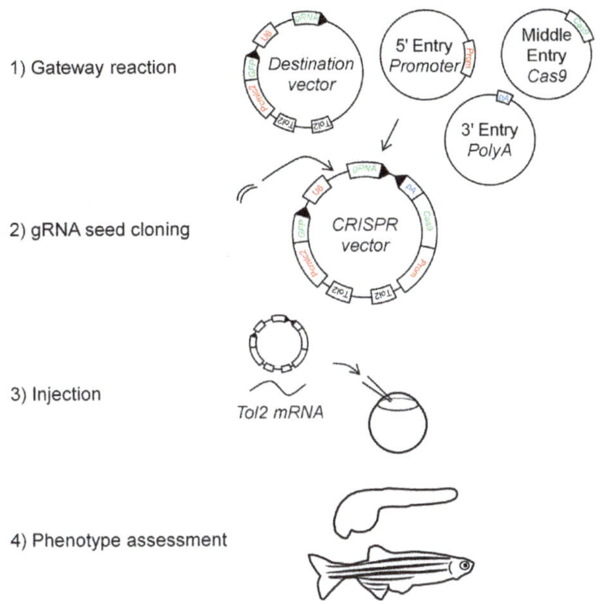

J. ABLAIN AND L.I. ZON, FIGURE 2

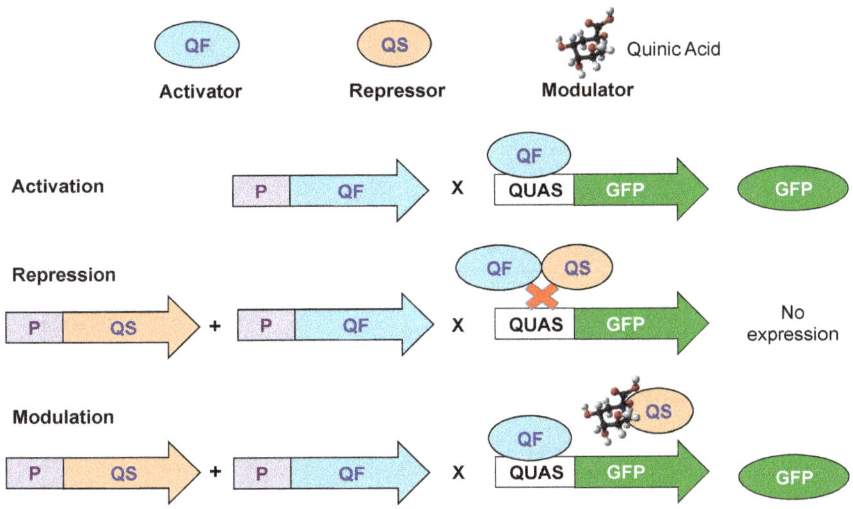

A. GHOSH AND M.E. HALPERN, FIGURE 1

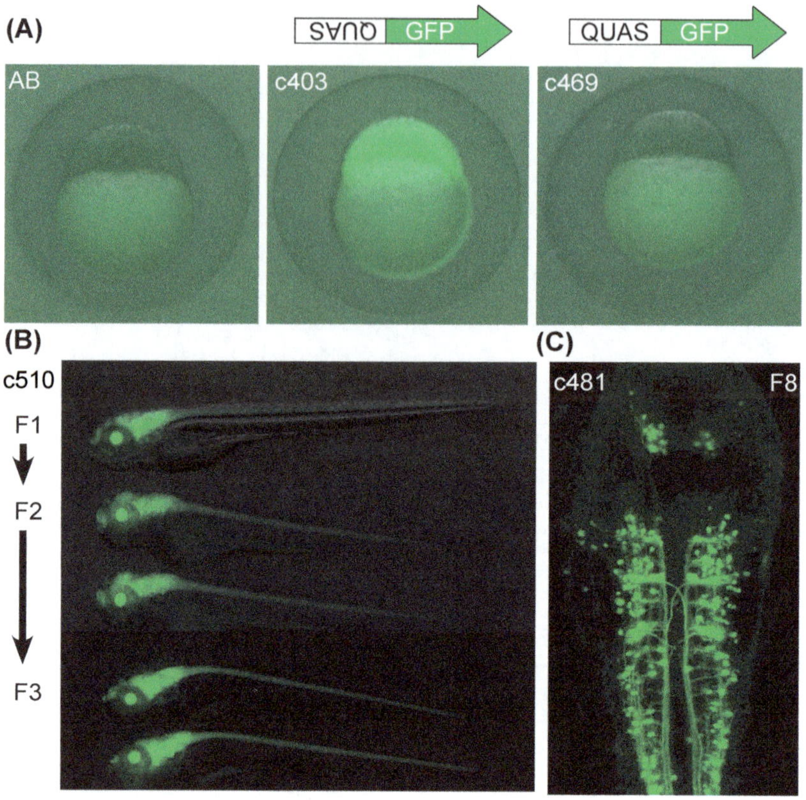

A. GHOSH AND M.E. HALPERN, FIGURE 3

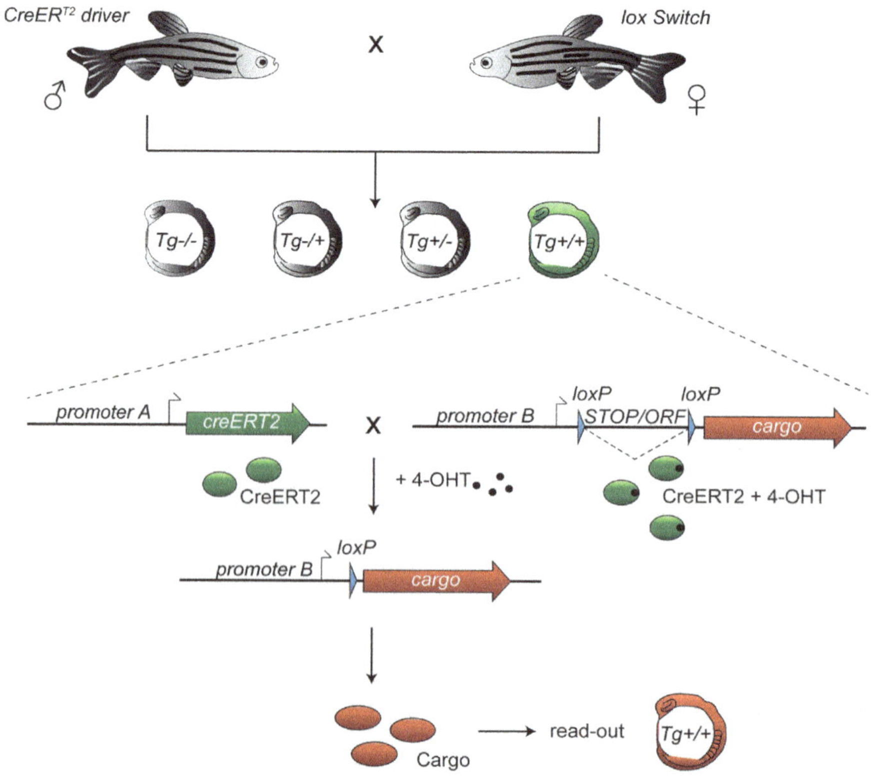

A. FELKER AND C. MOSIMANN, FIGURE 2

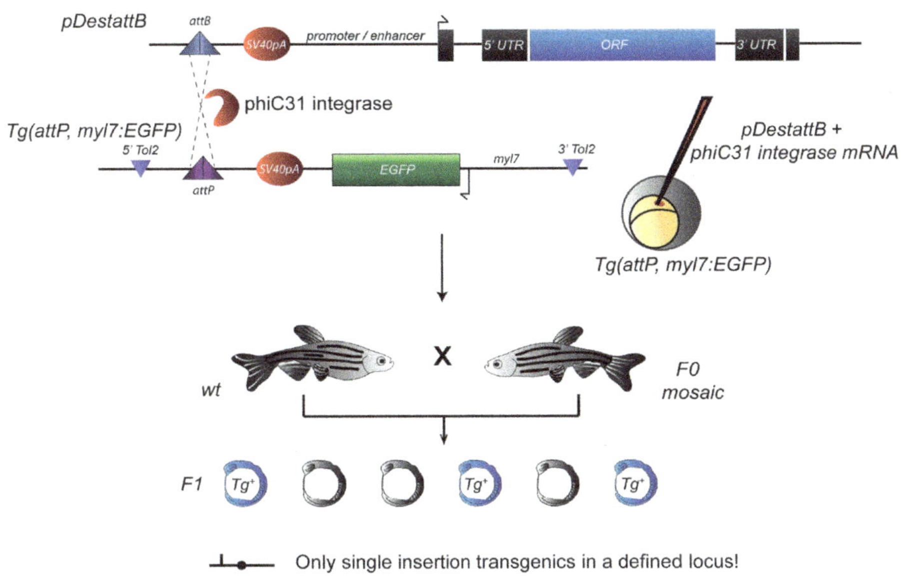

A. FELKER AND C. MOSIMANN, FIGURE 4

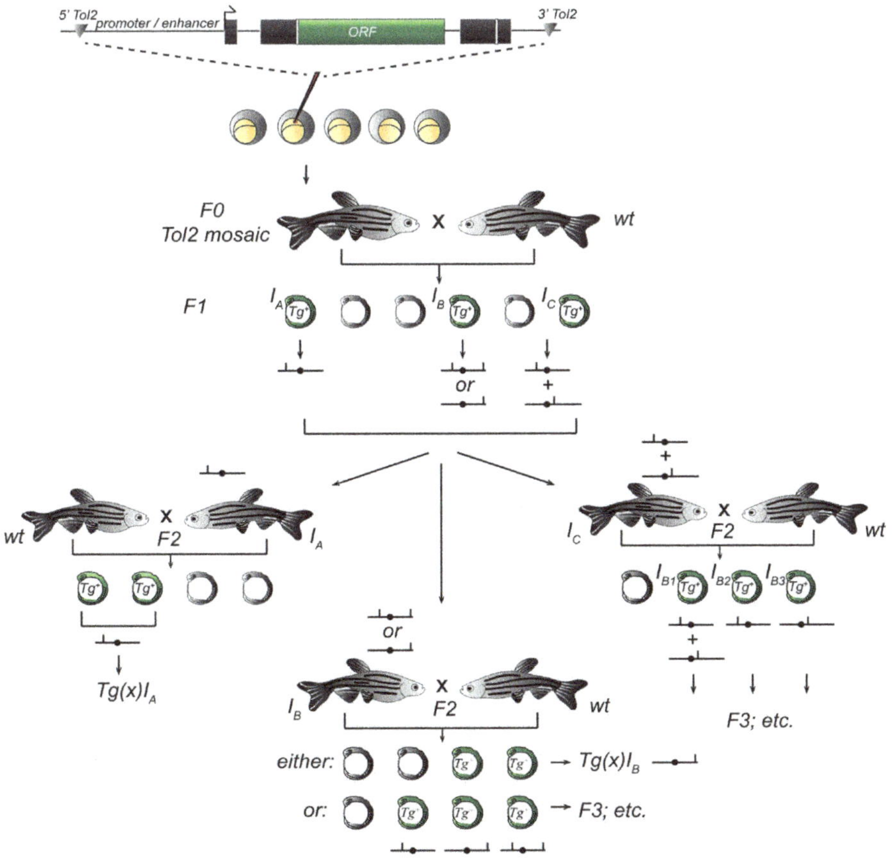

A. FELKER AND C. MOSIMANN, FIGURE 5

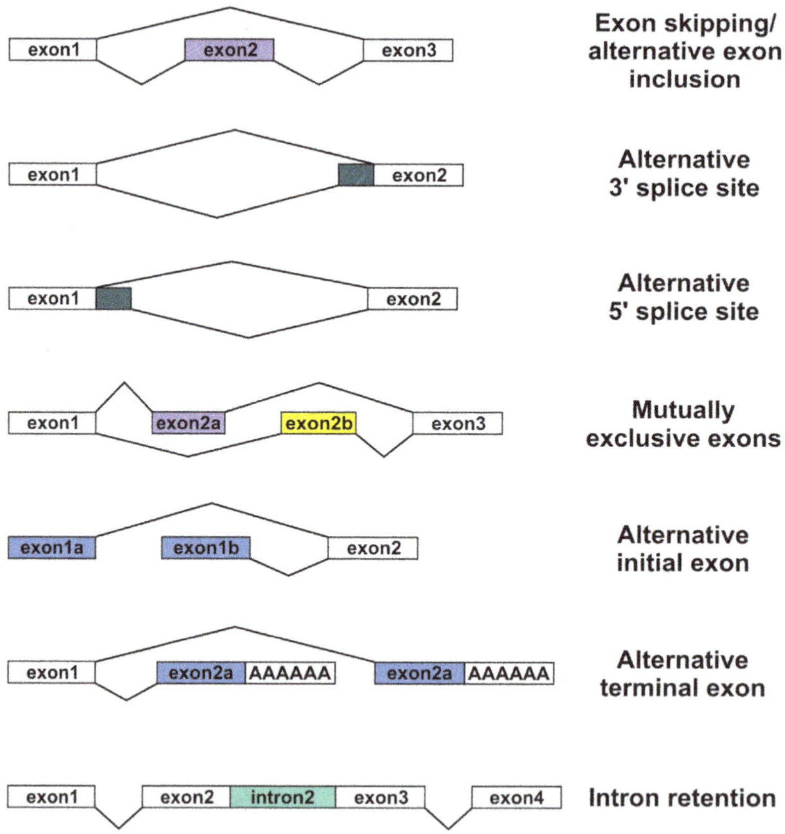

M.C. KEIGHTLEY *ET AL.*, FIGURE 1

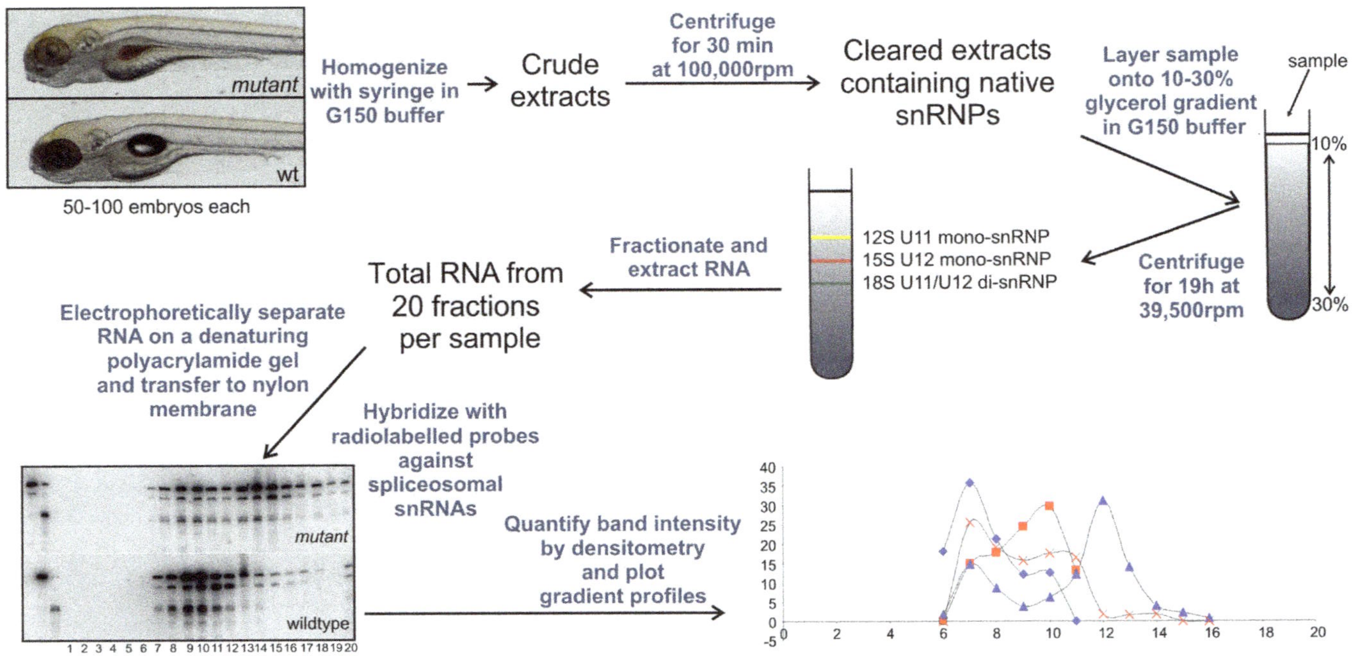

M.C. KEIGHTLEY *ET AL.*, FIGURE 5

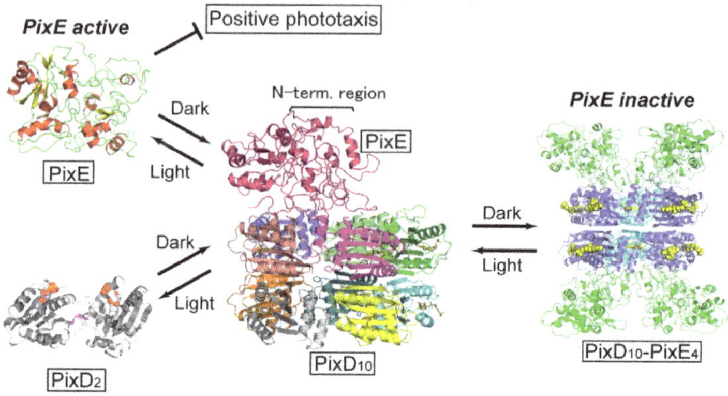

S. MASUDA AND M. TANAKA, FIGURE 1

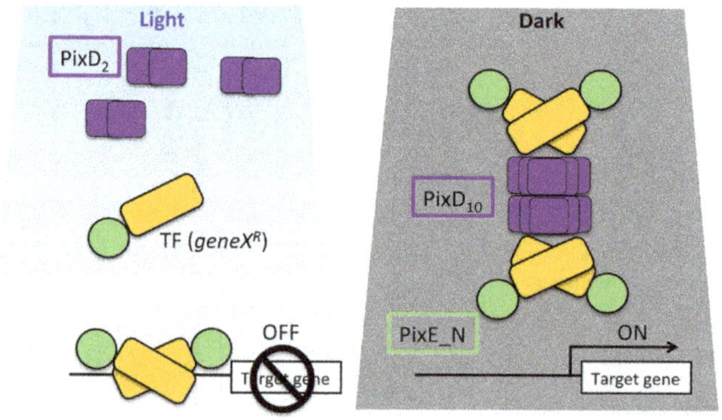

S. MASUDA AND M. TANAKA, FIGURE 2

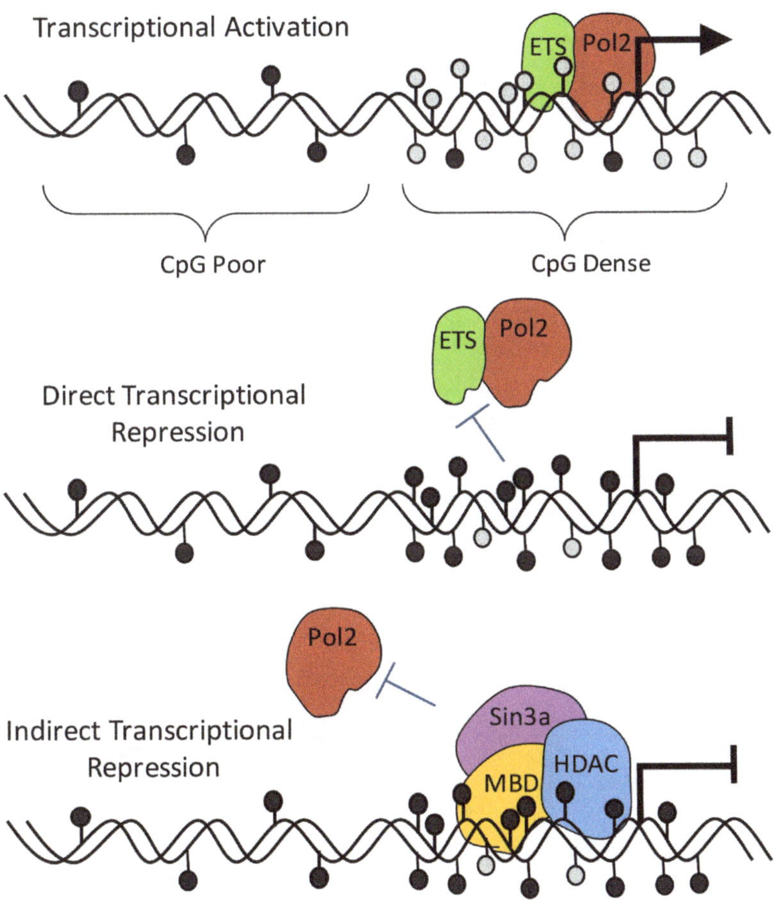

P.J. MURPHY AND B.R. CAIRNS, FIGURE 1

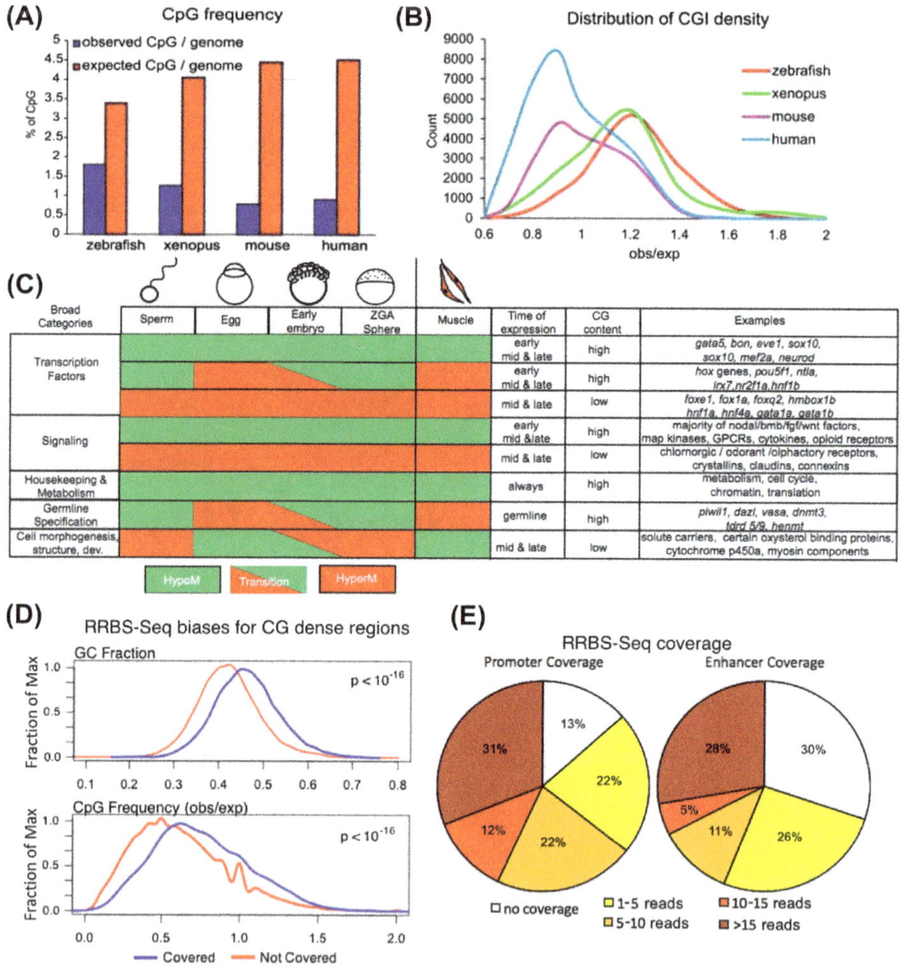

P.J. MURPHY AND B.R. CAIRNS, FIGURE 2

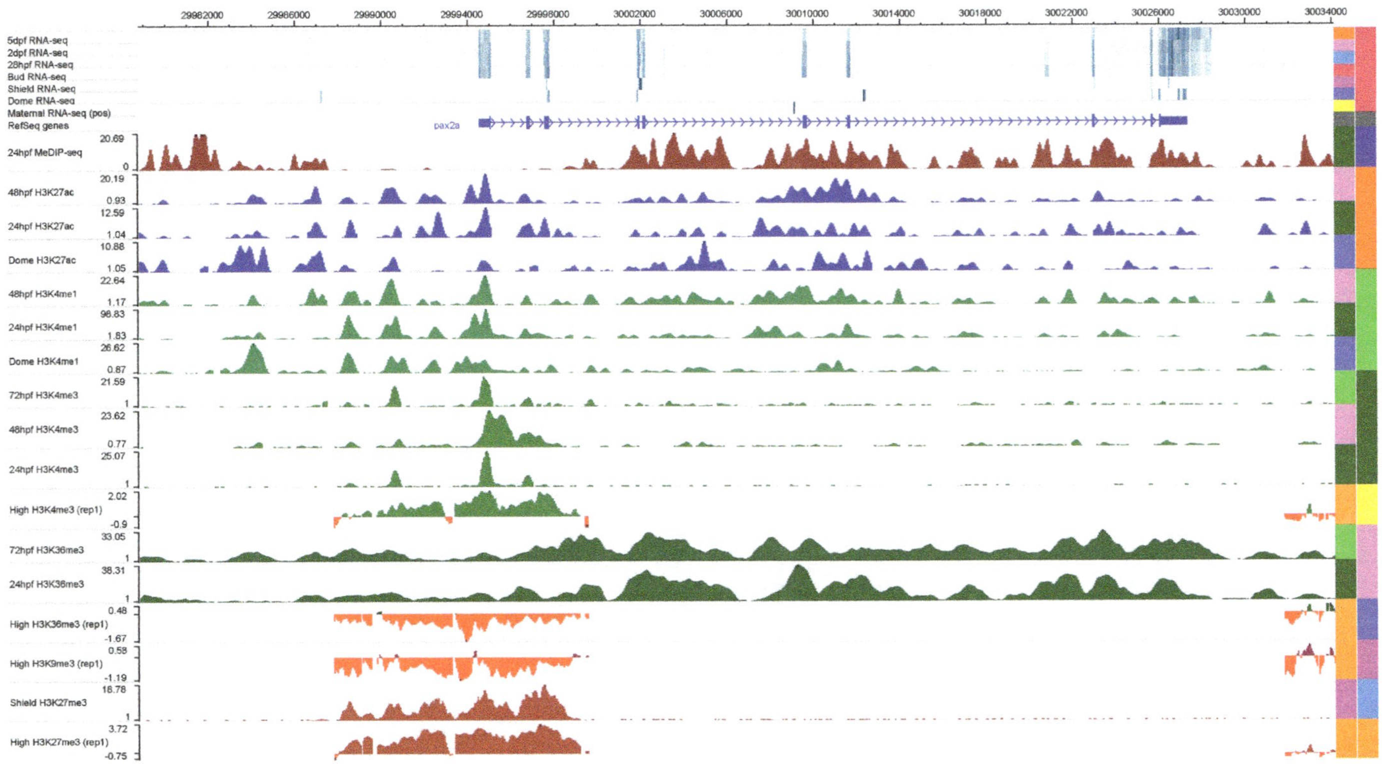

V.T. CUNLIFFE, FIGURE 2

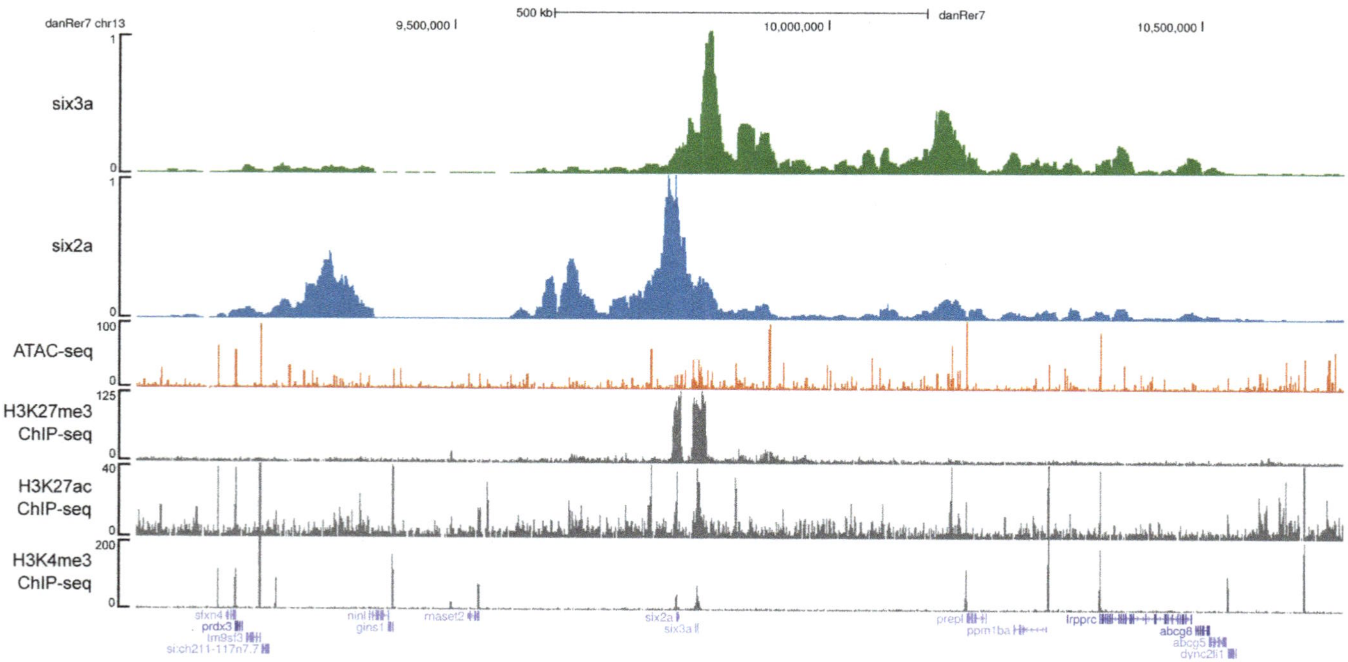

A. FERNÁNDEZ-MIÑÁN *ET AL.*, FIGURE 1

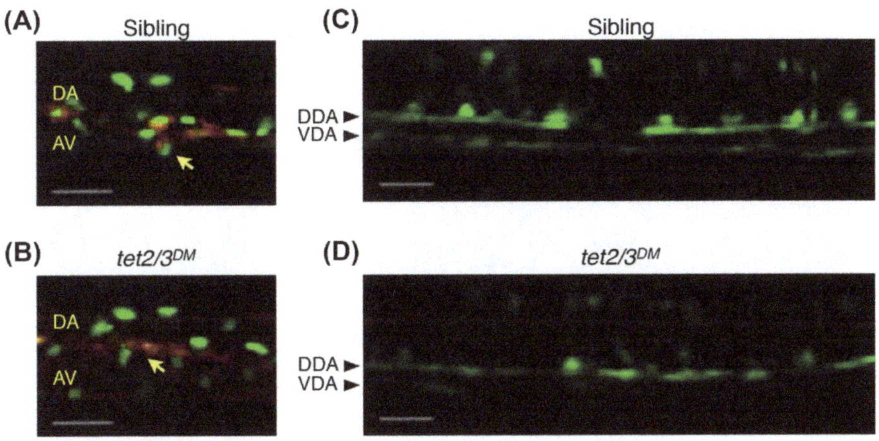

C. LI *ET AL.*, FIGURE 1

CPSIA information can be obtained
at www.ICGtesting.com
Printed in the USA
BVHW01*2010290618
520004BV00005B/4/P

9 780128 034743